Selected Titles in This Series

70 **Carmen Chicone and Yuri Latushkin,** Evolution semigroups in dynamical systems and differential equations, 1999
69 **C. T. C. Wall (A. A. Ranicki, Editor),** Surgery on compact manifolds, Second Edition, 1999
68 **David A. Cox and Sheldon Katz,** Mirror symmetry and algebraic geometry, 1999
67 **A. Borel and N. Wallach,** Continuous cohomology, discrete subgroups, and representations of reductive groups, Second Edition, 1999
66 **Yu. Ilyashenko and Weigu Li,** Nonlocal bifurcations, 1999
65 **Carl Faith,** Rings and things and a fine array of twentieth century associative algebra, 1999
64 **Rene A. Carmona and Boris Rozovskii, Editors,** Stochastic partial differential equations: Six perspectives, 1999
63 **Mark Hovey,** Model categories, 1999
62 **Vladimir I. Bogachev,** Gaussian measures, 1998
61 **W. Norrie Everitt and Lawrence Markus,** Boundary value problems and symplectic algebra for ordinary differential and quasi-differential operators, 1999
60 **Iain Raeburn and Dana P. Williams,** Morita equivalence and continuous-trace C^*-algebras, 1998
59 **Paul Howard and Jean E. Rubin,** Consequences of the axiom of choice, 1998
58 **Pavel I. Etingof, Igor B. Frenkel, and Alexander A. Kirillov, Jr.,** Lectures on representation theory and Knizhnik-Zamolodchikov equations, 1998
57 **Marc Levine,** Mixed motives, 1998
56 **Leonid I. Korogodski and Yan S. Soibelman,** Algebras of functions on quantum groups: Part I, 1998
55 **J. Scott Carter and Masahico Saito,** Knotted surfaces and their diagrams, 1998
54 **Casper Goffman, Togo Nishiura, and Daniel Waterman,** Homeomorphisms in analysis, 1997
53 **Andreas Kriegl and Peter W. Michor,** The convenient setting of global analysis, 1997
52 **V. A. Kozlov, V. G. Maz'ya, and J. Rossmann,** Elliptic boundary value problems in domains with point singularities, 1997
51 **Jan Malý and William P. Ziemer,** Fine regularity of solutions of elliptic partial differential equations, 1997
50 **Jon Aaronson,** An introduction to infinite ergodic theory, 1997
49 **R. E. Showalter,** Monotone operators in Banach space and nonlinear partial differential equations, 1997
48 **Paul-Jean Cahen and Jean-Luc Chabert,** Integer-valued polynomials, 1997
47 **A. D. Elmendorf, I. Kriz, M. A. Mandell, and J. P. May (with an appendix by M. Cole),** Rings, modules, and algebras in stable homotopy theory, 1997
46 **Stephen Lipscomb,** Symmetric inverse semigroups, 1996
45 **George M. Bergman and Adam O. Hausknecht,** Cogroups and co-rings in categories of associative rings, 1996
44 **J. Amorós, M. Burger, K. Corlette, D. Kotschick, and D. Toledo,** Fundamental groups of compact Kähler manifolds, 1996
43 **James E. Humphreys,** Conjugacy classes in semisimple algebraic groups, 1995
42 **Ralph Freese, Jaroslav Ježek, and J. B. Nation,** Free lattices, 1995
41 **Hal L. Smith,** Monotone dynamical systems: an introduction to the theory of competitive and cooperative systems, 1995
40.4 **Daniel Gorenstein, Richard Lyons, and Ronald Solomon,** The classification of the finite simple groups, number 4, 1999

(*Continued in the back of this publication*)

Evolution Semigroups in Dynamical Systems and Differential Equations

Mathematical
Surveys
and
Monographs

Volume 70

Evolution Semigroups in Dynamical Systems and Differential Equations

Carmen Chicone
Yuri Latushkin

American Mathematical Society

Editorial Board

Georgia M. Benkart Tudor Stefan Ratiu, Chair
Peter Landweber Michael Renardy

Supported by the NSF Grant DMS-9531811.
Supported by the NSF Grant DMS-9622105, by the Research Board and by the Research Council of the University of Missouri.

1991 *Mathematics Subject Classification.* Primary 47Dxx, 34Cxx; Secondary 47Bxx, 58Fxx.

ABSTRACT. In this book our main objective is to characterize asymptotic properties (stability, hyperbolicity, exponential dichotomy) of linear differential equations on Banach spaces and infinite dimensional dynamical systems in terms of spectral properties of a special type of associated semigroup that we call an *evolution semigroup*. We use methods from the theory of strongly continuous semigroups of linear operators, the theory of nonautonomous abstract Cauchy problems on Banach spaces, the theory of C^*- and Banach algebras, ergodic theory, the theory of hyperbolic dynamical systems and Lyapunov exponents. Applications to linear control theory, magnetohydrodynamics, and to the theory of transfer operators are given.

Library of Congress Cataloging-in-Publication Data
Chicone, Carmen Charles.
 Evolution semigroups in dynamical systems and differential equations / Carmen Chicone, Yuri Latushkin.
   p. cm. — (Mathematical surveys and monographs, ISSN 0076-5376 ; v. 70)
 Includes bibliographical references and index.
 ISBN 0-8218-1185-1
 1. Semigroups of operators. 2. Evolution equations. 3. Differentiable dynamical systems. 4. Differential equations. I. Latushkin, Yuri, 1956– II. Title. III. Series: Mathematical surveys and monographs ; no. 70.
QA329.C477 1999
515'.724–dc21   99-23729
                                                                                                       CIP

Copying and reprinting. Individual readers of this publication, and nonprofit libraries acting for them, are permitted to make fair use of the material, such as to copy a chapter for use in teaching or research. Permission is granted to quote brief passages from this publication in reviews, provided the customary acknowledgment of the source is given.

Republication, systematic copying, or multiple reproduction of any material in this publication is permitted only under license from the American Mathematical Society. Requests for such permission should be addressed to the Assistant to the Publisher, American Mathematical Society, P. O. Box 6248, Providence, Rhode Island 02940-6248. Requests can also be made by e-mail to reprint-permission@ams.org.

© 1999 by the American Mathematical Society. All rights reserved.
The American Mathematical Society retains all rights
except those granted to the United States Government.
Printed in the United States of America.
∞ The paper used in this book is acid-free and falls within the guidelines
established to ensure permanence and durability.
Visit the AMS home page at URL: http://www.ams.org/
10 9 8 7 6 5 4 3 2 1   04 03 02 01 00 99

To our families

Contents

Chapter 1.	Introduction	1
1.1.	Motivations and main results	2
1.2.	Historical comments	6
1.3.	Organization and content	8
Chapter 2.	Semigroups on Banach Spaces and Evolution Semigroups	21
2.1.	Introduction to semigroups	21
2.2.	Evolution semigroups and hyperbolicity	37
2.3.	Bibliography and remarks	53
Chapter 3.	Evolution Families and Howland Semigroups	57
3.1.	Evolution families and dichotomy	57
3.2.	Howland semigroups on the line	62
3.3.	Howland semigroups on the half line	73
3.4.	Bibliography and remarks	78
Chapter 4.	Characterizations of Dichotomy for Evolution Families	87
4.1.	Discrete dichotomies: an algebraic approach	87
4.2.	Green's function and evolution semigroups	103
4.3.	Dichotomy and solutions of nonhomogeneous equations	107
4.4.	Hyperbolicity and dissipativity	118
4.5.	Bibliography and remarks	126
Chapter 5.	Two Applications of Evolution Semigroups	131
5.1.	Control theory	131
5.2.	Persistence of dichotomy	155
5.3.	Bibliography and remarks	158
Chapter 6.	Linear Skew-Product Flows and Mather Evolution Semigroups	163
6.1.	Linear skew-product flows and dichotomy	163
6.2.	The Mather semigroup	173
6.3.	Sacker-Sell spectral theory	191
6.4.	Bibliography and remarks	200
Chapter 7.	Characterizations of Dichotomy for Linear Skew-Product Flows	205
7.1.	Pointwise dichotomies	205
7.2.	The Annular Hull Theorem	222
7.3.	Dichotomy, mild solutions, and Green's function	230
7.4.	Isomorphism Theorems	240
7.5.	Dichotomy and quadratic Lyapunov function	246
7.6.	Bibliography and remarks	256

Chapter 8. Evolution Operators and Exact Lyapunov Exponents		261
8.1. Oseledets' Theorem and linear skew-product flows		261
8.2. The kinematic dynamo operator		274
8.3. The Ruelle transfer operator		304
8.4. Bibliography and remarks		332
Bibliography		341
List of Notations		357
Index		359

CHAPTER 1

Introduction

The analytical theory of semigroups is a recent addition to the ever-growing list of mathematical disciplines. It was my good fortune to take an early interest in this discipline and to see it reach maturity. It has been a pleasant association: I hail a semi-group when I see one and I seem to see them everywhere! Friends have observed, however, that there are mathematical objects which are not semi-groups.

Einar Hille 1948 ([**Hil**, p. v]).

The subject matter of this book traces its origin to the classical results of J. Mather and R. Mañe on hyperbolic dynamical systems, and Y. Daleckij and M. Krein on nonautonomous Cauchy problems. In the same spirit as this pioneering work, our main objective is to characterize asymptotic properties (stability, hyperbolicity, exponential dichotomy) of linear differential equations on Banach spaces and infinite dimensional dynamical systems in terms of spectral properties of a special type of associated semigroup that we call an *evolution semigroup*.

Consider a discrete dynamical system; that is, a map on a metric space. The associated evolution operator is the product of a multiplication operator, and a composition (Koopman) operator that is induced by this map. The evolution operator acts on a space of vector functions defined on the metric space with values in a Banach space. Evolution operators are also known as weighted composition or weighted shift operators in general operator theory, as transfer operators in dynamical systems theory, and as push-forward operators in the theory of differentiable manifolds. If the underlying map is replaced by a flow, then it is natural to consider a semigroup of evolution operators, the *evolution semigroup*.

The concept of an evolution operator arises naturally in the study of finite and infinite dimensional dynamical systems and differential equations. Indeed, on a finite dimensional manifold, the flow of a smooth vector field induces an evolution semigroup corresponding to the "push-forward" operation on the space of sections of the tangent bundle of the manifold. The multiplication operator, in this case, is given by the differential of the flow. The infinitesimal generator of this evolution semigroup is Lie differentiation in the direction of the vector field that generates the flow. In infinite dimensions, the simplest example arises in the context of a well-posed autonomous linear differential equation on a Banach space. The underlying flow is now the translation flow on the real line (or the translation semiflow on the half line) and the multiplication operators are given by the elements of the semigroup generated by the linear operator that defines the differential equation. More generally, we will consider *nonautonomous* linear differential equations on Banach spaces; for instance, such differential equations arise as the linearizations of a nonlinear differential equation along its trajectories. In this case, there is a two-parameter family of solution operators, one for each current and initial value of time. For the nonautonomous case it is also possible to assign an evolution semigroup — a *one-parameter family* of operators — that carries all of the important "asymptotic information" of the original *two-parameter family* of solution operators. Moreover, in each case, we will show that the asymptotic behavior of the solutions is completely determined by the position in the complex plane of the spectrum of the assigned evolution operators and by the spectrum of the infinitesimal generator of the evolution semigroup.

These results require important new methods that we will explain in detail. Our exposition includes a systematic treatment of the recent results obtained by A. Baskakov, J. Campbell, S. Clark, M. Gundlach, Nguyen Van Minh, S. Montgomery-Smith, F. Räbiger, T. Randolph, R. Rau, A. Rhandi, R. Schnaubelt, A. Stepin, and the authors. Many applications of the theory are also discussed. In addition to these basic results we give a comprehensive survey of the vast literature related to the subject of the book.

The material in this book was used by the second author for the presentation of several special topic courses given at the University of Missouri and at the Georgia Institute of Technology. In fact, a major portion of the book was written while the second author was at the Georgia Institute of Technology during the Fall of 1996. He thanks the faculty of the School of Mathematics for their hospitality.

We thank our collaborators S. Montgomery-Smith, T. Randolph, R. Schnaubelt, A. Stepin, and R. Swanson for their many contributions. Many of their ideas are used in this book. Our special thanks go to L. Arnold, M. Gundlach, W. Liu, T. Randolph, R. Schnaubelt, Y. Tomilov, and, especially, F. Räbiger. They read and corrected various parts of the manuscript and gave many valuable suggestions.

We also thank Brenda Frazier for typing the manuscript.

1.1. Motivations and main results

This book has its origin in two famous theorems: the Lyapunov Stability Theorem for differential equations on Banach spaces and Mather's Hyperbolicity Theorem for hyperbolic diffeomorphisms on smooth manifolds.

The Lyapunov Stability Theorem states that if $A \in \mathcal{L}(X)$ is a bounded operator on a Banach space X and the spectrum of A, denoted hereafter by $\sigma(A)$, lies in the open left half of the complex plane, then the autonomous differential equation $\dot{x} = Ax$ on X is uniformly exponentially stable. In this case, the solution of the differential equation, starting at the point $x_0 \in X$, is given by $x(t) = e^{tA}x_0$ for $t \in \mathbb{R}$. The trivial solution, or the differential equation, is called uniformly exponentially stable if the function $t \mapsto \|e^{tA}\|$ tends to zero exponentially fast as $t \to \infty$. Equivalently, the spectrum $\sigma(e^{tA})$, for $t \neq 0$, is contained in the open unit disk $\mathbb{D}$. The central feature underlying the proof of the Lyapunov Stability Theorem is the fact that the following Spectral Mapping Theorem holds: If A is a *bounded* operator on a Banach space X, then, for the semigroup $\{e^{tA}\}_{t\geq 0}$,

$$(1.1) \qquad \sigma(e^{tA}) \setminus \{0\} = \exp(t\sigma(A)), \quad t > 0.$$

In fact, this theorem has many other consequences. For example, the condition $\sigma(A) \cap i\mathbb{R} = \emptyset$ is equivalent to the hyperbolicity of the semigroup $\{e^{tA}\}_{t\geq 0}$; that is, to the condition that $\sigma(e^{tA}) \cap \mathbb{T} = \emptyset$, for $t \neq 0$, where $\mathbb{T}$ denotes the unit circle. We will say that a strongly continuous semigroup has the *spectral mapping property* if the spectral condition (1.1) holds.

Mather's Hyperbolicity Theorem concerns diffeomorphisms on smooth finite dimensional manifolds. To state it, let φ denote a diffeomorphism of a compact manifold Θ, let $D\varphi$ denote the differential of φ, and let $C(\mathcal{T}\Theta)$ denote the space of continuous sections of the tangent bundle $\mathcal{T}\Theta$; that is, continuous maps $f : \Theta \to \mathcal{T}\Theta$ such that $f(\theta) \in \mathcal{T}_\theta\Theta$ where $\mathcal{T}_\theta\Theta$ denotes the tangent space of Θ at the point $\theta \in \Theta$. Consider the associated "push-forward" operator E defined on $C(\mathcal{T}\Theta)$ by

$$(Ef)(\theta) = D\varphi(\varphi^{-1}\theta)f(\varphi^{-1}\theta).$$

Mather's Theorem states that the diffeomorphism φ is hyperbolic (in other words, Anosov, or exponentially dichotomic) if and only if the operator E is hyperbolic; that is, $\sigma(E) \cap \mathbb{T} = \emptyset$.

Our starting point is the fact that Lyapunov's Stability Theorem fails for *unbounded* operators. For an operator A that generates a strongly continuous semigroup $\{e^{tA}\}_{t\geq 0}$ on a Banach space X, we will continue to write $x(t) = e^{tA}x_0$ for the solution of the differential equation $\dot{x} = Ax$. However, if A is *unbounded*, then the spectral mapping property generally *fails*—possibly in a very dramatic way. For instance, there are examples for which the spectrum of A is empty while the spectrum of e^{tA} is the closed unit disk.

Lyapunov's Theorem can be "saved" for unbounded operators. We will construct an operator Γ, related to the infinitesimal generator A, such that if $\sigma(\Gamma)$ lies in the open left half of the complex plane, then $\sigma(e^{tA})$ is contained in $\mathbb{D}$. In fact, the construction of the operator Γ is surprisingly simple: Consider a "super"-space $\mathcal{F}$ of functions $f : \mathbb{R} \to X$ where $\mathcal{F}$ is the space $L^p(\mathbb{R}; X)$, $1 \leq p < \infty$, or the space $\mathcal{F} = C_0(\mathbb{R}; X)$ of continuous functions vanishing at infinity. Define the semigroup $\{E^t\}_{t\geq 0}$ on $\mathcal{F}$ by

$$(1.2) \qquad (E^t f)(\theta) = e^{tA}f(\theta - t), \quad \theta \in \mathbb{R}.$$

The semigroup $\{E^t\}_{t\geq 0}$, called the *evolution semigroup associated with* $\{e^{tA}\}_{t\geq 0}$, is strongly continuous and its infinitesimal generator Γ is the closure of the operator $-d/dt + A$. It turns out that the generator Γ is the operator that saves Lyapunov's Theorem! In fact, the spectrum of Γ, relative to the space $\mathcal{F}$, lies in the open left half of the complex plane if and only if the spectrum of e^{tA} for each $t > 0$, relative to X, is contained in the open unit disk. Moreover, whether or not the spectral mapping property is valid for the semigroup $\{e^{tA}\}_{t\geq 0}$, it is always valid for the associated evolution semigroup $\{E^t\}_{t\geq 0}$; that is,

$$(1.3) \qquad \sigma(E^t) \setminus \{0\} = \exp(t\sigma(\Gamma)), \quad t > 0.$$

In addition, the semigroup $\{e^{tA}\}_{t\geq 0}$ is hyperbolic on X if and only if the semigroup $\{E^t\}_{t\geq 0}$ is hyperbolic on $\mathcal{F}$, or, equivalently, if and only if Γ is invertible on $\mathcal{F}$.

Consider a nonautonomous differential equation $\dot{x} = A(t)x$ on the Banach space X. Suppose that there exists a family $\{U(\theta, \tau) : \theta \geq \tau, \theta, \tau \in \mathbb{R}\}$ of solution operators with the initial condition $U(\tau, \tau) = I$ so that the solution of the differential equation with initial condition $x(\tau) = x_\tau$ is given by $x(t) = U(t, \tau)x_\tau$ for all $t \geq \tau$. In this book we will call $\{U(\theta, \tau)\}_{\theta \geq \tau}$ the *evolution family* associated with the differential equation. As is well-known, even for the case of a two-dimensional Banach space X, the set of spectra $\{\sigma(A(t)) : t \in \mathbb{R}\}$ does not generally determine the asymptotic behavior (stability, exponential dichotomy, etc.) of the solutions. However, the spectrum of the infinitesimal generator Γ of the evolution semigroup $\{E^t\}_{t \geq 0}$, defined on $\mathcal{F}$ by

$$(1.4) \qquad (E^t f)(\theta) = U(\theta, \theta - t)f(\theta - t), \quad \theta \in \mathbb{R}, \quad t \geq 0,$$

does determine the asymptotic behavior of the solutions for this nonautonomous equation. For instance, the generator Γ is invertible on $L^p(\mathbb{R}; X)$, or $C_0(\mathbb{R}; X)$, if and only if $\dot{x} = A(t)x$ has an exponential dichotomy on $\mathbb{R}$. Also, the spectral mapping property is valid for the semigroup $\{E^t\}_{t \geq 0}$.

How is the concept of evolution semigroups related to Mather's Hyperbolicity Theorem? To see the connection, let us define the translation flow $\{\varphi^t\}_{t \in \mathbb{R}}$ on $\Theta := \mathbb{R}$ by $\varphi^t \theta = \theta + t$. Also, using the evolution family $\{U(\theta, \tau)\}_{\theta \geq \tau}$, let us define an associated family $\{\Phi^t\}_{t \geq 0} = \{\Phi^t(\theta) : t \geq 0, \theta \in \Theta\}$ of operators on X by $\Phi^t(\theta) = U(\theta + t, \theta)$. Using the "semigroup properties" of the solutions of the differential equation $\dot{x} = A(t)x$, it follows that, for $t, s \geq 0$ and $\theta \in \mathbb{R}$, we have

$$\Phi^{t+s}(\theta) = \Phi^t(\varphi^s \theta)\Phi^s(\theta) \quad \text{and} \quad \Phi^0(\theta) = I.$$

A family $\{\Phi^t\}_{t \geq 0}$ of operators with these properties is called a *cocycle* over the flow $\{\varphi^t\}_{t \in \mathbb{R}}$. Note that, in this notation, the definition (1.4) of the evolution semigroup $\{E^t\}_{t \geq 0}$ can be recast as follows:

$$(1.5) \qquad (E^t f)(\theta) = \Phi^t(\varphi^{-t}\theta)f(\varphi^{-t}\theta), \quad t \geq 0, \quad \theta \in \Theta, \quad f \in C_0(\Theta; X).$$

For an arbitrary continuous flow $\{\varphi^t\}_{t \in \mathbb{R}}$ on a locally compact metric space Θ, let $\{\Phi^t\}_{t \geq 0}$ be a strongly continuous, exponentially bounded cocycle over $\{\varphi^t\}_{t \in \mathbb{R}}$ with values in $\mathcal{L}(X)$, and define the *linear skew-product flow* $\{\hat{\varphi}^t\}_{t \geq 0}$ as follows:

$$(1.6) \qquad \hat{\varphi}^t : \Theta \times X \to \Theta \times X : (\theta, x) \mapsto (\varphi^t \theta, \Phi^t(\theta)x).$$

By using formula (1.5), an *evolution semigroup* $\{E^t\}_{t \geq 0}$, on the space $C_0(\Theta; X)$ of continuous X-valued functions that vanish at infinity, can be associated with the cocycle $\{\Phi^t\}_{t \geq 0}$, or with the linear skew-product flow $\{\hat{\varphi}^t\}_{t \geq 0}$. Again, the spectral properties of the evolution operators in the semigroup $\{E^t\}_{t \geq 0}$ and the spectral properties of its infinitesimal generator Γ determine the asymptotic properties of the cocycle $\{\Phi^t\}_{t \geq 0}$.

Of course, the concept of a linear skew-product flow on the trivial bundle $\Theta \times X$ over Θ can be generalized to nontrivial vector bundles. For example, if $\{\varphi^t\}_{t \in \mathbb{R}}$ is a smooth flow on a finite dimensional smooth manifold Θ, then a linear skew-product flow $\{\hat{\varphi}^t\}_{t \in \mathbb{R}}$ is induced on the tangent bundle $\mathcal{T}\Theta$ by the cocycle given by the differential of the flow: $\Phi^t(\theta) := D\varphi^t(\theta)$.

We will prove three main results that hold for strongly continuous, exponentially bounded cocycles over flows on locally compact metric spaces in the Banach space setting. First, the linear skew-product flow $\{\hat{\varphi}^t\}_{t \geq 0}$ has an exponential dichotomy (is hyperbolic) if and only if the corresponding evolution semigroup

$\{E^t\}_{t\geq 0}$ is hyperbolic on $C_0(\Theta; X)$ (Dichotomy Theorem). Second, if the operator E^1 is hyperbolic, then the Riesz projection $\mathcal{P}$, that corresponds to the operator E^1 and the spectral set $\sigma(E^1) \cap \mathbb{D}$, is a multiplication operator of the form $(\mathcal{P}f)(\theta) = P(\theta)f(\theta)$, for $\theta \in \Theta$ and $f \in C_0(\Theta; X)$, where $P : \Theta \to \mathcal{L}(X)$ is a bounded, strongly continuous, projection-valued function (Spectral Projection Theorem). Moreover, the decomposition $X = \operatorname{Im} P(\theta) + \operatorname{Ker} P(\theta)$ gives invariant stable and unstable subbundles for the linear skew-product flow $\{\hat{\varphi}^t\}_{t\geq 0}$. Third, if the underlying flow $\{\varphi^t\}_{t\in\mathbb{R}}$ does not have open sets consisting of periodic orbits, then the spectral mapping property (1.3) is valid for the evolution semigroup $\{E^t\}_{t\geq 0}$ (Spectral Mapping Theorem). A related result, the Annular Hull Theorem, holds without the assumption on the periodic orbits of the base flow. As a result, the existence of an exponential dichotomy for the linear skew-product flow is equivalent to the following spectral condition on the generator of the evolution semigroup: $\sigma(\Gamma) \cap i\mathbb{R} = \emptyset$.

The main advantage of these results is that the theory of strongly continuous semigroups of linear operators can be used when dealing with questions concerning the existence of dichotomies for linear skew-product flows. We will discuss applications of this theory to several topics: persistence of exponential dichotomies, the existence of exponential dichotomies and Green's functions, the existence of exponential dichotomies and mild solutions, Sacker-Sell spectral theory, etc. We will mention here a few of these applications in more detail.

One of the important applications is related to a study of "pointwise" exponential dichotomies; that is, exponential dichotomies along individual orbits of the flow. To study this concept, we will introduce a special family of operators, called the *discrete evolution operators*. Fix $t > 0$, for example $t = 1$, and, for the cocycle $\{\Phi^t\}_{t\geq 0}$ and its corresponding evolution semigroup $\{E^t\}_{t\geq 0}$, let $E := E^1$ and $\Phi(\theta) := \Phi^1(\theta)$. Fix $\theta \in \Theta$, and consider the space $c_0(\mathbb{Z}; X)$ of *sequences* $(x_n)_{n\in\mathbb{Z}} \subset X$ with the property that $|x_n| \to 0$ as $n \to \pm\infty$. The associated discrete evolution operator $\pi_\theta(E)$ is given by

$$(1.7) \qquad \pi_\theta(E) : (x_n)_{n\in\mathbb{Z}} \mapsto \left(\Phi(\varphi^{n-1}\theta)x_{n-1}\right)_{n\in\mathbb{Z}}.$$

We will show that the hyperbolicity of the "continuous" evolution operator E is equivalent to the hyperbolicity of all discrete evolution operators uniformly for $\theta \in \Theta$. We will see that the hyperbolicity of $\pi_\theta(E)$ is equivalent to the exponential dichotomy of the linear skew-product flow along the trajectory of the base flow through θ. Thus, we will connect the existence of "global" and "pointwise" exponential dichotomies (Discrete Dichotomy Theorem).

As an important observation, we note that the operator E is a *weighted translation operator*; that is, $E = aV$ where the weight a is the multiplication operator given by $(af)(\theta) = \Phi(\varphi^{-1}\theta)f(\theta)$, and V is the translation operator defined by $(Vf)(\theta) = f(\varphi^{-1}\theta)$ for $\theta \in \Theta$ and $f \in C_0(\Theta; X)$. Likewise, $\pi_\theta(E)$ is a *weighted shift operator*. In effect $\pi_\theta(E) = dS$ where the weight $d = \operatorname{diag}\left(\Phi(\varphi^{n-1}\theta)\right)_{n\in\mathbb{Z}}$ is a diagonal operator, and $S : (x_n)_{n\in\mathbb{Z}} \mapsto (x_{n-1})_{n\in\mathbb{Z}}$ is the shift operator on $c_0(\mathbb{Z}; X)$. The map $\pi_\theta : E \to \pi_\theta(E)$, defined in display (1.7), gives rise to a representation in the space $c_0(\mathbb{Z}; X)$ of the algebra generated by the multiplication operators and translation operators. The study of this *weighted translation algebra* is another important theme of this book.

We will discuss the relationship between the existence of exponential dichotomy of linear skew-product flows and the Multiplicative Ergodic Theorem. For a compact metric space Θ and a Hilbert space X, if the family $\{\Phi(\theta) : \theta \in \Theta\}$ consists of compact operators, then, by the Multiplicative Ergodic Theorem, for each $x \in X$, there is an *exact* Oseledets-Lyapunov exponent

$$\lambda(\theta, x) = \lim_{n \to \infty} n^{-1} \log |\Phi^n(\theta) x|$$

for almost all $\theta \in \Theta$ with respect to each φ-ergodic measure ν. We will determine the spectrum and the spectral subspaces of E, and the dynamical spectrum and spectral subbundles for $\hat{\varphi}$ in terms of the Oseledets-Lyapunov exponents and the corresponding measurable subbundles.

As an application, we will give a description of the spectrum for the kinematic dynamo operator of magneto-hydrodynamics. Finally, we use our technique to study an important generalization of the evolution operator E; namely, the Ruelle-Perron-Frobenius transfer operator.

1.2. Historical comments

As we have mentioned above, the history of the subject matter of this book lies at the intersection of three main topics: dynamical systems, semigroup theory, and the theory of Banach algebras.

The "push-forward" semigroup generated by the Lie derivative is a classical object described in every book on the subject of differential manifolds, see for example Abraham, Marsden, and Ratiu [**AMR**]. As far as we know, the paper of Mather [**Ma**] was the first to establish a connection between the spectral properties of the "push-forward" operator defined by $Ef = D\varphi \circ f \circ \varphi^{-1}$ on $C(\mathcal{T}\Theta)$ and dynamical properties of the underlying map $\varphi : \Theta \to \Theta$. Mather proved that $\sigma(E) \cap \mathbb{T} = \emptyset$ if and only if the map φ is Anosov; that is, the tangent bundle can be decomposed as a direct sum of $D\varphi$-invariant subbundles, $\mathcal{T}\Theta = \mathcal{T}^s + \mathcal{T}^u$, and there are positive constants M and β such that

$$|D\varphi^n(\theta)x| \leq M e^{-n\beta} |x|, \quad x \in \mathcal{T}_\theta^s, \quad n \geq 0,$$
$$|D\varphi^n(\theta)x| \leq M e^{\beta n} |x|, \quad x \in \mathcal{T}_\theta^u, \quad n \leq 0, \quad \theta \in \Theta.$$

For a modern account of the theory of Anosov systems, see for example Katok and Hasselblatt [**KH**], or Pesin [**Ps**] and the references therein. Many different authors contributed to the theory of hyperbolic cocycles. Important contributions, related to the subject of this book, were made by Bronshtein [**Br**], Hirsh, Pugh, and Shub [**HPS**], Mañe [**Mn, Mn2**], and Otsuki [**Ot**]. Another group of authors, Churchill, Franks, and Selgrade [**CFS**], Franks and Robinson [**FRo**], Sacker and Sell [**SS**], and Selgrade [**Sg**] studied so-called quasi-Anosov systems, which are defined by the following property: the set $\{|D\varphi^n(\theta)x| : n \in \mathbb{Z}\}$ is unbounded for all nonzero $x \in \mathcal{T}_\theta\Theta$. It was proved that a diffeomorphism φ is quasi-Anosov if and only if $1 \notin \sigma_{ap}(E)$, see Mañe [**Mn2**], Chicone and Swanson [**CS, CS3**], and the literature cited therein. In a series of papers by Sacker and Sell [**SS2, SS3**] and Selgrade [**Sg**], linear skew-product flows were considered on arbitrary bundles and the now classical spectral theory was constructed (see also a later review by Johnson, Palmer, and Sell [**JPS**] and an infinite dimensional development by Sacker and Sell [**SS5**]). In the papers by Chicone and Swanson [**CS, CS2, CS3**] and in the papers by Johnson [**Jo, Jo2**], a connection is made between the Sacker-Sell spectral theory for linear skew-product flows and the spectral theory of evolution semigroups as defined in equation (1.5) for the finite dimensional case. These papers contain the first examples of spectral mapping theorems for evolution semigroups. We note that Johnson [**Jo2**] uses C^*-algebraic methods.

Motivated by a completely different circle of problems, Antonevich [**An2, An3**] used C^*-algebras of evolution operators (weighted translation algebras) to study hyperbolic extensions of dynamical systems, what we call linear skew-product flows. These methods are related to the theory of pseudo-differential operators, in particular, singular integral operators, with a "shift" (translation) of their arguments, see the work by Karlovich, Kravchenko, and Litvinchuk [**KLc, Lc**] and the bibliography therein. A part of this theory is concerned with the spectral properties of weighted translation operators; that is, evolution operators of the form $Ef = af \circ \varphi$ where a is a multiplication operator whose multiplier is an operator-valued function $a(\cdot)$ on Θ. A review of the relevant part of this theory is in the paper of Latushkin and Stepin [**LSt2**]. For more information on C^*- and Banach algebras generated by weighted translation operators, see the recent book of Antonevich and Lebedev [**AL2**], the paper of Karlovich [**Ka**], and the paper of Hadwin and Hoover [**HH3**]. The theory of weighted translation operators is connected to the general theory of Banach lattices (see Abramovich, Arenson, and Kitover [**AAK**]), functional-differential equations (see Kurbatov [**Ku**]), function theory (see Cowen and Shapiro [**Cw, Sh, Sh2**]), and operators on Hilbert spaces (see Nordgren [**No**]).

Evolution semigroups of operators for cocycles that are uniformly continuous in θ on Hilbert spaces are studied in the papers by Latushkin and Stepin [**LSt2, LSt3**]. In this work, C^*-algebraic methods, in the spirit of Antonevich's book, are used to construct the infinite dimensional Sacker-Sell spectral theory. In addition, generalizations of the spectral mapping theorems of Chicone and Swanson [**CS, CS2, CS3**] and Johnson [**Jo2**] are obtained. Moreover, the important connection with the Multiplicative Ergodic Theorem of Oseledets [**Os**] is studied. This connection was also studied by Johnson, Palmer, and Sell [**JPS**] in the context of finite dimensional linear skew-product flows.

This line of research has a natural intersection with another important and widely studied topic; namely, the theory of the transfer operator, also known as the Ruelle-Perron-Frobenius operator. This operator is a natural generalization of the evolution operator (1.5) to the case where the underlying dynamical system is not invertible. Indeed, for a covering map $\varphi : \Theta \to \Theta$ and a (matrix-valued) function Φ on Θ, the associated transfer operator $\mathcal{R}$ is defined on the space of (vector-valued) functions f on Θ by

$$(1.8) \qquad (\mathcal{R}f)(\theta) = \sum_{\eta \in \varphi^{-1}\theta} \Phi(\eta) f(\eta).$$

We will show that $\mathcal{R}$ shares many important properties with the operator $E := E^1$. The transfer operator is one of the central objects in statistical physics, dynamical systems, the theory of ζ-functions, etc. Therefore, it is the subject of extensive research by many authors; for example, V. Baladi, R. Bowen, P. Collet, N. Haydn, S. Isola, G. Keller, F. Ledrappier, R. de la Llave, D. Mayer, W. Parry, Ja. Pesin, M. Pollicott, D. Ruelle, Ja. Sinai, P. Walters, L.-S. Young, and many others, see the books [**Ba, Bo2, KS, My, PPo, Ru2, Ru7**] and the bibliographies therein.

The reader is perhaps surprised by the fact that neither strongly continuous semigroups nor differential equations on Banach spaces have been mentioned thus far in this historical sketch. In fact, the "link" between the theory of evolution semigroups, weighted translation operators, and the corresponding C^*- and Banach algebras as studied in dynamical system theory on the one hand, and their counterparts in differential equations on Banach spaces on the other hand, was established quite recently.

As far as we know, evolution semigroups of the form (1.4) were first studied by Howland [**Ho**] in connection with scattering theory. By a formal differentiation of equation (1.4), one can see that the generator Γ of the semigroup $\{E^t\}_{t \geq 0}$ is related to the operator $-d/dt + A(t)$. This latter operator was used already by Daleckij and Krein [**DK**, Lemma IV.3.3] and by Levitan and Zhikov [**LZ**, Chapt. 10] in their study of exponential dichotomy (see also the book by Massera and Schäffer [**MS2**]). A major contribution to the theory was made by Lumer [**Lu, Lu2, Lu3**] and Paquet [**Pq**]. The work by Howland was continued by Evans [**Ev**], and by Neidhardt [**Ne**]. Later, a systematic use of evolution semigroups in questions of dichotomy for nonautonomous differential equations on Banach spaces was initiated by R. Nagel in Tübingen. Rau [**Ra, Ra2, Ra3**] linked the study of Latushkin and Stepin [**LSt2**] and evolution semigroups as they appear in the theory of semigroups and abstract Cauchy problems. The reader will find many of his results in this book. His research, in turn, was developed further by Latushkin, Montgomery-Smith, and Randolph [**LM, LM2, LMR**]. In particular, the paper [**LM**] is foundational for this book. Räbiger, Rhandi, and Schnaubelt [**RS, RS2, RRS, Sc**] made a fundamental contribution to the subject. In fact, we will use many of the ideas introduced by these authors. We note that the notion of the evolution semigroup was also discovered and studied by Zabreiko and Nguyen Van Minh [**Mi, Mi2, ZM**]. An important step in the study of evolution semigroups of the type (1.2) was made by van Neerven [**vN**]. Also, important contributions were made by Baskakov [**Bs, Bs2, Bs3, Bs4**]. In fact, in the short paper [**Bs3**], he gives beautiful proofs of several results in this book.

One of the central objects of our study is the theory of exponential dichotomy. In addition to the classical sources Coppel [**Co**], Daleckij and Krein [**DK**], Hale [**Ha**], and Massera and Schäffer [**MS2**], we mention the contributions made by Chow, Hale, Henry, Leiva, Verduyn Lunel, Shen, Yi, and their students and collaborators, see [**CLe, CLe2, CLe3, Ha2, HL, He, SY2, Yi**] as well as the work of Palmer [**KP, Pa3, Pa4**], and the important papers [**BrGK, BG3, BG4, BGK**] by Bart, Ben-Artzi, Gohberg, and Kaashoek. Their research played a very important role in our understanding of the subject. We also mention connections of the theory to the fast dynamo problem (see the review by Arnold, Zel'dovich, Rasumaikin, and Sokolov [**AZRS**], and the papers by de la Llave [**dL**] and Friedlander and Vishik [**FV, FV2, Vs2, Vs3**]).

1.3. Organization and content

Chapter 2 is devoted to the study of evolution semigroups of the type defined in equation (1.2). Recall that evolution semigroups of this type are generated by a strongly continuous semigroup $\{e^{tA}\}_{t \geq 0}$ defined on a Banach space X. Thus, in effect, this chapter is concerned with the solution of the autonomous differential equation $\dot{x} = Ax$.

In Section 2.1 we give a brief account of well-known results on the asymptotic behavior of strongly continuous semigroups in a form that is convenient for our purposes. Detailed proofs and much more information can be found in Engel and Nagel [**EN**], Nagel [**Na**], Pazy [**Pz**], and van Neerven [**vN**]. After a discussion of our notation in Subsection 2.1.1, we consider, in Subsection 2.1.2, the spectral mapping property, as in formula (1.1), for various parts of the spectrum, and we also formulate the important Gearhart Spectral Mapping Theorem 2.10. This result asserts that, in the case of Hilbert spaces, the spectral mapping property as in display 1.1 is implied by the boundedness of the function $\lambda \mapsto \|(\lambda - A)^{-1}\|$ along the lines parallel to the imaginary axis of the complex plane. We introduce in Subsection 2.1.3 several characteristics that control the stability of semigroups: the spectral bound $s(A)$, the growth bound $\omega(A)$, and the abscissa of uniform boundedness $s_0(A)$. In Subsection 2.1.4 we present some basic results about hyperbolic semigroups; that is, semigroups $\{e^{tA}\}_{t\geq 0}$ such that $\sigma(e^{tA}) \cap \mathbb{T} = \emptyset$ for each $t > 0$. These semigroups correspond to exponentially dichotomic autonomous linear differential equations of the type $\dot{x} = Ax$. We will formulate a recent result by M. Kaashoek and S. M. Verduyn Lunel that gives conditions for a semigroup to be hyperbolic on a Banach space in terms of the resolvent of its generator. In the central part of this section, Subsection 2.1.5, we collect many examples of semigroups that violate the spectral mapping property. In addition to the examples that are due to W. Arendt, G. Greiner, S. Montgomery-Smith, V. Wrobel, M. Wolff, J. Voigt, and J. Zabczyk, we present two examples that have their origin in differential equations. One of these is the semigroup related to fractional integration; it goes back to E. Hille and R. S. Phillips but was also studied more recently by Henry [**He2**] and Arendt, El Mennaoui and Hieber [**AEMH**]. Another example is due to Renardy [**Rn**]. It provides a natural example of a hyperbolic partial differential equation—a first order perturbation of the two-dimensional wave equation—where the spectral mapping property does not hold for the corresponding semigroup.

In Section 2.2 we introduce and study three types of evolution semigroups, as defined by equation (1.2), which act on the space of p-integrable, or continuous, X-valued functions defined, respectively, on an interval, say $[0, 2\pi]$, on the half line $\mathbb{R}_+$, or on the entire real line $\mathbb{R}$. Consider, for the moment, the case of p-integrable functions. Heuristically, the evolution semigroup on $L^p([0, 2\pi]; X)$ is "responsible" for a single point in the resolvent set $\rho(e^{2\pi A})$, the evolution semigroup on $L^p(\mathbb{R}_+; X)$ is "responsible" for the stability of the semigroup $\{e^{tA}\}_{t\geq 0}$, and the evolution semigroup on $L^p(\mathbb{R}; X)$ is "responsible" for the hyperbolicity of the semigroup $\{e^{tA}\}_{t\geq 0}$. These three cases are considered, respectively, in Subsections 2.2.1, 2.2.3, and 2.2.2. In each case we will show that the spectral mapping property holds for the evolution semigroup $\{E^t\}_{t\geq 0}$ even when it fails for the underlying semigroup $\{e^{tA}\}_{t\geq 0}$. We will prove the following results: $1 \in \rho(e^{2\pi A})$ if and only if Γ is invertible for the $[0, 2\pi]$-case; $\{e^{tA}\}_{t\geq 0}$ is uniformly exponentially stable if and only if $s(\Gamma) < 0$ for the $\mathbb{R}_+$-case; $\{e^{tA}\}_{t\geq 0}$ is hyperbolic if and only if Γ is invertible for the $\mathbb{R}$-case; and $s(\Gamma) = \omega(A)$ for the $\mathbb{R}$-case. Using the evolution semigroup for the $[0, 2\pi]$-case, we will prove a general spectral mapping theorem for the semigroup $\{e^{tA}\}_{t\geq 0}$ on a Banach space, see Theorem 2.31. In fact, Theorem 2.31 is a direct generalization of the Gearhart Spectral Mapping Theorem mentioned above. In addition, we will give a formula for the norm of Γ^{-1} that is used later to compute an important characteristic, called the stability radius, that arises in control theory. Finally, for the $\mathbb{R}_+$-case, we will present an important theorem by R. Datko and J. van Neerven that gives a convolution formula for Γ^{-1}; this formula allows us to express one of the main objects of linear control theory, the input-output operator, in terms of Γ.

Chapter 3 deals with Howland semigroups; that is, with evolution semigroups of the type defined in (1.4) that are associated with a strongly continuous evolution family $\{U(\theta, \tau)\}_{\theta \geq \tau}$ on a Banach space X. In particular, this includes the study of nonautonomous differential equations of the form $\dot{x} = A(t)x$ on X. Our main emphasis in this chapter is the theory for evolution semigroups acting on the space $L^p(\mathbb{R}; X)$ or $L^p(\mathbb{R}_+; X)$; the corresponding theory for the space of continuous functions is covered in Chapter 6 in the more general setting of linear skew-product flows.

In Section 3.1 we discuss the notion of evolution families and their relation to nonautonomous differential equations of the type $\dot{x} = A(t)x$ as well as the notion of exponential dichotomy for the evolution families. In Subsection 3.1.1 we present the definitions and basic facts concerning abstract Cauchy problems and associated evolution families. In Subsection 3.1.2 we give the definition of exponential stability and exponential dichotomy for evolution families. In this book we do not address the deep question of the existence of an evolution family for particular classes of nonautonomous Cauchy problems; rather, we refer the reader to the literature on the subject, for example Fattorini [**Fa**], Pazy [**Pz**], and Tanabe [**Ta**]. One of the difficulties related to nonautonomous problems is that their associated evolution families are two-parameter families of operators. From this point of view, the evolution semigroup $\{E^t\}_{t \geq 0}$ as in (1.4) is a natural one-parameter semigroup related to the nonautonomous problem. Heuristically, the construction of the evolution semigroup $\{E^t\}_{t \geq 0}$ is a powerful way to "autonomize" a nonautonomous Cauchy problem: we replace the time dependent differential equation $\dot{x} = A(t)x$ on X by an autonomous differential equation $\dot{f} = \Gamma f$ on a "super"-space of X-valued functions defined on $\mathbb{R}$ or $\mathbb{R}_+$.

In Section 3.2 we study the evolution semigroup $\{E^t\}_{t \geq 0}$ as defined in (1.4) on $L^p(\mathbb{R}; X)$. In Subsection 3.2.1 we consider the evolution semigroup generated by the evolution family that solves the abstract Cauchy problem for $\dot{x} = A(t)x$. Following Latushkin, Montgomery-Smith, and Randolph [**LMR**] and Schnaubelt [**Sc**], we show that the generator Γ of $\{E^t\}_{t \geq 0}$ is the closure of the operator $f \mapsto -d/d\theta f + A(\cdot)f$. In Subsection 3.2.2 we prove the spectral mapping property (1.3) by reducing the proof to the autonomous case covered in Subsection 2.2.2 by means of a special "change-of-variables" technique that is frequently used in this book. Subsection 3.2.3 is devoted to the proof of the central result of the chapter; namely, Dichotomy Theorem 3.17. It states that the existence of an exponential dichotomy of an evolution family $\{U(\theta, \tau)\}_{\theta \geq \tau}$, for $\theta, \tau \in \mathbb{R}$, is equivalent to the hyperbolicity of the evolution semigroup $\{E^t\}_{t \geq 0}$, or, equivalently, to the invertibility of its generator Γ. This result follows from the Spectral Projection Theorem 3.14: Each spectral (Riesz) projection $\mathcal{P}$ for E^t is a multiplication operator whose multiplier is a bounded, strongly continuous, projection-valued function $P(\cdot)$; that is, $(\mathcal{P}f)(\theta) = P(\theta)f(\theta)$. The kernel and the range of $P(\theta)$ are the subspaces of the initial data for the exponentially decaying and growing solutions. The proof of the Spectral Projection Theorem, in turn, is based on two facts. The first is that $\mathcal{P}$ commutes with multiplication operators whose multipliers are scalar-valued functions. The second is an important regularization result by Räbiger and Schnaubelt [**RS**]: The range of the resolvent operator $R(\lambda; \Gamma)$ on $L^p(\mathbb{R}; X)$ is in the space of continuous functions on $\mathbb{R}$; and therefore, the domain $\mathcal{D}(\Gamma)$ of Γ on $L^p(\mathbb{R}; X)$ consists of continuous functions. For $\theta \in \Theta$, this allows us to define $P(\theta)x$ for x in a dense subset of X as follows: Choose an arbitrary function $f \in \mathcal{D}(\Gamma)$ such that

$f(\theta) = x$ and set $P(\theta)x = (\mathcal{P}f)(\theta)$. As a result, we obtain the following formula that relates the Bohl spectrum $\mathcal{B}$ for $\{U(\theta, \tau)\}_{\theta \geq \tau}$ to the spectrum of E^t and Γ:

$$\mathcal{B} = \sigma(\Gamma) \cap \mathbb{R} = t^{-1} \log |\sigma(E^t) \setminus \{0\}|,$$

see Daleckij and Krein [**DK**] for the definition of the Bohl spectrum.

In Section 3.3 we study a modification, $\{E_+^t\}_{t \geq 0}$, of the evolution semigroup (1.4) on $L^p(\mathbb{R}_+; X)$. Here, the invertibility of the generator of the evolution semigroup is equivalent to the stability of the evolution family $\{U(\theta, \tau)\}_{\theta \geq \tau \geq 0}$. In Subsection 3.3.1 we show that the evolution semigroup $\{E_+^t\}_{t \geq 0}$ has the spectral mapping property. Moreover, if $t > 0$, then the spectrum of the operator E_+^t is a disk centered at the origin, while the spectrum of the generator Γ_+ of the semigroup $\{E_+^t\}_{t \geq 0}$ is a half plane. In Subsection 3.3.2 our main goal is to discuss the theorem of R. Datko and J. van Neerven for the nonautonomous case in order to characterize the stability of $\{U(\theta, \tau)\}_{\theta \geq \tau}$ in terms of the boundedness of the "convolution" operator Γ_+^{-1}. This generalization is also needed for the applications to control theory in Chapter 5. We give two proofs of the nonautonomous Datko-van Neerven Theorem 3.26. One of them is taken from Van Minh, Räbiger and Schnaubelt [**MRS**] and is related to the mild integral equations studied in Section 4.3. Another proof reduces the nonautonomous case to the autonomous result by means of the "change-of-variables" technique mentioned above.

Chapter 4 provides several characterizations of the existence of an exponential dichotomy for evolution families. One of the main themes of this chapter is the study of the hyperbolicity of the discrete operators, in the family $\{\pi_\theta(E) : \theta \in \Theta\}$, that act on the space $\ell^p(\mathbb{Z}; X)$ of X-valued sequences, see display (1.7).

In Section 4.1 we study the discrete operators and related algebraic techniques. As we have mentioned above, the evolution operator $E := E^1$ as in display (1.4) is a weighted translation operator $E = aV$ where V is the translation operator defined by $(Vf)(\theta) = f(\theta - 1)$ and a is the multiplication operator corresponding to the function defined by $a(\theta) = U(\theta, \theta - 1)$ for $\theta \in \mathbb{R}$. The discrete operator, $\pi_\theta(E)$, $\theta \in \mathbb{R}$, is a weighted shift operator of the form $\pi_\theta(E) = \pi_\theta(a)S$ where $S : (x_n)_{n \in \mathbb{Z}} \mapsto (x_{n-1})_{n \in \mathbb{Z}}$ is the shift operator on $\ell^p(\mathbb{Z}; X)$ and $\pi_\theta(a) = \text{diag}(a(\theta + n))_{n \in \mathbb{Z}}$ is the weight. At this point we will have already proved that $\{U(\theta, \tau)\}_{\theta \geq \tau}$ has an exponential dichotomy if and only if each operator $\lambda - aV$, for $\lambda \in \mathbb{T}$, is invertible on $L^p(\mathbb{R}; X)$. As an additional result in this direction, we will prove that this happens if and only if each operator $\lambda - \pi_\theta(a)S$ is invertible on $\ell^p(\mathbb{Z}; X)$ and the norms of their inverses are bounded uniformly for $\theta \in \mathbb{R}$ and $\lambda \in \mathbb{T}$. In Subsection 4.1.1 we prove (Lemma 4.1) that the inverse of the operator $D = \lambda - \pi_\theta(a)S$ belongs to a certain algebra, $\mathfrak{C}$, of weighted shift operators on $\ell^p(\mathbb{Z}; X)$. The ideas behind the proof of this fact go back to Kurbatov [**Ku**]. Descriptively, the infinite matrix representing the operator $D = \lambda - \pi_\theta(a)S$ on $\ell^p(\mathbb{Z}; X)$ has two nonzero diagonals. For an arbitrarily $D \in \mathfrak{C}$, the corresponding matrix representation of D has infinitely many nonzero diagonals, and the sequence of the norms of these diagonal operators is summable. More generally, we also prove that the algebra $\mathfrak{C}$ is inverse-closed; that is, if $D \in \mathfrak{C}$ is invertible as an operator on $\ell^p(\mathbb{Z}; X)$, then $D^{-1} \in \mathfrak{C}$. This fact follows from a powerful generalization of the Wiener Theorem on Fourier series that is due to S. Bochner and R. S. Phillips. In particular, the hyperbolicity of $\pi_\theta(E)$ is an intrinsic algebraic property that does not depend on the choice of the space $\ell^p(\mathbb{Z}; X)$, $1 \leq p \leq \infty$, or the space $c_0(\mathbb{Z}; X)$. In Subsection 4.1.2 we use the fact that $\lambda - E$ belongs to a certain weighted translation algebra, $\mathfrak{B}$, that consists of operators on $L^p(\mathbb{R};X)$ of the form $b = \sum_{k=-\infty}^{\infty} a_k V^k$ where the a_k are multiplication operators whose multipliers are bounded, strongly continuous, operator-valued functions on $\mathbb{R}$ and $\Sigma_k \|a_k\| < \infty$. For $\theta \in \mathbb{R}$, we will consider the homomorphism $\pi_\theta : \mathfrak{B} \to \mathfrak{C}$ and show that if $\sup\{\|\pi_\theta(b)^{-1}\| : \theta \in \mathbb{R}\} < \infty$, then b is invertible on $L^p(\mathbb{R};X)$. This result uses the "inverse-closedness" of $\mathfrak{C}$, mentioned above. In Subsection 4.1.3 we show that the invertibility of the operator $\lambda - E$, for all $\lambda \in \mathbb{T}$, is equivalent to the uniform boundedness of $\|(\lambda - \pi_\theta(E))^{-1}\|$ for $\theta \in \mathbb{R}$ (Theorem 4.16). This gives a characterization of exponential dichotomy for the evolution family $\{U(\theta,\tau)\}_{\theta \geq \tau}$ in terms of the hyperbolicity of all discrete evolution operators in the family $\{\pi_\theta(E) : \theta \in \mathbb{R}\}$.

Section 4.2 is an exposition of the fact that the boundedness of Γ^{-1} is equivalent to the existence and uniqueness of the Green's function G_P for $\{U(\theta,\tau)\}_{\theta \geq \tau}$. In fact, $\Gamma^{-1} = -\mathbb{G}$ for the Green's operator $\mathbb{G}$ that is a "convolution" of f with the Green's function G_P. We prove this result first for the autonomous case; that is for semigroups (Subsection 4.2.1), and then, using evolution semigroups and our "autonomization" philosophy, we will translate the theorem to the nonautonomous case (Subsection 4.2.2).

Section 4.3 relates the existence of an exponential dichotomy for the evolution family $\{U(\theta,\tau)\}_{\theta \geq \tau}$ to the unique solvability of the mild form of the inhomogeneous differential equation $\dot{u} = A(t)u + g$. As we have mentioned, the existence of an exponential dichotomy for $\{U(\theta,\tau)\}_{\theta \geq \tau}$ is equivalent to the invertibility of the generator Γ on the space $\mathcal{F} = L^p(\mathbb{R};X)$ or $\mathcal{F} = C_0(\mathbb{R};X)$. In view of the "formula" $\Gamma = -d/dt + A(\cdot)$, this result can be recast as follows: $\{U(\theta,\tau)\}_{\theta \geq \tau}$ has an exponential dichotomy if and only if for each $g \in \mathcal{F}$ the equation $\dot{u} = A(t)u + g$ has a unique solution $u \in \mathcal{F}$. When the space $L^p(\mathbb{R};X)$, or $C_0(\mathbb{R};X)$, is replaced by $C_b(\mathbb{R};X)$, the space of bounded continuous functions, this result is in the spirit of a classical theorem of Perron [**Pn**], see the book by Daleckij and Krein [**DK**, IV.3] or Chapters 10 and 11 of the book by Levitan and Zhikov [**LZ**]. We prove a generalization of this classical result to the case of a strongly continuous evolution family $\{U(\theta,\tau)\}_{\theta \geq \tau}$ (with a proof different from that in [**LZ**]) and for the integral equation

$$(1.9) \qquad u(\theta) = U(\theta,\tau)u(\tau) + \int_\tau^\theta U(\theta,s)g(s)\,ds, \quad \theta \geq \tau.$$

In Subsection 4.3.1 we consider the above equation for $u, g \in C_b(\mathbb{R};X)$, while in Subsection 4.3.2 we take u and g from a certain scale of Banach spaces of exponentially bounded functions. This scale is frequently used in the theory of invariant manifolds, see Bates and Jones [**BJ**], Chow and Lu [**CLu**], Mallet-Paret and Sell [**MSe**], Vanderbauwhede, Van Gils, and Iooss [**VI**, **VG**], Wayne [**Way**], and Yi [**Yi**]. Thus, our result justifies a standard setting for this theory, see for example Condition (H) in [**VI**]. In Subsection 4.3.3 we return to the study of the family of discrete operators $\{\pi_\theta(E) : \theta \in \mathbb{R}\}$ and give a proof of a theorem of Baskakov [**Bs3**] (Theorem 4.37). This theorem implies, in particular, that the exponential dichotomy of the evolution family $\{U(\theta,\tau)\}_{\theta \geq \tau}$ is equivalent to the hyperbolicity of a *single* such discrete operator, for instance $\pi_0(E)$. In addition, this powerful result gives another, independent, proof of the Discrete Dichotomy Theorem 4.16. The main insight in the proof is the following connection between the generator Γ of the evolution semigroup $\{E^t\}_{t \geq 0}$ on the space $L^p(\mathbb{R};X)$, or $C_0(\mathbb{R};X)$, and the mild integral equation (1.9): The integral equation (1.9) holds for $u, g \in C_0(\mathbb{R};X)$

if and only if u is in the domain of Γ and $\Gamma u = -g$. This approach goes back to the classical results of Levitan and Zhikov [**LZ**], see also the work by Nguyen Van Minh, Räbiger, and Schnaubelt [**MRS**].

In Section 4.4 we give yet another characterization of the existence of exponential dichotomy for evolution families. We consider the case of a Hilbert space X and show that the existence of an exponential dichotomy is equivalent to the existence of a quadratic Lyapunov function. In other words, $\dot{x} = A(t)x$ has an exponential dichotomy if and only if there exists an indefinite scalar product such that the solution decays monotonically in the induced Lyapunov norm. This set of results generalizes the classical theory that can be found in the book by Daleckij and Krein [**DK**], the work of Massera and Schäffer [**MS2, MS**], Palmer [**Pa3, Pa4**], Kirchgraber [**KP**], and in many other places. Again, we prove this result in two steps: First, we prove the autonomous, or semigroup case, in Subsection 4.4.1. Then, in Subsection 4.4.2, we apply this result to the evolution semigroup corresponding to a given evolution family, in order to prove the nonautonomous case. We cite Nagel and Rhandi [**NR**] where similar results were obtained by a different method.

In Chapter 5 we will consider applications of evolution semigroups to control theory and to the problem of persistence of an exponential dichotomy.

In Section 5.1 we apply evolution semigroups to linear control theory. For this, we will consider systems of the form

$$(1.10) \quad x(t) = U(t,\tau)x_\tau + \int_\tau^t U(t,s)B(s)u(s)\,ds,$$

$$y(t) = C(t)x(t), \quad t \geq \tau \geq 0, \quad u(t) \in U, \quad y(t) \in Y,$$

where $\{U(t,\tau)\}_{t \geq \tau \geq 0}$ is a strongly continuous evolution family and both $B(\cdot)$ and $C(\cdot)$ are bounded, strongly continuous functions on $\mathbb{R}_+$ with values in the set of bounded operators. In particular, this is the mild form of the nonautonomous system

$$\dot{x}(t) = A(t)x(t) + B(t)u(t), \qquad y(t) = C(t)x(t),$$

for $t \geq 0$, where the operators in the family $\{A(t) : t \in \mathbb{R}_+\}$ are *not* assumed to be bounded. Our exposition is based on papers by Clark, Latushkin, Randolph, and Montgomery-Smith [**RLC, CLRM**]. In Subsection 5.1.1, we recall the basic definitions of internal, external and input-output stability, stabilizability, and detectability. Using the nonautonomous Datko–van Neerven's Theorem presented in Section 3.3, it follows that, for the internally stable systems, the input-output operator

$$(\mathbb{L}u)(t) := C(t)\int_0^t U(t,\tau)B(\tau)u(\tau)d\tau, \quad u \in L^p(\mathbb{R}_+;U),$$

can be decomposed as $\mathbb{L} = \mathcal{C} \cdot \Gamma_+^{-1} \cdot \mathcal{B}$ where Γ_+ is the generator of the evolution semigroup on $L^p(\mathbb{R}_+;X)$ corresponding to the evolution family $\{U(t,\tau)\}_{t \geq \tau \geq 0}$ and $\mathcal{B}$, respectively $\mathcal{C}$, is a multiplication operator with multiplier $B(\cdot)$, respectively $C(\cdot)$. Using this fact, we give a very short proof (Subsection 5.1.2) of the fact that system (1.10) is internally stable if and only if it is stabilizable, detectable, and input-output stable. For autonomous systems of type (1.10), we give new formulas for the norm of the input-output operator $\mathbb{L}$. These formulas generalize to the Banach space setting results of Rebarber [**Re2**] and Weiss [**Ws3**]. Also, they shed new light on the connections between the stability and frequency-domain approach, cf. [**Ad, Cu, HP3, JDP, Ja, JN, Re, Re2, Wi, Ws3**]. In particular, we give a characterization of internal stability for stabilizable and detectable autonomous systems in terms of the operators A, B, and C. In addition, we describe the class of autonomous systems on Banach spaces for which *external* stability implies internal stability under the assumptions of stabilizability and detectability. In Subsection 5.1.3 we deal with the problem of finding the Hinrichsen-Pritchard stability radius, r_{stab}, see [**HP2, HP3**]. For a stable system $\dot{x} = A(t)x$, it is defined to be the maximal size of the perturbation $\Delta(\cdot)$ such that the perturbed system $\dot{x} = [A(t) + B(t)\Delta(t)C(t)]x$ remains stable. We present the known estimate of r_{stab} in terms of $\|\mathbb{L}\|^{-1}$ and the norm of the transfer function, see [**HP3**]; however, our proofs make extensive use of the results on evolution semigroups from Chapters 2 and 3. We also give some new estimates for the stability radius, we discuss the dichotomy radius, and we consider the stability radius for a point in $\rho(e^{tA})$. Finally, we point out some radically new effects that appear in the Banach space setting and provide the corresponding counterexamples.

In Section 5.2 we discuss the persistence of exponential dichotomies. Since the existence of an exponential dichotomy is equivalent to the invertibility of a semigroup generator Γ, it is a spectral property. Since the spectrum "persists" under suitable perturbations, we might expect that exponential dichotomy does also. Indeed, in Subsection 5.2.1 we will give a one-line proof of the fact that if $\dot{x} = A(t)x$ has an exponential dichotomy and if a *bounded* perturbation $B(\cdot)$ is sufficiently small, then $\dot{x} = [A(t) + B(t)]x$ has an exponential dichotomy. Subsection 5.2.2 is an exposition of a much deeper recent result of F. Räbiger, R. Schnaubelt, A. Rhandi, and J. Voigt concerning unbounded perturbations that satisfy "Miyadera-type" conditions.

Chapter 6 is devoted to the study of linear skew-product flows $\{\hat{\varphi}^t\}_{t \geq 0}$ of the form given in display (1.6) where Θ is a locally compact metric space and $\{\Phi^t\}_{t \geq 0}$ is a strongly continuous exponentially bounded cocycle over a continuous flow $\{\varphi^t\}_{t \in \mathbb{R}}$ on Θ. Our main goal is to relate the exponential dichotomy for $\{\hat{\varphi}^t\}_{t \geq 0}$ to the spectral properties of the associated evolution semigroup $\{E^t\}_{t \geq 0}$ defined as in display (1.5) on the space $C_0(\Theta; X)$ of continuous X-valued functions that vanish at infinity. The results of this chapter remain valid for linear skew-product flows on nontrivial bundles where the semigroup and its generator act on the space of continuous sections of the bundle. This theory allows us to study variational equations of the form $\dot{x} = A(\varphi^t\theta)x$.

In Section 6.1 we define and give examples of cocycles and linear skew-product flows. A prototypical example is the cocycle corresponding to the solution operator $\Phi^t(\theta)$ for a variational equation of the form $\dot{x} = A(\varphi^t\theta)x$ where $\theta \in \Theta$. We stress the fact that, in this setting, the operators in the family $\{A(\theta) : \theta \in \Theta\}$ are *not* assumed to be bounded. Another natural situation occurs if Θ is a hull of the almost-periodic coefficient of the matrix differential equation $\dot{x} = \theta(t)x$ and $\{\varphi^t\}_{t \in \mathbb{R}}$ is the translation flow on Θ defined by $(\varphi^t\theta)(s) = \theta(s+t)$ where $s \in \mathbb{R}$. Infinite dimensional examples of this type are given in Subsection 6.1.1. In Subsection 6.1.2 we discuss the definition of an exponential dichotomy (versus hyperbolicity) for infinite dimensional linear skew-product flows, the Sacker-Sell spectrum, etc.

In Section 6.2 we begin the study of the Mather semigroup (1.5). In Subsection 6.2.1 we prove that the evolution semigroup (1.5) is strongly continuous on $C_0(\Theta; X)$ if and only if the linear skew-product flow (1.6) is continuous, and we give several examples where the generator Γ can be explicitly computed. In Subsection 6.2.2 we discuss properties of approximate eigenfunctions for the evolution operator $E := E^1$. By definition, $\lambda \in \sigma_{ap}(E)$ if there is a family of nonzero functions $\{g_\epsilon : \epsilon > 0\}$ in $C_0(\Theta; X)$ such that $\|(\lambda - E)g_\epsilon\| \leq O(\epsilon) \cdot \|g_\epsilon\|$ as $\epsilon \to 0$. For instance, we prove that if the aperiodic points of the base homeomorphism $\varphi := \varphi^1$ are dense, then $\sigma(E)$ is rotationally invariant. We call the construction required to prove this fact "Mather localization". The idea was introduced by Mather [**Ma**]; a similar approach is developed for scalar evolution operators by Arendt and Greiner [**AG**]. At any rate, starting with an approximate eigenfunction f for E, it is possible to construct another approximate eigenfunction, g, supported in a sufficiently long trajectory of a small set B that contains a point θ_0 where the norm of the vector $x_0 = f(\theta_0)$ in X is large. The function g is therefore "localized" along the trajectory through θ_0. In the finite dimensional case, this construction leads to the Mañe Lemma which is also discussed in Subsection 6.2.2. This lemma guarantees that, for each $e^\lambda \in \sigma_{ap}(E)$, there is a point θ_0 (the Mañe point) and a vector x_0 (the Mañe vector) such that $|x_0| = 1$ and $\sup\{e^{-t\lambda}\Phi^t(\theta_0)x_0 : t \in \mathbb{R}\} < \infty$. In Subsection 6.2.3 we prove the spectral mapping property (1.3) for the evolution semigroup $\{E^t\}_{t\geq 0}$ as in (1.5) under the assumption that the underlying flow $\{\varphi^t\}_{t\in\mathbb{R}}$ is aperiodic. The idea of the proof is to use the localized approximate eigenfunction for E to construct an approximate eigenfunction for Γ. In Subsection 6.2.4 we study Mather evolution semigroups on L^p-spaces; and, in particular, we prove the spectral mapping property in this setting.

In Section 6.3 we develop the spectral theory of linear skew-product flows (Sacker-Sell spectral theory) using evolution semigroups. We stress the generality of our strongly continuous, infinite dimensional, locally compact setting. Our main tool is again the Spectral Projection Theorem 6.38. It states that every Riesz spectral projection $\mathcal{P}$ for the evolution operator E is a multiplication operator of the form $\mathcal{P} = P(\cdot)$. The proof of this result, given in Subsection 6.3.1, is different from the proof for the case $\Theta = \mathbb{R}$ in Subsection 3.2.3. The Spectral Projection Theorem is used in Subsection 6.3.2 to prove the Dichotomy Theorem 6.41: The existence of an exponential dichotomy for the cocycle $\{\Phi^t\}_{t\geq 0}$ is equivalent to the hyperbolicity of the evolution operator E. Also, simple arguments from operator theory give (Subsection 6.3.3) the formula

$$\Sigma = \ln|\sigma(E^1)\setminus\{0\}| = \operatorname{Re}\sigma(\Gamma) \cap \mathbb{R}$$

for the Sacker-Sell dynamical spectrum Σ, the existence of spectral subbundles (cf. Sacker and Sell [**SS3**]), the characterization of Σ in terms of Lyapunov numbers (cf. [**JPS**]), and results on the normal and tangential spectra, cf. [**SS4**].

In Chapter 7 we continue the study of dichotomies for linear skew-product flows. The main tool here is the family of discrete evolution operators as defined in display (1.7).

In Section 7.1 we relate the existence of a "global" exponential dichotomy over all of Θ to the existence of "pointwise" exponential dichotomies along trajectories of the flow. Translated into a statement about operators, we relate the hyperbolicity of $E := E^1$, as in display (1.5), to the hyperbolicity of the discrete operators in the family $\{\pi_\theta(E) : \theta \in \Theta\}$ defined in display (1.7). In Subsection 7.1.1 we consider a convolution algebra $\hat{\mathfrak{B}}$, which, for the case of an aperiodic map φ, is isomorphic to the weighted translation algebra $\mathfrak{B}$ generated by the multiplication operators and the translation operator $f \mapsto f \circ \varphi^{-1}$. We represent the algebra $\hat{\mathfrak{B}}$ in the space $c_0(\mathbb{Z}; X)$ by operators of the type (1.7) and show that if $\sup\{\|[\lambda - \pi_\theta(E)]^{-1}\| : \lambda \in \mathbb{T}, \theta \in \Theta\} < \infty$, then the operator $\lambda - E$ is invertible on $C_0(\Theta; X)$. The technique used is borrowed from the theory of Wiener algebras (see Antonevich and Lebedev [**AL2**] for the C^*-algebra setting), but it is adapted to cover our results that are valid in Banach spaces. These results are the most important step in proving the Discrete Dichotomy Theorem 7.9, see Subsection 7.1.2. This theorem, roughly speaking, establishes that the existence of a global exponential dichotomy is equivalent to the existence of a pointwise exponential dichotomy along each orbit with dichotomy constants uniform in $\theta \in \Theta$. In terms of the spectrum of operators, these conditions are equivalent to the hyperbolicity of all operators in the family $\{\pi_\theta(E) : \theta \in \Theta\}$, uniformly for $\theta \in \Theta$, or to the hyperbolicity of E. In addition, we relate the existence of an exponential dichotomy to a family of evolution semigroups $\{\{\Pi_\theta^t\}_{t\geq 0} : \theta \in \Theta\}$. Each semigroup $\{\Pi_\theta^t\}_{t\geq 0}$ is of the type defined in display (1.4), but constructed from the evolution family $U_\theta(s,\tau) := \Phi^{s-\tau}(\varphi^\tau\theta)$ for $\theta \in \Theta$. Thus, we show, roughly speaking, that the existence of an exponential dichotomy on Θ is equivalent to the invertibility of the operator $f \mapsto -d/dt(f \circ \varphi^t(\theta))|_{t=0} + A(\theta)f(\theta)$ on $C_0(\Theta; X)$, and then we "reduce" the question of its invertibility to the question of the invertibility of all differential operators in the family $\{-d/d\tau + A(\varphi^\tau\theta) : \theta \in \Theta\}$ on $C_0(\mathbb{R}; X)$. In Subsection 7.1.3 we present an application of the Discrete Dichotomy Theorem 7.9 to prove an infinite dimensional variant of the Sacker-Sell Perturbation Theorem which gives the semicontinuity of the dynamical spectrum as a function of the compact invariant subsets in Θ.

In Section 7.2 we relate the hyperbolicity of the evolution semigroup $\{E^t\}_{t\geq 0}$ and the spectrum of its generator Γ. Simple examples show that the spectral mapping property (1.3) is not valid without the assumption that the flow $\{\varphi^t\}_{t\in\mathbb{R}}$ is aperiodic. Fortunately, for general flows there is a replacement; namely, the Annular Hull Theorem (Subsection 7.2.2). This theorem states that $\sigma(E^t) \subset \mathcal{AH}(e^{t\sigma(\Gamma)})$ where $\mathcal{AH}(S)$ denotes the union of all circles centered at the origin of the complex plane that intersect the set S. As a corollary of the Annular Hull Theorem, we have the fact that the stability and hyperbolicity of the semigroup $\{E^t\}_{t\geq 0}$ are determined by the position of the spectrum of its generator. For the proof of the Annular Hull Theorem we give formulas for the approximate eigenfunctions of Γ that are localized in "tubes" formed by the trajectories of the flow. The position of these flow-boxes is chosen using *Mañe sequences* (Subsection 7.2.1) which are natural infinite dimensional replacements for the Mañe points discussed above.

Section 7.3 further develops Section 4.2 and 4.3 for the situation of an exponential dichotomy of linear skew-product flows versus evolution families. In Subsection 7.3.1 we give a generalization of the Perron Theorem of Section 4.3. Here the dichotomy of the skew-product flow $\{\hat{\varphi}^t\}_{t\geq 0}$ is related to the existence and uniqueness, for each $g \in C_b(\Theta; X)$, of the solution $u \in C_b(\Theta; X)$ of the inhomogeneous equation

$$u(\varphi^t\theta) = \Phi^t(\theta)u(\theta) + \int_0^t \Phi^{t-s}(\varphi^s\theta)g(\varphi^s\theta)\,ds, \quad t \geq 0, \quad \theta \in \Theta.$$

To compare this result with the deep work of Sacker and Sell [**SS5**], recall that a weakly hyperbolic cocycle (see [**SS5**]) is one for which the homogeneous equation $u(\varphi^t\theta) = \Phi^t(\theta)u(\theta)$ has only the trivial solution (with certain additional assumptions on compactness of the operators in the family $\{\Phi^t\}_{t\geq 0}$). Weak hyperbolicity is equivalent to the absence of nontrivial bounded orbits for $\{\hat{\varphi}^t\}_{t\geq 0}$, cf. the Mañe Lemma 6.29 of Subsection 6.2.2. Among other results, R. Sacker and G. Sell show that a weakly hyperbolic cocycle has an exponential dichotomy provided that the stable sets have a fixed finite codimension. Thus, our Perron-type theorem gives a necessary and sufficient condition of a different type for cocycles to have an exponential dichotomy. In Subsection 7.3.2 we prove that the existence of an exponential dichotomy for $\{\hat{\varphi}^t\}_{t\geq 0}$ is equivalent to the existence and uniqueness of the corresponding Green's function and also that $\Gamma^{-1} = -\mathbb{G}$ for the Green's operator $\mathbb{G}$ on $C_0(\Theta; X)$. These results are a far-reaching generalization of theorems due to Mitropolskii, Samojlenko, Kulik, and other mathematicians from the Kiev school, cf. [**MSK, Sm**], which is well-known for its contributions to the theory of the existence of invariant tori for finite dimensional linear skew-product flows. These invariant tori, in our approach, correspond to solutions of the homogeneous equation $\Gamma u = 0$.

In Section 7.4 we specialize to the Hilbert space case. In Subsection 7.4.1 we give a general outline of the theory of weighted translation C^*-algebras and we explain their connection to evolution semigroups for cocycles that are continuous in θ in the uniform operator topology. This theory was developed by Antonevich, Karlovich, Lebedev, and others, see [**An3, AL2, Ka**] as well as the papers by Hadwin and Hoover [**HH, HH2, HH3**]. We do not give details; rather we just explain the general algebraic structure that underlies the family of discrete operators $\{\pi_\theta(E) : \theta \in \Theta\}$ defined in display (1.7). A very brief account of ideas related to general theorems about isomorphisms of C^*-dynamical systems is given in Subsection 7.4.2.

In Section 7.5 we discuss the connection between the existence of an exponential dichotomy for $\{\hat{\varphi}^t\}_{t\geq 0}$ and the existence of quadratic Lyapunov functions. In particular, this gives a generalization of the work of Lewowicz [**Lw**] and also Y. A. Mitropolskii and A. M. Samojlenko. Applications to Riccati, Hamiltonian, and Schrödinger equations, and to geodesic flows are given.

In Chapter 8 we collect the results connected to *exact* Oseledets-Lyapunov exponents and the Multiplicative Ergodic Theorem of V. Oseledets.

In Section 8.1 we describe a connection between this theorem and the Sacker-Sell spectral theory developed in Section 6.3 as well as a connection to the spectral theory for the evolution operators defined in display (1.5). In the finite dimensional case, important results in this direction were obtained by Johnson, Palmer, and Sell [**JPS**], but we follow Latushkin and Stepin [**LSt2, LSt3**] where the infinite dimensional case is considered. After recalling the Multiplicative Ergodic Theorem in Subsection 8.1.1, we show in Subsection 8.1.2 that the measurable spectral subbundles that appear in this theorem are the "footprints" of the continuous Sacker-Sell spectral subbundles that are produced by the spectral subspaces of the evolution operator. This allows us (Subsection 8.1.3) to compute the Sacker-Sell spectrum Σ, the spectrum of E, or the spectrum of Γ in terms of exact Oseledets-Lyapunov exponents. In particular, we show that the logarithm of the spectral radius of E (known also as the general Lyapunov exponent) can be computed as $\log r(E) = \sup\{\lambda_\nu^1 : \nu \in \text{Erg}(\Theta; \varphi)\}$ where λ_ν^1 is the largest Oseledets-Lyapunov exponent for the cocycle $\{\Phi^t\}_{t \in \mathbb{Z}^+}$ with respect to the φ-ergodic measure ν on Θ.

In Section 8.2 we apply our techniques to give a full description of the spectrum of one practically important example of an evolution semigroup: the group generated by the kinematic dynamo operator for an ideally conducting fluid. Suppose that $\Theta \subset \mathbb{R}^3$ and $\{\varphi^t\}_{t \in \mathbb{R}}$ is the flow generated by a divergence free vector field **v** on Θ, which we view as a steady-state solution of the nonlinear Euler equation. Also, consider the cocycle $\{\Phi^t\}_{t \in \mathbb{R}}$ defined by the differential of this flow as follows: $\Phi^t(\theta) = D\varphi^t(\theta)$. The kinematic dynamo operator, Γ_ϵ, is defined on an appropriate

space of divergence free vector fields by
$$\Gamma_\epsilon f = \epsilon \Delta f + \nabla \times (\mathbf{v} \times f)$$
where $\epsilon = 1/\operatorname{Re}_m$ and Re_m denotes the magnetic Reynolds number. The corresponding semigroup $\{e^{t\Gamma_\epsilon}\}_{t\geq 0}$ governs the evolution of induction of the Eulerian fluid; the case $\epsilon = 0$ corresponds to the ideally conducting fluid, see Friedlander and Vishik [**FV2, FV**], Moffatt [**Mo**], and Molchanov, Ruzmaikin, and Sokolov [**MRS**]. This semigroup is the subject of intensive study, especially in connection with the famous "fast" dynamo problem, see [**Arn, AZRS, BaCh, KY**]. The operator $\Gamma = \Gamma_0$ can be recast as the operator given by $\Gamma f = -\langle \mathbf{v}, \nabla \rangle f + \langle f, \nabla \rangle \mathbf{v}$; it generates a group as defined in (1.5). A study of its spectral properties on the space of continuous divergence free vector fields was initiated in the important paper by de la Llave [**dL**], see also [**dL2**]. In Subsection 8.2.1 we consider an n-dimensional manifold Θ and an evolution semigroup acting on the space of continuous divergence free sections of the tangent bundle of Θ as in display (1.5). We prove the spectral mapping property and annular hull property for the group $\{e^{t\Gamma}\}_{t\in\mathbb{R}}$. In particular, we show that the stability of this group is controlled by the spectral bound $s(\Gamma)$. The techniques that we use are a further (much more technical!) development of those in Subsections 6.2.3 and 7.2.2. In Subsection 8.2.2 we give a full description of the spectrum of E^t and Γ on the space of continuous divergence free vector fields. For instance, for an aperiodic flow $\{\varphi^t\}_{t\in\mathbb{R}}$, the spectrum of the operator Γ acting on the space of divergence free vector fields is just one vertical strip formed by filling the gaps in the spectrum of the operator Γ acting on the space of all continuous vector fields. Thus, the spectrum of Γ and E^t in this case can be completely characterized in terms of the Oseledets-Lyapunov exponent for the cocycle $\{\Phi^t\}_{t\in\mathbb{Z}}$. In Subsection 8.2.3 we briefly comment on connections with the fast dynamo problem and the work by S. Friedlander and M. Vishik. We also give a straightforward argument to show that the geodesic flow for a surface of constant negative curvature is produced by a fast dynamo.

In Section 8.3 we turn our attention to a study of the Ruelle transfer operator (1.8). Following [**CL, GL, LSt2**], we study the transfer operator on the space C^0 of continuous sections and the space $C^\mathbf{r}$ of smooth sections of an ℓ-dimensional bundle over an n-dimensional manifold Θ. The main case of interest is where the map $\varphi: \Theta \to \Theta$ is assumed to be expanding (small distances are increased by a factor $\rho > 1$). This setting is due to D. Ruelle. In the celebrated paper [**Ru4**], Ruelle proved the following estimate for the essential spectral radius of the transfer operator $\mathcal{R}$:
$$r_{\mathrm{ess}}(\mathcal{R}; C^\mathbf{r}) \leq \exp\Big(\sup\Big\{h_\nu + \int_\Theta \log \|\Phi(\theta)\| d\nu - \mathbf{r}\log\rho : \nu \in \mathrm{Erg}(\Theta, \varphi)\Big\}\Big).$$
Here h_ν is the entropy of φ with respect to an ergodic measure ν. We prove the following sharp estimate:
$$r_{\mathrm{ess}}(\mathcal{R}; C^\mathbf{r}) \leq \exp\big(\sup\big\{h_\nu + \lambda_\nu - \mathbf{r}\chi_\nu : \nu \in \mathrm{Erg}(\Theta,\varphi)\big\}\big),$$
and we show that, for $\mathbf{r} = 0$, the right hand side gives the spectral radius $r(\mathcal{R}; C^0)$. Here $\lambda_\nu = \lambda_\nu^1$ is the largest Oseledets-Lyapunov exponent for the cocycle $\{\Phi^t\}_{t\in\mathbb{Z}_+}$, while χ_ν is the smallest Oseledets-Lyapunov exponent for the cocycle $\{D\varphi^t\}_{t\in\mathbb{Z}_+}$. In Subsection 8.3.1 we describe the setting and recall the necessary definitions for the topological pressure, etc. In Subsection 8.3.2 we describe the spectral properties of the transfer operator on the space of continuous sections. The main tools here

are the Multiplicative Ergodic Theorem and a variant of the Mather localization described in Subsection 6.2.2. In Subsection 8.3.3 we relate the transfer operator $\mathcal{R}$ on $C^{\mathbf{r}}$ to another transfer operator, $\mathcal{K}$, acting on the space of continuous sections of a certain extended bundle. To this end, we develop a generalization of the Mather localization techniques for the case of the transfer operator. We apply the results of Subsection 8.3.2 to the operator $\mathcal{K}$ to derive the above estimate for $r_{\text{ess}}(\mathcal{R}; C^{\mathbf{r}})$.

CHAPTER 2

Semigroups on Banach Spaces and Evolution Semigroups

2.1. Introduction to semigroups

In this section we will collect some facts from operator theory and the theory of strongly continuous semigroups which will be used in this book. In particular, we will discuss the validity of the spectral mapping theorem in the context of semigroups; that is, we will discuss some conditions that imply that the spectrum of the semigroup can be obtained from the spectrum of the infinitesimal generator of the semigroup by exponentiation. There is a delicate relationship between the spectrum of the semigroup and the spectrum of its generator. In fact, we will give some examples of strongly continuous semigroups for which the spectral mapping theorem fails. In addition, we will define the spectral and growth bounds for a semigroup—the characteristics that determine the stability of a semigroup. Finally, we will give the definition and study the basic properties of hyperbolic semigroups; they are closely related to the study of dichotomy properties of the solutions of autonomous differential equations.

2.1.1. Basics. In this subsection we will recall the concept and the basic properties of strongly continuous semigroups of linear operators on Banach spaces.

2.1.1.1. *Resolvents, spectra, and projections.* Let X denote a complex Banach space with norm $|\cdot|_X$, and let $\mathcal{L}(X)$ denote the set of bounded linear operators on X.

A linear operator A on X with domain $\mathcal{D}(A)$ is called closed if its graph is a closed subset of $X \times X$. The resolvent set $\rho(A)$ of a closed operator A is the set of points $\lambda \in \mathbb{C}$ such that the resolvent operator $R(\lambda; A) := (\lambda - A)^{-1}$ exists, and, hence, is bounded on X. If $\lambda, \mu \in \rho(A)$, then we have the following resolvent identity:

(2.1) $$R(\lambda; A) - R(\mu; A) = (\mu - \lambda) R(\lambda; A) R(\mu; A).$$

The spectrum $\sigma(A)$ of A is defined to be the complement of the resolvent set in the complex plane $\mathbb{C}$. Several subsets of $\sigma(A)$ are classified as follows:

1. The point spectrum, denoted by $\sigma_p(A)$, is the set of all eigenvalues of A; that is, the set of all $\lambda \in \mathbb{C}$ such that $\ker(\lambda - A) \neq \{0\}$.
2. The continuous spectrum, denoted by $\sigma_c(A)$, is the set of all points $\lambda \notin \sigma_p(A)$ such that the range of the operator $\lambda - A$ is dense but not closed in X.
3. The residual spectrum, denoted by $\sigma_r(A)$, is the set of all points $\lambda \notin \sigma_p(A)$ such that the range of the operator $\lambda - A$ is not dense in X.
4. The approximate point spectrum, denoted by $\sigma_{\mathrm{ap}}(A)$, is the set of all points $\lambda \in \mathbb{C}$ such that either $\lambda \in \sigma_p(A)$ or the range of the operator $\lambda - A$ is not closed.

The spectrum can be decomposed as $\sigma(A) = \sigma_p(A) \cup \sigma_c(A) \cup \sigma_r(A)$.

For the approximate point spectrum we have that $\lambda \in \sigma_{\mathrm{ap}}(A)$ if and only if for every $\epsilon > 0$ there exists $x \in \mathcal{D}(A)$ such that $|(\lambda - A)x| < \epsilon |x|$. Equivalently, $\lambda \in \sigma_{\mathrm{ap}}(A)$ if and only if there exists a sequence $\{x_n\}_{n=1}^\infty \subset \mathcal{D}(A)$ with $|x_n| = 1$ and $\lim_{n \to \infty} |(\lambda - A)x_n| = 0$. If $\lambda \in \sigma_{\mathrm{ap}}(A)$ and if the sequence $\{x_n\}_{n=1}^\infty$ is defined as above, then λ is called an approximate eigenvalue and x_n is called an approximate eigenvector of the operator A.

We say that an operator A is *uniformly injective* on X if its lower bound
$$\|A\|_\bullet = \|A\|_{\bullet, X} := \inf\{|Ax| : x \in \mathcal{D}(A), |x| = 1\}$$
is positive.

A bounded linear operator P on X is called a projection if $P^2 = P$. We will always define the complementary projection Q to be the projection on X given by $Q = I - P$. This convention will also be used for projection-valued functions, $Q(\cdot) = I - P(\cdot)$.

Suppose that $\{A(\theta) : \theta \in \Theta\}$ is a family of operators on X indexed by θ in some set Θ; and, for each $\theta \in \Theta$, let $\mathcal{D}(A(\theta))$ denote the domain of the operator $A(\theta)$. We will use the corresponding calligraphic letter, $\mathcal{A}$, to denote the multiplication operator acting on a space of functions from Θ to X with the maximal domain. For instance, if $\Theta = \mathbb{R}$ and the space of functions is $L^p(\mathbb{R}; X)$, $1 \leq p \leq \infty$, then the domain of the operator $\mathcal{A}$ is defined as
$$\mathcal{D}(\mathcal{A}) = \{f \in L^p(\mathbb{R}; X) : f(\theta) \in \mathcal{D}(A(\theta)) \quad \text{a.a.} \quad \theta \in \mathbb{R}, \quad A(\cdot)f(\cdot) \in L^p(\mathbb{R}; X)\},$$
and $(\mathcal{A}f)(\theta) = A(\theta)f(\theta)$.

2.1.1.2. *Semigroups.* We will use the symbol $\mathbb{R}_+$ to denote the set $\{t \in \mathbb{R} : t \geq 0\}$. A family $\{T^t\}_{t \in \mathbb{R}_+}$, also denoted $\{T^t\}_{t \geq 0}$, of bounded linear operators on a Banach space X is called a *strongly continuous semigroup* if $T^0 = I$, $T^{t+s} = T^t T^s$ for all $t, s \in \mathbb{R}_+$, and $\lim_{t \downarrow 0} T^t x = x$ for every $x \in X$. If $\{T^t\}_{t \geq 0}$ is a strongly continuous semigroup, then there exist constants $\omega \geq 0$ and $M \geq 1$ such that $\|T^t\| \leq M e^{\omega t}$ for all $t \in \mathbb{R}_+$, see for example Pazy [**Pz**, p. 4]. In addition, the function $t \mapsto T^t x$ is continuous for every $x \in X$.

If $\{T^t\}_{t \geq 0}$ is a strongly continuous semigroup, then we define $\mathcal{D}$ to be the set of all $x \in X$ such that the limit $\lim_{t \downarrow 0} t^{-1}(T^t x - x)$ exists. The *infinitesimal generator* of the semigroup $\{T^t\}_{t \geq 0}$ is defined to be the operator A on X, with the domain $\mathcal{D}(A) := \mathcal{D}$, given by
$$Ax = \lim_{t \downarrow 0} \frac{T^t x - x}{t}, \quad x \in \mathcal{D}(A).$$
The name "infinitesimal generator" is used because we also have that
$$Ax = \left.\frac{dT^t x}{dt}\right|_{t=0}, \quad x \in \mathcal{D}(A).$$
We will often write $\{e^{tA}\}_{t \in \mathbb{R}_+}$, or $\{e^{tA}\}_{t \geq 0}$, to denote the semigroup generated by A.

If A denotes the infinitesimal generator of the strongly continuous semigroup $\{T^t\}_{t \geq 0}$ and $x \in \mathcal{D}(A)$, then the function $t \mapsto e^{tA} x$ is not only continuous, but also differentiable. In fact, if $x \in \mathcal{D}(A)$, then $e^{tA} x \in \mathcal{D}(A)$ with
$$\frac{d}{dt} e^{tA} x = A e^{tA} x = e^{tA} A x,$$
see [**Pz**, Theorem 2.4].

Consider the abstract Cauchy problem on the Banach space X given by

(2.2) $$\dot{x} = Ax, \quad x(0) = x_0.$$

If A is the infinitesimal generator of a strongly continuous semigroup and $x_0 \in X$, then the function $x(t) := e^{tA}x_0$ is called a *mild* solution of the differential equation (2.2). If $x_0 \in \mathcal{D}(A)$, then $x(t) = e^{tA}x_0$ is a *classical*; that is, differentiable, solution of the differential equation.

EXAMPLE 2.1. Let $X = BUC(\mathbb{R})$ be the space of bounded, uniformly continuous functions with the norm $|x|_X = \sup_{\theta \in \mathbb{R}} |x(\theta)|$, and consider the translation semigroup defined by $(T^t x)(\theta) = x(\theta + t)$, $\theta \in \mathbb{R}$. The infinitesimal generator of this semigroup is the operator given by $Ax = x'$ with domain

$$\mathcal{D}(A) = \{x \in X : x \text{ is differentiable with derivative in } X\}.$$

This fact is also valid when the Banach space X is the space of p-summable functions $L^p(\mathbb{R})$, $1 \leq p < \infty$, or the space $C_0(\mathbb{R})$ of continuous functions such that $\lim_{\theta \to \pm\infty} x(\theta) = 0$ with the *sup*-norm. $\diamond$

A semigroup $\{T^t\}_{t \geq 0}$ is called *uniformly continuous* if $t \mapsto T^t$ is continuous in the uniform operator topology. In fact, $\{T^t\}_{t \geq 0}$ is uniformly continuous provided that

$$\lim_{t \to 0^+} \|T^t - I\|_{\mathcal{L}(X)} = 0.$$

An operator A is the infinitesimal generator of a uniformly continuous semigroup if and only if $A \in \mathcal{L}(X)$. If this is the case, then $\{T^t\}_{t \geq 0}$ extends to a *group* of operators which can be represented as follows:

$$T^t = \sum_{n=0}^{\infty} \frac{(tA)^n}{n!} = e^{tA}, \quad t \in \mathbb{R},$$

where the series converges in the operator norm, see for example [**Pz**, p. 2].

The Hille-Yosida Theorem is the fundamental result in the theory of strongly continuous semigroups; it characterizes the generators of strongly continuous semigroups by means of the resolvent of the operator A, see for example [**Pz**, Theorem 5.3].

THEOREM 2.2 (Hille-Yosida). *A linear operator A is the infinitesimal generator of a strongly continuous semigroup $\{T^t\}_{t \in \mathbb{R}_+}$ satisfying $\|T^t\| \leq Me^{\omega t}$ if and only if*

(i) *A is closed and $\mathcal{D}(A)$ is dense in X; and*
(ii) *the resolvent set $\rho(A)$ contains the ray (ω, ∞) and the resolvent operator $R(\lambda; A) := (\lambda - A)^{-1}$ satisfies the estimates*

$$\|R(\lambda; A)^n\| \leq \frac{M}{(\lambda - \omega)^n},$$

for every $\lambda > \omega$ and every positive integer n.

We remark that, in the context of the Hille-Yosida Theorem, the half plane $\{\lambda : \operatorname{Re} \lambda > \omega\}$ belongs to $\rho(A)$, and the resolvent $R(\lambda; A)$ is given by the Laplace Transform Formula:

(2.3) $$R(\lambda; A)x = \int_0^{\infty} e^{-\lambda t} T^t x \, dt, \quad \operatorname{Re} \lambda > \omega, \quad x \in X.$$

If the operator A generates a strongly continuous semigroup, then $\sigma(A)$ belongs to the half plane $\{\lambda \in \mathbb{C} : \operatorname{Re}\lambda \leq \omega\}$.

REMARK 2.3. A strongly continuous semigroup $\{T^t\}_{t\in\mathbb{R}_+}$ can be extended to a strongly continuous group $\{T^t\}_{t\in\mathbb{R}}$ if and only if there exists a number $t_0 > 0$ such that T^{t_0} is an invertible operator. In that case A and $-A$ are generators of strongly continuous semigroups; and therefore, $\sigma(A)$ is contained in a vertical strip in $\mathbb{C}$, see Pazy [**Pz**, p. 24]. Also, each operator T^t is invertible. ◇

REMARK 2.4. We will often use a *rescaling* construction that we now describe. Let $\{T^t\}_{t\geq 0}$ be a strongly continuous semigroup with the infinitesimal generator A. If $\lambda \in \mathbb{C}$ and α is a positive real number, then the rescaled semigroup $\{e^{\lambda t}T^{\alpha t}\}_{t\geq 0}$ is a strongly continuous semigroup, its infinitesimal generator is $B = \alpha A + \lambda$, and $\mathcal{D}(B) = \mathcal{D}(A)$. Moreover, $\sigma(B) = \alpha\sigma(A) + \lambda$. This construction is convenient when it is desirable to "rescale" the spectrum. If, for example, $z = e^{-\lambda t_0} \in \sigma(T^{t_0})$ for some $\lambda \in \mathbb{C}$ and $t_0 > 0$, then $1 \in \sigma(e^{\lambda t_0}T^{t_0})$. Using this construction, we can work with the statement that "$1 \in \sigma(T^t)$", instead of "$z \in \sigma(T^t)$" with no loss of generality. ◇

REMARK 2.5. A linear subspace $\mathcal{D}_0 \subset \mathcal{D}(A)$ is called a *core* for A if A is the closure of its restriction on $\mathcal{D}_0$. In other words, the graph of A is the closure of the graph of $A|\mathcal{D}_0$; that is, $\mathcal{D}_0$ is dense in $\mathcal{D}(A)$ with respect to the graph-norm $|x|_A := |x| + |Ax|$ on $\mathcal{D}(A)$. If $\mathcal{D}_0$ is dense in $\mathcal{D}(A)$ with respect to the original norm on the Banach space and if $\mathcal{D}_0$ is $\{e^{tA}\}_{t\geq 0}$-invariant, then $\mathcal{D}_0$ is a core for A, see for example Nagel [**Na**, A-II.1.34].

If $\{e^{tA}\}_{t\geq 0}$ and $\{e^{tB}\}_{t\geq 0}$ are strongly continuous semigroups on X that commute; that is, $e^{tA}e^{tB} = e^{tB}e^{tA}$ for all $t \geq 0$, then $U^t := e^{tA}e^{tB}$ defines a strongly continuous semigroup $\{U^t\}_{t\geq 0}$ on X. If C denotes the infinitesimal generator of the semigroup $\{U^t\}_{t\geq 0}$, then $\mathcal{D}_0 := \mathcal{D}(A) \cap \mathcal{D}(B)$ is a core for C and $Cx = Ax + Bx$ for all $x \in \mathcal{D}_0$, see for example [**Na**, A-I.3.8]. ◇

2.1.2. Spectral mapping theorems. In this subsection we start the discussion of the spectral mapping property for strongly continuous semigroups. This property guarantees that the spectrum of the generator of a strongly continuous semigroups is mapped in a "correct" way to the spectrum of the semigroup under the exponential map.

2.1.2.1. *The general spectral mapping theorems on Banach spaces.* We will say that a strongly continuous semigroup $\{e^{tA}\}_{t\in\mathbb{R}_+}$ has the *spectral mapping property* if, for every $t > 0$,

$$(2.4) \qquad \sigma(e^{tA})\setminus\{0\} = e^{t\sigma(A)}.$$

If A is a bounded operator on the Banach space X, then, by Dunford's functional calculus, we even have the identity $\sigma(e^{tA}) = e^{t\sigma(A)}$ and therefore the group generated by A has the spectral mapping property. However, if A generates a semigroup that can not be continued to a group, then the operators e^{tA} are not invertible and the number 0 is in the spectrum of these operators. Since 0 does not belong to the range of the exponential function, it is not possible to have the spectral identity unless we subtract 0 from the spectrum of the operators in the semigroup as in equation (2.4).

The spectral mapping property does not hold in general for unbounded generators. Examples of strongly continuous semigroups for which the spectral mapping property fails will be given below. The next theorem shows that one inclusion required for the spectral mapping property is always true.

THEOREM 2.6 (Spectral Inclusion Theorem). *If $\{e^{tA}\}_{t\in\mathbb{R}_+}$ is a strongly continuous semigroup with infinitesimal generator A and $t \in \mathbb{R}_+$, then*
$$e^{t\sigma(A)} \subseteq \sigma(e^{tA}).$$

PROOF. Denote the semigroup by $\{T^t\}_{t\geq 0}$, and define
$$B_\lambda(t)x := \int_0^t e^{\lambda(t-s)} T^s x\, ds, \quad \lambda \in \mathbb{C}, \quad t > 0.$$
The following identities hold:
$$(\lambda - A)B_\lambda(t)x = e^{\lambda t}x - T^t x, \qquad x \in X;$$
$$B_\lambda(t)(\lambda - A)x = e^{\lambda t}x - T^t x, \qquad x \in \mathcal{D}(A),$$
see for example [**Pz**, Lemma I.2.2]. Let $e^{\lambda t} \in \rho(e^{tA})$. Using the fact that the operators $(e^{\lambda t} - T^t)^{-1}$ and $B_\lambda(t)$ commute, we have that
$$(\lambda - A)B_\lambda(t)(e^{\lambda t} - T^t)^{-1} x = x, \qquad x \in X,$$
$$(e^{\lambda T} - T^t)^{-1}B_\lambda(t)(\lambda - A)x = x, \qquad x \in \mathcal{D}(A),$$
$$B_\lambda(t)(e^{\lambda t} - T^t)^{-1}(\lambda - A)x = x, \qquad x \in \mathcal{D}(A).$$
Therefore, $\lambda \in \rho(A)$ with $R(\lambda; A) = B_\lambda(t)(e^{\lambda t} - T^t)^{-1}$, and $\rho(e^{tA}) \subset e^{t\rho(A)}$. □

Let us define some of the classes of strongly continuous semigroups that have the spectral mapping property. A semigroup $\{T^t\}_{t\geq 0}$ is called *eventually norm continuous* if there exists some $t_0 > 0$ such that the function $t \to T^t$ is norm-continuous for $t > t_0$; *eventually compact* (resp., *eventually differentiable*) if there exists some $t_0 > 0$ such that, for each $t > t_0$, the operator T^t is compact (resp., $t \mapsto T^t x$ is differentiable for each $x \in X$); and *analytic* if the map $t \to T^t \in \mathcal{L}(X)$ has an analytic extension to a sector containing the positive half line. The spectral mapping property for these classes of semigroups is specified in the following theorem. For a proof see, for example, Nagel [**Na**, A-III.6.7].

THEOREM 2.7. *The following classes of strongly continuous semigroups have the spectral mapping property:*
   (i) *eventually norm continuous semigroups;*
   (ii) *eventually compact semigroups;*
   (iii) *eventually differentiable semigroups;*
   (iv) *analytic semigroups;*
   (v) *uniformly continuous semigroups.*

The failure of the spectral mapping property is completely determined by the continuous spectrum. This is the content of the next two theorems, see [**Pz**, Theorems II.2.4-II.2.5].

THEOREM 2.8 (Spectral Mapping Theorem for the Point Spectrum). *If $\{e^{tA}\}_{t\in\mathbb{R}_+}$ is a strongly continuous semigroup with infinitesimal generator A, then*
$$e^{t\sigma_p(A)} \subset \sigma_p(e^{tA}) \subset e^{t\sigma_p(A)} \cup \{0\}.$$
More precisely, if $\lambda \in \sigma_p(A)$, then $e^{\lambda t} \in \sigma_p(e^{tA})$; and, if $e^{\lambda t} \in \sigma_p(e^{tA})$, then there is some integer k such that $\lambda_k := \lambda + 2\pi i k/t \in \sigma_p(A)$.

THEOREM 2.9 (Spectral Mapping Theorem for the Residual Spectrum). *Suppose that $\{e^{tA}\}_{t\in\mathbb{R}_+}$ is a strongly continuous semigroup with infinitesimal generator A. If $\lambda \in \sigma_r(A)$ and $\lambda_n := \lambda + 2\pi i n/t \notin \sigma_p(A)$ for all $n \in \mathbb{Z}$, then $e^{\lambda t} \in \sigma_r(e^{tA})$. If $e^{\lambda t} \in \sigma_r(e^{tA})$, then $\lambda_n = \lambda + 2\pi i n/t \notin \sigma_p(A)$ for all $n \in \mathbb{Z}$; and moreover, there is an integer k such that $\lambda_k \in \sigma_r(A)$.*

2.1.2.2. *Gearhart's spectral mapping theorem.* A semigroup has the spectral mapping property (2.4) if certain conditions are imposed on the growth of the norm of the resolvent of its generator. For a semigroup on a Hilbert space, it turns out that the resolvent $R(\cdot; A)$ of the generator must be bounded along vertical lines. This is the content of the next theorem due to L. Gearhart.

THEOREM 2.10 (Gearhart's Spectral Mapping Theorem). *Let A be the generator of a strongly continuous semigroup $\{T^t\}_{t\geq 0}$ on a Hilbert space X. The spectrum $\sigma(e^{tA}) \setminus \{0\}$ is equal to the set of all complex numbers of the form $e^{\lambda t}$ such that either $\lambda_k := \lambda + 2\pi i k/t$ is in the spectrum of A for some integer k, or the sequence $\{\|R(\lambda_k; A)\|\}_{k \in \mathbb{Z}}$ is unbounded.*

We point out that, by rescaling, the Gearhart Spectral Mapping Theorem is equivalent to the following statement: $1 \in \rho(e^{2\pi A})$ *if and only if*

$$(2.5) \qquad i\mathbb{Z} \subset \rho(A) \quad \text{and} \quad \sup_{k \in \mathbb{Z}} \|(ik - A)^{-1}\| < \infty.$$

We will show later, see page 42, how to derive Gearhart's Spectral Mapping Theorem as a special case of a more general result for semigroups on Banach spaces.

2.1.3. Growth and spectral bounds. If the operator A is the generator of a strongly continuous semigroup, then the "best possible" exponential growth bound for the mild solutions of the associated abstract Cauchy problem $\dot{x} = Ax$, $x(0) = x_0$ is called the growth bound of the semigroup. More precisely, we have the following definition.

DEFINITION 2.11. *Consider a strongly continuous semigroup $\{e^{tA}\}_{t \in \mathbb{R}_+}$ with infinitesimal generator A. The* growth bound *$\omega(A)$ of the semigroup, also denoted $\omega(\{e^{tA}\}_{t \in \mathbb{R}_+})$ or $\omega(e^{tA})$, is defined as follows:*

$$\omega(A) := \inf\{\omega \in \mathbb{R} : \text{ there exists a positive number } M = M(\omega)$$
$$\text{such that } \|e^{tA}\| \leq M e^{\omega t}, \quad \text{for } t \geq 0\}.$$

Using the Gelfand spectral radius formula

$$r(T) = \lim_{n \to \infty} \|T^n\|^{1/n}$$

for a bounded operator T, it follows easily that, for each $t_0 > 0$,

$$(2.6) \qquad \omega(A) = \frac{1}{t_0} \log r(T^{t_0}) = \lim_{t \to \infty} \frac{1}{t} \log \|T^t\|,$$

see for example van Neerven [**vN**, Prop.1.2.2]. In particular, for $t_0 = 1$ and $T := T^1$, we obtain the equation $\omega(A) = \log r(T)$.

Consider a strongly continuous semigroup $\{e^{tA}\}_{t \in \mathbb{R}_+}$ with infinitesimal generator A. The abstract Cauchy problem $\dot{x} = Ax$, $x(0) = x_0$ is called *uniformly exponentially stable* if all mild solutions decay exponentially to zero as $t \to +\infty$ uniformly with respect to initial conditions. In other words, $\|e^{tA}\| \to 0$ as $t \to \infty$. Thus, an abstract Cauchy problem is uniformly exponentially stable if and only if its growth bound $\omega(A)$ is negative. Likewise, the semigroup $\{e^{tA}\}_{t\in\mathbb{R}_+}$ is called *uniformly exponentially stable*, or just *stable*, if $\omega(A) < 0$.

A fundamental problem in the stability theory for autonomous differential equations of the form $\dot{x} = Ax$ is to determine if $\omega(A) < 0$ without constructing the semigroup $\{e^{tA}\}_{t\in\mathbb{R}_+}$; that is, without solving the differential equation. A classical result in this direction is the Lyapunov Stability Theorem: If A is a bounded operator, then the equation $\dot{x} = Ax$ is uniformly exponentially stable if and only if $\sigma(A)$ is contained in the open left half plane $\mathbb{C}_- = \{z \in \mathbb{C} : \operatorname{Re} z < 0\}$.

Let us consider an arbitrary semigroup generator A. In the following definition we introduce the *spectral bound* for the semigroup that is used to specify the position of $\sigma(A)$.

DEFINITION 2.12. Let A be the infinitesimal generator of a strongly continuous semigroup $\{e^{tA}\}_{t\in\mathbb{R}_+}$. The *spectral bound* $s(A)$, or $s(\{e^{tA}\}_{t\in\mathbb{R}_+})$, is defined by
$$s(A) := \sup\{\operatorname{Re} z : z \in \sigma(A)\}.$$

In terms of the spectral bound and the growth bound, Lyapunov's Theorem states that if $A \in \mathcal{L}(X)$ and $s(A) < 0$, then $\omega(A) < 0$. This result is a corollary of the following theorem: If $A \in \mathcal{L}(X)$, then
$$(2.7) \qquad s(A) = \omega(A).$$

Equation (2.7) does not hold, in general, for *unbounded* generators. This fact motivates some of the main results presented in this book as already outlined in the Introduction. For instance, we will show how it is possible to find an operator, Γ, such that $\omega(A) = s(\Gamma)$. In this regard, let us also note that the equality (2.7) is implied by the spectral mapping property (2.4). As a result, the spectral bound equals the growth bound for the various classes of semigroups listed in Theorem 2.7.

The growth bound $\omega(A)$ takes into account the exponential growth of the mild solutions of $\dot{x} = Ax$; to study the rate of growth of *classical* solutions, we define $\omega_1 = \omega_1(A)$ as follows:

(2.8) $\omega_1(A) := \inf\{\omega \in \mathbb{R} :$ there is a positive number $M = M(\omega)$

such that $|e^{tA}x| \leq Me^{\omega t}|x|_A$ for $t \geq 0$ and $x \in \mathcal{D}(A)\}$.

Here, $|x|_A = |x| + |Ax|$ denotes the graph norm of $x \in \mathcal{D}(A)$ and $\mathcal{D}(A)$ the domain of the operator A. In fact $s(A) \leq \omega_1(A) \leq \omega(A)$, see for example van Neerven [**vN**, Thm.1.2.3]. Also, $\omega_1(A)$ is equal to the infimum of all $\omega \in \mathbb{R}$ such that $\{\lambda : \operatorname{Re} \lambda > \omega\} \subset \rho(A)$ and
$$R(\lambda; A)x = \int_0^\infty e^{-\lambda t} e^{tA} x \, dt$$
for all such λ and all $x \in X$.

By the Laplace formula (2.3), it follows easily that $\sup\{\|R(\lambda; A)\| : \operatorname{Re} \lambda = \omega\}$ is finite uniformly for all $\omega \geq \omega'$ as long as $\omega' > \omega(A)$. However, the resolvent could be unbounded for $\operatorname{Re} \lambda = \omega$ with $\omega \in (s(A), \omega(A)]$. The abscissa of uniform boundedness of the resolvent, $s_0 = s_0(A)$, is defined by

(2.9)
$$s_0(A) := \inf\{\omega \in \mathbb{R} : \{\operatorname{Re} \lambda > \omega\} \subset \rho(A) \text{ and } \sup\{\|R(\lambda; A)\| : \operatorname{Re} \lambda > \omega\} < \infty\}.$$

It is true, see van Neerven [**vN**, p. 14], that
$$s(A) \leq \omega_1(A) \leq s_0(A) \leq \omega(A).$$

There are examples of semigroups such that all four numbers above are different, see [**vN**, p. 14]. However, as an application of Gearhart's Theorem 2.10, if A is a generator defined on a Hilbert space, then $s_0(A) = \omega(A)$. We refer to Example 2.23 below for a generator A defined on a Banach space such that $s_0(A) < \omega(A)$.

REMARK 2.13. Consider a strongly continuous semigroup $\{T^t\}_{t\geq 0}$ with the generator A on a Hilbert space X. Assume $\overline{\mathbb{C}_+} \subset \rho(A)$. Due to the analyticity of the resolvent,
$$\sup_{\operatorname{Re}\lambda \geq 0} \|(A-\lambda)^{-1}\| = \sup_{s\in\mathbb{R}} \|(A-is)^{-1}\|.$$
Therefore, Gearhart's Theorem 2.10 implies that $\{e^{tA}\}_{t\geq 0}$ is uniformly exponentially stable if and only if $\mathbb{C}_+ \subset \rho(A)$ and $\lambda \mapsto (A-\lambda)^{-1}$ is a bounded, analytic function on $\mathbb{C}_+$, that is, $\sup_{\operatorname{Re}\lambda > 0} \|(A-\lambda)^{-1}\| < \infty$. ◇

2.1.4. Hyperbolic semigroups. In this subsection we begin to study a concept of hyperbolicity which is more general than uniform exponential stability for semigroups. For differential equations $\dot{x} = Ax$ the corresponding notion is called an exponential dichotomy. Descriptively, a semigroup on X is hyperbolic if X can be decomposed as a direct sum of two subspaces (stable and unstable) such that the semigroup is uniformly exponentially stable for positive time on the stable subspace and uniformly exponentially stable for negative time on the unstable subspace.

An operator $T \in \mathcal{L}(X)$ is called *hyperbolic* if
$$\sigma(T) \cap \mathbb{T} = \emptyset \tag{2.10}$$
where $\mathbb{T} = \{z \in \mathbb{C} : |z| = 1\}$ denotes the unit circle. The spectral Riesz projection P for a hyperbolic operator T is given by the formula
$$P := \frac{1}{2\pi i} \int_{\mathbb{T}} (z-T)^{-1}\, dz.$$
This projection corresponds to the part of the spectrum of T contained in the open unit disk $\mathbb{D}$. We stress the important property that the projection P commutes with T. Using this fact, let us define the restriction $T_P = T|\operatorname{Im} P : \operatorname{Im} P \to \operatorname{Im} P$, and the restriction of T to the image of the complementary projection $T_Q = T|\operatorname{Im} Q : \operatorname{Im} Q \to \operatorname{Im} Q$. Because
$$\sigma(T_P) = \sigma(T) \cap \mathbb{D} \quad \text{and} \quad \sigma(T_Q) = \sigma(T) \cap (\mathbb{C}\setminus\overline{\mathbb{D}}),$$
the operator T_P has spectral radius $r(T_P) < 1$, the operator T_Q is invertible as an operator on $\operatorname{Im} Q$, and $r(T_Q^{-1}) < 1$. Hence, if T is hyperbolic, then there exist constants $\beta > 0$ and $M > 0$ such that, for all integers $n \geq 0$,
$$|T^n x| \leq M e^{-\beta n}|x|, \quad x \in \operatorname{Im} P, \quad \text{and} \quad |T^n x| \geq M^{-1} e^{\beta n}|x|, \quad x \in \operatorname{Im} Q.$$

A strongly continuous semigroup $\{T^t\}_{t\in\mathbb{R}_+}$ on X is called *hyperbolic* if there exists a direct sum decomposition $X = X^s \oplus X^u$ of $\{T^t\}_{t\geq 0}$-invariant closed subspaces X^s (stable subspace) and X^u (unstable subspace) such that, for the restrictions
$$T_s^t := T^t|X^s \quad \text{and} \quad T_u^t := T^t|X^u,$$
the following properties hold:
  (i) $\{T_s^t\}_{t\in\mathbb{R}_+}$ is uniformly exponentially stable; that is, $\omega(\{T_s^t\}_{t\in\mathbb{R}_+}) < 0$;
  (ii) $\{T_u^t\}_{t\in\mathbb{R}_+}$ extends to a strongly continuous *group* $\{T_u^t\}_{t\in\mathbb{R}}$ on X^u such that $\{T_u^{-t}\}_{t\in\mathbb{R}_+}$ is uniformly exponentially stable, that is, $\omega(\{T_u^{-t}\}_{t\in\mathbb{R}_+}) < 0$.

A projection P such that $\operatorname{Im} P = X^s$ and $\operatorname{Im} Q = X^u$, for $Q = I - P$, is called the *splitting projection* for $\{T^t\}_{t \in \mathbb{R}_+}$.

If A is a generator of a hyperbolic semigroup $\{T^t\}_{t \geq 0}$, then the differential equation $\dot{x} = Ax$ admits an exponential dichotomy with projection P such that $\operatorname{Im} P = X^s$ and $\operatorname{Im} Q = X^u$, where $Q = I - P$. In other words, the space of initial data X has a splitting $X = \operatorname{Im} P \oplus \operatorname{Im} Q$ such that the solutions $x(\cdot)$ starting in X^s (resp., in X^u) decay exponentially for $t > 0$ (resp., for $t < 0$) uniformly with respect to the initial data.

EXAMPLE 2.14. Let $A \in \mathcal{L}(X)$ and recall that A generates a group $\{e^{tA}\}_{t \in \mathbb{R}}$. Suppose that $\sigma(A) \cap i\mathbb{R} = \emptyset$, let $\sigma_s = \sigma(A) \cap \mathbb{C}_-$ and $\sigma_u = \sigma(A) \cap \mathbb{C}_+$, and assume that γ is a curve that surrounds σ_s and separates it from σ_u. The Riesz projection

$$P = \frac{1}{2\pi i} \int_\gamma (\lambda - A)^{-1} d\lambda$$

that corresponds to the spectral set σ_s together with its complementary projection $Q = I - P$ induces an A-invariant decomposition of X given by $X = X^s \oplus X^u$. Here $X^s = \operatorname{Im} P$ and $X^u = \operatorname{Im} Q$. This decomposition is also invariant for the corresponding hyperbolic group. Indeed, using Lyapunov's theorem,

$$\omega(A|\operatorname{Im} P) = s(A|\operatorname{Im} P) < 0, \quad \omega(-A|\operatorname{Im} Q) = s(-A|\operatorname{Im} Q) < 0,$$

where vertical bar denotes the restriction of an operator. $\diamond$

The following simple proposition gives a spectral characterization of hyperbolicity.

LEMMA 2.15. *For a strongly continuous semigroup $\{T^t\}_{t \geq 0}$ the following statements are equivalent:*

(a) $\{T^t\}_{t \geq 0}$ *is hyperbolic.*
(b) *There exists a number $t_0 > 0$ such that $\sigma(T^{t_0}) \cap \mathbb{T} = \emptyset$.*

Moreover, if either (a) or (b) holds, then $X = X^s \oplus X^u$ where $X^s = \operatorname{Im} P$, $X^u = \operatorname{Im} Q$, and

$$P = \frac{1}{2\pi i} \int_\mathbb{T} (\lambda - T^{t_0})^{-1} d\lambda$$

is the Riesz projection for T^{t_0} that corresponds to $\sigma(T^{t_0}) \cap \mathbb{D}$. Also, if $\{T^t\}_{t \geq 0}$ is hyperbolic, then $\sigma(T^t) \cap \mathbb{T} = \emptyset$ for every $t > 0$.

PROOF. If (a) holds with $X^s = \operatorname{Im} \hat{P}$ and $X^s = \operatorname{Im} \hat{Q}$ where $\hat{Q} = I - \hat{P}$, then $\sigma(T_s^{t_0}) \subset \mathbb{D}$ and $\sigma(T_u^{t_0}) \subset \mathbb{C} \setminus \overline{\mathbb{D}}$ for every $t_0 > 0$. Indeed, this follows from the inequalities

$$r(T_s^{t_0}) = e^{t_0 \omega(T_s^{t_0})} < 1, \quad r([T_u^{t_0}]^{-1}) = e^{t_0 \omega(T_u^{-t})} < 1.$$

Thus, (b) holds, the Riesz projection P is defined, and

$$Px = \frac{1}{2\pi i} \int_\mathbb{T} (\lambda - T_s^{t_0})^{-1} \hat{P} x \, d\lambda + \frac{1}{2\pi i} \int_\mathbb{T} (\lambda - T_u^{t_0})^{-1} \hat{Q} x \, d\lambda = \hat{P} x$$

by Cauchy's theorem.

If (b) holds, then $\{T^t\}_{t \geq 0}$ is hyperbolic with $X^s = \operatorname{Im} P$ and $X^u = \operatorname{Im} Q$ where $Q = I - P$ and P is the Riesz projection for T^{t_0}. In fact, it follows from the identity

$$PT^t x = \frac{1}{2\pi i} \int_\mathbb{T} (\lambda - T^{t_0})^{-1} T^t x \, d\lambda = \frac{1}{2\pi i} \int_\mathbb{T} T^t (\lambda - T^{t_0})^{-1} x \, d\lambda = T^t P x$$

that X^s and X^u are $\{T^t\}_{t\geq 0}$-invariant.

Using the fact that $r(T_s^{t_0}) < 1$, we have $\omega(T_s^{t_0}) < 0$. To define T_u^t for $t < 0$, we remark that $\sigma(T_u^{t_0}) \subset \mathbb{C}\setminus\overline{\mathbb{D}}$. In particular, for each integer $n \geq 1$, the operator $T_u^{t_0 n}$ is invertible. Also, we have that $r[(T_u^{t_0})^{-1}] < 1$. Define
$$T_u^{st_0} := [T_u^{t_0}]^{-1} T_u^{(1-s)t_0} \quad \text{for} \quad s \in [0,1].$$
For $t = nt_0 + \tau$ with $\tau \in [0, t_0)$, let $T_u^{-t} = [T_u^{t_0 n}]^{-1} T_u^{-\tau}$. Clearly, $\{T_u^t\}_{t\in\mathbb{R}}$ is a group such that
$$\omega(\{T_u^{-t}\}_{t\geq 0}) = \frac{1}{t_0} \log r(T_{u_0}^{-t_0}) < 0.$$
□

It is natural to ask whether the hyperbolicity of a semigroup can be expressed in terms of spectral properties of its generator. The following result in this direction is an application of the Gearhart Spectral Mapping Theorem 2.10.

THEOREM 2.16. *Let X be a Hilbert space. A strongly continuous semigroup $\{e^{tA}\}_{t\in\mathbb{R}_+}$ is hyperbolic if and only if there exists an open strip containing the imaginary axis on which the function $\lambda \mapsto \|R(\lambda; A)\|$ is uniformly bounded.*

If X is a Banach space, the spectral condition in Theorem 2.16 is necessary but not sufficient for the hyperbolicity of the semigroup (see Example 2.22 below). A necessary *and* sufficient condition in this setting was obtained by Kaashoek and Verduyn Lunel [**KL**]. To formulate this condition we need some additional notations.

If $f : \mathbb{R} \to X$ is an integrable function, then the *Cesàro mean* of the function $a \mapsto \int_{-a}^{a} f(s)\,ds$, for $a > 0$, is defined by
$$(C,1) - \int_{-\infty}^{\infty} f(s)\,ds = \lim_{N\to\infty} \int_{-N}^{N} f(s)\left(1 - \frac{|s|}{N}\right) ds$$
$$= \lim_{N\to\infty} \int_0^N \left(\frac{1}{N} \int_{-a}^{a} f(s)\,ds\right) da.$$
The Cesàro mean exists whenever is a certain type of "decay" for the function f at positive and negative infinity. Note that if the principal value
$$\text{PV} \int_{-\infty}^{\infty} f(s)\,ds := \lim_{a\to\infty} \int_{-a}^{a} f(s)\,ds$$
of the integral exists, then it is equal to the Cesàro mean.

Let $\mathcal{S}$ denote the Schwartz space of C^∞ functions $\phi : \mathbb{R} \to \mathbb{R}$ such that, for each pair of nonnegative integers k and q, there is a corresponding number $m_{kq}(\phi) > 0$ with
$$\sup_{\tau\in\mathbb{R}} |\tau^k \phi^{(q)}(\tau)| \leq m_{kq}(\phi). \tag{2.11}$$

In addition, let $\hat{\phi}$ denote the Fourier transform of $\phi \in \mathcal{S}$ and $\|\hat{\phi}\|_1$ its L^1-norm. Also, we let $\phi \mapsto \langle \psi, \phi\rangle$ denote the continuous linear functional (tempered distribution) defined by ψ on $\mathcal{S}$. Using the symbol X^* for the dual space of X, let $\langle x^*, x\rangle$ denote the value of the functional $x^* \in X^*$ on $x \in X$.

For each $x \in X$, $x^* \in X^*$, and $h \in \mathbb{R}$ such that $h + i\mathbb{R} \subset \rho(A)$, define a function $\psi(\cdot, h; x, x^*) : \mathbb{R} \to \mathbb{C}$ by
$$\psi(\tau, h; x, x^*) = \langle x^*, R(h+i\tau; A)x\rangle, \quad \tau \in \mathbb{R}. \tag{2.12}$$

THEOREM 2.17 ([**KL**]). *A strongly continuous semigroup $\{e^{tA}\}_{t\geq 0}$ on a Banach space X is hyperbolic if and only if both of the following conditions hold:*

1. *There exists an open strip containing the imaginary axis on which the function $\|R(\cdot;A)\|$ is uniformly bounded.*
2. *There exists a number $\gamma > 0$ such that*
   (a) *for each $x \in X$ and each h such that $0 < |h| < \gamma$, the Cesàro mean*
   $$(C,1) - \int_{-\infty}^{\infty} R(h+i\tau;A)x\,d\tau$$
   exists; and
   (b) *for each $x \in X$, $x^* \in X^*$, h such that $0 < |h| < \gamma$, and $\phi \in \mathcal{S}$, the function $\psi(\cdot,h;x,x^*)$ defined by equation (2.12) satisfies*
   $$|\langle \psi(\cdot,h;x,x^*),\phi\rangle| \leq K|x|\cdot|x^*|\cdot\|\hat{\phi}\|_1.$$

Moreover, the splitting projection P for the hyperbolic semigroup is given by the formula
$$Px = \frac{1}{2}x + \frac{1}{2\pi i}\,(C,1) - \int_{-\infty}^{\infty} R(h+i\tau;A)x\,d\tau.$$

The next corollary gives a "checkable" sufficient condition for a semigroup $\{e^{tA}\}_{t\geq 0}$ to be hyperbolic.

COROLLARY 2.18 ([**KL**]). *Let A be the infinitesimal generator of a strongly continuous semigroup $\{e^{tA}\}_{t\geq 0}$ on a Banach space X. Suppose that there is a number $\eta > \omega(\{e^{tA}\}_{t\geq 0})$ such that, for each $x \in X$ and each $x^* \in X^*$, the integrals*

$$(2.13) \qquad \int_{-\infty}^{\infty} |R(\eta+i\tau;A)x|^2\,d\tau \quad \text{and} \quad \int_{-\infty}^{\infty} |R(\eta+i\tau;A^*)x^*|_{X^*}^2\,d\tau,$$

are convergent. Then the semigroup $\{e^{tA}\}_{t\geq 0}$ is hyperbolic if and only if both of the following conditions hold:

1. *There exists a constant $\gamma > 0$ such that $\{z \in \mathbb{C} : |\operatorname{Re} z| < \gamma\} \subset \rho(A)$.*
2. $\sup\{\|R(z;A)\| : |\operatorname{Re} z| < \gamma\} < \infty.$

REMARK 2.19. Theorem 2.17 implies the following result (see van Neerven [**vN**, Cor. 4.6.12]): If there is an $\eta > \omega(A)$ such that the integrals (2.13) converge for each $x \in X$ and $x^* \in X^*$, then $\omega(A) = s_0(A)$ where $s_0(A)$ is the abscissa of uniform boundedness $s_0(A)$, see (2.9). If, in addition, the resolvent of A is uniformly bounded in the right half plane, then $\omega(A) < 0$, cf. Remark 2.13. ◇

2.1.5. Examples. In this subsection we will give several examples of strongly continuous semigroups for which the spectral mapping property (2.4) does not hold.

EXAMPLE 2.20. Consider the space $c_0 = c_0(\mathbb{Z}_+)$ of complex sequences that converge to zero and the operator A on c_0 given by $A\{x_n\} = \{inx_n\}$ with maximal domain. The spectrum of A is given by $\sigma(A) = \{in : n \in \mathbb{Z}_+\}$. If t is a positive number such that $2\pi/t$ is irrational, then $\mathbb{T} = \sigma(e^{tA}) = \overline{\{e^{int} : n \in \mathbb{Z}_+\}}$. Thus, we have a very simple example where the spectral mapping property does not hold, but where we do have $\sigma(e^{tA}) = \overline{\exp t\sigma(A)}$, for $t \geq 0$. ◇

EXAMPLE 2.21. For each integer $n \geq 1$, consider the (nilpotent) $n \times n$ matrix
$$A_n = \begin{bmatrix} 0 & 1 & \ldots & 0 \\ 0 & 0 & 1 \ldots & 0 \\ 0 & 0 & 0 \ldots & 1 \\ 0 & 0 & 0 \ldots & 0 \end{bmatrix}$$
on $\mathbb{C}^n$ and note that $\sigma(A_n) = \{0\}$. Also, let X be the Hilbert space $X = \oplus_{n \in \mathbb{N}} \mathbb{C}^n$ with ℓ^2-norm, and define the operator A on X by
$$Ax = \{A_n x_n + 2\pi i n x_n\}_{n \in \mathbb{N}}$$
with maximal domain in X. The operator A generates a semigroup $\{T^t\}_{t \geq 0}$ on X given by
$$T^t x = \{e^{2\pi i n t} e^{tA_n} x_n\}.$$
We claim its spectral and growth bounds do not coincide; in fact,
(2.14) $\qquad s(A) = 0 \quad \text{but} \quad \omega(A) = 1.$

Let $\lambda \in \mathbb{C}$ with $\operatorname{Re} \lambda > 0$, and note that if $R(\lambda; A)x = y$ for $y = (y_n)$, then
$$(\lambda - 2\pi i n - A_n)^{-1} x_n = y_n.$$
Moreover, in view of the fact that A_n is nilpotent, the resolvent $R(\lambda - 2\pi i n; A_n)$ can be represented as a finite sum. Also, we have that $\|A_n\| = 1$. Using these facts and an easy computation, we have the estimate
$$\|R(\lambda; A)\|^2 \leq \sum_{n=1}^{\infty} \|R(\lambda - 2\pi i n; A_n)\|^2 \leq \sum_{n=1}^{\infty} \frac{1}{|\lambda - 1 - 2\pi i n|^2} < \infty.$$
Thus, the spectrum of A lies in the closed left half plane. Because the set $\{2\pi i n : n \in \mathbb{Z}_+\}$ is in the point spectrum of A, we conclude that $s(A) = 0$.

To prove that $\omega(A) = 1$, set $x_n = n^{-\frac{1}{2}}(1, \ldots, 1) \in \mathbb{C}^n$, for $n \in \mathbb{N}$ and $x := (x_n)_{n \in \mathbb{N}}$. Clearly, $|x_n| = 1$. Moreover, using the nilpotence of A_n and standard estimates, it follows that
$$\lim_{n \to \infty} |e^{tA_n} x_n - e^t x_n| = 0,$$
and, therefore, $e^t \in \sigma_{\mathrm{ap}}(e^{tA})$. In particular,
$$\omega(A) = \log r(e^A) \geq 1.$$
Using the fact that $\|e^{tA}\| = e^t$, we also have the inequality $\log r(e^A) \leq 1$; and therefore, $\omega(A) = 1$, as required. $\diamond$

EXAMPLE 2.22. Consider the Banach space
$$X = C_0(\mathbb{R}_+) \cap L^1(\mathbb{R}_+, e^\tau d\tau), \quad \|x\|_X := \|x\|_\infty + \|x\|_1$$
where $\|x\|_\infty := \sup\{|x(\tau)| : \tau \in \mathbb{R}_+\}$ and $\|x\|_1 := \int_0^\infty |x(\tau)| e^\tau d\tau$, and define the operator A on X by
$$Ax = x', \quad \mathcal{D}(A) = \{x \in X : x \in C^1(\mathbb{R}_+), \quad x' \in X\}.$$
The operator A is the infinitesimal generator of the translation semigroup $\{T^t\}_{t \geq 0} = \{e^{tA}\}_{t \geq 0}$ given by $(T^t x)(\tau) = x(\tau + t)$. We claim that
(2.15) $\qquad s(A) = -1 \quad \text{but} \quad \omega(A) = 0.$

By a computation, we find that $\|T^t\|_{\mathcal{L}(X)} = 1$; and, as a result, $\omega(A) = 0$. For each $\lambda \in \mathbb{C}$, define $\epsilon_\lambda : \mathbb{R}_+ \to \mathbb{R}$ by $\tau \mapsto e^{\lambda \tau}$. For $\operatorname{Re} \lambda < -1$, we have that

$\epsilon_\lambda \in \mathcal{D}(A)$ and $A\epsilon_\lambda = \lambda\epsilon_\lambda$; and therefore, $s(A) \geq -1$. The fact that $s(A) = -1$ is an immediate consequence of the following proposition: If $\text{Re}\,\lambda > -1$, then

$$\tag{2.16} R(\lambda; A)x = \int_0^\infty e^{-\lambda t} T^t x \, dt.$$

Indeed, under the hypothesis, if $x \in X$, then $\|x\|_1 < \infty$, and it follows that there is a constant $M > 0$ such that $|x(t)| \leq Me^{-t}$; that is, $\|T^t x\|_1 \leq e^{-t}\|x\|_1$. Using these inequalities and a computation, it is easy to show that if $t \geq 0$, then the function $\tau \mapsto e^{-\lambda t} x(t + \tau)$ is in X. Again, using the same inequalities, it follows that the the improper integral in equation (2.16) is absolutely convergent in the norm of X and that $R(\lambda; A)$ is a bounded operator. Finally, it can be shown that

$$(\lambda - A)R(\lambda; A)x = x, \qquad R(\lambda; A)(\lambda - A)x = x.$$

Thus, λ is in the resolvent set, as required. $\diamond$

EXAMPLE 2.23. We will give an example of a strongly continuous semigroup such that $s_0(A) = \omega_1(A) < \omega(A) < 0$ where $\omega(A)$ is defined in Definition 2.11, $\omega_1(A)$ is defined in (2.8), and $s_0(A)$ in (2.9). For $1 \leq p < q < \infty$, define the Banach space $X = L^p(1, \infty) \cap L^q(1, \infty)$ with the norm $\|x\| = \max\{\|x\|_{L^p}, \|x\|_{L^q}\}$. Also, define a semigroup as follows:

$$(e^{tA}x)(\theta) = x(e^t\theta), \quad \theta > 1, \quad t \geq 0, \quad x \in X.$$

The infinitesimal generator of this semigroup is the operator A defined by

$$(Ax)(\theta) = \theta x'(\theta), \quad \mathcal{D}(A) = \{x \in X : \text{ the function } \theta \mapsto \theta x'(\theta) \text{ belongs to } X\};$$

and, for this semigroup,

$$s_0(A) = \omega_1(A) = s(A) = -\frac{1}{p} < -\frac{1}{q} = \omega(A),$$

see Arendt [**Ar**] or van Neerven [**vN**, Example 1.4.4]. We note that the first two equalities hold for every positive strongly continuous semigroup on a Banach lattice, see [**vN**, Thm. 1.4.1]. $\diamond$

EXAMPLE 2.24. Consider the semigroup $\{e^{tA}\}_{t\geq 0}$ defined by

$$(e^{tA}f)(\theta) = f(e^t\theta),$$

for $\theta \in (1, \infty)$, on the Sobolev space $H^1(1, \infty)$ with norm $\|f\|_{H^1} = (\|f\|_{L^2}^2 + \|f'\|_{L^2}^2)^{1/2}$. We claim that

$$s(A) \leq -\frac{1}{2} < \frac{1}{2} \leq \omega(A).$$

To prove this fact, let us first consider the semigroup $(T_2^t f)(\theta) = f(e^t\theta)$ defined on the space $L^2(1, \infty)$, and let A_2 denote its generator. Using the growth estimate

$$\|T_2^t f\|_{L^2} = \left(\int_{e^t}^\infty |f(\theta)|^2 \, d\theta\right)^{\frac{1}{2}} e^{-t/2} \leq e^{-t/2}\|f\|_{L^2}, \quad f \in L^2(1, \infty),$$

we have that $s(A_2) \leq \omega(A_2) \leq -1/2$. The semigroup $\{e^{tA}\}_{t\geq 0}$ on $H^1(1, \infty)$ is a restriction of the semigroup $\{T_2^t\}_{t\geq 0}$. Thus, A is the part of A_2 on $H^1(1, \infty)$. In fact, if $f \in L^2(1, \infty)$ and $g = R(0, A_2)f$, then

$$g(\theta) = \int_\theta^\infty \frac{f(\tau)}{\tau} \, d\tau, \quad g \in H^1(1, \infty), \quad g'(\theta) = -\frac{1}{\theta}f(\theta),$$

and, in particular, $\mathcal{D}(A_2) \subset H^1(1,\infty)$. Hence, $s(A) \leq s(A_2) \leq -1/2$.

To see that $\omega(A) \geq 1/2$, choose a function $f \in H^1(1,\infty)$ with $\operatorname{supp} f \subset (e^t, \infty)$, $\|f\|_{L^2} \leq 1$, and $\|f'\|_{L^2} = 1$. In particular, $\|f\|_{H^1} \leq \sqrt{2}$. Moreover, for $t > 0$, we have that

$$\|e^{tA}f\|_{H^1} \geq \|(e^{tA}f)'\|_{L^2} = \|e^t f'(e^t \cdot)\|_{L^2} = e^t \left(\int_1^\infty |f'(e^t\theta)|^2 d\theta\right)^{1/2}$$

$$= e^t e^{-\frac{t}{2}} \left(\int_{e^t}^\infty |f'(\theta)|^2 d\theta\right)^{1/2} = e^{t/2}\|f'\|_{L^2} = e^{t/2}.$$

$\diamondsuit$

EXAMPLE 2.25. We will give an example of a group $\{e^{tA}\}_{t\in\mathbb{R}}$ on the Hilbert space $L^2(0,1)$ such that $\sigma(A) = \emptyset$ but

(2.17) $\qquad \sigma(e^{tA}) = \sigma_{\mathrm{ap}}(e^{tA}) = \{z : e^{-|t|\pi/2} \leq |z| \leq e^{|t|\pi/2}\}, \quad t \neq 0.$

This example, originating in the book of Hille and Phillips [**HPh**, Sec. 23.16], is based on the concept of fractional integration. Later, Henry [**He2**] proved that equality (2.17) holds, and that the resolvent $R(\lambda, A)$ of the generator A in the Hille-Phillips example behaves as follows: For $\operatorname{Re}\lambda \in (-\pi/2, \pi/2)$, the norm $\|R(\lambda, A)\|$ is bounded as $\operatorname{Im}\lambda \to +\infty$ and unbounded as $\operatorname{Im}\lambda \to -\infty$. The Gearhart Spectral Mapping Theorem 2.10 explains the above formula for $\sigma(e^{tA})$. We will briefly sketch Henry's argument that is used to show (2.17). The estimates for the resolvent are more involved and are omitted. This semigroup is also studied by Arendt, El Mennaoui, and Hieber in [**AEMH**].

Let us define a semigroup J^t acting on $L^2(0,1)$ by

$$(J^t f)(y) = \int_0^y \frac{(y-s)^{t-1}}{\Gamma(t)} f(s)\,ds, \quad y \geq 0, \quad t \in \mathbb{C}, \quad \operatorname{Re} t \geq 0$$

where Γ denotes the Gamma function. Clearly, $J^t f$ can be written as a convolution

$$J^t f = K_t * f = \int_0^1 K_t(\cdot - s) f(s)\,ds$$

where $K_t(s) = s^{t-1}/\Gamma(t)$, for $s > 0$, and $K_t(s) = 0$, for $s \leq 0$. Using the estimate $\|J^t\|_{\mathcal{L}(L^2)} \leq \|K_t\|_{L^1}$, it follows that

$$\|J^t\|_{\mathcal{L}(L^2(0,1))} \leq \frac{1}{\operatorname{Re} t \cdot |\Gamma(t)|},$$

for $\operatorname{Re} t > 0$. Hence, the function $t \mapsto J^t f \in L^2(0,1)$ is continuous for $\operatorname{Re} t > 0$.

Let B denote the generator for $\{J^t\}_{t\geq 0}$. Because $\lim_{n\to\infty} |\Gamma(nt+1)|^{1/n} = \infty$, we conclude that the spectral radius $r(J^t) = 0$, for $t > 0$. Also, by the Spectral Inclusion Theorem 2.6, we have $e^{t\sigma(B)} \subset \sigma(J^t) = \{0\}$. Since the image of the exponential function does not contain zero, $\sigma(B) = \emptyset$.

In addition to the estimate given above, the more delicate estimate

(2.18) $\qquad \|J^t\|_{\mathcal{L}(L^2(0,1))} \leq \frac{e^{|\operatorname{Im} t|\pi/2}}{(1 - 2\operatorname{Re} t)^{1/2}}, \quad 0 \leq \operatorname{Re} t < 1/2$

is given in [**He2**, p. 63-64].

For $t = i\tau$, with $\operatorname{Re} t = 0$, consider $\{J^{i\tau}\}_{\tau\in\mathbb{R}}$, and let A denote the generator of this semigroup; that is, $J^{i\tau} = e^{\tau A}$ where $A = iB$. Then $\sigma(A) = \emptyset$, and we claim that (2.17) holds.

By (2.18), it follows that $r(J^{i\tau}) \leq e^{|\tau|\pi/2}$ and $\sigma(e^{tA})$ is contained in the set $\mathbb{A}$ defined by the right hand side of (2.17). To see that $\sigma(e^{tA})$ contains the set $\mathbb{A}$, we will construct approximate eigenfunctions for e^{tA} as follows: For $p \in \mathbb{C}$, with $\operatorname{Re} p > 0$, let
$$g_p(y) = y^p \sqrt{2\operatorname{Re} p + 1}, \quad \text{for } 0 \leq y \leq 1,$$
and note that $\|g_p\|_{L^2(0,1)} = 1$. Using Euler's Beta function
$$B(z_1, z_2) = \int_0^1 y^{z_1-1}(1-y)^{z_2-1}dy, \quad B(z_1, z_2) = \Gamma(z_1)\Gamma(z_2)/\Gamma(z_1+z_2),$$
it can be shown that
$$(2.19) \quad J^{i\tau}g_p(y) - \frac{\Gamma(p+1)}{\Gamma(p+1+i\tau)}g_p(y) = \frac{\Gamma(p+1)}{\Gamma(p+1+i\tau)}(y^{i\tau} - 1)g_p(y), \quad y \in (0,1).$$
Also, $\|(y^{i\tau} - 1)g_p\|_{L^2(0,1)} \to 0$ as $\operatorname{Re} p \to +\infty$.

Choose $p = r_n e^{i\theta}$, where $r_n = \exp(\alpha + 2\pi n/|\tau|)$, with α and $|\theta| < \pi/2$ fixed, and let $z = e^{-\tau(\theta+i\alpha)}$ denote a point in $\mathcal{A}$. Using some properties of the Gamma function, it can be shown that
$$\frac{\Gamma(p+1)}{\Gamma(p+1+i\tau)} = p^{-i\tau}(1 + O(|p|^{-1})) = e^{-\tau\theta - it\alpha}(1 + O(r_n^{-1}))$$
$$= z(1 + O(r_n^{-1})).$$

Therefore, by equation (2.19), we have $\|e^{tA}g_p - zg_p\| \to 0$ as $n \to \infty$, and (2.17) is proved. ◇

EXAMPLE 2.26. The following example is due to Renardy [**Rn**]. It shows that the strict inequality $s(A) < \omega(A)$ can hold for the generator of a semigroup that appears as a result of a first order perturbation of the two-dimensional wave equation.

Consider the solutions of the partial differential equation
$$u_{tt} = u_{ss} + u_{yy} + e^{iy}u_s$$
with periodic boundary conditions; that is, we consider solutions of the form $u = u(t, s, y)$ that are 2π-periodic in the space variables $(s, y) \in \mathbb{R}^2$ for $t \geq 0$. The above equation can be written as an evolution equation $\dot{x} = Ax$ for $x = (u, v)^T$, $v = u_t$ where
$$A = \begin{bmatrix} 0 & I \\ \Delta + e^{iy}d/ds & 0 \end{bmatrix}, \quad \Delta u = u_{ss} + u_{yy}, \quad \frac{d}{ds}u = u_s.$$

The operator A is the infinitesimal generator of a strongly continuous semigroup on the space $X = H^1 \times L^2$ of 2π-periodic functions, and the resolvent of A is a compact operator. Hence, $\sigma(A) = \sigma_p(A)$. We claim that $\sigma(A) \subset i\mathbb{R}$; and therefore, $s(A) = 0$. However, for this example, we also have that $\omega(A) \geq 1/2$.

To compute $\sigma(A)$, separate the variables to obtain $u(s, y) = w(y)e^{ins}$ and, in turn, the spectral problem
$$(\lambda^2 + n^2)w = w'' + ine^{iy}w.$$
Also, use the Fourier series representation $w = \sum_m w_m e^{imy}$ to find the recurrence relation
$$(\lambda^2 + n^2 + m^2)w_m = inw_{m-1}.$$

If $w_M \neq 0$ for some $M \in \mathbb{Z}$ and $\lambda^2 + n^2 + m^2 \neq 0$ for all $m \leq M$, then w_{M-1}, w_{M-2}, ... are all nonzero and $|w_m| \to \infty$ as $m \to -\infty$, contrary to the fact that the sequence $\{w_m\}$ is in ℓ^2. Thus, we conclude that $\lambda = \pm i\sqrt{n^2 + m^2}$ for some $m \in \mathbb{Z}$. It follows that $s(A) = 0$.

To prove the inequality $\omega(A) \geq 1/2$, it suffices to show that $\|R(\lambda; A)\|$ is unbounded on $\{\lambda : \operatorname{Re}\lambda = 1/2\}$. In fact, we will produce a sequence λ_n with $\operatorname{Re}\lambda_n \to 1/2$ and $x_n = (u_n, v_n) \in X$ such that $\|(A - \lambda)x_n\|_X / \|x_n\|_X$ approaches 0 as $n \to \infty$.

Select $\epsilon_n > 0$ such that $\epsilon_n \to 0$ and $\epsilon_n^2 n \to \infty$ as $n \to \infty$. Also, select a C^∞-function β with $\operatorname{supp}\beta \subset (-1, 1)$, and consider the function given by $\tilde{\alpha}_n(y) = \beta(y/\epsilon_n)$. For $\epsilon_n < \pi$ we have $\operatorname{sup}\tilde{\alpha}_n \subset (-\pi, \pi)$. Let α_n denote the 2π-periodic continuation of $\tilde{\alpha}_n$ and define
$$\lambda_n = (-n^2 + in)^{1/2}, \quad u_n(s, y) = e^{ins}\alpha_n(y), \quad v_n = \lambda_n u_n.$$
Now $\operatorname{Re}\lambda_n \to 1/2$ as $n \to \infty$, and we claim that
$$\|(A - \lambda_n)(u_n, v_n)\|_{H^1 \times L^2} / \|(u_n, v_n)\|_{H^1 \times L^2} \to 0 \quad \text{as} \quad n \to \infty.$$
To estimate $\|(u_n, v_n)\|_X$ from below, let us note that
$$\|u_n\|_{L^2}^2 = 2\pi \int_{-\pi}^{\pi} \beta^2(y/\epsilon_n)\,dy = 2\pi\epsilon_n \|\beta\|_{L^2}^2,$$
$$\|\nabla u_n\|_{L^2}^2 = \int_{-\pi}^{\pi}\int_{-\pi}^{\pi} [n^2\beta'^2(y/\epsilon_n) + \epsilon_n^{-2}\beta^2(y/\epsilon_n)]\,ds\,dy$$
$$= 2\pi(n^2\epsilon_n\|\beta\|_{L^2}^2 + \epsilon_n^{-1}\|\beta'\|_{L^2}^2).$$

By the choice of ϵ_n, the function $n \mapsto \|u_n\|_{H^1}$ is asymptotic to $n\epsilon_n^{1/2}$ as $n \to \infty$. On the other hand, for $v_n = \lambda_n u_n$, we have $(\lambda_n - A)(u_n, v_n) = (0, \lambda_n^2 u_n - (\Delta + e^{iy}d/ds)u_n)$; and, as a result,
$$\lambda_n^2 u_n - \frac{\partial^2 u_n}{\partial s^2} - \frac{\partial^2 u_n}{\partial y^2} - e^{iy}\frac{\partial u_n}{\partial s} = e^{ins}[\epsilon_n^{-2}\beta''(y/\epsilon_n) + in(e^{iy} - 1)\beta(y/\epsilon_n)].$$
The square of the L^2-norm of the last expression is bounded above by
$$2\pi\epsilon_n^{-3}\|\beta''\|_{L^2}^2 + 2\pi n^2\epsilon_n\|(e^{i\epsilon_n \cdot} - 1)\beta\|_{L^2}^2.$$
Thus, $\|(\lambda_n - A)x_n\|$ is asymptotic to $n\epsilon_n^{3/2}$ as $n \to \infty$. This proves our claim. $\diamond$

If A is the infinitesimal generator of a positive semigroup on $L^p(\Omega)$, then $s(A) = \omega(A)$ by a theorem of Weis, see [**We**]. This fact might lead to the conjecture that the hyperbolicity of a positive semigroup on $L^p(\Omega)$ is also determined by $\sigma(A)$. For example, we might conjecture that $(\xi + i\mathbb{R}) \cap \sigma(A) = \emptyset$ if and only if $e^{t\xi \mathbb{T}} \cap \sigma(e^{tA}) = \emptyset$ for all $\xi \in \mathbb{R}$. The following example due to Montgomery-Smith [**Mt**] shows that this is not true in general. In fact, there is a positive semigroup $\{e^{tA}\}_{t \geq 0}$ on an L^2-space such that $(1 + i\mathbb{R}) \cap \sigma(A) = \emptyset$ but $e^{2\pi} \in \sigma(e^{2\pi A})$.

EXAMPLE 2.27. Consider the matrices A_n defined in Example 2.21. One can check directly, see [**Mt**] for details, that e^{tA_n} is a positive operator (in the sense of lattices) and $\|e^{tA_n}\| \leq e^{t\|A_n\|} = e^t$. Consider the positive semigroup $\{e^{tB_n}\}_{t \geq 0}$ acting on $L^2([0, 2\pi])$ by
$$e^{tB_n} f(\theta) = (e^{4t} - 1)\frac{1}{2\pi}\int_0^{2\pi} f(\tau)\,d\tau + f(\theta + nt \pmod{2\pi}).$$

The infinitesimal generator of this semigroup is the closure of the operator B_n given by

$$(B_n f)(\theta) = \frac{4}{2\pi} \int_0^{2\pi} f(\tau)\, d\tau + n\frac{d}{d\theta} f(\theta),$$

where $d/d\theta$ is taken with the periodic boundary conditions. Note that $\|e^{tB_n}\| \leq e^{4t}$.

Next, consider the positive semigroup $\{e^{tC_n}\}_{t\geq 0} = \{e^{tB_n} \otimes e^{tA_n}_{t\geq 0}\}$ generated on the space

$$X_n = L^2([0, 2\pi]) \otimes \mathbb{C}^n = L^2([0, 2\pi] \times \{1, 2, \ldots, n\})$$

by $C_n = B_n \otimes I + I \otimes A_n$, and note that $\|e^{tC_n}\| \leq e^{5t}$.

Consider a typical element of X_n, say $f(\theta) = \sum_{k=-\infty}^{\infty} x_k e^{ik\theta}$ where $x_k \in \mathbb{C}^n$ and $\|f\|_{X_n}^2 = 2\pi \sum_k |x_k|_{\mathbb{C}^n}^2$. If $\lambda \neq 4$ and $\lambda \notin n\mathbb{Z}\setminus\{0\}$, then $\lambda \in \rho(C_n)$ and

$$(\lambda - C_n)^{-1} f(\theta) = (\lambda - 4 - A_n)^{-1} x_0 + \sum_{k \neq 0} (\lambda - ikn - A_n)^{-1} x_k e^{ik\theta}.$$

It follows that

$$\|(\lambda - C_n)^{-1}\| = \max\{\|(\lambda - 4 - C_n)^{-1}\|, \sup_{k \neq 0} \|(\lambda - ikn - C_n)^{-1}\|\}.$$

Therefore, if $\operatorname{Re}\lambda = 1$ and $|\lambda| \leq n - 2$, then $\|(\lambda - C_n)^{-1}\| \leq 1$. Also, if $\lambda = 1 + in$, then $\|(\lambda - C_n)^{-1}\| \geq n^{1/2}$.

Now consider the semigroup $\{e^{tA}\}_{t\geq 0} = \{\oplus_{n=1}^{\infty} e^{tC_n}\}_{t\geq 0}$ acting on the space

$$X = \oplus_{n=1}^{\infty} X_n = L^2(\bigvee_{n=1}^{\infty} ([0, 2\pi] \times \{1, 2, \ldots, n\})).$$

Note that $\|e^{tA}\| \leq e^{5t}$ and A is the closure of $\oplus_{n=1}^{\infty} C_n$. Thus, $\rho(A)$ is the set of all $\lambda \in \mathbb{C}$ such that

$$\|(\lambda - A)^{-1}\| = \sup_{n \geq 1} \|(\lambda - C_n)^{-1}\| < \infty$$

and, in particular, $\sigma(A) \subseteq \{z : |z - 4| \leq 1\} \bigcup i\mathbb{Z}\setminus\{0\}$. Thus, if $\operatorname{Re}\lambda = 1$, then $\lambda \in \rho(A)$. But, also

$$\sup_{\lambda \in 1 + i\mathbb{Z}} \|(\lambda - A)^{-1}\| = \infty.$$

By the Gearhart Spectral Mapping Theorem and a rescaling (see (2.5)), it follows that $e^{2\pi} \in \sigma(e^{2\pi A})$. $\diamond$

2.2. Evolution semigroups and hyperbolicity

Let X denote a Banach space and $\{e^{tA}\}_{t \geq 0}$ a strongly continuous semigroup on X. In this section we show that the hyperbolicity of the strongly continuous semigroup can be characterized in terms of an associated *evolution* semigroup, $\{E^t\}_{t \geq 0}$.

We will first consider the evolution semigroup $\{E^t\}_{t \geq 0}$, with infinitesimal generator Γ_{per}, defined on a space of X-valued periodic functions. The relationship between the spectra of e^{tA} and Γ_{per} is stated in Theorem 2.30. The main idea here is to construct an approximate eigenfunction for Γ_{per} whenever an approximate eigenvector for e^{tA} is given (Lemma 2.29). Theorem 2.30 allows us to extend the Gearhart Spectral Mapping Theorem 2.10 to the Banach-space setting (see Theorem 2.31).

For our characterization of hyperbolicity, we will consider an associated evolution semigroup $\{e^{t\Gamma}\}_{t \geq 0}$ acting on a space of X-valued functions defined on $\mathbb{R}$. In fact, we will show that the hyperbolicity of $\{e^{tA}\}_{t\geq 0}$ on X is equivalent to the hyperbolicity of $\{e^{t\Gamma}\}_{t\geq 0}$ or, equivalently, to the invertibility of Γ (see Theorem 2.39). We give two proofs of this result. The first proof uses a reduction to the case of an evolution semigroup defined on a space of periodic X-valued functions. The second proof (see Lemma 2.41) is more explicit. It gives a direct construction of approximate eigenfunctions for Γ whenever an approximate eigenvector for $\{E^t\}_{t\geq 0}$ is given.

Finally, we consider evolution semigroups $\{E_+^t\}_{t\geq 0}$ on spaces of functions defined on the positive half line $\mathbb{R}_+$. We will see that the invertibility of the corresponding generator Γ_+ is equivalent to the *stability* of the underlying semigroup. In addition, we will compute the norms of the operators $\Gamma_{\mathrm{per}}^{-1}$, Γ_+^{-1}, and Γ^{-1}.

2.2.1. Evolution semigroups on periodic functions. In this subsection we consider evolution semigroups on spaces of X-valued periodic functions. In Subsection 2.2.1.1 we describe a connection between the spectrum of the underlying strongly continuous semigroup on X and the spectrum of the corresponding evolution semigroup on the space of periodic functions. In Subsection 2.2.1.2 we prove a generalization of Gearhart's Spectral Mapping Theorem that holds for an arbitrary strongly continuous semigroup on a Banach space.

2.2.1.1. *The spectrum*. Let $\mathcal{F}_{\mathrm{per}}$ denote the space $L^p([0, 2\pi]; X)$, $1 \leq p < \infty$, or the space $C_{\mathrm{per}}([0, 2\pi]; X)$ that consists of all continuous functions $f : [0, 2\pi] \to X$ such that $f(0) = f(2\pi)$ with supremum norm. Of course, these spaces can also be viewed as spaces of 2π-periodic functions defined on the entire real line. The evolution semigroup $\{e^{t\Gamma_{\mathrm{per}}}\}_{t\geq 0}$ on $\mathcal{F}_{\mathrm{per}}$ is defined by

$$(2.20) \qquad (e^{t\Gamma_{\mathrm{per}}}f)(\theta) = e^{tA}f([\theta - t](\mathrm{mod}\,2\pi)), \qquad \theta \in [0, 2\pi].$$

For convenience, we consider 2π-periodic functions, but every other (nonzero) choice of the period would work just as well.

Note that the semigroup $\{e^{t\Gamma_{\mathrm{per}}}\}_{t\geq 0}$ can be represented as the product of two commuting semigroups, namely, the semigroups defined by $(V_{\mathrm{per}}^t f)(\theta) = f([\theta - t](\mathrm{mod}\,2\pi))$ and $(e^{tA}f)(\theta) = e^{tA}f(\theta)$. Hence, the generator Γ_{per} is the closure of the operator

$$(2.21) \qquad (\Gamma'_{\mathrm{per}}f)(\theta) = -\frac{d}{d\theta}f(\theta) + Af(\theta)$$

where Γ'_{per} is defined on the core $\mathcal{D}(\Gamma'_{\mathrm{per}})$ of Γ_{per} (see [**Na**, p. 24] and Remark 2.5 above). For the case $\mathcal{F}_{\mathrm{per}} = L^p([0, 2\pi]; X)$, let $\mathcal{D}(\Gamma'_{\mathrm{per}}) := \{f : [0, 2\pi] \to \mathcal{D}(A) : f \in \mathcal{F}_{\mathrm{per}}$ is absolutely continuous, $f(0) = f(2\pi)$, $\frac{d}{d\theta}f \in \mathcal{F}_{\mathrm{per}}$, and $\mathcal{A}f \in \mathcal{F}_{\mathrm{per}}\}$ while, for $\mathcal{F}_{\mathrm{per}} = C_{\mathrm{per}}([0, 2\pi]; X)$, let $\mathcal{D}(\Gamma'_{\mathrm{per}}) := \{f : [0, 2\pi] \to \mathcal{D}(A) : f \in \mathcal{F}_{\mathrm{per}}$ is differentiable, $\frac{d}{d\theta}f \in \mathcal{F}_{\mathrm{per}}$, and $\mathcal{A}f \in \mathcal{F}_{\mathrm{per}}\}$. The derivative $d/d\theta$ is taken in the strong sense in X where we have $\mathcal{D}(\Gamma'_{\mathrm{per}}) = \mathcal{D}(-d/d\theta) \cap \mathcal{D}(\mathcal{A})$.

Let us begin with a simple observation on a spectral symmetry for Γ.

PROPOSITION 2.28. *The spectrum $\sigma(\Gamma_{per})$ is invariant with respect to vertical translations by ik, $k \in \mathbb{Z}$.*

PROOF. Define $(L_k f)(\theta) = e^{ik\theta}f(\theta)$, $k \in \mathbb{Z}$. Then,

$$(e^{t\Gamma_{\mathrm{per}}}L_k f)(\cdot) = e^{ik(\cdot - t)}e^{tA}f(\cdot - t) = L_k e^{-ikt}(e^{t\Gamma_{\mathrm{per}}}f)(\cdot)$$

implies $L_k^{-1}(z - e^{t\Gamma_{\mathrm{per}}})L_k = z - e^{-ikt}e^{t\Gamma_{\mathrm{per}}}$ and $L_k^{-1}(z - \Gamma_{\mathrm{per}})L_k = z + ik - \Gamma_{\mathrm{per}}$. □

The next lemma is the heart of the proof of Theorem 2.30 that follows. Here we construct an approximate eigenfunction for Γ on $\mathcal{F}_{\text{per}}$ that corresponds to an approximate eigenvector for $e^{2\pi A}$ on X.

LEMMA 2.29. *If $1 \in \sigma_{ap}(e^{2\pi A})$, then $0 \in \sigma_{ap}(\Gamma_{per})$.*

PROOF. Fix $m \in \mathbb{N}$, $m \geq 2$. Since $1 \in \sigma_{\text{ap}}(e^{2\pi A})$, there is some $x \in X$ such that $|x|_X = 1$ and $|x - e^{2\pi A}x|_X < \frac{1}{m}$. Also, for this choice of x, we have $|e^{2\pi A}x|_X \geq 1 - \frac{1}{m}$.

Let $\alpha : [0, 2\pi] \to [0, 1]$ denote a smooth function with bounded derivative such that $\alpha(\theta) = 0$ for $\theta \in [0, \frac{2\pi}{3}]$, and $\alpha(\theta) = 1$ for $\theta \in [\frac{4\pi}{3}, 2\pi]$. Define $g : [0, 2\pi] \to X$ by

$$(2.22) \qquad g(\theta) = [1 - \alpha(\theta)]e^{(2\pi+\theta)A}x + \alpha(\theta)e^{\theta A}x.$$

Note that $g \in \mathcal{F}_{\text{per}}$, $g(0) = g(2\pi) = e^{2\pi A}x$, and

$$(e^{t\Gamma_{\text{per}}}g)(\theta) = [1 - \alpha(\theta - t(\text{mod } 2\pi))]e^{(2\pi+\theta)A}x + \alpha(\theta - t(\text{mod } 2\pi))e^{\theta A}x.$$

Since α is continuously differentiable with bounded derivative, it follows that $g \in \mathcal{D}(\Gamma_{\text{per}})$ and

$$(2.23) \qquad (\Gamma_{\text{per}}g)(\theta) = \alpha'(\theta)e^{\theta A}[e^{2\pi A}x - x].$$

Define

$$a = \max\{|\alpha'(\theta)| : \theta \in [0, 2\pi]\}, \qquad b = \max\{\|e^{\theta A}\| : \theta \in [0, 2\pi]\},$$

and note that if $\theta \in [0, 2\pi]$, then $|e^{2\pi A}x| = |e^{(2\pi-\theta)A}e^{\theta A}x| \leq b|e^{\theta A}x|$.

If $\mathcal{F}_{\text{per}} = L^p([0, 2\pi]; X)$, then

$$(2.24) \qquad \|\Gamma_{\text{per}}g\|_{L^p([0,2\pi];X)} \leq (2\pi)^{1/p}\frac{ab}{m}$$

and

$$\|g\|^p_{L^p([0,2\pi];X)} \geq \int_{\frac{4\pi}{3}}^{2\pi} |e^{\theta A}x|^p \, d\theta \geq \frac{2\pi}{3}b^{-p}|e^{2\pi A}x|^p \geq \frac{2\pi}{3}b^{-p}(1 - \frac{1}{m})^p.$$

In view of these inequalities, we have the estimate

$$(2.25) \qquad \|\Gamma_{\text{per}}g\|_{L^p([0,2\pi];X)} \leq (2\pi)^{1/p}\frac{ab}{m} \leq 3^{1/p}ab^2\|g\|_{L^p([0,2\pi];X)} \cdot \frac{1}{m-1}.$$

Since (2.25) holds for all $m \geq 2$, it follows that $0 \in \sigma_{\text{ap}}(\Gamma_{\text{per}})$.

If $\mathcal{F}_{\text{per}} = C_{\text{per}}([0, 2\pi]; X)$, then

$$(2.26) \quad \|\Gamma_{\text{per}}g\|_{C_{\text{per}}([0,2\pi],X)} \leq \frac{ab}{m}, \qquad \|g\|_{C([0,2\pi],X)} \geq |g(0)|_X = |e^{2\pi A}x| \geq 1 - \frac{1}{m}$$

and we have the estimate

$$(2.27) \qquad \|\Gamma_{\text{per}}g\|_{C_{\text{per}}([0,2\pi];X)} \leq \frac{ab}{m-1}\|g\|_{C([0,2\pi];X)}.$$

Since (2.27) holds for all $m \geq 2$, it follows that $0 \in \sigma_{\text{ap}}(\Gamma_{\text{per}})$. □

Recall that, for a strongly continuous semigroup $\{e^{tA}\}_{t \geq 0}$ on X, the condition $0 \in \rho(A)$ *does not* imply $1 \in \rho(e^{2\pi A})$ in general. Heuristically, if $1 \in \sigma_{\text{ap}}(e^{2\pi A})$, then the space X could be too "small" to contain an approximate eigenvector so that $0 \in \sigma(A)$. The following theorem shows that the space $\mathcal{F}_{\text{per}}$ is already "large" enough; that is, if $\Gamma = $ closure $(-d/d\theta + A)$, then the condition $0 \in \rho(\Gamma)$ on $\mathcal{F}_{\text{per}}$ *does imply* that $1 \in \rho(e^{2\pi A})$ on X.

THEOREM 2.30. *Let $\mathcal{F}_{per}$ denote one of the Banach spaces $L^p([0, 2\pi]; X)$, $1 \le p < \infty$, or $C_{per}([0, 2\pi]; X)$. The following statements are equivalent:*

1) $1 \in \rho(e^{2\pi A})$ in X.
2) $1 \in \rho(e^{2\pi \Gamma_{per}})$ in $\mathcal{F}_{per}$.
3) $0 \in \rho(\Gamma_{per})$ in $\mathcal{F}_{per}$.

PROOF. 1) $\Rightarrow$ 2). Since $(e^{2\pi \Gamma_{per}} f)(\theta) = e^{2\pi A} f(\theta)$, we have

$$\sigma(e^{2\pi A}; X) = \sigma(e^{2\pi \Gamma_{per}}; \mathcal{F}_{per}).$$

Also, note that $\sigma_r(e^{2\pi A}; X) = \sigma_r(e^{2\pi \Gamma_{per}}; \mathcal{F}_{per})$.

2) $\Rightarrow$ 3). This implication follows from the Spectral Inclusion Theorem 2.6: $e^{2\pi \sigma(\Gamma_{per})} \subset \sigma(e^{2\pi \Gamma_{per}})$.

3) $\Rightarrow$ 1). Suppose that $0 \in \rho(\Gamma_{per}; \mathcal{F}_{per})$, but

$$1 \in \sigma(e^{2\pi A}; X) = \sigma_{ap}(e^{2\pi A}; X) \cup \sigma_r(e^{2\pi A}; X).$$

We will derive a contradiction. By Lemma 2.29, we have $1 \in \sigma_r(e^{2\pi A}; X)$. Then $1 \in \sigma_r(e^{2\pi \Gamma_{per}}; \mathcal{F}_{per})$, and an application of the Spectral Mapping Theorem 2.9 for the Residual Spectrum shows that $ik \in \sigma_r(\Gamma_{per}; \mathcal{F}_{per})$ for some $k \in \mathbb{Z}$. Since, by Proposition 2.28, $\sigma(\Gamma_{per}; \mathcal{F}_{per})$ is invariant under translations by i, we have that $0 \in \sigma(\Gamma_{per}; \mathcal{F}_{per})$, contradicting statement 3). □

2.2.1.2. *A spectral mapping theorem.* In this section we will give a Banach space version of Gearhart's Theorem 2.10 as we formulated it in display (2.5).

THEOREM 2.31. *Let $\{e^{tA}\}_{t \ge 0}$ be a strongly continuous semigroup on a Banach space X, and let $\mathcal{F}_{per}$ denote one of the spaces $L^p([0, 2\pi]; X)$, $1 \le p < \infty$, or $C_{per}([0, 2\pi]; X)$. The following statements are equivalent:*

1) $1 \in \rho(e^{2\pi A})$.
2) $i\mathbb{Z} \subset \rho(A)$, *and there is a constant $C > 0$ such that*

(2.28) $$\|\sum_k (A - ik)^{-1} e^{ik\theta} x_k\|_{\mathcal{F}_{per}} \le C \|\sum_k e^{ik\theta} x_k\|_{\mathcal{F}_{per}}$$

for every finite sequence $\{x_k\} \subset X$.

PROOF. Consider the evolution semigroup $\{e^{t\Gamma_{per}}\}_{t \ge 0}$ on $\mathcal{F}_{per}$ defined by equation (2.20). Choose a finite sequence $\{x_k\} \subset X$, and assume that $(A - ik)^{-1}$ exists for all $k \in \mathbb{Z}$. Also, note that $(A - ik)^{-1} : X \to \mathcal{D}(A)$.

Define trigonometric polynomials $f, g \in \mathcal{F}_{per}$ by

(2.29) $$f(\theta) = \sum_k (A - ik)^{-1} e^{ik\theta} x_k, \quad g(\theta) = \sum_k e^{ik\theta} x_k, \quad \theta \in [0, 2\pi].$$

We claim that $\Gamma_{per} f = g$. Indeed, we have

$$(\Gamma_{per} f)(\theta) = \frac{d}{dt} e^{tA} f([\theta - t](\mathrm{mod}\, 2\pi))\big|_{t=0}$$
$$= \sum_k [A(A - ik)^{-1} e^{ik\theta} x_k - ik(A - ik)^{-1} e^{ik\theta} x_k] = g.$$

1) $\Rightarrow$ 2). If $1 \in \rho(e^{2\pi A})$, then the inclusion $i\mathbb{Z} \subset \rho(A)$ follows from the Spectral Inclusion Theorem 2.6: $e^{2\pi \sigma(A)} \subset \sigma(e^{2\pi A})$. By the equivalence of parts 1) and 3) of Theorem 2.30, the operator Γ_{per} has a bounded inverse Γ_{per}^{-1} on $\mathcal{F}_{per}$ provided that

$1 \in \rho(e^{2\pi A})$. Let $C := \|\Gamma_{\text{per}}^{-1}\|$, and consider the functions defined in display (2.29). We have $\|f\|_{\mathcal{F}_{\text{per}}} = \|\Gamma_{\text{per}}^{-1} g\|_{\mathcal{F}_{\text{per}}} \leq C\|g\|_{\mathcal{F}_{\text{per}}}$, and the inequality (2.28) is proved.

2) $\Rightarrow$ 1). First, we show that condition 2) implies $0 \notin \sigma_{\text{ap}}(\Gamma_{\text{per}})$. Indeed, the functions of type g in (2.29) are dense in $\mathcal{F}_{\text{per}}$. If we let $y_k = (A - ik)^{-1} x_k$, then the functions of type f are also dense in $\mathcal{D}(\Gamma)$. Thus, we see that inequality (2.28) implies $\|\Gamma_{\text{per}} f\|_{\mathcal{F}_{\text{per}}} = \|g\|_{\mathcal{F}_{\text{per}}} \geq C^{-1}\|f\|_{\mathcal{F}_{\text{per}}}$, and $0 \notin \sigma_{\text{ap}}(\Gamma_{\text{per}})$.

Assume that condition 2) holds, but $1 \in \sigma(e^{2\pi A}) = \sigma_r(e^{2\pi A}) \cup \sigma_{\text{ap}}(e^{2\pi A})$. If $1 \in \sigma_{\text{ap}}(e^{2\pi A})$ in X then, by Lemma 2.29, $0 \in \sigma_{\text{ap}}(\Gamma_{\text{per}})$, in contradiction to the results in the previous paragraph. On the other hand, if $1 \in \sigma_r(e^{2\pi A})$, then by the Spectrum Mapping Theorem 2.9 for the Residual Spectrum, we have that $ik \in \sigma_r(A)$ for some $k \in \mathbb{Z}$, in contradiction to the fact that $i\mathbb{Z} \subset \rho(A)$. □

REMARK 2.32. The implication 1) $\Rightarrow$ 2) in Theorem 2.31 can also be proved directly. Indeed, assuming 1), if we define $\phi(\theta) := (e^{2\pi A} - I)^{-1} e^{\theta A}$, $\theta \in [0, 2\pi]$, then the convolution operator

$$(Kf)(\theta) = \int_0^{2\pi} \phi(s) f(\theta - s)\, ds$$

is a bounded operator on $\mathcal{F}_{\text{per}}$. But, for each $x \in X$,

$$(A - ik)^{-1} x = \int_0^{2\pi} e^{-iks} (e^{2\pi A} - I)^{-1} e^{sA} x\, ds, \quad k \in \mathbb{Z},$$

are the Fourier coefficients of the function $\phi(\cdot)x : [0, 2\pi] \to X$. Apply K to the trigonometric polynomial g as in display (2.29). Inequality (2.28) can be viewed as the condition of boundedness of K, which implies statement 2). ◇

COROLLARY 2.33. *If $1 \in \rho(e^{2\pi A})$ on X, then Γ_{per} is invertible on the space $L^p = L^p([0, 2\pi]; X)$. Moreover, the norm of Γ_{per}^{-1} in $\mathcal{L}(L^p)$ is given by the formula*

$$(2.30) \quad \|\Gamma_{per}^{-1}\| = \sup_{\{x_k\} \subset \Lambda} \frac{\|\sum_k (A - ik)^{-1} x_k e^{ik\theta}\|_{L^p}}{\|\sum_k x_k e^{ik\theta}\|_{L^p}}$$

where Λ denotes the set of finite sequences $\{x_k\} \subset X$.

PROOF. The fact that $\Gamma_{\text{per}}^{-1} \in \mathcal{L}(L^p)$ follows from the implication 1) $\Rightarrow$ 3) of Theorem 2.30. Formula (2.30) follows from the fact that the set of trigonometric polynomials, g, defined as in display (2.29), is dense in $L^p([0, 2\pi]; X)$, and the observation that

$$\Gamma_{per}^{-1} g = \sum_k (A - ik)^{-1} e^{ik\theta} x_k.$$

□

REMARK 2.34. Theorem 2.31 and Corollary 2.33 imply that $1 \in \rho(e^{2\pi A})$ if and only if $i\mathbb{Z} \subset \rho(A)$ and

$$\|\Gamma_{per}\|_{\bullet} = \inf_{\{y_k\} \subset \Lambda_{\mathcal{D}}} \frac{\|\sum_k (A - ik) y_k e^{ik\theta}\|_{L^p}}{\|\sum_k y_k e^{ik\theta}\|_{L^p}} > 0$$

where $\Lambda_{\mathcal{D}}$ is the set of finite sequences $\{y_k\} \subset \mathcal{D}(A)$. To see this, observe the inequality $\|\Gamma_{per}^{-1}\| = \|\Gamma_{per}\|_{\bullet}^{-1}$ and apply formula (2.30) with $x_k = (A - ik) y_k$. ◇

Let us see that Theorem 2.31 is in fact a generalization of the Gearhart Spectral Mapping Theorem 2.10. Indeed, if X is a Hilbert space and $p = 2$, then an application of Parseval's identity yields the equations

$$\|\sum_k (A - ik)^{-1} e^{ik\theta} x_k\|_{L^2([0,2\pi];X)} = \left(2\pi \sum_k |(A-ik)^{-1} x_k|^2\right)^{1/2},$$

$$\|\sum_k e^{ik\theta} x_k\|_{L^2([0,2\pi];X)} = \left(2\pi \sum_k |x_k|^2\right)^{1/2}.$$

Clearly, the inequality (2.28) is equivalent to the condition $\sup\{\|(A-ik)^{-1}\| : k \in \mathbb{Z}\} < \infty$.

We conclude this subsection by giving four more statements equivalent to statements 1) or 2) in Theorem 2.31:

3) $i\mathbb{Z} \subset \rho(A)$ and there is a constant $C > 0$ such that
$$\|\Gamma_{\mathrm{per}} f\|_{L^1([0,2\pi];X)} \geq C^{-1} \|f\|_{C([0,2\pi];X)}$$
for all $f \in C_{\mathrm{per}}([0,2\pi]; X)$ such that $\Gamma_{\mathrm{per}} f \in L^1([0,2\pi]; X)$.

4) $i\mathbb{Z} \subset \rho(A)$ and there is a constant $C > 0$ such that
$$\|\Gamma_{\mathrm{per}} f\|_{C([0,2\pi];X)} \geq C^{-1} \|f\|_{L_1([0,2\pi];X)}$$
for all $f \in L^1([0,2\pi]; X)$ such that $\Gamma_{\mathrm{per}} f \in C_{\mathrm{per}}([0,2\pi]; X)$.

5) $i\mathbb{Z} \subset \rho(A)$ and there is a constant $C > 0$ such that
$$\|\sum_k (A - ik)^{-1} e^{ik\theta} x_k\|_{C([0,2\pi];X)} \leq C \|\sum_k e^{ik\theta} x_k\|_{L^1([0,2\pi];X)}$$
for any finite sequence $\{x_k\} \subset X$.

6) $i\mathbb{Z} \subset \rho(A)$ and there is a constant $C > 0$ such that
$$\|\sum_k (A - ik)^{-1} e^{ik\theta} x_k\|_{L^1([0,2\pi];X)} \leq C \|\sum_k e^{ik\theta} x_k\|_{C([0,2\pi];X)}$$
for every finite sequence $\{x_k\} \subset X$.

2.2.2. Evolution semigroups on the line. In this subsection we consider the evolution semigroup $\{E^t\}_{t\geq 0}$ given by

(2.31) $$(E^t f)(\theta) = e^{tA} f(\theta - t), \quad \theta \in \mathbb{R},$$

on the spaces $\mathcal{F} = L^p(\mathbb{R}; X)$, $1 \leq p < \infty$, and $\mathcal{F} = C_0(\mathbb{R}; X)$. Note, that $\{E^t\}_{t\geq 0}$ is a strongly continuous semigroup, and let Γ denote its generator.

The generator Γ of the evolution semigroup can be described explicitly. In fact, the evolution semigroup $\{E^t\}_{t\geq 0}$, with $E^t = e^{tA} \otimes V^t$, on $L^p(\mathbb{R}; X) = X \otimes L^p(\mathbb{R})$ is given by the tensor product of $\{e^{tA}\}_{t\geq 0}$ on X and the translation semigroup $(V^t f)(\theta) = f(\theta - t)$ on $L^p(\mathbb{R})$. By this fact, Γ is the closure of the operator $-\frac{d}{dt} + \mathcal{A}$, see for example Nagel [**Na**, p. 23–24]. We give more details in the following remark.

REMARK 2.35. Let A be the generator of a strongly continuous semigroup $\{e^{tA}\}_{t\in\mathbb{R}_+}$ on a Banach space X. Let $\mathcal{A}$ denote the operator of multiplication by A with maximal domain acting on the space $\mathcal{F} = L^p(\mathbb{R}; X)$ or the space $\mathcal{F} = C_0(\mathbb{R}; X)$; that is, $(\mathcal{A}f)(\theta) = Af(\theta)$ and $\mathcal{D}(\mathcal{A}) = \{f : f(\theta) \in \mathcal{D}(A)$ (a.e. for $\mathcal{F} = L^p(\mathbb{R}; X)$) such that $Af(\cdot) \in \mathcal{F}\}$. Let us consider two semigroups on $\mathcal{F}$; namely, $\{e^{tA}\}_{t\geq 0}$ and $\{V^t\}_{t\in\mathbb{R}}$ where $(V^t f)(\theta) = f(\theta - t)$ is the translation semigroup with the generator $-d/dt$. For $\mathcal{F} = C_0(\mathbb{R}; X)$, we have

$\mathcal{D}(-d/dt) = \{f \in C^1(\mathbb{R}; X) : f' \in C_0(\mathbb{R}; X)\}$ while, for $\mathcal{F} = L^p(\mathbb{R}; X)$, we have $\mathcal{D}(-d/dt) = \{f$ is absolutely continuous with $f' \in L^p\}$.

Note that the semigroups $\{V^t\}_{t\geq 0}$ and $\{e^{t\mathcal{A}}\}_{t\geq 0}$ commute. For the operator $\Gamma' := -d/dt + \mathcal{A}$, with $\mathcal{D}(\Gamma') := \mathcal{D}(-d/dt) \cap \mathcal{D}(\mathcal{A})$, and the evolution semigroup $\{E^t\}_{t\geq 0}$ with the generator Γ, the set $\mathcal{D}(\Gamma')$ is a core for Γ (see Remark 2.5) and Γ is the closure of Γ'. Symbolically, we write $\Gamma = -d/dt + \mathcal{A}$. ◇

Let us discuss some symmetry properties of the spectra of Γ and E^t.

PROPOSITION 2.36. (a) *The spectrum $\sigma(e^{t\Gamma})$, for $t > 0$, is invariant with respect to rotations centered at the origin of the complex plane.*
(b) *The sets $\sigma(\Gamma)$, $\sigma_{ap}(\Gamma)$, $\sigma(\Gamma)$, and $\sigma_r(\Gamma)$ are invariant under translations parallel to $i\mathbb{R}$.*

PROOF. For each $\xi \in \mathbb{R}$, define the operator $(L_\xi f)(\theta) = e^{i\xi\theta}f(\theta)$. Then, cf. Proposition 2.28, the identities
$$e^{t\Gamma}e^{i\xi\cdot}f(\cdot) = e^{i\xi\cdot}e^{-i\xi t}e^{t\Gamma}f(\cdot), \quad \Gamma e^{i\xi\cdot}f(\cdot) = e^{i\xi\cdot}(\Gamma - i\xi)f(\cdot)$$
imply $L_\xi^{-1} e^{t\Gamma} L_\xi = e^{-i\xi t}e^{t\Gamma}$ and $L_\xi^{-1}\Gamma L_\xi = -i\xi + \Gamma$. □

Our main objective is to relate the hyperbolicity of the semigroups $\{e^{t\mathcal{A}}\}_{t\geq 0}$ and $\{E^t\}_{t\geq 0}$. We start with two lemmas.

Let $\mathcal{F}_s$ denote one of the spaces $\mathcal{F}_s = \ell^p(\mathbb{Z}; X)$, $1 \leq p < \infty$, or $\mathcal{F}_s = c_0(\mathbb{Z}; X)$ (the sequences $(x_n)_{n\in\mathbb{Z}}$ such that $x_n \to 0$ as $n \to \pm\infty$). Let S denote the shift operator on $\mathcal{F}_s$ given by $S : (x_n) \mapsto (x_{n-1})$. For an operator a on X, we will denote by D_a the diagonal operator $D_a = \text{diag}\{a\}_{n\in\mathbb{Z}}$ on $\mathcal{F}_s$ defined by $D_a : (x_n)_{n\in\mathbb{Z}} \mapsto (ax_n)_{n\in\mathbb{Z}}$. Note that $\sigma(a) = \sigma(D_a)$. Also, as in Proposition 2.36, one can use the operator $L_\xi = \text{diag}(e^{-in\xi})_{n\in\mathbb{Z}}$ on $\mathcal{F}_s$ to show that $\sigma(D_a S)$ is rotationally invariant.

LEMMA 2.37. *The following statements are equivalent:*
1) $\sigma(a) \cap \mathbb{T} = \emptyset$.
2) $\sigma(D_a S) \cap \mathbb{T} = \emptyset$.

Moreover,
  a) *if $\|I - D_a S\|_\bullet > 0$, then $\|I - a\|_\bullet > 0$; and*
  b) *if $I - D_a S$ is surjective, then $I - a$ has dense range.*

PROOF. We will give the proof for the space $\mathcal{F}_s = \ell^p = \ell^p(\mathbb{Z}; X)$; the proof for $\mathcal{F}_s = c_0(\mathbb{Z}; X)$ is similar.

1) $\Rightarrow$ 2). Since $SD_a = D_a S$, we have
$$\sigma(D_a S) \subset \sigma(S) \cdot \sigma(D_a) = \mathbb{T} \cdot \sigma(a).$$

2) $\Rightarrow$ 1). As above, we have the inclusion
$$\sigma(a) = \sigma(D_a) = \sigma(D_a S S^{-1}) \subset \sigma(D_a S) \cdot \sigma(S^{-1}) = \sigma(D_a S) \cdot \mathbb{T},$$
as required.

To prove statement a), assume that $\|I - D_a S\|_\bullet > 0$ in ℓ^p, but, for each $\epsilon > 0$, there is a vector $x \in X$ such that $|x|_X = 1$ and $|x - ax|_X < \epsilon$. Also, fix $\gamma > 0$ such that $\left|1 - e^{\pm\gamma}\right| < \epsilon$. If $v = (x_n) \in \ell^p$ denotes the sequence given by $x_n = e^{-\gamma|n|}x$, for $n \in \mathbb{Z}$, then
$$I - D_a S : (x_n)_{n\in\mathbb{Z}} \mapsto \left(e^{-\gamma|n|}(x - ax) + (e^{-\gamma|n|} - e^{-\gamma|n-1|})ax\right)_{n\in\mathbb{Z}}.$$

However, by a direct calculation, it follows that
$$\|(I - D_a S)(x_n)\|_{\ell^p} \leq (1 + \|a\|) \cdot \epsilon \cdot \|(x_n)\|_{\ell^p},$$
in contradiction.

To prove statement b), consider, for each $x \in X$, the sequence $(u_n) \in \ell^p$ defined by $u_0 = x$, and $u_n = 0$ for $n \neq 0$. By our hypotheses, there is a sequence $v = (x_n) \in \ell^p$ such that $(I - D_a S)(x_n) = (u_n)$; that is, $x_n - a x_{n-1} = u_n$ for $n \in \mathbb{Z}$. But then, for $k \in \mathbb{N}$, we have the equality
$$x = \sum_{n=-k}^{k}(x_n - a x_{n-1}) = x_k - a x_{-k-1} + (y_k - a y_k) \quad \text{where} \quad y_k = \sum_{n=-k}^{k-1} x_n.$$
The sequence $y_k - a y_k$ in $\operatorname{Im}(I-a)$ converges to x because $x_k \to 0$ and $a x_{-k-1} \to 0$ as $k \to \infty$. $\square$

We will also need a "continuous" version of the previous lemma. To state it, let $\{V^t\}$ be the translation group on $\mathcal{F} = L^p(\mathbb{R}; X)$ or $\mathcal{F} = C_0(\mathbb{R}; X)$ defined as $(V^t u)(\theta) = u(\theta - t)$. Also, for each operator $B \in \mathcal{L}(X)$, let $\mathcal{B} \in \mathcal{L}(C_0(\mathbb{R}; X))$ denote the multiplication operator defined by $(\mathcal{B}u)(\theta) = Bu(\theta)$, $\theta \in \mathbb{R}$.

LEMMA 2.38. *The following statements are equivalent:*
1) $\sigma(B) \cap \mathbb{T} = \emptyset$.
2) $\sigma(\mathcal{B}V^t) \cap \mathbb{T} = \emptyset$ *for some (hence all)* $t > 0$.

Moreover,
   a) *if* $\|I - \mathcal{B}V^t\|_\bullet > 0$, *then* $\|I - B\|_\bullet > 0$; *and*
   b) *if* $I - \mathcal{B}V^t$ *is surjective, then* $I - B$ *has dense range.*

PROOF. This time let us consider the space $\mathcal{F} = C_0(\mathbb{R}; X)$, the proof for p-summable functions is similar. The equivalence 1) $\Leftrightarrow$ 2) follows exactly as in the proof of the equivalence 1) $\Leftrightarrow$ 2) in Lemma 2.37.

To prove statement a), let us assume that $\|I - \mathcal{B}V^t\|_\bullet > 0$, but for each $\epsilon > 0$ there is a vector $x \in X$ such that $|x|_X = 1$ and $|x - Bx|_X < \epsilon$. Fix $\gamma > 0$ such that $|1 - e^{\pm\gamma}| < \epsilon$, and define the function $v \in C_0(\mathbb{R}; X)$ by $v(\tau) = e^{-\gamma|\tau|} x$, $\tau \in \mathbb{R}$. As in the proof of statement a) in Lemma 2.37, a direct calculation yields
$$\|[I - \mathcal{B}V^t] v\|_{C_0(\mathbb{R};X)} = O(\epsilon),$$
contradicting our assumption.

To prove statement b), consider, for each $x \in X$, the function $u \in C_0(\mathbb{R}; X)$ defined by $u(\tau) = \alpha(\tau) x$, where $\alpha : \mathbb{R} \to [0, 1]$ is a continuous bump-function supported in $(-t/2, t/2)$ with $\alpha(0) = 1$. Since $I - \mathcal{B}V^t$ is surjective, there is a function $v \in C_0(\mathbb{R}; X)$ such that $v(\tau) - Bv(\tau - t) = u(\tau)$. The rest of the proof goes through as in Lemma 2.37, part b). $\square$

THEOREM 2.39. *Let* $\mathcal{F}$ *denote one of the spaces* $L^p(\mathbb{R}; X)$, $1 \leq p < \infty$, *or* $C_0(\mathbb{R}; X)$. *For* $t > 0$, *the following statements are equivalent:*
1) $\sigma(e^{tA}) \cap \mathbb{T} = \emptyset$ *in* X.
2) $\sigma(e^{t\Gamma}) \cap \mathbb{T} = \emptyset$ *in* $\mathcal{F}$.
3) $0 \in \rho(\Gamma)$ *in* $\mathcal{F}$.

PROOF. The implication 2) $\Rightarrow$ 3) follows from the Spectral Inclusion Theorem 2.6 for $\{e^{t\Gamma}\}_{t \geq 0}$.

2.2. EVOLUTION SEMIGROUPS AND HYPERBOLICITY

We will prove the implication 3) $\Rightarrow$ 2) for the space $\mathcal{F} = L^p(\mathbb{R}; X)$; the proof for $\mathcal{F} = C_0(\mathbb{R}; X)$ is similar.

The spectrum $\sigma(e^{t\Gamma})$ is invariant under all rotations centered at origin. Thus, it suffices to prove that condition 3) implies that $1 \in \rho(e^{t\Gamma})$. We will consider only the case $t = 2\pi$. However, the proof is essentially the same for arbitrary t. We will use a "change-of-variables" trick.

The idea of the proof is to apply the implication 3) $\Rightarrow$ 1) of Theorem 2.30 to show that if $0 \in \rho(\tilde{\Gamma})$, then $1 \in \rho(e^{2\pi\Gamma})$. Here, the operator $\tilde{\Gamma} = -\frac{d}{ds} - \frac{d}{d\theta} + \mathcal{A}$ acts on $L^p([0, 2\pi] \times \mathbb{R}; X)$, $s \in [0, 2\pi]$, $\theta \in \mathbb{R}$. Indeed, by formula (2.21), we have that $\Gamma = -\frac{d}{d\theta} + \mathcal{A}$. Hence, $\tilde{\Gamma}$, defined on $L^p([0, 2\pi]; L^p(\mathbb{R}; X))$, is the generator of the evolution semigroup corresponding to the semigroup $\{e^{t\Gamma}\}_{t \geq 0}$ on $L^p(\mathbb{R}; X)$. Finally, the change of variables $u = [s - \theta](\text{mod} 2\pi)$, $v = \theta$ is used to show that $\rho(\tilde{\Gamma}) = \rho(-\frac{d}{dv} + \mathcal{A}) = \rho(\Gamma)$.

Let us now make this argument precise. Consider the semigroups given by

$$(e^{t\tilde{\Gamma}}h)(s, \theta) = e^{t\mathcal{A}}h([s-t](\text{mod} 2\pi), \theta - t), \quad t > 0, \quad s \in [0, 2\pi], \quad \theta \in \mathbb{R},$$
$$(e^{tG}h)(s, \theta) = e^{t\mathcal{A}}h(s, \theta - t), \quad t > 0, \quad s \in [0, 2\pi], \quad \theta \in \mathbb{R},$$

and an invertible isometry J that acts on the space

$$L^p([0, 2\pi] \times \mathbb{R}; X) = L^p([0, 2\pi]; L^p(\mathbb{R}; X))$$

by $(Jh)(s, \theta) = h([s + \theta](\text{mod} 2\pi), \theta)$. Since the operator e^{tG}, which acts on the space $L^p([0, 2\pi]; L^p(\mathbb{R}; X))$, is a multiplication operator with multiplier the operator $e^{t\Gamma}$ in $L^p(\mathbb{R}; X)$, we have $\sigma(e^{tG}) = \sigma(e^{t\Gamma})$ and $\sigma(G; L^p([0, 2\pi]; L^p(\mathbb{R}; X))) = \sigma(\Gamma; L^p(\mathbb{R}; X))$. Also, it follows from the equality

$$\left(Je^{t\tilde{\Gamma}}h\right)(s, \theta) = e^{t\mathcal{A}}h([s + \theta - t](\text{mod} 2\pi), \theta - t) = \left(e^{tG}Jh\right)(s, \theta)$$

that $Je^{t\tilde{\Gamma}} = e^{tG}J$ and $J\tilde{\Gamma} = GJ$. Therefore, we have the identities

$$\sigma(e^{tG}) = \sigma(e^{t\tilde{\Gamma}}) \text{ and } \sigma(G) = \sigma(\tilde{\Gamma})$$

in $L^p([0, 2\pi]; L^p(\mathbb{R}; X))$. Thus, statement 3) implies $0 \in \rho(G)$ and $0 \in \rho(\tilde{\Gamma})$.

The semigroup $\{e^{t\tilde{\Gamma}}\}_{t \geq 0}$ acts on $L^p([0, 2\pi]; L^p(\mathbb{R}; X))$ by the rule

$$(e^{t\tilde{\Gamma}}f)(s) = e^{t\Gamma}f([s-t](\text{mod} 2\pi)),$$

where $f(s) = h(s, \cdot) \in L^p(\mathbb{R}; X)$ for almost all $s \in [0, 2\pi]$. Hence, the semigroup $\{e^{t\tilde{\Gamma}}\}_{t \geq 0}$ defined on the space $L^p([0, 2\pi]; L^p(\mathbb{R}; X))$ is the evolution semigroup for the semigroup $\{e^{t\Gamma}\}_{t \geq 0}$ on $L^p(\mathbb{R}; X)$. By the implication 3) $\Rightarrow$ 1) of Theorem 2.30 where we replace A in that theorem by Γ, let us conclude that $1 \in \rho(e^{2\pi\Gamma})$ on $L^p(\mathbb{R}; X)$. Thus, the equivalence 2) $\Leftrightarrow$ 3) is proved.

The equivalence 1) $\Leftrightarrow$ 2) follows from Lemma 2.38 with $B = e^{t\mathcal{A}}$.

We will give yet another proof 1) $\Leftrightarrow$ 2) for $L^p(\mathbb{R}; X)$ that uses Lemma 2.37. Define $a = e^{2\pi\mathcal{A}}$ and the operator $(V^{2\pi}f)(\theta) = f(\theta - 2\pi)$ on $L^p(\mathbb{R}; X)$, and note that $e^{2\pi\Gamma} = aV^{2\pi}$. Consider the invertible isometry

$$j : L^p(\mathbb{R}; X) \to \ell^p(\mathbb{Z}; L^p([0, 2\pi]; X)) : f \mapsto (f_n)$$

where $f_n(s) := f(s + 2\pi n)$, for $n \in \mathbb{Z}$ and $s \in [0, 2\pi)$, and let $S : (f_n) \mapsto (f_{n-1})$ denote the shift operator on $\ell^p(\mathbb{Z}; L^p([0, 2\pi]; X))$. Then, $jaV^{2\pi} = D_aSj$ and $\sigma(aV^{2\pi}) = \sigma(D_aS)$. Therefore, statement 2) is equivalent to the identity

$\sigma(D_a S) \cap \mathbb{T} = \emptyset$. By Lemma 2.37 this is in turn equivalent to $\sigma(a) \cap \mathbb{T} = \emptyset$ in $L^p(\mathbb{R}; X)$. $\square$

The equivalence 2) $\Leftrightarrow$ 3) of Theorem 2.39 together with the spectral symmetry described in Proposition 2.28 imply that the evolution semigroup $\{E^t\}_{t\geq 0}$ enjoys the spectral mapping property on $\mathcal{F}$.

COROLLARY 2.40. *Let $\mathcal{F}$ denote one of the spaces $L^p(\mathbb{R}; X)$, $1 \leq p < \infty$ or $C_0(\mathbb{R}; X)$. If $\{E^t\}_{t\geq 0}$ is the evolution semigroup on $\mathcal{F}$ defined in equation (2.31), then*
$$\sigma(E^t) \setminus \{0\} = \exp t\sigma(\Gamma), \quad t > 0.$$
Moreover, the following formula holds for the spectral and growth bounds:
$$\omega(\Gamma) = s(\Gamma) = \omega(A).$$

The last statement follows from the equivalence 1) $\Leftrightarrow$ 2) of Theorem 2.39 and a simple rescaling. We stress that the evolution semigroup $\{E^t\}_{t\geq 0}$ has the spectral mapping property on $\mathcal{F}$ even if the underlying semigroup $\{e^{tA}\}_{t\geq 0}$ does not have the spectral mapping property on the Banach space X.

We will now give an alternate proof of the implication 3) $\Rightarrow$ 2) of Theorem 2.39 to show that it can be proved directly by constructing an approximate eigenfunction g for Γ as in the proof of Lemma 2.29. Using rescaling, the Spectral Mapping Theorem 2.9 for the Residual Spectrum, and the Spectral Mapping Theorem 2.8 for the Point Spectrum, it suffices to prove the implication
$$1 \in \sigma_{\mathrm{ap}}(E^1) \quad \Rightarrow \quad 0 \in \sigma_{\mathrm{ap}}(\Gamma).$$
This is the content of the following Lemma 2.41. We note that the construction of the approximate eigenfunction g for Γ used in Lemma 2.41 will be generalized later in Chapters 6–8, see for example Theorem 6.37 and Lemmas 7.23–7.24. Also, we have the implication $1 \in \sigma_{\mathrm{ap}}(e^{tA}) \Rightarrow 1 \in \sigma_{\mathrm{ap}}(E^t)$, see statement b) of Lemma 2.38. Thus, by Lemma 2.41, we also have the implication $\sigma_{\mathrm{ap}}(e^{tA}) \cap \mathbb{T} \neq \emptyset \Rightarrow i\mathbb{R} \subset \sigma_{\mathrm{ap}}(\Gamma)$.

LEMMA 2.41. *If $\sigma_{ap}(E^t) \cap \mathbb{T} \neq \emptyset$ for some $t \neq 0$, then $\sigma_{ap}(\Gamma) \supset i\mathbb{R}$ in $\mathcal{F}$.*

PROOF. It suffices to prove that the condition $1 \in \sigma_{\mathrm{ap}}(E^t)$ implies $0 \in \sigma_{\mathrm{ap}}(\Gamma)$, see Proposition 2.36. We will consider only the case $t = 1$. The proof remains the same for arbitrary t. Let us assume that $1 \in \sigma_{\mathrm{ap}}(E)$ where $E := E^1$. We will show that $0 \in \sigma_{\mathrm{ap}}(\Gamma)$. The proof will be given for $\mathcal{F} = L^p(\mathbb{R}; X)$; the space $\mathcal{F} = C_0(\mathbb{R}; X)$ can be considered similarly (see for example Lemma 7.23).

Note that
$$\|f\|_{L^p(\mathbb{R};X)}^p = \sum_{n=-\infty}^{\infty} \int_n^{n+1} |f(\theta)|^p \, d\theta.$$

Since $1 \in \sigma_{\mathrm{ap}}(E)$ on $L^p(\mathbb{R}; X)$, if $\varepsilon \in (0, 1/2)$, then there exists some $f \in L^p(\mathbb{R}; X)$ such that $\|Ef - f\|_{L^p} \leq \varepsilon \|f\|_{L^p}$. Therefore, for some $\theta_0 \in \mathbb{R}$, the sequence $\bar{x} = (x_n)_{n \in \mathbb{Z}}$ with $x_n = f(n + \theta_0)$ belongs to $\ell^p(\mathbb{Z}; X)$; and, moreover, we have the inequality
$$\sum_{n=-\infty}^{\infty} |e^A f(n + \theta_0 - 1) - f(n + \theta_0)|^p \leq \varepsilon^p \sum_{n=-\infty}^{\infty} |f(n + \theta_0)|^p.$$

For $a = e^A$, using the notations in Lemma 2.37, the last inequality becomes $\|D_a S\bar{x} - \bar{x}\|_{\ell^p} \leq \varepsilon \|\bar{x}\|_{\ell^p}$. Since $\varepsilon < 1/2$, the triangle inequality gives $\|D_a S\bar{x} - \bar{x}\| \leq 2\varepsilon \|D_a S\bar{x}\|$; or, equivalently,

$$(2.32) \qquad \sum_{n=-\infty}^{\infty} |e^A f(n+\theta_0-1) - f(n+\theta_0)|^p \leq 2^p \varepsilon^p \sum_{n=-\infty}^{\infty} |e^A f(n+\theta_0)|^p.$$

Since the sequence $(e^A f(n+\theta_0))_{n \in \mathbb{Z}}$ belongs to $\ell^p(\mathbb{Z}; X)$, there exists a natural number $N > 1/\varepsilon$ such that

$$(2.33) \qquad \frac{1}{2} \sum_{n=-\infty}^{\infty} |e^A f(n+\theta_0)|^p \leq \sum_{n=-N}^{N-1} |e^A f(n+\theta_0)|^p.$$

Let $\alpha : [0,1] \to [0,1]$ denote a smooth function with bounded derivative such that $\alpha(\theta) = 0$ for $\theta \in [0, \frac{1}{3}]$, and $\alpha(\theta) = 1$ for $\theta \in [\frac{2}{3}, 1]$, cf. the proof of Lemma 2.29. Also, let $\gamma : \mathbb{R} \to [0,1]$ denote a smooth function with $\operatorname{supp}\gamma \subset (-2N, 2N)$ such that $\gamma(\theta) = 1$ for $\theta \in [-N, N]$, and $\|\gamma'\|_\infty \leq c\varepsilon$. Here and below c denotes various absolute constants.

For $\theta = n + \theta_0 + \tau$ where $n \in \mathbb{Z}$ and $\tau \in [0,1)$, we define (cf. formula (2.22)):

$$(2.34) \qquad g(\theta) = \gamma(\theta)\Big[(1-\alpha(\tau))e^{(\tau+1)A}f(n+\theta_0-1) + \alpha(\tau)e^{\tau A}f(n+\theta_0)\Big].$$

Note that the inequality $|e^A x| \leq |e^{(1-\tau)A} e^{\tau A} x| \leq c |e^{\tau A} x|$, for $x \in X$, and the choice of α, γ, and N as in (2.33) imply the following lower estimate:

$$\|g\|_{L^p}^p = \sum_{n=-\infty}^{\infty} \int_0^1 |g(n+\theta_0+\tau)|^p \, d\tau \geq \sum_{n=-N}^{N-1} \int_{2/3}^{1} |e^{\tau A} f(n+\theta_0)|^p$$

$$\geq c \sum_{n=-N}^{N-1} |e^A f(n+\theta_0)|^p \geq c \sum_{n=-\infty}^{\infty} |e^A f(n+\theta_0)|^p.$$

As in the proof of Lemma 2.29, $g \in \mathcal{D}(\Gamma)$ and

$$(\Gamma g)(\theta) = e^{\tau A}\Big[\big(\gamma(\theta)\alpha'(\theta) + \gamma'(\theta)\alpha(\theta)\big)[e^A f(n+\theta_0-1) - f(n+\theta_0)]$$
$$- \gamma'(\theta) e^A f(n+\theta_0-1)\Big].$$

Using (2.32) and the lower estimate for $\|g\|_{L^p}$, it follows that $\|\Gamma g\|_{L^p} \leq c\varepsilon \|g\|_{L^p}$. $\square$

2.2.3. Evolution semigroups on the half line. In this subsection we will study evolution semigroups acting on certain spaces of X-valued functions defined on the half line $\mathbb{R}_+$. Let $\mathcal{F}_+$ denote one of the spaces $L^p(\mathbb{R}_+; X)$, $1 \leq p < \infty$, or the space $C_{00}(\mathbb{R}_+; X)$ of continuous functions $f : \mathbb{R}_+ \to X$ such that $f(0) = 0$ and $\lim_{\tau \to +\infty} f(\tau) = 0$ with the supremum norm. For a strongly continuous semigroup $\{e^{tA}\}_{t \geq 0}$ on X, we define the associated evolution semigroup $\{E_+^t\}_{t \geq 0}$ on $\mathcal{F}_+$ by

$$(2.35) \qquad (E_+^t f)(\theta) = \begin{cases} e^{tA} f(\theta - t), & \text{if } \theta \geq t; \\ 0, & \text{if } 0 \leq \theta \leq t. \end{cases}$$

As we will see, this evolution semigroup can be used to determine the stability of the underlying semigroup $\{e^{tA}\}_{t \geq 0}$.

The semigroup $\{E_+^t\}_{t \geq 0}$ is strongly continuous on $\mathcal{F}_+$. Thus, it has an infinitesimal generator Γ_+. As in Remark 2.35, it follows that Γ_+ is the closure of the operator $\Gamma'_+ = -\frac{d}{dt} + \mathcal{A}$ with domain $\mathcal{D}(\Gamma'_+) = \mathcal{D}_0(d/dt) \cap \mathcal{D}(\mathcal{A})$ where $\mathcal{D}_0(d/dt)$ is the domain of the operator of differentiation with zero boundary condition at zero on $\mathcal{F}_+$.

2.2.3.1. *The autonomous Datko-van Neerven Theorem.* In this section we will describe some simple spectral properties of the evolution semigroup $\{E_+^t\}_{t\geq 0}$, formulate the Spectral Mapping Theorem for $\{E_+^t\}_{t\geq 0}$, and state an important result by R. Datko and J. van Neerven that gives a formula for Γ_+^{-1} and a characterization of the stability for $\{e^{tA}\}_{t\geq 0}$ in terms of a convolution operator on $\mathcal{F}_+$.

For each $f \in \mathcal{F}_+$ and $\xi \in \mathbb{R}$, we have that

$$E_+^t e^{i\xi \cdot} f = e^{i\xi \cdot} e^{-i\xi t} E_+^t f \quad \text{and} \quad \Gamma_+ e^{i\xi \cdot} = e^{i\xi \cdot}(\Gamma_+ - i\xi).$$

Therefore, $\sigma(E_+^t)$ is invariant under rotations around the origin, and $\sigma(\Gamma_+)$ is invariant under translations along $i\mathbb{R}$, cf. Propositions 2.28 and 2.36. Also, the following Spectral Mapping Theorem holds (cf. Corollary 2.40).

THEOREM 2.42. *Let $\{e^{tA}\}_{t\geq 0}$ be a strongly continuous semigroup on X and $\{E_+^t\}_{t\geq 0}$ the evolution semigroup on $\mathcal{F}_+$ defined in (2.35). Then, $\sigma(\Gamma_+)$ is a half plane, $\sigma(E_+^t)$, for $t > 0$, is a disc centered at the origin, and the spectral mapping property is valid for $\{E_+^t\}_{t\geq 0}$; that is,*

$$\sigma(E_+^t)\setminus\{0\} = \exp(t\sigma(\Gamma_+)), \quad t > 0.$$

We will not prove Theorem 2.42 here since it is a corollary of the more general Theorem 3.22 proved below. Theorem 3.22 is stated for evolution semigroups on $\mathcal{F}_+$ induced by *evolution families* $\{U(\theta, \tau)\}_{\theta \geq \tau}$, see Definition 3.1. At this point, we simply note that a semigroup $\{e^{tA}\}_{t\geq 0}$ can be viewed as an evolution family $U(\theta, \tau) := e^{(\theta - \tau)A}$. We stress that the spectral mapping property holds for the evolution semigroup $\{E_+^t\}_{t\geq 0}$ even when it does not hold for the underlying semigroup $\{e^{tA}\}_{t\geq 0}$.

By Theorem 2.30, the evolution semigroup $\{E_{\text{per}}^t\}_{t\geq 0}$, defined in display (2.20) for the various spaces $\mathcal{F}_{\text{per}}$ of periodic functions, can be used to determine if a single point belongs to the resolvent set $\rho(e^{tA})$. Also, by Theorem 2.39, the evolution semigroup $\{E^t\}_{t\geq 0}$ defined in (2.31) on the spaces $\mathcal{F}$ of functions from $\mathbb{R}$ to X can be used to determine if $\{e^{tA}\}_{t\geq 0}$ is hyperbolic. The following result shows that the evolution semigroup (2.35) on $\mathcal{F}_+$ can be used to determine the stability of the semigroup $\{e^{tA}\}_{t\geq 0}$ on X.

PROPOSITION 2.43. *A strongly continuous semigroup $\{e^{tA}\}_{t\geq 0}$ is stable on X if and only if the evolution semigroup $\{E_+^t\}_{t\geq 0}$ is stable on $\mathcal{F}_+$, equivalently $s(\Gamma_+) < 0$.*

Again, we postpone the proof because Proposition 2.43 is a corollary of Theorem 3.23, given below.

To formulate the result of R. Datko and J. van Neerven, which gives a characterization of uniform exponential stability for the semigroup $\{e^{tA}\}_{t\geq 0}$, let us define the following convolution operator on $\mathcal{F}_+$:

(2.36) $$(\mathbb{G}f)(\theta) := \int_0^\theta e^{\tau A} f(\theta - \tau)\, d\tau, \quad \theta \geq 0, \quad f \in \mathcal{F}_+.$$

Clearly, if $\{E_+^t\}_{t\geq 0}$ is stable, then, using formula (2.3), we have that

(2.37) $$\mathbb{G}f = \int_0^\infty E_+^t f \, dt = -\Gamma_+^{-1} f, \qquad f \in \mathcal{F}_+.$$

THEOREM 2.44 (Datko-van Neerven Theorem). *Let $\{e^{tA}\}_{t\geq 0}$ be a strongly continuous semigroup on X and $1 \leq p < \infty$. The following statements are equivalent:*

(i) $\omega(A) < 0$.
(ii) $\mathbb{G}f \in L^p(\mathbb{R}_+ X)$ for each $f \in L^p(\mathbb{R}_+; X)$.
(iii) $\mathbb{G}f \in C_0(\mathbb{R}_+; X)$ for each $f \in C_0(\mathbb{R}_+; X)$.

REMARK 2.45. We stress the fact that condition (ii) of Theorem (2.44) is equivalent to the boundedness of $\mathbb{G}$ on $L^p(\mathbb{R}_+; X)$. In view of the Closed Graph Theorem, it suffices to show that the map $f \mapsto \mathbb{G}f$ is a closed operator on $L^p(\mathbb{R}_+; X)$. For this, note that if $f_n \to f$ and $\mathbb{G}f_n \to g$ in $L^p(\mathbb{R}_+; X)$, then $(\mathbb{G}f_n)(s) \to (\mathbb{G}f)(s)$ for each $s \in \mathbb{R}$. Also, note that every norm-convergent sequence in $L^p(\mathbb{R}_+; X)$ contains a subsequence that converges pointwise almost everywhere. Thus, $(\mathbb{G}f_{n_k})(s) \to g(s)$ for almost all s, and $\mathbb{G}f = g$, as claimed. We also note that condition (iii) is equivalent to the boundedness of $\mathbb{G}$ on $C_0(\mathbb{R}_+; X)$. This follows from the Uniform Boundedness Principle applied to the operators $\mathbb{G}_\theta : f \mapsto \int_0^\theta e^{\tau A} f(\theta - \tau) \, d\tau$. $\diamond$

We refer the reader to Datko [**Da2**] and van Neerven [**vN**, **vN2**] for different proofs of Theorem 2.44. We will use the implication (ii) $\Rightarrow$ (i) of this theorem later to derive the analogous result in the nonautonomous case, see the second proof of Theorem 3.26 in Subsection 3.3.2.

Next, we will give a proof of the implication (ii) $\Rightarrow$ (i) of Theorem 2.44 that is based on the classical Datko-Pazy Theorem, see for example [**Pz**, p. 116].

PROOF. Assume statement (ii) in Theorem 2.44 holds; that is, $\mathbb{G}$ is a bounded operator on $L^p(\mathbb{R}_+; X)$. By the Datko-Pazy Theorem, to prove $\omega(A) < 0$ we need to check that the integral $\int_0^\infty |e^{\theta A} x|^p \, d\theta$ is finite for each $x \in X$. To this end, fix $x \in X$ and define $f \in L^p(\mathbb{R}_+; X)$ by $f(\theta) = e^{\theta A} x$ for $\theta \in [0,1]$, and $f(\theta) = 0$ otherwise. By the assumption,

$$\|\mathbb{G}f\|_{L^p}^p \leq \|\mathbb{G}\|^p \|f\|_{L^p}^p = \|\mathbb{G}\|^p \int_0^1 |e^{\theta A} x|^p \, d\theta.$$

Moreover, for $\theta > 1$, we have the equality

$$(\mathbb{G}f)(\theta) = \int_0^\theta e^{(\theta-\tau)A} f(\tau) \, d\tau = \int_0^1 e^{(\theta-\tau)A} e^{\tau A} x \, d\tau = e^{\theta A} x;$$

and therefore,

$$\int_0^\infty |e^{\theta A} x|^p \, d\theta = \int_0^1 |e^{\theta A} x|^p \, d\theta + \int_1^\infty |(\mathbb{G}f)(\theta)|^p \, d\theta \leq (1 + \|\mathbb{G}\|) \int_0^1 |e^{\theta A} x|^p \, d\theta$$

is finite, as required. $\square$

The following corollary is obtained by combining Theorems 2.42 and 2.44.

COROLLARY 2.46. *A strongly continuous semigroup $\{e^{tA}\}_{t\geq 0}$ is stable if and only if the generator Γ_+ of the evolution semigroup $\{E_+^t\}_{t\geq 0}$ defined on $\mathcal{F}_+$ in (2.35) is invertible. If this is the case, then formula (2.37) holds for Γ_+^{-1}.*

2.2.3.2. *Formulas for the norm of the inverse of the generator.* In this section we will obtain a formula for the norm of the inverse of the infinitesimal generator of the evolution semigroup (2.35) on $L^p(\mathbb{R}_+; X)$. Theorem 2.49 states formulas for the norms of Γ_+^{-1} and Γ^{-1} on $L^p(\mathbb{R}_+; X)$ and $L^p(\mathbb{R}; X)$, respectively, that are similar to the formula (2.30) in Corollary 2.33 for $\|\Gamma_{\text{per}}^{-1}\|$. These results will be used in Chapter 5 in our applications to control theory.

Recall that formula (2.30) is obtained from the formula

$$\|\Gamma_{\text{per}}^{-1}\|_{\mathcal{L}(L^p([0,2\pi];X))} = \sup_f \frac{\|f\|_{L^p([0,2\pi];X)}}{\|\Gamma_{\text{per}}f\|_{L^p([0,2\pi];X)}} = \sup_{\{x_k\}\in\Lambda} \frac{\|f\|_{L^p([0,2\pi];X)}}{\|g\|_{L^p([0,2\pi];X)}}.$$

Here, Λ is the set of all finite sequences in X, and the trigonometric polynomials $f = f_{\{x_k\}}$ and $g = g_{\{x_k\}}$ are defined using $\{x_k\} \in \Lambda$ by the following formulas:

$$(2.38) \quad f(\theta) = \sum_k (A - ik)^{-1} x_k e^{ik\theta}, \quad g(\theta) = \sum_k x_k e^{ik\theta}, \quad \theta \in [0, 2\pi], \quad \{x_k\} \in \Lambda.$$

As we have seen on page 40, $\Gamma_{\text{per}} f = g$. Note that the set $\mathfrak{F}_{\text{per}}$ of polynomials of the form f in (2.38) is dense in $\mathcal{D}(\Gamma_{\text{per}})$, while the set $\mathfrak{G}_{\text{per}}$ of polynomials of the form g in (2.38) is dense in $L^p([0, 2\pi]; X)$.

Our strategy for computing $\|\Gamma^{-1}\|_{\mathcal{L}(L^p(\mathbb{R};X))}$ and $\|\Gamma_+^{-1}\|_{\mathcal{L}(L^p(\mathbb{R}_+;X))}$ is the following. Instead of the set Λ in (2.38), the Schwartz class $\mathcal{S} = \mathcal{S}(\mathbb{R}; X)$ of rapidly decaying X-valued functions (cf. (2.11)) on $\mathbb{R}$ is used to define a dense subset $\mathfrak{F}$ in $\mathcal{D}(\Gamma)$ and a dense subset $\mathfrak{G}$ in $L^p(\mathbb{R}; X)$ such that, for $f \in \mathfrak{F}$ and $g \in \mathfrak{G}$ we have the equality $\Gamma f = g$. We will show that the norm of Γ^{-1} is given by

$$\|\Gamma^{-1}\|_{\mathcal{L}(L^p(\mathbb{R};X))} = \sup_{f \in \mathfrak{F}} \frac{\|f\|_{L^p(\mathbb{R};X)}}{\|\Gamma f\|_{L^p(\mathbb{R};X)}} = \sup_{f \in \mathfrak{F}} \frac{\|f\|_{L^p(\mathbb{R};X)}}{\|g\|_{L^p(\mathbb{R};X)}}.$$

Finally, we will obtain a formula for the norm of Γ_+^{-1} using the following proposition.

PROPOSITION 2.47. *If Γ_+ has a bounded inverse on $L^p(\mathbb{R}_+; X)$, then Γ has a bounded inverse on $L^p(\mathbb{R}; X)$ and*

$$\|\Gamma_+^{-1}\|_{\mathcal{L}(L^p(\mathbb{R}_+;X))} = \|\Gamma^{-1}\|_{\mathcal{L}(L^p(\mathbb{R};X))}.$$

PROOF. By Corollary 2.46, if $\Gamma_+^{-1} \in \mathcal{L}(L^p(\mathbb{R}_+; X))$, then $\omega(A) < 0$, and the semigroup $\{e^{tA}\}$ is, trivially, hyperbolic. By the implication 1) $\Rightarrow$ 3) of Theorem 2.39, we have that $\Gamma^{-1} \in \mathcal{L}(L^p(\mathbb{R}; X))$.

Recall, see Remark 2.35, that Γ is the closure of the operator $\Gamma' = -d/dt + \mathcal{A}$ with $\mathcal{D}(\Gamma') = \mathcal{D}(-d/dt) \cap \mathcal{D}(\mathcal{A})$. Similarly, Γ_+ is the closure of $\Gamma'_+ = -d/dt + \mathcal{A}$ on $L^p(\mathbb{R}_+; X)$ with $\mathcal{D}(\Gamma'_+) = \mathcal{D}_0(-d/dt) \cap \mathcal{D}(\mathcal{A})$ where the subscript 0 is used to remind us that the derivative operator d/dt is taken with the zero boundary condition at $t = 0$.

We claim that $\|\Gamma\|_{\bullet, L^p(\mathbb{R};X)} \leq \|\Gamma_+\|_{\bullet, L^p(\mathbb{R}_+;X)}$. Indeed, for a given $\epsilon > 0$, take $f_+ \in \mathcal{D}(\Gamma'_+)$ such that

$$\|\Gamma_+ f_+\|_{L^p(\mathbb{R}_+;X)} \leq \|\Gamma_+\|_{\bullet, L^p(\mathbb{R}_+;X)} + \epsilon.$$

Extend f_+ to $\mathbb{R}$ by setting $f = f_+$ on $\mathbb{R}_+$ and $f = 0$ otherwise. Then, $f \in \mathcal{D}(\Gamma')$. Also, $(\Gamma f)(\theta) = -f'(\theta) + Af(\theta) = (\Gamma_+ f)(\theta)$ for $\theta \in \mathbb{R}_+$, and $(\Gamma f)(\theta) = 0$ otherwise.

Thus,
$$\|\Gamma\|_{\bullet, L^p(\mathbb{R};X)} \leq \|\Gamma f\|_{L^p(\mathbb{R};X)} = \|\Gamma_+ f\|_{L^p(\mathbb{R}_+;X)} \leq \|\Gamma_+ f\|_{\bullet, L^p(\mathbb{R}_+;X)} + \epsilon,$$
as required.

To see that $\|\Gamma\|_{\bullet, L^p(\mathbb{R};X)} \geq \|\Gamma_+\|_{\bullet, L^p(\mathbb{R}_+;X)}$, let $\epsilon > 0$ be given, and choose $f \in \mathcal{D}(\Gamma')$ with compact support such that $\|f\|_{L^p(\mathbb{R};X)} = 1$ and $\|\Gamma\|_{\bullet, L^p(\mathbb{R};X)} \geq \|\Gamma f\|_{L^p(\mathbb{R};X)} - \epsilon$. Next, choose $\tau \in \mathbb{R}$ such that $\theta \mapsto f_\tau(\theta) := f(\theta - \tau)$ is a function in the space $L^p(\mathbb{R}; X)$ with $\operatorname{supp} f_\tau \subset \mathbb{R}_+$. If f_τ^+ denotes the element of $L^p(\mathbb{R}_+; X)$ that coincides with f_τ on $\mathbb{R}_+$, then $\|f_\tau\|_{L^p(\mathbb{R};X)} = \|f_\tau^+\|_{L^p(\mathbb{R}_+;X)}$. Also,
$$(\Gamma_+ f_\tau^+)(\theta) = -\frac{d}{dt} f(\theta - \tau) + A f(\theta - \tau) = (\Gamma f)(\theta - \tau) = (\Gamma f)_\tau(\theta)$$
for $\theta \geq 0$, and $(\Gamma f)_\tau(\theta) = 0$ for $\theta < 0$. Therefore,
$$\begin{aligned}
\|\Gamma f\|_{L^p(\mathbb{R};X)} - \epsilon &= \|(\Gamma f)_\tau\|_{L^p(\mathbb{R};X)} - \epsilon \\
&= \|\Gamma_+ f_\tau^+\|_{L^p(\mathbb{R}_+;X)} - \epsilon \geq \|\Gamma_+\|_{\bullet, L^p(\mathbb{R}_+;X)} - \epsilon.
\end{aligned}$$
$\square$

We will define the $\mathbb{R}$-analogs of the trigonometric polynomials (2.38). Suppose that the semigroup $\{e^{tA}\}_{t \geq 0}$ is hyperbolic. Then, $\sigma(A) \cap i\mathbb{R} = \emptyset$ and, moreover, $\sup\{\|(A - is)^{-1}\| : s \in \mathbb{R}\} < \infty$. Given $v \in \mathcal{S}(\mathbb{R}; X)$ where $\mathcal{S}(\mathbb{R}; X)$ is the Schwartz class, let f_v and g_v denote the functions
$$f_v(\theta) = \frac{1}{2\pi} \int_\mathbb{R} (A - is)^{-1} v(s) e^{i\theta s} ds, \quad g_v(\theta) = \frac{1}{2\pi} \int_\mathbb{R} v(s) e^{i\theta s} ds, \quad \theta \in \mathbb{R},$$
and define the sets
$$\mathfrak{F} := \{f_v : v \in \mathcal{S}(\mathbb{R}; X)\}, \quad \mathfrak{G} := \{g_v : v \in \mathcal{S}(\mathbb{R}; X)\}.$$

PROPOSITION 2.48. *If $\sup_{s \in \mathbb{R}} \|(A - is)^{-1}\| < \infty$, then*
(i) $\mathfrak{G}$ *is dense in $L^p(\mathbb{R}; X)$;*
(ii) $\mathfrak{G}$ *consists of differentiable functions;*
(iii) $\mathfrak{F}$ *is dense in $\mathcal{D}(\Gamma')$; and*
(iv) *if $v \in \mathcal{S}(\mathbb{R}; X)$, then $\Gamma f_v = g_v$.*

PROOF. Let us denote the Fourier transform of $g \in L^1(\mathbb{R}; X)$ by
$$\hat{g}(\theta) = \frac{1}{2\pi} \int_\mathbb{R} e^{-is\theta} g(s) ds.$$
Note that $\mathfrak{G} = \{g : \mathbb{R} \to X : \hat{g} = v \text{ for some } v \in \mathcal{S}(\mathbb{R}, X)\}$. Thus, $\mathfrak{G}$ contains the set $\{g \in L^1(\mathbb{R}; X) : \hat{g} \in \mathcal{S}(\mathbb{R}; X)\}$. Since the latter set is dense in $L^p(\mathbb{R}; X)$, property (i) follows.

Property (ii) holds because, for each $v \in \mathcal{S}(\mathbb{R}; X)$, the inverse Fourier transform $g_v = \check{v}$ is again in $\mathcal{S}(\mathbb{R}; X)$.

For property (iii), note that if $v \in \mathcal{S}(\mathbb{R}; X)$, then the function w given by $w(s) = (A - is)^{-1} v(s)$, for $s \in \mathbb{R}$, is also in $\mathcal{S}(\mathbb{R}; X)$. Hence f_v is differentiable with derivative
$$f_v'(\theta) = \frac{1}{2\pi} \int_\mathbb{R} is(A - is)^{-1} v(s) e^{i\theta s} ds = \frac{1}{2\pi} \int_\mathbb{R} is w(s) e^{i\theta s} ds,$$

and it follows that $f_v' \in L^p(\mathbb{R}; X)$. Since the function $s \mapsto A(A - is)^{-1}v(s)e^{i\theta s}$ is integrable and A is a closed operator, we have

$$f_v(\theta) \in \mathcal{D}(A) \quad \text{and} \quad Af_v(\theta) = \frac{1}{2\pi} \int_{\mathbb{R}} A(A - is)^{-1}v(s)e^{is\theta}\, ds, \quad \theta \in \mathbb{R}.$$

Using $v(s) = (A - is)\phi(s)x$ where ϕ is a scalar-valued Schwartz function and $x \in \mathcal{D}(A)$, we conclude that the set $\mathfrak{F}$ is dense in $\mathcal{D}(\Gamma') = \mathcal{D}(d/dt) \cap \mathcal{D}(\mathcal{A})$.

The proof of property (iv) is simply a computation:

$$(\Gamma f_v)(\theta) = \frac{1}{2\pi} \int_{\mathbb{R}} [-is(A - is)^{-1}v(s)e^{is\theta} + A(A - is)^{-1}v(s)e^{is\theta}]\, ds$$

(2.39)
$$= \frac{1}{2\pi} \int_{\mathbb{R}} (A - is)(A - is)^{-1}v(s)e^{is\theta}\, ds = g_v(\theta).$$

$\square$

Define
$$\Lambda_\mathcal{D} := \{w \in \mathcal{S}(\mathbb{R}; X) : w(s) \in \mathcal{D}(A) \quad \text{for} \quad s \in \mathbb{R}\},$$
and suppose that Γ, respectively Γ_+, denotes the generator of the evolution semigroup $\{E^t\}_{t \geq 0}$, respectively $\{E_+^t\}_{t \geq 0}$, that is defined by formula (2.31), respectively formula (2.35), on $L^p(\mathbb{R}; X)$, respectively $L^p(\mathbb{R}_+; X)$.

THEOREM 2.49. (i) If $\sigma(A) \cap i\mathbb{R} = \emptyset$ and $\sup_{s \in \mathbb{R}} \|(A - is)^{-1}\| < \infty$, then

$$\|\Gamma\|_{\bullet, L^p(\mathbb{R}; X)} = \inf_{w \in \Lambda_\mathcal{D}} \frac{\|\int_{\mathbb{R}} (A - is)w(s)e^{is(\cdot)}ds\|_{L^p(\mathbb{R}; X)}}{\|\int_{\mathbb{R}} w(s)e^{is(\cdot)}ds\|_{L^p(\mathbb{R}; X)}}.$$

(ii) Γ is invertible on $L^p(\mathbb{R}; X)$ if and only if $\{e^{tA}\}_{t \geq 0}$ is hyperbolic; and, if this is the case, then

$$\|\Gamma^{-1}\|_{\mathcal{L}(L^p(\mathbb{R}; X))} = \sup_{v \in \mathcal{S}(\mathbb{R}; X)} \frac{\|\int_{\mathbb{R}} (A - is)^{-1}v(s)e^{is(\cdot)}ds\|_{L^p(\mathbb{R}; X)}}{\|\int_{\mathbb{R}} v(s)e^{is(\cdot)}ds\|_{L^p(\mathbb{R}; X)}}.$$

(iii) Γ_+ is invertible on $L^p(\mathbb{R}_+; X)$ if and only if $\{e^{tA}\}_{t \geq 0}$ is uniformly exponentially stable; and, if this is the case, then

$$\|\Gamma_+^{-1}\|_{\mathcal{L}(L^p(\mathbb{R}_+; X))} = \|\Gamma^{-1}\|_{\mathcal{L}(L^p(\mathbb{R}; X))}.$$

PROOF. To show (i), let $v \in \mathcal{S}(\mathbb{R}; X)$ and note that $s \mapsto w(s) = (A - is)^{-1}v(s)$ defines a function w in $\Lambda_\mathcal{D}$. We have that

$$g_v(\tau) = \frac{1}{2\pi} \int_{\mathbb{R}} (A - is)(A - is)^{-1}v(s)e^{is\tau}\, ds$$
$$= \frac{1}{2\pi} \int_{\mathbb{R}} (A - is)w(s)e^{is\tau}\, ds$$

and
$$f_v(\tau) = \frac{1}{2\pi} \int_{\mathbb{R}} w(s)e^{is\tau}\, ds.$$

Also, by Proposition 2.48,

$$\|\Gamma\|_{\bullet, L^p(\mathbb{R}; X)} = \inf_{f_v \in \mathfrak{F}} \frac{\|\Gamma f_v\|_{L^p(\mathbb{R}; X)}}{\|f_v\|_{L^p(\mathbb{R}; X)}} = \inf_{v \in \mathcal{S}(\mathbb{R}; X)} \frac{\|g_v\|}{\|f_v\|}$$
$$= \inf_{w \in \Lambda_\mathcal{D}} \frac{\|\int_{\mathbb{R}} (A - is)w(s)e^{is(\cdot)}\, ds\|}{\|\int_{\mathbb{R}} w(s)e^{is(\cdot)}\, ds\|}.$$

To prove statement (ii), note that

$$\|\Gamma^{-1}\|_{\mathcal{L}(L^p(\mathbb{R};X))} = \|\Gamma\|_{\bullet, L^p(\mathbb{R};X)}^{-1} = \left[\inf_{v \in \mathcal{S}(\mathbb{R};X)} \frac{\|\Gamma f_v\|}{\|f_v\|}\right]^{-1} = \sup_{v \in \mathcal{S}(\mathbb{R};X)} \frac{\|f_v\|}{\|g_v\|},$$

and use Theorem 2.39.

Assertion (iii) follows from Proposition 2.47 and Corollary 2.46. □

2.3. Bibliography and remarks

Introduction to Semigroups. Of the many excellent books on semigroups and their applications, we suggest Davies [**Dv**], deLaubenfels [**De**], Engel and Nagel [**EN**], Goldstein [**Go2**], Nagel [**Na**], Pazy [**Pz**], and Webb [**Wb**].

Basics. The books by Nagel [**Na**] and by van Neerven [**vN**] give a systematic account of the asymptotic behavior of semigroups. See also the web site http://ma1serv. mathematik. uni-karlsruhe. de/evolve-l.

Spectral mapping theorems. Theorems 2.6, 2.8 and 2.9 are taken from Pazy [**Pz**]; Theorem 2.7 from Nagel [**Na**].

Theorem 2.10 is due to Gearhart [**Ge**] for the case of contraction semigroups. It has been modified and generalized in various directions by G. Greiner, I. Herbst, F. Huang, and J. Prüss. A beautiful proof of this result, as well as a generalization to Banach spaces, is due to G. Greiner, see for example Nagel [**Na**] and van Neerven [**vN**, Section 2.2]. A similar theorem asserting that boundedness of the resolvent of the generator on the closed right half plane implies stability on a Hilbert space is known in the control theory literature as Huang's Theorem, see [**Hu**], Theorem 5.1.5 in the book by Curtain and Zwart [**CZ**], and the paper of Weiss [**Ws**].

Theorem 2.10 (without the contractivity assumption) as well as (2.5) was proved by Prüss [**Pr**]. One of the main results in this paper may be stated as follows:

THEOREM 2.50. *If X is a Banach space, then $1 \in \rho(e^{2\pi A})$ if and only if, for each continuous 2π-periodic function $f : [0, 2\pi] \to X$, there exists a unique continuous 2π-periodic solution u of the mild equation*

$$u(t) = e^{At}u(0) + \int_0^t e^{A(t-s)} f(s)\, ds, \quad t \in [0, 2\pi].$$

Gearhart's Theorem 2.10 is obtained in [**Pr**] from Theorem 2.50. In Theorem 4.28 below we give a nonautonomous version of Theorem 2.50. A simple proof of Gearhart's Theorem is given by Howland in [**Ho2**]; the argument (for the Hilbert space setting) is close to our proof of Theorem 2.31.

Spectral mapping theorems for integrated semigroups are proved by Day [**Dy**] and Greiner and Müller [**GM**]. See [**Na**] for a discussion of the *weak spectral mapping property;* namely, the equality $\sigma(e^{tA}) = \overline{\exp t\sigma(A)}$, for all $t \geq 0$. This property also does not hold for all strongly continuous semigroups. For weak spectral mapping theorems see [**Na, vN**], Arendt and Greiner [**AG**], Greiner and Schwarz [**GS**], and Vu Quoc Phong and Lyubich [**LP2**].

We mention papers by Wrobel [**Wr, Wr2**] and by Weiss [**We, WW**], see also Martinez and Mazon [**MM**] and several papers by Renardy [**Rn, Rn2, Rn3**]. For some recent advances, see the work by S. Huang; for example, [**Hn**].

Recently, Nagel and Poland [**NP**] introduced the *critical spectrum*, denoted $\sigma_{\mathrm{crit}}(e^{tA})$, for a strongly continuous semigroup $\{e^{tA}\}_{t \geq 0}$. The critical spectrum gives (in an optimal way) the spectral mapping theorem $\sigma(e^{tA}) = e^{t\sigma(A)} \cup \sigma_{\text{crit}}(e^{tA})$. See also related papers by Blake [**Bl**].

Growth and spectral bounds. See Nagel [**Na**], van Neerven [**vN**], and an excellent recent review by Arendt [**Ar2**]. We refer to Remark 5.9 below for a *formula* for the growth bound $\omega(A)$ given in terms similar to Theorem 2.49; that is, in fact, in terms of Fourier multipliers.

There is a vast literature on *individual stability* and related questions, see for example [**vN**], Batty, van Neerven, and Räbiger [**BvNR, BvNR2**], Vũ Quoc Phong and Ruess [**PR**], Arendt and Batty [**AB2**], Batty and Chill [**BC**], and, of course, Arendt and Batty [**AB**], and Lyubich and Vũ Quoc Phong [**LP**].

Hyperbolic semigroups. For the case of bounded generators (Example 2.14), see the book by Daleckij and Krein [**DK**]. Lemma 2.15 is taken from the preliminary version of [**vN**]. Theorem 2.16 is a direct application of Gearhart's Theorem, see for example Kaashoek and Verduyn Lunel [**KL**]. Theorem 2.17 and Corollary 2.18 are due to Kaashoek and Verduyn Lunel [**KL**]. See [**vN**, Section 4.6] for a further development of Kaashoek-Lunel theory. The paper [**KL**] contains several examples — delay equations and one-dimensional hyperbolic PDEs — where the conditions of Theorem 2.17 and Corollary 2.18 can be effectively checked.

Examples. Example 2.21 is due to Zabczyk [**Za**]. To our knowledge it is the first example such that $s(A) < \omega(A)$. We follow the exposition in [**Na**]. See also [**vN**, Example 1.2.4] for the proof that $s(A) < \omega_1(A)$ in Zabczyk's example. Example 2.22 is due to Greiner, Voigt, and Wolff [**GVW**], but we follow the exposition in [**Na**]. Examples 2.23 and 2.24 are due to Arendt [**Ar**]. See Davies [**Dv**, p. 44] for an example of a strongly continuous group on ℓ^2 with compact resolvent and $s(A) < \omega(A)$. See Wolff [**Wo**] for an example of a positive group with $s(A) < \omega(A)$. However, note that $s(A) = \omega(A)$ for positive strongly continuous semigroups on $C_0(\Theta)$ (Batty-Davies), on $L^1(\Theta)$ (Derndinger-Greiner), and on $L^2(\Theta)$ (Greiner-Nagel), see [**Na**]. Using Theorem 2.39 and Corollary 2.40, Weis [**We**] showed this result for any $L^p(\Theta)$, $1 \leq p < \infty$, see also Montgomery-Smith [**Mt**] and [**We3**] for shorter proofs. Example 2.25 is due to Hille and Phillips [**HPh**, Sec. 23.16]; our exposition is taken from Henry [**He2**]. This semigroup was studied by Arendt, El Mennaoui, and Hieber [**AEMH**] for the spaces $L^p(0,1)$, $1 < p < \infty$, where it was shown that A has a compact resolvent. Example 2.26 is due to Renardy [**Rn**]. This result is especially exciting, because $s(A) = \omega(A)$ for generators that correspond to *one-dimensional* hyperbolic equations, see [**KL**] and Neves, Ribeiro, and Lopes [**NRL**]. Example 2.27 is due to Montgomery-Smith [**Mt**]; it shows that the existence of a dichotomy for a semigroup is not controlled by the spectrum of its infinitesimal generator. However, we stress again that $\omega(A) = s(A)$ for every positive semigroup on L^2, see [**Na**]. Thus, the stability of a semigroup is controlled by the spectrum of its generator.

Evolution semigroups and hyperbolicity. In this chapter we considered the evolution semigroups that are generated by a strongly continuous semigroup $\{e^{tA}\}_{t \geq 0}$ on X, or by an autonomous differential equation $\dot{x} = Ax$. In Chapters 3 and 6 we will define evolution semigroups of type (2.20) and (2.31) for the more general (nonautonomous) situation that corresponds to a differential equation $\dot{x} = A(t)x$ or to a variational equation $\dot{x} = A(\varphi^t \theta)x$. Therefore, we postpone the detailed bibliographical remarks about evolution semigroups.

Evolution semigroups on periodic functions. This subsection is based on the papers by Latushkin and Montgomery-Smith [**LM, LM2**], see also the book by van

Neerven [**vN**]. Propositions 2.28 and 2.36 are the simplest (autonomous) versions of the general phenomenon of rotational invariance of the spectrum of evolution operators, see Lemma 6.28 in Subsection 6.2.2. Lemma 2.29 goes back to the papers by Chicone and Swanson [**CS, CS3**] where the situation of finite dimensional linear skew-product flows is considered. These papers already contain the main idea of "interpolation" of an approximate eigenvector, see the definition of the approximate eigenfunction g for Γ_{per} in (2.22). Formula (2.22), in fact, explains the reason why the evolution semigroups have the spectral mapping property in $\mathcal{F}$ even when the underlying semigroup $\{e^{tA}\}_{t\geq 0}$ does not have this property in X. Indeed, the "super"-space $\mathcal{F}$ has many more potential approximate eigenfunctions for Γ than there are in the corresponding space X for the generator A.

Theorem 2.30 and Theorem 2.31 are taken from [**LM, LM2**]. Another Banach-space version of Gearhart's Theorem is proved by G. Greiner, see [**Na**] or [**vN**]. The idea to use the discrete Fourier transform, which is the main idea for the proof, was first used for the Hilbert space setting by Howland in [**Ho2**]. Formula (2.30) is important by itself. An analogue of this formula for the case of the real line is used in Subsection 2.2.3 to compute the norms of the inverses to the generators of evolution semigroups. In turn, this result is used in Section 5.1 to obtain an application to control theory. Remark 2.32 corresponds to the Datko-van Neerven Theorem 2.44 that uses convolution operators to characterize the stability of $\{e^{tA}\}_{t\geq 0}$. We mention an important paper by Vu Quoc Phong and Schüler [**PS**] where, in particular, yet another proof of Theorem 2.31 is given. Their proof is based on the solvability of an operator equation of Lyapunov type.

Evolution semigroups are related to L^p-maximal regularity, in particular, to the question of when the operator $-d/dt + \mathcal{A}$ is closed, cf. Remark 2.35. For the literature on maximal regularity see the excellent review by Dore [**Do**], the books by Lunardi [**Ln**] and Prüss [**Pr2**], the papers by Clement and Guerre-Delabriere [**CG**] and by Lancien and Le Merdy [**LlM**], and the bibliographies therein. Recall, that a differential equation $\dot{u} = Au + f$ has L^p-*maximal regularity* on $[0, 2\pi]$ if for each $f \in L^p([0, 2\pi]; X)$ there exists a unique solution $u \in W_1^p([0, 2\pi]; X) \cap L^p([0, 2\pi]; \mathcal{D}(A))$ such that $u(0) = u(2\pi)$. Here W_1^p is the Sobolev space and $\mathcal{D}(A)$ is equipped with the graph-norm. Maximal regularity is equivalent to the fact the operator $\mathcal{A}(-d/dt + \mathcal{A})^{-1}$, or $d/dt(-d/dt + \mathcal{A})^{-1}$, is bounded on $L^p([0, 2\pi]; X)$. Maximal regularity implies certain restrictions on the generator A (it should generate an analytic semigroup) and on the Banach space X (it should have the UMD-property). Similarly to the proof of Theorem 2.31, one can give the following characterization: Under the assumption that $i\mathbb{Z} \subset \rho(A)$, the equation $\dot{x} = Ax$ has maximal regularity if and only if there is a constant $C > 0$ such that

$$\|\sum_k k(A - ik)^{-1} e^{ik\theta} x_k\|_{L^p([0,2\pi];X)} \leq C\|\sum_k e^{ik\theta} x_k\|_{L^p([0,2\pi];X)}$$

for every finite sequence $\{x_k\} \subset X$. Parseval's identity immediately implies a classical result by De Simon: L^2-maximal regularity holds provided X is a Hilbert space, see for example [**Do**].

Evolution semigroups on the real line. This subsection is based on the papers [**LM, LM2**]. Lemma 2.37 and its continuous counterpart Lemma 2.38 are obtained from the fact that $\ell^p(\mathbb{Z}; X)$ and $L^p(\mathbb{R}; X)$ can be represented as tensor products of the scalar spaces $\ell^p(\mathbb{Z})$ and $L^p(\mathbb{R})$ with X. In this tensor product, the evolution semigroup is a tensor product of the translation semigroup and the underlying semigroup on X. The short proof of these lemmas was suggested by F. Räbiger. Theorem 2.39 is taken from [**LM, LM2**]. There are several proofs available for this result, see Räbiger and Schnaubelt [**RS, RS2, Sc**]. Our second proof of this theorem, based on Lemma 2.41, is due to S. Montgomery-Smith and Y. Latushkin (unpublished). Corollary 2.40 summarizes the results in Theorem 2.39 in a different way. It shows that the generator Γ of the evolution semigroup is the correct "replacement" for an unbounded generator A in the classical Lyapunov Theorem, see the Introduction. Corollary 2.40 was used by Weis [**We**] to obtain an affirmative answer to the following long-standing open question: Is $s(A) = \omega(A)$ for every generator A of a positive semigroup on $L^p(\Omega, \mu)$?

Evolution semigroups on the half line. Theorem 2.42 and Proposition 2.43 are taken from the paper by van Neerven [**vN2**], see also [**vN**]. Theorem 2.44 is due to Datko [**Da2**] and van Neerven [**vN2**]. There is a connection between Theorem 2.44 and a result on the Green's function to be found Chapter 4 of this book. Theorem 2.49 and its proof are taken from [**CLRM**] where the formulas for the norm of Γ_{per}^{-1} are given and used to estimate the stability radius, see Section 5.1 for applications to control theory.

CHAPTER 3

Evolution Families and Howland Semigroups

In this chapter we will study evolution families $\{U(\theta,\tau)\}_{\theta \geq \tau}$ on a Banach space X and the corresponding evolution semigroups $\{E^t\}_{t \geq 0}$ defined on "super-spaces" of X-valued functions on $\mathbb{R}$ or $\mathbb{R}_+$. Evolution families arise naturally from the solutions of nonautonomous differential equations on Banach spaces.

We will prove the Spectral Mapping Theorem, Theorem 3.13, for the evolution semigroups. The main result of this chapter is the Dichotomy Theorem 3.17. It states that an evolution family has an exponential dichotomy if and only if the corresponding evolution semigroup on $\mathbb{R}$ is hyperbolic. The difficult part of this theorem is to show that each spectral projection for the evolution operator E^t is a multiplication operator defined by a strongly continuous projection-valued operator function. We call this result the Spectral Projection Theorem, see Theorem 3.14 below. The proof of the Spectral Projection Theorem is based on the following two facts: the Commutation Lemma 3.15, which states that a spectral projection for E^t commutes with the multiplication operators defined by scalar functions, and the Räbiger-Schnaubelt Lemma 3.16, which states that the domain of the infinitesimal generator Γ of the semigroup E^t has a certain regularity property. Finally, we will characterize the stability of an evolution family in terms of the stability of the corresponding evolution semigroup on the half line.

3.1. Evolution families and dichotomy

In this section we give some preliminary definitions and results about strongly continuous evolution families. Also, we discuss the definitions of exponential stability and dichotomy for evolution families.

3.1.1. Abstract Cauchy problems and their evolution families. Unlike an autonomous differential equation $\dot{x} = Ax$ on a Banach space X where the solution may be given by a one-parameter semigroup $\{e^{tA}\}_{t \geq 0}$, the solution of a nonautonomous differential equation $\dot{x} = A(t)x$ may be given by means of a two-parameter family $\{U(\theta,\tau)\}_{\theta \geq \tau}$ of bounded operators on X which we will call an *evolution family*. Of course, it may be very difficult to prove the existence of an evolution family for a given differential equation on an infinite dimensional Banach space. In fact, existence is proved only for some special classes of nonautonomous differential equations. We will not pursue in this book the problem of the existence of evolution families for differential equations. Instead, we will always assume that an evolution family is given and then explore its properties.

DEFINITION 3.1. A family of operators $\{U(\theta,\tau)\}_{\theta \geq \tau} \subset \mathcal{L}(X)$, with $\theta,\tau \in \mathbb{R}$ or $\theta,\tau \in \mathbb{R}_+$, is called an *evolution family* if
  (i) $U(\theta,\tau) = U(\theta,s)U(s,\tau)$ and $U(\theta,\theta) = I$ for all $\theta \geq s \geq \tau$; and
  (ii) for each $x \in X$, the function $(\theta,\tau) \mapsto U(\theta,\tau)x$ is continuous for $\theta \geq \tau$.

An evolution family $\{U(\theta,\tau)\}_{\theta\geq\tau}$ is called *exponentially bounded* if, in addition,

(iii) there exist constants $M \geq 1$ and $\omega > 0$ such that
$$\|U(\theta,\tau)\| \leq Me^{\omega(\theta-\tau)}, \quad \theta \geq \tau.$$

Some authors, see for example Datko [**Da2**] and Fattorini [**Fa**], consider evolution families for $0 \leq \tau \leq \theta$ only. However, a canonical extension for all $\theta \geq \tau \in \mathbb{R}$ is obtained by setting
$$\tilde{U}(\theta,\tau) = U(\max\{\theta,0\}, \max\{\tau,0\}).$$

If A is the infinitesimal generator of a strongly continuous semigroup on X, then $U(\theta,\tau) = e^{(\theta-\tau)A}$, for $\theta \geq \tau$, is a strongly continuous, exponentially bounded evolution family. However, generally, properties of evolution families are quite different from the properties of strongly continuous semigroups. We mention a few of the differences.

Instead of a one-parameter family of operators defining a semigroup, an evolution family is a *two-parameter* family of operators. To overcome this problem, we will associate a strongly continuous semigroup, called an evolution semigroup, to each evolution family. It turns out that many properties of the evolution family can be obtained from the properties of this associated semigroup.

If $\{e^{tA}\}_{t\geq 0}$ is a strongly continuous semigroup, then the continuity of the function $t \mapsto e^{tA}x$ for all $x \in X$ implies its differentiability for $x \in \mathcal{D}(A)$. In contrast, strong continuity of evolution families does not imply their differentiability. For example, consider a continuous function $p : \mathbb{R} \to [1/2, 1]$ and define an evolution family on $X = \mathbb{R}$ by $U(\theta,\tau) = \frac{p(\theta)}{p(\tau)}$, $\theta \geq \tau$. With an appropriate choice of p this evolution family is strongly continuous, but not differentiable.

On a reflexive space X, the adjoint operation does not preserve the strong continuity of evolution families. For example, let u denote a function from $\mathbb{R}$ to the set of bounded invertible operators on X. If u and u^{-1}, $u^{-1}(\tau) = [u(\tau)]^{-1}$, are bounded and strongly continuous, but the adjoint function $u^* : \mathbb{R} \to \mathcal{L}(X^*)$ is not strongly continuous, then $U(\theta,\tau) = u(\theta)u(\tau)^{-1}$ defines a strongly continuous evolution family on X such that the adjoint evolution family $\tilde{U}(\theta,\tau) = U(\theta,\tau)^*$ is *not* strongly continuous on X^*. As a concrete example, let $X = L^2(\mathbb{R}_+)$, consider the semigroup of left translations on X given by $(\tilde{V}^t f)(\tau) = f(\tau + t)$, for $\tau \in \mathbb{R}$, and define $u(\theta) = I - \frac{1}{2}\tilde{V}^{1/|\theta|}$ for $\theta \neq 0$ and $u(0) = I$.

Evolution families appear as solutions for abstract Cauchy problems of the form

(ACP) $\quad \dot{x}(t) = A(t)x(t), \quad x(\tau) = x_\tau, \quad x_\tau \in \mathcal{D}(A(\tau)), \quad t \geq \tau, \quad t, \tau \in \mathbb{R}$

where the domain $\mathcal{D}(A(\tau))$ of the operator $A(\tau)$ is assumed to be dense in X.

DEFINITION 3.2. An evolution family $\{U(\theta,\tau)\}_{\theta\geq\tau}$ is said to *solve the abstract Cauchy problem* (ACP) if, for each $\tau \in \mathbb{R}$, there exists a dense subset $Y_\tau \subseteq \mathcal{D}(A(\tau))$ such that, for each $x_\tau \in Y_\tau$, the function $x(\cdot) = U(\cdot,\tau)x_\tau$ given by $x(t) := U(t,\tau)x_\tau$, for $t \geq \tau$, is differentiable, $x(t) \in \mathcal{D}(A(t))$, and (ACP) holds. The solution $x(\cdot)$ is called the *classical solution* of (ACP). The abstract Cauchy problem (ACP) is called *well-posed* if there is an evolution family that solves (ACP).

This definition of well-posedness is quite general; we follow Schnaubelt [**Sc**], see also Engel and Nagel [**EN**]. More restrictive definitions can be given if we require that $Y_\tau = \mathcal{D}(A(\tau))$ for each $\tau \in \mathbb{R}$ or even that $Y_\tau = \mathcal{D}(A(\tau)) = \mathcal{D}$ is independent of τ, see Fattorini [**Fa**].

Since the definition of the evolution family requires merely that $(\theta, \tau) \mapsto U(\theta, \tau)$, $\theta \geq \tau$, is *strongly* continuous, the operators $A(t)$ in (ACP) can be unbounded.

There are several difficulties that distinguish the case of autonomous equations $\dot{x} = Ax$ from the nonautonomous abstract Cauchy problems (ACP). For example, there is no general existence theorem analogous to the Hille-Yosida Theorem 2.2 for the nonautonomous case. This theorem gives a characterization of the infinitesimal generators of strongly continuous semigroups; and therefore, it characterizes all well-posed autonomous abstract Cauchy problems.

The conditions required to obtain an evolution family that solves the abstract Cauchy problem (ACP) where $A(t)$ is a family of *bounded* operators are well understood. For instance, if the function $t \mapsto A(t) \in \mathcal{L}(X)$ is bounded and continuous for $t \in \mathbb{R}$, then $U(\theta, \tau)$ is defined for all $(\theta, \tau) \in \mathbb{R}^2$, each such operator is invertible, and the function given by $(\theta, \tau) \mapsto U(\theta, \tau)$ is $\|\cdot\|_{\mathcal{L}(X)}$-continuous, see Daleckij and Krein [**DK**] or Fattorini [**Fa**] for details.

Generally, when the operators $A(t)$ are unbounded, it is a very delicate matter to prove that an abstract Cauchy problem is well posed. We refer the reader to the sources cited in the bibliographical remarks at the end of this chapter.

We remark that the well-posedness of (ACP) in the sense of classical solutions can be destroyed by a bounded and continuous perturbation. For instance, consider a strongly continuous semigroup $\{e^{tA_0}\}_{t \geq 0}$ on X, let $B(t) \in \mathcal{L}(X)$ for $t \geq 0$, and define $A(t) = A_0 + B(t)$ with $\mathcal{D}(A(t)) = \mathcal{D}(A_0)$. The equation $\dot{x} = A(t)X$ may not have a differentiable solution for all initial conditions $x(0) = x_0 \in \mathcal{D}(A_0)$, even if $t \mapsto B(t)$ is continuous, see for example Phillips [**Ph**].

Consequently, it is important to consider solutions that exist only in a "mild sense". For this, suppose that $\{U(\theta, \tau)\}_{\theta \geq \tau}$ is the evolution family that solves the abstract Cauchy problem (ACP) with $Y_\tau = \mathcal{D}(A(\tau))$. Consider an initial value problem for the nonautonomous inhomogeneous differential equation

(3.1) $$\dot{x}(t) = A(t)x(t) + g(t), \quad x(\tau) = x, \quad \tau \in \mathbb{R},$$

where g is a locally integrable X-valued function on $\mathbb{R}$. For some $x \in \mathcal{D}(A(\tau))$, the initial value problem may have a classical solution $x(\cdot) \in C^1([\tau, \infty); X)$ with $x(t) \in \mathcal{D}(A(t))$ for $t \geq \tau$. Since (ACP) is well-posed, we have that

(3.2) $$\frac{\partial}{\partial t} U(t, \tau)x = A(t)U(t, \tau)x, \quad \frac{\partial}{\partial \tau} U(t, \tau)x = -U(t, \tau)A(\tau)x.$$

Indeed, the second equality in display (3.2) follows from the first and the identity

$$h^{-1}\Big[U(t, \tau + h) - U(t, \tau)\Big]x = U(t, \tau + h) \cdot h^{-1}\Big[I - U(\tau + h, \tau)\Big]x, \quad h > 0.$$

Now (3.1) and (3.2) imply that the function $\tau \mapsto U(t, \tau)x(\tau)$ is differentiable with $\frac{\partial}{\partial \tau} U(t, \tau)x(\tau) = U(t, \tau)g(\tau)$. Therefore, our classical solution satisfies the following mild integral equation (see more details in Pazy [**Pz**, p. 129]):

(3.3) $$x(t) = U(t, \tau)x(\tau) + \int_\tau^t U(t, s)g(s)\, ds, \quad t \geq \tau.$$

Thus, it is natural to give the following definition: A continuous function $t \mapsto x(t)$ with values in X is called a *mild solution* of (3.1) with initial value $x(\tau) = x \in \mathcal{D}(A(\tau))$ if (3.3) holds. We will consider mild solutions in Chapter 4.

We conclude this subsection with a brief description of an example of a well-posed nonautonomous Cauchy problem. We refer to Pazy [**Pz**], Section 7.6, and Tanabe [**Ta**], Chapter 5, for the basic theory of parabolic differential equations with variable coefficients defined on a bounded domain $\Omega \subset \mathbb{R}^n$ and for the proof of the result that we now outline.

EXAMPLE 3.3. We mention here an example of a class of well-posed PDEs that can be written in the form $\dot{x} = A(t)x$ where the operators $A(t)$ are unbounded, and where it is possible to prove the existence of an evolution family that solves the corresponding (ACP).

Consider the differential operator
$$\mathcal{A}(t,y) := -\Sigma_{|\alpha| \leq 2m} a_\alpha(t,y) D^\alpha$$
where $\alpha = (\alpha_1, \ldots, \alpha_n) \in \mathbb{N}^n$ is a multi-index, $D^\alpha := D^{\alpha_1} \cdot \ldots \cdot D^{\alpha_n}$, and $D^j := \frac{\partial}{\partial y_j}$. The coefficients $a_\alpha(t, \cdot)$ are assumed to be smooth functions in $\overline{\Omega}$ that satisfy certain conditions (see [**Pz**, p. 226]). In particular, the operators $-\mathcal{A}(t,y)$, for $t \geq 0$, are uniformly strongly elliptic in Ω; that is, there is some constant $c > 0$ such that
$$(-1)^m \operatorname{Re} \sum_{|\alpha|=2m} a_\alpha(t,y)\xi^\alpha \geq c|\xi|^{2m}$$
for all $y \in \overline{\Omega}$, $0 \leq t \leq T$, and $\xi \in \mathbb{R}^n$.

Consider the Cauchy problem given by

(3.4)
$$\begin{aligned}
\frac{\partial x}{\partial t} &= \mathcal{A}(t,y)x, \quad (t,y) \in [0,T] \times \Omega, \\
D^\alpha x(t,y) &= 0 \quad \text{for} \quad |\alpha| < m, \quad (t,y) \in [0,T] \times \partial\Omega, \\
x(0,y) &= x_0(y), \quad y \in \Omega.
\end{aligned}$$

There is a family of unbounded linear operators $A(t)$ on $X = L^p(\Omega)$, $1 < p < \infty$, associated with the operators $\mathcal{A}(t,y)$. The operators $A(t)$ are given by $A(t)x = \mathcal{A}(t,\cdot)x$, for $x \in \mathcal{D}$, where $\mathcal{D} := \mathcal{D}(A(t)) = W_{2m}^p(\Omega) \cap W_{m,0}^p(\Omega)$ is the indicated intersection of Sobolev spaces. Given $x_0 \in \mathcal{D}$, the solution to the abstract Cauchy problem $\dot{x} = A(t)x$, $x(0) = x_0$ is defined to be the generalized solution of (3.4). In other words, the solution $x(\cdot)$ is such that $x(t) \in \mathcal{D}$ for $t > 0$, the derivative dx/dt exists in L^p and is continuous on $(0,T]$, and the solution $x(\cdot)$ is continuous on $[0,T]$, see [**Pz**, p. 226]. It can be proved, see Theorem 5.6.1 and Lemma 7.6.1 in [**Pz**], that there exists an evolution family $\{U(\theta,\tau)\}$ that solves the corresponding abstract Cauchy problem. ◇

3.1.2. Stability and dichotomy of evolution families. The definition of a uniformly exponentially stable evolution family is similar to the one given for semigroups.

DEFINITION 3.4. An evolution family $\{U(\theta,\tau)\}_{\theta \geq \tau}$ is called *uniformly exponentially stable*, or just *stable*, if its *growth bound*, defined by
$$\omega(U) := \inf\{\omega : \text{there exists } M = M(\omega) \text{ such that }$$
$$\|U(\theta,\tau)\|_{\mathcal{L}(X)} \leq Me^{\omega(\theta-\tau)} \text{ for all } \theta \geq \tau\},$$
is negative.

The following example shows that the growth bound of an evolution family that solves an abstract Cauchy problem cannot be characterized in terms of the spectra of the operators $A(t)$ even for a nonautonomous differential equation in the Banach space $\mathbb{R}^2$.

EXAMPLE 3.5. Let

$$A_0 = \begin{bmatrix} -1 & -5 \\ 0 & -1 \end{bmatrix}, \quad W(t) = \begin{bmatrix} \cos t & \sin t \\ -\sin t & \cos t \end{bmatrix}, \quad A(t) = W^{-1}(t) A_0 W(t),$$

and note that $\sigma(A(t)) = \sigma(A_0) = \{-1\}$ for $t \in \mathbb{R}$, an indication of stability. The time dependent change of variables given by $z(t) = W(t)x(t)$ transforms the nonautonomous differential equation $\dot{x} = A(t)x$ to the autonomous differential equation $\dot{z} = Bz$ where

$$B = A_0 + \dot{W}(t)W^{-1}(t) = \begin{bmatrix} -1 & -4 \\ -1 & -1 \end{bmatrix}.$$

But, since $\sigma(B) = \{-2, 2\}$, the trivial solution of the differential equation $\dot{z} = Bz$ is unstable. Clearly, this also implies that the trivial solution of $\dot{x} = A(t)x$ is unstable. $\diamond$

We now give the definition of exponential dichotomy for a (strongly continuous) evolution family. To do this, let us first agree that if $P : \mathbb{R} \to \mathcal{L}(X)$ is a projection valued function, then the function whose values are the complementary projections is denoted by $Q(\theta) = I - P(\theta)$ for each $\theta \in \mathbb{R}$. If, for all $\theta \geq \tau$, we have $P(\theta)U(\theta, \tau) = U(\theta, \tau)P(\tau)$, then we denote by

$$U_P(\theta, \tau) := P(\theta)U(\theta, \tau)P(\tau), \quad U_Q(\theta, \tau) := Q(\theta)U(\theta, \tau)Q(\tau),$$

the restrictions of the operator $U(\theta, \tau)$ on $\operatorname{Im} P(\tau)$ and $\operatorname{Im} Q(\tau)$, respectively. We stress that $U_P(\theta, \tau)$ is an operator from $\operatorname{Im} P(\tau)$ to $\operatorname{Im} P(\theta)$ while $U_Q(\theta, \tau)$ maps $\operatorname{Im} Q(\tau)$ to $\operatorname{Im} Q(\theta)$.

DEFINITION 3.6. An evolution family $\{U(\theta, \tau)\}_{\theta \geq \tau}$ is said to have an *exponential dichotomy* (with constants $M > 0$ and $\beta > 0$) if there exists a projection-valued function $P : \mathbb{R} \to \mathcal{L}(X)$ such that, for each $x \in X$, the function $\theta \mapsto P(\theta)x$ is continuous and bounded, and, for all $\theta \geq \tau$, the following conditions hold:

(i) $P(\theta)U(\theta, \tau) = U(\theta, \tau)P(\tau)$.
(ii) $U_Q(\theta, \tau)$ is invertible as an operator from $\operatorname{Im} Q(\tau)$ to $\operatorname{Im} Q(\theta)$.
(iii) $\|U_P(\theta, \tau)\| \leq M e^{-\beta(\theta - \tau)}$.
(iv) $\|[U_Q(\theta, \tau)]^{-1}\| \leq M e^{-\beta(\theta - \tau)}$.

REMARK 3.7. If $\{U(\theta, \tau)\}_{\theta \geq \tau \geq 0}$ is an exponentially bounded, strongly continuous evolution family, then the boundedness and the strong continuity of $\theta \mapsto P(\theta)$ follows from conditions (i)–(iv) of Definition 3.6. For a proof of this fact, see Nguyen Van Minh, Räbiger and Schnaubelt [**MRS**, Lemma 4.2], and Daleckij and Krein [**DK**], Lemma IV.1.1 and Lemma IV.3.2. $\diamond$

REMARK 3.8. 1. If, for each $\theta \geq \tau$, the operator $U(\theta, \tau)$ is invertible, then U can be extended to all of $\mathbb{R}^2$ by using the definition $U(\tau, \theta) := [U(\theta, \tau)]^{-1}$. Also, in this case, condition (ii) holds for all $(\theta, \tau) \in \mathbb{R}^2$.

2. If the evolution family $\{U(\theta, \tau)\}_{(\theta,\tau)\in\mathbb{R}^2}$ consists of invertible operators, then, for each $\theta \in \mathbb{R}$, we have that $P(\theta) = U(\theta, 0)P(0)U(0, \theta)$. Hence, the conditions (ii)–(iv) can be replaced by the equivalent conditions

$$\|U(\theta, 0)P(0)U(0, \tau)\| \leq Me^{-\beta(\theta-\tau)}, \quad \theta \geq \tau,$$
$$\|U(\theta, 0)Q(0)U(0, \tau)\| \leq Me^{\beta(\theta-\tau)}, \quad \theta \leq \tau.$$

3. Sometimes conditions (ii)–(iv) are replaced by
(iii') $|U_P(\theta, \tau)x| \leq Me^{-\beta(\theta-\tau)}|x|$ for $x \in \operatorname{Im} P(\tau)$,
(iv') $|U_Q(\theta, \tau)x| \geq M^{-1}e^{\beta(\theta-\tau)}|x|$ for $x \in \operatorname{Im} Q(\tau)$.

Note that the inequality (iv') implies that the operator $U_Q(\theta, \tau)$ is uniformly injective from $\operatorname{Im} Q(\tau)$ to $\operatorname{Im} Q(\theta)$. However, to obtain condition (ii), the inequality (iv') must be augmented by the following condition:

(v') $\operatorname{Im} U_Q(\theta, \tau)$ is dense in $\operatorname{Im} Q(\theta)$.

We note that condition (iv') implies condition (v') provided that $\dim \operatorname{Im} Q(\theta) < \infty$. Hence, the set of conditions (iii')–(iv') is equivalent to the set (ii)–(iv) provided that $\dim \operatorname{Im} Q(\theta) < \infty$ for all $\theta \in \mathbb{R}$. This is indeed the case if the operators $U(\theta, \tau)$ are all compact. Also, the conditions (iii')-(iv') are equivalent to the set (iii)-(iv) for invertible evolution families $\{U(\theta, \tau)\}_{(\theta,\tau)\in\mathbb{R}^2}$. $\diamond$

Let us note that if the evolution family $\{U(\theta, \tau)\}_{\theta \geq \tau}$ solves the abstract Cauchy problem (ACP) and $\lambda \in \mathbb{C}$, then the evolution family $\{e^{-\lambda(\theta-\tau)}U(\theta, \tau)\}_{\theta \geq \tau}$ solves the abstract Cauchy problem for the equation $\dot{x} = [A(t) - \lambda]x$.

DEFINITION 3.9. The *Bohl spectrum* $\mathcal{B} = \mathcal{B}(\mathcal{U})$ of an evolution family $\mathcal{U} = \{U(\theta, \tau)\}_{\theta \geq \tau}$ is defined by

$$\mathcal{B} = \{\lambda \in \mathbb{R} : \{e^{-\lambda(\theta-\tau)}U(\theta, \tau)\}_{\theta \geq \tau} \text{ does not have an exponential dichotomy}\}.$$

DEFINITION 3.10. The *dichotomy bound* $\beta(\mathcal{U})$ of an exponentially dichotomic evolution family $\mathcal{U} = \{U(\theta, \tau)\}_{\theta \geq \tau}$ is defined to be

$$\beta(\mathcal{U}) := \sup\{\beta > 0 : \{U(\theta, \tau)\}_{\theta \geq \tau} \text{ has exponential dichotomy}$$
$$\text{with constants } \beta \text{ and } M = M(\beta)\}.$$

If the Bohl spectrum has a spectral gap containing zero, then the dichotomy bound $\beta(\mathcal{U})$ measures its width.

In case $A(\cdot)$ is a periodic operator-valued function, the abstract Cauchy problem (ACP) can be analyzed using "Floquet theory", see Daleckij and Krein [**DK**] for the case of bounded operators and Daners and Koch Medina [**DKo**] for the general case. In the next section we will show how semigroup theory can be used to treat the general nonperiodic case.

3.2. Howland semigroups on the line

For an evolution family $\{U(\theta, \tau)\}_{\theta \geq \tau}$ on a Banach space X, let us define an associated evolution semigroup $\{E^t\}_{t \geq 0}$ on $L^p(\mathbb{R}; X)$, $1 \leq p < \infty$, or on $C_0(\mathbb{R}; X)$ as follows:

(3.5) $$(E^t f)(\theta) = U(\theta - t, \theta)f(\theta - t), \quad t \geq 0, \quad \theta \in \mathbb{R}.$$

This semigroup is called the *Howland evolution semigroup* on the real line.

3.2.1. The generator of the evolution semigroup.
The definition (3.5) of the Howland evolution semigroup is based on the classical idea of defining "time" to be a new variable in order to make a nonautonomous Cauchy problem autonomous. In particular, the nonautonomous differential equation $\dot{x} = A(t)x$ is equivalent to the autonomous system

$$\dot{x} = A(\tau)x, \qquad \dot{\tau} = 1.$$

However, this associated autonomous system is nonlinear. The advantage of using the definition (3.5) is that one obtains an autonomous, one-parameter family of *linear* operators, the evolution semigroup.

Note that the evolution semigroup (3.5) is a multiplicative perturbation of the semigroup of translations defined by $(V^t f)(\theta) = f(\theta - t)$. Thus, E^t is a weighted translation operator that can be written as $E^t = aV^t$ where a denotes the multiplication operator $(af)(\theta) = a(\theta)f(\theta)$ associated with the operator-valued function $a(\theta) = U(\theta, \theta - t)$. We will make extensive use of the representation $E^t = aV^t$ in Chapter 4.

PROPOSITION 3.11. *If $\{U(\theta, \tau)\}_{\theta \geq \tau}$ is an exponentially bounded evolution family on a Banach space X, then the semigroup $\{E^t\}_{t \geq 0}$, as defined in (3.5), is a strongly continuous semigroup on $L^p(\mathbb{R}; X)$, $1 \leq p < \infty$, and on $C_0(\mathbb{R}; X)$.*

We omit the elementary proof of the last proposition, but see Theorem 6.20 below. In view of the proposition, the semigroup $\{E^t\}_{t \geq 0}$ has an infinitesimal generator that we denote by Γ. We will also use the notation $E^t = e^{t\Gamma}$.

Suppose that $\{U(\theta, \tau)\}_{\theta \geq \tau}$ solves the abstract Cauchy problem (ACP). We will describe the generator Γ of the evolution semigroup $\{E^t\}$ as follows. Consider the space $C_0(\mathbb{R}; X)$ (respectively, $L^p(\mathbb{R}; X)$, $1 \leq p < \infty$) and define the operator Γ' by

$$(3.6) \qquad (\Gamma' f)(\theta) = -\frac{df}{d\theta} + A(\theta)f(\theta), \quad \theta \in \mathbb{R},$$

with domain the set of differentiable (respectively, absolutely continuous) functions f such that $f(\theta) \in \mathcal{D}(A(\theta))$ and $\Gamma' f \in C_0(\mathbb{R}; X)$ (respectively, $L^p(\mathbb{R}; X)$). In other words, $\mathcal{D}(\Gamma') = \mathcal{D}(-d/d\theta) \cap \mathcal{D}(\mathcal{A})$ where $\mathcal{A}$ is the multiplication operator with maximal domain, given by $(\mathcal{A}f)(\theta) = A(\theta)f(\theta)$, see page 22.

We will identify a dense subset $\mathcal{D}_\Gamma \subset \mathcal{D}(\Gamma')$ of $C_0(\mathbb{R}; X)$ (or $L^p(\mathbb{R}; X)$) such that $\mathcal{D}_\Gamma \subset \mathcal{D}(\Gamma)$ and such that Γ is the closure of $(\Gamma', \mathcal{D}_\Gamma)$. In particular, the operator Γ is given by $\Gamma f = -f' + \mathcal{A}f$ on the dense set $\mathcal{D}_\Gamma$.

Let $C_c^1(\mathbb{R})$ denote the set of smooth functions $\alpha: \mathbb{R} \to \mathbb{R}$ with compact support. Fix $\tau \in \mathbb{R}$ and $x_\tau \in Y_\tau \subseteq \mathcal{D}(A(\tau))$, see Definition 3.2 where the sets Y_τ are described. For each $\alpha \in C_c^1(\mathbb{R})$ such that $\operatorname{supp} \alpha \subset (\tau, \infty)$, define the function $f = f_{\alpha, \tau, x_\tau}$ by

$$(3.7) \qquad f(\theta) = \alpha(\theta) U(\theta, \tau) x_\tau \quad \text{for} \quad \theta > \tau, \quad \text{and} \quad f(\theta) = 0 \quad \text{for} \quad \theta \leq \tau.$$

Let $\mathcal{D}_\Gamma$ denote the linear span of all functions f_{α, τ, x_τ} given as in (3.7) for $\tau \in \mathbb{R}$, $x_\tau \in Y_\tau \subseteq \mathcal{D}(A(\tau))$, and α in the class defined above.

THEOREM 3.12. *Assume that the exponentially bounded, strongly continuous evolution family $\{U(\theta, \tau)\}_{\theta \geq \tau}$ solves the abstract Cauchy problem (ACP). The generator Γ of the associated evolution semigroup (3.5) is the closure of the restriction to the set $\mathcal{D}_\Gamma$ of the operator Γ', as defined in equation (3.6).*

PROOF. We will give the proof for $C_0(\mathbb{R}; X)$. Similar arguments can be used to prove the theorem for the space $L^p(\mathbb{R}; X)$, $1 \leq p < \infty$.

Fix $f = f_{\alpha,\tau,x_\tau}$ as defined in (3.7). We claim that $f \in \mathcal{D}(\Gamma)$, $f \in \mathcal{D}(\Gamma')$, and $\Gamma f = \Gamma' f$. Indeed, note that

$$(E^t f)(\theta) = \alpha(\theta - t)U(\theta, \theta - t)U(\theta - t, \tau)x_\tau = \alpha(\theta - t)U(\theta, \tau)x_\tau$$

for $\theta - t > \tau$, and $(E^t f)(\theta) = 0$ otherwise. Hence,

$$(\Gamma f)(\theta) = \frac{d}{dt}(E^t f)(\theta)\bigg|_{t=0} = -\alpha'(\theta)U(\theta, \tau)x_\tau, \quad \theta \in \mathbb{R}.$$

On the other hand, since the function $\theta \mapsto U(\theta, \tau)x_\tau$ satisfies the abstract Cauchy problem (ACP), we also have

$$\frac{df}{d\theta} = \alpha'(\theta)U(\theta, \tau)x_\tau + \alpha(\theta)\frac{d}{d\theta}(U(\theta, \tau)x_\tau) = \alpha'(\theta)U(\theta, \tau)x_\tau + \alpha(\theta)A(\theta)U(\theta, \tau)x_\tau.$$

To complete the proof, we must show that the set of linear combinations of functions f as in (3.7) are dense in $C_0(\mathbb{R}; X)$. First, observe that the set of finite sums of the form $\sum \beta_j x_j$ where $x_j \in X$ and $\beta_j \in C_c^1(\mathbb{R})$, is dense in $C_0(\mathbb{R}; X)$. Next, suppose that $g \in C_0(\mathbb{R}; X)$ is given by $g = \beta x$ for some $x \in X$ and some $\beta \in C_c^1(\mathbb{R})$ with $\operatorname{supp} \beta \subset [a, b]$. We will show that g can be approximated by a sum of functions of the form of f as in (3.7).

Fix $\epsilon > 0$. For every $\theta_0 \in \operatorname{supp} \beta$, the map $(\theta, \tau) \mapsto U(\theta, \tau)x$ is continuous at the point (θ_0, θ_0). Hence, there are points $\tau_0 \leq \tau_0'$ such that the interval $I_0 = I(\theta_0) := (\tau_0, \tau_0')$ contains θ_0, and $|U(\theta, \tau_0)x - x| \leq \epsilon/2$ for $\theta \in I_0$. Thus, we obtain an open covering of $\operatorname{supp} \beta$ formed by the collection of intervals of the form $I(\theta_0)$ with $\theta_0 \in \operatorname{supp} \beta$. By the compactness of $\operatorname{supp} \beta$, there is a finite subcovering that we denote by $\{I_j\}_{j=1}^n$. For each $I_j := (\tau_j, \tau_j')$, we have that $|U(\theta, \tau_j)x - x| \leq \epsilon/2$ for $\theta \in I_j$, $j = 1, \ldots, n$.

Choose a smooth partition of unity $\{\gamma_j\}_{j=1}^n$ subordinate to the cover $\{I_j\}_{j=1}^n$; that is, smooth functions $\gamma_j : \mathbb{R} \to [0, 1]$ such that

$$\sum_{j=1}^n \gamma_j(\theta) = 1 \text{ for } \theta \in \operatorname{supp} \beta, \text{ and } \operatorname{supp} \gamma_j \subset I_j, \quad j = 1, \ldots, n.$$

Using the fact that, for each such j, the set $Y_{\tau_j} \subseteq \mathcal{D}(A(\tau_j))$ is dense in X, there is some $x_j \in Y_{\tau_j} \subseteq \mathcal{D}(A(\tau_j))$ such that

$$|x - x_j| \leq \frac{\epsilon}{2} M^{-1} e^{-\omega(b-a)}$$

where M and ω are as in part (iii) of Definition 3.1.

Define the function

$$h(\theta) = \sum_{j=1}^n \beta(\theta)\gamma_j(\theta)U(\theta, \tau_j)x_j, \quad \theta \in \mathbb{R},$$

and note that since $\operatorname{supp}\gamma_j \subset (\tau_j, \tau_j')$, the function h is a sum of functions f as defined in (3.7). Also, for each $\theta \in \mathbb{R}$, we have that

$$|g(\theta) - h(\theta)| = |\beta(\theta)\sum_{j=1}^{n}\gamma_j(\theta)x - \sum_{j=1}^{n}\beta(\theta)\gamma_j(\theta)U(\theta, \tau_j)x_j|$$

$$\leq \|\beta\|_\infty \Big(\sum_{j=1}^{n}\gamma_j(\theta)|x - U(\theta, \tau_j)x_j| + \sum_{j=1}^{n}\gamma_j(\theta)\|U(\theta, \tau_j)\||x - x_j|\Big)$$

$$\leq \|\beta\|_\infty \Big(\frac{\epsilon}{2}\sum_{j=1}^{n}\gamma_j(\theta) + \sum_{j=1}^{n}\gamma_j(\theta)Me^{\omega(b-a)}|x - x_j|\Big) \leq \epsilon\|\beta\|_\infty,$$

as required. $\square$

3.2.2. The spectral mapping theorem for evolution semigroups. Consider an exponentially bounded, strongly continuous evolution family $\{U(\theta, \tau)\}_{\theta \geq \tau}$ on a Banach space X, and let $(E^t f)(\theta) = U(\theta, \theta - t)f(\theta - t)$ denote the associated evolution semigroup on $\mathcal{F} = L^p(\mathbb{R}; X)$, $1 \leq p < \infty$, or $\mathcal{F} = C_0(\mathbb{R}; X)$.

THEOREM 3.13. *Let Γ be the infinitesimal generator of the evolution semigroup $\{E^t\}_{t \geq 0}$. The spectrum $\sigma(\Gamma)$ is invariant under translations along the imaginary axis, and the spectrum $\sigma(e^{t\Gamma})$, for $t > 0$, is invariant under rotations centered at the origin. Moreover, the following statements are equivalent:*

(1) $0 \in \rho(\Gamma)$.
(2) $\sigma(e^{t\Gamma}) \cap \mathbb{T} = \emptyset$, $\quad t > 0$.

In particular, the evolution semigroup has the spectral mapping property; that is, $\sigma(e^{t\Gamma}) \setminus \{0\} = e^{t\sigma(\Gamma)}$ for $t > 0$.

PROOF. We will give the proof for the space $L^p(\mathbb{R}; X)$; the arguments for $C_0(\mathbb{R}; X)$ are similar.

For each $\xi \in \mathbb{R}$, define the operator L_ξ on $L^p(\mathbb{R}; X)$ by $(L_\xi f)(\theta) = e^{i\xi\theta}f(\theta)$. Then, as in Proposition 2.36, we have that

(3.8) $$L_\xi^{-1}e^{t\Gamma}L_\xi = e^{-i\xi t}e^{t\Gamma} \quad \text{and} \quad L_\xi^{-1}\Gamma L_\xi = -i\xi + \Gamma.$$

The rotational invariance of $\sigma(E^t)$ and the translational invariance of $\sigma(\Gamma)$ follow at once.

The implication (2)$\Rightarrow$(1) follows from the Spectral Inclusion Theorem 2.6.

(1)$\Rightarrow$(2). The idea is the same "change-of-variables" trick as in the proof of the implication 3) $\Rightarrow$ 2) in Theorem 2.39. Heuristically, if the equation $\dot{x} = A(t)x$ is solved by a smooth evolution family, then the generator Γ is given by $(\Gamma f)(\theta) = -\frac{df}{d\theta} + A(\theta)f(\theta)$, $\theta \in \mathbb{R}$, see Theorem 3.12 above. Consider the evolution semigroup $\{\tilde{E}^t\}_{t \geq 0}$ on $L^p(\mathbb{R}; L^p(\mathbb{R}; X))$ that corresponds to the semigroup $\{E^t\}_{t \geq 0}$ on $L^p(\mathbb{R}; X)$ with generator Γ as in (2.31). The generator of $\{\tilde{E}^t\}_{t \geq 0}$ is given by $(\tilde{\Gamma}F)(\tau) = -\frac{dF}{d\tau} + \Gamma F(\tau)$. That is, for $F(\tau) = f(\tau, \cdot)$, we have $(\tilde{\Gamma}f)(\tau, \theta) = -\frac{\partial f}{\partial \tau} - \frac{\partial f}{\partial \theta} + A(\theta)f(\tau, \theta)$. Consider the change of variables defined by $u = \tau - \theta$ and $v = \theta$. Then, for each function f, the corresponding function given by $h(u, v) = f(u+v, v)$ in the new variables is such that $(\Gamma h)(u, v) = -\frac{\partial h}{\partial v} + A(v)h(u, v) = (\tilde{\Gamma}f)(\tau, \theta)$. By this identity, we have $\rho(\tilde{\Gamma}) = \rho\left(-\frac{d}{dv} + A(\cdot)\right) = \rho(\Gamma)$. Using this fact that $0 \in \rho(\Gamma) = \rho(\tilde{\Gamma})$, the implication (3)$\Rightarrow$(1) of Theorem 2.39 shows that $\sigma(e^{t\Gamma}) \cap \mathbb{T} = \emptyset$. We will make this argument precise.

Consider two semigroups $\{\tilde{E}^t\}_{t\geq 0}$ and $\{\mathcal{E}^t\}_{t\geq 0}$ on the Banach space $L^p(\mathbb{R} \times \mathbb{R}; X) = L^p(\mathbb{R}; L^p(\mathbb{R}; X))$ defined by

$$(\tilde{E}^t h)(\tau, \theta) = U(\theta, \theta - t) h(\tau - t, \theta - t), \quad (\tau, \theta) \in \mathbb{R}^2,$$
$$(\mathcal{E}^t h)(\tau, \theta) = U(\theta, \theta - t) h(\tau, \theta - t), \quad t \geq 0,$$

and let $\tilde{\Gamma}$ and G denote their respective generators. The semigroup $\{\tilde{E}^t\}_{t\geq 0}$ is the evolution semigroup on $L^p(\mathbb{R}; Y)$ associated as in (2.31) with the semigroup $\{e^{t\Gamma}\}_{t\geq 0}$ defined on $Y = L^p(\mathbb{R}; X)$. Indeed, for a function $F(\tau) = h(\tau, \cdot) \in L^p(\mathbb{R}; X)$, the semigroup $\{\tilde{E}^t\}_{t\geq 0}$ is given by $(\tilde{E}^t F)(\tau) = e^{t\Gamma} F(\tau - t)$, $\tau \in \mathbb{R}$. By the implication (3)$\Rightarrow$(1) of Theorem 2.39, if $0 \in \rho(\tilde{\Gamma})$, then statement (2) of the theorem holds.

The semigroup $\{\mathcal{E}^t\}_{t\geq 0}$ is a semigroup of multiplication operators whose multipliers are given by $E^t = e^{t\Gamma}$. In fact, we have $(\mathcal{E}^t F)(\tau) = e^{t\Gamma} F(\tau)$ and, similarly, $(GF)(\tau) = \Gamma F(\tau)$ for $\tau \in \mathbb{R}$. Thus, G has a bounded inverse G^{-1} on $L^p(\mathbb{R}; Y)$ provided that Γ has a bounded inverse on $Y = L^p(\mathbb{R}; X)$. Therefore, statement (1) implies $0 \in \rho(G)$.

Consider the isometry J on $L^p(\mathbb{R} \times \mathbb{R}; X)$ given by $(Jh)(\tau, \theta) = h(\tau + \theta, \theta)$ and note that

$$(\mathcal{E}^t J h)(\tau, \theta) = U(\theta, \theta - t) h(\tau + \theta - t, \theta - t) = (J \tilde{E}^t h)(\tau, \theta).$$

Hence,

$$GJh = J\tilde{\Gamma}h, \quad h \in \mathcal{D}(\tilde{\Gamma}) \quad \text{and} \quad J^{-1} G h = \tilde{\Gamma} J^{-1} h, \quad h \in \mathcal{D}(G);$$

and therefore, $\rho(G) = \rho(\tilde{\Gamma})$ and statement (1) of the theorem implies $0 \in \rho(\tilde{\Gamma})$. □

3.2.3. Spectral Projection and Dichotomy Theorems. In this subsection we will prove that exponential dichotomy for an exponentially bounded evolution family $\{U(\theta, \tau)\}_{\theta \geq \tau}$ on a Banach space X is equivalent to the hyperbolicity of the corresponding evolution semigroup $(E^t f)(\theta) = U(\theta, \theta - t) f(\theta - t)$, $\theta \in \mathbb{R}$, defined on $L^p(\mathbb{R}; X)$, $1 \leq p < \infty$, or to the invertibility of its generator Γ. Similar results for the space $C_0(\mathbb{R}; X)$ are proved in Chapter 6.

Our main tool for the proof of these results is the Spectral Projection Theorem. To state this theorem, recall that $\mathcal{L}_s(X)$ denotes the space of bounded operators on the Banach space X equipped with strong operator topology, and $C_b(\mathbb{R}; \mathcal{L}_s(X))$ denotes the space of bounded continuous functions from $\mathbb{R}$ to $\mathcal{L}_s(X)$.

THEOREM 3.14 (Spectral Projection Theorem). *If $t > 0$ and $\mathcal{P}$ is a spectral projection for the operator E^t, then there is a projection-valued function $P \in C_b(\mathbb{R}; \mathcal{L}_s(X))$ such that $(\mathcal{P} f)(\theta) = P(\theta) f(\theta)$, $f \in L^p(\mathbb{R}; X)$, $\theta \in \mathbb{R}$.*

Recall that $\sigma(E^t)$ is rotationally invariant. Thus, when dealing with a disjoint decomposition of $\sigma(E^t)$, we may assume, by rescaling, that $\sigma(E^t) \cap \mathbb{T} = \emptyset$. Also, without loss of generality, it suffices to assume that $t = 1$. In view of these facts, we can restrict to the case where $E := E^1 = e^{\Gamma}$ is a hyperbolic operator and $\mathcal{P}$ is the Riesz projection that corresponds to the spectral set $\sigma(E) \cap \mathbb{D}$. As usual, we set $\mathcal{Q} = I - \mathcal{P}$. To construct $P(\cdot) \in C_b(\mathbb{R}; \mathcal{L}_s(X))$ such that $(\mathcal{P} f)(\theta) = P(\theta) f(\theta)$ for $f \in L^p(\mathbb{R}; X)$ and $\theta \in \mathbb{R}$, we will need two lemmas.

For each scalar function $\chi(\cdot) : \mathbb{R} \to \mathbb{R}$ let us also denote by $\chi := \chi I$ the multiplication operator given by $(\chi f)(\theta) = \chi(\theta) f(\theta)$, $\theta \in \mathbb{R}$.

LEMMA 3.15 (Commutation Lemma). *If $\chi \in L^\infty(\mathbb{R}; \mathbb{R})$, then*

(3.9) $$\chi \mathcal{P} = \mathcal{P}\chi.$$

PROOF. The decomposition $L^p(\mathbb{R}; X) = \operatorname{Im}\mathcal{P} \oplus \operatorname{Im}\mathcal{Q}$ is E-invariant. Define
$$E_P := \mathcal{P}E\mathcal{P} = E|\operatorname{Im}\mathcal{P}, \quad E_Q := \mathcal{Q}E\mathcal{Q} = E|\operatorname{Im}\mathcal{Q},$$
and note that $\sigma(E_P) \subset \mathbb{D}$ and E_Q is invertible with $\sigma(E_Q^{-1}) \subset \mathbb{D}$. Hence, there exist constants $\omega > 0$ and $M > 0$ such that, for all $n \in \mathbb{N}$, the following inequalities hold:

(3.10) $\quad\quad\quad\quad \|E_P^n f\|_{L^p} \leq M e^{-\omega n} \|f\|_{L^p}, \quad f \in \operatorname{Im}\mathcal{P},$

(3.11) $\quad\quad\quad\quad \|E_Q^n f\|_{L^p} \geq M^{-1} e^{\omega n} \|f\|_{L^p}, \quad f \in \operatorname{Im}\mathcal{Q}.$

We show first that $\operatorname{Im}\mathcal{P} = \{f \in L^p(\mathbb{R}; X) : E^n f \to 0 \text{ as } n \to \infty\}$. Indeed, if $f \in \operatorname{Im}\mathcal{P}$, then $\lim_{n\to\infty} E^n f = 0$ by (3.10). If $\lim_{n\to\infty} E^n f = 0$ for $f = \mathcal{P}f + \mathcal{Q}f$, then, by inequality (3.11),
$$\|\mathcal{Q}f\| \leq Me^{-\omega n}\|E_Q^n \mathcal{Q}f\| \leq Me^{-\omega n}\{\|E^n f\| + \|E_P^n f\|\}.$$
By passing to the limit as $n \to \infty$, we see that $\|\mathcal{Q}f\| = 0$; and therefore, $f \in \operatorname{Im}\mathcal{P}$.

Fix $\chi \in L^\infty(\mathbb{R}; \mathbb{R})$, and note that $(E^n \chi f)(\theta) = \chi(\theta - n)(E^n f)(\theta)$, $\theta \in \mathbb{R}$. Hence, for $f \in \operatorname{Im}\mathcal{P}$, we have the inequality
$$\|E^n \chi f\|_{L^p} \leq \|\chi\|_\infty \|E^n f\|_{L^p}.$$
As the right hand side of this inequality converges to zero as $n \to \infty$, it follows that $\chi f \in \operatorname{Im}\mathcal{P}$.

Thus, to prove equality (3.9), it suffices to show that if $f \in \operatorname{Im}\mathcal{Q}$, then $\chi f \in \operatorname{Im}\mathcal{Q}$. For this, fix $f \in \operatorname{Im}\mathcal{Q}$ and recall that E_Q is invertible on $\operatorname{Im}\mathcal{Q}$. For each integer $n \geq 0$, define functions $f_n := E_Q^{-n} f \in \operatorname{Im}\mathcal{Q}$ and $g_n(\theta) := \chi(\theta + n) f_n(\theta)$, $\theta \in \mathbb{R}$. If we decompose g_n as $g_n = \mathcal{P}g_n + \mathcal{Q}g_n$, then, using the fact that E and $\mathcal{P}$ commute, we have
$$\chi f = E^n g_n, \quad \mathcal{P}\chi f = E_P^n \mathcal{P}g_n, \quad \mathcal{Q}\chi f = E_Q^n \mathcal{Q}g_n.$$
From the inequalities (3.10) and (3.11), it follows that
$$\|\mathcal{P}\chi f\| \leq Me^{-\omega n}\|\mathcal{P}g_n\| \leq Me^{-\omega n}\|\mathcal{P}\| \cdot \|\chi\|_\infty \cdot \|f_n\|$$
$$\leq M^2 e^{-2\omega n}\|\mathcal{P}\| \cdot \|\chi\|_\infty \cdot \|f\|.$$
Passing to the limit as $n \to \infty$, we see that $\|\mathcal{P}\chi f\| = 0$. Thus, $\chi f \in \operatorname{Im}\mathcal{Q}$. □

The regularization property of the resolvent $R(\lambda; \Gamma) = (\lambda - \Gamma)^{-1}$ recorded in the Räbiger-Schnaubelt Lemma 3.16 below is a key fact that is used in the proof of the existence of an operator $P(\cdot) \in C_b(\mathbb{R}; X)$ such that $(\mathcal{P}f)(\theta) = P(\theta)f(\theta)$. This lemma states that the image of the space $L^p(\mathbb{R}; X)$, $1 \leq p < \infty$, under the resolvent operator or, equivalently, the domain of the generator Γ of the evolution semigroup $\{E^t\}_{t \geq 0}$ on the space $L^p(\mathbb{R}; X)$, belongs to the space $C_0(\mathbb{R}; X)$ of *continuous* functions.

Consider the space $\mathcal{G}_p := L^p(\mathbb{R}; X) \cap C_0(\mathbb{R}; X)$ endowed with the norm
$$\|f\|_{\mathcal{G}_p} := \max\{\|f\|_p, \|f\|_\infty\}.$$
If $1 \leq p < \infty$, then we let $\Gamma = \Gamma_p$ denote the generator of the evolution semigroup $\{E^t\}_{t \geq 0}$ on the space $L^p(\mathbb{R}; X)$. In addition, we let $\Gamma = \Gamma_\infty$ denote the generator of the evolution semigroup $\{E^t\}_{t \geq 0}$ on the space $C_0(\mathbb{R}; X)$.

LEMMA 3.16 (Räbiger-Schnaubelt). *If $\lambda \in \rho(\Gamma)$, then*
$$R(\lambda; \Gamma) : L^p(\mathbb{R}; X) \to \mathcal{G}_p, \quad 1 \leq p < \infty,$$
is a bounded operator. In particular, $\mathcal{D}(\Gamma) \subseteq \mathcal{G}_p \subseteq C_0(\mathbb{R}; X)$, and $\mathcal{D}(\Gamma)$ is dense in the spaces $\mathcal{G}_p$ and $C_0(\mathbb{R}; X)$.

PROOF. The idea of the proof is to use the fact that the Laplace transform gives the same formula for the resolvent operator on both $L^p(\mathbb{R}; X)$ and $C_0(\mathbb{R}; X)$.

Recall that the evolution family is exponentially bounded. Thus, there are constants $M \geq 1$ and $\omega \in \mathbb{R}$ such that $\|U(\theta, \tau)\|_{\mathcal{L}(X)} \leq M e^{\omega(\theta - \tau)}$ for $\theta \geq \tau$. Using the definition of the associated evolution semigroup, we then have that
$$\|E^t\|_{\mathcal{L}(L^p(\mathbb{R}; X))} \leq M e^{\omega t} \quad \text{and} \quad \|E^t\|_{\mathcal{L}(C_0(\mathbb{R}; X))} \leq M e^{\omega t}$$
for all $t \geq 0$. If $\mu > \omega$, then by the Laplace Transform Formula (2.3) we have that $\mu \in \rho(\Gamma_p) \cap \rho(\Gamma_\infty)$ and
$$R(\mu; \Gamma) = \int_0^\infty e^{-\mu t} e^{t\Gamma} \, dt, \quad \text{on} \quad L^p(\mathbb{R}; X) \quad \text{and} \quad C_0(\mathbb{R}; X).$$

If $f \in \mathcal{G}_p = L^p(\mathbb{R}; X) \cap C_0(\mathbb{R}; X)$, $1 \leq p < \infty$, then $R(\mu; \Gamma_p) f = R(\mu; \Gamma_\infty) f$. Hence, $R(\mu; \Gamma_p)$ maps $\mathcal{G}_p$ to $\mathcal{G}_p$. Moreover, using Hölder's inequality, we obtain

$$|(R(\mu; \Gamma_p)f)(\theta)| = \Big| \int_0^\infty e^{-\mu t}(e^{t\Gamma_\infty} f)(\theta) \, dt \Big| \leq \int_0^\infty e^{-\mu t} |U(\theta, \theta - t) f(\theta - t)| \, dt$$
$$\leq M \int_0^\infty e^{(\omega - \mu)t} |f(\theta - t)| \, dt \leq M \Big(\int_0^\infty e^{(\omega - \mu)qt} \, dt \Big)^{\frac{1}{q}} \Big(\int_0^\infty |f(\theta - t)|^p \, dt \Big)^{\frac{1}{p}}$$
$$\leq MC \|f\|_{L^p}$$

for some constant $C > 0$ and all $\theta \in \mathbb{R}$. As a result, there is a constant $M_1 > 0$ such that $\|R(\mu; \Gamma_p) f\|_\infty \leq M_1 \|f\|_{L^p}$. Note that $\mathcal{G}_p = L^p(\mathbb{R}; X) \cap C_0(\mathbb{R}; X)$ is dense in $L^p(\mathbb{R}; X)$. Hence, $\|R(\mu; \Gamma_p) f\|_\infty \leq M_1 \|f\|_{L^p}$ for all $f \in L^p(\mathbb{R}; X)$. This shows that if $\mu > \omega$, then $R(\mu; \Gamma) : L^p(\mathbb{R}; X) \to \mathcal{G}_p$ is a bounded operator.

For each $\lambda \in \rho(\Gamma)$, an application of the resolvent identity (2.1) yields the following operator identity on $L^p(\mathbb{R}; X)$:
$$R(\lambda; \Gamma) \left[R(\mu; \Gamma) - \frac{1}{\mu - \lambda} \right] = \frac{1}{\lambda - \mu} R(\mu; \Gamma).$$

We claim that the operator $[R(\mu; \Gamma) - (\mu - \lambda)^{-1}]^{-1}$ exists and is bounded as an operator on $L^p(\mathbb{R}; X)$. Indeed, using the fact that the maps $R(\mu; \Gamma) : L^p(\mathbb{R}; X) \to \mathcal{D}(\Gamma)$ and $\lambda - \Gamma : \mathcal{D}(\Gamma) \to L^p(\mathbb{R}; X)$ are injective and surjective, it follows that the operator
$$R(\mu; \Gamma) - \frac{1}{\mu - \lambda} = \left[I - \frac{1}{\mu - \lambda}(\mu - \Gamma) \right] R(\mu; \Gamma) = \frac{1}{\lambda - \mu}(\lambda - \Gamma) R(\mu; \Gamma)$$
is both injective and surjective. These facts together with the Inverse Mapping Theorem prove the claim. In view of this result, the operator
$$R(\lambda; \Gamma) = \frac{1}{\lambda - \mu} R(\mu; \Gamma) \left[R(\mu; \Gamma) - \frac{1}{\mu - \lambda} \right]^{-1}$$
is bounded from $L^p(\mathbb{R}; X)$ to $\mathcal{G}_p$.

To finish the proof, let us note that $\mathcal{D}(\Gamma) = \operatorname{Im} R(\lambda; \Gamma)$ on $L^p(\mathbb{R}; X)$. Since $R(\lambda; \Gamma)$ maps $L^p(\mathbb{R}; X)$ to $\mathcal{G}_p$, we have the inclusion $\mathcal{D}(\Gamma_p) \subseteq \mathcal{G}_p$. To see that $\mathcal{D}(\Gamma_p)$ is dense in $\mathcal{G}_p \subset C_0(\mathbb{R}; X)$, consider the evolution semigroup $\{e^{t\Gamma_{\mathcal{G}_p}}\}_{t \geq 0}$

induced by $\{E^t\}_{t\geq 0}$ on $\mathcal{G}_p$. Its generator $\Gamma_{\mathcal{G}_p}$ is a part of Γ_p in $\mathcal{G}_p$. In particular, $\mathcal{D}(\Gamma_{\mathcal{G}_p}) \subseteq \mathcal{D}(\Gamma_p) \subset \mathcal{G}_p \subset C_0(\mathbb{R}; X)$, and $\mathcal{G}_p$ is dense in $C_0(\mathbb{R}; X)$. Thus, the domain $\mathcal{D}(\Gamma_{\mathcal{G}_p}) \subseteq \mathcal{D}(\Gamma_p)$ is dense in the spaces $\mathcal{G}_p$ and $C_0(\mathbb{R}; X)$. $\square$

We are now ready to prove the Spectral Projection Theorem 3.14.

PROOF. We must construct a function $P(\cdot) : \mathbb{R} \to \mathcal{L}(X)$ such that the Riesz projection $\mathcal{P}$ corresponding to the hyperbolic operator $E := E^1$ and the spectral set $\sigma(E) \cap \mathbb{D}$ is given by $(\mathcal{P}f)(\theta) = P(\theta)f(\theta)$.

Let us define $P(\cdot)$ as follows. Consider a function $f \in \mathcal{D}(\Gamma)$ for Γ acting on $L^p(\mathbb{R}; X)$. By the Räbiger-Schnaubelt Lemma 3.16, $f \in C_0(\mathbb{R}; X)$. Moreover, since $f \in \mathcal{D}(\Gamma)$, there is a $g \in L^p(\mathbb{R}; X)$ such that $f = R(\lambda; \Gamma)g$. Note that $\mathcal{P}$ and $R(\lambda; \Gamma)$ commute. Thus, we have

$$\mathcal{P}f = \mathcal{P}R(\lambda; \Gamma)g = R(\lambda; \Gamma)\mathcal{P}g.$$

As a result, $\mathcal{P}f \in \mathcal{D}(\Gamma)$ and, by the Räbiger-Schnaubelt Lemma 3.16, $\mathcal{P}f \in C_0(\mathbb{R}; X)$.

Fix $\tau \in \mathbb{R}$. By the Commutation Lemma 3.15, we have that $\mathcal{P}\chi_{(\tau,\tau+\epsilon)} = \chi_{(\tau,\tau+\epsilon)}\mathcal{P}$ for the characteristic function $\chi_{(\tau,\tau+\epsilon)}$ of the interval $(\tau, \tau+\epsilon)$ where $\epsilon > 0$. Using the fact that both $(\mathcal{P}f)(\cdot)$ and $f(\cdot)$ are continuous at τ, we have the equality

$$|(\mathcal{P}f)(\tau)|^p = \lim_{\epsilon \to 0^+} \frac{1}{\epsilon} \int_\tau^{\tau+\epsilon} |(\mathcal{P}f)(s)|^p \, ds = \lim_{\epsilon \to 0^+} \frac{1}{\epsilon} \|\chi_{(\tau,\tau+\epsilon)}\mathcal{P}f\|^p_{L^p(\mathbb{R};X)}$$

$$\leq \|\mathcal{P}\|^p_{\mathcal{L}(L_p(\mathbb{R};X))} \lim_{\epsilon \to 0^+} \frac{1}{\epsilon} \int_\tau^{\tau+\epsilon} |f(s)|^p \, ds = \|\mathcal{P}\|^p_{\mathcal{L}(L^p(\mathbb{R};X))} |f(\tau)|^p;$$

and therefore,

(3.12) $\qquad |(\mathcal{P}f)(\tau)| \leq \|\mathcal{P}\|_{\mathcal{L}(L^p(\mathbb{R};X))} |f(\tau)|, \quad \tau \in \mathbb{R}.$

Define $X_\tau := \{f(\tau) : f \in \mathcal{D}(\Gamma)\}$. Also, define an operator $P(\tau) : X_\tau \to X_\tau$ by the following rule: For each $x \in X_\tau$, choose a function $f \in \mathcal{D}(\Gamma)$ such that $f(\tau) = x$, and then set

(3.13) $\qquad\qquad P(\tau)x = (\mathcal{P}f)(\tau).$

In view of the inequality (3.12), if $f_1(\tau) = f_2(\tau) = x$, then $P_{f_1}(\tau)x = P_{f_2}(\tau)x$. Thus, $P(\tau)$ is a well-defined projection. By the Räbiger-Schnaubelt Lemma 3.16, the set $\mathcal{D}(\Gamma)$ is dense in $C_0(\mathbb{R}; X)$. Thus, for each $\tau \in \mathbb{R}$, the set X_τ is dense in X. By continuity, $P(\tau)$ can be extended to all of X with $\|P(\tau)\|_{\mathcal{L}(X)} \leq \|\mathcal{P}\|_{\mathcal{L}(L^p(\mathbb{R};X))}$.

Finally, let us show that the function $\tau \mapsto P(\tau)x$ is continuous for each $x \in X$. Let $\epsilon > 0$ be given. For $y \in X$ such that $|x - y| < \epsilon$, choose $f \in \mathcal{D}(\Gamma)$ such that $f(\tau) = y$. In view of the fact that both f and $\mathcal{P}f$ are in $C_0(\mathbb{R}; X)$, there exists $\delta > 0$ so that $|f(\tau) - f(s)| < \epsilon$ and $|(\mathcal{P}f)(\tau) - (\mathcal{P}f)(s)| < \epsilon$ whenever $|\tau - s| < \delta$. The following computation completes the proof:

$$|P(\tau)x - P(s)x| \leq |P(\tau)x - P(\tau)y| + |P(\tau)y - P(s)f(s)|$$
$$+ |P(s)f(s) - P(s)y| + |P(s)y - P(s)x|$$
$$\leq \|P(\tau)\|_{\mathcal{L}(X)} |x - y| + |(\mathcal{P}f)(\tau) - (\mathcal{P}f)(s)|$$
$$+ \|P(s)\|_{\mathcal{L}(X)} |f(s) - f(\tau)| + \|P(s)\|_{\mathcal{L}(X)} |y - x|$$
$$\leq O(\epsilon).$$

$\square$

The main results of this section are summarized in the following theorem. It states necessary and sufficient conditions for an evolution family to have an exponential dichotomy (see Definition 3.6) in terms of the spectrum of an associated evolution semigroup.

THEOREM 3.17 (Dichotomy Theorem). *Let $\{U(\theta,\tau)\}_{\theta\geq\tau}$ be a strongly continuous, exponentially bounded evolution family on a Banach space X, let $\{E^t\}_{t\geq 0}$ be the corresponding evolution semigroup given by $(E^t f)(\theta) = U(\theta, \theta-t)f(\theta-t)$ on $L^p(\mathbb{R};X)$, $1 \leq p < \infty$, and let Γ denote its infinitesimal generator. The following statements are equivalent:*

(1) *$\{U(\theta,\tau)\}_{\theta\geq\tau}$ has an exponential dichotomy on X.*
(2) *$\sigma(E^t) \cap \mathbb{T} = \emptyset$ for $t > 0$.*
(3) *$0 \in \rho(\Gamma)$.*

Moreover, the Riesz projection $\mathcal{P}$ that corresponds to the hyperbolic operator E^1 and the spectral subset $\sigma(E^1) \cap \mathbb{D}$ is related to the dichotomy projection $P : \mathbb{R} \to \mathcal{L}_s(X)$, as in Definition 3.6, by the formula $(\mathcal{P}f)(\theta) = P(\theta)f(\theta)$ where $\theta \in \mathbb{R}$ and $f \in L^p(\mathbb{R};X)$.

The Dichotomy Theorem also holds if the space $L^p(\mathbb{R};X)$ is replaced by $C_0(\mathbb{R};X)$, see the Dichotomy Theorem 6.41 in Chapter 6.

PROOF. The equivalence (2)⇔(3) is proved in Theorem 3.13.

(1)⇔(2). We will use the following auxiliary result. Assume that $\{U(\theta,\tau)\}_{\theta\geq\tau}$ has an exponential dichotomy with the dichotomy projection $P(\cdot)$. By part (ii) in Definition 3.6, we have that $U_Q(\theta,\tau) = U(\theta,\tau)|\operatorname{Im} Q(\tau)$ is invertible as an operator from $\operatorname{Im} Q(\tau)$ to $\operatorname{Im} Q(\theta)$ for $\theta \geq \tau$. Let $\mathcal{F}_\infty^u$ denote the subspace of functions $f \in C_0(\mathbb{R};X)$ such that $f(\theta) \in \operatorname{Im} Q(\theta)$ for each $\theta \in \mathbb{R}$. Also, for each $t \geq 0$, define the operator R_t on $\mathcal{F}_\infty^u$ by
$$(R_t f)(\theta) = [U_Q(\theta+t, \theta)]^{-1} f(\theta+t).$$

LEMMA 3.18. *The operator R_t is bounded on the space $\mathcal{F}_\infty^u$. Moreover,*
$$\|R_t f\|_\infty \leq M e^{-\beta t} \|f\|_\infty$$
where M and β are the dichotomy constants as in Definition 3.6.

PROOF. By parts (i) and (iv) in Definition 3.6, if $\theta \in \mathbb{R}$ and $t \geq 0$, then $Q(\theta)U(\theta,\theta-t) = U(\theta,\theta-t)Q(\theta-t)$ and $\|[U_Q(\theta,\theta-t)]^{-1}\| \leq Me^{-\beta t}$. Also, if $f \in \mathcal{F}_\infty^u$, then $f(\theta) = Q(\theta)f(\theta)$ for all $\theta \in \mathbb{R}$. Using these facts, for $\theta, \tau \in \mathbb{R}$, we have the inequalities
$$|(R_t f)(\tau-t) - (R_t f)(\theta-t)| = |[U_Q(\tau,\tau-t)]^{-1}f(\tau) - [U_Q(\theta,\theta-t)]^{-1}f(\theta)|$$
$$\leq |U_Q^{-1}(\tau,\tau-t)f(\tau)$$
$$- U_Q^{-1}(\tau,\tau-t)U_Q(\tau,\tau-t)Q(\tau-t)U_Q^{-1}(\theta,\theta-t)f(\theta)|$$
$$+ |Q(\tau-t)U_Q^{-1}(\theta,\theta-t)f(\theta) - Q(\theta-t)U_Q^{-1}(\theta,\theta-t)f(\theta)|$$
$$\leq Me^{-\beta t}|f(\tau) - Q(\tau)U_Q(\tau,\tau-t)U_Q^{-1}(\theta,\theta-t)f(\theta)|$$
$$+ |(Q(\tau-t) - Q(\theta-t))U_Q^{-1}(\theta,\theta-t)f(\theta)|.$$

Because $Q(\cdot)$ is strongly continuous, it follows that
$$\lim_{\tau\to\theta} f(\tau) = f(\theta) = Q(\theta)f(\theta) = \lim_{\tau\to\theta} Q(\tau)U(\tau,\tau-t)U_Q^{-1}(\theta,\theta-t)f(\theta).$$

Therefore, $(R_t f)(\cdot)$ is continuous. Also, we have the estimate

$$\|R_t f\|_\infty = \sup_{\theta \in \mathbb{R}} |U_Q^{-1}(\theta + t, \theta) f(\theta + t)| \le M e^{-\beta t} \|f\|_\infty.$$

Finally, let us note that

$$(R_t f)(\theta) = U_Q^{-1}(\theta + t, \theta) f(\theta + t) = Q(\theta) R_t f(\theta) \in \operatorname{Im} Q(\theta);$$

and therefore, $R_t f \in \mathcal{F}_\infty^u$. $\square$

Using the previous lemma, we will prove the implication (1)⇒(2) stated in Theorem 3.17. Suppose that the evolution family $\{U(\theta, \tau)\}_{\theta \ge \tau}$ has a dichotomy, and let $P(\cdot) \in C_b(\mathbb{R}; \mathcal{L}_s(X))$ denote the corresponding projection. Define the operator $\mathcal{P}$ by $(\mathcal{P}f)(\theta) = P(\theta)f(\theta)$ on the space $L^p(\mathbb{R}; X)$. Clearly, we have $E^t \mathcal{P} = \mathcal{P} E^t$ since $P(\theta) U(\theta, \tau) = U(\theta, \tau) P(\tau)$, $\theta \ge \tau$, by part (i) of Definition 3.6. Let $\{E_P^t\}_{t \ge 0}$ and $\{E_Q^t\}_{t \ge 0}$ denote the induced semigroups on $\operatorname{Im} \mathcal{P}$ and $\operatorname{Im} \mathcal{Q}$. They are given, respectively, by

$$(E_P^t f)(\theta) = U_P(\theta, \theta - t) f(\theta - t), \quad f \in \operatorname{Im} \mathcal{P}, \quad \theta \in \mathbb{R},$$
$$(E_Q^t f)(\theta) = U_Q(\theta, \theta - t) f(\theta - t), \quad f \in \operatorname{Im} \mathcal{Q}, \quad \theta \in \mathbb{R}.$$

By part (iii) in Definition 3.6, we have the inequality $\|E_P^t\| \le M e^{-\beta t}$; and, as a result, $\sigma(E_P^t) \subset \mathbb{D}$. By statement (ii) in Definition 3.6, the operator E_Q^t is invertible on $\operatorname{Im} \mathcal{Q}$, and, by Lemma 3.18, $(E_Q^t)^{-1} = R_t$ and $\|(E_Q^t)^{-1}\| \le M e^{-\beta t}$. Thus, we have the inclusion $\sigma(E_Q^t) \subset \mathbb{C} \setminus \overline{\mathbb{D}}$, and the hyperbolicity of $\{E^t\}_{t \ge 0}$ is proved.

We will prove the implication (2) ⇒ (1) in Theorem 3.17. Assume that $\{E^t\}_{t \ge 0}$ is hyperbolic. Let $\mathcal{P}$ denote the spectral projection corresponding to $E := E^1$ and the spectral set $\sigma(E) \cap \mathbb{D}$. By the Spectral Projection Theorem 3.14, it follows that there is a projection-valued function $P(\cdot) \in C_b(\mathbb{R}; \mathcal{L}_s(X))$ such that $(\mathcal{P}f)(\theta) = P(\theta)f(\theta)$. We will show that the function P satisfies conditions (i)–(iv) of Definition 3.6.

Define the operators $E_P^t := E^t|\operatorname{Im}\mathcal{P}$ and $E_Q^t := E^t|\operatorname{Im}\mathcal{Q}$. Clearly, condition (i) of Definition 3.6; that is, the identity $P(\theta) U(\theta, \tau) = U(\theta, \tau) P(\tau)$, $\theta \ge \tau$, follows from the fact that $E^t \mathcal{P} = \mathcal{P} E^t$. Indeed, we have

$$(E^t \mathcal{P} f)(\theta) = U(\theta, \theta - t) P(\theta - t) f(\theta - t) = P(\theta) U(\theta, \theta - t) f(\theta - t) = (\mathcal{P} E^t f)(\theta).$$

Also, condition (iii); that is, the estimate $\|U_P(\theta, \tau)\| \le M e^{-\beta(\theta - \tau)}$, follows from the fact that the spectral radius $r(E_P^t) < 1$.

Next, we show that there exist positive constants M and β such that

(3.14) $\qquad |U_Q(\theta, \tau) x| \ge M^{-1} e^{\beta(\theta - \tau)} |x|, \quad x \in \operatorname{Im} Q(\tau).$

Since $\sigma([E_Q^t]^{-1}) \subset \mathbb{D}$, there are positive constants M and β such that

(3.15) $\qquad \|E^t \mathcal{Q} f\|_{L^p(\mathbb{R}; X)} \ge M^{-1} e^{\beta t} \|\mathcal{Q} f\|_{L^p(\mathbb{R}; X)}.$

Fix $\theta > \tau \in \mathbb{R}$ and $x \in X$. Choose $f \in C_0(\mathbb{R}; X)$ such that f has compact support and $f(\tau) = x$. Note that $E^t \mathcal{Q} f(\cdot) \in C_0(\mathbb{R}; X)$. Also, by the Commutation Lemma 3.15, the identity $\mathcal{Q}\chi = \chi \mathcal{Q}$ holds for every characteristic function $\chi \in$

$L^\infty(\mathbb{R};\mathbb{R})$. For $t = \theta - \tau$, we have that

$$\begin{aligned}
|U(\theta,\tau)Q(\tau)x|^p &= |(E^t\mathcal{Q}f)(\tau+t)|^p \\
&= \lim_{\epsilon\to 0^+} \frac{1}{\epsilon}\int_{\tau+t}^{\tau+t+\epsilon} |(E^t\mathcal{Q}f)(s)|^p\, ds \quad (\text{ by the continuity of } E^t\mathcal{Q}f) \\
&= \lim_{\epsilon\to 0^+} \frac{1}{\epsilon}\|\chi_{[\tau+t,\tau+t+\epsilon]}E^t\mathcal{Q}f\|_{L^p(\mathbb{R};\chi)}^p = \lim_{\epsilon\to 0^+}\frac{1}{\epsilon}\|E^t\mathcal{Q}\chi_{[\tau,\tau+\epsilon]}f\|_{L^p(\mathbb{R};X)}^p \\
&\geq M^{-p}e^{p\beta t}\lim_{\epsilon\to 0^+}\frac{1}{\epsilon}\|\mathcal{Q}\chi_{[\tau,\tau+\epsilon]}f\|_{L^p}^p \quad (\text{by (3.15)}) \\
&= M^{-p}e^{p\beta t}\lim_{\epsilon\to 0^+}\frac{1}{\epsilon}\int_\tau^{\tau+\epsilon}|(\mathcal{Q}f)(s)|^p\, ds = M^{-p}e^{p\beta t}|Q(\tau)x|^p,
\end{aligned}$$

and the inequality (3.14) is proved.

It remains to show that the operator $U_Q(\theta,\tau): \operatorname{Im} Q(\tau) \to \operatorname{Im} Q(\theta)$, $\theta \geq \tau$, is invertible. By (3.14), it suffices to prove the surjectivity of this operator. By (2) in Theorem 3.17, if $t > 0$, then the operator E_Q^t is invertible on $\operatorname{Im}\mathcal{Q}$. For sufficiently large $\operatorname{Re}\lambda$, the resolvent $R(\lambda;\Gamma)$ can be represented using the Laplace Transform Formula (2.3). Therefore, since each operator E^t commutes with $\mathcal{Q}$, so does $R(\lambda;\Gamma)$. By the Räbiger-Schnaubelt Lemma 3.16 the range of $R(\lambda;\Gamma)$ on $L^p(\mathbb{R};X)$ is dense in $C_0(\mathbb{R};X)$. Note that $\mathcal{Q}$ is bounded on $C_0(\mathbb{R};X)$. Therefore, if $t = \theta - \tau \geq 0$, then

$$\begin{aligned}
\mathcal{Q}R(\lambda;\Gamma)L^p(\mathbb{R};X) &= R(\lambda;\Gamma)\mathcal{Q}L^p(\mathbb{R};X) = R(\lambda;\Gamma)E_Q^t\mathcal{Q}L^p(\mathbb{R};X) \\
&= E^t\mathcal{Q}R(\lambda;\Gamma)L^p(\mathbb{R};X)
\end{aligned}$$

is dense in $\mathcal{Q}C_0(\mathbb{R};X)$. Thus,

$$\begin{aligned}
Q(\theta)X &= \{f(\theta) : f \in \mathcal{Q}C_0(\mathbb{R};X) \\
&= \operatorname{closure}\{f(\theta) : f \in E^t\mathcal{Q}R(\lambda;\Gamma)L^p(\mathbb{R};X)\} \\
&= \operatorname{closure}\{U(\theta,\theta-t)Q(\theta-t)g(\theta-t) : g \in R(\lambda;\Gamma)L^p(\mathbb{R};X)\} \\
&= U(\theta,\tau)Q(\tau)X;
\end{aligned}$$

and hence, $U_Q(\theta,\tau)$ is surjective. □

REMARK 3.19. Returning to the definition of the Bohl spectrum for the evolution family $\{U(\theta,\tau)\}_{\theta\geq\tau}$, see Definition 3.9, we conclude that

$$(3.16) \qquad \mathcal{B} = \sigma(\Gamma) \cap \mathbb{R} = \frac{1}{t}\log|\sigma(E^t)\setminus\{0\}|.$$

Each spectral component (connected subset) of $\mathcal{B}$ generates a vertical strip (or a half plane) in $\sigma(\Gamma)$ corresponding to an annulus in $\sigma(e^{t\Gamma})$. By the Spectral Projection Theorem 3.14, each annulus in $\sigma(e^{t\Gamma})$ has an associated Riesz projection $\mathcal{P}$ that is of the from $\mathcal{P} = P(\cdot)$. Thus, we obtain the existence of U-invariant subbundles in $\mathbb{R}\times X$ that correspond to spectral components of $\mathcal{B}$. In particular, the evolution family $\{U(\theta,\tau)\}_{\theta\geq\tau}$ is stable, see Definition 3.4, if and only if $\omega(\Gamma) = s(\Gamma) < 0$. Also, the dichotomy bound, $\beta(\mathcal{U})$, see Definition 3.10, is equal to the distance from $0 \notin \sigma(\Gamma)$ to $\sigma(\Gamma)$. ◇

REMARK 3.20. The Spectral Projection Theorem 3.14 and the Dichotomy Theorem 3.17 are valid if we replace the space $L^p(\mathbb{R};X)$ by $C_0(\mathbb{R};X)$. For the proofs, see the Spectral Projection Theorem 6.38 and the Dichotomy Theorem 6.41 in Chapter 6. ◇

3.3. Howland semigroups on the half line

In this section we will study an evolution semigroup $\{E_+^t\}_{t\geq 0}$ defined on a "super"-space $\mathcal{F}_+$ of functions $f : \mathbb{R}_+ \to X$ that is associated with an exponentially bounded evolution family $\{U(\theta, \tau)\}_{\theta \geq \tau \geq 0}$. As we will see, the uniform exponential stability of the evolution family is equivalent to the stability of the evolution semigroup or the invertibility of its generator Γ_+. In addition, we will prove the Spectral Mapping Theorem for $\{E_+^t\}_{t\geq 0}$, and give a formula for the inverse operator Γ_+^{-1} in terms of an integral ("convolution") operator.

Let X denote a Banach space and let $\mathcal{F}_+$ denote one of the spaces $L^p(\mathbb{R}_+; X)$, $1 \leq p < \infty$, or the space $C_{00}(\mathbb{R}_+; X)$ of continuous functions $f : \mathbb{R}_+ \to X$ such that $f(0) = 0$ and $\lim_{t \to +\infty} f(t) = 0$ with the supremum norm. Let $\{U(\theta, \tau)\}_{\theta \geq \tau \geq 0}$ denote a strongly continuous, exponentially bounded evolution family on X, see Definition 3.1. Throughout this section we will assume that the evolution family $\{U(\theta, \tau)\}_{\theta \geq \tau}$ is defined for $\theta, \tau \in \mathbb{R}_+$. Define the associated evolution semigroup $\{E_+^t\}_{t\geq 0}$ on $\mathcal{F}_+$ by

$$(3.17) \quad (E_+^t f)(\theta) = \begin{cases} U(\theta, \theta - t) f(\theta - t), & \theta \geq t, \\ 0, & 0 \leq \theta < t. \end{cases}$$

This evolution semigroup is strongly continuous. Thus, it has an infinitesimal generator Γ_+, and $E_+^t = e^{t\Gamma_+}$, $t \geq 0$.

3.3.1. The spectral mapping theorem for evolution semigroups.
Consider the evolution semigroup $\{E_+^t\}_{t \geq 0}$ defined on the space $\mathcal{F}_+$ by formula (3.17). We begin with the observation concerning the spectral symmetry for $\{E_+^t\}_{t \geq 0}$ and Γ_+, cf. Proposition 2.36 and Theorem 3.13.

PROPOSITION 3.21. *The spectrum $\sigma(\Gamma_+)$ is invariant under translations along $i\mathbb{R}$ and the spectrum $\sigma(E_+^t)$ is invariant under rotations about zero. Moreover, for each $t > 0$ the spectrum $\sigma(E_+^t)$ is a disk centered at zero.*

PROOF. For each $\xi \in \mathbb{R}$, define an invertible operator on $\mathcal{F}_+$ by $(L_\xi f)(\theta) = e^{i\xi\theta} f(\theta)$. The required spectral symmetry is a consequence of the fact that

$$(3.18) \quad E_+^t L_\xi = e^{-i\xi t} L_\xi E_+^t \quad \text{and} \quad \Gamma_+ L_\xi = L_\xi (\Gamma_+ - i\xi).$$

To prove that, for $t_0 > 0$, the spectrum $\sigma(E_+^{t_0})$ is a disk, observe that if $z \in \rho(E_+^{t_0})$, then the circle γ with the radius $|z|$ also belongs to $\rho(E_+^{t_0})$. Let $\mathcal{F}_+^u$ denote the spectral subspace for $E_+^{t_0}$ such that $0 \notin \sigma(E_+^{t_0}|\mathcal{F}_+^u)$, for the restriction $E_+^{t_0}$ on $\mathcal{F}_+^u$), and note that $\mathcal{F}_+^u = \operatorname{Im}(I - \mathcal{P})$ for the Riesz projection $\mathcal{P}$ given as an integral of $R(\lambda; E_+^{t_0})$ over γ. Since $E_+^{t_0}|\mathcal{F}_+^u$ is invertible, the strongly continuous semigroup $\{E_+^t|\mathcal{F}_+^u\}_{t \geq 0}$ can be extended to a group, see Remark 2.3. Therefore, $E_+^t|\mathcal{F}_+^u$ is invertible for each $t > 0$. Fix $f \in \mathcal{F}_+^u$, and let $g = \left(E_+^t|\mathcal{F}_+^u\right)^{-1} f$, $t > 0$. Then $f(\theta) = (E_+^t g)(\theta) = 0$ for $0 \leq \theta < t$ by the definition of E_+^t. Since t is arbitrary, $\mathcal{F}_+^u = \{0\}$; and, therefore, $|z|$ is larger than the spectral radius $r(E_+^{t_0})$. □

THEOREM 3.22. *Let $\mathcal{F}_+$ denote the space $C_{00}(\mathbb{R}_+; X)$ or $L^p(\mathbb{R}; X)$, $1 \leq p < \infty$. The spectrum $\sigma(\Gamma_+)$ is a half plane, the spectrum $\sigma(E_+^t)$ is a disk centered at the origin, and the spectral mapping property holds for $\{E_+^t\}_{t \geq 0}$:*

$$(3.19) \quad \exp t\sigma(\Gamma_+) = \sigma(E_+^t) \setminus \{0\}, \quad t > 0.$$

PROOF. The proofs for the two spaces $\mathcal{F}_+ = C_{00}(\mathbb{R}_+; X)$ and $\mathcal{F}_+ = L^p(\mathbb{R}; X)$ are similar. We will only give the proof for the first case.

The inclusion $\exp t\sigma(\Gamma_+) \subset \sigma(E_+^t)\setminus\{0\}$ follows from the Spectral Inclusion Theorem 2.6. In view of the spectral symmetry, it suffices to show that $\sigma(E_+^t) \cap \mathbb{T} = \emptyset$ whenever $0 \in \rho(\Gamma_+)$. To this end, we employ the "change-of-variables" technique used in the proofs of the implication 3) $\Rightarrow$ 2) of Theorem 2.39 and the implication (1)$\Rightarrow$(2) of Theorem 3.13. In particular, we will exploit the known relationships between the strongly continuous semigroup $\{E_+^t\}_{t\geq 0}$ on $\mathcal{F}_+$ and the evolution semigroup it induces on the space of functions on $\mathbb{R}$ with values in $\mathcal{F}_+$.

Consider two semigroups $\{\tilde{E}^t\}_{t\geq 0}$ and $\{\mathcal{E}^t\}_{t\geq 0}$ with generators $\tilde{\Gamma}$ and G, respectively, acting on the space $C_0(\mathbb{R}; C_{00}(\mathbb{R}_+; X))$ by the following rules:

$$(\tilde{E}^t h)(\tau, \theta) = \begin{cases} U(\theta, \theta - t) h(\tau - t, \theta - t) & \text{for } \theta \geq t, \\ 0 & \text{for } 0 \leq \theta < t, \end{cases}$$

$$(\mathcal{E}^t h)(\tau, \theta) = \begin{cases} U(\theta, \theta - t) h(\tau, \theta - t) & \text{for } \theta \geq t, \\ 0, & \text{for } 0 \leq \theta \leq t \end{cases}$$

where $\tau \in \mathbb{R}$ and $h(\tau, \cdot) \in C_{00}(\mathbb{R}_+; X)$. Note that $\{\tilde{E}^t\}_{t\geq 0}$ is the evolution semigroup as in formula (2.31) that is associated with the semigroup $\{E_+^t\}_{t\geq 0}$; in fact, $(\tilde{E}^t F)(\tau) = E_+^t F(\tau - t)$ for $F(\tau) = h(\tau, \cdot)$. Also, the operator $\mathcal{E}^t$ is the multiplication operator given by $(\mathcal{E}^t F)(\tau) = E_+^t F(\tau)$. Similarly, the generator G of the semigroup $\{\mathcal{E}^t\}_{t\geq 0}$ satisfies $(GF)(\tau) = \Gamma_+(F(\tau))$ where $F(\tau) \in \mathcal{D}(\Gamma_+)$ for $\tau \in \mathbb{R}$. In particular, if $0 \in \rho(\Gamma_+)$ on $\mathcal{F}_+$, then $(G^{-1} F)(\tau) = \Gamma_+^{-1}(F(\tau))$. Hence, $0 \in \rho(G)$.

Let J denote the isometry on $C_0(\mathbb{R}; C_{00}(\mathbb{R}_+; X))$ given by $(Jh)(\tau, \theta) = h(\tau + \theta, \theta)$ for $\tau \in \mathbb{R}$ and $\theta \in \mathbb{R}_+$, and note that we have the identity

$$(\mathcal{E}^t Jh)(\tau, \theta) = (J\tilde{E}^t h)(\tau, \theta), \quad \tau \in \mathbb{R}, \quad \theta \in \mathbb{R}_+.$$

It follows that $GJF = J\tilde{\Gamma} F$ for $F \in \mathcal{D}(\tilde{\Gamma})$, and $J^{-1}GF = \tilde{\Gamma}J^{-1}F$ for $F \in \mathcal{D}(G)$. Consequently, $\sigma(G) = \sigma(\tilde{\Gamma})$ on $C_0(\mathbb{R}; C_{00}(\mathbb{R}_+; X))$. In particular, $0 \in \rho(\tilde{\Gamma})$. An application of the implication 3)$\Rightarrow$1) of Theorem 2.39 to the semigroup $\{E_+^t\}_{t\geq 0}$ on the space $\mathcal{F}_+$ shows that $\sigma(E_+^t) \cap \mathbb{T} = \emptyset$ for $t > 0$.

Proposition 3.21 and the spectral mapping property (3.19) imply $\sigma(\Gamma_+)$ is a half plane. $\square$

3.3.2. Characterization of stability.
In this subsection we prove the nonautonomous versions of Proposition 2.43 and the Datko-van Neerven Theorem 2.44.

The hyperbolicity of the evolution semigroup $\{E^t\}_{t\geq 0}$ on $\mathcal{F}$ defined as in (3.5) is equivalent to the existence of an exponential dichotomy for $\{U(\theta, \tau)\}_{\theta \geq \tau}$ (Theorem 3.17). As the following simple result shows, the stability of the evolution semigroup $\{E_+^t\}_{t\geq 0}$ on $\mathcal{F}_+$, as defined in (3.17), is equivalent to the stability of the evolution family $\{U(\theta, \tau)\}_{\theta \geq \tau \geq 0}$. Let us recall that $\omega(\Gamma_+) = s(\Gamma_+)$ by the spectral mapping property (3.19).

THEOREM 3.23. *The evolution family $\{U(\theta, \tau)\}_{\theta \geq \tau \geq 0}$ is uniformly exponentially stable on X if and only if the spectral bound $s(\Gamma_+)$ of the evolution semigroup defined on $\mathcal{F}_+$ by (3.17) is negative.*

PROOF. Let $\mathcal{F}_+ = C_{00}(\mathbb{R}_+; X)$. If $\{U(\theta, \tau)\}_{\theta \geq \tau}$ is uniformly exponentially stable, then there exist constants $M > 1$ and $\beta > 0$ such that $\|U(\theta, \tau)\|_{\mathcal{L}(X)} \leq$

$Me^{-\beta(\theta-\tau)}$, for $\theta \geq \tau$. Hence, for $\theta \geq 0$ and $f \in C_{00}(\mathbb{R}_+; X)$, we have the inequality

$$\|E_+^t f\|_{C_{00}(\mathbb{R}_+;X)} = \sup_{\theta \geq 0} |E_+^t f(\theta)| = \sup_{\theta \geq t} |U(\theta, \theta - t)f(\theta - t)|$$
$$\leq \sup_{\theta \geq t} \|U(\theta, \theta - t)\|_{\mathcal{L}(X)} |f(\theta - t)| \leq Me^{-\beta t} \|f\|_{C_{00}(\mathbb{R}_+;X)}.$$

Conversely, assume that there exist constants $M > 1$ and $\alpha > 0$ such that $\|E_+^t\| \leq Me^{-\alpha t}$ for $t \geq 0$. Also, let $x \in X$ be such that $|x| = 1$. For fixed $\theta > \tau > 0$, choose $f \in C_{00}(\mathbb{R}_+; X)$ such that $\|f\|_{C_{00}(\mathbb{R}_+;X)} = 1$ and $f(\tau) = x$. Then,

$$|U(\theta, \tau)x| = |U(\theta, \tau)f(\tau)| = |(E_+^{\theta-\tau} f)(\theta)|$$
$$\leq \sup_{s \geq 0} |(E_+^{\theta-\tau} f)(s)| = \|E_+^{\theta-\tau} f\|_{C_{00}(\mathbb{R}_+;X)} \leq Me^{-\alpha(\theta-\tau)},$$

as required. A similar argument works for $\mathcal{F}_+ = L^p(\mathbb{R}_+; X)$. $\square$

The following classical result by Datko [**Da2**] can be derived as a corollary of Theorem 3.23.

COROLLARY 3.24. *Suppose that $1 \leq p < \infty$. An exponentially bounded evolution family $\{U(\theta, \tau)\}_{\theta \geq \tau \geq 0}$ is uniformly exponentially stable if and only if, for each $x \in X$, there exists a constant $M = M(x)$ such that*

$$(3.20) \qquad \int_\tau^\infty |U(s, \tau)x|^p \, ds < M(x),$$

uniformly for $\tau \geq 0$.

PROOF. The "only if" part is trivial. For the "if" part, consider the family $\{S_\tau : \tau \geq 0\}$ of linear operators $S_\tau : X \to L^p(\mathbb{R}_+, X)$ defined by $(S_\tau x)(s) = U(s + \tau, \tau)x$. The operator S_τ is closed. Hence, it is bounded. Using (3.20) and the Uniform Boundedness Principle, $M := \sup_{\tau \geq 0} \|S_\tau\| < \infty$. Hence, for each $x \in X$,

$$\int_\tau^\infty |U(s, \tau)x|^p \, ds = \int_0^\infty |U(s + \tau, \tau)x|^p \, ds = \|S_\tau x\|_{L^p}^p \leq M^p |x|^p.$$

By Theorem 3.23, it suffices to show that this inequality implies the stability of the semigroup $\{E_+^t\}_{t \geq 0}$ on $L^p(\mathbb{R}_+; X)$. In turn, by the Datko-Pazy Theorem [**Pz**, Thm. 4.4.1], the semigroup $\{E_+^t\}$ is stable if

$$\int_0^\infty \|E_+^t f\|_{L^p(\mathbb{R}_+;X)}^p dt \leq M(f) \quad \text{for each} \quad f \in L^p(\mathbb{R}_+; X).$$

In fact, we have that

$$\int_0^\infty \|E_+^t f\|_{L^p(\mathbb{R}_+;X)}^p \, dt = \int_0^\infty \int_t^\infty |U(s, s-t)f(s-t)|^p \, ds dt$$
$$= \int_0^\infty \left(\int_0^\infty |U(s+t, s)f(s)|^p dt\right) ds$$
$$\leq \int_0^\infty M^p |f(s)|^p \, ds \leq M^p \|f\|_{L^p(\mathbb{R}_+;X)}^p.$$

$\square$

Our next goal is to give the nonautonomous version of the Datko-van Neerven Theorem 2.44 which characterizes stability in terms of a convolution operator, see (2.36). For an exponentially bounded, strongly continuous evolution family $\{U(\theta,\tau)\}_{\theta\geq\tau\geq0}$, we define the associated ("convolution") operator $\mathbb{G}$ on $\mathcal{F}_+$ as follows:

$$(\mathbb{G}f)(\theta) = \int_0^\theta U(\theta, \theta - t)f(\theta - t)\,dt = \int_0^\theta U(\theta, t)f(t)\,dt, \quad \theta \geq 0.$$

Using the definition (3.17) of the evolution semigroup $\{E_+^t\}_{t\geq0}$, we have that

$$(3.21) \qquad (\mathbb{G}f)(\theta) = \int_0^\infty (E_+^t f)(\theta)\,dt, \quad \theta \geq 0.$$

By the Laplace Transform Formula (2.3), we have that $\mathbb{G} = -\Gamma_+^{-1}$ whenever the semigroup $\{E_+^t\}_{t\geq0}$ (or, by Theorem 3.23, the evolution family $\{U(\theta,\tau)\}_{\theta\geq\tau}$) is uniformly exponentially stable. The following proposition shows that if $\mathbb{G}$ is a bounded operator, then $-\mathbb{G}$ is the inverse of Γ_+.

PROPOSITION 3.25. *Let $\{E_+^t\}_{t\geq0}$ be the evolution semigroup defined by the formula (3.17) on the space $\mathcal{F}_+ = L^p(\mathbb{R}_+; X)$ or $\mathcal{F}_+ = C_{00}(\mathbb{R}_+; X)$, and let Γ_+ denote its generator. Assume that $u, f \in \mathcal{F}_+$. The following statements are equivalent:*

(a) $u \in \mathcal{D}(\Gamma_+)$ *and* $\Gamma_+ u = -f$.
(b) $u(\theta) = \int\limits_0^\theta U(\theta, \tau) f(\tau)\,d\tau, \theta \geq 0$.

PROOF. Assume that (a) holds. By elementary properties of strongly continuous semigroups, see Pazy [**Pz**, p. 4], if $t > 0$, then

$$E_+^t u - u = \int_0^t E_+^\tau \Gamma_+ u\,d\tau = -\int_0^t E_+^\tau f\,d\tau.$$

Fix $\theta \geq 0$ and consider $t > \theta$. Using the definition of E_+^t, observe that $(E_+^t u)(\theta) - u(\theta) = -u(\theta)$ and

$$\int_0^t (E_+^\tau f)\theta)\,d\tau = \int_0^\theta U(\theta, \theta - \tau)f(\theta - \tau)\,d\tau = \int_0^\theta U(\theta, \tau)f(\tau)\,d\tau.$$

Thus, we have proved statement (b).

Assume that statement (b) holds. The definition of E_+^t gives

$$\int_0^t (E_+^\tau f)(\theta)\,d\tau = \int_{\max\{0,\theta-t\}}^\theta U(\theta, \tau) f(\tau)\,d\tau, \quad \theta \geq 0, \quad t > 0.$$

Using (b), the last integral is equal to $(E_+^t u)(\theta) - u(\theta)$. Hence, the equality

$$E_+^t u - u = -\int_0^t E_+^\tau f\,d\tau, \quad t > 0,$$

implies statement (a). □

THEOREM 3.26. *For an evolution family $\{U(\theta,\tau)\}_{\theta\geq\tau\geq0}$ on the Banach space X, the following statements are equivalent:*

(i) $\{U(\theta,\tau)\}_{\theta \geq \tau \geq 0}$ *is uniformly exponentially stable.*
(ii) $\mathbb{G}$ *is a bounded operator on* $L^p(\mathbb{R}_+; X)$.
(iii) $\mathbb{G}$ *is a bounded operator on* $C_0(\mathbb{R}_+; X)$.

Before proceeding with the proof, note that statement (ii) is equivalent to the statement: $\mathbb{G}f \in L^p(\mathbb{R}_+; X)$ for each $f \in L^p(\mathbb{R}_+; X)$. This fact is obtained using the argument in Remark 2.45.

PROOF. By Theorem 3.23, statement (i) implies $\{E_+^t\}_{t\geq 0}$ is stable, and formula (3.21) implies statements (ii) and (iii).

We will prove (iii) $\Rightarrow$ (i); the implication (ii) $\Rightarrow$ (i) is proved similarly. Since $(\mathbb{G}f)(0) = 0$, the operator $\mathbb{G}$ is bounded on $C_{00}(\mathbb{R}_+; X)$ by statement (iii). By Proposition 3.25, the operator Γ_+ on $C_{00}(\mathbb{R}_+; X)$ is injective and $\Gamma_+ u = -f$ is equivalent to $u = \mathbb{G}f$. Thus, Γ_+ is invertible with $\Gamma_+^{-1} = -\mathbb{G}$. By Theorem 3.22, we have $s(\Gamma_+) < 0$. An application of Theorem 3.23 gives statement (i). $\square$

Theorem 3.26 makes explicit, in the case of the half line $\mathbb{R}_+$, the relationship between the stability of an evolution family $\{U(\theta,\tau)\}_{\theta \geq \tau}$ and the generator, Γ_+, of the corresponding evolution semigroup (3.17). Indeed, as shown above, stability is equivalent to the boundedness of $\mathbb{G}$, in which case $\mathbb{G} = -\Gamma_+^{-1}$. Combining Theorems 3.22 and 3.26 yields the following corollary.

COROLLARY 3.27. *Let* $\{U(\theta,\tau)\}_{\theta \geq \tau \geq 0}$ *be an exponentially bounded evolution family on* X *and* Γ_+ *the generator of the induced evolution semigroup* (3.17) *on* $\mathcal{F}_+ = L^p(\mathbb{R}_+; X)$, $1 \leq p < \infty$, *or on* $\mathcal{F}_+ = C_{00}(\mathbb{R}_+; X)$. *The following statements are equivalent:*

(i) $\{U(\theta,\tau)\}_{\theta \geq \tau}$ *is uniformly exponentially stable.*
(ii) Γ_+ *is invertible with* $\Gamma_+^{-1} = -\mathbb{G}$.
(iii) *the spectral bound* $s(\Gamma_+) < 0$.

Finally, we will give yet another proof of the implication (ii) $\Rightarrow$ (i) of Theorem 3.26 based on the "change-of-variables" technique, as in the proof of Theorem 3.22, and the implication (ii) $\Rightarrow$ (i) of the autonomous Datko-van Neerven Theorem 2.44. The operator $\mathcal{G}_*$ that appears in this proof is used in Subsection 5.1.3.2 for a discussion of the transfer function for nonautonomous control systems.

PROOF. Consider the multiplication operator $\mathcal{G}$ on the space
$$L^p(\mathbb{R}; L^p(\mathbb{R}_+; X)) = L^p(\mathbb{R} \times \mathbb{R}_+; X)$$
given by multiplication by $\mathbb{G}$. More precisely, the operator $\mathcal{G}$ is defined by
$$(\mathcal{G}h)(\tau,\theta) = \int_0^\theta U(\theta, \theta - t) h(\tau, \theta - t)\, dt, \quad \theta \in \mathbb{R}_+, \quad \tau \in \mathbb{R}.$$

In view of statement (ii) in Theorem 3.26, this operator is bounded. For the isometry J defined on the space $L^p(\mathbb{R}, L^p(\mathbb{R}_+; X))$ by $(Jh)(\tau,\theta) = h(\tau + \theta, \theta)$, we have

(3.22) $\quad (J^{-1}\mathcal{G}Jh)(\tau, M\theta) = \displaystyle\int_0^\theta U(\theta, \theta - t) h(\tau - t, \theta - t)\, dt.$

Let $\mathbb{G}_*$ denote the operator defined on $L^p(\mathbb{R}; L^p(\mathbb{R}_+; X))$ by

(3.23) $\quad (\mathbb{G}_* H)(\tau) = \displaystyle\int_0^\infty E_+^t H(\tau - t)\, dt$

where $\{E_+^t\}_{t\geq 0}$ is the evolution semigroup (3.17) on $L^p(\mathbb{R}_+, X)$ induced by the evolution family $\{U(\theta, \tau)\}_{\theta \geq \tau}$. Using the definition (3.17), observe that if $H(\tau) = h(\tau, \cdot) \in L^p(\mathbb{R}_+; X)$ for $\tau \in \mathbb{R}$, then

$$(3.24) \quad (\mathbb{G}_* h)(\tau, \theta) = \int_0^\theta U(\theta, \theta - t) h(\tau - t, \theta - t) \, dt, \quad \theta \in \mathbb{R}_+, \quad \tau \in \mathbb{R}.$$

By equation (3.22), the operator $\mathbb{G}_* = J^{-1} \mathcal{G} J$ is bounded on $L^p(\mathbb{R}; L^p(\mathbb{R}_+; X))$.

Each function $H_+ \in L^p(\mathbb{R}_+; L^p(\mathbb{R}_+; X))$ is an $L^p(\mathbb{R}_+; X)$-valued function on the half line $\mathbb{R}_+$. Extend each such H_+ to a function $H \in L^p(\mathbb{R}; L^p(\mathbb{R}_+; X))$ by setting $H(\tau) = H_+(\tau)$ for $\tau \geq 0$, and $H(\tau) = 0$ for $\tau < 0$. Note that $\mathbb{G}_* H \in L^p(\mathbb{R}; L^p(\mathbb{R}_+; X))$ because $\mathbb{G}_*$ is bounded on $L^p(\mathbb{R}; L^p(\mathbb{R}_+; X))$. Also, consider the function $F_+ : \mathbb{R}_+ \to L^p(\mathbb{R}_+; X)$ defined by

$$F_+(\tau) = \int_0^\tau E_+^t H_+(\tau - t) \, dt = \int_0^\infty E_+^t H_+(\tau - t) \, dt, \quad \tau \in \mathbb{R}_+.$$

To complete the proof, it suffices to show that $F_+ \in L^p(\mathbb{R}_+; L^p(\mathbb{R}_+; X))$. Indeed, the operator $H_+ \mapsto F_+$ is the convolution operator as in (2.36) defined for the semigroup $\{E_+^t\}_{t \geq 0}$ instead of $\{e^{tA}\}_{t\geq 0}$. By the autonomous Datko-van Neerven Theorem 2.44, applied to $\{E_+^t\}_{t\geq 0}$ on $L^p(\mathbb{R}_+; X)$, we conclude that the claim "$F_+ \in L^p(\mathbb{R}_+; L^p(\mathbb{R}_+; X))$ for each $H_+ \in L^p(\mathbb{R}_+; L^p(\mathbb{R}_+; X))$" implies the semigroup $\{E_+^t\}_{t\geq 0}$ is stable on $L^p(\mathbb{R}_+; X)$. Then, by Theorem 3.23, the evolution family $\{U(\theta, s)\}_{\theta \geq s}$ is stable.

To prove that $F_+ \in L^p(\mathbb{R}_+; L^p(\mathbb{R}_+; X))$, apply formula (3.24) with $h(\tau, \theta) = h_+(\tau, \theta)$ for $\tau \geq 0$, and $h(\tau, \theta) = 0$ for $\tau < 0$ where $h_+(\tau, \cdot) = H_+(\tau)$ and $\theta \in \mathbb{R}_+$. This gives

$$(\mathbb{G}_* h)(\tau, \theta) = \int_0^{\min\{\tau, \theta\}} U(\theta, \theta - t) h_+(\tau - t, \theta - t) \, dt \quad \text{for} \quad \tau, \theta \in \mathbb{R}_+,$$

and $(\mathbb{G}_* h)(\tau, \theta) = 0$ for $\tau < 0$ and $\theta \in \mathbb{R}_+$. Thus, the function

$$\tau \mapsto (\mathbb{G}_* h)(\tau, \cdot) = (\mathbb{G}_* H)(\tau) \in L^p(\mathbb{R}_+; X)$$

is in the space $L^p(\mathbb{R}_+; L^p(\mathbb{R}_+; X))$. On the other hand, for $F_+(\tau) = f_+(\tau, \cdot) \in L^p(\mathbb{R}_+; X)$, we have that

$$f_+(\tau, \theta) = \int_0^{\min\{\tau, \theta\}} U(\theta, \theta - t) h_+(\tau - t, \theta - t) \, dt, \quad \tau, \theta \in \mathbb{R}_+.$$

Thus, $\tau \mapsto f_+(\tau, \cdot) = (\mathbb{G}_* h)(\tau, \cdot)$ is a function from $L^p(\mathbb{R}_+; L^p(\mathbb{R}_+; X))$, and the claim is proved. $\square$

3.4. Bibliography and remarks

To our knowledge, Howland [**Ho**] was the first to introduce evolution semigroups on vector-valued function spaces. Motivated by a study of scattering theory for time-dependent Hamiltonians $H(t)$, he considers a self-adjoint operator $K = -id/dt + H(t)$ on $L^2(\mathbb{R}; X)$ where X is a Hilbert space. Note that the group generated by iK on $\mathcal{F} = L^2(\mathbb{R}; X)$ is of the form given in formula (3.5). Howland proves, in particular, that a strongly continuous, bounded group $\{E^t\}_{t \in \mathbb{R}}$ on $\mathcal{F}$ is an evolution group if and only if there exists a bounded, measurable, operator-valued function $u : \mathbb{R} \to \mathcal{L}(X)$ such that $E^t = \mathcal{U} V^t \mathcal{U}^{-1}$ where $(\mathcal{U} f)(\theta) = u(\theta) f(\theta)$ and $(V^t f)(\theta) = f(\theta - t)$ is the group of translations. Another characterization, proved in [**Ho**], states that $\{e^{t\Gamma}\}_{t \in \mathbb{R}}$ is an evolution group on $\mathcal{F}$ if and only if, for each

$\alpha \in C_c^1(\mathbb{R})$ (the space of compactly supported and continuously differentiable functions), $M_\alpha \mathcal{D}(\Gamma) \subset \mathcal{D}(\Gamma)$ and $\Gamma M_\alpha f - M_\alpha \Gamma f = M_{\alpha'} f$ for each $f \in \mathcal{D}(\Gamma)$ where M_α is the operator of multiplication by α, cf. Theorem 3.30. In addition, he studies perturbations of evolution groups and applications of the theory to time-dependent Schrödinger equations.

We also mention a paper by Lovelady [**Lo**] where the evolution semigroup on $C_0(\mathbb{R}; X)$ is associated to a strongly continuous evolution family. A characterization of evolution semigroups is given, evolution semigroups are applied to produce formulas for evolution families associated with a nonautonomous Cauchy problem, and some results on perturbations of evolution semigroups are proved. See also the paper by Elliot [**El**] and the references cited on p. 459 in Fattorini [**Fa**].

The line of Howland's research was continued by Evans [**Ev**] who proved similar results for bounded evolution groups on $\mathcal{F} = C_0(\mathbb{R}; X)$ or $L^p(\mathbb{R}; X)$ where X is a separable Banach space. Among other results, the condition of separability was removed by Neidhardt [**Ne**] who also considered evolution *semi*groups acting on the space $L^p([0, T]; X)$ or $C_{00}([0, T]; X)$. A one-to-one correspondence between evolution semigroups and evolution families was also established.

Important contributions to the problem of the characterization of evolution semigroups were made by Lumer [**Lu, Lu2, Lu3**] and Paquet [**Pq**] in connection to applications for parabolic equations with time-dependent domains, see also the recent paper by Lumer and Schnaubelt [**LSc**].

A further study of evolution semigroups was initiated by R. Nagel in the Tübingen group, see [**Na2**]. In the dissertation of Rau [**Ra**], see also the important papers [**Ra2, Ra3**], the connection between the hyperbolicity of evolution semigroups on $C_0(\mathbb{R}; X)$ or $L^p(\mathbb{R}; X)$ and the dichotomy of evolution families is studied. In particular, the Spectral Mapping Theorem for the Hilbert space case is proved using Gearhart's Spectral Mapping Theorem 2.10, see also Latushkin and Stepin [**LSt2**]. Moreover, the Spectral Projection Theorem 3.14 and Dichotomy Theorem 3.17 are proved for $\mathcal{F} = C_0(\mathbb{R}; X)$. Generalizations of these proofs are used in Chapter 6 of this book.

Rau's research was continued by Latushkin, Montgomery-Smith and Randolph [**LM, LM2, LMR, LR**] and, in Tübingen, by Hutter, Nagel, Nickel Räbiger, Rhandi, and Schnaubelt, see [**Ht, Na2, NR, NR2, Ni, Rh, RS, RS2, RRS, Sc**].

We remark that evolution semigroups were independently discovered and studied also by Zabreiko and Nguyen Van Minh, see [**ZM**] and [**Mi, Mi2, Mi3**], see also Appel, Lakshmikantham, Nguyen Van Minh, Zabreiko [**ALMZ**] and Aulbach and Nguyen Van Minh [**AM**].

Evolution semigroups are also studied in very interesting papers by Baskakov [**Bs, Bs2, Bs3, Bs4**]. He develops further the techniques of Levitan and Zhikov [**LZ**]. See also related work by Tyurin [**Ty**].

See also an important paper by Dorroh and Neuberger [**DN**].

Evolution families and dichotomy. *Abstract Cauchy problems and their evolution families.* The material in this section is standard. It can be found, for example, in Coppel [**Co**], Daners and Koch Medina [**DKo**], Daleckij and Krein [**DK**], Fattorini [**Fa**], Hale [**Ha2**], Henry [**He**], Pazy [**Pz**], or Tanabe [**Ta**].

Stability and dichotomy of evolution families. Modifications of Example 3.5 appear in many places, for example in Coppel [**Co**], Hale [**Ha2**], and Nagel [**Na2**].

See the classical sources by Daleckij and Krein [**DK**] and Massera and Schäffer [**MS2**] for the definition of dichotomy for the evolution family generated by the differential equation $\dot{x} = A(t)x$ where $t \mapsto A(t)$ is a bounded, continuous function with values in $\mathcal{L}(X)$. Our Definition 3.6 is a standard (see, e.g., [**He**]) modification of the usual definition of dichotomy from [**DK, MS2**]. We stress again the importance of condition (ii) in Definition 3.6 of a dichotomy. Under additional assumptions on compactness of the operators $U(\theta, \tau)$, $\theta \geq \tau$, the invertibility of the evolution family on the unstable fibers is automatic. In our general setting this is a part of the definition of a dichotomy.

The literature on exponential dichotomy is vast. Besides the classical books [**Co, DK, Ha2, MS2**], we mention important contributions by Palmer [**Pa, Pa2, Pa3, Pa4**] and the book by Kirchgraber and Palmer [**KP**]. There are several unpublished sets of notes by Henry [**He3**]. The classical series of papers by R. Sacker and G. Sell on this subject will be mentioned in Chapters 6 and 7, see [**SS2, SS3, SS4, SS5, Sa, Sa2**]. We also mention the following papers on dichotomy: Coppel [**Co2**], Kurzweil and Papaschinopoulos [**KPp**], Megan and Latcu [**ML**], Preda and Megan [**PM**], Rodrigues and Ruas-Filho [**RR**], Volevich and Shirikyan [**VS**], Zeng [**Ze**], Zhang [**Zh**], and Willems [**Wm**].

Dichotomy theory for the equation $\dot{x} = A(t)x$, where each $A(t)$ is an unbounded operator on an infinite dimensional space, is studied by Chow and Leiva [**CLe, CLe2, CLe3, CLe4**]. We mention also the work by Lin [**Li**], Johnson and Nerurkar [**JN**], Aulbach and Garay [**AGa**], Laederich [**La**], Vinograd [**Vn**], Elaydi and Hajen [**EH**], and Levenshtam [**Lv**]. Recently, dichotomy was used by Shen and Yi in their study of automorphic dynamics, see [**SY, SY2**]. An important series of papers relating dichotomies to the invertibility of certain differential operators was written by Ben-Artzi, Gohberg, and Kaashoek, see [**BGK, BG3, BG4**]. These authors give applications of dichotomies in [**BG**], [**BG2**], [**BGK**], see also [**BrGK**] and other publications of these authors.

Usually, see for example [**DK**], the Bohl spectrum is defined in terms of general exponents; for instance,

$$\omega_0(\{U(\theta, \tau)\}) := \limsup \frac{1}{\theta - \tau} \ln \|U(\theta, \tau)\| \quad \text{as} \quad \theta - \tau \to \infty \quad \text{and} \quad \theta \to \infty.$$

Our definition is related to the dynamical, or Sacker-Sell spectrum, cf. [**SS3**] and [**CLe**].

Howland semigroups on the line. There are several ways to assign "evolution semigroups" to an evolution family $\{U(\theta, \tau)\}_{\theta \geq \tau}$ on a Banach space X that are different from the choice made in this book. For example, we can define

$$E_{op}^t F(\theta) = U(\theta, \theta - t) F(\theta - t), \quad \theta \in \mathbb{R}, \quad t \geq 0,$$

for *operator-valued* functions $F \in C_0(\mathbb{R}; \mathcal{L}(X))$. Also, instead of using functions defined on $\mathbb{R}$, we can use functions on $\mathbb{R}_+$ together with left translations to define

$$(E_{\text{left}}^t f)(\theta) = U(\theta, \theta + t) f(\theta + t), \quad \theta \in \mathbb{R}, \quad t \geq 0.$$

A disadvantage of this last formula is the requirement that $\{U(\theta, t)\}$ is defined for $\theta \leq t$, cf. (3.17).

In this book, the evolution semigroup associated with a strongly continuous, exponentially bounded evolution family $\{U(\theta, \tau)\}_{\theta \geq \tau}$ on a Banach space X is defined on a "super"-space $\mathcal{F}$ by the rule $(E^t f)(\theta) = U(\theta, \theta - t) f(\theta - t)$. A natural question arises: What characterizes an evolution semigroup? More precisely, which semigroups on $\mathcal{F}$ are evolution semigroups generated by some evolution family? This question goes back to Howland [**Ho**], Lumer [**Lu3**], and

[**Ev, Ne, Pq**]. Very general results in this direction were obtained by Räbiger, Rhandi, and Schnaubelt [**RRS, Sc**]. We state here a simplified version of a theorem in [**RRS**].

THEOREM 3.28. *Suppose that $\{e^{t\Gamma}\}_{t\geq 0}$ is a strongly continuous semigroup on $L^p(\mathbb{R};X)$ for which there exists a point $\lambda \in \rho(\Gamma)$ such that the resolvent operator $R(\lambda;G)$ maps $L^p(\mathbb{R};X)$ to $C_0(\mathbb{R};X)$ as a continuous operator with dense range. The following statements are equivalent:*

(a) $\{e^{t\Gamma}\}_{t\geq 0}$ *is an evolution semigroup; that is, there exists a strongly continuous, exponentially bounded evolution family $\{U(\theta,\tau)\}_{\theta \geq \tau}$ such that*
$$(e^{t\Gamma}f)(\theta) = U(\theta, \theta-t)f(\theta-t) \quad \text{for all} \quad f \in L^p(\mathbb{R};X), \quad \theta \in \mathbb{R}.$$
(b) $[e^{t\Gamma}(\alpha f)](\theta) = \alpha(\theta-t)(e^{t\Gamma}f)(\theta)$, $\theta \in \mathbb{R}$, $t \geq 0$, *for each $\alpha \in C_c^1(\mathbb{R})$ and all $f \in L^p(\mathbb{R};X)$.*
(c) $\Gamma(\alpha f) = -\alpha' f + \alpha \Gamma f$ *for each $f \in \mathcal{D}(\Gamma)$ and $\alpha \in C_c^1(\mathbb{R})$.*

The implication (a) $\Rightarrow$ (c) follows from the definition of the generator Γ. The implication (c)$\Rightarrow$(b) follows from the fact that $u(t)(\cdot) = \alpha(\cdot - t)(e^{t\Gamma}f)(\cdot)$ solves the abstract Cauchy problem $\frac{d}{dt}u(t) = \Gamma u(t)$. The main idea in the proof of the implication (b)$\Rightarrow$(a) is to check that the operator $M_\theta = e^{\theta\Gamma}V^{-\theta}$, $V^\theta f(\cdot) = f(\cdot - \theta)$ is a multiplication operator with multiplier the operator-function $M_\theta = U(\theta, \cdot)$. The structure of this operator is a consequence of the fact that $M_\theta(\alpha f) = \alpha M_\theta f$.

We mention that results of this type were applied in [**RRS**] to prove the existence of an evolution family for a class of nonautonomous heat equations. This requires an application of a perturbation theorem by Voigt [**Vo**]. Subsection 5.2.2 contains a discussion for perturbation theorems for evolution semigroups, see also Räbiger, Rhandi, Schnaubelt, and Voigt [**RRSV**] for further advances.

Generator of the evolution semigroup. Theorem 3.12 first appeared in Latushkin and Montgomery-Smith [**LMR**]; the corrected version given here is due to Schnaubelt [**Sc**]; but the proof is taken from [**LMR**]. See Nickel [**Ni**] for further developments.

Some sufficient conditions for the invertibility of the operator $d/dt - A(t)$ acting on $L^2(\mathbb{R};\mathcal{H})$ where H is a Hilbert space and $A : \mathbb{R} \to \mathcal{L}(\mathcal{H})$ is a continuous, bounded function are obtained by Baskakov and Yurgelas [**BYu**], see also further developments for $L^p(\mathbb{R};X)$ in [**Bs**].

Theorem 3.12 addresses the case where an abstract Cauchy problem is known to have a corresponding evolution family. As we have mentioned in Subsection 3.1.1, it is of course of great interest to know under which condition an abstract Cauchy problem has a solution given by an evolution family. One of the ways to attack this problem is to find conditions under which there exists an evolution semigroup (induced by an evolution family) that is related to the the abstract Cauchy problem. Following [**Na2**], we give a typical result in this direction that is due to Nagel and Rhandi [**NR2**]. We stress that much more general results are currently available; our exposition is intended to outline the simplest possible situation where the evolution semigroups can be used to show the existence of a corresponding evolution family.

Since the evolution semigroup $\{E^t\}_{t\geq 0}$ is a multiplicative perturbation of the translation semigroup $\{V^t\}_{t\geq 0}$, one might expect that its generator Γ is an additive perturbation of the generator of the translation semigroup. Let $\Gamma_0 = -d/d\theta$ denote the generator of the translation semigroup $(V^t f)(\theta) = f(\theta - t)$ on the space $\mathcal{F} :=$

$C_{UB}(\mathbb{R}; X)$ of uniformly continuous, bounded X-valued functions on $\mathbb{R}$. Every $A(\cdot) \in C_{UB}(\mathbb{R}; \mathcal{L}(X))$ defines a bounded linear operator $\mathcal{A}$ on $C_{UB}(\mathbb{R}; X)$ by the formula $(\mathcal{A}f)(\theta) = A(\theta)f(\theta)$, $\theta \in \mathbb{R}$.

PROPOSITION 3.29 ([**Na2**, **NR2**]). *If $A(\cdot) \in C_{UB}(\mathbb{R}; \mathcal{L}(X))$, then there exists a strongly continuous evolution family $\{U(\theta, \tau)\}_{\theta \geq \tau}$ such that $\Gamma = \Gamma_0 + \mathcal{A}$ with domain $\mathcal{D}(\Gamma) = \mathcal{D}(\Gamma_0)$ is the generator of the corresponding evolution semigroup on $\mathcal{F}$.*

To outline the proof, we remark that Γ is a generator by the Bounded Perturbation Theorem. By the Dyson-Phillips expansion (see Goldstein [**Go2**, Thm. I.6.5]), the semigroup $\{E^t\}_{t \geq 0}$ with generator Γ is given by $E^t = \sum_{n=0}^{\infty} E_n^t$ where

$$E_n^t f = \int_0^t E_0^{t-s} \mathcal{A} E_{n-1}^s f \, ds$$

for each integer $n \geq 1$. If E_n^t is of the form $(E_n^t f)(\theta) = U_n(\theta, \theta - t)f(\theta - t)$, then, for $\theta \in \mathbb{R}$ and some function $U_n(\cdot, \cdot - t) \in C_{UB}(\mathbb{R}; \mathcal{L}(\mathcal{F}))$, we have

$$(E_{n+1}^t f)(\theta) = \int_0^t E_0^{t-s} A(\theta) U_n(\theta, \theta - s) f(\theta - s) \, ds$$

$$= \int_0^t A(\theta - t - s) U_n(\theta - t - s, \theta - t) \, ds \cdot f(\theta - t)$$

$$=: U_{n+1}(\theta, \theta - t) f(\theta - t).$$

Since $\sum_{n=0}^{\infty} E_n^t$ converges in $\mathcal{L}(\mathcal{F})$, it follows that $U(\theta, \tau) := \sum_{n=0}^{\infty} U_n(\theta, \tau)$ is well defined and $(E^t f)(\theta) = U(\theta, \theta - t) f(\theta - t)$ for each $f \in \mathcal{F}$. Finally, using the fact that $\{E^t\}_{t \geq 0}$ is strongly continuous, it follows that $\{U(\theta, \tau)\}_{\theta \geq \tau}$ is an evolution family, and the proof is completed.

For a fixed $x \in X$, let $\mathbb{I}_x(\theta) = x$, $\theta \in \mathbb{R}$. Since $E^t \mathbb{I}_x$ is in $\mathcal{D}(\Gamma)$, by differentiation we see that the evolution family obtained in the previous proposition solves the abstract Cauchy problem for $A(\cdot)$. Thus, we have proved that a nonautonomous Cauchy problem is well-posed by using the perturbation theory for evolution semigroups. An application of the Chernoff product formula to the operators $W(t) \in \mathcal{L}(\mathcal{F})$ defined by

$$(W^t f)(\theta) = \exp\left(\int_{\theta - t}^{\theta} A(s) \, ds\right) f(\theta - t)$$

yields the following representation for the evolution family in Proposition 3.29:

$$U(\theta, \theta - t) x = \lim_{n \to \infty} \prod_{k=0}^{n-1} \exp\left(\int_{\frac{kt}{n}}^{\frac{(k+1)t}{n}} A(\theta - s) \, ds\right) x, \quad x \in X,$$

see Goldstein [**Go2**, Thm. I.8.4].

The Spectral Mapping Theorem for evolution semigroups. Theorem 3.13 and its "indirect" proof, based on the "change-of-variable" trick, cf. Theorem 2.39, is taken from Latushkin and Montgomery-Smith [**LM**]. For the case of a group acting in a Hilbert space, Rau [**Ra**, Prop. 1.7] gave a proof of this result based on the Gearhart Spectral Mapping Theorem 2.10, see also Latushkin and Stepin [**LSt2**]. A general result of this type (with a different direct proof) valid for a wide class of "super"-spaces $\mathcal{F}$ is due to Räbiger and Schnaubelt [**RS2**].

For the case $\mathcal{F} = L^p(\mathbb{R}; X)$, the following direct proof of the Spectral Mapping Theorem can be found in Latushkin and Montgomery-Smith [**LM3**]. To prove the equality $\sigma(e^{t\Gamma})\setminus\{0\} = e^{t\sigma(\Gamma)}$ directly (see the proofs of Theorem 2.39 in Chapter 2), the usual method is to assume that $1 \in \sigma_{\mathrm{ap}}(E^1)$ and then show that $0 \in \sigma_{\mathrm{ap}}(\Gamma)$ by constructing an approximate eigenfunction $g \in \mathcal{D}(\Gamma)$. If $\|f\|_{L^p} = 1$ and $\|E^1 f - f\|_{L^p} < \epsilon$, then there exists $\theta_0 \in [0,1]$ such that

$$\sum_{n=-\infty}^{\infty} |U(n+\theta_0, n+\theta_0-1)f(n+\theta_0-1) - f(n+\theta_0)|^p < \epsilon^p \sum_{n=-\infty}^{\infty} |f(n+\theta_0)|^p.$$

By constructing a smooth function $\alpha : [0,1] \to [0,1]$ such that $\alpha(\theta) = 0$ for $\theta \in [0, 1/3]$, and $\alpha(\theta) = 1$ for $\theta \in [2/3, 1]$, as in the proof of Lemma 2.29 in Chapter 2, g is defined by the formula (cf. (2.34))

$$g(n + \theta_0 + \tau) = \chi(\theta)[(1-\alpha(\tau)]U(n+\theta_0+\tau, n+\theta_0-1)f(n+\theta_0-1) \\ + \alpha(\tau)U(n+\theta_0+\tau, n+\theta_0)f(n+\theta_0)$$

where $0 \leq \tau \leq 1$, $n \in \mathbb{Z}$, and $\chi : \mathbb{R} \to [0,1]$ is a bump-function with the following properties: $\chi(0) = 1$, $\chi(t) = 0$ for $|t| > 1/\epsilon$, and $|\chi'(t)| \leq 2\epsilon$.

Spectral Projection and Dichotomy Theorems. The Spectral Projection Theorem 3.14 goes back to Antonevich [**An2, An3**] for the case of linear skew-product flows, see Chapter 6 and Latushkin and Stepin [**LSt2**]. For *invertible* evolution families, Rau [**Ra**] proved the Spectral Projection Theorem 3.14 and the Dichotomy Theorem 3.17 on the space $C_0(\mathbb{R}; X)$. The general theorem presented in this book follows the proof given by Räbiger and Schnaubelt in [**RS**]. A different proof was given by Latushkin and Montgomery-Smith [**LM**] for separable Banach spaces (this latter condition, in fact, is not needed in [**LM**]). The main idea of [**LM**] is to use discrete operators $\pi_\theta(E)$, see Section 4.1 below.

The idea underlying the Commutation Lemma 3.15 goes back to Mather [**Ma**] (see also Antonevich [**An2**]) where an analogue of the space $C_0(\mathbb{R}; X)$ is used. Lemma 3.15 is taken from [**LM**], see [**LSt2**] and [**Ra**] for earlier versions. The beautiful regularity result recorded as Lemma 3.16 is due to Räbiger and Schnaubelt [**RS**]. It shows that, in the proof of the Spectral Projection Theorem 3.14, it suffices to use *continuous* functions instead of L^p-functions in $\mathcal{D}(\Gamma)$ to define the projection-valued function $P(\cdot)$ by (3.13). The proof of Lemma 3.16 as well as the equivalence (1)⇔(2) in the Dichotomy Theorem 3.17 are taken from [**RS**].

Formula (3.16) is a particular case of a more general relation involving the Sacker-Sell spectrum, see Corollary 6.45 in Chapter 6. This formula shows that the Bohl spectrum; that is, the "spectrum of general exponents", see [**DK**], is indeed a "spectrum".

We mention again papers by Van Minh and Zabreiko [**Mi, ZM**]. In particular, [**Mi**] contains the Spectral Mapping Theorem and the Dichotomy Theorem for the case where each $A(t)$ is a bounded operator.

This line of research was continued by A. G. Baskakov. In [**Bs2**], Baskakov considers the evolution semigroup $\{E^t\}_{t \geq 0}$ on $L^p(\mathbb{R}; X)$ or $C_0(\mathbb{R}; X)$ generated by a strongly continuous evolution family $\{U(\theta, \tau)\}_{\theta \geq \tau}$. Among other results, the Spectral Mapping Theorem for the evolution semigroup and the Dichotomy Theorem are proved in [**Bs2**]. This work was continued in [**Bs3, Bs4**], see more comments on page 128.

Howland semigroups on the half line. Our definition (3.17) of the evolution semigroup $\{E_+^t\}_{t \geq 0}$ on the spaces $\mathcal{F}_+ = C_{00}(\mathbb{R}_+; X)$ or $L^p(\mathbb{R}_+; X)$ is a direct generalization of (2.35). To our knowledge, these semigroups were first studied by

Rau [**Ra**]. The following result characterizes evolution semigroups on the half line, see Schnaubelt [**Sc**].

THEOREM 3.30. *Let* $\{E_+^t\}_{t\geq 0}$ *denote a strongly continuous semigroup on the space* $\mathcal{F}_+ = L^p(\mathbb{R}_+; X)$ *with infinitesimal generator* Γ_+. *The following statements are equivalent:*
  (i) *The semigroup* $\{E_+^t\}_{t\geq 0}$ *is an* evolution semigroup *induced by an evolution family* $\{U(\theta, \tau)\}_{\theta \geq \tau \geq 0}$.
  (ii) *There exists a core* $\mathcal{D}$ *for* Γ_+ *such that for all* $\alpha \in C_0^1(\mathbb{R}_+)$ *and* $f \in \mathcal{D}$ *it follows that* $\alpha f \in \mathcal{D}(\Gamma_+)$ *and* $\Gamma_+(\alpha f) = -\alpha' f + \alpha \Gamma_+ f$. *Moreover, there exists* $\lambda \in \rho(\Gamma_+)$ *such that the resolvent* $R(\lambda; \Gamma_+) \colon \mathcal{F}_+ \to C_{00}(\mathbb{R}_+; X)$ *is continuous with dense range.*

The Spectral Mapping Theorem. Proposition 3.21 is due to Rau [**Ra**]. Note that (3.18) implies $\|\Gamma_+^{-1}\| = \|(\Gamma_+ - i\xi)^{-1}\|$ for $\xi \in \mathbb{R}$. If $p = 2$ and X is a Hilbert space then Gearhart's Theorem 2.10 implies the Spectral Mapping Theorem for $\{E_+^t\}_{t\geq 0}$. This idea was used by Latushkin and Stepin [**LSt2**] and Rau [**Ra**] in their proofs of the spectral mapping property for the Hilbert space setting. Theorem 3.22 and its "indirect" proof, based on the "change-of-variable" trick is taken from a paper by Clark, Latushkin, Randolph and Montgomery-Smith [**CLRM**]. "Direct" proofs can be found in van Neerven, Van Minh, Räbiger, Rau, and Schnaubelt [**RS2, vN2, MRS, Ra**].

Characterization of stability. An early version of Theorem 3.23 is due to Rau [**Ra**] for the space $L^2(\mathbb{R}_+; X)$ where X is a Hilbert space. He also noticed that the following statements are equivalent: $\{E_+^t\}_{t\geq 0}$ is stable; $\{E_+^t\}_{t\geq 0}$ is hyperbolic; $I - E_+^1$ is invertible; and Γ_+ is invertible. See Datko [**Da2**] for another characterization of stability given in Corollary 3.24. The elegant proof of Corollary 3.24, given above, is due to R. Schnaubelt (private communication).

Proposition 3.25 is due to Van Minh, Räbiger and Schnaubelt [**MRS**]. It relates the generator of the evolution semigroup and solutions of the mild inhomogeneous differential equation $\dot{u} = A(t)u + f$. See Proposition 4.32 for the line case. Our proof of Theorem 3.26, the nonautonomous version of the Datko-van Neerven Theorem 2.44, is an immediate application of this result. The second proof of the implication (ii) $\Rightarrow$ (i) in Theorem 3.26 is taken from [**CLRM**]. Note that the proof by van Neerven [**vN2**] of Theorem 2.44 can be generalized directly to give our Theorem 3.26. A similar result was proved independently by Buse [**Bu**] using a different idea. A characterization of dichotomy for evolution families in terms of the boundedness of the corresponding "convolution" (Green's) operator $\mathbb{G}$ acting in the space of functions on $\mathbb{R}$ is considered later in Chapter 4. Corollary 3.27 is taken from [**CLRM**].

The current "state-of-the-art" results on the connections between exponential *dichotomy on the half line* and evolution semigroups can be found in a paper by Van Minh, Räbiger and Schnaubelt [**MRS**]. We will state some results of these authors.

Consider the evolution semigroup $\{E_+^t\}_{t\geq 0}$ on $C_{00}(\mathbb{R}; X)$, and let Γ_+ denote its generator. Also define a second evolution semigroup $\{T_+^t\}_{t\geq 0}$ on the space $C_0(\mathbb{R}; X)$ by

$$(T_+^t f)(\theta) = \begin{cases} U(\theta, \theta - t) f(\theta - t), & \theta \geq t \\ U(\theta, 0) f(0), & 0 \leq \theta \leq t, \end{cases}$$

and let G_+ denote its generator. Note, that $\mathcal{D}(\Gamma_+) = \mathcal{D}(G_+) \cap C_{00}(\mathbb{R}_+; X)$. Consider the inhomogeneous equation

$$(3.25) \qquad u(\theta) = U(\theta, \tau)u(\tau) + \int_\tau^\theta U(\theta, s)f(s)\, ds, \quad \theta \geq \tau \geq 0.$$

If $u, f \in C_0(\mathbb{R}_+; X)$ satisfy this equation, we set $Ju = f$, cf. Proposition 3.25 and a discussion on page 115. This defines an operator on $C_0(\mathbb{R}_+; X)$ with the domain

$$\mathcal{D}(J) = \{u \in C_0(\mathbb{R}_+; X) : \text{there exists } f \in C_0(\mathbb{R}_+; X)$$

such that u and f satisfy (3.25) $\}$.

Define the stable subspace of initial data $X_0 := \{x \in X : \lim_{t \to 0} U(t, 0)x = 0\}$. The result by Van Minh, Räbiger, and Schnaubelt from [**MRS**] is formulated in the next theorem.

THEOREM 3.31. *The following statements are equivalent:*
(i) $\{U(\theta, \tau)\}_{\theta \geq \tau \geq 0}$ *has an exponential dichotomy on* $\mathbb{R}_+$.
(ii) $\operatorname{Im} G_+ = C_{00}(\mathbb{R}_+; X)$ *and* X_0 *is complemented in* X.
(iii) J *is surjective and* X_0 *is complemented in* X.

The condition "X_0 is complemented in X" can be eliminated. For this, let us consider restrictions of T_+^t and G_+ to another space. For a closed linear subspace Z in X, let

$$C_{0,Z}(\mathbb{R}_+; X) = \{f \in C_0(\mathbb{R}_+; X) : f(0) \in Z\}.$$

Denote by $G_{+,Z}$ the part of G_+ in $C_{0,Z}(\mathbb{R}_+; X)$, and let J_Z denote the restriction of J to the space $C_{0,Z}(\mathbb{R}_+; X)$. The following statements are equivalent, see [**MRS**]:

(i') $\{U(\theta, \tau)\}_{\theta \geq \tau \geq 0}$ has an exponential dichotomy on $\mathbb{R}_+$ with $\operatorname{Ker} P(0) = Z$.
(ii') $G_{+,Z} : \mathcal{D}(G_{+,Z}) \subseteq C_{0,Z}(\mathbb{R}_+; X) \to C_{00}(\mathbb{R}_+; X)$ is invertible.
(iii') $J_Z : \mathcal{D}(J_Z) \subseteq C_{0,Z}(\mathbb{R}_+; X) \to C_0(\mathbb{R}_+; X)$ is invertible.

CHAPTER 4

Characterizations of Dichotomy for Evolution Families

In this chapter we give four different characterizations of exponential dichotomy (Definition 3.6) for a strongly continuous, exponentially bounded evolution family $\{U(\theta, \tau)\}_{\theta \geq \tau}$, $\theta, \tau \in \mathbb{R}$, (Definition 3.1) on a Banach space X. In Chapter 3, the existence of an exponential dichotomy for $\{U(\theta, \tau)\}_{\theta \geq \tau}$ is characterized in terms of the hyperbolicity of the evolution semigroup $\{E^t\}_{t \geq 0}$ on $L^p(\mathbb{R}; X)$ given by $(E^t f)(\theta) = U(\theta, \theta - \tau) f(\theta - \tau)$, $\theta \in \mathbb{R}$ or, equivalently, in terms of the invertibility of the infinitesimal generator Γ of this evolution semigroup, see the Dichotomy Theorem 3.17. The main result of Section 4.1 (Discrete Dichotomy Theorem 4.16) characterizes the existence of an exponential dichotomy for $\{U(\theta, \tau)\}_{\theta \geq \tau}$ in terms of the hyperbolicity of a family of operators $\pi_\theta(E)$, for $\theta \in \mathbb{R}$ and $E := E^1$, that act on the space $\ell^p(\mathbb{Z}; X)$ of X-valued sequences. The operators $\pi_\theta(E)$ are weighted shift operators, while E is a weighted translation operator. To relate E and $\pi_\theta(E)$, we will study certain Banach algebras of weighted translations and weighted shifts (Subsection 4.1.1). The results of this subsection on algebras of weighted shift operators will be also used is Chapter 7 for a treatment of exponential dichotomy for linear skew-product flows. Section 4.2 continues with the theme of convolution-type results started in Subsection 3.3.2. In particular, we will show (Theorem 4.25) that $\{U(\theta, \tau)\}_{\theta \geq \tau}$ has an exponential dichotomy if and only if a certain integral operator $\mathbb{G}$, called the Green's operator, is bounded on $L^p(\mathbb{R}; X)$. We will also show that $\mathbb{G} = -\Gamma^{-1}$. In Section 4.3 we will prove some Perron-type results (Theorem 4.28) that relate the existence of an exponential dichotomy to the existence of a unique bounded, continuous solution of a nonhomogeneous equation with bounded and continuous right-hand side. In addition, we present an important result by A. G. Baskakov (Theorem 4.37) that gives a different proof and, at the same time, improves the Discrete Dichotomy Theorem 4.16. Finally, in Section 4.4 we consider the case where X is a Hilbert space and determine the existence of exponential dichotomies in terms of the existence of quadratic Lyapunov functions for the evolution family.

4.1. Discrete dichotomies: an algebraic approach

Let $\{U(\theta, \tau)\}_{\theta \geq \tau}$, $\theta, \tau \in \mathbb{R}$, be a strongly continuous, exponentially bounded evolution family on X. In this section we will use the fact that each evolution operator E^t on $L^p(\mathbb{R}; X)$, given by $(E^t f)(\theta) = U(\theta, \theta - t) f(\theta - t)$, $\theta \in \mathbb{R}$, is a *weighted translation* operator. Indeed, fix t, for example at $t = 1$, define the translation operator V by $(Vf)(\theta) = f(\theta - 1)$, let a denote the strongly continuous, bounded, operator-valued function $\theta \mapsto U(\theta, \theta - 1)$, and define the multiplication

operator in $L^p(\mathbb{R}; X)$, also denoted by a, given by
$$(af)(\theta) = a(\theta)f(\theta) = U(\theta, \theta - 1)f(\theta).$$

Then, the evolution operator $E := E^1$ is expressible in the form $E = aV$; that is, E is a weighted translation operator. We will associate with E a family of *weighted shift* operators $\pi_\theta(E)$ acting on the space $\ell^p(\mathbb{Z}; X)$ as follows: For each fixed $\theta \in \mathbb{R}$, define $\pi_\theta(E) = \pi_\theta(a)S$ where $S : (x_n) \mapsto (x_{n-1})$ is the shift operator on $\ell^p(\mathbb{Z}; X)$ and $\pi_\theta(a) : (x_n) \mapsto (a(\theta + n)x_n)$ is a diagonal operator given by $\pi_\theta(a) = \mathrm{diag}(a(\theta + n))_{n \in \mathbb{Z}}$.

By the Dichotomy Theorem 3.17 and the circular symmetry of the spectrum $\sigma(E)$ (Theorem 3.13), the existence of a dichotomy for $\{U(\theta, \tau)\}_{\theta \geq \tau}$ is equivalent to the invertibility of the operator $I - E$ on $L^p(\mathbb{R}; X)$. We will relate the invertibility of $I - E$ on $L^p(\mathbb{R}; X)$ to the invertibility of all the operators on $\ell^p(\mathbb{Z}; X)$ in the family $\{I - \pi_\theta(E) : \theta \in \mathbb{R}\}$. To this end, we will immerse the operator E in a certain algebra $\mathfrak{B}$ of operators on $L^p(\mathbb{R}; X)$ and the family of operators $\pi_\theta(E)$, $\theta \in \mathbb{R}$, in a certain algebra $\mathfrak{C}$ of operators on $\ell^p(\mathbb{Z}; X)$ so that $\pi_\theta : \mathfrak{B} \to \mathfrak{C}$ is a homomorphism.

The main result of this section, the Discrete Dichotomy Theorem 4.16, states that E is hyperbolic on $L^p(\mathbb{R}; X)$ if and only if each weighted shift operator $\pi_\theta(E)$ is hyperbolic on $\ell^p(\mathbb{Z}; X)$, and $\|R(\lambda; \pi_\theta(E))\|$ is uniformly bounded for $\lambda \in \mathbb{T}$ and $\theta \in \mathbb{R}$. In addition, we will derive some facts of independent interest concerning the invertibility of operators in the algebras $\mathfrak{C}$ and $\mathfrak{B}$. These results will be used in Section 7.1. Also, they are used in Proposition 4.18 where we give a new, algebraic proof of the Spectral Projection Theorem 3.14 which is based on the observation that the resolvent operators $R(\lambda; E)$ belong to the algebra $\mathfrak{B}$.

4.1.1. Algebras of weighted shifts.
In this subsection we will study the invertibility of the operators in the *weighted shift algebra* $\mathfrak{C}$ of operators on $\ell^p(\mathbb{Z}; X)$.

For a sequence of operators $(d^{(n)})_{n \in \mathbb{Z}} \subset \mathcal{L}(X)$, let $d = \mathrm{diag}(d^{(n)})_{n \in \mathbb{Z}}$ denote the diagonal operator $d : (x_n)_{n \in \mathbb{Z}} \mapsto (d^{(n)}x_n)_{n \in \mathbb{Z}}$ on $\ell^p(\mathbb{Z}; X)$, $1 \leq p \leq \infty$. Let $\mathfrak{D}$ denote the Banach algebra of bounded diagonal operators on $\ell^p(\mathbb{Z}; X)$ with the norm
$$\|d\|_\mathfrak{D} = \|d\|_{\mathcal{L}(\ell^p(\mathbb{Z}; X))} = \sup_{n \in \mathbb{Z}} \|d^{(n)}\|_{\mathcal{L}(X)}.$$

Also, let $S : (x_n)_{n \in \mathbb{Z}} \mapsto (x_{n-1})_{n \in \mathbb{Z}}$ denote the shift operator on $\ell^p(\mathbb{Z}; X)$ and $\mathfrak{C}$ the *weighted shift algebra* of operators on $\ell^p(\mathbb{Z}; X)$ represented by absolutely convergence series, in powers of the operator S, with coefficients from $\mathfrak{D}$; that is,

$$\mathfrak{C} = \left\{ D = \sum_{k=-\infty}^{\infty} d_k S^k \in \mathcal{L}(\ell^p(\mathbb{Z}; X)) : d_k \in \mathfrak{D} \text{ and } \|D\|_\mathfrak{C} := \sum_{k=-\infty}^{\infty} \|d_k\|_\mathfrak{D} < \infty \right\}.$$

As we will show later, see Corollary 4.8, the algebra $\mathfrak{C}$ is inverse-closed in $\mathcal{L}(\ell^p(\mathbb{Z}; X))$; that is, if $D \in \mathfrak{C}$ is invertible as an operator on $\ell^p(\mathbb{Z}; X)$, then its inverse D^{-1} belongs to $\mathfrak{C}$. Thus, the invertibility of an operator $D \in \mathfrak{C}$ is an intrinsic algebraic property; for example, it does not depend on the particular ℓ^p-space where D is acting, etc. An operator $D \in \mathfrak{C}$ may be viewed as an infinite "matrix" such that the norms of its diagonals form an absolutely summable sequence. Thus, we will show that the diagonals of the "matrix" $D^{-1} \in \mathfrak{C}$ have the same ℓ^1-summability property.

Let us begin by discussing the case of a two-diagonal operator D. Although this is the simplest nontrivial case, this is the case of primary interest because it corresponds to the operator $I - E$.

LEMMA 4.1. *If $D = I - dS$, $d \in \mathfrak{D}$, is invertible in $\mathcal{L}(\ell^p(\mathbb{Z}; X))$, $1 \leq p \leq \infty$, then $C := D^{-1}$ belongs to $\mathfrak{C}$; that is, it has the form*

$$C = \sum_{k=-\infty}^{\infty} C_k S^k \quad \text{where} \quad C_k \in \mathfrak{D} \quad \text{and} \quad \sum_{k=-\infty}^{\infty} \|C_k\|_{\mathfrak{D}} < \infty.$$

PROOF. For each $x \in X$ and each $l \in \mathbb{Z}$, define a sequence $e_l \otimes x \in \ell^p(\mathbb{Z}; X)$, $e_l \otimes x = (x_n)_{n \in \mathbb{Z}}$, by setting $x_n = x$ for $n = l$, and $x_n = 0$ for $n \neq l$. If $C \in \mathcal{L}(\ell^p(\mathbb{Z}; X))$, then C has the matrix representation $C = [C_{nl}]_{n,l \in \mathbb{Z}}$ obtained by defining

$$C_{nl} x := (C(e_l \otimes x))_n, \quad n, l \in \mathbb{Z}, \quad \text{where} \quad C_{nl} \in \mathcal{L}(X).$$

Our proof that $C = D^{-1} \in \mathfrak{C}$ proceeds in three steps. We first show that, for some $\gamma > 0$, there is some $R > 0$ such that

(4.1) $$\|C_{n,0}\|_{\mathcal{L}(X)} \leq R e^{-\gamma |n|}, \quad n \in \mathbb{Z}.$$

This inequality is then used to prove the inequality

(4.2) $$\|C_{nl}\|_{\mathcal{L}(X)} \leq R e^{-\gamma |n-l|}, \quad n, l \in \mathbb{Z}.$$

Finally, the lemma is derived from inequality (4.2).

First step. Choose $\gamma > 0$ and define the Banach space

$$\ell^p_\gamma := \{(x_n)_{n \in \mathbb{Z}} \in \ell^p(\mathbb{Z}; X) : (e^{\gamma |n|} x_n)_{n \in \mathbb{Z}} \in \ell^p(\mathbb{Z}; X)\}, \quad 1 \leq p \leq \infty,$$

with the norm

$$\|(x_n)_{n \in \mathbb{Z}}\|_{\ell^p_\gamma} := \|(e^{\gamma |n|} x_n)_{n \in \mathbb{Z}}\|_{\ell^p(\mathbb{Z}; X)}.$$

The operator $J_\gamma : \ell^p(\mathbb{Z}, X) \to \ell^p_\gamma$, defined by $J_\gamma : (x_n)_{n \in \mathbb{Z}} \mapsto (e^{-\gamma |n|} x_n)_{n \in \mathbb{Z}}$, is an isometry. Also, the operator $D = I - dS$ maps ℓ^p_γ into ℓ^p_γ. Hence, the operator $D(\gamma)$, given by

$$D(\gamma) = J_\gamma^{-1} D J_\gamma, \quad D(\gamma) = I - \operatorname{diag}(e^{\gamma(|n|-|n-1|)} d^{(n)})_{n \in \mathbb{Z}} S,$$

is a bounded operator on $\ell^p(\mathbb{Z}; X)$. Note that

(4.3) $$\|D(\gamma) - D\|_{\mathcal{L}(\ell^p(\mathbb{Z}; X))} \leq \max\{|e^{\pm \gamma} - 1|\} \sup_{n \in \mathbb{Z}} \|d^{(n)}\|_{\mathcal{L}(X)};$$

and therefore, $\lim_{\gamma \to 0} D(\gamma) = D$ in $\mathcal{L}(\ell^p(\mathbb{Z}; X))$. Since D is invertible in $\mathcal{L}(\ell^p(\mathbb{Z}; X))$ by our assumption, there is some $\gamma > 0$ such that $D(\gamma)$ has a bounded inverse on $\ell^p(\mathbb{Z}; X)$ with norm $R := \|D(\gamma)^{-1}\|_{\mathcal{L}(\ell^p(\mathbb{Z}; X))}$.

Fix $x \in X$, define $\bar{x} := e_0 \otimes x$, and note that $J_\gamma \bar{x} = \bar{x}$. Since $D(\gamma)^{-1}$ is bounded on $\ell^p(\mathbb{Z}; X)$ and $\bar{x} \in \ell^p(\mathbb{Z}; X)$, it follows that

$$[J_\gamma^{-1} D J_\gamma] D(\gamma)^{-1} \bar{x} = \bar{x} \quad \text{and} \quad D J_\gamma D(\gamma)^{-1} \bar{x} = J_\gamma \bar{x} = \bar{x}.$$

Let us also note that $\bar{y} := J_\gamma D(\gamma)^{-1} \bar{x}$ belongs to ℓ^p_γ because $D(\gamma)^{-1} \bar{x} \in \ell^p(\mathbb{Z}; X)$ and J_γ maps $\ell^p(\mathbb{Z}; X)$ on ℓ^p_γ. Since $\ell^p_\gamma \subset \ell^p(\mathbb{Z}; X)$, we conclude that $\bar{y} \in \ell^p(\mathbb{Z}; X)$ and $D\bar{y} = \bar{x}$ as elements of $\ell^p(\mathbb{Z}; X)$.

By assumption, D is invertible on $\ell^p(\mathbb{Z}; X)$ with $C = D^{-1}$. Therefore, $\bar{y} = C\bar{x}$, or $J_\gamma D(\gamma)^{-1} \bar{x} = C\bar{x}$. Since $C\bar{x} = \bar{y} \in \ell^p_\gamma$, we can apply the operator $J_\gamma^{-1} : \ell^p_\gamma \to \ell^p(\mathbb{Z}; X)$ to the last identity to obtain the equality

$$D(\gamma)^{-1} \bar{x} = J_\gamma^{-1} C \bar{x} = J_\gamma^{-1} C J_\gamma \bar{x}.$$

As a result,
$$D(\gamma)^{-1}\bar{x} = J_\gamma^{-1}CJ_\gamma\bar{x} = J_\gamma^{-1}C(e_0 \otimes x) = J_\gamma^{-1}(C_{n0}x)_{n\in\mathbb{Z}} = (e^{\gamma|n|}C_{n0}x)_{n\in\mathbb{Z}}.$$

Hence, for $1 \leq p < \infty$, if $n \in \mathbb{Z}$, then
$$|e^{\gamma|n|}C_{n0}x|^p \leq \sum_{n=-\infty}^{\infty} |e^{\gamma|n|}C_{n0}x|^p = \|D(\gamma)^{-1}\bar{x}\|^p_{\ell^p(\mathbb{Z};X)} \leq R^p\|\bar{x}\|^p_{\ell^p(\mathbb{Z},X)} = R^p|x|^p.$$

If, on the other hand, $p = \infty$, then
$$|e^{\gamma|n|}C_{n0}x| \leq \sup_{n\in\mathbb{Z}}|e^{\gamma|n|}C_{n0}x| = \|D(\gamma)^{-1}\bar{x}\|_{\ell^\infty(\mathbb{Z};X)} \leq R|x|.$$

This finishes the proof of the inequality (4.1).

Second step. Fix $n, l \in \mathbb{Z}$, and define the operators $\tilde{D} = S^{-l}DS^l$ and $\tilde{C} = \tilde{D}^{-1}$. For each finitely supported sequence of the form $\bar{x} = (x_k)_{k\in\mathbb{Z}} = \sum_j e_j \otimes x_j$, it follows that
$$\tilde{C}\bar{x} = S^{-l}CS^l(x_k)_{k\in\mathbb{Z}} = S^{-l}\Big(\sum_{j=-\infty}^{\infty} C_{kj}x_{j-l}\Big)_{k\in\mathbb{Z}} = \Big(\sum_{j=-\infty}^{\infty} C_{k+l,j+l}x_j\Big)_{k\in\mathbb{Z}}.$$

Therefore, for $k, j \in \mathbb{Z}$, we have $\tilde{C}_{kj} = C_{k+l,j+l}$ or, equivalently, $C_{nl} = \tilde{C}_{n-l,0}$. An application of the inequality (4.1) to $\tilde{C} = [\tilde{C}_{kj}]$ results in the estimate
$$\|C_{nl}\|_{\mathcal{L}(X)} = \|\tilde{C}_{n-l,0}\|_{\mathcal{L}(X)} \leq Re^{-\gamma|n-l|},$$
which is just the required inequality (4.2).

Third step. To finish the proof of the lemma, let us note that a formal inverse for D is given by
$$D^{-1} = C = [C_{ij}] = \sum_{k=-\infty}^{\infty} \text{diag}(C_{n,n-k})_{n\in\mathbb{Z}}S^k = \sum_{k=-\infty}^{\infty} C_k S^k$$
where $C_k = \text{diag}(C_{n,n-k})_{n\in\mathbb{Z}} \in \mathfrak{D}$. Use the estimate (4.2) to see that
$$(4.4) \quad \sum_{k=-\infty}^{\infty} \|C_k\|_{\mathcal{L}(\ell^p(\mathbb{Z};X))} = \sum_{k=-\infty}^{\infty} \sup_{n\in\mathbb{Z}} \|C_{n-n-k}\|_{\mathcal{L}(X)} \leq R\sum_{k=-\infty}^{\infty} e^{-\gamma|k|} < \infty.$$

Hence, $D^{-1} \in \mathfrak{C}$, as required. □

REMARK 4.2. Using the proof of Lemma 4.1 we have the following additional result: If $D \in \mathfrak{C}$ is a finite-diagonal operator that is invertible in $\mathcal{L}(\ell^p(\mathbb{Z};X))$, then $D^{-1} \in \mathfrak{C}$. Indeed, let us consider an arbitrary *finite* sum $D = \sum_{k=-N}^{N} d_k S^k$, $d_k \in \mathfrak{D}$, and observe that D leaves ℓ^p_γ invariant. The proof of the lemma shows that the inverse $C = D^{-1}$ belongs to $\mathfrak{C}$; that is, $C = \sum_{k=-\infty}^{\infty} C_k S^k$ with $\sum_k \|C_k\|_{\mathcal{L}(\ell^p(\mathbb{Z};X))} < \infty$. ◇

Lemma 4.1 follows from the more general Corollary 4.8 given below. However, our proof of Lemma 4.1 gives some additional information needed in the next Lemma 4.3. Indeed, our objective is to study the invertibility of the operators in the family $I - \pi_\theta(E) = I - d_\theta S$ where $\theta \in \mathbb{R}$ and $d_\theta = \text{diag}(U(\theta+n, \theta+n-1))_{n\in\mathbb{Z}}$. Here d_θ depends on the parameter $\theta \in \mathbb{R}$ and the function $\theta \mapsto \|d_\theta\|_\mathfrak{D}$ is bounded. Thus, we will be dealing with *families* of operators $D_\theta \in \mathfrak{C}$, parameterized by points of $\mathbb{R}$. To study the invertibility of D_θ we will need the following variant of Lemma 4.1.

LEMMA 4.3. *Suppose that $d_\theta \in \mathfrak{D}$, $\theta \in \mathbb{R}$, is a family of operators such that each operator $D_\theta := I - d_\theta S$ is invertible on $\ell^p(\mathbb{Z}; X)$ and, in addition,*

$$\sup_{\theta \in \mathbb{R}} \|d_\theta\|_{\mathcal{L}(\ell^p(\mathbb{Z},X))} < \infty \quad \text{and} \quad \sup_{\theta \in \mathbb{R}} \|D_\theta^{-1}\|_{\mathcal{L}(\ell^p(\mathbb{Z};X))} < \infty.$$

Then, $C(\theta) := D_\theta^{-1}$ has the form $C(\theta) = \sum_{k=-\infty}^{\infty} C_k(\theta) S^k$ where $C_k(\theta) \in \mathfrak{D}$ and

(4.5) $$\sum_{k=-\infty}^{\infty} \sup_{\theta \in \mathbb{R}} \|C_k(\theta)\|_{\mathcal{L}(\ell^p(\mathbb{Z};X))} < \infty.$$

PROOF. Since, for $d_\theta = \mathrm{diag}(d_\theta^{(n)})_{n \in \mathbb{Z}}$, we have the bound

$$\sup_{\theta \in \mathbb{R}} \|d_\theta\|_{\mathcal{L}(\ell^p(\mathbb{Z};X))} = \sup_{\theta \in \mathbb{R}} \sup_{n \in \mathbb{Z}} \|d_\theta^{(n)}\|_{\mathcal{L}(X)} < \infty,$$

the estimate for $\|D_\theta(\gamma) - D_\theta\|_{\mathcal{L}(\ell^p(\mathbb{Z};X))}$, as in (4.3), is uniform in $\theta \in \mathbb{R}$. Hence, we can fix $\gamma > 0$, independent of $\theta \in \mathbb{R}$, such that $D_\theta(\gamma)$ has a bounded inverse on $\ell^p(\mathbb{Z}; X)$ for each $\theta \in \mathbb{R}$. Moreover, since

$$D_\theta(\gamma)^{-1} = [(D_\theta(\gamma) - D_\theta) + D_\theta]^{-1} = [I - D_\theta^{-1}(D_\theta - D_\theta(\gamma))]^{-1} D_\theta^{-1},$$

we have that

$$R := \sup_{\theta \in \mathbb{R}} \|D_\theta(\gamma)^{-1}\|_{\mathcal{L}(\ell^p(\mathbb{Z};X))} < \infty$$

by the hypothesis of the lemma. Thus, for $C(\theta) = [C_{nl}(\theta)]_{n,l \in \mathbb{Z}}$, the estimate

$$\sup_{\theta \in \mathbb{R}} \|C_{nl}(\theta)\|_{\mathcal{L}(X)} \leq R e^{-\gamma |n-l|}, \quad n, l \in \mathbb{Z},$$

similar to inequality (4.2), together with exactly the same estimates as in display (4.4), implies the required bound (4.5). $\square$

REMARK 4.4. Recall the two-term operator $D = I - dS$ defined in Lemma 4.1. The spectrum $\sigma(dS)$ is rotationally invariant. Indeed, as in the proof of Proposition 2.28, for each $\xi \in \mathbb{R}$, define $L_\xi : (x_n)_{n \in \mathbb{Z}} \mapsto (e^{in\xi} x_n)_{n \in \mathbb{Z}}$ on $\ell^p(\mathbb{Z}; X)$, and note that $dS L_\xi = e^{-i\xi} L_\xi dS$. Thus, if $D = I - dS$ is invertible on $\ell^p(\mathbb{Z}; X)$, then $\lambda - dS$ is invertible on $\ell^p(\mathbb{Z}; X)$ for every $\lambda \in \mathbb{T}$. Moreover, by Lemma 4.1, the resolvent operator $R(\lambda; dS)$ belongs to $\mathfrak{C}$ for every $\lambda \in \mathbb{T}$. $\diamond$

Let $c_0(\mathbb{Z}; X)$ denote the subspace of sequences $(x_n)_{n \in \mathbb{Z}} \in \ell^\infty(\mathbb{Z}; X)$ such that $\lim_{n \to \pm\infty} |x_n| = 0$. We remark that Lemmas 4.1–4.3 are proved exactly the same way if the space $\ell^\infty(\mathbb{Z}; X)$ is replaced by $c_0(\mathbb{Z}; X)$.

LEMMA 4.5. *If $D = I - dS$, $d \in \mathfrak{D}$, is invertible on $\ell^\infty(\mathbb{Z}; X)$, then $C = D^{-1}$ maps $c_0(\mathbb{Z}; X)$ to $c_0(\mathbb{Z}; X)$.*

PROOF. Fix $\bar{x} = (x_l)_{l \in \mathbb{Z}} \in c_0(\mathbb{Z}; X)$. For each $n \in \mathbb{Z}$, the estimate

(4.6) $$|(C\bar{x})_n| = \left| \sum_{l=-\infty}^{\infty} C_{nl} x_l \right| \leq R \sum_{l=-\infty}^{\infty} e^{-\gamma|n-l|} |x_l|$$

follows from the inequality (4.2). Also, because

$$\sum_{l=-\infty}^{\infty} e^{-\gamma|n-l|} |x_l| \leq \|\bar{x}\|_\infty \sum_{l=-\infty}^{\infty} e^{-\gamma|l|},$$

the sum on the right hand side of inequality (4.6) converges uniformly for $n \in \mathbb{Z}$. By passing to the limit as $n \to \pm\infty$ in (4.6), it follows that $\lim_{n \to \pm\infty} |(C\bar{x})_n| = 0$. $\square$

REMARK 4.6. Lemma 4.1 implies that the invertibility of an operator $D = I - dS$, for $d \in \mathfrak{D}$, does not depend upon the choice of the space $\ell^p(\mathbb{Z}; X)$, $1 \le p \le \infty$, or $c_0(\mathbb{Z}; X)$; that is, if D is invertible on one of the spaces $\ell^{p_0}(\mathbb{Z}; X)$ or $c_0(\mathbb{Z}; X)$ for some fixed p_0, $1 \le p_0 \le \infty$, then D is invertible on each of the spaces $\ell^p(\mathbb{Z}; X)$, $1 \le p \le \infty$, and $c_0(\mathbb{Z}; X)$. The following similar statement holds in the situation of Lemma 4.3: If, for a family of operators $d_\theta \in \mathfrak{D}$, $d_\theta = \mathrm{diag}(d_\theta^{(n)})_{n \in \mathbb{Z}}$, $\theta \in \mathbb{R}$, the operator $D_\theta = I - d_\theta S$ is invertible on one of the spaces $Y = \ell^{p_0}(\mathbb{Z}; X)$ or $Y = c_0(\mathbb{Z}; X)$ and

$$(4.7) \qquad \sup_{\theta \in \mathbb{R}} \|d_\theta\|_{\mathcal{L}(Y)} < \infty, \qquad \sup_{\theta \in \mathbb{R}} \|D_\theta^{-1}\|_{\mathcal{L}(Y)} < \infty,$$

then $C(\theta) = D_\theta^{-1}$ is the same operator for all spaces $\ell^p(\mathbb{Z}; X)$, $1 \le p \le \infty$ and $c_0(\mathbb{Z}; X)$. Also, $C(\theta)$ has the form $C(\theta) = \sum_{k=-\infty}^{\infty} C_k(\theta) S^k$. The norm of $C_k(\theta) \in \mathfrak{D}$ as an operator on either one of the spaces is just $\|C_k(\theta)\|_\mathfrak{D} = \sup_{n \in \mathbb{Z}} \|C_k^{(n)}(\theta)\|_{\mathcal{L}(X)}$. In particular, this norm does not depend on the choice of the space $\ell^p(\mathbb{Z}; X)$, $1 \le p \le \infty$, or $c_0(\mathbb{Z}; X)$. For each of these spaces, we have the norm estimate

$$(4.8) \qquad \|C(\theta)\|_{\mathfrak{C}} \le \sum_{k=-\infty}^{\infty} \sup_{\theta \in \mathbb{R}} \|C_k(\theta)\|_\mathfrak{D} < \infty, \quad \theta \in \mathbb{R}.$$

◇

We will use a theorem of Bochner and Phillips [**BPh**], see also Gohberg and Leiterer [**GL**], that is a very general version of the Wiener Theorem on absolutely convergent Fourier series. To formulate this result, consider a noncommutative Banach algebra $\mathcal{A}$ with unity and a function $F: \mathbb{T} \to \mathcal{A}$. We will say that F is represented by an absolutely convergent Fourier series if $F(z) = \sum_{k=-\infty}^{\infty} A_k z^k$ where $A_k \in \mathcal{A}$, $|z| = 1$, and $\sum_k \|A_k\|_{\mathcal{A}} < \infty$.

THEOREM 4.7 (Bochner-Phillips). *Suppose that F is represented by an absolutely convergent Fourier series and, for each $z \in \mathbb{T}$, the element $F(z) \in \mathcal{A}$ is invertible in $\mathcal{A}$. Then, the function $G: \mathbb{T} \to \mathcal{A}$ defined by $G(z) = [F(z)]^{-1}$ is represented by an absolutely convergent Fourier series.*

COROLLARY 4.8. *The algebra $\mathfrak{C}$ is inverse-closed; that is, if $D \in \mathfrak{C}$ is an invertible operator on $\ell^p(\mathbb{Z}; X)$, then its inverse $C = D^{-1}$ belongs to $\mathfrak{C}$.*

PROOF. For each operator $D = \sum_{k=-\infty}^{\infty} d_k S^k \in \mathfrak{C}$ that is invertible in $\mathcal{L}(\ell^p(\mathbb{Z}; X))$ and each $t \in [0, 2\pi)$, define the operators

$$D(t) = \sum_{k=-\infty}^{\infty} d_k S^k e^{ikt} \in \mathfrak{C} \quad \text{and} \quad W(t) = \mathrm{diag}(e^{-int})_{n \in \mathbb{Z}}.$$

Since, for each $t \in [0, 2\pi)$, we have

$$(4.9) \qquad D(t) = W(t)^{-1} D W(t),$$

the operator $D(t)$ is invertible in $\mathcal{L}(\ell^p(\mathbb{Z}; X))$. An application of the Bochner-Phillips Theorem 4.7 shows that the function $t \mapsto D(t)^{-1}$ can be represented as an absolutely convergent Fourier series. In other words, there is a sequence of operators $\{C_k\}_{k\in\mathbb{Z}}$ in $\mathcal{L}(\ell^p(\mathbb{Z};X))$ such that

$$D(t)^{-1} = \sum_{k=-\infty}^{\infty} C_k e^{ikt} \quad \text{and} \quad \sum_{k=-\infty}^{\infty} \|C_k\|_{\mathcal{L}(\ell^p(\mathbb{Z};X))} < \infty.$$

We claim that the identity (4.9) implies $C_k S^{-k}$ is a diagonal operator for each $k \in \mathbb{Z}$. Indeed, since we are dealing with absolutely convergent series, (4.9) implies

$$C_k e^{ikt} = \text{diag}(e^{int})_{n\in\mathbb{Z}} C_k \, \text{diag}(e^{-int})_{n\in\mathbb{Z}} \quad \text{for each} \quad k \in \mathbb{Z}.$$

Then, for $C_k S^{-k} = [a_{nl}]_{n,l\in\mathbb{Z}}$ and all $t \in [0, 2\pi)$, it follows that

$$C_k S^{-k} = \text{diag}(e^{int}) C_k S^{-k} \text{diag}(e^{-int}) \quad \text{or} \quad a_{nl} = e^{i(n-l)t} a_{nl}.$$

Hence, $a_{nl} = 0$ for $n \neq l$, and the claim is proved.

As a result, if we set $\tilde{C}_k = C_k S^{-k} \in \mathfrak{D}$, then

$$D^{-1} = D(0)^{-1} = \sum_{k=-\infty}^{\infty} C_k = \sum_{k=-\infty}^{\infty} \tilde{C}_k S^k, \quad \sum_{k=-\infty}^{\infty} \|\tilde{C}_k\| = \sum_{k=-\infty}^{\infty} \|C_k\| < \infty,$$

and $D^{-1} \in \mathfrak{C}$. $\square$

4.1.2. Algebras of weighted translations. In this subsection we will study the invertibility of elements of the *weighted translation algebra* $\mathfrak{B}$ defined in display (4.10) below. The main technique we use involves representing each weighted translation operator b on $L^p(\mathbb{R};X)$ by a family of weighted shift operators $\pi_\theta(b)$, $\theta \in \mathbb{R}$ on $\ell^p(\mathbb{Z};X)$. Suppose that $b = I - aV \in \mathfrak{B}$, each operator $\pi_\theta(b)$ is invertible, and the family of norms $\{\|[\pi_\theta(b)]^{-1}\|_{\mathcal{L}(\ell^p)} : \theta \in \mathbb{R}\}$ is uniformly bounded. Then, we will see that b is invertible with $b^{-1} \in \mathfrak{B}$ (Theorem 4.9). The converse result is "almost" true (see Lemma 4.11). The correct formulation of a partial converse allows us to identify a subalgebra $\mathfrak{B}_* \subset \mathfrak{B}$ that is inverse-closed (Corollary 4.14).

Define $\mathfrak{A}$ to be the subalgebra of $\mathcal{L}(L^p(\mathbb{R};X))$, $1 \leq p < \infty$, whose elements are multiplication operators given by $(af)(\theta) = a(\theta)f(\theta)$, $\theta \in \mathbb{R}$, where the multiplier a is a strongly continuous, bounded function in $C_b(\mathbb{R};\mathcal{L}_s(X))$. Also, let $\mathfrak{B}$ denote the subalgebra of operators on $L^p(\mathbb{R};X)$ defined by

$$\mathfrak{B} = \Big\{ b = \sum_{k=-\infty}^{\infty} a_k V^k \in \mathcal{L}(L^p(\mathbb{R};X)), \quad a_k \in \mathfrak{A} \quad \text{such that}$$

(4.10)
$$\|b\|_{\mathfrak{B}} := \sum_{k=-\infty}^{\infty} \|a_k\|_{\mathcal{L}(L^p(\mathbb{R};X))} < \infty \Big\}.$$

For each $a \in \mathfrak{A}$ and for each $\theta \in \mathbb{R}$, we define operators on $\ell^p(\mathbb{Z};X)$ as follows:

$$\pi_\theta(a) = \text{diag}(a(\theta+n))_{n\in\mathbb{Z}} \quad \text{and} \quad \pi_\theta(V) = S$$

where S is the shift operator. Clearly, π_θ can be extended, by linearity, to a bounded homomorphism $\pi_\theta : \mathfrak{B} \to \mathfrak{C}$ given by

$$\pi_\theta(b) = \sum_{k=-\infty}^{\infty} \pi_\theta(a_k) S^k \in \mathfrak{C} \quad \text{for} \quad b = \sum_{k=-\infty}^{\infty} a_k V^k \in \mathfrak{B}.$$

In particular, if $b = I - aV$, then $\pi_\theta(b) = I - d_\theta S$ with

$$d_\theta = \pi_\theta(a) = \text{diag}(a(\theta+n))_{n\in\mathbb{Z}}.$$

The operators $\pi_\theta(b)$ are called *discrete operators*.

In the next result we concentrate on the two-term operator $b = I - aV$. See Proposition 4.10 for a similar result that holds for arbitrary $b \in \mathfrak{B}$.

THEOREM 4.9. *Suppose that $a \in \mathfrak{A}$ and $b := I - aV$. If $\pi_\theta(b)$ is invertible in $\mathcal{L}(\ell^p(\mathbb{Z}; X))$, $1 \leq p < \infty$, for each $\theta \in \mathbb{R}$ and there exists a constant $B > 0$ such that*

$$\|\pi_\theta(b)^{-1}\|_{\mathcal{L}(\ell^p(\mathbb{Z};X))} \leq B \quad \text{for all} \quad \theta \in \mathbb{R}, \tag{4.11}$$

then b is invertible in $\mathcal{L}(L^p(\mathbb{R}; X))$. Moreover, its inverse b^{-1} belongs to $\mathfrak{B}$; that is, for some $a_k \in \mathfrak{A}$,

$$b^{-1} = \sum_{k=-\infty}^{\infty} a_k V^k \quad \text{with} \quad \sum_{k=-\infty}^{\infty} \|a_k\|_{\mathcal{L}(\ell^p(\mathbb{Z};X))} < \infty.$$

PROOF. Define $d_\theta := \pi_\theta(a)$ and $D_\theta := \pi_\theta(b)$ for $\theta \in \mathbb{R}$. By Lemma 4.3, the operator $\pi_\theta(b)^{-1} \in \mathfrak{C}$ for each $\theta \in \mathbb{R}$. Moreover,

$$\pi_\theta(b)^{-1} = \sum_{k=-\infty}^{\infty} C_k(\theta) S^k, \quad C_k(\theta) \in \mathfrak{D},$$

$$\text{and} \quad \sum_{k=-\infty}^{\infty} \sup_{\theta \in \mathbb{R}} \|C_k(\theta)\|_{\mathcal{L}(\ell^p(\mathbb{Z};X))} < \infty. \tag{4.12}$$

We claim that if $C_k(\theta) = \operatorname{diag}(C_k^{(l)}(\theta))_{l \in \mathbb{Z}}$, then

$$C_k^{(n)}(\theta) = C_k^{(0)}(\theta + n) \tag{4.13}$$

for all $n \in \mathbb{Z}$ and $\theta \in \mathbb{R}$. Indeed, for each $n \in \mathbb{Z}$ and $\theta \in \mathbb{R}$, we have the equalities

$$\sum_{k=-\infty}^{\infty} \operatorname{diag}(C_k^{(l)}(\theta+n))_{l \in \mathbb{Z}} S^k = \sum_{k=-\infty}^{\infty} C_k(\theta+n) S^k = [\pi_{\theta+n}(b)]^{-1}$$

$$= \pi_\theta(V^{-n} b V^n)^{-1} = S^{-n} \pi_\theta(b)^{-1} S^n = S^{-n} \Big(\sum_{k=-\infty}^{\infty} C_k(\theta) S^k\Big) S^n$$

$$= S^{-n}\Big(\sum_{k=-\infty}^{\infty} \operatorname{diag}(C_k^{(l)}(\theta))_{l \in \mathbb{Z}} S^k\Big) S^n = \sum_{k=-\infty}^{\infty} \operatorname{diag}(C_k^{(l+n)}(\theta))_{l \in \mathbb{Z}} S^k.$$

Since all these series converge absolutely, we conclude that

$$C_k^{(l)}(\theta + n) = C_k^{(l+n)}(\theta), \quad k, l \in \mathbb{Z}.$$

This proves the claim (4.13) in case $l = 0$.

Our second observation is that, for each $k \in \mathbb{Z}$, the function $\theta \mapsto C_k^{(0)}(\theta)$ is strongly continuous. Indeed, let $x \in X$ and define $\bar{x} = e_{-k} \otimes x$; that is, $\bar{x} = (x_l)_{l \in \mathbb{Z}}$ with $x_{-k} = x$ and $x_l = 0$ for $l \neq -k$. Then, for each fixed $\theta_0 \in \mathbb{R}$ and every $\theta \in \mathbb{R}$, we have the estimate

$$\left|[C_k^{(0)}(\theta) - C_k^{(0)}(\theta_0)]x_{-k}\right|^p = \left|\sum_{j=-\infty}^{\infty}[C_j^{(0)}(\theta) - C_j^{(0)}(\theta_0)]x_{-j}\right|^p$$

$$\leq \sum_{n=-\infty}^{\infty}\left|\sum_{j=-\infty}^{\infty}\left[C_j^{(n)}(\theta) - C_j^{(n)}(\theta_0)\right]x_{n-j}\right|^p$$

$$= \left\|\sum_{j=-\infty}^{\infty}[C_j(\theta) - C_j(\theta_0)]S^j\bar{x}\right\|^p_{\ell^p(\mathbb{Z};X)} = \|[\pi_\theta(b)^{-1} - \pi_{\theta_0}(b)^{-1}]\bar{x}\|^p_{\ell^p(\mathbb{Z};X)}.$$

As a result, with the aid of the estimate (4.11), it follows that

$$|[C_k^{(0)}(\theta) - C_k^{(0)}(\theta_0)]x|^p \leq \|\pi_\theta(b)^{-1}[\pi_{\theta_0}(b) - \pi_\theta(b)]\pi_{\theta_0}(b)^{-1}\bar{x}\|^p_{\ell^p(\mathbb{Z};X)}$$
$$\leq B\|[\pi_{\theta_0}(b) - \pi_\theta(b)]\bar{y}\|^p_{\ell^p(\mathbb{Z},X)}$$

for $\bar{y} = \pi_{\theta_0}(b)^{-1}\bar{x}$. Since the function from $\mathbb{R}$ to $\mathcal{L}_s(X)$ defined by $\theta \mapsto a(\theta)$ is continuous, so is the function from $\mathbb{R}$ to $\mathcal{L}_s(\ell^p(\mathbb{Z},X))$ defined by $\theta \mapsto \pi_\theta(b)$. In particular, for θ_0 fixed, $\|[\pi_{\theta_0}(b) - \pi_\theta(b)]\bar{y}\|_{\ell^p(\mathbb{Z},X)} \to 0$ as $\theta \to \theta_0$. This proves that the function from $\mathbb{R}$ to $\mathcal{L}_s(X)$ defined by $\theta \mapsto C_k^{(0)}(\theta)$ is continuous.

Next, using (4.12), note that

$$\sup_{\theta\in\mathbb{R}}\|C_k^{(0)}(\theta)\|_{\mathcal{L}(X)} \leq \sum_{k=-\infty}^{\infty}\sup_{\theta\in\mathbb{R}}\|C_k^{(0)}(\theta)\|_{\mathcal{L}(X)} \leq \sum_{k=-\infty}^{\infty}\sup_{\theta\in\mathbb{R}}\|C_k(\theta)\|_{\mathcal{L}(\ell^p(\mathbb{Z};X))} < \infty;$$

and therefore, the function from $\mathbb{R}$ to $\mathcal{L}(X)$ defined by $\theta \mapsto C_k^{(0)}(\theta)$ is bounded.

Now, for $\theta \in \mathbb{R}$, we define $a_k(\theta) = C_k^{(0)}(\theta)$ and $d = \sum_{k=-\infty}^{\infty} a_k V^k$ on $L^p(\mathbb{R};X)$. Clearly, from what we have discussed so far, it follows that $a_k \in \mathfrak{A}$ and $d \in \mathfrak{B}$. We claim that $d = b^{-1}$.

For $d \in \mathfrak{B}$ formula (4.13) implies

$$\pi_\theta(d) = \sum_{k=-\infty}^{\infty}\text{diag}(C_k^{(0)}(\theta+n))_{n\in\mathbb{Z}}S^k = \sum_{k=-\infty}^{\infty}\text{diag}(C_k^{(n)}(\theta))_{n\in\mathbb{Z}}S^k$$

$$= \sum_{k=-\infty}^{\infty}C_k(\theta)S^k = \pi_\theta(b)^{-1}.$$

Since π_θ is a homomorphism, we have

$$\pi_\theta(bd - I) = \pi_\theta(b)\pi_\theta(d) - I = 0 = \pi_\theta(db - I), \quad \theta \in \mathbb{R}.$$

Note that there is a representation of the element $bd - I \in \mathfrak{B}$ given by

$$bd - I = \sum_{k=-\infty}^{\infty}e_k V^k \quad \text{where} \quad e_k \in \mathfrak{A} \quad \text{and} \quad \sum_{k=-\infty}^{\infty}\|e_k\|_{\mathcal{L}(\ell^p(\mathbb{Z};X))} < \infty.$$

For this representation we have

$$0 = \pi_\theta(bd - I) = \pi_\theta\left(\sum_{k=-\infty}^{\infty}e_k V^k\right) = \sum_{k=-\infty}^{\infty}\pi_\theta(e_k)S^k.$$

Since $\sum_{k=-\infty}^{\infty} \|\pi_\theta(e_k)\|_{\mathcal{L}(\ell^p(\mathbb{Z};X))} < \infty$, it follows that $\pi_\theta(e_k) = \mathrm{diag}(e_k(\theta+n))_{n \in \mathbb{Z}} = 0$ for all $k \in \mathbb{Z}$, and $e_k(\theta) = 0$ for all $\theta \in \mathbb{R}$ and $k \in \mathbb{Z}$. Thus, we have proved that $bd - I = \sum_{k=-\infty}^{\infty} e_k V^k = 0$, as required. $\square$

We will use the same arguments, as in the proof of Theorem 4.9 and Corollary 4.8, to prove the next proposition.

PROPOSITION 4.10. *For each $b \in \mathfrak{B}$, if the operator $\pi_\theta(b)$ is invertible on $\ell^p(\mathbb{Z};X)$ for each $\theta \in \mathbb{R}$ and $\sup_{\theta \in \mathbb{R}} \|[\pi_\theta(b)]^{-1}\|_{\mathcal{L}(\ell^p(\mathbb{Z};X))} < \infty$, then b is invertible in $\mathcal{L}(L^p(\mathbb{R};X))$ and $b^{-1} \in \mathfrak{B}$.*

PROOF. Let $\mathcal{F} := C_0(\mathbb{R}; \ell^p(\mathbb{Z};X))$ and suppose that $\mathcal{A} \subset \mathcal{L}(\mathcal{F})$ is the algebra of multiplication operators whose multipliers are bounded, strongly continuous functions on $\mathbb{R}$ with values in $\mathcal{L}(\ell^p(\mathbb{Z};X))$. By the assumptions, $\mathcal{A}$ contains the operators $\mathcal{D}$ and $\mathcal{D}^{-1}$ defined by $(\mathcal{D}f)(\theta) = \pi_\theta(b)f(\theta)$ and $(\mathcal{D}^{-1}f)(\theta) = [\pi_\theta(b)]^{-1}f(\theta)$, $f \in \mathcal{F}$, $\theta \in \mathbb{R}$. As in the proof of Corollary 4.8, we let $\mathcal{W}(t) \in \mathcal{L}(\mathcal{F})$ denote the multiplication operator $(\mathcal{W}(t)f)(\theta) = W(t)f(\theta)$ where $W(t) = \mathrm{diag}(e^{-int})_{n \in \mathbb{Z}}$, and we define $\mathcal{D}(t) := \mathcal{W}(t)^{-1}\mathcal{D}\mathcal{W}(t)$, $t \in [0, 2\pi)$. Then, the function $t \mapsto \mathcal{D}(t) \in \mathcal{A}$ is represented by the absolutely convergent series $\sum_{k=-\infty}^{\infty} \mathcal{B}_k e^{ikt}$ where $\mathcal{B}_k \in \mathcal{L}(\mathcal{F})$ is the multiplication operator defined by $(\mathcal{B}_k f)(\theta) = B_k(\theta) f(\theta)$ with the multiplier $B_k(\theta) = \mathrm{diag}(a_k(\theta+n))_{n \in \mathbb{Z}} S^k$. By Bochner-Phillips Theorem 4.7,

(4.14)
$$\mathcal{D}(t)^{-1} = \sum_{k=-\infty}^{\infty} \mathcal{C}_k e^{ikt} \quad \text{with} \quad \sum_{k=-\infty}^{\infty} \|\mathcal{C}_k\|_{\mathcal{L}(\mathcal{F})} = \sum_{k=-\infty}^{\infty} \sup_{\theta \in \mathbb{R}} \|C_k(\theta)\|_{\mathcal{L}(\ell^p(\mathbb{Z};X))} < \infty$$

where $\mathcal{C}_k \in \mathcal{A}$ are multiplication operators with operator-valued multipliers $C_k(\cdot)$. The same argument as in the proof of Corollary 4.8 shows that $C_k(\theta)S^{-k}$ is a diagonal operator for each $\theta \in \mathbb{R}$ and $k \in \mathbb{Z}$. Also, using the representation in display (4.14), we have the representation given in display (4.12). Hence, the result follows as in the proof of Theorem 4.9. $\square$

For a bounded operator A on a Banach space Y recall the notation $\|A\|_{\bullet,Y} := \inf\{|Ax| : |x| = 1\}$. Also, recall that A is called uniformly injective if $\|A\|_{\bullet,Y} > 0$

Let us consider the converse of Theorem 4.9. We ask the following question: If $b \in \mathfrak{B}$ is a uniformly injective operator on $L^p(\mathbb{R};X)$, is each operator in the family $\pi_\theta(b)$, for $\theta \in \mathbb{R}$, uniformly injective on $\ell^p(\mathbb{Z};X)$? Moreover, is there a positive uniform lower bound over $\theta \in \mathbb{R}$? The next lemma answers these questions.

LEMMA 4.11. *If $1 \leq p < \infty$ and an operator $b \in \mathfrak{B}$ is uniformly injective on $L^p(\mathbb{R};X)$, then $\pi_\theta(b)$, for each $\theta \in \mathbb{R}$, is uniformly injective on $\ell^p(\mathbb{Z};X)$. Moreover,*

$$\|\pi_\theta(b)\|_{\bullet,\ell^p(\mathbb{Z};X)} \geq \|b\|_{\bullet,L^p(\mathbb{R};X)} \quad \text{for all} \quad \theta \in \mathbb{R}.$$

Note that Lemma 4.11 is applicable to the operator $b = I - aV \in \mathfrak{B}$.

PROOF. We will prove that if $\bar{x} \in \ell^p(\mathbb{Z};X)$, then, for every $\theta \in \mathbb{R}$,

(4.15) $\qquad \|b\|_{\bullet,L^p(\mathbb{R};X)} \|\bar{x}\|_{\ell^p(\mathbb{Z};X)} \leq \|\pi_\theta(b)\bar{x}\|_{\ell^p(\mathbb{Z};X)}.$

REMARK 4.12. It suffices to prove the inequality (4.15) for $\theta = 0$. Indeed, let us consider the translation group $(V^t f)(\theta) = f(\theta - t)$ on $L^p(\mathbb{R}; X)$. Clearly, if $b = \sum a_k V^k$ and $\theta \in \mathbb{R}$, then $V^{-\theta} b V^\theta = \sum a_k(\cdot + \theta) V^k$. Hence,

$$\pi_\theta(b) = \pi_0(V^{-\theta} b V^\theta) = \pi_0(\tilde{b}), \quad \tilde{b} = \sum_{k=-\infty}^{\infty} a_k(\cdot + \theta) V^k,$$

and it follows that if the inequality (4.15) holds for $\tilde{b}$ at $\theta = 0$, then the desired result also holds for b. $\diamond$

To prove inequality (4.15), let us fix a sequence $\bar{x} = (x_n)_{n \in \mathbb{Z}} \in \ell^p(\mathbb{Z}; X)$ and $\epsilon > 0$. Recall that if $b = \sum_{k=-\infty}^{\infty} a_k V^k \in \mathfrak{B}$, then $\sum_{k=-\infty}^{\infty} \|a_k\|_\infty < \infty$. Choose a natural number $N > 1/\epsilon$ such that

(4.16) $\quad \sum_{|k|>N} \|a_k\|_\infty < \epsilon/(4 \sup_{n \in \mathbb{Z}} |x_n|) \quad \text{and} \quad \sup_{|n|>N} |x_n| < \epsilon/(4 \sum_{k=-\infty}^{\infty} \|a_k\|_\infty).$

Recall that, for each $x \in X$ and each $k \in \mathbb{Z}$, the function from $\mathbb{R}$ to X given by $\theta \mapsto a_k(\theta) x$ is continuous, and choose $\delta \in (0, 1)$ such that

(4.17) $\quad \sum_{|k| \leq N} \left| [a_k(\theta + n) - a_k(n)] x_{n-k} \right| < \epsilon/2 \quad \text{for} \quad \theta \in [0, \delta] \quad \text{and} \quad |n| \leq 2N.$

Also, define $f \in L^p(\mathbb{R}; X)$ as follows: $f(\theta) = x_n$ for $|n| \leq N$ and $\theta \in [n, n + \delta]$, and $f(\theta) = 0$ otherwise. In particular, $\|f\|_\infty \leq \sup_{n \in \mathbb{Z}} |x_n|$ and $f(\theta + n - k) = 0$ for $|n - k| > N$, $n, k \in \mathbb{Z}$, and $\theta \in [0, \delta]$.

Let us note that

(4.18)
$$\|bf\|_{L^p}^p \geq \|b\|_{\bullet, L^p}^p \|f\|_{L^p}^p = \|b\|_{\bullet, L^p}^p \sum_{n=-N}^{N} \int_n^{n+\delta} |x_n|^p d\theta = \delta \|b\|_{\bullet, L^p}^p \sum_{n=-N}^{N} |x_n|^p.$$

On the other hand, if we use the definition of f and (4.16), we have the estimate

$$\|bf\|_{L^p}^p = \sum_{n=-\infty}^{\infty} \int_0^\delta \left| \sum_{k=-\infty}^{\infty} a_k(\theta + n) f(\theta + n - k) \right|^p d\theta$$

$$\leq \sum_{|n| \leq 2N} \int_0^\delta \left(\left| \sum_{|k| \leq N} a_k(\theta + n) f(\theta + n - k) \right| + \epsilon/4 \right)^p d\theta$$

$$\leq \sum_{|n| \leq 2N} \int_0^\delta \left(\left| \sum_{\substack{|k| \leq N \\ |n-k| \leq N}} a_k(n) x_{n-k} \right| \right.$$

$$\left. + \sum_{|k| \leq N} \left| [a_k(\theta + n) - a_k(n)] x_{n-k} \right| + \epsilon/4 \right)^p d\theta.$$

Now (4.16) and (4.17) imply

$$\|bf\|_{L^p}^p \leq \delta \sum_{|n|\leq 2N} \Big(\Big|\sum_{\substack{|k|\leq N \\ |n-k|\leq N}} a_k(n)x_{n-k}\Big| + 3\epsilon/4\Big)^p$$

$$\leq \delta \sum_{|n|\leq 2N} \Big(\Big|\sum_{|k|\leq N} a_k(n)x_{n-k}\Big| + \sum_{\substack{|k|\leq N \\ |n-k|>N}} \|a_k\|\|x_{n-k}\| + 3\epsilon/4\Big)^p$$

$$\leq \delta \sum_{|n|\leq 2N} \Big(\Big|\sum_{|k|\leq N} a_k(n)x_{n-k}\Big| + \epsilon\Big)^p.$$

Combining this inequality with the inequality (4.18), we conclude that

$$\|b\|_{\bullet,L^p} \sum_{|n|\leq N} |x_n|^p \leq \sum_{|n|\leq 2N} \Big(\Big|\sum_{|k|\leq N} a_k(n)x_{n-k}\Big| + \epsilon\Big)^p.$$

Thus, if $\epsilon \to 0$ and $N \to \infty$, then

$$\|b\|_{\bullet,L^p}\|\bar{x}\|_{\ell^p} \leq \Big(\sum_{n=-\infty}^{\infty}\Big|\sum_{k=-\infty}^{\infty} a_k(n)x_{n-k}\Big|^p\Big)^{1/p} = \|\pi_0(b)\bar{x}\|_{\ell^p}.$$

$\square$

Suppose that $1 < p < \infty$. Let us define $\mathfrak{A}_*$ to be the set of all $a \in \mathfrak{A}$ such that, for the adjoint operator $a^*(\theta) := [a(\theta)]^*$, the function $\theta \mapsto a^*(\theta)$ belongs to the space $C_b(\mathbb{R}; \mathcal{L}_s(X^*))$. Here X^* denotes the dual space. Also, let $(\mathfrak{B}_*, \|\cdot\|_{\mathfrak{B}_*})$ denote the Banach algebra of all operators in $\mathcal{L}(L^p(\mathbb{R}, X))$ of the form

$$b = \sum_{k=-\infty}^{\infty} a_k V^k \quad \text{where} \quad a_k \in \mathfrak{A}_* \quad \text{and} \quad \|b\|_{\mathfrak{B}_*} := \sum_{k=-\infty}^{\infty} \|a_k\|_{\mathcal{L}(L^p(\mathbb{R};X))} < \infty.$$

The following lemma is analogous to Lemma 4.11. Recall that $L^q(\mathbb{R}; X^*)$ is isometrically embedded in $L^p(\mathbb{R}; X)^*$ for $p^{-1} + q^{-1} = 1$ (see for example Diestel and Uhl [**DU**]).

LEMMA 4.13. *Suppose that $b = \sum_{k=-\infty}^{\infty} a_k V^k \in \mathfrak{B}_*$. If the operator b^* is uniformly injective on $L^p(\mathbb{R}; X)^*$, then $\pi_\theta(b^*)$ is uniformly injective on $\ell^q(\mathbb{Z}; X^*)$. Moreover, $\|\pi_\theta(b^*)\|_{\bullet,\ell^q(\mathbb{Z};X^*)} \geq \|b^*\|_{\bullet,L^p(\mathbb{R};X)^*}$ for all $\theta \in \mathbb{R}$.*

PROOF. As in the proof of Lemma 4.11, let us check that the inequality

$$\|b^*\|_{\bullet,L^p(\mathbb{R};X)^*}\|\bar{y}\|_{\ell^q(\mathbb{Z};X^*)} \leq \|\pi_\theta(b^*)\bar{y}\|_{\ell^q(\mathbb{Z};X^*)}$$

holds for all $\theta \in \mathbb{R}$ and $\bar{y} \in \ell^q(\mathbb{Z}; X^*)$. Also, as before, it suffices to consider $\theta = 0$.

Let $\epsilon > 0$ be given, fix $\bar{y} = (y_n)_{n\in\mathbb{Z}} \in \ell^q(\mathbb{Z}; X^*)$, and choose a natural number $N > 1/\epsilon$ so that

$$\sum_{|k|>N} \|a_k^*\|_\infty < \epsilon/(4\sup_n |y_n|) \quad \text{and} \quad \sup_{|n|>N} |y_n| \leq \epsilon/(4\sum_{k=-\infty}^{\infty} \|a_k^*\|_\infty).$$

Because $b \in \mathfrak{B}_*$, the function $\theta \mapsto a_k^*(\theta)y$ is continuous for each $y \in X^*$ and $k \in \mathbb{Z}$. Hence, there is some $\delta \in (0,1)$ such that

$$\sum_{|k|\leq N} \Big|[a_k^*(\theta+n+k) - a_k^*(n+k)]y_{n+k}\Big| < \epsilon/2 \quad \text{for} \quad \theta \in [0,\delta] \quad \text{and} \quad |n| \leq 2N.$$

Let us define $g \in L^q(\mathbb{R}; X^*)$ by $g(\theta) = y_n$ for $|n| \leq N$ and $\theta \in [n, n+\delta]$, and $g(\theta) = 0$ otherwise. Since $L^q(\mathbb{R}; X^*)$ is isometrically embedded in $L^p(\mathbb{R}; X)^*$, we have
$$\|g\|^q_{L^p(\mathbb{R};X)^*} = \|g\|^q_{L^q(\mathbb{R};X^*)} = \delta \sum_{|n| \leq N} |y_n|^q_{X^*}.$$

Also, since the operator $b^* = \sum V^{-k} a_k^* = \sum a_k^*(\cdot + k) V^{-k}$ maps $L^q(\mathbb{R}; X^*)$ into itself, it follows that

(4.19)
$$\delta \|b^*\|^q_{\bullet, L^{p*}} \sum_{|n| \leq N} |y_n|^q = \|b^*\|^q_{\bullet, L^{p*}} \|g\|^q_{L^p(\mathbb{R};X)^*} \leq \|b^* g\|^q_{L^p(\mathbb{R};X)^*} = \|b^* g\|^q_{L^q(\mathbb{R};X^*)}.$$

Arguing as in Lemma 4.11, we have the inequality
$$\|b^* g\|^q_{L^q(\mathbb{R};X^*)} \leq \delta \sum_{|n| \leq 2N} \left(\left| \sum_{|k| \leq N} a_k^*(n+k) y_{n+k} \right| + \epsilon \right)^q.$$

This inequality, together with the estimate (4.19), yields
$$\|b^*\|^q_{\bullet, L^{p*}} \sum_{|n| \leq N} |y_n|^q \leq \sum_{|n| \leq 2N} \left(\sum_{|k| \leq N} |a_k^*(n+k) y_{n+k}| + \epsilon \right)^q.$$

Passing to the limit as $\epsilon \to 0$ and $N \to \infty$ it follows that
$$\|b^*\|_{\bullet, L^p(\mathbb{R};X)^*} \|(y_n)_{n \in \mathbb{Z}}\|_{\ell^q(\mathbb{Z};X^*)} \leq \|\pi_0(b^*)((y_n)_{n \in \mathbb{Z}})\|_{\ell^q(\mathbb{Z};X^*)}.$$
□

COROLLARY 4.14. *The algebra $\mathfrak{B}_*$ is inverse-closed in $\mathcal{L}(L^p(\mathbb{R}; X))$. Moreover, an operator $b = \sum a_k V^k \in \mathfrak{B}_*$ is invertible in $\mathcal{L}(L^p(\mathbb{R}; X))$ if and only if each operator $\pi_\theta(b)$, for $\theta \in \mathbb{R}$, is invertible in $\mathcal{L}(\ell^p(\mathbb{Z}; X))$ and*
$$\sup_{\theta \in \mathbb{R}} \|\pi_\theta(b)^{-1}\|_{\mathcal{L}(\ell^p(\mathbb{Z};X))} < \infty.$$

PROOF. If $b = \sum a_k V^k \in \mathfrak{B}$ is invertible in $\mathcal{L}(L^p(\mathbb{R}; X))$, then, by Lemma 4.11, each operator in the family $\pi_\theta(b)$, for $\theta \in \mathbb{R}$, is uniformly injective and these operators also have a positive uniform lower bound; that is, $\inf_\theta \|\pi_\theta(b)\|_{\bullet, \ell^p(\mathbb{Z};X)} > 0$. On the other hand, since b^* is invertible in $\mathcal{L}(L^p(\mathbb{R}; X)^*)$, by Lemma 4.13 each operator $\pi_\theta(b)^* = \pi_\theta(b^*)$ on the space $\ell^q(\mathbb{Z}; X^*)$ is uniformly injective. Hence, each operator $\pi_\theta(b)$ and $\pi_\theta(b^*)$ is invertible. Moreover, we have the bounds

(4.20) $\quad \sup_{\theta \in \mathbb{R}} \|\pi_\theta(b)^{-1}\|_{\mathcal{L}(\ell^p(\mathbb{Z};X))} < \infty \quad$ and $\quad \sup_{\theta \in \mathbb{R}} \|\pi_\theta(b^*)^{-1}\|_{\mathcal{L}(\ell^q(\mathbb{Z};X^*))} < \infty.$

By Theorem 4.9 and Proposition 4.10, the first inequality in (4.20) implies $b^{-1} \in \mathfrak{B}$ on $L^p(\mathbb{R}; X)$; that is, $b^{-1} = \sum a_k V^k$ and each function $a_k : \mathbb{R} \to \mathcal{L}_s(X)$ is bounded and continuous. By Theorem 4.9 and Proposition 4.10 applied to the operator b^* on $L^q(\mathbb{R}; X^*)$, the second inequality in (4.20) implies $[b^*]^{-1} \in \mathfrak{B}$ on $L^q(\mathbb{R}; X^*)$; that is, each function $a_k^* : \mathbb{R} \to \mathcal{L}_s(X^*)$ is bounded and continuous. It follows that $b^{-1} \in \mathfrak{B}_*$. □

REMARK 4.15. If $b \in \mathfrak{B}$, $\theta \in \mathbb{R}$, and $j \in \mathbb{Z}$, then $S^j \pi_\theta(b) S^{-j} = \pi_{\theta-j}(b)$. Hence, if I is an interval of unit length; for example, $I = [-1/2, 1/2]$, then
$$\sup_{\theta \in \mathbb{R}} \|\pi_\theta(b)^{-1}\|_{\mathcal{L}(\ell^p(\mathbb{Z};X))} = \sup_{\theta \in I} \|\pi_\theta(b)^{-1}\|_{\mathcal{L}(\ell^p(\mathbb{Z};X))}.$$

As a consequence, the condition "for all $\theta \in \mathbb{R}$" in this subsection (see Theorem 4.9, Lemma 4.11, and Corollary 4.14) can be replaced by the formally weaker condition "for all $\theta \in [-1/2, 1/2]$". ◇

4.1.3. Dichotomy and discrete operators. We will apply some of the results of the previous subsection to evolution operators. As usual, let $\{U(\theta, \tau)\}_{\theta \geq \tau}$ be a strongly continuous, exponentially bounded evolution family on a Banach space X and $\{E^t\}_{t \geq 0}$, defined by $(E^t f)(\theta) = U(\theta, \theta - t) f(\theta - t)$, the corresponding evolution semigroup on $L^p(\mathbb{R}; X)$. Also, let $a(\theta) = U(\theta, \theta - 1)$ for $\theta \in \mathbb{R}$. Observe that $a \in \mathfrak{A}$; that is, $a(\cdot) \in C_b(\mathbb{R}; \mathcal{L}_s(X))$, and $E^1 = aV$ where $(Vf)(\theta) = f(\theta - 1)$. Finally, let us define $E := E^1$.

THEOREM 4.16 (Discrete Dichotomy Theorem). *Suppose that $1 \leq p < \infty$. The following statements are equivalent:*

(1) $\{U(\theta, \tau)\}_{\theta \geq \tau}$ *has an exponential dichotomy on X.*
(2) *The operator $I - E$ is invertible on $L^p(\mathbb{R}; X)$.*
(3) *For each $\theta \in \mathbb{R}$, the operator $I - \pi_\theta(E)$ is invertible on $\ell^p(\mathbb{Z}; X)$ and*

$$(4.21) \qquad \sup_{\theta \in \mathbb{R}} \|[I - \pi_\theta(E)]^{-1}\|_{\mathcal{L}(\ell^p(\mathbb{Z}; X))} < \infty.$$

If one of the equivalent statements is true, then the operators $\pi_\theta(E)$, for $\theta \in \mathbb{R}$, and E are hyperbolic. Moreover, the dichotomy projection $P(\cdot)$ for $\{U(\theta, \tau)\}_{\theta \geq \tau}$, the Riesz projection $\mathcal{P}$ that corresponds to the operator E and the spectral set $\sigma(E) \cap \mathbb{D}$, and the Riesz projection P_θ that correspond to the operator $\pi_\theta(E)$ and the spectral set $\sigma(\pi_\theta(E)) \cap \mathbb{D}$, for $\theta \in \mathbb{R}$, are related by the following formulas:

$$(4.22) \qquad (\mathcal{P}f)(\theta) = P(\theta) f(\theta), \quad P_\theta = \operatorname{diag}(P(\theta + n))_{n \in \mathbb{Z}}.$$

Also, the supremum over $\mathbb{R}$ in (4.21) can be replaced by the supremum over an interval $I \subset \mathbb{R}$ of unit length.

See Remark 4.19 where condition (4.21) is relaxed. Also, see Subsection 4.3.3 where we show that, in fact, condition (3) holds provided only the *single* discrete operator $I - \pi_0(E)$ is invertible.

REMARK 4.17. The space $\ell^p(\mathbb{Z}; X)$ in (3) can be replaced by the space $c_0(\mathbb{Z}; X)$. This fact follows from the result in Remark 4.6 applied to the families of operators $d_\theta = \pi_\theta(E)$ and $D_\theta = I - \pi_\theta(E)$ for $\theta \in \mathbb{R}$. Indeed, assume that statement (3) of the theorem holds for one of the spaces $Y = l^p(\mathbb{Z}; X)$, or $Y = c_0(\mathbb{Z}; X)$. Then, the bounds in display (4.7) of Remark 4.6 hold on Y. On the space Y, we have $[I - \pi_\theta(E)]^{-1} \in \mathfrak{C}$; that is, the operator $C(\theta) = [I - \pi_\theta(E)]^{-1}$ has the form $C(\theta) = \sum_{k=-\infty}^{\infty} C_k(\theta) S^k$, $\theta \in \mathbb{R}$, where $C_k(\theta) \in \mathfrak{D}$. Moreover, the same operator $C(\theta)$ is the inverse for $I - \pi_\theta(E)$ on $\ell^p(\mathbb{Z}; X)$, $1 \leq p \leq \infty$, and $c_0(\mathbb{Z}; X)$. By the bound (4.8), the norm of this inverse for each of these spaces is such that

$$\sup_{\theta \in \mathbb{R}} \|[I - \pi_\theta(E)]^{-1}\| = \sup_{\theta \in \mathbb{R}} \|\sum_{k=-\infty}^{\infty} C_k(\theta) S^k\| \leq \sum_{k=-\infty}^{\infty} \|C_k(\theta)\|_{\mathfrak{D}} < \infty.$$

Thus, statement (3) either holds on $\ell^p(\mathbb{Z}; X)$, for $1 \leq p \leq \infty$, and on $c_0(\mathbb{Z}; X)$ or it holds for none of these spaces. ◇

Next, we give the proof of the Discrete Dichotomy Theorem 4.16.

PROOF. Recall that if $\theta \in \mathbb{R}$, then $\sigma(E)$ and $\sigma(\pi_\theta(E))$ are invariant with respect to rotations centered at zero, see Theorem 3.13 and Remark 4.4, respectively. Thus, E (respectively, $\pi_\theta(E)$) is hyperbolic if and only if $I - E$ (respectively, $I - \pi_\theta(E)$) is invertible.

The implication (1)$\Rightarrow$(2) is stated in the Dichotomy Theorem 3.17 where its proof is a relatively simple observation. Also, the implication (2)$\Rightarrow$(1) is stated in Theorem 3.17, but the proof is much more difficult. In fact, the main part of the proof is a separate result; namely, the Spectral Projection Theorem 3.14. By this result, condition (2) implies the Riesz projection $\mathcal{P}$ for E is a multiplication operator $\mathcal{P} = P(\cdot)$ for some $P(\cdot) \in C_b(\mathbb{R}; \mathcal{L}_s(X))$. (In Proposition 4.18 we will give an algebraic proof of this result.)

(2)$\Rightarrow$(3). By Lemma 4.11, condition (2) implies that each operator in the family $I - \pi_\theta(E)$ is uniformly injective and that there is a positive uniform lower bound; that is, $\inf_{\theta \in \mathbb{R}} \|I - \pi_\theta(E)\|_\bullet > 0$. Thus, to prove statement (3) it suffices to show that $I - \pi_\theta(E)$ is surjective on $\ell^p(\mathbb{Z}; X)$ for each $\theta \in \mathbb{R}$. In fact, by the same argument as in Remark 4.12, it suffices to show that $I - \pi_0(E)$ is surjective.

Fix $\bar{x} = (x_n)_{n \in \mathbb{Z}} \in \ell^p(\mathbb{Z}; X)$, and let
$$f(\tau) := U(\tau, n-1)x_{n-1}, \quad \tau \in \left[n - \tfrac{1}{2}, n + \tfrac{1}{2}\right], \quad n \in \mathbb{Z}.$$
Since $\{U(\theta, \tau)\}_{\theta \geq \tau}$ is exponentially bounded, $f \in L^p(\mathbb{R}; X)$. Since $I - E$ is invertible by statement (2), there exists $g \in L^p(\mathbb{R}; X)$ such that
$$(4.23) \qquad g(\tau) - U(\tau, \tau - 1)g(\tau - 1) = f(\tau)$$
for almost all $\tau \in \mathbb{R}$. Also, since
$$\|g\|_{L^p(\mathbb{R};X)}^p = \sum_{n=-\infty}^{\infty} \int_{n-\frac{1}{2}}^{n+\frac{1}{2}} |g(s)|^p ds = \int_{-\frac{1}{2}}^{\frac{1}{2}} \left(\sum_{n=-\infty}^{\infty} |g(s+n)|^p \right) ds < \infty,$$
the sequence $(g(s+n))_{n \in \mathbb{Z}}$ belongs to $\ell^p(\mathbb{Z}; X)$ for all $s \in \Omega$ where $\Omega \subset \left(-\tfrac{1}{2}, \tfrac{1}{2}\right)$ is some subset of full measure.

For each $s \in \Omega$, define $h_s : \mathbb{R} \to X$ as follows:
$$h_s(\tau) = \begin{cases} g(\tau), & \text{if } n - \tfrac{1}{2} \leq \tau \leq n + s, \\ U(\tau, n+s)g(n+s), & \text{if } n + s \leq \tau < n + \tfrac{1}{2}, \end{cases} \quad n \in \mathbb{Z}.$$
Since $\{U(\theta, \tau)\}_{\theta \geq \tau}$ is exponentially bounded and $(g(s+n))_{n \in \mathbb{Z}} \in \ell^p(\mathbb{Z}, X)$, it follows that $h_s \in L^p(\mathbb{R}; X)$ for each $s \in \Omega$.

For each $s \in \Omega$, the function h_s is a solution of equation (4.23). Indeed, for $\tau \in [n - \tfrac{1}{2}, n + s]$, we have that $g(\tau) = h_s(\tau)$ and (4.23) is satisfied. On the other hand, if $\tau \in [n + s, n + 1/2)$, then
$$h_s(\tau) - U(\tau, \tau - 1)h_s(\tau - 1) = U(\tau, n+s)[g(n+s)$$
$$-U(n+s, n-1+s)g(n-1+s)]$$
$$= U(\tau, n+s)f(n+s) = U(\tau, n+s)U(n+s, n-1)x_{n-1} = f(\tau).$$
Let us also note that since equation (4.23) has only one solution in $L^p(\mathbb{R}; X)$, $g = h_s$ for all $s \in \Omega$.

Suppose that $s \in \Omega$ and $s < 0$. The values $h_s(\tau)$ are defined for $\tau = n$, $n \in \mathbb{Z}$. Moreover, the sequence $(h_s(n))_{n \in \mathbb{Z}}$ belongs to $\ell^p(\mathbb{Z}; X)$. Indeed, $h_s(n) = U(n, n+s)g(n+s)$, the sequence $(g(n+s))_{n \in \mathbb{Z}}$ belongs to $\ell^p(\mathbb{Z}; X)$, and, because $\{U(\theta, \tau)\}_{\theta \geq \tau}$ is exponentially bounded, $\|U(n, n+s)\|_{\mathcal{L}(X)} \leq Me^{-\omega s}$.

Define $y_n := h_s(n) + x_n$ and $\bar{y} := (y_n)_{n \in \mathbb{Z}}$. Since $g = h_s(\cdot)$ satisfies equation (4.23) for $\tau = n$, $n \in \mathbb{Z}$, we have that

$$([I - \pi_0(E)]\bar{y})_n = y_n - U(n, n-1)y_{n-1}$$
$$= h_s(n) + x_n - U(n, n-1)h_s(n-1) - U(n, n-1)x_{n-1}$$
$$= h_s(n) + x_n - U(n, n-1)h_s(n-1) - f(n) = x_n.$$

Thus, $I - \pi_\theta(E)$ is surjective on $\ell^p(\mathbb{Z}; X)$. This proves the implication (2)⇒(3).

The implication (3)⇒(2) follows directly from Theorem 4.9.

It remains to prove the statements in Theorem 4.16 related to the Riesz and dichotomy projections. First, as a corollary of Theorem 4.9, we conclude that $R(\lambda; E) \in \mathfrak{B}$ for each $\lambda \in \mathbb{T}$ provided that statement (3) holds. By the following proposition, this fact by itself implies the conclusion of the Spectral Projection Theorem 3.14.

PROPOSITION 4.18. *If $R(\lambda; E) \in \mathfrak{B}$ for each $\lambda \in \mathbb{T}$, then $\mathcal{P} \in \mathfrak{A}$.*

PROOF. To prove the proposition, we note that $f : \lambda \mapsto \lambda I - E = \lambda - aV$ is an absolutely convergent Fourier series in λ with values in $\mathfrak{B}$. Since $[f(\lambda)]^{-1} \in \mathfrak{B}$ for each $\lambda \in \mathbb{T}$, by the Bochner-Phillips generalization of Wiener's theorem; namely Theorem 4.7, we have

$$[f(\lambda)]^{-1} = R(\lambda; E) = \sum_{n=-\infty}^{\infty} b_n \lambda^n, \quad \text{where} \quad \sum_{n=-\infty}^{\infty} \|b_n\|_{\mathfrak{B}} < \infty, \quad \lambda \in \mathbb{T}, \quad b_n \in \mathfrak{B}.$$

As a result, we have

$$\mathcal{P} = \frac{1}{2\pi i} \int_{\mathbb{T}} R(\lambda; E) \, d\lambda = \frac{1}{2\pi i} \int_{\mathbb{T}} \sum_{n=-\infty}^{\infty} b_n \lambda^n \, d\lambda = b_{-1} \in \mathfrak{B};$$

and therefore,

$$\mathcal{P} = \sum_{k=-\infty}^{\infty} a_k V^k \quad \text{where} \quad a_k \in \mathfrak{A} \quad \text{and} \quad \sum_{k=-\infty}^{\infty} \|a_k\|_{\mathfrak{A}} < \infty.$$

By the Commutation Lemma 3.15, $\mathcal{P}\chi = \chi\mathcal{P}$ for every $\chi \in L^\infty(\mathbb{R}; \mathbb{R})$. Since the series for $\mathcal{P}$ converges absolutely, the identity

$$\chi \mathcal{P} - \mathcal{P} \chi = \sum_{k=-\infty}^{\infty} a_k(\cdot)[\chi(\cdot) - \chi(\cdot - k)]V^k = 0$$

implies $a_k(\cdot)[\chi(\cdot) - \chi(\cdot - k)] = 0$ for each $k \in \mathbb{Z}$. If we fix $k \neq 0$, $\theta \in \mathbb{R}$, and pick χ such that $\chi(\theta) \neq \chi(\theta - k)$, then $a_k(\theta) = 0$. Thus, $a_k = 0$ for all $k \neq 0$, and $\mathcal{P} = a_0 \in \mathfrak{A}$. □

We are ready to verify the statements of Theorem 4.16 that are related to the Riesz and dichotomy projections.

By Proposition 4.18, $\mathcal{P} = P(\cdot)$ for some $P(\cdot) \in C_b(\mathbb{R}; (\mathcal{L}_s(X)))$, Hence, by an argument as in the proof of the implication (2) ⇒ (1) of the Dichotomy Theorem 3.17, we conclude that $P(\cdot)$ is the dichotomy projection.

To prove that $P_\theta = \text{diag}(P(\theta + n))_{n \in \mathbb{Z}}$, we use the fact that π_θ is a homomorphism:

$$P_\theta = \frac{1}{2\pi i} \int_{\mathbb{T}} [\lambda - \pi_\theta(E)]^{-1} \, d\lambda = \frac{1}{2\pi i} \int_{\mathbb{T}} \pi_\theta([\lambda - E]^{-1}) \, d\lambda = \pi_\theta(\mathcal{P}).$$

The last statement in Theorem 4.16 follows from Remark 4.15. □

REMARK 4.19. If $\{U(\theta,\tau)\}_{(\theta,\tau)\in\mathbb{R}^2}$ is an invertible, exponentially bounded evolution family with $\|U(\theta,\tau)\| \leq Me^{\omega|t-s|}$, then condition (3) in the Discrete Dichotomy Theorem 4.16 can be replaced by the following condition:

(3′) $I - \pi_{\theta_0}(E)$ is invertible for some $\theta_0 \in \mathbb{R}$.

Indeed, for $\theta_1, \theta_2 \in \mathbb{R}$, let us define an invertible operator
$$K = \operatorname{diag}(U(\theta_1+n, \theta_2+n))_{n\in\mathbb{Z}} \quad \text{on} \quad \ell^p(\mathbb{Z}; X).$$
By the intertwining $K\pi_{\theta_2}(E) = \pi_{\theta_1}(E)K$, we have the desired result. ◇

4.2. Green's function and evolution semigroups

In this section we will study a connection between the existence of an exponential dichotomy for a strongly continuous evolution family $\{U(\theta,\tau)\}_{\theta\geq\tau}$ and the existence of a Green's operator $\mathbb{G}$. The Green's operator gives a formula for the inverse of the generator Γ of the evolution semigroup $\{E^t\}_{t\geq 0}$ defined on $L^p(\mathbb{R}; X)$ by the formula $(E^t f)(\theta) = U(\theta, \theta-t)f(\theta-t)$, $\theta \in \mathbb{R}$.

By the Dichotomy Theorem 3.17, $\{U(\theta,\tau)\}_{\theta\geq\tau}$ has a dichotomy if and only if Γ has a bounded inverse on $L^p(\mathbb{R}; X)$ or, in other words, if and only if the nonhomogeneous equation $\Gamma u = f$ has a unique solution for each choice of f. It is expected, see statement (ii) in Corollary 3.27, that this solution u is given by $u = G_P * f$ where G_P, the Green's function, is the kernel of a certain integral operator $\mathbb{G}$. In other words, it is expected that the existence of an exponential dichotomy for $\{U(\theta,\tau)\}_{\theta\geq\tau}$ is equivalent to the boundedness of $\mathbb{G}$ on $L^p(\mathbb{R}; X)$. This expectation is made rigorous by using "change-of-variables" techniques as in Theorem 3.26 or Theorem 3.13. We will consider in Subsection 4.2.1 the case of a hyperbolic semigroup $\{T^t\}_{t\geq 0}$ on a Banach space X, and in Subsection 4.2.2 we will apply the results for general semigroups to the evolution semigroup $\{E^t\}_{t\geq 0}$ on $L^p(\mathbb{R}; X)$.

4.2.1. Semigroup case. Let $\{T^t\}_{t\geq 0}$ be a strongly continuous semigroup on a Banach space X with infinitesimal generator A. We say that a projection $P \in \mathcal{L}(X)$ is a *splitting projection* for $\{T^t\}_{t\geq 0}$ if $PT^t = T^t P$ and, for the corresponding restrictions $T^t_P : \operatorname{Im} P \to \operatorname{Im} P$ and $T^t_Q : \operatorname{Im} Q \to \operatorname{Im} Q$ where $Q = I - P$, the operator T^t_Q is invertible as an operator on $\operatorname{Im} Q$. If P is a splitting projection, we define $G_P : \mathbb{R} \setminus \{0\} \to \mathcal{L}(X)$ by
$$G_P(\tau) = T^\tau_P P \quad \text{for} \quad \tau > 0, \quad \text{and} \quad G_P(\tau) = -T^\tau_Q Q \quad \text{for} \quad \tau < 0.$$

DEFINITION 4.20. Suppose that $1 \leq p < \infty$. The function G_P is called the *Green's function* for the semigroup $\{T^t\}_{t\geq 0}$ if the operator $\mathbb{G}$, called the *Green's operator*, defined by
$$(\mathbb{G}f)(\theta) = \int_{-\infty}^{\infty} G_P(\theta-\tau)f(\tau)\, d\tau, \quad f \in L^p(\mathbb{R}; X),$$
is bounded on $L^p(\mathbb{R}; X)$.

Clearly, $\mathbb{G}$ is a convolution operator $\mathbb{G}f = G_P * f$; see a similar definition for the case of $L^p(\mathbb{R}_+; X)$ in display (2.36).

Let V^τ denote the translation operator $(V^\tau f)(\theta) = f(\theta - \tau)$ on $L^p(\mathbb{R}; X)$. For each $\tau \in \mathbb{R} \setminus \{0\}$, the operator $G_P(\tau)$ can be viewed as the multiplication operator

on $L^p(\mathbb{R}; X)$ that is given by $(G_P(\tau)f)(\theta) = G_P(\tau)f(\theta)$. With this interpretation, we have that
$$\mathbb{G}f = \int_{-\infty}^{\infty} G_P(\tau)V^\tau f\, d\tau, \quad f \in L^p(\mathbb{R}; X).$$
Also, for the splitting projection P, let us define the operator $\tilde{G}$ on X by
$$\tilde{G}x = \int_0^\infty T_Q^{-t} x\, dt - \int_0^\infty T_P^t x\, dt. \tag{4.24}$$

PROPOSITION 4.21. *Suppose that A is the generator of the semigroup $\{T^t\}_{t \geq 0}$ and P is a splitting projection. If the operator $\tilde{G}$ defined in equation (4.24) is bounded on X, then $0 \in \rho(A)$ and $A^{-1} = \tilde{G}$.*

PROOF. We will show that $\tilde{G}$ is a right inverse for A and $\text{Ker } A = \{0\}$.
First, let $x \in X$ and note that
$$T^t \tilde{G}x - \tilde{G}x = \int_{-\infty}^t T_Q^\tau x\, d\tau - \int_t^\infty T_P^\tau x\, d\tau - \int_{-\infty}^0 T_Q^\tau x\, d\tau + \int_0^\infty T_Q^\tau x\, d\tau$$
$$= \int_0^t T_Q^\tau x\, d\tau + \int_0^t T_P^\tau x\, d\tau.$$

Thus, $t^{-1}(T^t \tilde{G}x - \tilde{G}x)$ converges to $Qx + Px = x$ as $t \to 0$. Therefore, $\tilde{G}x \in \mathcal{D}(A)$ and $A\tilde{G}x = x$.

To see that $\text{Ker } A = \{0\}$, let $x \in \mathcal{D}(A)$ and $Ax = 0$. The functions $t \mapsto T_P^t x$ and $t \mapsto T_Q^{-t} x$, defined for $t \in \mathbb{R}_+$, are both constant. Indeed, we have the equations
$$\frac{d}{dt} T_P^t x = \frac{d}{dt} P T^t x = P T^t A x = 0,$$
$$\frac{d}{dt} T_Q^{-t} x = \frac{d}{dt}([QT^t Q]^{-1} x) = [QT^t Q]^{-1} Q A x = 0.$$
Thus, for $t > 0$, we have that $T_P^t x \equiv T_P^0 x = Px$ and $T_Q^{-t} x = Qx$; and therefore,
$$\tilde{G}x = \int_0^\infty T_Q^{-\tau} x\, d\tau - \int_0^\infty T_P^\tau x\, d\tau = \int_0^\infty (Qx - Px)\, d\tau.$$
But, $\tilde{G}x \in X$ because $\tilde{G} \in \mathcal{L}(X)$ by assumption. Hence, $Qx = Px$, and as a result, $x = 0$. □

Let us now consider the evolution semigroup $\{E^t\}_{t \geq 0}$ that is associated with the strongly continuous semigroup $\{T^t\}_{t \geq 0}$. This evolution semigroup is defined on $L^p(\mathbb{R}; X)$ by $(E^t f)(\theta) = T^t f(\theta - t)$, see (2.31). As usual, we let Γ denote the infinitesimal generator of $\{E^t\}_{t \geq 0}$.

THEOREM 4.22. *A strongly continuous semigroup $\{T^t\}_{t \geq 0}$ is hyperbolic on X if and only if there exists a unique Green's function G_P for $\{T^t\}_{t \geq 0}$. Moreover, if this is the case, then $\mathbb{G} = -\Gamma^{-1}$ on $L^p(\mathbb{R}; X)$.*

PROOF. "If"-part. We will prove the following proposition: *If there is a Green's function G_P for $\{T^t\}_{t \geq 0}$, then $\Gamma^{-1} \in \mathcal{L}(L^p(\mathbb{R}; X))$.*

Recall the splitting projection P from Definition 4.20 and define a projection $\mathcal{P}$ on $L^p(\mathbb{R}; X)$ by $(\mathcal{P}f)(\theta) = Pf(\theta)$, $\theta \in \mathbb{R}$. Also, define the complementary projection $\mathcal{Q} = I - \mathcal{P}$. Construct $\tilde{G}_E$ as in (4.24), but for the semigroup $\{E^t\}_{t\geq 0}$ on $L^p(\mathbb{R}; X)$ instead of $\{T^t\}_{t\geq 0}$ on X, to obtain

$$\tilde{G}_E f = \int_0^\infty (E_Q^\tau)^{-1} f \, d\tau - \int_0^\infty E_P^\tau f \, d\tau$$

where $E_P^t = \mathcal{P} E^t \mathcal{P}$ and $E_Q^t = \mathcal{Q} E^t \mathcal{Q}$ are the corresponding restrictions.

Since, as in the proof of the Dichotomy Theorem 3.17,

$$(E_Q^\tau)^{-1} = (T^\tau V^\tau Q)^{-1} = (T^\tau Q V^\tau)^{-1} = V^{-\tau}(T^\tau Q)^{-1} = (T_Q^\tau)^{-1} V^{-\tau}$$

and $E^\tau \mathcal{P} = T^\tau V^\tau \mathcal{P} = T^\tau \mathcal{P} V^\tau$, it follows that the inverse $(E_Q^\tau)^{-1}$ exists for each $\tau > 0$. Also, we have the identity $\tilde{G}_E f = -\mathbb{G} f$, a fact that is verified by the following computation:

$$(\tilde{G}_E f)(\theta) = \int_0^\infty [T_Q^\tau]^{-1} f(\theta + \tau) \, d\tau - \int_0^\infty T_P^\tau f(\theta - \tau) \, d\tau$$
$$= \int_\theta^\infty (T_Q^{\theta-\tau} f)(\tau) \, \tau - \int_{-\infty}^\theta (T_P^{\theta-\tau} f)(\tau) \, d\tau$$
$$= -\int_{-\infty}^\infty G_P(\theta - \tau) f(\tau) \, d\tau = -(\mathbb{G} f)(\theta), \quad \theta \in \mathbb{R}.$$

Since $\mathbb{G} \in \mathcal{L}(L^p(\mathbb{R}; X))$ by assumption, it follows from the identity that $\tilde{G}_E \in \mathcal{L}(L^p(\mathbb{R}; X))$. Now apply Proposition 4.21 above to the semigroup $\{E^t\}_{t\geq 0}$. Then Γ is invertible with $\Gamma^{-1} = \tilde{G}_E = -\mathbb{G}$. By the implication (3) $\Rightarrow$ (1) of Theorem 2.39, the semigroup $\{T^t\}_{t\geq 0}$ is hyperbolic on X with the hyperbolicity projection P equal to the Riesz projection corresponding to T^1 and the spectral set $\sigma(T^1) \cap \mathbb{D}$.

"Only if" part. Conversely, suppose that $\{T^t\}_{t\geq 0}$ is hyperbolic; that is, $\sigma(T^t) \cap \mathbb{T} = \emptyset$. Let P denote the Riesz projection corresponding to T^1 and the spectral set $\sigma(T^1) \cap \mathbb{D}$. Recall that $T^t P = P T^t$, $\sigma(T_P^t) \subset \mathbb{D}$, $\sigma((T_Q^t)^{-1}) \subset \mathbb{D}$, and there exist positive constants M and β such that $\|T_P^t\| \leq M e^{-\beta t}$ and $\|T_Q^{-t}\| \leq M e^{-\beta t}$ for $t > 0$. Therefore, $\|G_P(\tau)\|_{\mathcal{L}(X)} \leq M e^{-\beta |\tau|}$, for $\tau \in \mathbb{R}$, and

$$\|\mathbb{G} f\|_{L^p(\mathbb{R}; X)} \leq \int_{-\infty}^\infty \|G_P(\tau)\|_{\mathcal{L}(X)} \|f\|_{L^p(\mathbb{R}; X)} \, d\tau \leq \frac{2M}{\beta} \|f\|_{L^p(\mathbb{R}; X)}.$$

This prove the existence of the Green function.

To prove the uniqueness of G_P, suppose that $G_{P'}$ is also a Green's function for $\{T^t\}_{t\geq 0}$. By the "If"-part of the theorem, P' is the hyperbolic projection for $\{T^t\}_{t\geq 0}$, that is, P' is the Riesz projection that corresponds to T^1 and the spectral set $\sigma(T^1) \cap \mathbb{D}$. Thus $P = P'$. □

REMARK 4.23. Since Theorem 2.39 holds also for the space $C_0(\mathbb{R}; X)$ instead of $L^p(\mathbb{R}; X)$, the results of this subsection are valid if one replaces $L^p(\mathbb{R}; X)$ by $C_0(\mathbb{R}; X)$. That is, $\{T^t\}_{t\geq 0}$ is hyperbolic on X if and only if there exists a unique splitting projection P such that the operator $\mathbb{G}$ from Definition 4.20 is a bounded operator on $C_0(\mathbb{R}; X)$. ◇

4.2.2. Evolution families. For a strongly continuous, exponentially bounded evolution family $\{U(\theta, \tau)\}_{\theta \geq \tau}$, a *splitting projection* is an operator valued function $P(\cdot) \in C_b(\mathbb{R}, \mathcal{L}_s(X))$ such that statements (i) and (ii) of the Definition 3.6 of an exponential dichotomy hold; that is, $U(\theta, \tau) P(\tau) = P(\theta) U(\theta, \tau)$ for $\theta \geq \tau$ and the restriction $U_Q(\theta, \tau) = U(\theta, \tau)|\operatorname{Im} Q(\tau)$ is an invertible operator from $\operatorname{Im} Q(\tau)$

to $\operatorname{Im} Q(\theta)$. For the splitting projection $P(\cdot)$, define $G_P(\theta, \tau) = U_P(\theta, \tau)P(\tau)$ for $\theta > \tau$, and $G_P(\theta, \tau) = -[U_Q(\tau, \theta)]^{-1}Q(\tau)$ for $\theta < \tau$.

DEFINITION 4.24. Suppose that $1 \leq p < \infty$. The function G_P is called the *Green's function* for the evolution family $\{U(\theta, \tau)\}_{\theta \geq \tau}$ if the operator $\mathbb{G}$, defined on $L^p(\mathbb{R}, X)$ by

$$(4.25) \qquad (\mathbb{G}f)(\theta) = \int_{-\infty}^{\infty} G_P(\theta, \tau)f(\tau)\, d\tau, \quad f \in L^p(\mathbb{R}; X),$$

is bounded. The operator $\mathbb{G}$ is called the *Green's operator*.

Let Γ be the generator of the evolution semigroup $(E^t f)(\theta) = U(\theta, \theta - t)f(\theta - t)$ on $L^p(\mathbb{R}; X)$.

THEOREM 4.25. *An exponentially bounded, strongly continuous evolution family $\{U(\theta, \tau)\}_{\theta \geq \tau}$ has an exponential dichotomy on X if and only if there exists a unique Green's function G_P for $\{U(\theta, \tau)\}_{\theta \geq \tau}$. Moreover, if this is the case, then $\mathbb{G} = -\Gamma^{-1}$ on $L^p(\mathbb{R}; X)$.*

PROOF. By the Dichotomy Theorem 3.17, the evolution family $\{U(\theta, \tau)\}_{\theta \geq \tau}$ has an exponential dichotomy if and only if the evolution semigroup $\{E^t\}_{t \geq 0}$ is hyperbolic on $L^p(\mathbb{R}; X)$. By Theorem 4.22 applied to the evolution semigroup, we have that $\{E^t\}_{t \geq 0}$ is hyperbolic if and only if there exists a unique splitting projection $\mathcal{P} \in \mathcal{L}(L^p(\mathbb{R}; X))$ for $\{E^t\}_{t \geq 0}$ such that, for the Green's function on $L^p(\mathbb{R}; X)$ defined by $G_\mathcal{P}(\tau) = E_P^\tau$ for $\tau \geq 0$, and $G_\mathcal{P}(\tau) = -E_Q^\tau$ for $\tau \leq 0$, the corresponding Green's operator

$$(\hat{\mathbb{G}}F)(\tau) = \int_{-\infty}^{\infty} G_\mathcal{P}(\tau - s)F(s)\, ds, \quad F \in L^p(\mathbb{R}; L^p(\mathbb{R}; X))$$

is bounded on $L^p(\mathbb{R}; L^p(\mathbb{R}; X))$. We will prove that the existence of a unique Green's function $G_\mathcal{P}$ for $\{E^t\}_{t \geq 0}$ is equivalent to the existence of the Green's function G_P for $\{U(\theta, \tau)\}_{\theta \geq \tau}$.

To prove the desired equivalence, let us consider the isometry

$$\mathcal{L}(L^p(\mathbb{R}; X)) \to \mathcal{L}(L^p(\mathbb{R}; L^p(\mathbb{R}; X))) : K \mapsto I \otimes K$$

where, if $F(\tau) = h(\tau, \cdot) \in L^p(\mathbb{R}; X)$ for almost all $\tau \in \mathbb{R}$, then $[(I \otimes K)h](\tau, \theta) = Kh(\tau, \theta)$. We note that

$$(\hat{\mathbb{G}}h)(\tau, \cdot) = \int_{-\infty}^{\infty} G_\mathcal{P}(\tau - s)h(s, \cdot)\, ds,$$

or, more explicitly,

$$(\hat{\mathbb{G}}h)(\tau, \theta) = \int_{-\infty}^{\tau} U_P(\theta, \theta - (\tau - s))P(\theta - (\tau - s))h(s, \theta - (\tau - s))\, ds$$
$$- \int_{\tau}^{\infty} U_Q(\theta, \theta - (\tau - s))Q(\theta - (\tau - s))h(s, \theta - (\tau - s))\, ds.$$

As in the proof of Theorem 2.39, if J is the isometric isomorphism defined on the space $L^p(\mathbb{R}; L^p(\mathbb{R}; X))$ by $(Jh)(\tau, \theta) = h(\tau + \theta, \theta)$, then $J\hat{\mathbb{G}}J^{-1} = I \otimes \mathbb{G}$. Thus, we have that $\hat{\mathbb{G}} \in \mathcal{L}(L^p(\mathbb{R}; L^p(\mathbb{R}; X)))$ if and only if $\mathbb{G} \in \mathcal{L}(L^p(\mathbb{R}; X))$. □

REMARK 4.26. The results of this section are valid if we replace the space $L^p(\mathbb{R};X)$ by $C_0(\mathbb{R};X)$, see Corollary 7.27. With the aid of Proposition 4.32, we will see below that the invertibility of Γ, or the existence of the Green's function, is equivalent to the existence for each $g \in L^p(\mathbb{R};X)$ of a unique *mild* solution $u \in L^p(\mathbb{R};X)$ for the equation $\dot{u} = A(t)u + g$. Similar results hold if one replaces $L^p(\mathbb{R};X)$ by $C_0(\mathbb{R};X)$. ◇

4.3. Dichotomy and solutions of nonhomogeneous equations

In this section we present generalizations of classical theorems that go back to Perron [**Pn**], A. D. Majzel', Daleckij and Krein [**DK**], and Levitan and Zhikov [**LZ**]. The classical result states that an evolution family, generated by the abstract Cauchy problem for the differential equation $\dot{x} = A(t)x$ where $A : \mathbb{R} \to \mathcal{L}(X)$ is a bounded and continuous function, has an exponential dichotomy if and only if, for each bounded, continuous function g on $\mathbb{R}$, there is a unique bounded, continuous solution u of the nonhomogeneous equation $\dot{u} = A(t)u + g(t)$. We will extend this result to the case of a strongly continuous evolution family. In particular, our extension includes evolution families generated by the abstract Cauchy problem where the operators in the family $\{A(t) : t \in \mathbb{R}\}$ are, generally, unbounded.

As we have mentioned in Remarks 3.20 and 4.26, our previous results are valid if the space $L^p(\mathbb{R};X)$, $1 \le p < \infty$, is replaced by the space $C_0(\mathbb{R};X)$ consisting of the continuous functions that vanish at infinity. In particular, the existence of an exponential dichotomy for an evolution family $\{U(\theta,\tau)\}_{\theta \ge \tau}$ is equivalent to the existence of a unique solution $u \in C_0(\mathbb{R};X)$ for the equation $\Gamma u = -g$ for each $g \in C_0(\mathbb{R};X)$. If the evolution family solves an abstract Cauchy problem for $\dot{x} = A(t)x$, then, see Theorem 3.12, the generator Γ is a closure of the operator $\Gamma' = -d/dt + \mathcal{A}$ with $\mathcal{D}(\Gamma') = \mathcal{D}(-d/dt) \cap \mathcal{D}(\mathcal{A})$. Note that the solutions u of the equation $\Gamma'u = -g$ are the *classical* solutions of the differential equation $\dot{u} = A(t)u + g$, see Definition 3.2. We will see in Proposition 4.32 that the solutions to the equation $\Gamma u = -g$ on $C_0(\mathbb{R};X)$ are exactly the *mild* solutions of the differential equation $\dot{u} = A(t)u + g$. In particular, mild solutions are classical provided that the operator Γ' is already closed. Thus, we have the following result: Dichotomy is equivalent to the existence, for each $g \in C_0(\mathbb{R};X)$, of a unique mild solution $u \in C_0(\mathbb{R};X)$ of the corresponding nonhomogeneous differential equation.

In Subsection 4.3.1 we will replace the space $C_0(\mathbb{R};X)$ in this last result by the space $C_b(\mathbb{R};X)$ of bounded, continuous functions, and in Subsection 4.3.2 we will replace the space $C_0(\mathbb{R};X)$ by certain spaces $\mathcal{F}_\alpha$ consisting of functions with exponential growth. Since the evolution semigroup $\{E^t\}_{t \ge 0}$ is *not* a strongly continuous semigroup on $C_b(\mathbb{R};X)$, characterization of dichotomy in terms of bounded, continuous solutions requires some additional work. Finally, in Subsection 4.3.3 we will significantly improve the Discrete Dichotomy Theorem 4.16 and show (Baskakov's Theorem 4.37) that dichotomy is equivalent to the invertibility of the single discrete operator $I - \pi_0(E)$.

4.3.1. Bounded solutions.
In this subsection we relate the exponential dichotomy for a given strongly continuous, exponentially bounded evolution family to the following Condition (M):

Condition (M). *For every $g \in C_b(\mathbb{R}; X)$, there exists a unique function $u \in C_b(\mathbb{R}; X)$ such that*

$$(4.26) \qquad u(\theta) = U(\theta, \tau)u(\tau) + \int_\tau^\theta U(\theta, s)g(s)\,ds, \qquad \theta \geq \tau.$$

REMARK 4.27. Condition (M) states that, for each $g \in C_b(\mathbb{R}; X)$, there is a unique solution $u \in C_b(\mathbb{R}; X)$ of the mild integral equation (4.26). Thus, if we define $Gg = u$, we obtain an operator on $C_b(\mathbb{R}; X)$. By the Closed Graph Theorem, the operator G is bounded. Indeed, if $g_n \to g$ and $Gg_n := u_n \to u$ in $C_b(\mathbb{R}; X)$, then, for each $\theta \in \mathbb{R}$, we have that

$$u(\theta) = \lim_{n\to\infty} u_n(\theta) = \lim_{n\to\infty} \left(U(\theta, \tau)u_n(\tau) + \int_\tau^\theta U(\theta, s)g_n(s)\,ds \right)$$

$$= U(\theta, \tau)u(\tau) + \int_\tau^\theta U(\theta, s)g(s)\,ds;$$

that is, $u = Gg$. In the next theorem we will prove that if $\{U(\theta, \tau)\}_{\theta \geq \tau}$ has an exponential dichotomy, then G is equal to the Green's operator $\mathbb{G}$. $\diamond$

THEOREM 4.28. *A strongly continuous, exponentially bounded evolution family $\{U(\theta, \tau)\}_{\theta \geq \tau}$ on X has an exponential dichotomy if and only if Condition (M) is satisfied.*

PROOF. **"Only if" part.** If $\{U(\theta, \tau)\}_{\theta \geq \tau}$ has an exponential dichotomy, then, by Theorem 4.25, there is a Green's function G_P associated with the dichotomy projection $P(\cdot)$. Let us use it to define the Green's operator $\mathbb{G}$ as in the equation (4.25); that is,

$$(\mathbb{G}u)(\theta) = \int_{-\infty}^\theta U_P(\theta, \tau)u(\tau)\,d\tau - \int_\theta^\infty U_Q^{-1}(\tau, \theta)u(\tau)\,d\tau.$$

We note that $\mathbb{G}$ is bounded on $C_b(\mathbb{R}; X)$; in fact, $\|\mathbb{G}\|_{\mathcal{L}(C_b(\mathbb{R};X))} \leq 2M/\beta$ where M and β are the positive constants as in the Definition 3.6 of exponential dichotomy. The existence of a solution $u \in C_b(\mathbb{R}; X)$ of the equation (4.26) for each $g \in C_b(\mathbb{R}; X)$ follows from the next proposition.

PROPOSITION 4.29. *If the Green's operator $\mathbb{G}$, defined in display (4.25), is bounded on $C_b(\mathbb{R}; X)$, then, for each $g \in C_b(\mathbb{R}; X)$, there exists a solution $u \in C_b(\mathbb{R}; X)$ of equation (4.26).*

PROOF. If $g \in C_b(\mathbb{R}; X)$ and $u := \mathbb{G}g$, then, for $\theta \geq \tau$, the equation (4.26) holds. Indeed, we have that

$$u(\theta) - U(\theta, \tau)u(\tau) = (\mathbb{G}g)(\theta) - U(\theta, \tau)(\mathbb{G}g)(\tau)$$

$$= \int_{-\infty}^\theta P(\theta)U(\theta, s)P(s)g(s)\,ds - U(\theta, \tau)\int_{-\infty}^\tau P(\tau)U(\tau, s)P(s)g(s)\,ds$$

$$- \int_\theta^\infty U_Q^{-1}(s, \theta)Q(s)g(s)\,ds + U(\theta, \tau)\int_\tau^\infty U_Q^{-1}(s, \tau)Q(s)g(s)\,ds$$

$$= \int_\tau^\theta P(\theta)U(\theta, s)P(s)g(s)\,ds - \int_\theta^\infty U_Q^{-1}(s, \theta)Q(s)g(s)\,ds$$

$$+ \int_\tau^\theta U(\theta, s)U(s, \tau)U_Q^{-1}(s, \tau)Q(s)g(s)$$

$$+ \int_\theta^\infty U(\theta,\tau)[U_Q(s,\theta)U_Q(\theta,\tau)]^{-1}Q(s)g(s)\,ds$$
$$= \int_\tau^\theta P(\theta)U(\theta,s)P(s)g(s)\,ds + \int_\tau^\theta U(\theta,s)Q(s)g(s)\,ds = \int_\tau^\theta U(\theta,s)g(s)\,ds,$$
and Proposition 4.29 is proved. $\square$

To prove the uniqueness of our solution of equation (4.26), let $g = 0$ and suppose that there exists $u \in C_b(\mathbb{R};X)$ such that $u(\theta) = U(\theta,\tau)u(\tau)$, $\theta \geq \tau$. Since $\{U(\theta,\tau)\}_{\theta \geq \tau}$ has an exponential dichotomy, we have
$$P(\theta)u(\theta) = U_P(\theta,\tau)P(\tau)u(\tau) \quad \text{and} \quad Q(\theta)u(\theta) = U_Q(\theta,\tau)Q(\tau)u(\tau), \quad \theta \geq \tau.$$
Since $|u(\cdot)|$ is bounded, estimate (iii) in the Definition 3.6 of an exponential dichotomy implies
$$|P(\theta)u(\theta)| \leq Me^{-\beta(\theta-\tau)}|u(\tau)|.$$
By passing to the limit as $\tau \to -\infty$, it follows that $P(\theta)u(\theta) = 0$ for all $\theta \in \mathbb{R}$. On the other hand, estimate (iv) in Definition 3.6 implies
$$|Q(\tau)u(\tau)| = |[U_Q(\theta,\tau)]^{-1}Q(\theta)u(\theta)| \leq Me^{-\beta(\theta-\tau)}|u(\theta)|.$$
Hence, by passing to the limit as $\theta \to \infty$, we have that $Q(\tau)u(\tau) = 0$ for all $\tau \in \mathbb{R}$. Thus, $u = 0$, as required.

"If" part. To prove that Condition (M) implies the existence of an exponential dichotomy for $\{U(\theta,\tau)\}_{\theta \geq \tau}$, we will use the implication $(3) \Rightarrow (1)$ of the Discrete Dichotomy Theorem 4.16, see also Remark 4.17. To this end, we will consider the family of operators $\pi_\theta(E)$, for $\theta \in \mathbb{R}$, defined on $c_0(\mathbb{Z};X)$ and $\ell^\infty(\mathbb{Z};X)$ by
$$\pi_\theta(E) : (x_n)_{n \in \mathbb{Z}} \mapsto (U(\theta+n,\theta+n-1)x_{n-1})_{n \in \mathbb{Z}}.$$
Condition (M) will be used in Lemmas 4.30 and 4.31 to show that, for each $\theta \in \mathbb{R}$, the inverse $(I - \pi_\theta(E))^{-1}$ exists and the norms of these inverse operators are uniformly bounded on $\ell^\infty(\mathbb{Z};X)$. Next, by Lemma 4.5 and Remark 4.6, this fact implies that each operator $I - \pi_\theta(E)$, for $\theta \in \mathbb{R}$, is invertible on $c_0(\mathbb{Z};X)$ and there is a constant $B > 0$ such that
$$(4.27) \qquad \|(I - \pi_\theta(E))^{-1}\|_{\mathcal{L}(c_0(\mathbb{Z};X))} \leq B \quad \text{for all} \quad \theta \in \mathbb{R}.$$
Finally, by the Discrete Dichotomy Theorem 4.16 and Remark 4.17, condition (4.27) implies the evolution family $\{U(\theta,\tau)\}_{\theta \geq \tau}$ has an exponential dichotomy. $\square$

LEMMA 4.30. *If Condition (M) holds, then each operator $I - \pi_\theta(E)$, for $\theta \in \mathbb{R}$, is uniformly injective on $\ell^\infty(\mathbb{Z};X)$ or $c_0(\mathbb{Z};X)$. Moreover, there is a positive uniform lower bound over $\theta \in \mathbb{R}$.*

PROOF. We will prove the result for $\ell^\infty(\mathbb{Z};X)$; the result for $c_0(\mathbb{Z};X)$ is an immediate corollary. Suppose that
$$(4.28) \qquad \inf_{\theta \in \mathbb{R}} \|I - \pi_\theta(E)\|_{\bullet,\ell^\infty(\mathbb{Z};X)} = 0.$$
Then, for each $\epsilon > 0$, there is some $\theta \in \mathbb{R}$ and $\bar{x} \in \ell^\infty(\mathbb{Z};X)$ such that
$$\|\bar{x}\|_{\ell^\infty(\mathbb{Z};X)} = 1 \quad \text{and} \quad \|[I - \pi_\theta(E)]\bar{x}\|_{\ell^\infty(\mathbb{Z};X)} < \epsilon.$$
Fix $n \in \mathbb{N}$ and note that there is some $\theta = \theta(n) \in \mathbb{R}$ and $\bar{x} = \bar{x}(n) \in \ell^\infty(\mathbb{Z};X)$ such that
$$(4.29) \qquad 1/2 < \|[\pi_\theta(E)]^k \bar{x}\|_{\ell^\infty(\mathbb{Z};X)} \leq 2 \quad \text{for} \quad k = 0, 1, \ldots, 2n.$$

Also, note that if $\bar{x} = (x_m)_{m \in \mathbb{Z}}$, then
$$[\pi_\theta(E)]^k \bar{x} = (U(\theta + m, \theta + m - k)x_{m-k})_{m \in \mathbb{Z}}.$$
Thus, we can choose $l \in \mathbb{Z}$, $l = l(n)$, such that condition (4.29) with $k = n$ implies
$$1/2 \leq |U(\theta + l, \theta + l - n)x_{l-n}|.$$
Consider the generator Γ of the evolution semigroup $\{E^t\}_{t \geq 0}$ on the space $C_0(\mathbb{R}; X)$. We will use Condition (M) to derive a contradiction to the assumption (4.28) by constructing a sequence of functions $\{u_n\}_{n=1}^\infty \subset \mathcal{D}(\Gamma) \subset C_0(\mathbb{R}; X)$ such that $u_n = -G\Gamma u_n$ and the sequence $\{\Gamma u_n\}_{n=1}^\infty$ converges to 0, but the sequence $\{u_n\}_{n=1}^\infty$ does *not* converge to 0. Here, the operator G is, by Condition (M), the *bounded* operator that recovers u for a given g in (4.26), see Remark 4.27.

To construct u_n, let us first define $\theta_n := \theta + l$ and $y_n := x_{l-n}$. Choose a smooth bump-function $\alpha_n : \mathbb{R} \to \mathbb{R}$ with the following properties:
(i) $\operatorname{supp}(\alpha_n) \subseteq (\theta_n - n, \theta_n + n)$;
(ii) $0 \leq \alpha_n \leq 1$ with $\alpha_n(\theta_n) = 1$;
(iii) $\|\alpha_n'\|_\infty \leq 2/n$.

Also, define the function $u_n \in C_0(\mathbb{R}; X)$ as follows:
$$u_n(\tau) = \begin{cases} \alpha_n(\tau) U(\tau, \theta_n - n) y_n & \text{for } \tau \geq \theta_n - n, \\ 0 & \text{for } \tau < \theta_n - n. \end{cases}$$

Let us note that
$$(4.30) \qquad \|u_n\|_\infty \geq |u_n(\theta_n)| = |U(\theta_n, \theta_n - n)y_n| \geq 1/2.$$

Since $u_n \in C_0(\mathbb{R}; X)$, we can use the strongly continuous evolution semigroup $\{E^t\}_{t \geq 0}$ and its generator Γ on $C_0(\mathbb{R}; X)$. Let us note that
$$(E^t u_n)(\tau) = U(\tau, \tau - t) \alpha_n(\tau - t) U(\tau - t, \theta_n - n) y_n = \alpha_n(\tau - t) U(\tau, \theta_n - n) y_n$$
for $\tau - t \geq \theta_n - n$. Hence, $u_n \in \mathcal{D}(\Gamma)$ and, by differentiation, we have that $(\Gamma u_n)(\tau) = -\alpha_n'(\tau) U(\tau, \theta_n - n) y_n$ for $\tau \geq \theta_n - n$ and $(\Gamma f_n)(\tau) = 0$ for $\tau < \theta_n - n$. Using this result, the condition (4.29), property (iii) and the fact that $\{U(\theta, \tau)\}_{\theta \geq \tau}$ is exponentially bounded, there is constant $c > 0$ such that
$$(4.31) \qquad \|\Gamma u_n\|_\infty \leq 2cn^{-1} \sup_{0 \leq k \leq 2n} |U(k + \theta_n - n, \theta_n - n)y_n| \leq 4cn^{-1}.$$

Next, we show that Condition (M) implies $u_n = -G\Gamma u_n$. Indeed, for $\theta \geq \tau \geq \theta_n - n$, we have the identities
$$u_n(\theta) = \alpha_n(\theta) U(\theta, \theta_n - n) y_n$$
$$= \alpha_n(\tau) U(\theta, \theta_n - n) y_n + \alpha_n(\theta) U(\theta, \theta_n - n) y_n - \alpha_n(\tau) U(\theta, \theta_n - n) y_n$$
$$= U(\theta, \tau)[\alpha_n(\tau) U(\tau, \theta_n - n) y_n] + U(\theta, \theta_n - n) y_n \int_\tau^\theta \alpha'(s)\, ds$$
$$= U(\theta, \tau)[\alpha_n(\tau) U(\tau, \theta_n - n) y_n] + \int_\tau^\theta U(\theta, s)[\alpha'(s) U(s, \theta_n - n) y_n]\, ds$$
$$= U(\theta, \tau) u_n(\tau) - \int_\tau^\theta U(\theta, s)(\Gamma u_n)(s)\, ds,$$
and similar identities for the other cases.

Since G is a bounded operator by Condition (M), see Remark 4.27, we conclude from the inequality (4.31) that the sequence with elements $u_n = -G(\Gamma u_n)$ converges to zero in $C_0(\mathbb{R}; X)$, in contradiction to inequality (4.30). $\square$

LEMMA 4.31. *If Condition (M) holds, then, for each $\theta \in \mathbb{R}$, the operator $I - \pi_\theta(E)$ from $\ell^\infty(\mathbb{Z}; X)$ to $\ell^\infty(\mathbb{Z}; X)$ is surjective.*

PROOF. Let $\bar{x} = (x_n)_{n \in \mathbb{Z}} \in \ell^\infty(\mathbb{Z}; X)$ and fix $\theta \in \mathbb{R}$. It suffices to find $\bar{y} = (y_n)_{n \in \mathbb{Z}} \in \ell^\infty(\mathbb{Z}; X)$ such that

(4.32) $\qquad y_n - U(\theta + n, \theta + n - 1)y_{n-1} = U(\theta + n, \theta + n - 1)x_{n-1}, \quad n \in \mathbb{Z};$

that is, $[I - \pi_\theta(E)]\bar{y} = \pi_\theta(E)\bar{x}$. If this is the case, then

$$[I - \pi_\theta(E)](\bar{y} + \bar{x}) = \pi_\theta(E)\bar{x} + \bar{x} - \pi_\theta(E)\bar{x} = \bar{x}.$$

To obtain $\bar{y}$ such that (4.32) holds, we claim that there exists a function $g \in C_b(\mathbb{R}; X)$ such that

(4.33) $\qquad U(\theta + n, \theta + n - 1)x_{n-1} = \int_{\theta+n-1}^{\theta+n} U(\theta + n, s)g(s)\, ds, \quad n \in \mathbb{Z}.$

Since Condition (M) holds, it then follows that there exists a unique $u \in C_b(\mathbb{R}; X)$ such that

$$u(\theta + n) = U(\theta + n, \theta + n - 1)u(\theta + n - 1) + \int_{\theta+n-1}^{\theta+n} U(\theta + n, s)g(s)\, ds, \quad n \in \mathbb{Z}.$$

By setting $y_n = u(\theta + n)$, we have that $\bar{y} \in \ell^\infty(\mathbb{Z}; X)$ satisfies (4.32), as required.

To construct g as in (4.33), choose $\alpha \in C([0, 1])$ such that $\alpha(0) = 0$, $\alpha(1) = 1$, and $\int_0^1 \alpha(s)\, ds = 1$; and, for $\tau \in [\theta + n - 1, \theta + n)$, define

$$g(\tau) = \alpha(\tau - \theta - n + 1)U(\tau, \theta + n - 1)x_{n-1}$$
$$+ [1 - \alpha(\tau - \theta - n + 1)]U(\tau, \theta + n - 2)x_{n-2}.$$

It follows that

$$\int_{\theta+n-1}^{\theta+n} U(\theta + n, s)g(s)\, ds = \int_0^1 \alpha(s)\, ds \cdot U(\theta + n, \theta + n - 1)x_{n-1}$$
$$+ [1 - \int_0^1 \alpha(s)\, ds] \cdot U(\theta + n, \theta + n - 2)x_{n-2}$$

and (4.33) is satisfied. $\square$

4.3.2. Exponentially bounded solutions. In this section we will show that Condition (M) in Theorem 4.28 can be replaced by any one of three conditions formulated below. To state these conditions, we will use the scale of function spaces $\mathcal{F}_\alpha$, for $\alpha > 0$, defined by

$$\mathcal{F}_\alpha := \{f \in C(\mathbb{R}; X): \quad e^{-\alpha|\cdot|}f(\cdot) \in C_b(\mathbb{R}; X)\}.$$

In other words, $\mathcal{F}_\alpha$ is the space of continuous, exponentially bounded functions with exponent α. These spaces are important in the theory of center manifolds, see Remark 4.34 below.

Condition ($\mathbf{M}_{C_0}$). For every $g \in C_0(\mathbb{R}; X)$, the integral equation (4.26) has a unique solution $u \in C_0(\mathbb{R}; X)$.

Condition ($\mathbf{M}_{L^p}$). For every $g \in L^p(\mathbb{R}; X)$, $1 \leq p < \infty$, the integral equation (4.26) has a unique solution $u \in L^p(\mathbb{R}; X)$.

Condition ($M_{\mathcal{F}_\alpha}$). *For every $g \in \mathcal{F}_\alpha$, the integral equation (4.26) has a unique solution $u \in \mathcal{F}_\alpha$.*

We start with a variant of Proposition 3.25 for the line case.

PROPOSITION 4.32. *Let $\mathcal{F}$ denote one of the spaces $C_0(\mathbb{R}; X)$ or $L^p(\mathbb{R}; X)$, $1 \leq p < \infty$, and Γ the generator of the evolution semigroup $\{E^t\}_{t \geq 0}$ on $\mathcal{F}$. Also, assume that $u, g \in \mathcal{F}$. The following statements are equivalent:*
   (i) *$u \in \mathcal{D}(\Gamma)$ and $\Gamma u = -g$.*
   (ii) *u is a solution of equation (4.26) that corresponds to g.*

PROOF. Assume that (i) holds. By an elementary property of strongly continuous semigroups, see Pazy [**Pz**, p. 4],

$$(4.34) \qquad E^t u - u = \int_0^t E^s \Gamma u \, ds = -\int_0^t E^s g \, ds, \quad t \geq 0.$$

Also, $(E^t u)(\theta) = U(\theta, \theta - t)u(\theta - t)$ by the definition of the evolution semigroup. Statement (ii) is verified by substituting this formula for E^t into (4.34).

Assume that statement (ii) holds. For $t \geq 0$, $\theta - t \geq \tau$, and u, a solution of equation (4.26), we have the identity

$$(E^t u)(\theta) = U(\theta, \theta - t)\Big[U(\theta - t, \tau)u(\tau) + \int_\tau^{\theta - t} U(\theta - t, s)g(s)\, ds\Big]$$
$$= U(\theta, \tau)u(\tau) + \int_\tau^{\theta - t} U(\theta, s)g(s)\, ds;$$

and therefore,

$$t^{-1}\Big[(E^t u)(\theta) - u(\theta)\Big] = t^{-1}\Big[U(\theta, \tau)u(\tau) + \int_\tau^{\theta - t} U(\theta, s)g(s)\, ds$$
$$- \Big(U(\theta, \tau)u(\tau) + \int_\tau^\theta U(\theta, s)g(s)\, ds\Big)\Big]$$
$$= -t^{-1}\int_{\theta - t}^\theta U(\theta, s)f(s)\, ds = -t^{-1}\int_0^t U(\theta, \theta - s)g(\theta - s)\, ds.$$

By Theorem 4.2 in [**Ne**], it follows that

$$t^{-1}(E^t u - u) = -t^{-1} \int_0^t E^s g\, ds.$$

Hence, $u \in \mathcal{D}(\Gamma)$ and $\Gamma u = -g$. □

THEOREM 4.33. *Let $\{U(\theta, \tau)\}_{\theta \geq \tau}$ be an exponentially bounded evolution family on X.*
(a) *The following statements are equivalent:*
   (i) *$\{U(\theta, \tau)\}_{\theta \geq \tau}$ has an exponential dichotomy.*
   (ii) *Condition (M) holds.*
   (iii) *Condition (M_{C_0}) holds.*
   (iv) *Condition (M_{L^p}) holds.*
(b) *The operator G defined by Conditions (M), (M_{C_0}), or (M_{L^p}) as in Remark 4.27, is equal to the Green's operator $\mathbb{G}$ as defined in display (4.25). Moreover, on the space $C_0(\mathbb{R}; X)$, or $L^p(\mathbb{R}; X)$ with $1 \leq p < \infty$, we have that $G = -\Gamma^{-1}$ where Γ denotes the generator of the evolution semigroup $\{E^t\}_{t \geq 0}$ on $C_0(\mathbb{R}; X)$ or $L^p(\mathbb{R}; X)$.*

4.3. DICHOTOMY AND SOLUTIONS OF NONHOMOGENEOUS EQUATIONS

PROOF. If $\{U(\theta,\tau)\}_{\theta \geq \tau}$ has an exponential dichotomy, then the Green's operator $\mathbb{G}$ is defined on $L^p(\mathbb{R};X)$ and $C_0(\mathbb{R};X)$, and, by Theorem 4.25, we have that $\mathbb{G} = -\Gamma^{-1}$. Also, by the Räbiger-Schnaubelt Lemma 3.16, $\mathbb{G}$ maps $L^p(\mathbb{R};X)$ into $C_0(\mathbb{R};X)$. Using the same argument as in the proof of the "only if" part of Theorem 4.28, it is easy to see that (M_{C_0}) and (M_{L^p}) hold, and $G = \mathbb{G}$.

By Theorem 4.28, we conclude that Condition (M) is equivalent to (i).

To see that Condition (M_{C_0}), resp. (M_{L^p}), yields the exponential dichotomy for $\{U(\theta,\tau)\}_{\theta \geq \tau}$, define G using Condition (M_{C_0}), resp. (M_{L^p}), as in Remark 4.27. By the Dichotomy Theorem 3.17 and Remark 3.20, it suffices to show that Condition (M_{L^p}), resp. (M_{C^0}), implies the invertibility of Γ on $L^p(\mathbb{R};X)$, resp. $C_0(\mathbb{R};X)$.

Let $g \in L^p(\mathbb{R};X)$ or $C_0(\mathbb{R};X)$. Then, for $u = Gg$, Proposition 4.32 implies $u \in \mathcal{D}(\Gamma)$ and $\Gamma G g = -g$. Thus, Γ is right invertible. To prove, that Γ is left invertible, let us suppose that $u \in \mathcal{D}(\Gamma)$ and $\Gamma u = 0$. Then, for $t > 0$ and $n \in \mathbb{N}$ large enough, we have that

$$\left(\frac{n}{t} - \Gamma\right) u = \frac{n}{t} u, \quad \text{or} \quad u = \frac{n}{t}\left(\frac{n}{t} - \Gamma\right)^{-1} u,$$

and, as a result,

$$u = \lim_{n \to \infty} \left[\frac{n}{t}\left(\frac{n}{t} - \Gamma\right)^{-1}\right]^n u = E^t u, \quad t > 0.$$

Thus, $u(\theta) = U(\theta,\tau)u(\tau)$ for $\theta \geq \tau$; and therefore, equation (4.26) holds for $g = 0$. It follows that $u = 0$ and Γ is invertible. □

Our next goal is to prove that the existence of a dichotomy is equivalent to Condition $(M_{\mathcal{F}_\alpha})$.

REMARK 4.34. Before proceeding, we pause to motivate our use of the scale of Banach spaces $\mathcal{F}_\alpha$. These spaces arise in center manifold theory (see, e.g., [**DK, VI**]). For example, in [**VI**] a semilinear differential equation of the form $\dot{y} = By + F(y)$ on a Banach space Y is considered where the existence of a B-invariant decomposition $Y = Z \oplus X$ with restrictions $A = B|X$ and $C = B|Z$ is assumed. Here, Z represents the "central" part in the sense that $\sigma(C) \subset i\mathbb{R}$, and X represents the "hyperbolic" part. The hyperbolicity condition, Condition (H) in [**VI**], is given in terms of the nonhomogeneous equation $\dot{x} = Ax + g$. This condition can be reformulated as follows: For every $g \in \mathcal{F}_\alpha$, there exists a unique solution $x \in \mathcal{F}_\alpha$ of the equation $\dot{x} = Ax + g$. This solution is given by $x = Gg$ for some operator $G \in \mathcal{L}(\mathcal{F}_\alpha)$ with the property that $\|G\|_{\mathcal{L}(\mathcal{F}_\alpha)} \leq \rho(\alpha)$ for some continuous function $\rho: [0,\beta) \to \mathbb{R}_+$.

We will see that Condition (H) is equivalent to the fact that the evolution family $\{e^{(\theta-\tau)A}\}_{\theta \geq \tau}$ has an exponential dichotomy. This result is contained in the next theorem. The theorem is valid for every evolution family (nonautonomous equation) and it requires only the existence of mild solutions. ◇

THEOREM 4.35. *Let $\{U(\theta,\tau)\}_{\theta \geq \tau}$ be an exponentially bounded evolution family on X.*
(a) The following statements are equivalent:
   (i) *$\{U(\theta,\tau)\}_{\theta \geq \tau}$ has an exponential dichotomy.*
   (ii) *There is some $\beta' > 0$ such that if $\alpha \in [0,\beta')$ then condition $(M_{\mathcal{F}_\alpha})$ holds for $\{U(\theta,\tau)\}_{\theta \geq \tau}$.*

(b) *If either (i) or (ii) holds, then the dichotomy bound $\beta = \beta(\mathcal{U})$ for the evolution family $\{U(\theta,\tau)\}_{\theta \geq \tau}$ can be estimated as follows:*

(4.35)
$$\beta \leq \sup\{\beta' > 0 \colon \text{ for each } \alpha \in [0, \beta') \text{ condition } (M_{\mathcal{F}_\alpha}) \text{ holds for } \{U(\theta,\tau)\}_{\theta \geq \tau}\}.$$

In addition, for each $\alpha > 0$ and each $g \in \mathcal{F}_\alpha$, the solution of the integral equation (4.26) is given by $u = Gg$ where $G \in \mathcal{L}(\mathcal{F}_\alpha)$ is equal to the Green's operator $\mathbb{G}$ on $\mathcal{F}_\alpha$ as defined in (4.25). Also, there is a continuous function $\rho : [0, \beta) \to \mathbb{R}_+$ such that $\|G\|_{\mathcal{L}(\mathcal{F}_\alpha)} \leq \rho(\alpha)$.

PROOF. If $\alpha = 0$, then condition $(M_{\mathcal{F}_\alpha})$ is the same as condition (M). Thus, statement (ii) implies statement (i) by Theorem 4.28.

To prove that statement (i) implies statement (ii), let us suppose that the evolution family $\{U(\theta,\tau)\}_{\theta \geq \tau}$ has an exponential dichotomy and let β denote the dichotomy bound. If $\beta' \in (0, \beta)$ then $\{U(\theta,\tau)\}_{\theta \geq \tau}$ has an exponential dichotomy with constants β' and $M = M(\beta')$. If $\alpha \in [0, \beta')$, then the evolution family $\{U_\alpha(\theta,\tau)\}_{\theta \geq \tau}$ defined by

$$U_\alpha(\theta,\tau) = e^{-\alpha(|\theta|-|\tau|)}U(\theta,\tau), \quad \theta \geq \tau,$$

has an exponential dichotomy with constants $M(\beta')$ and $\beta' - \alpha$. By Theorem 4.28, condition (M) holds for $\{U_\alpha(\theta,\tau)\}_{\theta \geq \tau}$.

We will prove that condition $(M_{\mathcal{F}_\alpha})$ holds for $\{U(\theta,\tau)\}_{\theta \geq \tau}$. In fact, we have the following result:

CLAIM. *The condition $(M_{\mathcal{F}_\alpha})$ holds for $\{U(\theta,\tau)\}_{\theta \geq \tau}$ if and only if (M) holds for $\{U_\alpha(\theta,\tau)\}_{\theta \geq \tau}$.*

Proof of Claim. Note that if condition (M) holds for $\{U_\alpha(\theta,\tau)\}_{\theta \geq \tau}$, then there is a bounded operator G_α on $C_b(\mathbb{R}; X)$ defined by the rule $G_\alpha g = u$. Let us also define the operator $J_\alpha : \mathcal{F}_\alpha \to C_b(\mathbb{R}; X)$ by $(J_\alpha f)(\theta) = e^{-\alpha|\theta|}f(\theta)$, $\theta \in \mathbb{R}$. Similarly, if $\{U(\theta,\tau)\}_{\theta \geq \tau}$ satisfies condition $(M_{\mathcal{F}_\alpha})$, then there is a a bounded operator $G \in \mathcal{L}(\mathcal{F}_\alpha)$ defined by the rule $Gg = u$. We have that $G_\alpha = J_\alpha G J_\alpha^{-1}$, and therefore condition (M) holds for $\{U_\alpha(\theta,\tau)\}_{\theta \geq \tau}$ if and only if $G_\alpha \in \mathcal{L}(C_b(\mathbb{R}; X))$ if and only if $G \in \mathcal{L}(\mathcal{F}_\alpha)$ if and only if $(M_{\mathcal{F}_\alpha})$ holds for $\{U(\theta,\tau)\}_{\theta \geq \tau}$. This proves the claim.

To finish the proof of Theorem 4.35, let us note first that since $\beta = \sup \beta'$, we have the inequality (4.35). Next, let $0 \leq \alpha < \beta' < \beta$ and, by condition $(M_{\mathcal{F}_\alpha})$ for the evolution family $\{U(\theta,\tau)\}_{\theta \geq \tau}$, let $G \in \mathcal{L}(\mathcal{F}_\alpha)$ be the operator defined by $Gu = g$. Since $G_\alpha = \mathbb{G}_\alpha$, where $\mathbb{G}_\alpha$ is the Green's operator for the dichotomic evolution family $\{U_\alpha(\theta,\tau)\}_{\theta \geq \tau}$, and $G_\alpha = J_\alpha^{-1} G J_\alpha$ as in the proof of the claim, we have the following estimate:

$$\|G\|_{\mathcal{L}(\mathcal{F}_\alpha)} = \|G_\alpha\|_{\mathcal{L}(C_b(\mathbb{R};X))} = \|\mathbb{G}_\alpha\|_{\mathcal{L}(C_b(\mathbb{R};X))} \leq \frac{2M(\beta')}{\beta' - \alpha}.$$

$\square$

EXAMPLE 4.36. It is possible that the condition $(M_{\mathcal{F}_\alpha})$ holds for all $\alpha \in (0, \beta)$, but the evolution family $\{U(\theta,\tau)\}_{\theta \geq \tau}$ has no exponential dichotomy. Moreover, the inequality in (4.35) may be strict. Indeed, for $X = \mathbb{C}$ and $0 \leq \epsilon \leq 1$, we have the strongly continuous evolution family defined by

$$U^{(\epsilon)}(\theta,\tau) = \begin{cases} e^{\epsilon(\theta-\tau)}, & 0 \geq \theta \geq \tau, \\ e^{\theta}e^{-\epsilon\tau}, & \theta \geq 0 \geq \tau, \\ e^{\theta-\tau}, & \theta \geq \tau > 0; \end{cases}$$

and, for $\alpha \geq 0$, we have the family

$$U_\alpha^{(\epsilon)}(\theta,\tau) = e^{-\alpha|\theta|}e^{\alpha|\tau|}U^{(\epsilon)}(\theta,\tau) = \begin{cases} e^{(\alpha+\epsilon)(\theta-\tau)}, & 0 > \theta \geq \tau, \\ e^{(1-\alpha)\theta}e^{-(\epsilon+\alpha)\tau}, & \theta \geq 0 \geq \tau, \\ e^{(1-\alpha)(\theta-\tau)}, & \theta \geq \tau > 0. \end{cases}$$

Consequently, $\{U^{(0)}(\theta,\tau)\}_{\theta\geq\tau}$ has no dichotomy, but the condition $(M_{\mathcal{F}_\alpha})$ holds for $\{U^{(0)}(\theta,\tau)\}_{\theta\geq\tau}$ and each $\alpha \in (0,1)$. Moreover, for $\epsilon > 0$ the dichotomy bound for $\{U^{(\epsilon)}(\theta,\tau)\}_{\theta\geq\tau}$ is equal to ϵ, but the condition $(M_{\mathcal{F}_\alpha})$ again holds for $\{U^{(\epsilon)}(\theta,\tau)\}_{\theta\geq\tau}$ and each $\alpha \in (0,1)$. ◇

4.3.3. Baskakov's Theorem. In this subsection we return to the theme of discrete operators, $\pi_\theta(E)$, see the Discrete Dichotomy Theorem 4.16 and the algebraic discussion in Section 4.1. We will present a theorem due to Baskakov [**Bs3**] to show that condition (3) in the Discrete Dichotomy Theorem 4.16, which requires the uniform boundedness of $\|(I - \pi_\theta(E))^{-1}\|$ for *all* $\theta \in \mathbb{R}$, is equivalent to the invertibility of the *single* operator $I - \pi_0(E)$. Moreover, this theorem gives a formula that directly relates the inverses for the operators Γ and $I - \pi_0(E)$.

Recall that the discrete operator $I - \pi_0(E)$ acts on one of the spaces $\mathcal{F}_s(\mathbb{Z};X) = \ell^p(\mathbb{Z};X)$, $1 \leq p \leq \infty$ or $\mathcal{F}_s(\mathbb{Z};X) = c_0(\mathbb{Z};X)$ by the rule

(4.36) $\qquad I - \pi_0(E) : (x_n)_{n\in\mathbb{Z}} \mapsto (x_n - U(n, n-1)x_{n-1})_{n\in\mathbb{Z}}.$

Here, as usual, $\{U(\theta,\tau)\}_{\theta\geq\tau}$, is a strongly continuous, exponentially bounded evolution family.

Let $\mathcal{F}$ be one of the spaces $L^p(\mathbb{R};X)$, $1 \leq p < \infty$, $C_b(\mathbb{R};X)$, or $C_0(\mathbb{R};X)$. We will define an operator Γ on $\mathcal{F}$ that is associated with the mild integral equation (4.26) as follows: A function u is in the domain of Γ on $\mathcal{F}$ if there exists a function $g \in \mathcal{F}$ such that u is a solution of the corresponding mild integral equation (4.26). In this case, we define $\Gamma u = -g$.

Note that the notation Γ for this operator is consistent: Indeed, by Proposition 4.32, if $\mathcal{F} = L^p(\mathbb{R};X)$, $1 \leq p < \infty$, or $\mathcal{F} = C_0(\mathbb{R};X)$, then the operator Γ just defined is the generator of the evolution semigroup $\{E^t\}_{t\geq 0}$ on $\mathcal{F}$. Also, the operator Γ is invertible on $\mathcal{F} = L^p(\mathbb{R};X)$, $\mathcal{F} = C_b(\mathbb{R};X)$, or $\mathcal{F} = C_0(\mathbb{R};X)$ if and only if Condition (M_{L^p}), (M), or (M_{C_0}), respectively, is satisfied. If this is the case, then $\Gamma^{-1} = -G$ where G is the operator defined as in Remark 4.27 for the corresponding condition. We also stress the fact that, by Theorem 4.33, the operator Γ is invertible on each space $\mathcal{F}$ if and only if the evolution family $\{U(\theta,\tau)\}_{\theta\geq\tau}$ has an exponential dichotomy.

For $\mathcal{F} = L^p(\mathbb{R};X)$, $1 \leq p < \infty$, $\mathcal{F} = C_b(\mathbb{R};X)$, or $\mathcal{F} = C_0(\mathbb{E};X)$, let us define, respectively, $\mathcal{F}_s = \ell^p(\mathbb{Z};X)$, $1 \leq p < \infty$, $\mathcal{F}_s = \ell^\infty(\mathbb{Z};X)$, or $\mathcal{F}_s = c_0(\mathbb{Z};X)$. Fix a smooth function $\alpha : \mathbb{R} \to \mathbb{R}$ with period 1 such that $\alpha(0) = \alpha(1) = 0$ and $\int_0^1 \alpha(s)\,ds = 1$. Also, define the bounded operator $L : \mathcal{F}_s \to \mathcal{F}$ by the formula

$(L\bar{x})(\theta) = \alpha(\theta)U(\theta, n-1)x_{n-1} \quad \text{for} \quad \theta \in [n-1, n], \quad \text{and} \quad \bar{x} = (x_n), \quad n \in \mathbb{Z}.$

THEOREM 4.37 (Baskakov's Theorem). *The following statements are equivalent:*

(a) *The operator Γ defined above by means of the mild integral equation (4.26) is invertible on $\mathcal{F}$.*
(b) *The discrete operator $I - \pi_0(E)$ defined in (4.36) is invertible on $\mathcal{F}_s$.*

Moreover, if (a) or (b) holds and $\bar{x} = (x_n)_{n \in \mathbb{Z}} \in \mathcal{F}_s$, then

$$(4.37) \quad [I - \pi_0(E)]^{-1} \bar{x} = \bar{x} + \bar{z} \quad \text{where} \quad \bar{z} = (z_n), \quad z_n := (\Gamma^{-1} L \bar{x})(n), \quad n \in \mathbb{Z}.$$

PROOF. (a) $\Rightarrow$ (b). If Γ is injective on $\mathcal{F}$, then $I - \pi_0(E)$ is injective on $\mathcal{F}_s$. Indeed, assume that $\bar{x} \in \text{Ker } [I - \pi_0(E)]$; that is, $x_n = U(n, n-1) x_{n-1}$, and define the function $u(\theta) = U(\theta, n) x_n$ for $\theta \in [n, n+1]$, $n \in \mathbb{Z}$. Then, $u(\theta) = U(\theta, \tau) u(\tau)$ for all $\theta \geq \tau$ and, by Proposition 4.32, we have $\Gamma u = 0$, in contradiction.

If Γ is surjective on $\mathcal{F}$, then $I - \pi_0(E)$ is surjective on $\mathcal{F}_s$. Indeed, fix any $\bar{x} = (x_n)_{n \in \mathbb{Z}}$ and consider $L\bar{x} \in \mathcal{F}$. Since Γ is surjective on $\mathcal{F}$, there is some $u \in \mathcal{D}(\Gamma)$ such that $\Gamma u = -L\bar{x}$. Note that if $\mathcal{F} = L^p(\mathbb{R}; X)$, $1 \leq p < \infty$, then $u = -\Gamma^{-1} L\bar{x} \in C_0(\mathbb{R}; X)$ by the Räbiger-Schnaubelt Lemma 3.16. For each space $\mathcal{F}$, by Proposition 4.32 and the definition of L, we have

$$u(n) = U(n, n-1) u(n-1) + \int_{n-1}^{n} U(n, s)(L\bar{x})(s)\, ds$$

$$= U(n, n-1) u(n-1) + \int_{n-1}^{n} U(n, s) \alpha(s) U(s, n-1) x_{n-1}\, ds$$

$$= U(n, n-1) u(n-1) + U(n, n-1) x_{n-1} \int_{n-1}^{n} \alpha(s)\, ds.$$

Therefore, if $\bar{z} = (z_n)_{n \in \mathbb{Z}}$ with $z_n = u(n) = (\Gamma^{-1} L \bar{x})(n)$, then $[I - \pi_0(E)] \bar{z} = \pi_0(E) \bar{x}$ and $[I - \pi_0(E)](\bar{z} + \bar{x}) = \bar{x}$. Thus, $I - \pi_0(E)$ is surjective on $\mathcal{F}_s$ and formula (4.37) holds.

(b) $\Rightarrow$ (a). Assume that $I - \pi_0(E)$ is invertible on $\mathcal{F}_s$. Since the invertibility of this operator does not depend on the choice of the space ℓ^p or c_0, see Remark 4.6, we may assume that $I - \pi_0(E)$ is invertible on $c_0(\mathbb{Z}; X)$. We will show that Γ is invertible on $C_0(\mathbb{R}; X)$. Then, by the Dichotomy Theorem 3.17, the evolution family $\{U(\theta, \tau)\}_{\theta \geq \tau}$ has an exponential dichotomy and, as a result, Γ is invertible on $L^p(\mathbb{R}; X)$ or $C_b(\mathbb{R}; X)$ by Theorem 4.33.

First, if $I - \pi_0(E)$ is injective on $c_0(\mathbb{Z}; X)$, then Γ is injective on $C_0(\mathbb{R}; X)$. Indeed, suppose that $\Gamma u = 0$; that is, $u(\theta) = U(\theta, \tau) u(\tau)$ for $\theta \geq \tau$. Then, the sequence $(u(n))_{n \in \mathbb{Z}}$ is in $c_0(\mathbb{Z}; X)$ and belongs to Ker $(I - \pi_0(E))$. Therefore, for each $n \in \mathbb{Z}$, we have $u(n) = 0$, and $u(\theta) = U(\theta, n) u(n) = 0$ for $\theta \in [n, n+1]$.

Next, if $I - \pi_0(E)$ is surjective on $c_0(\mathbb{Z}; X)$, then Γ is surjective on $C_0(\mathbb{R}; X)$. Indeed, fix $g \in C_0(\mathbb{R}; X)$ and let

$$y_n = \int_{n-1}^{n} U(n, s) g(s)\, ds \quad \text{for} \quad n \in \mathbb{Z}.$$

4.3. DICHOTOMY AND SOLUTIONS OF NONHOMOGENEOUS EQUATIONS

Since $\{U(\theta,\tau)\}_{\theta\geq\tau}$ is exponentially bounded, we have $(y_n)_{n\in\mathbb{Z}} \in c_0(\mathbb{Z};X)$. Hence, there exists $(x_n)_{n\in\mathbb{Z}} \in c_0(\mathbb{Z};X)$ such that $(I - \pi_0(E))(x_n) = (y_n)$; that is,

$$(4.38) \qquad x_n = U(n,n-1)x_{n-1} + \int_{n-1}^{n} U(n,s)g(s)\,ds, \quad n \in \mathbb{Z}.$$

Define a function $u \in C_0(\mathbb{R};X)$ by

$$u(\theta) = U(\theta,n)x_n + \int_{n}^{\theta} U(\theta,s)g(s)\,ds \quad \text{for} \quad \theta \in [n,n+1].$$

We claim that $\Gamma u = -g$; that is, equation (4.26) holds for this u.

Indeed, for $n+1 \geq \theta \geq \tau \geq n$, we have the identity

$$U(\theta,\tau)u(\tau) + \int_{\tau}^{\theta} U(\theta,s)g(s)\,ds = U(\theta,\tau)\Big[U(\tau,n)x_n + \int_{n}^{\tau} U(\tau,s)g(s)\,ds\Big]$$

$$+ \int_{\tau}^{\theta} U(\theta,s)g(s)\,ds = U(\theta,n)x_n + \int_{n}^{\theta} U(\theta,s)g(s)\,ds = u(\theta).$$

Also, for $m+1 \geq \theta \geq m \geq n \geq \tau \geq n-1$, $m,n \in \mathbb{Z}$, we have

$$U(\theta,\tau)u(\tau) + \int_{\tau}^{\theta} U(\theta,s)g(s)\,ds = U(\theta,\tau)\Big[U(\tau,n-1)x_{n-1} + \int_{n-1}^{\tau} U(\tau,s)g(s)\,ds\Big]$$

$$+ \int_{\tau}^{\theta} U(\theta,s)g(s)\,ds = U(\theta,n-1)x_{n-1} + \int_{n-1}^{\theta} U(\theta,s)g(s)\,ds.$$

On the other hand, using (4.38), we have the identity

$$u(\theta) = U(\theta,m)x_m + \int_{m}^{\theta} U(\theta,s)g(s)\,ds$$

$$= U(\theta,m)\Big[U(m,m-1)x_{m-1} + \int_{m-1}^{m} U(m,s)g(s)\,ds\Big] + \int_{m}^{\theta} U(\theta,s)g(s)\,ds$$

$$= U(\theta,m-1)x_{m-1} + \int_{m-1}^{\theta} U(\theta,s)g(s)\,ds$$

$$= U(\theta,n-1)x_{n-1} + \int_{n-1}^{\theta} U(\theta,s)g(s)\,ds,$$

and (4.26) holds. □

4.4. Hyperbolicity and dissipativity

In this section we describe a characterization of exponential dichotomy in terms of the existence of quadratic Lyapunov functions for strongly continuous semigroups (Subsection 4.4.1) and strongly continuous, exponentially bounded evolution families (Subsection 4.4.2).

Throughout this section we let $\mathcal{H}$ be a Hilbert space with the scalar product $\langle \cdot, \cdot \rangle$. We say that a bounded operator $A \in \mathcal{L}(\mathcal{H})$ is *uniformly negative* and write $A << 0$ if there exists a positive constant δ such that $\langle Ax, x \rangle \leq -\delta |x|^2$ for all $x \in \mathcal{H}$. We say that A is uniformly positive and write $A >> 0$ if $-A << 0$. The real part of an operator $A \in \mathcal{L}(\mathcal{H})$ is defined to be Re $A := (A+A^*)/2$. An operator $A \in \mathcal{L}(\mathcal{H})$ is called *dissipative* if

$$\langle (\text{Re } A)x, x \rangle = \text{Re } \langle Ax, x \rangle \leq 0, \quad x \in \mathcal{H},$$

and *uniformly dissipative* if Re $A << 0$.

If Re $A << 0$, then because

$$\frac{d}{dt}|e^{tA}x|^2 = \text{Re } \langle Ae^{tA}x, e^{tA}x \rangle \leq -\delta |e^{tA}x|^2, \quad x \in \mathcal{H},$$

the norm of the solution $x(t) = e^{tA}x$ of the differential equation $\dot{x} = Ax$ decays to zero *monotonically*. Moreover, if Re $A << 0$, then the spectral bound $s(A)$ is negative; that is, $\sigma(A) \subset \mathbb{C}_-$. Conversely, if A is bounded and $s(A) < 0$, then the norm of the solution of the differential equation $\dot{x} = Ax$ decays to zero. However, as the simple example provided by the operator $A = \begin{bmatrix} -1 & 4 \\ 0 & -1 \end{bmatrix}$ on $\mathcal{H} = \mathbb{R}^2$ shows, condition $s(A) < 0$ does not, generally, imply that Re $A << 0$ in the original scalar product. Thus, the decay of the norm *is not necessarily monotone*. In fact, by the Generalized Lyapunov Theorem, $s(A) < 0$ if and only if there exists a uniformly positive operator $W >> 0$ such that WA is uniformly dissipative (see, e.g., Daleckij and Krein [**DK**], Theorem I.5.1). As a result, if the spectral bound $s(A) < 0$, then the solutions of the homogeneous equation $\dot{x} = Ax$ do decay monotonically to zero in the norm induced by the new scalar product

$$\langle x, y \rangle_W := \langle Wx, y \rangle.$$

In other words, the new scalar product defines a quadratic Lyapunov function $l : t \mapsto \langle x(t), x(t) \rangle_W$ for the differential equation $\dot{x} = Ax$.

More generally, a bounded operator $A \in \mathcal{L}(\mathcal{H})$ generates a hyperbolic semigroup $\{e^{tA}\}_{t \geq 0}$; equivalently, $\sigma(A) \cap i\mathbb{R} = \emptyset$, if and only if the operator A is W-uniformly dissipative, Re $(WA) << 0$, for some (indefinite) self-adjoint operator $W \in \mathcal{L}(\mathcal{H})$ (see for example [**DK**], Theorem I.7.1). We stress the fact that W is not positive, but the indefinite scalar product $\langle x, y \rangle_W := \langle Wx, y \rangle$ still defines a quadratic Lyapunov function such that if $t \mapsto x(t)$ is a solution of the differential equation $\dot{x} = Ax$, then the function $l(t) := \langle x(t), x(t) \rangle_W$ decays monotonically to zero.

In this section, we will generalize the results about quadratic Lyapunov functions to the case of unbounded operators A (Subsection 4.4.1) and nonautonomous abstract Cauchy problems (Subsection 4.4.2).

4.4.1. Hyperbolic semigroups. Let A denote the infinitesimal generator of a strongly continuous semigroup $\{T^t\}_{t \geq 0}$ on a Hilbert space $\mathcal{H}$. We say that A is *uniformly W-dissipative* for some bounded, self-adjoint operator $W \in \mathcal{L}(\mathcal{H})$ if there is a constant $\delta > 0$ such that

(4.39) $\quad\quad\quad\quad \operatorname{Re}\langle Ax, Wx\rangle \leq -\delta|x|^2 \quad$ for each $\quad x \in \mathcal{D}(A).$

THEOREM 4.38. *The strongly continuous semigroup $\{T^t\}_{t\geq 0}$, $T^t = e^{tA}$, is hyperbolic if and only if there exist self-adjoint operators W and V in $\mathcal{L}(\mathcal{H})$ such that A is uniformly W-dissipative and A^* is uniformly V-dissipative.*

The generators of contraction semigroups are characterized by the classical theorem of Lumer and Phillips, see for example [**Pz**, p. 14]. Recall that a semigroup $\{e^{tA}\}_{t\geq 0}$ is called a *contraction semigroup* if $\|e^{tA}\| \leq 1$ for $t \geq 0$. For a linear operator A with dense domain, the Lumer-Phillips Theorem states:

(a) If A is dissipative and there exists $\xi_0 > 0$ such that the operator $\xi_0 - A$ is surjective, then A is the infinitesimal generator of a strongly continuous contraction semigroup.
(b) If A is the generator of a strongly continuous contraction semigroup, then $\xi - A$ is surjective for all $\xi > 0$ and A is dissipative.

Clearly, if A is uniformly dissipative, then the growth bound $\omega(A)$ of the semigroup $\{e^{tA}\}_{t\geq 0}$ is negative. The following result gives a characterization of hyperbolic semigroups.

COROLLARY 4.39. *If A is uniformly W-dissipative for some self-adjoint operator W and there is some $\xi_0 \in \mathbb{R}$ such that the operator $i\xi_0 - A$ is surjective, then the strongly continuous semigroup $\{T^t\}_{t\geq 0}$, $T^t = e^{tA}$, is hyperbolic. Conversely, if the semigroup is hyperbolic, then A is uniformly W-dissipative and $i\xi - A$ is surjective for each $\xi \in \mathbb{R}$.*

The next result allows one to obtain the operators W and V as the solutions of linear Riccati equations. As usual, for a hyperbolic operator $T = e^A$, we let P denote the Riesz projection that corresponds to T and the spectral set $\sigma(T) \cap \mathbb{D}$. The operator $Q := I - P$ is the complementary projection. Also, T_P^t for $t \geq 0$, and T_Q^t for $t \in \mathbb{R}$ denote the corresponding restrictions. For each operator $H \in \mathcal{L}(\mathcal{H})$, we define $H_{PQ} = P^*HP + Q^*HQ$. Recall that a subspace $\mathcal{H}' \subset \mathcal{H}$ is called *uniformly W-positive* if there is an $\alpha > 0$ such that $\langle x, x\rangle_W \geq \alpha\langle x, x\rangle$ for all $x \in \mathcal{H}$.

THEOREM 4.40. *The semigroup $\{T^t\}_{t\geq 0}$, $T^t = e^{tA}$, is hyperbolic if and only if there exist bounded operators $R \ll 0$ and $R_* \ll 0$ such that the operator equations*

(4.40) $\quad\quad\quad\quad (WA + A^*W)x = Rx, \quad x \in \mathcal{D}(A),$
(4.41) $\quad\quad\quad\quad (VA^* + AV)x = R_*x, \quad x \in \mathcal{D}(A^*),$

have, respectively, self-adjoint solutions $W : \mathcal{D}(A) \to \mathcal{D}(A^*)$ and $V : \mathcal{D}(A^*) \to \mathcal{D}(A)$. In addition, if $\{T^t\}_{t\geq 0}$ is hyperbolic then W and V are invertible. Also, if $H \gg 0$ is a bounded operator, the operator R in equation (4.40) is given by $R = -H_{PQ}$, and the operator R_* in equation (4.41) is given by $R_* = -H_{P^*Q^*}$, then solution operators W and V are given by the formulas

(4.42) $\quad\quad\quad\quad W = \int_0^\infty \{T_P^{t*}HT_P^t - T_Q^{-t*}HT_Q^{-t}\}\,dt,$

(4.43) $\quad\quad\quad\quad V = \int_0^\infty \{T_P^t HT_P^{t*} - T_Q^{-t}HT_Q^{-t*}\}\,dt.$

With this choice of W and V the subspace $\operatorname{Im} P$ is uniformly W-positive and $\operatorname{Im} Q$ is uniformly W-negative, while the subspace $\operatorname{Im} P^*$ is uniformly V-positive and $\operatorname{Im} Q^*$ is uniformly V-negative. Also, $\operatorname{Im} P$ and $\operatorname{Im} Q$ are W-orthogonal while $\operatorname{Im} P^*$ and $\operatorname{Im} Q^*$ are V-orthogonal.

It is convenient to give the proofs of Theorems 4.38–4.40 and Corollary 4.39 simultaneously.

PROOF. **Sufficiency.** Assume there is an operator $R \ll 0$ with $\langle Rx, x \rangle \leq -2\delta|x|^2$ such that equation (4.40) has a solution W with the property $W : \mathcal{D}(A) \to \mathcal{D}(A^*)$. Then,

$$(4.44) \qquad 2\operatorname{Re} \langle Ax, Wx \rangle = \langle (WA + A^*W)x, x \rangle \leq -2\delta|x|^2, \quad x \in \mathcal{D}(A).$$

Thus, A is uniformly W-dissipative. Also, since $\langle Wx, x \rangle \in \mathbb{R}$ for all $x \in \mathcal{H}$, the inequality (4.39) implies that if $\xi \in \mathbb{R}$ and $x \in \mathcal{D}(A)$, then

$$\delta|x|^2 \leq |\operatorname{Re} \langle Ax, Wx \rangle| = |\operatorname{Re} \langle (A - i\xi)x, Wx \rangle|$$
$$\leq |\langle (A - i\xi)x, Wx \rangle| \leq |(A - i\xi)x| \cdot \|W\|_{\mathcal{L}(\mathcal{H})}|x|.$$

Hence, $\|A - i\xi\|_\bullet \geq \delta \|W\|^{-1}$, and $A - i\xi$ is uniformly injective for all $\xi \in \mathbb{R}$. Similarly, for $x \in \mathcal{D}(A^*)$ and $\xi \in \mathbb{R}$, if A^* is V-dissipative, then $|(A^* - i\xi)x| \geq \delta \|V\|^{-1}|x|$. Hence, the operator $A - i\xi$ is invertible for each $\xi \in \mathbb{R}$, and the resolvent of A is uniformly bounded along $i\mathbb{R}$, that is $\|R(i\xi; A)\|_{\mathcal{L}(\mathcal{H})} \leq \delta^{-1} \|W\|_{\mathcal{L}(\mathcal{H})}$. By the Gearhart Spectral Mapping Theorem 2.10, we have $\sigma(T^t) \cap \mathbb{T} = \emptyset$.

To complete the proof of sufficiency for Corollary 4.39, note that, as we have seen before, if A is W-dissipative and $\xi \in \mathbb{R}$, then there is some $\delta > 0$ such that the inequality $|(A - i\xi)x| \geq \delta \|W\|^{-1}|x|$, $x \in \mathcal{D}(A)$, holds. Hence, $\sigma_{\text{ap}}(A) \cap i\mathbb{R} = \emptyset$. Since the boundary of $\sigma(A)$ is in $\sigma_{\text{ap}}(A)$, either $\sigma(A) \cap i\mathbb{R} = \emptyset$ or, for each $\xi \in \mathbb{R}$, the operator $i\xi - A$ is not surjective. Since $i\xi - A$ is surjective for $\xi = \xi_0$, we conclude that $\sigma(A) \cap i\mathbb{R} = \emptyset$. Also, $\|R(i\xi; A)\| \leq \delta^{-1} \|W\|$ whenever $\xi \in \mathbb{R}$; and therefore, $\sigma(T^t) \cap \mathbb{T} = \emptyset$ by the Gearhart Spectral Mapping Theorem 2.10. $\square$

PROOF. **Necessity.** Suppose that $\{T^t\}_{t \geq 0}$ is hyperbolic. By the Spectral Inclusion Theorem 2.6, it follows that $\sigma(A) \cap i\mathbb{R} = \emptyset$ and the second condition in Corollary 4.39 holds.

Fix $H \gg 0$. We will show that the operator W defined as in equation (4.42) is bounded, $W : \mathcal{D}(A) \to \mathcal{D}(A^*)$, and W satisfies the equation (4.40) when $R = -H_{PQ}$. It then follows that A is W-dissipative by (4.44), as required in Theorem 4.38 and Corollary 4.39.

Since the spectral radii of both T_P^1 and $(T_Q^1)^{-1}$ are less than one, the operators W_P and W_Q given by

$$W_P = \int_0^\infty T_P^{t*} H T_P^t \, dt \quad \text{and} \quad W_Q = -\int_0^\infty T_Q^{-t*} H T_Q^{-t} \, dt$$

are bounded.

LEMMA 4.41. *If $y \in \mathcal{D}(A)$, then $W_P y \in \mathcal{D}(A^*)$ and $A^* W_P y = -P^* H P y - W_P A y$.*

PROOF. We need to check that

$$I_t := t^{-1} \left(T_P^{t*} W_P y - W_P y \right) + P^* H P y + W_P A y$$

converges to zero as $t \to 0$. Since $T_P^{t*}W_Py = \int_t^\infty T_P^{s*}HT_P^{(s-t)}y\,ds$, we have that

$$I_t = P^*HP^*y - t^{-1}\int_0^t T_P^{s*}HT_P^s y\,ds + \int_0^t T_P^{s*}HT_P^s Ay\,ds$$
$$+ \int_t^\infty \left[t^{-1}\{T_P^{s*}HT_P^{(s-t)}y - T_P^{s*}HT_P^s y\} + T_P^{s*}HT_P^s Ay\right]ds.$$

Since the last integral is equal to

$$T_P^{t*}\int_0^\infty T_P^{s*}HT_P^s\,ds\left[t^{-1}(y - T_P^t y) + T_P^t Ay\right],$$

it follows that $I_t \to 0$. □

Similarly, if $y \in \mathcal{D}(A)$, then $W_Q y \in \mathcal{D}(A^*)$ and $A^*W_Q y = -Q^*HQy - W_Q Ay$. Hence, the operator W as in (4.42) satisfies equation (4.40) with $R = -H_{PQ}$.

As $\{T^t\}_{t \geq 0}$ is assumed hyperbolic, $\sigma(A) \cap i\mathbb{R} = \emptyset$ and A is invertible. This implies that every self-adjoint solution W of equation (4.40) is an invertible operator. Indeed, since the operator $R << 0$ in (4.40) is bounded, there is some number $\delta > 0$ such that if $x \in \mathcal{D}(A)$, then

$$2\mathrm{Re}\,\langle A^{-1}x, Wx\rangle = \langle (A^*W + WA)A^{-1}x, A^{-1}x\rangle$$
$$= \langle RA^{-1}x, A^{-1}x\rangle \leq -2\delta\|A^{-1}\|^2|x|^2.$$

It follows that

$$\delta\|A^{-1}\||x|^2 \leq |\mathrm{Re}\,\langle A^{-1}x, Wx\rangle| \leq |\langle A^{-1}x, Wx\rangle| \leq \|A^{-1}\| \cdot |x| \cdot |Wx|.$$

Hence,

(4.45) $$|Wx| \geq \delta\|A^{-1}\||x|$$

and, as a result, the self-adjoint operator W is invertible.

To prove that $\mathrm{Im}\,P$ is uniformly W-positive, define $T := T^1$, and note that

$$T^*WT - W = \int_1^\infty T_P^{t*}HT_P^t\,dt - \int_0^\infty T_P^{t*}HT_P^t\,dt$$
$$- \int_{-1}^\infty T_Q^{-t*}HT_Q^{-t}\,dt + \int_0^\infty T_Q^{-t*}HT_Q^{-t}\,dt$$

(4.46) $$= -\int_0^1 (T_P^{t*}HT_P^t + T_Q^{t*}HT_Q^t)\,dt = -\int_0^1 T^{s*}H_{PQ}T^s\,ds << 0.$$

Also, for each $x \in \mathrm{Im}\,P$, we have that $\langle Wx, x\rangle = \langle (WP + P^*W - W)x, x\rangle$. Next, by the integral formula for the Riesz projection P for T, we have

$$WP + P^*W - W = \frac{1}{2\pi}\int_0^{2\pi}(e^{-i\tau} - T^*)^{-1}(W - T^*WT)(e^{i\tau} - T)^{-1}\,d\tau.$$

Hence, since $W - T^*WT$ is uniformly positive, it follows that $WP + P^*W - W$ is uniformly positive. Similarly, $\mathrm{Im}\,Q$ is uniformly W-negative. Obviously, $\mathrm{Im}\,P$ and $\mathrm{Im}\,Q$ are W-orthogonal; that is, $\langle Wx, y\rangle = 0$ for $x \in \mathrm{Im}\,P$ and $y \in \mathrm{Im}\,Q$.

Also, since $\{T^{t*}\}_{t \geq 0}$ is hyperbolic whenever $\{T^t\}_{t \geq 0}$ is hyperbolic, all statements about the operator V in Theorem 4.38-4.40 can be obtained by applying the arguments above to the strongly continuous semigroup $\{T^{t*}\}_{t \geq 0}$ on the Hilbert space $\mathcal{H}$. □

REMARK 4.42. 1. If the operator W that satisfies equation (4.40) is invertible, and $W^{-1}: \mathcal{D}(A^*) \to \mathcal{D}(A)$, then $V = W^{-1}$ is a solution of (4.41) with $R_* = W^{-1}RW^{-1}$.

2. If $\{T^t\}_{t\in\mathbb{R}}$ is a hyperbolic strongly continuous group, then, in addition to the fact that W defined by equation (4.42) is invertible, see (4.45), we claim that $W^{-1}: \mathcal{D}(A^*) \to \mathcal{D}(A)$. Indeed, using Lemma 4.41, we have the identity

$$\frac{d}{dt}(T^{t*}W_P T^t x)|_{t=0} = -P^*HPx, \qquad x \in \operatorname{Im} P \cap \mathcal{D}(A).$$

It follows that if $x \in \operatorname{Im} P \cap \mathcal{D}(A)$, then

$$\frac{d}{dt}(T^t W_P^{-1} T^{t*} x)\Big|_{t=0} = \frac{d}{dt}(T^{-t*}W_P T^{-t})^{-1}x\Big|_{t=0} = -W_P^{-1} P^* H P W_P^{-1} x.$$

Hence, $W_P^{-1}: \mathcal{D}(A^*) \to \mathcal{D}(A)$ and

$$(AW_P^{-1} + W_P^{-1}A^*)x = -W_P^{-1}P^*HPW^{-1}Px, \qquad x \in \mathcal{D}(A^*),$$

and similarly, for W_Q. We thus obtain the following fact.

COROLLARY 4.43. *A strongly continuous group $\{T^t\}_{t\in\mathbb{R}}$ is hyperbolic if and only if there exists an invertible operator W such that A is W-dissipative and A^* is W^{-1}-dissipative.*

3. The operator W from Theorems 4.38-4.40 defines the Lyapunov function $l(x,y) = \langle Wx, y\rangle$ for the differential equation $\dot{x} = Ax$. That is, for each solution $x(t) = e^{tA}x(0)$ of the differential equation, the function $t \mapsto l(x(t), x(t))$ decreases monotonically. Similarly, V defines a Lyapunov function for the differential equation $\dot{x} = A^*x$.

4. Generally, the condition of V-dissipativity of A^* in Theorem 4.38 cannot be dropped, see Example 3.2 in Massera and Schäffer [**MS**]. $\diamond$

Later, in Section 7.5, we will need the following facts about the *group* $\{T^t\}_{t\in\mathbb{R}}$.

LEMMA 4.44. *If the generator A of a strongly continuous group $\{T^t\}_{t\in\mathbb{R}}$ is W-dissipative, then T^t, $t > 0$, is a W-contraction; that is, $T^{t*}WT^t << W$.*

PROOF. Suppose that $x \in \mathcal{D}(A)$ is such that $|x| = 1$, and define

$$g(\tau) := \langle WT^\tau x, T^\tau x\rangle - \langle Wx, x\rangle, \quad \tau \in [0,t].$$

By a calculation, we find that

$$g'(\tau) = \langle WAT^\tau x, T^\tau x\rangle + \langle WT^\tau x, AT^\tau x\rangle = 2\operatorname{Re}\langle AT^\tau x, WT^\tau x\rangle.$$

Since $\{T^t\}_{t\in\mathbb{R}}$ is a group, there is some $\beta > 0$ such that $|T^\tau x| \geq \beta$ for $\tau \in [0,t]$. Let $g_1(\tau) := g(\tau) + \delta\beta^2\tau$. Since A is W-dissipative, $g_1'(\tau) \leq -2\delta|T^\tau x|^2 + \delta\beta^2 \leq 0$. Thus, g_1 decreases, $0 = g_1(0) > g_1(t)$, and $g(t) \leq -\delta\beta^2 t$. $\square$

COROLLARY 4.45. *A strongly continuous group $\{T^t\}_{t\in\mathbb{R}}$ is hyperbolic if and only if there exists an invertible self-adjoint operator $W \in \mathcal{L}(\mathcal{H})$ such that*

(4.47) $\qquad T^*WT << W \quad \text{and} \quad TW^{-1}T^* << W^{-1}, \quad T = T^1.$

PROOF. As shown in [**DK**, Lemma I.7.3], the conditions (4.47) imply $\sigma(T) \cap \mathbb{T} = \emptyset$. If $\{T^t\}_{t\geq 0}$ is hyperbolic, then, by Corollary 4.43, the operator A is W-dissipative and A^* is W^{-1}-dissipative. By Lemma 4.44, we have (4.47). $\square$

4.4.2. Dichotomic evolution families. In this subsection we will apply Theorem 4.38 and Theorem 4.40 to the evolution semigroup $\{E^t\}_{t\geq 0}$ on $L^2(\mathbb{R};\mathcal{H})$ that is generated by the strongly continuous evolution family $\{U(\theta,\tau)\}_{\theta\geq \tau}$. This allows us to characterize the dichotomy of the evolution family in terms of the dissipativity of the generator Γ of the evolution semigroup.

We will assume that both $\{U(\theta,\tau)\}_{\theta\geq \tau}$ and $\{U(\theta,\tau)^*\}_{\theta\geq \tau}$ are exponentially bounded, strongly continuous evolution families on a Hilbert space $\mathcal{H}$.

THEOREM 4.46. *The evolution family $\{U(\theta,\tau)\}_{\theta\geq \tau}$ has a dichotomy if and only if there exist self-adjoint operators $W, V \in \mathcal{L}(L^2(\mathbb{R};\mathcal{H}))$ such that the generator Γ of the evolution semigroup $\{E^t\}_{t\geq 0}$ on $L^2(\mathbb{R};\mathcal{H})$ defined by $(E^t f)(\theta) = U(\theta, \theta - t) f(\theta - t)$ is uniformly W-dissipative and the operator Γ^* is uniformly V-dissipative on $L^2(\mathbb{R};\mathcal{H})$:*

(4.48) $$\operatorname{Re}\langle \Gamma f, Wf\rangle_{L^2} \leq -\delta\|f\|_{L^2}^2, \qquad f \in \mathcal{D}(\Gamma),$$

(4.49) $$\operatorname{Re}\langle \Gamma^* f, Vf\rangle_{L^2} \leq -\delta\|f\|_{L^2}^2, \qquad f \in \mathcal{D}(\Gamma^*).$$

The operators W and V also enjoy all properties listed in Theorem 4.40. Moreover, if $\{U(\theta,\tau)\}_{\theta\geq \tau}$ has an exponential dichotomy and $H(\cdot) \in C_b(\mathbb{R};\mathcal{L}(\mathcal{H}))$ is such that $H(\theta) << 0$ uniformly for $\theta \in \mathbb{R}$, then solutions W and V of the operator equations

$$(W\Gamma + \Gamma^* W)f = -H_{PQ}f, \qquad f \in \mathcal{D}(\Gamma),$$
$$(V\Gamma^* + \Gamma V)f = -H_{P^*Q^*}f, \qquad f \in \mathcal{D}(\Gamma^*),$$

are given by the multiplication operators associated with the multipliers $W(\cdot), V(\cdot) \in C_b(\mathbb{R};\mathcal{L}(\mathcal{H}))$ defined by

(4.50) $$W(\theta) = \int_0^\infty \{U_P^*(\theta+\tau,\theta)H(\theta+\tau)U_P(\theta+\tau,\theta) \\ - U_Q^{-1*}(\theta, \theta-\tau)H(\theta-\tau)U_Q^{-1}(\theta, \theta-\tau)\}\,d\tau,$$

(4.51) $$V(\theta) = \int_0^\infty \{U_P(\theta,\theta-\tau)H(\theta-\tau)U_P^*(\theta,\theta-\tau) \\ - U_Q^{-1}(\theta+\tau,\theta)H(\theta+\tau)U_Q^{-1*}(\theta+\tau,\theta)\}\,d\tau.$$

PROOF. The theorem follows from Theorem 4.38 and Theorem 4.40 above, and the following facts: $\{U(\theta,\tau)\}_{\theta\geq \tau}$ has an exponential dichotomy if and only if $\{E^t\}_{t\geq 0}$ is hyperbolic (see the Dichotomy Theorem 3.17) and the Riesz projection $\mathcal{P}$ for E^1 has the form $\mathcal{P} = \mathcal{P}(\cdot)$ for some $P(\cdot) \in C_b(\mathbb{R};\mathcal{L}_s(\mathcal{H}))$ (see the Spectral Projection Theorem 3.14). Formulas (4.50) – (4.51) follow from (4.42) – (4.43). □

If the evolution family $\{U(\theta,\tau)\}_{\theta\geq \tau}$ solves an abstract Cauchy problem for the equation

(4.52) $$\dot{x} = A(t)x, \qquad t \in \mathbb{R},$$

then Γ can be computed "explicitly", see Theorem 3.12. We will give a variant of Theorem 4.46 for the following situation.

Let $\{A(t)\}_{t\in\mathbb{R}}$ in equation (4.52) be a stable family of generators of strongly continuous semigroups on $\mathcal{H}$; that is, each operator $A(t)$ generates a strongly continuous semigroup on $\mathcal{H}$, and there exist real numbers $M \geq 1$ and β such that, for each $T > 0$, all $\lambda > \beta$, and all partitions $0 \leq t_1 \leq t_2 \leq \ldots \leq t_k \leq T$ where k is a positive integer,

$$\|(A(t_k) - \lambda)^{-1}(A(t_{k-1}) - \lambda)^{-1} \cdot \ldots \cdot (A(t_1) - \lambda)^{-1}\| \leq M(\lambda - \beta)^{-k},$$

see [**Ta**, p.93].

In addition, assume that the domain $\mathcal{D}(A(t)) = \mathcal{D}$ is independent of the choice of $t \in \mathbb{R}$ and, for each $x \in \mathcal{D}$, the function from $\mathbb{R}$ to $\mathcal{H}$ given by $\theta \mapsto A(\theta)x$ is continuously differentiable.

Under these assumptions (see [**Ta**, p.102]), there exists an evolution family $\{U(\theta, \tau)\}_{\theta \geq \tau}$ that solves an abstract Cauchy problem for (4.52). Moreover, $U(\theta, \tau) : \mathcal{D} \to \mathcal{D}$, each function $\theta \mapsto U(\theta, \tau)x$ for $\tau \in \mathbb{R}$, and each function $\tau \mapsto U(\theta, \tau)x$ for $\theta \in \mathbb{R}$ is (strongly) continuously differentiable for each $x \in \mathcal{D}$, and such that

$$\frac{d}{d\theta}U(\theta, \tau)x = A(\theta)U(\theta, \tau)x, \quad \frac{d}{d\tau}U(\theta, \tau)x = -U(\theta, \tau)A(\tau)x, \quad \theta \geq \tau.$$

Also, we impose the same set of assumptions on the operators $A(t)^*$. That is, we assume that $\{A(t)^*\}_{t \in \mathbb{R}}$ is a stable family of generators, the domain $\mathcal{D}(A(t)^*) = \mathcal{D}^*$ is independent of the choice of $t \in \mathbb{R}$, and for each $y \in \mathcal{D}^*$ the function from $\mathbb{R}$ to $\mathcal{H}$ given by $\theta \mapsto A(\theta)^* y$ is continuously differentiable. As above, $\{U(\theta, \tau)^*\}_{\theta \geq \tau}$ is a strongly continuous evolution family and $U(\theta, \tau)^* : \mathcal{D}^* \to \mathcal{D}^*$.

Consider the associated evolution semigroups $\{E^t\}_{t \geq 0}$ and $\{E^{t*}\}_{t \geq 0}$. By Theorem 3.12, their generators Γ and Γ^* are the closures of the operators Γ' and $\Gamma^{*\prime}$, respectively, given by the following formulas:

$$(\Gamma' f)(\theta) = -\frac{d}{d\theta}f + A(\theta)f(\theta), \qquad (\Gamma^{*\prime} f)(\theta) = \frac{df}{d\theta} + A^*(\theta)f(\theta)$$

with $\mathcal{D}(\Gamma') = \mathcal{D}(-d/d\theta) \cap \mathcal{D}(\mathcal{A})$ and $\mathcal{D}(\Gamma^{*\prime}) = \mathcal{D}(d/d\theta) \cap \mathcal{D}(\mathcal{A}^*)$, see a related discussion on page 63.

THEOREM 4.47. *Under the assumptions stated above, the evolution family $\{U(\theta, \tau)\}_{\theta \geq \tau}$ has an exponential dichotomy if and only if there exist bounded functions $W : \mathbb{R} \to \mathcal{L}(\mathcal{H})$ and $V : \mathbb{R} \to \mathcal{L}(\mathcal{H})$ with self-adjoint values that satisfy the following conditions:*

(i) *For each $x, y \in \mathcal{H}$, the functions $\theta \mapsto \langle W(\theta)x, y\rangle$ and $\theta \mapsto \langle V(\theta)x, y\rangle$ are continuous.*

(ii) *For each $x \in \mathcal{D}$ and $y \in \mathcal{D}^*$, the functions $\theta \mapsto \langle W(\theta)x, x\rangle$ and $\theta \mapsto \langle V(\theta)y, y\rangle$ are continuously differentiable.*

(iii) *There is a number $\delta > 0$ such that, for all $\theta \in \mathbb{R}$,*

$$\frac{d}{d\theta}\langle W(\theta)x, x\rangle + 2\mathrm{Re}\,\langle A(\theta)x, W(\theta)x\rangle \leq -2\delta|x|^2, \qquad x \in \mathcal{D},$$

$$\frac{d}{d\theta}\langle V(\theta)y, y\rangle + 2\mathrm{Re}\,\langle A(\theta)y, V(\theta)y\rangle \leq -2\delta|y|^2, \qquad y \in \mathcal{D}^*.$$

PROOF. **Necessity.** Suppose that $W(\cdot)$ and $V(\cdot)$ satisfy (i)-(iii). Consider the bounded, self-adjoint multiplication operators on $L^2(\mathbb{R}; \mathcal{H})$ corresponding to the functions W and V. We claim that W satisfies condition (4.48) of Theorem 4.46 (condition (4.49) for V can be proved similarly). To show this, fix $f \in \mathcal{D}(\Gamma')$. Since the function $\theta \mapsto \|W(\theta)\|$ is bounded, conditions (i) and (ii) yield the following "Leibniz rule":

$$\frac{d}{d\theta}\langle W(\theta)f(\theta), f(\theta)\rangle = \frac{d}{dt}\langle W(\theta + t)f(\theta), f(\theta)\rangle\Big|_{t=0} + 2\mathrm{Re}\,\langle \frac{d}{d\theta}f(\theta), W(\theta)f(\theta)\rangle.$$

Since $f \in L^2(\mathbb{R}; \mathcal{H})$, we find that

$$2\mathrm{Re}\,\langle \frac{d}{d\theta}f, Wf\rangle_{L^2(\mathbb{R};\mathcal{H})} = -\int_{\mathbb{R}} \frac{d}{dt}\langle W(\theta + t)f(\theta), f(\theta)\rangle\Big|_{t=0}\,d\theta.$$

Now, since $\Gamma' = -\frac{d}{d\theta} + \mathcal{A}$, the first inequality in (iii) gives

$$2\operatorname{Re} \langle \Gamma' f, Wf \rangle_{L^2(\mathbb{R};\mathcal{H})} = -2\operatorname{Re} \langle \frac{d}{d\theta} f, Wf \rangle_{L^2(\mathbb{R};\mathcal{H})} + 2\operatorname{Re} \langle \mathcal{A}f, Wf \rangle_{L^2(\mathbb{R};\mathcal{H})}$$
$$= \int_{\mathbb{R}} \left\{ \frac{d}{dt} \langle W(\theta + t) f(\theta), f(\theta) \rangle \Big|_{t=0} + 2\operatorname{Re} \langle A(\theta) f(\theta), W(\theta) f(\theta) \rangle \right\} d\theta$$
$$\leq -2\delta \|f\|^2_{L^2(\mathbb{R};X)},$$

and (4.48) is proved. $\square$

PROOF. **Sufficiency.** Suppose that the evolution family $\{U(\theta, \tau)\}_{\theta \geq \tau}$ has an exponential dichotomy. We will show that $W(\cdot)$ from (4.50) satisfies (i)-(iii), similarly for $V(\cdot)$. Split $W(\cdot)$ from (4.50) as $W(\theta) = W_P(\theta) + W_Q(\theta)$ where $W_P(\theta)$ is equal to

$$\int_0^\infty U_P^*(\theta + \tau, \theta) H(\theta + \tau) U_P(\theta + \tau, \theta) \, d\tau = \int_\theta^\infty U_P^*(\tau, \theta) H(\tau) U_P(\tau, \theta) \, d\tau.$$

Since the function $\theta \mapsto \|H(\theta)\|_{\mathcal{L}(\mathcal{H})}$ is bounded, the exponential dichotomy for $\{U(\theta, \tau)\}_{\theta \geq \tau}$ implies the convergence of the last integral. Moreover, for each $x, y \in \mathcal{H}$, the function

$$\theta \mapsto \langle W_P(\theta) x, y \rangle = \int_\theta^\infty \langle H(\tau) U_P(\tau, \theta) x, U_P(\tau, \theta) y \rangle \, d\tau$$

is continuous since $\theta \mapsto U_P(\tau, \theta) x$ is continuous and the last integral converges absolutely and uniformly for $\theta \in \mathbb{R}$. For each $x \in \mathcal{D}$, by the definition (4.50), we have that

$$\frac{d}{d\theta} \langle W_P(\theta) x, x \rangle = -\langle H(\theta) U_P(\theta, \theta) x, U_P(\theta, \theta) x \rangle$$
$$- \int_\theta^\infty \big\{ \langle H(\tau) U_P(\tau, \theta) A(\theta) x, U_P(\tau, \theta) x \rangle$$
$$+ \langle H(\tau) U_P(\tau, \theta) x, U_P(\tau, \theta) A(\theta) x \rangle \big\} d\tau$$
$$= -\langle P^*(\theta) H(\theta) P(\theta) x, x \rangle - 2\operatorname{Re} \langle A(\theta) x, W_P(\theta) x \rangle.$$

The same argument can be applied for $W_Q(\theta)$. As a result,

$$\frac{d}{d\theta} \langle W(\theta) x, x \rangle = -\langle H_{PQ}(\theta) x, x \rangle - 2\operatorname{Re} \langle A(\theta) x, W(\theta) x \rangle,$$

and (i)–(iii) hold for W. $\square$

Let us specialize to the case of a *norm* continuous evolution family. In particular, let us consider the propagator $\{U(\theta, \tau)\}_{(\theta, \tau) \in \mathbb{R}^2}$ for a differential equation $\dot{x} = A(t)x$ defined by a bounded, norm-continuous, operator valued function $A : \mathbb{R} \to \mathcal{L}(\mathcal{H})$. Recall that, for the derivatives taken in $\mathcal{L}(\mathcal{H})$, we have the following formulas:

$$\frac{d}{d\theta} U(\theta, \tau) = A(\theta) U(\theta, \tau), \quad \frac{d}{d\tau} U(\theta, \tau) = -U(\theta, \tau) A(\tau), \quad (\theta, \tau) \in \mathbb{R}^2,$$

see for example [**DK**].

COROLLARY 4.48. *The existence of an exponential dichotomy for the equation* $\dot{x} = A(t)x$ *with* $A \in C_b(\mathbb{R}; \mathcal{L}(\mathcal{H}))$ *is equivalent to the existence of a bounded, norm differentiable function* $W : \mathbb{R} \to \mathcal{L}(\mathcal{H})$ *with self-adjoint and invertible values such that the function from* $\mathbb{R}$ *to* $\mathcal{L}(\mathcal{H})$ *defined by* $\theta \mapsto [W(\theta)]^{-1}$ *is bounded and*

$$\hat{W}(\theta) := \frac{dW}{d\theta}(\theta) + A^*(\theta)W(\theta) + W(\theta)A(\theta)$$

is uniformly negative ($\hat{W}(\theta) << 0$) *in* $\mathcal{H}$ *uniformly for* $\theta \in \mathbb{R}$.

PROOF. By Theorem 4.46, the existence of an exponential dichotomy for the evolution family $\{U(\theta, \tau)\}_{(\theta,\tau) \in \mathbb{R}^2}$ is equivalent to the existence of multiplication operators $W = W(\cdot)$ and $V = V(\cdot)$ that satisfy the inequalities (4.48) and (4.49). By Corollary 4.43, since $\{E^t\}_{t \in \mathbb{R}}$, $E^t = e^{t\Gamma}$, is a *group*, we can assume with no loss of generality that $V = W^{-1}$. Recall that Γ is the closure of the operator $\Gamma' = -\frac{d}{d\theta} + \mathcal{A}$, see Theorem 3.12. Using Γ', we see that condition (4.48) is equivalent to the statement that $\hat{W} << 0$ uniformly for $\theta \in \mathbb{R}$. Indeed, if W is the multiplication operator defined by $W(\cdot)$ in $L^2(\mathbb{R}; \mathcal{H})$, then $\frac{d}{d\theta} \cdot W = \frac{dW}{d\theta} + W \frac{d}{d\theta}$ where $\frac{dW}{d\theta}$ is the multiplication operator with multiplier $\frac{dW}{d\theta}(\cdot)$ in $L^2(\mathbb{R}; \mathcal{H})$. Also, we have

$$\frac{d}{d\theta} W^{-1} = -W^{-1} \frac{dW}{d\theta} W^{-1} + W^{-1} \frac{d}{d\theta}.$$

Thus, with $V = W^{-1}$, the inequality (4.49) is equivalent to

$$W^{-1}\Big(\frac{d}{d\theta} + A^*\Big) + \Big(-\frac{d}{d\theta} + A\Big)W^{-1} = W^{-1}\Big(\frac{dW}{d\theta} + A^*W + WA\Big)W^{-1}$$
$$= W^{-1}\hat{W}W^{-1} = -H_{P^*Q^*} << 0.$$

□

4.5. Bibliography and remarks

Discrete dichotomies: an algebraic approach. The connection between the existence of a "global" dichotomy for an evolution family $\{U(\theta, \tau)\}_{\theta \geq \tau}$, as defined in Definition 3.6, and the existence of a "discrete" dichotomy, equivalent to the hyperbolicity of the operator $\pi_\theta(E)$ on $\ell^p(\mathbb{Z}; X)$, will be developed further in Chapter 7 for the situation of linear skew-product flows.

Using a different method, the connection between "global" and "discrete" dichotomies was studied by Henry [**He, He3**], and Chow and Leiva [**CLe2**]. The main ingredient in their study is the "discrete" Green's function (in our setting, the operator $[I - \pi_\theta(E)]^{-1}$) and its relation to the Green's function (see Definition 4.24) for the dichotomic evolution family $\{U(\theta, \tau)\}_{\theta \geq \tau}$. See also Baskakov [**Bs3**].

Another type of "discrete" dichotomy and the related weighted shift operators is studied by Ben-Artzi and Gohberg, see [**BG, BG2, BG3**]. We take a more "algebraic" approach. Our exposition follows the treatment by Latushkin and Randolph [**LR**].

Algebras of weighted shifts. The algebra $\mathfrak{C}$ appears in the study of C^*-algebras of weighted composition (translation) operators, see the work by Antonevich and Lebedev [**An2, An3, AL2**], Karlovich [**Ka**], Kurbatov [**Ku**], Semenyuta [**Sy**], and a review by Karlovich, Kravchenko and Litvinchuk [**KKL**], and the bibliography therein.

The main idea of the proof of Lemma 4.1 taken from [**LR**], goes back to Kurbatov [**Ku**]. Of course, Lemma 4.1 also follows from Corollary 4.8, a result that is based on the general Bochner-Phillips Theorem 4.7. Our direct proof shows that the norms of the diagonals of $C = D^{-1}$ for the two-diagonal operator D decay exponentially. Note that the space ℓ^p_γ of weighted ℓ^p-sequences, used in the proof, can be thought of as a discrete analogue of the scale of exponentially bounded functions $\mathcal{F}_\alpha$, see Subsection 4.3.2. The proof of Lemma 4.1 that uses the space ℓ^p_γ does not work for an arbitrary ("infinite-diagonal") $b \in \mathfrak{B}$. This fact was pointed out in an interesting paper by Marchesi [**Mr**]. Lemma 4.3 also follows from Proposition 4.10, another result based on the Bochner-Phillips Theorem 4.7.

Corollary 4.8 records an important feature of weighted shift algebras: the fact that they are inverse-closed. Since $\mathfrak{C}$ is an intrinsic algebraic property, the inverse D^{-1} for a $D \in \mathfrak{C}$ is the same operator for $\ell^p(\mathbb{Z}; X)$ or $c_0(\mathbb{Z}; X)$. More information about inverse-closed algebras can be found in Antonevich [**An3**], Latushkin and Stepin [**LSt2**], Shubin [**Sb, Sb2**], and the literature cited above. Generalizations and applications of the Bochner-Phillips Theorem 4.7 can be found in papers by Allan [**Al**] and Gohberg and Leiterer [**GL**].

Algebras of weighted translations. The algebra $\mathfrak{B}$ was introduced and studied in [**LR**] as the Banach-space analogue of the algebra $\mathfrak{B}$ (see Chapter 7) of weighted translations studied by Antonevich [**An2**]. All results of this subsection are taken from [**LR**]. We stress that the multiplication operators a are assumed to be merely *strongly* continuous, while in the previous related study ([**An2, An3, KKL**]) they are always assumed to be continuous in norm. As a result, we can not prove that $\mathfrak{B}$ is inverse-closed. Instead, we prove that a slightly smaller algebra $\mathfrak{B}_*$ is inverse-closed (see Corollary 4.14).

Dichotomy and discrete operators. Theorem 4.16 is taken from [**LR**], see also Latushkin and Montgomery-Smith [**LM**]. The idea of the proof of Proposition 4.18 goes back to Antonevich [**An2**]; similar propositions were used by Latushkin, Stepin, Montgomery-Smith and Randolph, see [**LSt2, Lt2**] and [**LM, LMR**]. In a sense, this proposition is the core of the applications of our algebraic approach to the study of dichotomy. Indeed, an algebraic fact; namely, $(\lambda - E)^{-1} \in \mathfrak{B}$, together with the Commutation Lemma 3.15 implies the conclusion of the Spectral Projection Theorem 3.14.

Green's function and evolution semigroups. This material is taken from [**LR**]. See the work by Mitropolskij, Samojlenko, and Kulik [**MSK, Sm**] for more information about Green's function and dichotomy. Theorem 4.22 is in the spirit of the Datko-van Neerven Theorem 2.44. See also Bart, Gohberg, and Kaashoek [**BrGK**, Proposition 2.1] concerning the uniqueness of the splitting projection. The proof of Theorem 4.25 is based on the "change-of-variable" trick, cf. Chapter 3.

Dichotomy and solutions of nonhomogeneous equations. This section is based on Latushkin, Randolph, and Schnaubelt [**LRS**], see also Van Minh, Räbiger and Schnaubelt [**MRS**] and the comments following Chapter 3 on dichotomy for the half line and mild solutions.

Bounded solutions. Theorem 4.28 is a direct generalization of a classical theorem of Perron [**Pn**], see Daleckij and Krein [**DK**, Thm. IV.3.3'] and Massera and Schäffer [**MS2**], for the case of differentiable evolution families (or bounded $A(t)$). Recent results for the finite dimensional case are obtained by Palmer [**Pa3**] and Ben-Artzi and Gohberg [**BG3**].

For the case where the operators in the family $\{A(t) : t \in \mathbb{R}\}$ are, generally, unbounded, a result of this type concerning classical solutions of $\dot{x} = A(t)x + g(t)$, obtained by a completely different method, and applications of this result are contained in the book [**LZ**] by Levitan and Zhikov, see Chapters 10 and 11. The

"L^p-theorems" of this type can be found in Dore [**Do**]. Results of this type for nonautonomous equations on the half line are considered (under some additional assumptions) in Rodrigues and Ruas-Filho [**RR**], and a certain class of nonautonomous parabolic equations on the half line is considered by Zhang [**Zh**]. Also, see [**MRS**] and the comments following Chapter 3 for the generalization of Theorem 4.28 to the case of the half line. Finally, Theorem 4.28 is a direct generalization of a result by Prüss [**Pr**] for the autonomous case that is recorded as Theorem 2.50.

Exponentially bounded solutions. Proposition 4.32 can be found in [**MRS**]. Theorems 4.33 and 4.35 are taken from [**LRS**]. A different proof of the equivalence (i) ⇔ (iii) in Theorem 4.35 is given in Levitan and Zhikov [**LZ**, Chap. 10].

We give a short argument due to F. Räbiger [private communication] that allows one to derive the implication "Condition (M) implies dichotomy" of Theorem 4.28 without using the discrete operators $\pi_\theta(E)$; that is, without the Discrete Dichotomy Theorem 4.16. By Theorem 4.33, Condition (M_{C_0}) implies dichotomy. So, we will assume that Condition (M) holds and derive Condition (M_{C_0}).

Consider the bounded operator G on $C_b(\mathbb{R}; X)$ defined as in Remark 4.27 using Condition (M). Let $\mathcal{F}$ be the maximal closed subspace of $C_b(\mathbb{R}; X)$ which is invariant with respect to the evolution semigroup $\{E^t\}_{t\geq 0}$ and on which $\{E^t\}_{t\geq 0}$ is strongly continuous. Let $\Gamma_\mathcal{F}$ denote the generator of the restricted semigroup $\{E^t|\mathcal{F}\}_{t\geq 0}$. As in Theorem 3.28, it is easy to see that if $u \in \mathcal{D}(\Gamma_\mathcal{F})$ and $\alpha \in C_c^1(\mathbb{R})$, then $\alpha u \in \mathcal{D}(\Gamma_\mathcal{F})$ and $\Gamma_\mathcal{F}(\alpha u) = -\alpha' u + \alpha \Gamma_\mathcal{F} u$. Take $g \in C_b(\mathbb{R}; X)$ with compact support and let $u = Gg \in C_b(\mathbb{R}; X)$ be the solution of the integral equation (4.26). We want to prove that $u \in C_0(\mathbb{R}; X)$. As in Proposition 4.32, we have $u \in \mathcal{D}(\Gamma_\mathcal{F})$ and $\Gamma_\mathcal{F} u = -g$. Choose $\alpha_n \in C_c^1(\mathbb{R})$ such that $\|\alpha'\|_\infty \to 0$ as $n \to \infty$ and $\alpha_n g = g$. Then, $\Gamma_\mathcal{F}(\alpha_n u) = -\alpha_n' u + \alpha_n \Gamma_\mathcal{F} u = -\alpha_n' u - g$. In particular, $\alpha_n u = G(-\alpha_n' u - g)$ is the *unique* solution to the integral equation (4.26) that corresponds to $-\alpha_n' u - g$. Note that $\alpha_n' u + g \to g$ in $C_b(\mathbb{R}; X)$. Since G is bounded,

$$\alpha_n u = G\Gamma_\mathcal{F}(\alpha_n u) = G(\alpha_n' u + g) \to Gg = u \quad \text{in} \quad C_b(\mathbb{R}; X) \quad \text{as} \quad n \to \infty.$$

Since each $\alpha_n u$ has compact support, $u \in C_0(\mathbb{R}; X)$ and the proof is complete.

Conditions (M), (M_{C_0}), and (M_{L^p}) are related to the classical question of regular admissibility. Recall that a closed subspace $\mathcal{M}$ of the space of bounded, uniformly continuous functions $C_{UB}(\mathbb{R}; X)$ is called *regularly admissible* if for each $g \in \mathcal{M}$ there exists a unique solution $u \in \mathcal{M}$ of the mild equation (4.26). In addition to the literature cited above we mention the work of Coppel [**Co**], Datko [**Da**], Hale and Verduyn Lunel [**HL**], Prüss [**Pr2**]; see also an important paper by Vu Quoc Phong and Schüler [**PS**] for a new approach to this question that involves Lyapunov equations.

The elegant proof of the Baskakov's Theorem 4.37 is taken from [**Bs3**, Thm. 3], cf. also Lemmas 4.30–4.31. Using results of Levitan and Zhikov [**LZ**, Chap. 10], Baskakov [**Bs3**, Thm. 4] also shows that Theorem 4.37 implies the conclusions of the Dichotomy Theorem 3.17. The Spectral Mapping Theorem for $\{E^t\}_{t\geq 0}$ on $\mathcal{F} = C_0(\mathbb{R}; X)$ or $L^p(\mathbb{R}; X)$ is derived in [**Bs3**, Thm. 2] from the following formula:

$$((E^t - I)^{-1} f)(\theta) = -\sum_{k \in \mathbb{Z}} G_P(\theta, \theta + kt) f(\theta + kt), \quad t > 0, \quad \theta \in \mathbb{R},$$

that holds provided Γ is invertible. Here G_P is the Green's function for the evolution family $\{U(\theta,\tau)\}_{\theta\geq\tau}$. Using G_P, one can also show that

$$(I - \pi_0(E))^{-1} : (x_n)_{n\in\mathbb{Z}} \mapsto \left(\sum_{k\in\mathbb{Z}} G_P(n,k)x_n\right)_{n\in\mathbb{Z}}.$$

In other words, one can obtain the discrete Green's function for the operator $I - \pi_0(E)$, see also Henry [**He**, Thm. 7.6.5]. Finally, the paper [**Bs4**] contains a treatment of the half line case.

Hyperbolicity and dissipativity. The material of this section is based on the papers by Chicone and Latushkin [**ChLt, ChLt2**], see Subsection 7.5 for related results on the dichotomy of linear skew-product flows. A beautiful explanation of the second Lyapunov method (quadratic Lyapunov function), from the point of view of the change of scalar product, can be found in Sections I.5 and II.2 of Daleckij and Krein [**DK**] for the case of differentiable evolution families (for bounded $A(t)$ in the abstract Cauchy problem).

Hyperbolic semigroups. The Lumer-Phillips Theorem can be found, for example, in Pazy [**Pz**, p.14]. Theorems 4.38–4.40 are direct generalizations of the classical Generalized Lyapunov Theorem (see [**DK**, Thm. I.5.1]) for the case of differential equations $\dot{x} = Ax$ defined by an *unbounded* operator A. We stress the fact that, for the case of an unbounded generator A of a strongly continuous *semigroup*, one has to deal with *two* Lyapunov equations, see (4.40)–(4.41). See Goldstein [**Go**], Pandolfi [**Pf**], and Pazy [**Pz2**] for the Lyapunov equation with unbounded A.

Dichotomic evolution families. Arguments in this subsection again illustrate our general philosophy of "autonomization" of nonautonomous problems by passing from evolution families to associated evolution semigroups, see for example Theorem 4.46. Theorems 4.46–4.47 are taken from [**ChLt2**]. See Problems I.15–I.17 and IV.15–IV.16 in [**DK**] for a discussion related to Corollary 4.48. See Nagel and Rhandi [**NR, Rh**] for an another approach to the proofs for the results of this subsection. They associate yet another evolution semigroup on $\mathcal{L}(X)$ to the evolution family $\{U(\theta,\tau)\}_{\theta\geq\tau}$ and thereby define an action of $\{U(\theta,\tau)\}_{\theta\geq\tau}$ on a certain operator algebra.

CHAPTER 5

Two Applications of Evolution Semigroups

In this chapter we discuss applications of evolution semigroups to linear control theory and to the problem of persistence of exponential dichotomy under perturbations.

In Section 5.1 the main tool is the Howland evolution semigroup on the half line. As we have seen in Chapters 2 and 3, the stability of this semigroup is equivalent to the stability of the underlying evolution family. We will use the evolution semigroup to study several different notions of stability (internal, external, input-output stability) that appear in linear control theory. In addition, we will show how to use the evolution semigroup to give explicit formulas for the estimates of the stability radius of a control system.

Section 5.2 contains several results that ensure the persistence of dichotomy of evolution families under perturbations. Since the main object of this section is the exponential *dichotomy*, we will use the Howland evolution semigroup on the entire line. Our persistence results for the dichotomy of evolution families are obtained from the results on perturbation of evolution semigroups and their generators.

5.1. Control theory

In this section we apply the results of previous chapters to linear control theory. The main observation is that the inverse Γ_+^{-1} of the generator of the evolution semigroup $\{E_+^t\}_{t \geq 0}$ on $L^p(\mathbb{R}_+; X)$, $1 \leq p < \infty$, given by (3.17), is closely related to the input-output operator, one of the main objects in control theory. Using this observation, we will study the input-output operator in a general setting.

In Subsection 5.1.1 we give a brief account of basic concepts from linear control theory. In Subsection 5.1.2 we study the relationships between various notions of stability: internal, external, and input-output stability. In particular, we give a very short proof of the fact (Theorem 5.3) that a nonautonomous control system on a Banach space is internally stable if and only if it is stabilizable, detectable, and input-output stable. Also, we give an explicit formula (Theorem 5.8) for the norm of the input-output operator for an autonomous control system in a Banach space setting. Subsection 5.1.3 is devoted to a discussion of the stability radius; that is the size of the smallest perturbation under which a stable system loses its stability. In particular, we give sharp estimates for the stability radius (Theorem 5.15). We also discuss some new effects (Examples 5.20 and 5.22) that appear for control systems in Banach spaces that are not seen in the Hilbert space setting.

5.1.1. Some basic notions in control theory. In this subsection we give a brief account of the basic ideas of linear control theory that will be used in this chapter. For more detailed information, the reader should consult the sources in control theory listed in the bibliographical notes at the end of this chapter.

Throughout this section we assume that the evolution family is defined for $\theta \geq \tau \geq 0$, as in the case of the half line, see Section 3.3.

5.1.1.1. *A setting for control systems.* Let X, U, and Y be complex Banach spaces and consider a strongly continuous, exponentially bounded evolution family $\{U(\theta,\tau)\}_{\theta \geq \tau \geq 0}$. In particular, one can view $\{U(\theta,\tau)\}_{\theta \geq \tau \geq 0}$ as the propagator for a nonautonomous system $\dot{x} = A(t)x$, $t \geq 0$, where the operators in the family $\{A(t) : t \geq 0\}$ on X are generally unbounded. This differential equation is called the *nominal system* and X is called the *state space*. Also, let $B(t) : U \to X$ and $C(t) : X \to Y$ be bounded operators such that $B(\cdot) \in L^\infty(\mathbb{R}_+; \mathcal{L}_s(U, X))$ and $C(\cdot) \in L^\infty(\mathbb{R}_+; \mathcal{L}_s(X, Y))$. We will study the linear nonautonomous (time-varying) control system

$$(5.1) \quad \begin{aligned} x(t) &= U(t,\tau)x(\tau) + \int_\tau^t U(t,s)B(s)u(s)\,ds, \\ y(t) &= C(t)x, \quad t \geq \tau \geq 0. \end{aligned}$$

Here, U is called the *control space* and Y the *output space*. Let us note that the control system (5.1) is the mild form of the classical nonautonomous linear control system

$$(5.2) \quad \begin{aligned} \dot{x}(t) &= A(t)x(t) + B(t)u(t), \quad x(\tau) = x_\tau \in \mathcal{D}(A(\tau)), \\ y(t) &= C(t)x(t), \quad t \geq \tau \geq 0. \end{aligned}$$

Also, for the autonomous case, where $U(\theta,\tau) = e^{(\theta-\tau)A}$, $\theta \geq \tau \geq 0$, is given by a strongly continuous semigroup on X generated by A and the operators $B(t) \equiv B$ and $C(t) \equiv C$ are both operators independent of t, the mild control system (5.1) has the form

$$(5.3) \quad \begin{aligned} x(t) &= e^{tA}x_0 + \int_0^t e^{(t-s)A}Bu(s)\,ds, \\ y(t) &= Cx(t), \quad t \geq 0. \end{aligned}$$

The *input-output operator* $\mathbb{L}$ for the nonautonomous system (5.1) is defined on functions $u : \mathbb{R}_+ \to U$ by the rule

$$(5.4) \quad (\mathbb{L}u)(\theta) = C(\theta)\int_0^\theta U(\theta,s)B(s)u(s)\,ds, \quad \theta \in \mathbb{R}_+.$$

We will consider $\mathbb{L}$ as an operator from $L^p(\mathbb{R}_+; U)$ to $L^p(\mathbb{R}_+; Y)$, $1 \leq p < \infty$, although *a priori* $\mathbb{L}$ is not a bounded operator between these two spaces.

Consider the autonomous system (5.3), and for $\lambda \in \rho(A)$, define the *transfer function* H by $H(\lambda) = C(\lambda - A)^{-1}B$. The input-output operator $\mathbb{L}$ and the transfer function H are related as follows: If $\{e^{tA}\}_{t\geq 0}$ is stable, $p = 2$, and X, U, and Y are Hilbert spaces, then H is the unique bounded, analytic $\mathcal{L}(U, Y)$-valued function defined on $\mathbb{C}_+ = \{\lambda \in \mathbb{C} : \operatorname{Re} \lambda > 0\}$ such that

$$\widehat{\mathbb{L}u}(\lambda) = H(\lambda)\hat{u}(\lambda), \quad \lambda \in \mathbb{C}_+,$$

whenever $u \in L^2(\mathbb{R}_+; U)$ and "hat" denotes the Laplace transform (see, e.g., Weiss [**Ws3**]).

There are several notions of stability for the control systems (5.1) and (5.3).

DEFINITION 5.1. (i) A nonautonomous system (5.1) is called *internally stable* if the evolution family $\{U(\theta,\tau)\}_{\theta \geq \tau \geq 0}$ is uniformly exponentially stable (in the sense of Definition 3.4).

(ii) A nonautonomous system (5.1) is called *input-output stable* if the corresponding input-output operator $\mathbb{L}$ is a bounded operator from $L^p(\mathbb{R}_+; U)$ to $L^p(\mathbb{R}_+; Y)$, $1 \leq p < \infty$.

(iii) An autonomous system (5.3) is called *externally stable* if the transfer function $H(\lambda) = C(A - \lambda)^{-1} B$ is a bounded, analytic function of λ in the right half plane $\mathbb{C}_+ = \{\lambda \in \mathbb{C} : \operatorname{Re} \lambda > 0\}$.

Let us note that by the analyticity of the resolvent,
$$\sup_{\lambda \in \mathbb{C}_+} \|C(A - \lambda)^{-1} B\| = \sup_{s \in \mathbb{R}} \|C(A - is)^{-1} B\|$$
for every externally stable autonomous system. As we will see, internal stability implies input-output stability and external stability. Moreover, if X, U, and Y are Hilbert spaces and $p = 2$, then external stability is equivalent to input-output stability, see for example Weiss [**Ws2, Ws3**] and Curtain, Logemann, Townley, and Zwart [**CLTZ**].

The next definition is used in Subsection 5.1.2 to describe the assumptions under which input-output stability implies internal stability.

DEFINITION 5.2. (a) The nonautonomous system (5.1) is said to be *stabilizable* if there exists $F \in L^\infty(\mathbb{R}_+; \mathcal{L}_s(X, U))$ and a corresponding uniformly exponentially stable evolution family $\{U_{BF}(\theta, \tau)\}_{\theta \geq \tau \geq 0}$ such that

(5.5) $$U_{BF}(\theta, \tau) x = U(\theta, \tau) x + \int_\tau^\theta U(\theta, s) B(s) F(s) U_{BF}(s, \tau) x \, ds$$

for all $\theta \geq \tau \geq 0$ and $x \in X$.

(b) The nonautonomous system (5.1) is said to be *detectable* if there exists $K \in L^\infty(\mathbb{R}_+; \mathcal{L}_s(Y, X))$ and a corresponding uniformly exponentially stable evolution family $\{U_{KC}(\theta, \tau)\}_{\theta \geq \tau \geq 0}$ such that

(5.6) $$U_{KC}(\theta, \tau) x = U(\theta, \tau) x + \int_\tau^\theta U_{KC}(\theta, s) K(s) C(s) U(s, \tau) \, ds$$

for all $\theta \geq \tau \geq 0$ and $x \in X$.

An autonomous control system is called *stabilizable* if there is an operator $F \in \mathcal{L}(X, U)$ such that $A + BF$ generates a uniformly exponentially stable semigroup; that is, $\omega(A + BF) < 0$. It is called *detectable* if there is an operator $K \in \mathcal{L}(Y, X)$ such that $A + KC$ generates a uniformly exponentially stable semigroup.

5.1.1.2. *The input-output operator and evolution semigroups.* We will describe an important relationship between linear control systems and evolution semigroups.

Suppose that $\{U(\theta, \tau)\}_{\theta \geq \tau \geq 0}$ is the evolution family in (5.1), and consider the Howland evolution semigroup $\{E_+^t\}_{t \geq 0}$ defined on $L^p(\mathbb{R}_+; X)$ by the formula (3.17); that is, the evolution semigroup given by

(5.7) $(E_+^t f)(\theta) = U(\theta, \theta - t) f(\theta - t)$ for $\theta \geq t$, and $(E_+^t f)(\theta) = 0$ for $0 \leq \theta < t$.

Also, consider the corresponding Green's operator $\mathbb{G}$ defined in display (3.21):

$$(\mathbb{G} f)(\theta) = \int_0^\infty (E_+^t f)(\theta) \, dt = \int_0^\theta U(\theta, \theta - t) f(\theta - t) \, dt$$

(5.8) $$= \int_0^\theta U(\theta, t) f(t) \, dt, \quad \theta \geq 0.$$

Using the functions $B(\cdot)$ and $C(\cdot)$ in (5.1), define on $L^p(\mathbb{R}_+;U)$ and $L^p(\mathbb{R}_+;X)$, respectively, the multiplication operators $\mathcal{B}$ and $\mathcal{C}$ as follows: $(\mathcal{B}u)(\theta) = B(\theta)u(\theta)$ and $(\mathcal{C}f)(\theta) = C(\theta)f(\theta)$, $\theta \in \mathbb{R}_+$. By the definition (5.4) of the input-output operator $\mathbb{L}$ and equation (5.8), we have that $\mathbb{L} = \mathcal{C}\mathbb{G}\mathcal{B}$.

By Theorem 3.26, the evolution family $\{U(\theta,\tau)\}_{\theta \geq \tau \geq 0}$ is (internally) stable if and only if $\mathbb{G}$ is a bounded operator in $L^p(\mathbb{R}_+;X)$; and, if this is the case, then $\mathbb{G} = -\Gamma_+^{-1}$ where Γ_+ is the generator of the semigroup $\{E_+^t\}_{t \geq 0}$ (cf. Corollary 3.27). Thus, if system (5.1) is internally stable, then

$$\mathbb{L} = \mathcal{C}\mathbb{G}\mathcal{B} = -\mathcal{C}\Gamma_+^{-1}\mathcal{B}.$$

Moreover, since $\mathcal{C}$ and $\mathcal{B}$ are bounded, it is clear that internal stability implies input-output stability. For the transfer function H, it is known that $\sup_{\lambda \in \mathbb{C}_+} \|H(\lambda)\| \leq \|\mathbb{L}\|$ in the Banach space setting and equality holds in this estimate in the Hilbert space setting if $p = 2$, see Weiss [**Ws3**] or Theorem 2.3 and Remark 2.4 in [**Ws2**]. Thus, it follows that external stability is implied by input-output stability in the Banach space setting, and these notions are equivalent in the Hilbert space setting if $p = 2$.

5.1.2. Stability and the input-output operator. In this subsection we relate the notions of internal and input-output stability for the general nonautonomous system (5.1) in the Banach space setting. The main result of this subsection is the following fact.

THEOREM 5.3. *The system* (5.1) *is internally stable if and only if it is stabilizable, detectable, and input-output stable.*

Theorem 5.3 appears as part of Theorem 5.4 below. We will also consider the autonomous system (5.3) and give an explicit formula for the norm of its input-output operator in terms of the operators A, B, and C. It turns out that the quantity $\|\mathbb{L}\|^{-1}$ gives a lower estimate for the stability radius of the system, see Subsection 5.1.3. Finally, using the stabilizability and detectability, we give explicit conditions in terms of A, B, and C that imply the internal stability of the autonomous system.

5.1.2.1. *Internal versus input-output stability: nonautonomous case.* We will use Theorem 3.26 to characterize uniform exponential stability in terms of the operator $\mathbb{G}$ as in (5.8) and prove the following theorem by a straightforward manipulation of the appropriate operators.

THEOREM 5.4. *The following statements are equivalent for a strongly continuous, exponentially bounded evolution family* $\mathcal{U} = \{U(\theta,\tau)\}_{\theta \geq \tau \geq 0}$ *on a Banach space* X.

(i) $\mathcal{U}$ *is uniformly exponentially stable on* X *(that is, the system* (5.1) *is internally stable).*
(ii) $\mathbb{G}$ *is a bounded operator on* $L^p(\mathbb{R}_+;X)$.
(iii) *System* (5.1) *is stabilizable and* $\mathbb{G}\mathcal{B}$ *is a bounded operator from* $L^p(\mathbb{R}_+;U)$ *to* $L^p(\mathbb{R}_+;X)$.
(iv) *System* (5.1) *is detectable and* $\mathcal{C}\mathbb{G}$ *is a bounded operator from* $L^p(\mathbb{R}_+;X)$ *to* $L^p(\mathbb{R}_+;Y)$.
(v) *System* (5.1) *is stabilizable and detectable, and* $\mathbb{L} = \mathcal{C}\mathbb{G}\mathcal{B}$ *is a bounded operator from* $L^p(\mathbb{R}_+;U)$ *to* $L^p(\mathbb{R}_+;Y)$ *(that is, the system* (5.1) *is input-output stable).*

PROOF. The equivalence of statements (i) and (ii) is the equivalence of statements (i) and (ii) in the nonautonomous Datko-van Neerven Theorem 3.26.

To see that statement (ii) implies statements (iii), (iv), and (v), note that $\mathcal{B}$ and $\mathcal{C}$ are bounded; and therefore, $\mathbb{L}$ is bounded whenever $\mathbb{G}$ is bounded. Hence, statement (ii), the uniform exponential stability of $\mathcal{U}$, together with the boundedness of $B(\cdot)$, $C(\cdot)$, $F(\cdot)$, and $K(\cdot)$, ensure the existence of the evolution families $\{U_{BF}(\theta,\tau)\}_{\theta\geq\tau}$ and $\{U_{KC}(\theta,\tau)\}_{\theta\geq\tau}$ as solutions of the integral equations in Definition 5.2. This proves statements (iii), (iv), and (v).

To prove the implication (iii)$\Rightarrow$(ii), first note that the assumption of stabilizability ensures the existence of a uniformly exponentially stable evolution family $\mathcal{U}_{BF} = \{U_{BF}(\theta,\tau)\}_{\theta\geq\tau\geq 0}$ satisfying equation (5.5) for some $F \in L^\infty(\mathbb{R}_+; \mathcal{L}_s(X,U))$. Given this uniformly exponentially stable evolution family, we define the operator $\mathbb{G}_{BF}$ by

$$(5.9) \qquad \mathbb{G}_{BF}f(\tau) := \int_0^\tau U_{BF}(\tau,t)f(t)\,dt = \int_0^\infty (E_{BF}^t f)(\tau)\,dt$$

where $\{E_{BF}^t\}_{t\geq 0}$ is the semigroup induced by the evolution family $\mathcal{U}_{BF}$ as described in equation (5.7). $\mathbb{G}_{BF}$ is a bounded operator on $L^p(\mathbb{R}_+; X)$ by the equivalence of statements (i) and (ii).

For $f \in L^p(\mathbb{R}_+; X)$ and $\tau \in \mathbb{R}_+$, put $x = f(\tau)$ in equation (5.5). Then, set $\xi = s - \tau$ to obtain

$$U_{BF}(\theta,\tau)f(\tau) = U(\theta,\tau)f(\tau)$$
$$+ \int_0^{\theta-\tau} U(\theta,\xi+\tau)B(\xi+\tau)F(\xi+\tau)U_{BF}(\xi+\tau,\tau)f(\tau)\,d\xi.$$

From this equation and the definition of the semigroups $\{E_+^t\}$ and $\{E_{BF}^t\}$, we obtain the equality

$$(E_{BF}^{\theta-\tau}f)(\theta) = (E_+^{\theta-\tau}f)(\theta) + \int_0^{\theta-\tau}(E_+^{\theta-\tau-\xi}\mathcal{B}\mathcal{F}E_{BF}^\xi f)(\theta)\,d\xi;$$

and hence, for $0 \leq r$ and $0 \leq \sigma$, we have

$$(E_{BF}^r f)(\sigma) = (E_+^r f)(\sigma) + \int_0^r (E_+^{r-\xi}\mathcal{B}\mathcal{F}E_{BF}^\xi f)(\sigma)\,d\xi.$$

Integrate from 0 to ∞ to obtain

$$(\mathbb{G}_{BF}f)(\sigma) = (\mathbb{G}f)(\sigma) + \int_0^\infty \int_0^r (E_+^{r-\xi}\mathcal{B}\mathcal{F}E_{BF}^\xi f)(\sigma)\,d\xi\,dr.$$

Then, set $r = \zeta + \eta$ and $\xi = \eta$ to obtain

$$(5.10) \qquad \begin{aligned}(\mathbb{G}_{BF}f)(\sigma) &= (\mathbb{G}f)(\sigma) + \int_0^\infty \int_0^\infty (E_+^\zeta \mathcal{B}\mathcal{F}E_{BF}^\eta f)(\sigma)\,d\eta\,d\zeta \\ &= (\mathbb{G}f)(\sigma) + (\mathbb{G}\mathcal{B}\mathcal{F}\mathbb{G}_{BF}f)(\sigma).\end{aligned}$$

In other words, $\mathbb{G}_{BF} = \mathbb{G} + (\mathbb{G}\mathcal{B})\mathcal{F}\mathbb{G}_{BF}$. The fact that $\mathbb{G}$ is bounded now follows from the boundedness of $\mathbb{G}\mathcal{B}$, and the boundedness of $\mathbb{G}_{BF}$ and $\mathcal{F}$.

To prove the implication (iv)$\Rightarrow$(ii), first note that the assumption of detectability ensures the existence of a uniformly exponentially stable evolution family $\mathcal{U}_{KC} = \{U_{KC}(\theta,t)\}_{\theta\geq\tau\geq 0}$ satisfying equation (5.6) for some $K \in L^\infty(\mathbb{R}_+; \mathcal{L}_s(Y,X))$. Given this uniformly exponentially stable evolution family, the operator $\mathbb{G}_{KC}$, defined in a manner analogous to $\mathbb{G}_{BF}$ in equation (5.9), is a bounded operator on $L^p(\mathbb{R}_+; X)$.

A derivation beginning with equation (5.6), and similar to that which gave equation (5.10), now gives $\mathbb{G}_{KC} = \mathbb{G} + \mathbb{G}_{KC}\mathcal{K}(\mathcal{C}\mathbb{G})$. This equation, together with the assumed boundedness of $\mathbb{G}_{KC}$, $\mathcal{K}$, and $\mathcal{C}\mathbb{G}$, gives the boundedness of $\mathbb{G}$.

Finally, to prove the implication (v)$\Rightarrow$(ii), again note that the assumption of detectability yields a uniformly exponentially stable evolution family $\mathcal{U}_{KC}$ and an associated bounded operator $\mathbb{G}_{KC}$. For $u \in L^p(\mathbb{R}; U)$ and $\tau \in \mathbb{R}_+$, put $x = B(\tau)u(\tau)$ in equation (5.6). A calculation similar to that which gave equation (5.10) now gives $\mathbb{G}_{KC}\mathcal{B} = \mathbb{G}\mathcal{B} + \mathbb{G}_{KC}\mathcal{K}\mathcal{C}\mathbb{G}\mathcal{B}$. Since $\mathbb{L} = \mathcal{C}\mathbb{G}\mathcal{B}$, $\mathcal{K}$, and $\mathbb{G}_{KC}$ are bounded, so is $\mathbb{G}\mathcal{B}$. By the boundedness of $\mathbb{G}\mathcal{B}$, the assumption of stabilizability, and the equivalence of statements (iii) and (ii), the operator $\mathbb{G}$ is bounded. $\square$

5.1.2.2. *The norm of the input-output operator.* Here we compute the norm of the input-output operator $\mathbb{L}$ for the autonomous system (5.3) using constructions similar to those in Theorem 2.49. Our objective is to provide explicit formulas in terms of the operators A, B, and C.

Recall that $\mathcal{S}(\mathbb{R}; X)$ denotes the Schwartz class of rapidly decaying X-valued functions and assume $\sup_{s \in \mathbb{R}} \|(A - is)^{-1}\| < \infty$. As in Proposition 2.48, let us define, for each $v \in \mathcal{S}(\mathbb{R}; X)$, the functions

$$f_v(\theta) = \frac{1}{2\pi} \int_{\mathbb{R}} (A - is)^{-1} v(s) e^{i\theta s}\, ds, \quad g_v(\theta) = \frac{1}{2\pi} \int_{\mathbb{R}} v(s) e^{i\theta s}\, ds, \quad \theta \in \mathbb{R}.$$

We will need the following analogue of Proposition 2.48, a proposition that was used to derive a formula for the norm of the inverse of the generator of the evolution semigroup.

PROPOSITION 5.5. *The set $\mathfrak{G}_U = \{g_u : u \in \mathcal{S}(\mathbb{R}, U)\}$ is dense in $L^p(\mathbb{R}; U)$. If $u \in \mathcal{S}(\mathbb{R}, U)$ and $B \in \mathcal{L}(U, X)$, then $Bu \in \mathcal{S}(\mathbb{R}, X)$, $f_{Bu} \in \mathcal{D}(\Gamma)$, and $\Gamma f_{Bu} = \mathcal{B} g_u$.*

PROOF. The proof of the first statement is similar to the proof of the implication (i)$\Rightarrow$(ii) in Proposition 2.48. The proof of the second statement is similar to the proof of statement (iii) in Proposition 2.48. It uses the properties of Schwartz functions and the following computation as in display (2.39):

$$\Gamma f_{Bu}(\tau) = g_{Bu}(\tau) = \frac{1}{2\pi} \int_{\mathbb{R}} Bu(s) e^{is\tau}\, ds = B \frac{1}{2\pi} \int_{\mathbb{R}} u(s) e^{is\tau}\, ds.$$

$\square$

By Theorem 2.39 and Corollary 2.46, the semigroup $\{e^{tA}\}_{t\geq 0}$ is hyperbolic (resp., stable) on X if and only if Γ (resp., Γ_+) is invertible on $L^p(\mathbb{R}; X)$ (resp., $L^p(\mathbb{R}_+; X)$), $1 \leq p < \infty$. Recall that for the autonomous system the operators $\mathcal{B}$ and $\mathcal{C}$ are defined by $(\mathcal{B}u)(s) = Bu(s)$ and $(\mathcal{C}u)(s) = Cu(s)$ for $s \in \mathbb{R}$ or $s \in \mathbb{R}_+$.

THEOREM 5.6. *If Γ is invertible on $L^p(\mathbb{R}; X)$, then the norm of $\mathcal{C}\Gamma^{-1}\mathcal{B}$, as an operator from $L^p(\mathbb{R}; U)$ to $L^p(\mathbb{R}; Y)$, is given by the formula*

$$(5.11) \qquad \|\mathcal{C}\Gamma^{-1}\mathcal{B}\| = \sup_{u \in \mathcal{S}(\mathbb{R}, U)} \frac{\|\int_{\mathbb{R}} C(A - is)^{-1} Bu(s) e^{is(\cdot)}\, ds\|_{L^p(\mathbb{R}; Y)}}{\|\int_{\mathbb{R}} u(s) e^{is(\cdot)}\, ds\|_{L^p(\mathbb{R}; U)}}.$$

If Γ_+ is invertible on $L^p(\mathbb{R}_+; X)$, then the norm of the input-output operator $\mathbb{L} = \mathcal{C}\Gamma_+^{-1}\mathcal{B}$, as an operator from $L^p(\mathbb{R}_+; U)$ to $L^p(\mathbb{R}_+; Y)$, is given by

$$(5.12) \qquad \|\mathbb{L}\| = \|\mathcal{C}\Gamma^{-1}\mathcal{B}\|.$$

5.1. CONTROL THEORY

If, in addition, U and Y are Hilbert spaces and $p = 2$, then

(5.13) $$\|\mathbb{L}\| = \sup_{s \in \mathbb{R}} \|C(A - is)^{-1} B\|_{\mathcal{L}(U,Y)}.$$

PROOF. For $u \in \mathcal{S}(\mathbb{R}, U)$, consider the functions f_{Bu} and g_u. Proposition 5.5 gives $f_{Bu} = \Gamma^{-1} \mathcal{B} g_u$ and

$$\|\mathcal{C}\Gamma^{-1}\mathcal{B}\| = \sup_{g_u \in \mathfrak{G}_U} \frac{\|\mathcal{C}\Gamma^{-1}\mathcal{B}g_u\|_{L^p(\mathbb{R};Y)}}{\|g_u\|_{L^p(\mathbb{R};U)}} = \sup_{g_u \in \mathfrak{G}_U} \frac{\|\mathcal{C}f_{Bu}\|}{\|g_u\|}$$

$$= \sup_{u \in \mathcal{S}(\mathbb{R},U)} \frac{\|\int_{\mathbb{R}} C(A-is)^{-1} Bu(s) e^{is(\cdot)} \, ds\|_{L^p(\mathbb{R};Y)}}{\|\int_{\mathbb{R}} u(s) e^{is(\cdot)} \, ds\|_{L^p(\mathbb{R};U)}},$$

which proves (5.11).

Now, if Γ_+ is invertible on $L^p(\mathbb{R}_+; X)$, then $\{e^{tA}\}_{t \geq 0}$ is uniformly exponentially stable by Corollary 2.46. Hence, Γ is invertible on $L^p(\mathbb{R}; X)$. Moreover, for the case of the *stable* semigroup $\{e^{tA}\}_{t \geq 0}$, the formula for Γ^{-1} has the form

$$(\Gamma^{-1} f)(\theta) = \int_0^\infty e^{sA} f(\theta - s) \, ds = \int_{-\infty}^0 e^{-sA} f(\theta + s) \, ds$$

$$= \int_{-\infty}^\theta e^{(\theta - s)A} f(s) \, ds, \quad f \in L^p(\mathbb{R}; X).$$

If $\operatorname{supp} f \subseteq (0, \infty)$, then

(5.14) $$(\Gamma^{-1} f)(\theta) = \int_{-\infty}^\theta e^{(\theta-s)A} f(s) \, ds = \int_0^\theta e^{(\theta-s)A} f(s) \, ds.$$

For a function $h \in L^p(\mathbb{R}_+; X)$, define an extension $\tilde{h} \in L^p(\mathbb{R}; X)$ by $\tilde{h}(\theta) = h(\theta)$ for $\theta \geq 0$, and $\tilde{h}(\theta) = 0$ for $\theta < 0$. Then, by the identity (5.14), we have $\Gamma^{-1}\tilde{h} = (\Gamma_+^{-1} h)^\sim$. In particular, for $u \in L^p(\mathbb{R}_+; U)$, we have

$$\widetilde{\mathbb{L} u} = \widetilde{\mathcal{C}\Gamma_+^{-1}\mathcal{B} u} = \mathcal{C}\Gamma^{-1}\mathcal{B}\tilde{u}.$$

Hence,

$$\|\mathbb{L} u\|_{L^p(\mathbb{R}_+;Y)} = \|\widetilde{\mathbb{L} u}\|_{L^p(\mathbb{R};Y)} = \|\mathcal{C}\Gamma^{-1}\mathcal{B}\tilde{u}\|_{L^p(\mathbb{R};Y)}$$
$$\leq \|\mathcal{C}\Gamma^{-1}\mathcal{B}\| \cdot \|\tilde{u}\|_{L^p(\mathbb{R};U)} = \|\mathcal{C}\Gamma^{-1}\mathcal{B}\| \cdot \|u\|_{L^p(\mathbb{R}_+;U)},$$

and $\|\mathbb{L}\| \leq \|\mathcal{C}\Gamma^{-1}\mathcal{B}\|$.

To prove that equality holds in (5.12), let $\epsilon > 0$ and choose $u \in L^p(\mathbb{R}; U)$ such that $\|u\| = 1$ and

$$\|\mathcal{C}\Gamma^{-1}\mathcal{B} u\|_{L^p(\mathbb{R};Y)} \geq \|\mathcal{C}\Gamma^{-1}\mathcal{B}\| - \epsilon.$$

Without loss of generality, u may be assumed to have compact support. Now choose r such that $\operatorname{supp} u(\cdot - r) \subseteq (0, \infty)$ and set $w(\cdot) := u(\cdot - r)$. Then, $w \in L^p(\mathbb{R}; U)$ and $\operatorname{supp} w \subseteq (0, \infty)$. Let $\overline{w}$ denote the element of $L^p(\mathbb{R}_+; U)$ that coincides with w on $\mathbb{R}_+$. As in equality (5.14), we have

$$\mathcal{C}\Gamma^{-1}\mathcal{B} w(\theta) = C \int_0^\theta e^{(\theta-s)A} Bw(s) \, ds = C \int_{-\infty}^\theta e^{(\theta-s)A} Bw(s) \, ds.$$

Since $\|\overline{w}\|_{L^p(\mathbb{R}_+;U)} = \|w\|_{L^p(\mathbb{R};U)} = \|u\|_{L^p(\mathbb{R};U)} = 1$, it follows that

$$\|\mathbb{L}\| \geq \|\mathbb{L}\overline{w}\|_{L^p(\mathbb{R}_+;Y)} = \|\widetilde{\mathbb{L}\overline{w}}\|_{L^p(\mathbb{R};Y)} = \|\mathbb{L}\widetilde{\overline{w}}\|_{L^p(\mathbb{R};Y)} = \|\mathcal{C}\Gamma^{-1}\mathcal{B}w\|_{L^p(\mathbb{R};Y)}$$

$$= \|C\int_{-\infty}^{\cdot} e^{(\cdot-\tau)A}Bu(\tau)\,d\tau\|_{L^p(\mathbb{R};Y)} = \|\mathcal{C}\Gamma^{-1}\mathcal{B}u\|_{L^p(\mathbb{R};Y)} \geq \|\mathcal{C}\Gamma^{-1}\mathcal{B}\| - \epsilon.$$

This confirms (5.12).

To see (5.13), apply Plancherel's formula in the right hand side of equality (5.11) to obtain

(5.15) $$\|\mathcal{C}\Gamma^{-1}\mathcal{B}\| = \sup_{u\in\mathcal{S}(\mathbb{R};U)} \frac{\left(\int_\mathbb{R} |C(A-is)^{-1}Bu(s)|_Y^2\,ds\right)^{1/2}}{\left(\int_\mathbb{R} |u(s)|_U^2\,ds\right)^{1/2}}.$$

The last expression is less than or equal to $\sup_{s\in\mathbb{R}} \|C(A-is)^{-1}B\|_{\mathcal{L}(U,Y)}$, and so "≤" in (5.13) holds. To prove "≥" in (5.13), fix $\epsilon > 0$, select $s_0 \in \mathbb{R}$ and $u_0 \in U$ such that $|u_0| = 1$ and $|C(A-is_0)^{-1}Bu_0| \geq \sup_{s\in\mathbb{R}} \|C(A-is)^{-1}B\| - \epsilon$. Choose $\alpha \in C_c^1(\mathbb{R})$ such that $\|\alpha\|_{L^2} = 1$ and, for all $s \in \text{supp}\,\alpha$, one has

$$\|C[(A-is)^{-1} - (A-is_0)^{-1}]B\| \leq \epsilon.$$

Also, define $u(s) := \alpha(s)u_0$, $h(s) := C(A-is)^{-1}Bu(s)$,

$$h_1(s) := C[(A-is)^{-1} - (A-is_0)^{-1}]Bu(s),$$

and $h_0(s) := C(A-is_0)^{-1}Bu(s)$ for $s \in \mathbb{R}$. We have that $\|u\|_{L^2} = 1$ and the right hand side of equality (5.15) is estimated from below by

$$\|h\|_{L^2} = \|h_0 + h_1\|_{L^2} \geq \|h_0\|_{L^2} - \|h_1\|_{L^2}$$
$$\geq |C(A-is_0)^{-1}Bu_0| - \epsilon \geq \sup_{s\in\mathbb{R}} \|C(A-is)^{-1}B\| - 2\epsilon.$$

□

5.1.2.3. *Internal versus input-output stability: autonomous case.* In this section we prove a theorem that refines Theorems 5.3 and 5.4. In particular, we will formulate the conditions that imply internal stability in terms of the operators A, B, and C.

Let us note that, in the Hilbert space setting, the words "input-output stable" in Theorem 5.3 can be replaced by "externally stable", see Proposition 5.7 below. As motivation, let us review some known properties of autonomous systems. Let $\{e^{tA}\}_{t\geq 0}$ be a strongly continuous semigroup generated by A on X, and let $H_+^\infty(\mathcal{L}(X))$ denote the space of operator-valued functions $G: \mathbb{C}_+ \to \mathcal{L}(X)$ which are analytic on $\mathbb{C}_+$ with $\sup_{\lambda\in\mathbb{C}_+} \|G(\lambda)\| < \infty$. If X is a Hilbert space, then the analyticity of the resolvent together with the Gearhart Spectral Mapping Theorem 2.10, see Remark 2.13, imply $\{e^{tA}\}_{t\geq 0}$ is uniformly exponentially stable if and only if $G: \lambda \mapsto (\lambda - A)^{-1}$ is an element of $H_+^\infty(\mathcal{L}(X))$, see also Curtain and Zwart [**CZ**], Theorem 5.1.5. In other words, this is a consequence of the fact that if X is a Hilbert space, then $s_0(A) = \omega(A)$ where s_0 is the abscissa of uniform boundedness of the resolvent, see (2.9). Extending these ideas to address system (5.3), let us consider $H(\lambda) = C(\lambda - A)^{-1}B$. We have the following proposition, see the paper [**Re2**] by Rebarber and Theorem 5.8 of Curtain, Logemann, Townley, and Zwart [**CLTZ**] for a more general result of this type.

PROPOSITION 5.7. *If U and Y are Hilbert spaces, then system (5.3) is internally stable if and only if it is stabilizable, detectable, and externally stable (that is, $H(\cdot) \in H_+^\infty(\mathcal{L}(U,Y))$).*

If X is a Banach space, then strict inequality $s_0(A) < \omega(A)$ can hold. Thus, uniform exponential stability of the semigroup $\{e^{tA}\}_{t \geq 0}$ is no longer valid if we have $G \in H_+^\infty(\mathcal{L}(X))$ where $G(\lambda) = (\lambda - A)^{-1}$. Similarly, in Banach spaces, the conditions of external stability, stabilizability, and detectability are *not* sufficient to ensure the internal stability; that is, $H \in H_+^\infty(\mathcal{L}(X))$ where $H(\lambda) = C(\lambda - A)^{-1}B$, does not imply internal stability. Indeed, let A generate a semigroup for which $s_0(A) < \omega(A) = 0$ (see Example 5.20). Then, the system (5.3) with $B = I$ and $C = I$ is trivially stabilizable, detectable, and externally stable. But since $\omega(A) = 0$, it is not internally stable. Since Proposition 5.7, concerning external stability, fails for Banach-space systems (5.3), we will prove the following extension of this proposition.

Let $A_\alpha := A - \alpha$ denote the generator of the rescaled semigroup $\{e^{-\alpha t} e^{tA}\}_{t \geq 0}$.

THEOREM 5.8. *Let $\{e^{tA}\}_{t \geq 0}$ be a strongly continuous semigroup on a Banach space X generated by A. Let U and Y be Banach spaces and assume $B \in \mathcal{L}(U, X)$ and $C \in \mathcal{L}(X, Y)$. Then, the following statements are equivalent:*

(1) $\{e^{tA}\}_{t \geq 0}$ *is uniformly exponentially stable.*

(2) $\mathbb{G}$ *is a bounded operator on $L^p(\mathbb{R}_+; X)$.*

(3) $\sigma(A) \cap \overline{\mathbb{C}}_+ = \emptyset$ *and* $\displaystyle\sup_{v \in \mathcal{S}(\mathbb{R}, X)} \frac{\|\int_\mathbb{R} (A_\alpha - is)^{-1} v(s) e^{is(\cdot)} ds\|_{L^p(\mathbb{R}; X)}}{\|\int_\mathbb{R} v(s) e^{is(\cdot)} ds\|_{L^p(\mathbb{R}; X)}} < \infty$ *for all $\alpha \geq 0$.*

(4) $\sigma(A) \cap \overline{\mathbb{C}}_+ = \emptyset$, $\displaystyle\sup_{u \in \mathcal{S}(\mathbb{R}, U)} \frac{\|\int_\mathbb{R} (A_\alpha - is)^{-1} Bu(s) e^{is(\cdot)} ds\|_{L^p(\mathbb{R}; X)}}{\|\int_\mathbb{R} u(s) e^{is(\cdot)} ds\|_{L^p(\mathbb{R}; U)}} < \infty$ *for all $\alpha \geq 0$, and (5.3) is stabilizable.*

(5) $\sigma(A) \cap \overline{\mathbb{C}}_+ = \emptyset$, $\displaystyle\sup_{v \in \mathcal{S}(\mathbb{R}, X)} \frac{\|\int_\mathbb{R} C(A_\alpha - is)^{-1} v(s) e^{is(\cdot)} ds\|_{L^p(\mathbb{R}; Y)}}{\|\int_\mathbb{R} v(s) e^{is(\cdot)} ds\|_{L^p(\mathbb{R}; X)}} < \infty$ *for all $\alpha \geq 0$ and (5.3) is detectable.*

(6) $\sigma(A) \cap \overline{\mathbb{C}}_+ = \emptyset$, $\displaystyle\sup_{u \in \mathcal{S}(\mathbb{R}, U)} \frac{\|\int_\mathbb{R} C(A_\alpha - is)^{-1} Bu(s) e^{is(\cdot)} ds\|_{L^p(\mathbb{R}; Y)}}{\|\int_\mathbb{R} u(s) e^{is(\cdot)} ds\|_{L^p(\mathbb{R}; U)}} < \infty$ *for all $\alpha \geq 0$ and (5.3) is both stabilizable and detectable.*

Moreover, if $\{e^{tA}\}_{t \geq 0}$ is uniformly exponentially stable, then the norm of the input-output operator $\mathbb{L} = \mathcal{C}\mathbb{G}\mathcal{B}$, as an operator from $L^p(\mathbb{R}_+; U)$ to $L^p(\mathbb{R}_+; Y)$, is equal to

$$\sup_{u \in \mathcal{S}(\mathbb{R}, U)} \frac{\|\int_\mathbb{R} C(A - is)^{-1} Bu(s) e^{is(\cdot)} ds\|_{L^p(\mathbb{R}; Y)}}{\|\int_\mathbb{R} u(s) e^{is(\cdot)} ds\|_{L^p(\mathbb{R}; U)}}.$$

PROOF. First, note that the equivalence of statements (1) and (2) follows from the autonomous Datko–van Neerven Theorem 2.44. Also, the implication (1)$\Rightarrow$(6) and the last statement of the theorem follow from Theorem 5.6. The uniform exponential stability of $\{e^{tA}\}_{t \geq 0}$ implies the invertibility of Γ (Theorem 2.39), and so statement (3) follows from statement (1) by Theorem 2.49(ii).

To prove the implication (3)⇒(1), begin by setting $\alpha = 0$. We wish to use the properties of $\mathfrak{F}$ as in Proposition 2.48. We begin by observing that if the expression in statement (3) is finite, then $\sup_{s \in \mathbb{R}} \|(A - is)^{-1}\| < \infty$. Indeed, if this were not the case, then there would exist sequences $\{s_n\}_{n=1}^{\infty} \subset \mathbb{R}$ and $\{x_n\}_{n=1}^{\infty} \subset \mathcal{D}(A)$ with $|x_n| = 1$ such that $|(A - is_n)x_n| \to 0$ as $n \to \infty$. We will construct a sequence of scalar Schwartz functions $\{\beta_n\}_{n=1}^{\infty} \subset \mathcal{S}(\mathbb{R})$ such that

$$(5.16) \qquad \lim_{n \to \infty} \frac{\|\int_{\mathbb{R}} \beta_n(s)(is_n - is)e^{is(\cdot)}\,ds\|_{L^p(\mathbb{R})}}{\|\int_{\mathbb{R}} \beta_n(s)e^{is(\cdot)}\,ds\|_{L^p(\mathbb{R})}} = 0.$$

Without loss of generality, assume that $s_n = 0$ in (5.16) and choose a bump-function $\beta_0(\cdot)$ with support in $(-1, 1)$ such that $\beta_0(0) = 1$. Then, set $\beta_n(s) := n\beta_0(ns)$. Let us denote the inverse Fourier transform by the symbol "ˇ". Then, $\check{\beta}_n(\tau) = \check{\beta}_0(\tau/n)$. Also, for $\alpha_n(s) = s\beta_n(s)$, we have the identity $\check{\alpha}_n(\tau) = \check{\alpha}_0(\tau/n)/n$. Hence,

$$\frac{\|\int_{\mathbb{R}} \beta_n(s)s e^{is(\cdot)}\,ds\|_{L^p(\mathbb{R})}^p}{\|\int_{\mathbb{R}} \beta_n(s)e^{is(\cdot)}\,ds\|_{L^p(\mathbb{R})}^p} = \frac{\|\check{\alpha}_n\|^p}{\|\check{\beta}_n\|^p} = \frac{\left(\frac{1}{n}\right)^p \|\check{\alpha}_0\|^p}{\|\check{\beta}_0\|^p} \to 0 \quad \text{as } n \to \infty.$$

The function defined by $v_n(s) := \beta_n(s)(A - is)x_n$ is in $\mathcal{S}(\mathbb{R}, X)$ with the additional property that $(A - is)^{-1}v_n(s) = \beta_n(s)x_n$. Thus, we have the inequality

$$\frac{\|\int_{\mathbb{R}} (A - is)^{-1}v_n(s)e^{is(\cdot)}\,ds\|}{\|\int_{\mathbb{R}} v_n(s)e^{is(\cdot)}\,ds\|} = \frac{\|\int_{\mathbb{R}} \beta_n(s)x_n e^{is(\cdot)}\,ds\|}{\|\int_{\mathbb{R}} \beta_n(s)(A - is)x_n e^{is(\cdot)}\,ds\|}$$

$$= \frac{\|\int_{\mathbb{R}} \beta_n(s)x_n e^{is(\cdot)}\,ds\|}{\|\int_{\mathbb{R}} \beta_n(s)(A - is_n)x_n e^{is(\cdot)}\,ds + \beta_n(s)(is_n - is)x_n e^{is(\cdot)}\,ds\|}$$

$$\geq \frac{\|\int_{\mathbb{R}} \beta_n(s)x_n e^{is(\cdot)}\,ds\|}{|(A - is_n)x_n| \|\int_{\mathbb{R}} \beta_n(s)e^{is(\cdot)}\,ds\| + \|\int_{\mathbb{R}} \beta_n(s)(is_n - is)x_n e^{is(\cdot)}\,ds\|}$$

$$= \left(|(A - is_n)x_n| + \frac{\|\int_{\mathbb{R}} \beta_n(s)(is_n - is)e^{is(\cdot)}\,ds\|}{\|\int_{\mathbb{R}} \beta_n(s)e^{is(\cdot)}\,ds\|}\right)^{-1}.$$

By the choice of s_n, x_n, and β_n, the last expression converges to ∞ as $n \to \infty$, in contradiction to statement (3). Hence, if the expression in statement (3) is finite for $\alpha = 0$, then $\sup_{s \in \mathbb{R}} \|(A - is)^{-1}\| < \infty$.

Now, by parts (iii)–(iv) of Proposition 2.48 and Theorem 2.49, we have that

$$\|\Gamma\|_{\bullet} = \inf_{v \in \mathcal{S}(\mathbb{R}, X)} \frac{\|\Gamma f_v\|}{\|f_v\|} = \inf_{v \in \mathcal{S}(\mathbb{R}, X)} \frac{\|g_v\|}{\|f_v\|} = \left(\sup_{v \in \mathcal{S}(\mathbb{R}, X)} \frac{\|f_v\|}{\|g_v\|}\right)^{-1} > 0;$$

and therefore, $0 \notin \sigma_{\mathrm{ap}}(\Gamma)$. Hence, by Lemma 2.38 and Lemma 2.41, it follows that $\sigma_{\mathrm{ap}}(e^{tA}) \cap \mathbb{T} = \emptyset$. On the other hand, since $\sigma(A) \cap i\mathbb{R} = \emptyset$, by an application of the Residual Spectral Mapping Theorem 2.9, we find that

$$\sigma(e^{tA}) \cap \mathbb{T} = \left[\sigma_{\mathrm{ap}}(e^{tA}) \cup \sigma_r(e^{tA})\right] \cap \mathbb{T} = \emptyset.$$

The same argument holds for every $\alpha \geq 0$. As a result, $\{e^{tA_\alpha}\}_{t \geq 0}$ is hyperbolic for each $\alpha \geq 0$, and therefore $\{e^{tA}\}_{t \geq 0}$ is uniformly exponentially stable.

So far we have proved that statements (1)–(3) are equivalent, and statement (1) implies statement (6). The proof will be complete once we show the implications (6)⇒(4)⇒(3) and (6)⇒(5)⇒(3).

To prove the implication (6)⇒(4), begin by setting $\alpha = 0$. Since the autonomous control system (5.3) is detectable, there exists $K \in \mathcal{L}(Y, X)$ such that the operator $A + KC$ generates a uniformly exponentially stable semigroup. By the implication (1)⇒(3) for the semigroup $\{e^{t(A+KC)}\}_{t\geq 0}$, it follows that

$$M_1 := \sup_{v \in \mathcal{S}(\mathbb{R},X)} \frac{\|\int_{\mathbb{R}} (A + KC - is)^{-1} v(s) e^{is(\cdot)}\, ds\|}{\|\int_{\mathbb{R}} v(s) e^{is(\cdot)}\, ds\|}$$

is finite. Thus, we have that

(5.17)
$$\sup_{u \in \mathcal{S}(\mathbb{R},U)} \frac{\|\int_{\mathbb{R}} (A + KC - is)^{-1} Bu(s) e^{is(\cdot)}\, ds\|}{\|\int_{\mathbb{R}} u(s) e^{is(\cdot)}\, ds\|}$$
$$= \sup_{u \in \mathcal{S}(\mathbb{R},U)} \frac{\|\int_{\mathbb{R}} (A + KC - is)^{-1} Bu(s) e^{is(\cdot)}\, ds\|}{\|\int_{\mathbb{R}} Bu(s) e^{is(\cdot)}\, ds\|} \cdot \frac{\|B \int_{\mathbb{R}} u(s) e^{is(\cdot)}\, ds\|}{\|\int_{\mathbb{R}} u(s) e^{is(\cdot)}\, ds\|} \leq M_1 \|B\|.$$

By the hypothesis in statement (6),

$$M_2 := \sup_{u \in \mathcal{S}(\mathbb{R},U)} \frac{\|\int_{\mathbb{R}} C(A - is)^{-1} Bu(s) e^{is(\cdot)}\, ds\|}{\|\int_{\mathbb{R}} u(s) e^{is(\cdot)}\, ds\|}$$

is finite. For $u \in \mathcal{S}(\mathbb{R},U)$, let $w(s) = KC(A - is)^{-1} Bu(s)$, $s \in \mathbb{R}$, and note that

(5.18)
$$\frac{\|\int_{\mathbb{R}} (A + KC - is)^{-1} KC(A - is)^{-1} Bu(s) e^{is(\cdot)}\, ds\|}{\|\int_{\mathbb{R}} u(s) e^{is(\cdot)}\, ds\|}$$

$$= \frac{\|\int_{\mathbb{R}} (A + KC - is)^{-1} w(s) e^{is(\cdot)}\, ds\|}{\|\int_{\mathbb{R}} w(s) e^{is(\cdot)}\, ds\|} \cdot \frac{\|K \int_{\mathbb{R}} C(A - is)^{-1} Bu(s) e^{is(\cdot)}\, ds\|}{\|\int_{\mathbb{R}} u(s) e^{is(\cdot)}\, ds\|}$$
$$\leq M_1 \|K\| M_2.$$

Finally, since
$$(A - is)^{-1} B = (A + KC - is)^{-1} B$$
$$+ (A + KC - is)^{-1} KC(A - is)^{-1} B,$$

it follows from (5.17) and (5.18) that

$$\sup_{u \in \mathcal{S}(\mathbb{R},U)} \frac{\|\int_{\mathbb{R}} (A - is)^{-1} Bu(s) e^{is(\cdot)}\, ds\|}{\|\int_{\mathbb{R}} u(s) e^{is(\cdot)}\, ds\|} \leq M_1 \|B\| + M_1 \|K\| M_2.$$

This argument holds for all $\alpha \geq 0$, so the implication (6)⇒(4) follows.

To prove the implication (4)⇒(3), we again argue only in the case $\alpha = 0$. Since (5.3) is stabilizable, there exists $F \in B(X,U)$ such that $A + BF$ generates a uniformly exponentially stable semigroup. By the implication (1)⇒(3) for the semigroup $\{e^{t(A+BF)}\}_{t\geq 0}$, it follows that the quantity

$$M_3 := \sup_{v \in \mathcal{S}(\mathbb{R},X)} \frac{\|\int_{\mathbb{R}} (A + BF - is)^{-1} v(s) e^{is(\cdot)}\, ds\|}{\|\int_{\mathbb{R}} v(s) e^{is(\cdot)}\, ds\|}$$

is finite. By the hypothesis (4),

$$M_4 := \sup_{u \in \mathcal{S}(\mathbb{R},U)} \frac{\|\int_{\mathbb{R}} (A - is)^{-1} Bu(s) e^{is(\cdot)}\, ds\|}{\|\int_{\mathbb{R}} u(s) e^{is(\cdot)}\, ds\|}$$

is finite. Hence, if $v \in \mathcal{S}(\mathbb{R}, X)$ and $w(s) = F(A + BF - is)^{-1}v(s)$ for $s \in \mathbb{R}$, then

$$(5.19) \quad \sup_{v \in \mathcal{S}(\mathbb{R},X)} \frac{\|\int_{\mathbb{R}} (A - is)^{-1} BF(A + BF - is)^{-1} v(s) e^{is(\cdot)} \, ds\|}{\|\int_{\mathbb{R}} v(s) e^{is(\cdot)} \, ds\|}$$

$$= \frac{\|\int_{\mathbb{R}} (A - is)^{-1} Bw(s) e^{is(\cdot)} \, ds\|}{\|\int_{\mathbb{R}} w(s) e^{is(\cdot)} \, ds\|} \cdot \frac{\|F \int_{\mathbb{R}} (A + BF - is)^{-1} v(s) e^{is(\cdot)} \, ds\|}{\|\int_{\mathbb{R}} v(s) e^{is(\cdot)} \, ds\|}$$

$$\leq M_4 \|F\| M_3.$$

Since
$$(A - is)^{-1} = (A + BF - is)^{-1} + (A - is)^{-1} BF(A + BF - is)^{-1},$$

it follows from (5.19) that

$$\sup_{v \in \mathcal{S}(\mathbb{R},X)} \frac{\|\int_{\mathbb{R}} (A - is)^{-1} v(s) e^{is(\cdot)} \, ds\|}{\|\int_{\mathbb{R}} v(s) e^{is(\cdot)} \, ds\|} \leq M_3 + M_4 \|F\| M_3.$$

Thus, statement (3) follows from statement (4). Similar arguments prove the implications (6)⇒(5) and (5)⇒(3). □

REMARK 5.9. From the equivalence of statements (1) and (3), it follows that the growth bound of a semigroup on a Banach space is given by

$$(5.20) \quad \omega(A) = \inf \left\{ \alpha > s(A) : \sup_{v \in \mathcal{S}(\mathbb{R},X)} \frac{\|\int_{\mathbb{R}} (A_\alpha - is)^{-1} v(s) e^{is(\cdot)} \, ds\|}{\|\int_{\mathbb{R}} v(s) e^{is(\cdot)} \, ds\|} < \infty \right\}$$

where $s(A)$ is the spectral bound. This is a natural generalization of the formula for the growth bound for a semigroup on a Hilbert space as provided by Gearhart's Theorem 2.10 where the bound is

$$\omega(A) = s_0(A) = \inf \left\{ \alpha > s(A) : \sup_{s \in \mathbb{R}} \|(A_\alpha - is)^{-1}\| < \infty \right\}.$$

◇

As we have mentioned on page 139, for stabilizable and detectable systems, Proposition 5.7 ensures that external stability implies internal stability in the Hilbert space setting. This implication does not hold in general for Banach spaces because of a possible failure of the Gearhart Spectral Mapping Theorem: on a Banach space the inequality $s_0(A) \leq \omega(A)$ might be *strict*. The following result shows that this is the only reason why Proposition 5.7 does not hold in the Banach space setting.

THEOREM 5.10. *Suppose that $\{e^{tA}\}_{t \geq 0}$ is a strongly continuous semigroup on a Banach space X with the property that $s_0(A) = \omega(A)$, and assume that (5.3) is stabilizable and detectable. If $\overline{\mathbb{C}}_+ \subset \rho(A)$ and $M := \sup_{s \in \mathbb{R}} \|C(A - is)^{-1} B\| < \infty$, then $\{e^{tA}\}_{t \geq 0}$ is uniformly exponentially stable. In other words, external stability implies internal stability.*

PROOF. Choose operators $F \in \mathcal{L}(X, U)$ and $K \in \mathcal{L}(Y, X)$ such that the semigroups generated by $A + BF$ and $A + KC$ are uniformly exponentially stable. Then, $s_0(A + BF) < 0$ and $s_0(A + KC) < 0$; and therefore,

$$M_1 := \sup_{s \in \mathbb{R}} \|(A + BF - is)^{-1}\| \quad \text{and} \quad M_2 := \sup_{s \in \mathbb{R}} \|(A + KC - is)^{-1}\|$$

are both finite. Since
$$(A - is)^{-1}B = (A + KC - is)^{-1}B + (A + KC - is)^{-1}KC(A - is)^{-1}B,$$
it follows that
$$M_3 := \sup_{s \in \mathbb{R}} \|(A - is)^{-1}B\| \leq M_2\|B\| + M_2\|K\|M.$$
Also,
$$(A - is)^{-1} = (A + BF - is)^{-1} + (A - is)^{-1}BF(A + BF - is)^{-1},$$
and so
$$\sup_{s \in \mathbb{R}} \|(A - is)^{-1}\| \leq M_1 + M_3\|F\|M_1.$$
Therefore, $\omega(A) = s_0(A) < 0$. $\square$

The following result, based on Corollary 2.18, describes a particular situation in which $s_0(A) = \omega(A)$, see Kaashoek and Verduyn Lunel [**KL**].

COROLLARY 5.11. *Suppose that A is the generator of a strongly continuous semigroup $\{e^{tA}\}_{t \geq 0}$ on a Banach space X with adjoint space X^* and $\eta > \omega(A)$ is a number such that*

(5.21) $$\int_{-\infty}^{\infty} |(\eta + i\tau - A)^{-1}x|_X^2 \, d\tau < \infty \quad \text{for each} \quad x \in X,$$

and

(5.22) $$\int_{-\infty}^{\infty} |(\eta + i\tau - A^*)^{-1}x^*|_{X^*}^2 \, d\tau < \infty \quad \text{for each} \quad x^* \in X^*.$$

Then, the system (5.3) is internally stable if and only if it is stabilizable, detectable, and externally stable.

PROOF. By Corollary 2.18 (see [**KL**] and also Corollary 4.6.12 in van Neerven [**vN**]), the conditions (5.21)–(5.22) imply $s_0(A) = \omega(A)$. Then, Theorem 5.10 gives the result. $\square$

5.1.3. Stability radius. The goal of this subsection is to use the previous results in this chapter to study the (complex) stability radius of a uniformly exponentially stable system. Loosely speaking, the stability radius is a measure of the size of the smallest operator under which the additively perturbed system loses uniform exponential stability. We also discuss the notion of the transfer function for nonautonomous systems, and we conclude with counterexamples that demonstrate new effects in the study of the stability radius in the Banach space setting.

5.1.3.1. *General estimates.* In this section we give estimates for the stability radius of general nonautonomous systems on Banach spaces. The perturbations considered here are additive "structured" perturbations of output feedback type. More precisely, let U and Y be Banach spaces and $\Delta(t) : Y \to U$ a disturbance operator with $\Delta(\cdot) \in L^\infty(\mathbb{R}_+; \mathcal{L}_s(Y, U))$. The operators $B(t) : U \to X$ and $C(t) : X \to Y$ describe the structure of the perturbation in the following sense: If $u(t) = \Delta(t)y(t)$ is viewed as a feedback for the system (5.2), then the nominal system $\dot{x}(t) = A(t)x(t)$ is subject to the structured perturbation as follows:

(5.23) $$\dot{x}(t) = [A(t) + B(t)\Delta(t)C(t)]x(t), \quad t \geq 0.$$

In other words, in this section B and C do not represent input and output operators; rather, they describe the structure of the uncertainty of the system.

The control systems considered throughout this section are not assumed to have differentiable solutions. Thus, system (5.23) is to be interpreted in the mild sense as in display (5.1).

For the definition of the stability radius, let us suppose that $\mathcal{U} = \{U(\theta, \tau)\}_{\theta \geq \tau \geq 0}$ is a uniformly exponentially stable evolution family on X. Let $\tilde{\Delta}$ denote the multiplication operator with multiplier $\Delta(\cdot)$. Set $\mathcal{D} = \mathcal{B}\tilde{\Delta}\mathcal{C}$ and let $\mathcal{U}_\Delta = \{U_\Delta(t, \tau)\}_{t \geq \tau \geq 0}$ denote the evolution family corresponding to solutions of the perturbed equation; that is, $\mathcal{U}_\Delta$ satisfies the equation

$$U_\Delta(t,\tau)x = U(t,\tau)x + \int_\tau^t U(t,s)B(s)\Delta(s)C(s)U_\Delta(s,\tau)x\,ds, \quad x \in X, t \geq \tau \geq 0.$$

The (complex) *stability radius* for $\mathcal{U}$ with respect to the perturbation structure $(B(\cdot), C(\cdot))$ is the quantity

$r_{\text{stab}}(\mathcal{U}, B, C)$
$\quad := \sup\{r \geq 0 : \|\Delta(\cdot)\|_\infty \leq r \Rightarrow \mathcal{U}_\Delta$ is uniformly exponentially stable$\}$.

This definition applies to both nonautonomous and autonomous systems. In the autonomous case, the notation $r_{\text{stab}}(A, B, C)$ will be used to distinguish the case in which all the operators except $\Delta(t)$ are independent of t. We will have occasion to consider the *constant stability radius* that is defined for the case where $\Delta(t) \equiv \Delta$ is independent of t. In this case, we will write $rc_{\text{stab}}(A, B, C)$ or $rc_{\text{stab}}(\mathcal{U}, B, C)$ depending on the context.

Since, by Corollary 3.27, the stability of the evolution family $\{U_\Delta(\theta, \tau)\}_{\theta \geq \tau \geq 0}$ is determined by the invertibility of the operator $\Gamma_+ + \mathcal{D}$, it is clear that

(5.24) $\quad r_{\text{stab}}(\mathcal{U}, B, C) = \sup\{r \geq 0 : \|\Delta(\cdot)\|_\infty \leq r \Rightarrow \Gamma_+ + \mathcal{B}\tilde{\Delta}\mathcal{C}$ is invertible$\}$.

It is well known that, for autonomous systems where U and Y are Hilbert spaces and $p = 2$, the stability radius can be expressed in terms of the norm of the input-output operator or the transfer function as follows:

(5.25) $\quad \dfrac{1}{\|\mathbb{L}\|_{\mathcal{L}(L^2)}} = r_{\text{stab}}(A, B, C) = \dfrac{1}{\sup_{s \in \mathbb{R}} \|C(A - is)^{-1}B\|};$

see for example Hinrichsen and Pritchard [**HP3**, Theorem 3.5]. For nonautonomous equations, a scalar example given by Hinrichsen, Ilchmann, and Pritchard [**HIP**, Exm. 4.4] shows that, in general, the strict inequality $1/\|\mathbb{L}\| < r_{\text{stab}}(\mathcal{U}, B, C)$ may hold. Moreover, even for autonomous systems, as we will show in Examples 5.20 and 5.22, the equalities in (5.25) may not hold in the Banach space setting or in the case where $p \neq 2$. Below, see page 151, we will consider autonomous control systems and prove the following theorem.

THEOREM 5.12. *In the Banach space setting, the following inequalities hold for every internally stable autonomous system:*

(5.26) $\quad \dfrac{1}{\|\mathbb{L}\|_{\mathcal{L}(L^p)}} \leq r_{stab}(A, B, C) \leq \dfrac{1}{\sup_{s \in \mathbb{R}} \|C(A - is)^{-1}B\|}, \quad 1 \leq p < \infty.$

The lower bound for the stability radius in Theorem 5.12 holds for general nonautonomous systems. We will prove this inequality in a very direct way by using the structure of the operator $\mathbb{L} = \mathcal{C}\mathbb{G}\mathcal{B}$. The same result is proved by Hinrichsen and Pritchard [**HP3**, Theorem 3.2] using a different approach.

THEOREM 5.13. *Let $\{U(\theta,\tau)\}_{\theta \geq \tau \geq 0}$ be a uniformly exponentially stable evolution family and Γ_+ the generator of the corresponding evolution semigroup $\{E_+^t\}_{t \geq 0}$ on $L^p(\mathbb{R}_+; X)$. If*
$$B(\cdot) \in L^\infty(\mathbb{R}_+; \mathcal{L}_s(U, X)) \quad \text{and} \quad C(\cdot) \in L^\infty(\mathbb{R}_+; \mathcal{L}_s(X, Y)),$$
then $\mathbb{L}$ is a bounded operator from $L^p(\mathbb{R}_+; U)$ to $L^p(\mathbb{R}_+; Y)$, $1 \leq p < \infty$, the formula
$$\mathbb{L} = \mathcal{C}\mathbb{G}\mathcal{B} = -\mathcal{C}\Gamma_+^{-1}\mathcal{B}$$
holds, and

(5.27) $$\frac{1}{\|\mathbb{L}\|} \leq r_{stab}(\mathcal{U}, B, C).$$

In the "unstructured" case, if $U = Y = X$ and $B = C = I$, then
$$\mathbb{L} = -\Gamma_+^{-1}, \quad \text{and} \quad \frac{1}{\|\Gamma_+^{-1}\|} \leq r_{stab}(\mathcal{U}, I, I) \leq \frac{1}{r(\Gamma_+^{-1})}$$
where $r(\cdot)$ denotes the spectral radius.

PROOF. Since $\mathcal{U}$ is uniformly exponentially stable, Γ_+ is invertible and $\Gamma_+^{-1} = -\mathbb{G}$ by Corollary 3.27. The required formula for $\mathbb{L}$ follows from equation (5.8).

Set $H := \Gamma_+^{-1}\mathcal{B}\tilde{\Delta}$. To prove the inequality (5.27), let $\Delta(\cdot) \in L^\infty(\mathbb{R}_+; \mathcal{L}_s(Y, U))$ and suppose that $\|\Delta(\cdot)\|_\infty < 1/\|\mathbb{L}\|$. In follows that $\|\mathbb{L}\tilde{\Delta}\| < 1$; and therefore, the operator $I - \mathbb{L}\tilde{\Delta} = I + \mathcal{C}\Gamma_+^{-1}\mathcal{B}\tilde{\Delta}$ is invertible on $L^p(\mathbb{R}_+; Y)$. Since $I + \mathcal{C}H$ is invertible on $L^p(\mathbb{R}_+; Y)$, we conclude that $I + H\mathcal{C}$ is invertible on $L^p(\mathbb{R}_+; X)$ with inverse $I - H(I + \mathcal{C}H)^{-1}\mathcal{C}$. But, since
$$\Gamma_+ + \mathcal{B}\tilde{\Delta}\mathcal{C} = \Gamma_+(I + \Gamma_+^{-1}\mathcal{B}\tilde{\Delta}\mathcal{C}) = \Gamma_+(I + H\mathcal{C}),$$
the operator $\Gamma_+ + \mathcal{B}\tilde{\Delta}\mathcal{C}$ is invertible. It follows from the expression (5.24) that $1/\|\mathbb{L}\| \leq r_{\text{stab}}(\mathcal{U}, B, C)$.

For the last assertion of the theorem, suppose that $r_{\text{stab}}(\mathcal{U}, I, I) > 1/r(\Gamma_+^{-1})$. In this case, there is a number λ such that $|\lambda| = r(\Gamma_+^{-1})$ and $\lambda + \Gamma_+^{-1}$ is not invertible. By setting $\Delta = \lambda^{-1}$ we have that $\|\Delta\| = |\lambda|^{-1} < r_{\text{stab}}(\mathcal{U}, I, I)$. Hence, $\Gamma_+ + \Delta = \Delta(\lambda + \Gamma_+^{-1})\Gamma_+$ is invertible, a contradiction. □

5.1.3.2. *The transfer function for nonautonomous systems.* In this section we consider a time-varying version of equation (5.25) and observe that the concept of a transfer function, also called the frequency-response function, arises naturally for nonautonomous control systems. We will assume here that X, U, and Y are Hilbert spaces and $p = 2$.

Let $\{U(t,\tau)\}_{t \geq \tau \geq 0}$ be a uniformly exponentially stable evolution family and $\{E_+^t\}_{t \geq 0}$ the induced evolution semigroup with generator Γ_+ on $L^2(\mathbb{R}_+; X)$. Recall, that $\mathcal{B}$ and $\mathcal{C}$ denote multiplication operators, with respective multipliers $B(\cdot)$ and $C(\cdot)$, that act on the spaces $L^2(\mathbb{R}_+; U)$ and $L^2(\mathbb{R}_+; X)$, respectively. Let $\tilde{\mathcal{B}}$ and $\tilde{\mathcal{C}}$ denote the multiplication operators induced by $\mathcal{B}$ and $\mathcal{C}$, respectively; that is, $(\tilde{\mathcal{B}}\mathbf{u})(t) = \mathcal{B}(\mathbf{u}(t))$ for $\mathbf{u} \in L^2(\mathbb{R}_+; L^2(\mathbb{R}_+; U))$, and $(\tilde{\mathcal{C}}\mathbf{v})(t) = \mathcal{C}(\mathbf{v}(t))$ for $\mathbf{v} \in L^2(\mathbb{R}_+; L^2(\mathbb{R}_+; X))$. Also, recall the definition (5.8) of the Green's operator $\mathbb{G}$

associated with the evolution family $\{U(\theta,\tau)\}_{\theta \geq \tau \geq 0}$ on $L^2(\mathbb{R}_+; X)$, and define $\mathbb{G}_*$ on $L^2(\mathbb{R}; L^2(\mathbb{R}_+; X))$ as in equations (3.23)–(3.24); that is,

$$(5.28) \qquad \mathbb{G}_* h(\tau, \theta) = \int_0^\theta U(\theta, \theta - t) h(\tau - t, \theta - t)\, dt, \quad \tau \in \mathbb{R}, \quad \theta \in \mathbb{R}_+.$$

Note that the operator $\mathbb{L}_* := \tilde{\mathcal{C}} \mathbb{G}_* \tilde{\mathcal{B}}$ may be viewed (formally) as an input-output operator for the "autonomized" system $\dot{f} = \Gamma_+ f + \mathcal{B}\mathbf{u}$, $g = \mathcal{C}f$ where the state space is $L^2(\mathbb{R}_+; X)$. Thus, using the fact that in Hilbert space the inequalities in (5.26) are equalities, see (5.25), we have that

$$\frac{1}{\|\mathbb{L}_*\|} = r_{\text{stab}}(\Gamma_+, \mathcal{B}, \mathcal{C}) = \frac{1}{\sup_{s \in \mathbb{R}} \|\mathcal{C}(\Gamma_+ - is)^{-1}\mathcal{B}\|}.$$

Because of the similarity identities for Γ_+, see (3.18), it follows that

$$\|\mathbb{L}_*\| = \|\mathcal{C}(\Gamma_+ - is)^{-1}\mathcal{B}\| = \|\mathcal{C}\Gamma_+^{-1}\mathcal{B}\| = \|\mathbb{L}\|;$$

and therefore the stability radius for the evolution semigroup is equal to $1/\|\mathbb{L}\|$.

In view of the nonautonomous scalar example from [**HIP**] mentioned above for which $1/\|\mathbb{L}\| < r_{\text{stab}}(\mathcal{U}, B, C)$, we see that, even though the evolution semigroup (or its generator) completely determines the uniform exponential stability of a system, it does not provide a formula for the stability radius.

The appearance of the operator $\mathcal{C}(\Gamma_+ - is)^{-1}\mathcal{B}$ suggests that the transfer function for time-varying systems arises naturally when viewed in the context of evolution semigroups. Indeed, several authors have considered such a concept. In particular, the work of Ball, Gohberg, and Kaashoek [**BaGK**] seems to be the most comprehensive in providing a system-theoretic input-output interpretation for the value of such a transfer function at a point. Their interpretation justifies the term *frequency response* function for a time-varying finite dimensional systems with "time-varying complex exponential inputs". However, our remarks below concerning the frequency response for time-varying infinite dimensional systems will be restricted to inputs of the form $u(t) = u_0 e^{\lambda t}$.

For motivation, consider the input-output operator $\mathbb{L}$ associated with an autonomous system (5.3) where the nominal system is uniformly exponentially stable. The *transfer function* of $\mathbb{L}$ is the unique, bounded, analytic $\mathcal{L}(U, Y)$-valued function H, defined on $\mathbb{C}_+ = \{\lambda \in \mathbb{C} : \operatorname{Re} \lambda > 0\}$ such that, for each $u \in L^2(\mathbb{R}_+; U)$, we have

$$(\widehat{\mathbb{L}u})(\lambda) = H(\lambda)\hat{u}(\lambda), \qquad \lambda \in \mathbb{C}_+,$$

where the symbol " $\widehat{}$ " denotes the Laplace transform (see for example Weiss [**Ws3**]). In this autonomous setting, A generates a uniformly exponentially stable strongly continuous semigroup, and $\mathbb{L} = \mathcal{C}\mathbb{G}\mathcal{B}$ where $\mathbb{G}$ is the operator of convolution with the semigroup operators e^{tA} (see (2.36)). Standard arguments show that $(\widehat{\mathbb{L}u})(\lambda) = C(\lambda - A)^{-1} B \hat{u}(\lambda)$; that is, $H(\lambda) = C(\lambda - A)^{-1} B$.

Let $\mathbb{L}$ denote the input-output operator for the nonautonomous system (5.1). We wish to identify the transfer function of $\mathbb{L}$ as the Laplace transform of an appropriate operator. We are guided by the fact that, just as $(\lambda - A)^{-1}$ may be expressed as the Laplace transform of the semigroup generated by A, the operator $(\lambda - \Gamma_+)^{-1}$ is the Laplace transform of the evolution semigroup. For nonautonomous systems, $\mathbb{L}$ is again given by $\mathcal{C}\mathbb{G}\mathcal{B}$; but, in this case, the operator $\mathbb{G}$ as in display (5.8) is not, generally, a convolution operator. Instead, recall that the operator $\mathbb{G}_*$ in display

(5.28) is the operator of convolution with the evolution semigroup $\{E_+^t\}_{t\geq 0}$. As noted above, the operator $\mathbb{L}_* := \tilde{\mathcal{C}}\mathbb{G}_*\tilde{\mathcal{B}}$ may be viewed as an input-output operator for an autonomous system (where the state space is $L^2(\mathbb{R}_+; X)$). Therefore, the autonomous theory applies directly to show that, for $\mathbf{u} \in L^2(\mathbb{R}_+; L^2(\mathbb{R}_+; U))$,

$$\widehat{(\mathbb{L}_*\mathbf{u})}(\lambda) = \mathcal{C}(\lambda - \Gamma_+)^{-1}\mathcal{B}\hat{\mathbf{u}}(\lambda).$$

In other words, the transfer function for $\mathbb{L}_*$ is $\mathcal{C}(\lambda - \Gamma_+)^{-1}\mathcal{B}$ where

$$\mathcal{C}(\lambda - \Gamma_+)^{-1}\mathcal{B}u = \mathcal{C}\int_0^\infty e^{-\lambda\tau}E_+^\tau \mathcal{B}u\,d\tau, \qquad u \in L^2(\mathbb{R}_+;U).$$

Moreover, if we evaluate these expressions at $t \in \mathbb{R}_+$, then

(5.29) $$[\mathcal{C}(\lambda - \Gamma_+)^{-1}\mathcal{B}u](t) = \int_0^t C(t)U(t,\tau)B(\tau)u(\tau)e^{-\lambda(t-\tau)}\,d\tau.$$

It is natural to call $\mathcal{C}(\lambda - \Gamma_+)^{-1}\mathcal{B}$ the transfer function for the nonautonomous system. Also, the following argument indicates that the right hand side of equation (5.29) gives a natural frequency response function for the nonautonomous system. Observe that the definition of the transfer function for autonomous systems can be extended to allow for a class of "Laplace transformable" functions that are in $L^2_{loc}(\mathbb{R}_+; U)$ (see for example Weiss [**Ws3**]). This class includes constant functions of the form $v_0(t) = u_0$, $t \geq 0$, where $u_0 \in U$. If a periodic input signal of the form $u(t) = u_0 e^{i\omega t}$, $t \geq 0$ (for some $u_0 \in U$ and $\omega \in \mathbb{R}$), is fed into an autonomous system with an initial condition $x(0) = x_0$, then, by the definition of the input-output operator, we have that

$$(\mathbb{L}u)(t) = C(i\omega - A)^{-1}Bu_0 \cdot e^{i\omega t} - Ce^{tA}x_0, \quad (\mathbb{L}u)(t) = C\int_0^t e^{(t-s)A}Bu(s)\,ds.$$

Thus, the output

$$y(t; u(\cdot), x_0) = (\mathbb{L}u)(t) + Ce^{tA}x_0 = C(i\omega - A)^{-1}Bu_0 \cdot e^{i\omega t}$$

has the same frequency as the input. This is why the function $C(i\omega - A)^{-1}B$ is sometimes called the frequency response function. Recall that the semigroup $\{e^{tA}\}_{t\geq 0}$ is stable; and therefore, $\lim_{t\to\infty}|Ce^{tA}x_0| = 0$. On the other hand, consider $v(t) = u_0$ and (formally!) apply $\mathcal{C}(i\omega - \Gamma_+)^{-1}\mathcal{B}$ to this v. If $x_0 = (i\omega - A)^{-1}Bu_0$, then a calculation based on the Laplace formula (2.3) for the semigroup $\{E_+^t\}_{t\geq 0}$ yields the identity

$$\left[\mathcal{C}(i\omega - \Gamma_+)^{-1}\mathcal{B}u_0\right](t) = C(i\omega - A)^{-1}Bu_0 - Ce^{tA}x_0 \cdot e^{-i\omega t}.$$

Let us consider the nonautonomous case. By equation (5.29), the "expression" $[\mathcal{C}(i\omega - \Gamma_+)^{-1}\mathcal{B}u_0](t)$ coincides with the frequency response function for the time-varying systems defined in [**BaGK**, Corollary 3.2] by the formula

$$\int_0^t C(t)U(t,\tau)B(\tau)u_0 e^{i\omega(\tau-t)}\,d\tau.$$

Also, as noted in this reference, the result of our derivation agrees with the Arveson frequency response function that appears in [**SK**].

5.1.3.3. *Autonomous systems.* In this section we will prove the inequalities in display (5.26) in case X, U, and Y are Banach spaces. We will also introduce two more "stability radii": the pointwise stability radius and the dichotomy radius.

We begin by stating a generalization of the Spectral Mapping Theorem 2.31. Recall from Section 2.2.1.2 that $\mathcal{F}_{\text{per}}$ denotes the Banach space $L^p([0, 2\pi]; X)$, $1 \leq p < \infty$. Also, for the strongly continuous semigroup $\{e^{tA}\}_{t\geq 0}$ on X, let $\{E_{\text{per}}^t\}_{t\geq 0}$ denote the evolution semigroup defined on $\mathcal{F}_{\text{per}}$ by the rule $E_{\text{per}}^t f(\theta) = e^{tA} f([\theta - t](\text{mod } 2\pi))$, and let its generator be denoted by Γ_{per}. The symbol Λ will be used to denote the set of all finite sequences $\{v_k\}_{k=-N}^N$ in X or $\mathcal{D}(A)$, or $\{u_k\}_{k=-N}^N$ in U.

THEOREM 5.14. *Suppose that A generates a strongly continuous semigroup $\{e^{tA}\}_{t\geq 0}$ on X, the operator $B \in \mathcal{L}(U, X)$, the operator $C \in \mathcal{L}(X, Y)$, and $\Delta \in \mathcal{L}(Y, U)$, and $\{e^{t(A+B\Delta C)}\}_{t\geq 0}$ is the strongly continuous semigroup generated by $A + B\Delta C$. The following statements are equivalent:*

(i) $1 \in \rho(e^{2\pi(A+B\Delta C)})$.

(ii) $i\mathbb{Z} \subset \rho(A + B\Delta C)$ and
$$\sup_{\{u_k\}\in\Lambda} \frac{\|\sum_k (A - ik + B\Delta C)^{-1} u_k e^{ik(\cdot)}\|_{\mathcal{F}_{\text{per}}}}{\|\sum_k u_k e^{ik(\cdot)}\|_{\mathcal{F}_{\text{per}}}} < \infty.$$

(iii) $i\mathbb{Z} \subset \rho(A + B\Delta C)$ and
$$\inf_{\{v_k\}\in\Lambda} \frac{\|\sum_k (A - ik + B\Delta C) v_k e^{ik(\cdot)}\|_{\mathcal{F}_{\text{per}}}}{\|\sum_k v_k e^{ik(\cdot)}\|_{\mathcal{F}_{\text{per}}}} > 0.$$

In addition, if $1 \in \rho(e^{2\pi A})$, then Γ_{per} is invertible and

(5.30) $$\|\mathcal{C}\Gamma_{\text{per}}^{-1}\mathcal{B}\| = \sup_{\{u_k\}\in\Lambda} \frac{\|\sum_k C(A - ik)^{-1} B u_k e^{ik(\cdot)}\|_{L^p([0,2\pi];Y)}}{\|\sum_k u_k e^{ik(\cdot)}\|_{L^p([0,2\pi];U)}}$$

where $\mathcal{C}\Gamma_{\text{per}}^{-1}\mathcal{B} \in \mathcal{L}(L^p([0, 2\pi]; U), L^p([0, 2\pi]; Y))$.

PROOF. The equivalence of statements (i)–(iii) is proved in Theorem 2.31. For the last statement, let $\{u_k\}$ be a finite set in U and consider the functions f and g defined by
$$f(\theta) = \sum_k (A - ik)^{-1} B u_k e^{ik\theta}, \qquad g(\theta) = \sum_k B u_k e^{ik\theta}.$$

Note that $f = \Gamma_{\text{per}}^{-1} g$. Indeed, we have
$$(\Gamma_{\text{per}} f)(\theta) = \frac{d}{dt}\bigg|_{t=0} e^{tA} f([\theta - t] \text{mod} 2\pi)$$
$$= \sum_k [A(A - ik)^{-1} B u_k e^{ik\theta} - ik(A - ik)^{-1} B u_k e^{ik\theta}] = g(\theta).$$

Also, for functions of the form $h(\theta) = \sum_k u_k e^{ik\theta}$ where $\{u_k\}$ is a finite set in U, we have that $\mathcal{C}\Gamma_+^{-1}\mathcal{B} h = \sum_k C(A - ik)^{-1} B u_k e^{ik(\cdot)}$. By taking the supremum over all such functions, it follows that

$$\|\mathcal{C}\Gamma_{\text{per}}^{-1}\mathcal{B}\| = \sup_h \frac{\|\mathcal{C}\Gamma_{\text{per}}^{-1}\mathcal{B}h\|}{\|h\|}$$

$$= \sup_{\{u_k\}\in\Lambda} \frac{\|\sum_k C(A-ik)^{-1}Bu_k e^{ik(\cdot)}\|_{L^p([0,2\pi];Y)}}{\|\sum_k u_k e^{ik(\cdot)}\|_{L^p([0,2\pi];U)}}.$$

$\square$

In view of the previous theorem, let us introduce the following "pointwise" variant of the constant stability radius: For $t_0 > 0$ and $\lambda \in \rho(e^{t_0 A})$, we define

$$rc_{\text{stab}}^\lambda(e^{t_0 A}, B, C) := \sup\{r > 0 : \|\Delta\|_{\mathcal{L}(Y,U)} \leq r \Rightarrow \lambda \in \rho(e^{t_0(A+B\Delta C)})\}.$$

By rescaling, the study of this quantity can be reduced to the case where $\lambda = 1$ and $t_0 = 2\pi$. Indeed,

$$rc_{\text{stab}}^\lambda(e^{t_0 A}, B, C) = \frac{2\pi}{t_0} rc_{\text{stab}}^\lambda(e^{2\pi A'}, B, C) \quad \text{where} \quad A' = \frac{t_0}{2\pi} A.$$

Also, write $\lambda = |\lambda|e^{i\theta}$ ($\theta \in \mathbb{R}$) and note that

$$rc_{\text{stab}}^\lambda(e^{2\pi A}, B, C) = rc_{\text{stab}}^1(e^{2\pi A''}, B, C) \quad \text{for} \quad A'' = A - \frac{1}{2\pi}(\ln|\lambda| + i\theta).$$

Therefore, we have that

$$rc_{\text{stab}}^\lambda(e^{t_0 A}, B, C) = \frac{2\pi}{t_0} rc_{\text{stab}}^1(e^{2\pi A'''}, B, C) \quad \text{for} \quad A''' = \frac{1}{2\pi}(t_0 A - \ln|\lambda| - i\theta).$$

In the following theorem we estimate the quantity $rc_{\text{stab}}^1(e^{2\pi A}, B, C)$.

THEOREM 5.15. *Suppose that $\{e^{tA}\}_{t\geq 0}$ is a strongly continuous semigroup generated by A on X, the number $1 \in \rho(e^{2\pi A})$, and Γ_{per} is the generator of the induced evolution semigroup on $\mathcal{F}_{per}$. If $B \in \mathcal{L}(U,X)$ and $C \in \mathcal{L}(X,Y)$, then*

(5.31) $$\frac{1}{\|\mathcal{C}\Gamma_{per}^{-1}\mathcal{B}\|} \leq rc_{\text{stab}}^1(e^{2\pi A}, B, C) \leq \frac{1}{\sup_{k\in\mathbb{Z}}\|C(A-ik)^{-1}B\|}.$$

In addition, if U and Y are Hilbert spaces and $p = 2$, then the inequalities in (5.31) can be replaced by equalities.

PROOF. The proof of the first inequality is similar to the proof of Theorem 5.13. For the second inequality, let $\epsilon > 0$ be given. Choose $\overline{u} \in U$ with $|\overline{u}| = 1$, and $k_0 \in \mathbb{Z}$ such that

$$|C(A-ik_0)^{-1}B\overline{u}|_Y \geq \sup_{k\in\mathbb{Z}}\|C(A-ik)^{-1}B\| - \epsilon > 0.$$

By the Hahn-Banach Theorem, there is some $y^* \in Y^*$ with $|y^*| \leq 1$ such that

$$\left\langle y^*, \frac{C(A-ik_0)^{-1}B\overline{u}}{|C(A-ik_0)^{-1}B\overline{u}|_Y} \right\rangle = 1.$$

Define $\Delta \in \mathcal{L}(Y,U)$ by

$$\Delta y = -\frac{\langle y^*, y\rangle}{|C(A-ik_0)^{-1}B\overline{u}|_Y}\overline{u}, \quad y \in Y,$$

and note that

(5.32) $$\Delta C(A-ik_0)^{-1}B\overline{u} = -\frac{\langle y^*, C(A-ik_0)^{-1}B\overline{u}\rangle}{|C(A-ik_0)^{-1}B\overline{u}|_Y}\overline{u} = -\overline{u}$$

and

(5.33) $$\|\Delta\| \leq \frac{1}{|C(A-ik_0)^{-1}B\overline{u}|_Y} \leq \frac{1}{\sup_{k\in\mathbb{Z}}|C(A-ik_0)^{-1}B\overline{u}|_Y - \epsilon}.$$

Also, set $\overline{v} := (A-ik_0)^{-1}B\overline{u}$ in X. By equation (5.32), we have $\Delta C\overline{v} = -\overline{u}$. Hence,

$$(A - ik_0 + B\Delta C)\overline{v} = (A - ik_0)\overline{v} + B\Delta C\overline{v} = B\overline{u} - B\overline{u} = 0;$$

and therefore,

$$\inf_{\{v_k\}\in\Lambda} \frac{\|\sum_k(A-ik+B\Delta C)v_k e^{ik(\cdot)}\|_{\mathcal{F}_{\text{per}}}}{\|\sum_k u_k e^{ik(\cdot)}\|_{\mathcal{F}_{\text{per}}}} \leq \frac{\|(A-ik_0+B\Delta C)\overline{v}e^{ik_0(\cdot)}\|_{\mathcal{F}_{\text{per}}}}{\|\overline{v}e^{ik_0(\cdot)}\|_{\mathcal{F}_{\text{per}}}} = 0.$$

By Theorem 5.14, $1 \notin \rho(e^{2\pi(A+B\Delta C)})$, and it follows that $rc_{\text{stab}}^1(e^{2\pi A}, B, C) \leq \|\Delta\|$.

To finish the proof, suppose that

$$rc_{\text{stab}}^1(e^{2\pi A}, B, C) > \frac{1}{\sup_{k\in\mathbb{Z}}\|C(A-ik)^{-1}B\|}.$$

If

$$r := \frac{1}{\sup_{k\in\mathbb{Z}}|C(A-ik_0)^{-1}B\overline{u}|_Y - \epsilon}$$

and $\epsilon > 0$ is sufficiently small, then

$$\frac{1}{\sup_{k\in\mathbb{Z}}|C(A-ik_0)^{-1}B\overline{u}|_Y} < r < rc_{\text{stab}}^1(e^{2\pi A}, B, C).$$

By the inequalities (5.33), we have $\|\Delta\| \leq r < rc_{\text{stab}}^1(e^{2\pi A}, B, C)$, in contradiction.

For the last statement of the theorem, note that if Parseval's formula is applied in equation (5.30), then

(5.34) $$\|\mathcal{C}\Gamma_{\text{per}}^{-1}\mathcal{B}\| = \sup_{\{u_k\}\in\Lambda} \frac{\left(\sum_k |C(A-ik)^{-1}Bu_k|_Y^2\right)^{1/2}}{\left(\sum_k |u_k|_U^2\right)^{1/2}} \leq \sup_{k\in\mathbb{Z}}\|C(A-ik)^{-1}B\|.$$

Hence, both equalities hold in display (5.31). □

Let us consider a "hyperbolic" variant of the constant stability radius that is defined as follows: For a hyperbolic semigroup $\{e^{tA}\}_{t\geq 0}$ and operators B and C, the *constant dichotomy radius* is

$$rc_{\text{dich}}(A, B, C) := \sup\{r \geq 0 : \|\Delta\|_{\mathcal{L}(Y,U)} \leq r \text{ implies}$$
$$\sigma(e^{t(A+B\Delta C)}) \cap \mathbb{T} = \emptyset \text{ for all } t > 0\}.$$

The dichotomy radius measures the size of the smallest Δ for which the perturbed equation $\dot{x} = [A + B\Delta C]x$ loses its exponential dichotomy.

If $\{e^{tA}\}_{t\geq 0}$ is a semigroup and $\xi \in [0,1]$, then we let $A_\xi := A - i\xi$ denote the infinitesimal generator of the rescaled semigroup $\{e^{-i\xi t}e^{tA}\}_{t\geq 0}$. The pointwise stability radius can be related to the dichotomy radius as follows.

LEMMA 5.16. *If* $\{e^{tA}\}_{t\geq 0}$ *is a hyperbolic semigroup, then*

$$rc_{dich}(A, B, C) = \inf_{\xi\in[0,1]} rc_{stab}^1(e^{2\pi A_\xi}, B, C).$$

PROOF. Denote the left-hand side of the equality in the statement of the lemma by α and the right-hand side by β. Suppose that $r < \beta$. If $\xi \in [0,1]$ and $\|\Delta\| \leq r$, then $1 \in \rho(e^{2\pi(A_\xi + B\Delta C)})$. Thus, we have $e^{i\xi 2\pi} \in \rho(e^{2\pi(A+B\Delta C)})$ for all $\xi \in [0,1]$ and it follows that $e^{is} \in \rho(e^{2\pi(A+B\Delta C)})$ for all $s \in \mathbb{R}$. In other words, $\sigma(e^{2\pi(A+B\Delta C)}) \cap \mathbb{T} = \emptyset$. This proves the inequality $r \leq \alpha$; and therefore, $\beta \leq \alpha$.

Suppose that $r < \alpha$. If $\|\Delta\| \leq r$, then $\sigma(e^{t(A+B\Delta C)}) \cap \mathbb{T} = \emptyset$, and so $e^{i\xi t} \in \rho(e^{t(A+B\Delta C)})$ for all $\xi \in [0,1]$ and $t \in \mathbb{R}$; that is, $1 \in \rho(e^{t(A_\xi + B\Delta C)})$. But then, $r \leq \beta$ and so $\alpha \leq \beta$. □

Under the additional assumption that the semigroup $\{e^{tA}\}_{t\geq 0}$ is uniformly exponentially stable (that is, hyperbolic with a trivial dichotomy projection $P = I$), Lemma 5.16 gives a formula for the constant *stability* radius. Indeed, the following simple proposition holds.

PROPOSITION 5.17. *If $\{e^{tA}\}_{t\geq 0}$ is a uniformly exponentially stable semigroup, then $rc_{dich}(A,B,C) = rc_{stab}(A,B,C)$.*

PROOF. Denote the left-hand side of the equation in the statement of the proposition by α and the right-hand side by β. Suppose that $r < \beta$ and Δ is such that $\|\Delta\| \leq r$. By the definition of the constant stability radius, we have $\omega(A + B\Delta C) < 0$ and, in particular, $\sigma(e^{t(A+B\Delta C)}) \cap \mathbb{T} = \emptyset$. Thus, $r \leq \alpha$; and therefore, $\beta \leq \alpha$.

Suppose that $\beta < r < \alpha$ for some r. By the definition of the stability radius β, there is some operator Δ with $\|\Delta\| \in (\beta, r)$ such that the semigroup $\{e^{t(A+B\Delta C)}\}_{t\geq 0}$ is not stable.

For each $\tau \in [0,1]$ we have that $\|\tau\Delta\| \leq r < \alpha$. By the definition of the dichotomy radius α, it follows that the semigroup $\{e^{t(A+\tau B\Delta C)}\}_{t\geq 0}$ is hyperbolic for each $\tau \in [0,1]$. Consider its dichotomy projection

$$P(\tau) = (2\pi i)^{-1} \int_{\mathbb{T}} \left(\lambda - e^{A+\tau B\Delta C}\right)^{-1} d\lambda;$$

that is, the Riesz projection corresponding to the part of $\sigma(e^{A+\tau B\Delta C})$ located inside of the open unit disk. The function $\tau \mapsto P(\tau)$ is norm continuous. Indeed, since the bounded perturbation $\tau B\Delta C$ of the generator A is continuous in τ, the operators $e^{t(A+\tau B\Delta C)}$, $t \geq 0$, depend continuously on τ (see for example [**Pz**, Corollary 3.1.3]), and this implies the continuity of $P(\cdot)$ (see for example [**DK**, Theorem I.2.2]).

By our assumption, $\{e^{tA}\}_{t\geq 0}$ is uniformly exponentially stable. In particular, we have that $P(0) = I$. Also, $P(1) \neq I$ since the semigroup $\{e^{t(A+B\Delta C)}\}_{t\geq 0}$ with $\|\Delta\| \leq r < \alpha$ is hyperbolic but not stable. Since either $\|I - P(\tau)\| = 0$ or $\|I - P(\tau)\| \geq 1$, this contradicts the continuity of $\|P(\cdot)\|$. □

The inequality claimed in display (5.26) in Theorem 5.12 can now be proved.

PROOF. Because $r_{\text{stab}}(A,B,C) \leq rc_{\text{stab}}(A,B,C)$, we have the following inequalities

$$\frac{1}{\|\mathbb{L}\|} \leq r_{\text{stab}}(A,B,C) \leq rc_{\text{stab}}(A,B,C)$$
$$= rc_{\text{dich}}(A,B,C) \quad \text{(Proposition 5.17)}$$
$$= \inf_{\xi \in [0,1]} rc^1_{\text{stab}}(e^{2\pi A_\xi}, B, C) \quad \text{(Lemma 5.16)}$$

$$\le \inf_{\xi\in[0,1]} \frac{1}{\sup_{k\in\mathbb{Z}} \|C(A_\xi - ik)^{-1}B\|} \qquad \text{(Theorem 5.15)}$$
$$= \frac{1}{\sup_{s\in\mathbb{R}} \|C(A - is)^{-1}B\|}$$

We will need the following corollary that holds for *bounded* generators A.

COROLLARY 5.18. *If $A \in \mathcal{L}(X)$ generates a (uniformly continuous) stable semigroup on a Banach space X, then*

$$(5.35) \qquad rc_{stab}(A, B, C) = \frac{1}{\sup_{s\in\mathbb{R}} \|C(A - is)^{-1}B\|}.$$

PROOF. By Theorem 5.12, it suffices to prove only the inequality "$\ge$". Fix Δ with $\|\Delta\|$ strictly less than the right hand side of equation (5.35). Since $A + B\Delta C \in \mathcal{L}(X)$, it suffices to show that the operator

$$A + B\Delta C - \lambda = (A - \lambda)(I + (A - \lambda)^{-1} B\Delta C)$$

is invertible for each λ with $\operatorname{Re} \lambda \ge 0$. By the analyticity of the resolvent, we have the inequality $\sup_{\operatorname{Re} \lambda \ge 0} \|C(A-\lambda)^{-1}B\| \le \sup_{s\in\mathbb{R}} \|C(A-is)^{-1}B\|$. Thus, because the inequalities

$$\|\Delta\| < \frac{1}{\sup_{s\in\mathbb{R}} \|C(A - is)^{-1}B\|} \le \frac{1}{\sup_{\operatorname{Re} \lambda \ge 0} \|C(A-\lambda)^{-1}B\|} \le \frac{1}{\|C(A-\lambda)^{-1}B\|}$$

hold for each λ with $\operatorname{Re} \lambda \ge 0$, the operator $I + C(A-\lambda)^{-1} B\Delta$ is invertible. Therefore, using the same argument as in the proof of Theorem 5.13, the operator $I + (A - \lambda)^{-1} B\Delta C$ is invertible. □

If $\{e^{tA}\}_{t\ge 0}$ is uniformly exponentially stable, then the inequalities in display (5.26) give upper and lower bounds on the stability radius in terms of the operators $\mathbb{L}$ and $C(A-is)^{-1}B$, respectively. Moreover, by Theorem 5.6, we have an explicit expression for $\|\mathbb{L}\|$ in terms of integrals involving $C(A-is)^{-1}B$. Let us observe here that a lower bound for the constant stability radius can be expressed by a similar formula involving sums. For this, let $\xi \in [0,1]$ and define

$$S_\xi := \sup_{\{u_k\}\in\Lambda} \frac{\|\sum_k C(A - i\xi - ik)^{-1} Bu_k e^{ik(\cdot)}\|_{L^p([0,2\pi];Y)}}{\|\sum_k u_k e^{ik(\cdot)}\|_{L^p([0,2\pi];U)}}.$$

The number S_ξ is computed as in the right hand side of equation (5.30) with A replaced by $A_\xi = A - i\xi$.

COROLLARY 5.19. *If $\{e^{tA}\}_{t\ge 0}$ is a uniformly exponentially stable semigroup generated by A, then*

$$\frac{1}{\sup_{\xi\in[0,1]} S_\xi} \le rc_{stab}(A, B, C) \le \frac{1}{\sup_{s\in\mathbb{R}} \|C(A-is)^{-1}B\|}.$$

PROOF. Fix $\xi \in [0,1]$ and let $\Gamma_{\text{per},\xi}$ denote the generator on $L^p([0,2\pi];X)$ of the evolution semigroup induced by $\{e^{tA_\xi}\}_{t\ge 0}$. By Theorem 5.14, we have $\|\mathcal{C}\Gamma_{\text{per},\xi}^{-1}\mathcal{B}\| = S_\xi$, and, by Theorem 5.15,

$$\frac{1}{S_\xi} \le rc_{stab}^1(e^{2\pi A_\xi}, B, C) \le \frac{1}{\sup_{k\in\mathbb{Z}} \|C(A_\xi - ik)^{-1}B\|}.$$

By Proposition 5.17, if we take the infimum over $\xi \in [0,1]$, then

$$\frac{1}{\sup_{\xi \in [0,1]} S_\xi} \le \inf_{\xi \in [0,1]} rc^1_{\text{stab}}(e^{2\pi A_\xi}, B, C) = rc_{\text{stab}}(A,B,C)$$

$$\le \inf_{\xi \in [0,1]} \frac{1}{\sup_{k \in \mathbb{Z}} \|C(A_\xi - ik)^{-1} B\|} = \frac{1}{\sup_{s \in \mathbb{R}} \|C(A - is)^{-1} B\|}.$$

□

5.1.3.4. *Two counterexamples.* In contrast to the Hilbert space setting, the following examples show that the inequalities (5.26) can not, in general, be replaced by equalities in Banach spaces.

We begin with an example where the second inequality in (5.26) is strict.

EXAMPLE 5.20. W. Arendt constructed a (positive) strongly continuous semigroup $\{e^{tA}\}_{t \ge 0}$ on a Banach space X such that $s_0(A) < \omega(A) < 0$, see in Example 2.23 of Chapter 2, or for example [**vN**, Example 1.4.5]. Here, $s_0(A)$ is the abscissa of uniform boundedness of the resolvent, see (2.9), and $\omega(A)$ is the growth bound. Let α be a number such that $0 \le \alpha \le -\omega(A)$, consider a rescaled semigroup generated by $A + \alpha$, and let $\Gamma_{A+\alpha}$ denote the generator of the induced evolution semigroup on $L^p(\mathbb{R}_+; X)$. The following relationships hold: If $0 \le \alpha < -\omega(A)$, then

$$s_0(A+\alpha) = s_0(A) + \alpha < \omega(A) + \alpha = \omega(A+\alpha) < 0;$$

and if $\alpha_0 := -\omega(A)$, then $s_0(A+\alpha_0) < \omega(A+\alpha_0) = 0$. In particular, $s_0(A+\alpha) < 0$ for all $\alpha \in [0, \alpha_0]$; and therefore,

$$M := \sup_{\alpha \in [0, \alpha_0]} \sup_{s \in \mathbb{R}} \|(A + \alpha - is)^{-1}\| < \infty.$$

By Corollary 3.27, $\omega(A+\alpha) < 0$ if and only if $\|\Gamma_{A+\alpha}^{-1}\| < \infty$. Also, since $\omega(A+\alpha) \to 0$ as $\alpha \to \alpha_0$, we conclude that $\|\Gamma_{A+\alpha}^{-1}\| \to \infty$ as $\alpha \to \alpha_0$. Since the function given by $\alpha \mapsto \|\Gamma_{A+\alpha}^{-1}\|$ is continuous on the interval $[0, \alpha_0)$, there is some $\alpha_1 \in [0, \alpha_0)$ such that $\|\Gamma_{A+\alpha_1}^{-1}\| > M$; and therefore, the following inequality is strict:

$$\frac{1}{\|\Gamma_{A+\alpha_1}^{-1}\|} < \frac{1}{\sup_{s \in \mathbb{R}} \|(A + \alpha_1 - is)^{-1}\|}.$$

Also, we claim that there exists $\alpha_2 \in [0, \alpha_0)$ such that the following inequality is strict:

$$rc_{\text{stab}}(A + \alpha_2, I, I) < \frac{1}{\sup_{s \in \mathbb{R}} \|(A + \alpha_2 - is)^{-1}\|}.$$

To prove this fact, let us suppose that if $\alpha \in [0, \alpha_0)$, then $rc_{\text{stab}}(A + \alpha, I, I) \ge 1/(2M)$. Again, because $\omega(A+\alpha) \to 0$ as $\alpha \to \alpha_0$, there is some $\alpha \in [0, \alpha_0)$ such that $|\omega(A+\alpha)| < 1/(2M)$. Let $\Delta = \omega(A+\alpha)I$. Since $\|\Delta\| = |\omega(A+\alpha)|$, if we use the definition of the stability radius, then we have

$$0 > \omega(A + \alpha + \Delta) = \omega(A+\alpha) - \omega(A+\alpha) = 0,$$

in contradiction. Thus, there exists $\alpha_2 \in [0, \alpha_0)$ such that

$$rc_{\text{stab}}(A + \alpha_2, I, I) \le \frac{1}{2M} < \frac{1}{M} \le \frac{1}{\sup_{s \in \mathbb{R}} \|(A + \alpha_2 - is)^{-1}\|},$$

as claimed. ◇

The previous example shows that the second inequality in (5.26) can be strict as a result of the Banach-space pathologies related to the failure of the Gearhart Theorem 2.10. The example given below shows that the first inequality in (5.26) can be strict for $p = 1$ because Parseval's formula does not apply (see (5.34) as in the proof of Theorem 5.15). In particular, the requirement that $p = 2$ for equation (5.25) is as important as the fact that the spaces U and Y are Hilbert spaces.

To obtain the desired example, let us first find a formula for the norm of the input-output operator on $L^1(\mathbb{R}_+; X)$.

PROPOSITION 5.21. *Suppose that $\{e^{tA}\}_{t \geq 0}$ is a uniformly exponentially stable strongly continuous semigroup on a Banach space X. If $\mathbb{L} = \Gamma_+^{-1}$ denotes the input-output operator on $L^1(\mathbb{R}_+; X)$, then*

$$(5.36) \qquad \|\Gamma_+^{-1}\|_{\mathcal{L}(L^1(\mathbb{R}_+;X))} = \sup_{|x|=1} \int_0^\infty \left|e^{tA} x\right| dt.$$

PROOF. Recall displays (2.36) and (2.37), and note that the operator defined by

$$(\Gamma_+^{-1} f)(\theta) = -\int_0^\theta e^{\tau A} f(\theta - \tau) \, d\tau, \quad \theta \in \mathbb{R}_+, \quad f \in L^1(\mathbb{R}_+; X),$$

is a convolution operator. Choose a sequence of positive functions $\{\delta_n\}_{n=1}^\infty \subset L^1(\mathbb{R}_+; \mathbb{R})$ with $\|\delta_n\|_{L^1} = 1$ such that

$$\|g * \delta_n - g\|_{L^1(\mathbb{R}_+;X)} \to 0 \quad \text{as} \quad n \to \infty \quad \text{for each} \quad g \in L^1(\mathbb{R}_+; X).$$

Also, fix some $x \in X$ with $|x| = 1$, define $f = \delta_n x \in L^1(\mathbb{R}_+; X)$, and note that

$$(\Gamma_+^{-1} f)(\theta) = -\int_0^\theta e^{\tau A} x \, \delta_n(\theta - \tau) \, d\tau = -(g * \delta_n)(\theta) \quad \text{for} \quad g(\theta) = e^{\theta A} x, \quad \theta \in \mathbb{R}_+.$$

This implies "$\geq$" in (5.36). To prove "$\leq$", choose functions $\alpha_i \in L^1(\mathbb{R}_+; \mathbb{R})$ with disjoint supports and vectors $x_i \in X$ with $|x_i| = 1$ for $i = 1, \ldots, N$. Consider $f = \sum_{i=1}^N \alpha_i x_i$. It follows that $\|f\|_{L^1(\mathbb{R}_+;X)} = \sum_i \|\alpha_i\|_{L^1}$ and, for $f_i(\theta) = e^{\theta A} x_i$, we have the identity

$$(\Gamma_+^{-1} f)(\theta) = -\int_0^\theta \sum_i e^{\tau A} x_i \alpha_i(\theta - \tau) \, d\tau = -\sum_i (f_i * \alpha_i)(\theta).$$

Using Young's inequality,

$$\|\Gamma_+^{-1} f\|_{L^1(\mathbb{R}_+;X)} \leq \sum_i \|f_i * \alpha_i\|_{L^1(\mathbb{R}_+;X)} \leq \sum_i \|f_i\|_{L^1(\mathbb{R}_+;X)} \|\alpha_i\|_{L^1}$$

$$\leq \sup_{|x|=1} \int_0^\infty \left|e^{\theta A} x\right| d\theta \sum_i \|\alpha_i\|_{L^1}.$$

□

EXAMPLE 5.22. Consider the Banach space $X = \mathbb{C}^2$ with the ℓ_1 norm and the matrix
$$A = \begin{pmatrix} -1 & 1 \\ -1 & -1 \end{pmatrix}.$$
Let us note that
$$e^{tA} = \begin{pmatrix} e^{-t}\cos t & e^{-t}\sin t \\ -e^{-t}\sin t & e^{-t}\cos t \end{pmatrix},$$
$$(A - is)^{-1} = \frac{1}{(1+is)^2 + 1}\begin{pmatrix} -1-is & -1 \\ 1 & -1-is \end{pmatrix}.$$
Since the extreme points of the unit ball of X are $e^{i\theta}e_1$ and $e^{i\theta}e_2$ ($\theta \in \mathbb{R}$) where e_1 and e_2 are the unit vectors of $\mathbb{C}^2$, it follows that
$$\|(A - is)^{-1}\| = \frac{|1+is|+1}{|(1+is)^2+1|}.$$
Moreover, by a numerical computation, we have the approximation
$$\sup_{s\in\mathbb{R}} \|(A-is)^{-1}\| \approx 1.087494476.$$
By Corollary 5.18, the reciprocal of the last expression is equal to $rc_{\text{stab}}(A, I, I)$. On the other hand, using Proposition 5.21,
$$\|\mathbb{L}\| = \|\Gamma_A^{-1}\| = \sup_{|x|=1} \int_0^\infty |e^{tA}x|\,dt$$
$$= \int_0^\infty |e^{-t}\cos t| + |e^{-t}\sin t|\,dt \approx 1.262434309.$$
Therefore, for this example, the first inequality in (5.26) is strict. $\diamond$

5.2. Persistence of dichotomy

By the Dichotomy Theorem 3.17, the existence of a dichotomy for a strongly continuous, exponentially bounded evolution family $\{U(\theta, \tau)\}_{\theta \geq \tau}$ is a spectral property. Hence, it should persist under small perturbations. The case of bounded perturbations is considered in Subsection 5.2.1. Also, we briefly indicate a class of unbounded perturbations that preserve the dichotomy in Subsection 5.2.2. In this section we consider the line case where evolution families are defined for $\infty > \theta \geq \tau > -\infty$.

5.2.1. Bounded perturbations. In this subsection we show that the dichotomy persists under small bounded perturbations. In view of the Dichotomy Theorem 3.17, this result is not surprising. Indeed, the proof of the following theorem is quite short (cf. corresponding results in Daleckij and Krein [**DK**, Sect.IV.5] or Massera and Schäffer [**MS2**, Ch.7]).

THEOREM 5.23. *If a strongly continuous, exponentially bounded evolution family $\{U(\theta, \tau)\}_{\theta \geq \tau}$ has an exponential dichotomy on X, then, for each $t > 0$, there exists an $\epsilon > 0$ such that $\{U_1(\theta, \tau)\}_{\theta \geq \tau}$ has an exponential dichotomy whenever $\{U_1(\theta, \tau)\}_{\theta \geq \tau}$ is a strongly continuous, exponentially bounded evolution family such that*
$$\sup_{\theta \in \mathbb{R}} \|U_1(\theta + t, \theta) - U(\theta + t, \theta)\|_{\mathcal{L}(X)} \geq \epsilon.$$

PROOF. For $(E^t f)(\theta) = U(\theta, \theta - t)f(\theta - t)$ and $(E_1^t f)(\theta) = U_1(\theta, \theta - t)f(\theta - t)$ on $L^p(\mathbb{R}; X)$, we have the estimate

$$\|E_1^t f - E^t f\|_{L^p}^p = \int_\mathbb{R} |U_1(\theta, \theta - t)f(\theta - t) - U(\theta, \theta - t)f(\theta - t)|^p d\theta$$

$$= \int_\mathbb{R} |[U_1(\theta + t, \theta) - U(\theta + t, \theta)]f(\theta)|^p d\theta \leq \epsilon^p \|f\|_{L^p}^p;$$

and therefore, $\|E_1^t - E^t\|_{\mathcal{L}(L^p(\mathbb{R};X))} \leq \epsilon$.

Since $\sigma(E^t) \cap \mathbb{T} = \emptyset$ by the equivalence (1) $\Leftrightarrow$ (2) in the Dichotomy Theorem 3.17, the semicontinuity of the spectrum implies that $\sigma(E_1^t) \cap \mathbb{T} = \emptyset$ for sufficiently small ϵ. Hence, $\{U_1(\theta, \tau)\}_{\theta \geq \tau}$ has an exponential dichotomy. □

Theorem 5.23 asserts that a dichotomy persists under small perturbation of the evolution family. The next theorem shows a similar result for additive perturbations in the corresponding abstract Cauchy problem.

THEOREM 5.24. *Suppose that the evolution family* $\{U(\theta, \tau)\}_{\theta \geq \tau}$ *solves the abstract Cauchy problem*

(ACP) $\qquad \dot{x} = A(t)x(t), \quad x(\tau) = x_\tau, \quad x_\tau \in \mathcal{D}(A(\tau)), \quad t \geq \tau, \quad t, \tau \in \mathbb{R},$

and has an exponential dichotomy. Then, there is an $\epsilon > 0$ *with the following property: For every*

$$B \in C_b(\mathbb{R}; \mathcal{L}(X)) \quad \text{with} \quad \|B\|_\infty := \sup_{\theta \in \mathbb{R}} \|B(\theta)\|_{\mathcal{L}(X)} \geq \epsilon$$

and such that the perturbed equation $\dot{x} = [A(t) + B(t)]x(t)$ *is well-posed, the corresponding evolution family* $\{U_1(\theta, \tau)\}_{\theta \geq \tau}$ *that solves the abstract Cauchy problem for the perturbed equation also has an exponential dichotomy.*

PROOF. The evolution family $\{U_1(\theta, \tau)\}_{\theta \geq \tau}$ is a solution of the integral equation

(5.37) $\qquad U_1(\theta, \tau)x = U(\theta, \tau)x + \int_\tau^\theta U(\theta, x)B(s)U_1(s, \tau)x \, ds, \quad x \in X.$

Also, let Γ, respectively, Γ_1, denote the generator of the evolution semigroup, associated with the evolution family $\{U(\theta, \tau)\}_{\theta \geq \tau}$, respectively $\{U_1(\theta, \tau)\}_{\theta \geq \tau}$.

Consider the operator $\Gamma + \mathcal{B}$, where, as usual, $(\mathcal{B}f)(\theta) = B(\theta)f(\theta)$, $\theta \in \mathbb{R}$. Since $\mathcal{B}$ is a bounded operator, the operator $\Gamma + \mathcal{B}$ generates a strongly continuous semigroup $\{T^t\}_{t \geq 0}$ which is the unique semigroup that satisfies the equation

$$T^t f = E^t f + \int_0^t E^{t-s} \mathcal{B} T^s f \, ds, \quad E^t = e^{t\Gamma}, \quad t \geq 0,$$

see for example Pazy [**Pz**, Proposition 3.1.2]. We claim that $\{e^{t\Gamma_1}\}_{t \geq 0}$ also satisfies this equation. Once we prove this fact, it follows that $e^{t\Gamma_1} = T^t$ and $\Gamma_1 = \Gamma + \mathcal{B}$.

By the implication (1)$\Rightarrow$(3) of Theorem 3.17, we have that $0 \in \rho(\Gamma)$. Hence, if $\epsilon \geq \|B\|_\infty = \|\mathcal{B}\|$ is sufficiently small, then $0 \in \rho(\Gamma + \mathcal{B}) = \rho(\Gamma_1)$. By the implication (3)$\Rightarrow$(1) of the Dichotomy Theorem 3.17, we conclude that $\{U_1(\theta, \tau)\}_{\theta \geq \tau}$ has an exponential dichotomy.

To prove the claim, use identity (5.37) with $\tau = \theta - t$ and $x = f(\theta - t)$ to obtain the equality

$$(e^{t\Gamma_1} f)(\theta) = (E^t f)(\theta) + \int_0^t U(\theta, \theta - (t-s)) B(\theta - (t-s))(e^{s\Gamma_1} f)(\theta - (t-s))\, ds$$

$$= (E^t f)(\theta) + \int_0^t (E^{t-s} B e^{s\Gamma_1} f)(\theta)\, ds, \quad \theta \in \Theta.$$

$\square$

5.2.2. Miyadera-type perturbations. In this subsection we will prove the persistence of an exponential dichotomy for an evolution family $\{U(\theta, \tau)\}_{\theta \geq \tau}$ that solves an abstract Cauchy problem (ACP) relative to the class of perturbations, denoted $B(\cdot)$, that satisfy the so-called Miyadera condition.

For linear operators $B(\theta)$ on X, we let $\mathcal{B}$ denote the operator $(\mathcal{B}f)(\theta) = B(\theta)f(\theta)$, $\theta \in \mathbb{R}$, on $L^p(\mathbb{R}; X)$ with the maximal domain, see page 22. Recall that, by Theorem 3.17, the generator Γ of the evolution semigroup $\{E^t\}_{t \geq 0}$, associated with an exponentially dichotomic evolution family $\{U(\theta, \tau)\}_{\theta \geq \tau}$, is invertible on $L^p(\mathbb{R}; X)$, $1 \leq p < \infty$. The next result is a direct application of the Dichotomy Theorem 3.17.

THEOREM 5.25. *Suppose that the evolution family $\{U(\theta, \tau)\}_{\theta \geq \tau}$ solves (ACP) and has an exponential dichotomy. Also, assume that the operator $\mathcal{B}$, considered on the domain $\mathcal{D}(\Gamma) \cap \mathcal{D}(\mathcal{B})$, has an extension $\hat{\mathcal{B}}$ on $\mathcal{D}(\Gamma)$ such that the operator $\Gamma_1 := \Gamma + \hat{\mathcal{B}}$ with $\mathcal{D}(\Gamma_1) = \mathcal{D}(\Gamma)$ generates an evolution semigroup associated with the evolution family $\{U_1(\theta, \tau)\}_{\theta \geq \tau}$ that solves the abstract Cauchy problem for the differential equation $\dot{x} = [A(t) + B(t)]x(t)$. If for some constants a and b we have*

(5.38) $\quad \|\hat{\mathcal{B}} f\| \leq a \|f\| + b\|\Gamma f\| \quad \text{for} \quad f \in \mathcal{D}(\Gamma) \quad \text{and} \quad a\|\Gamma^{-1}\| + b < 1,$

then the perturbed evolution family $\{U_1(\theta, \tau)\}_{\theta \geq \tau}$ has an exponential dichotomy.

PROOF. By Theorem IV.1.16 in Kato [**Ka**], we have that Γ_1 is invertible on $L^p(\mathbb{R}; X)$. Since, by the hypotheses, Γ_1 is the generator for the evolution semigroup associated with $\{U_1(\theta, \tau)\}_{\theta \geq \tau}$, the result follows from the implication (3)$\Rightarrow$(1) of the Dichotomy Theorem 3.17. $\square$

Our next goal is to describe a class of perturbations, $B(\cdot)$, for which the hypotheses of Theorem 5.25 can be verified. The following exposition is adapted from a paper by Räbiger, Rhandi, Schnaubelt, and Voigt [**RRSV**] and [**Sc**].

Recall that $\{U(\theta, \tau)\}_{\theta \geq \tau}$ is assumed to be an exponentially bounded evolution family and $p \in [1, \infty)$. Let us consider the following assumptions for a family of linear operators $B(\theta)$, for $\theta \in \mathbb{R}$, (see Schnaubelt [**Sc**, p. 41]):

(i) There is a dense subspace Y of X such that $U(\theta, \tau) Y \subseteq Y$ for all $\theta \geq \tau$ and $Y \subseteq \mathcal{D}(B(\theta))$ for almost all $\theta \in \mathbb{R}$. Moreover, the operator $B(\theta)$ is closed for almost all $\theta \in \mathbb{R}$.

(ii) For each $\tau \in \mathbb{R}$ and $y \in Y$, the function $\theta \mapsto B(\theta) U(\theta, \tau) y$ belongs to the space $L^p_{\text{loc}}([\tau, \infty); X)$.

(iii) There exist constants $\alpha > 0$ and $\beta \geq 0$ such that $\alpha^{1/q} \beta < 1$, $p^{-1} + q^{-1} = 1$, and

$$\int_0^\alpha |B(\tau + s) U(\tau + s, s) y|^p\, ds \leq \beta^p |y|^p$$

for all $y \in Y$ and almost all $\tau \in \mathbb{R}$.

Conditions (i)–(iii) hold for many examples; for instance, they are clearly satisfied for $B(\cdot) \in L^\infty(\mathbb{R}; \mathcal{L}_s(X))$. For an important example of a family of unbounded operators $B(\theta)$ that satisfy these conditions see Example 2.16 in [**Sc**].

Let Γ be the generator of the evolution semigroup on $L^p(\mathbb{R}; X)$ induced by an evolution family $\{U(\theta, \tau)\}_{\theta \geq \tau}$. The following theorem is a simplified version of a result proved in [**RRSV**], see also [**Sc**].

THEOREM 5.26. *If the family of operators $B(\theta)$ satisfies (i)–(iii), then the following statements hold:*

(a) *The multiplication operator $\mathcal{B}$ with multiplier $B(\cdot)$ on $L^p(\mathbb{R}; X)$ has a unique extension $\hat{\mathcal{B}}$ to $\mathcal{D}(\Gamma)$. This extension is Γ-bounded; that is, $\|\hat{\mathcal{B}}f\| \leq a\|f\| + b\|\Gamma f\|$, $f \in \mathcal{D}(\Gamma)$, for some constants a and b. Moreover, the operator $\Gamma_1 := \Gamma + \hat{\mathcal{B}}$ with $\mathcal{D}(\Gamma_1) = \mathcal{D}(\Gamma)$ generates an evolution semigroup on $L^p(\mathbb{R}; X)$ induced by an exponentially bounded evolution family $\{U_1(\theta, \tau)\}_{\theta \geq \tau}$.*

(b) *The evolution family $\{U_1(\theta, \tau)\}_{\theta \geq \tau}$ in statement (a) is the only exponentially bounded evolution family on X that satisfies the equation*

$$U_1(\tau + t, \tau)f(\tau) = U(\tau + t, \tau)f(\tau) + \int_\tau^{\tau+t} U_1(\tau + t, s)B(s)U(s, \tau)f(\tau)\, ds$$

for all $f \in \mathcal{D}_\Gamma$ and $t \geq 0$. Here $\mathcal{D}_\Gamma$ is the core for the operator Γ defined in Theorem 3.12.

(c) *The domain $\mathcal{D}(\Gamma)$ is a subset of $\mathcal{D}(\mathcal{B})$ and $\mathcal{B}f = \hat{\mathcal{B}}f$ for $f \in \mathcal{D}(\Gamma)$.*

Moreover, if $f \in \mathcal{D}(\Gamma)$ and $t \geq 0$, then

$$U_1(\tau + t, \tau)f(\tau) = U(\tau + t, \tau)f(\tau) + \int_\tau^{\tau+t} U(\tau + t, s)B(s)U_1(s, \tau)f(\tau)\, ds.$$

5.3. Bibliography and remarks

Control theory. The main results in this section are taken from the papers by Clark, Latushkin, Montgomery-Smith, and Randolph [**RLC, CLRM**].

Some basic notions in control theory. There are several excellent books on the subject of infinite dimensional control theory; for example, Bensoussan, Da Prato, Delfour, and Mitter [**BPDM**], Curtain and Pritchard [**CP**], Curtain and Zwart [**CZ**], van Keulen [**vK**], and Zabczyk [**Za2**].

Stability and the input-output operator. Theorem 5.3 generalizes well-known finite dimensional results to the Banach space setting. A version of this result for finite dimensional time-varying systems is proved by Anderson in [**Ad**]. The fact that Theorem 5.3 actually *extends* the Hilbert-space Proposition 5.7 follows from the fact that the Banach-space inequality $\sup_{\lambda \in \mathbb{C}_+} \|H(\lambda)\| \leq \|\mathcal{L}\|$, which relates the operators that define external and input-output stability, is an *equality* for Hilbert-space systems (see Theorem 2.3 and Remark 2.4 by Weiss [**Ws2**], and also [**Ws3**]).

Theorem 5.4 is taken from [**CLRM**]. Proposition 5.7 goes back to Jacobson and Nett [**JN**]. Its proof can be found in Rebarber [**Re2**] or Curtain, Logemann, Townley, and Zwart [**CLTZ**, Thm.5.8]. It should be pointed out that these authors allow for unboundedness of the operators B and C. Such "regular" systems (see [**Ws3**]), and their time-varying generalizations might be addressed by combining the techniques of the present section (including the characterization of generators of evolution semigroups as found in Räbiger, Rhandi, Schnaubelt, and Voigt [**RRSV**]) with those of Hinrichsen and Pritchard [**HP3**] and Jacob, Dragon, and Pritchard [**JDP**]. See also Curtain [**Cu**], [**CLTZ**], and Lasiecka and Triggiani [**LT**].

Theorem 5.6 is taken from [**CLRM**]. This result is a direct generalization of Theorem 2.49 for the case of autonomous control systems. Theorem 5.8 is also taken from [**CLRM**]. Note that the expressions (see the right hand side of (5.11)) which appear in these theorems seem to be the correct Banach space replacements for the classical Hilbert space formulas for the norm of the transfer function. Theorem 5.10 from [**CLRM**] is a natural generalization of Gearhart's Theorem, a result known in the control literature as Huang's Theorem, see Huang [**Hu**], and also Nagel [**Na**], Prüss [**Pr**], van Neerven [**vN**], and Engel and Nagel [**EN**, Ex. V.1.13]. In fact, formula (5.20) states that $\omega(A) < 0$ if and only if the function $s \mapsto (A_\alpha - is)^{-1}$, $\alpha > 0$, is an L^p-Fourier multiplier, see for example Weis [**We2**] for further developments of this theme.

Stability radius. The stability radius is an important concept for linear systems theory. It was introduced by Hinrichsen and Pritchard as the basis for a state-space approach to studying robustness of linear time-invariant [**HP**] and time-varying systems, see Hinrichsen, Ilchmann, and Pritchard [**HIP**], Pritchard and Townley [**PT**], and [**HP, HP3**]. An excellent review of this topic is given in [**HP2**]. The paper [**HIP**] contains a study of nonautonomous finite dimensional situation, see also recent results by Jacob [**Ja1**].

Theorems 5.12, 5.13, and 5.14 are taken from [**CLRM**]. The idea of the proof of Theorem 5.15 goes back to [**HP3**], see also further developments by Fischer and van Neerven [**FvN**]. The examples in Section 5.1.3.4 are taken from [**CLRM**].

We cite the following finite dimensional result (see [**HP, HP2**]) that characterizes the stability radius in terms of the solutions of the parameterized Riccati equation

$$(\text{ARE}_\rho) \qquad A^*X + XA - \rho^2 C^*C - XBB^*X = 0.$$

THEOREM 5.27. *Suppose that X, Y, and U are finite dimensional, and $\sigma(A) \subset \mathbb{C}_-$.*

(i) *If $\rho \in [0, r_{stab}(A, B, C))$, then there exists a negative semi definite solution P_ρ of (ARE_ρ) such that P_ρ is stabilizing, i.e. $\sigma(A - BB^*P_\rho) \subset \mathbb{C}_-$. The solution P_ρ is unique among all Hermitian solutions with this property.*
(ii) *If $\rho = r_{stab}(A, B, C) < \infty$, then there exists a negative semi definite solution P_ρ of (ARE_ρ) satisfying $\sigma(A - BB^*P_\rho) \subset \overline{\mathbb{C}}_-$ and $\sigma(A - BB^*P_\rho) \cap i\mathbb{R} \neq \emptyset$.*
(iii) *If $\rho > r_{stab}(A, B, C)$, then there does not exist a Hermitian solution of (ARE_ρ).*

A systematic recent study of various stability radii is given by Fischer and van Neerven [**FvN**]. Using the methods utilized in the proof of Theorem 5.15, they prove the following result.

THEOREM 5.28. *Suppose that A is a closed operator on a Banach space X, $\lambda \in \rho(A)$, $B \in \mathcal{L}(U, X)$, and $C \in \mathcal{L}(X, Y)$ where U and Y are also Banach spaces. The pointwise constant stability radius*

$$rc_{stab}^\lambda := \sup\{r \geq 0 : \lambda \in \rho(A + B\Delta C) \text{ for all } \Delta \in \mathcal{L}(Y, U) \text{ with } \|\Delta\| \leq r\}$$

can be calculated by the formula

$$rc_{stab}^\lambda = \|C(\lambda - A)^{-1}B\|^{-1}.$$

Several other stability radii are studied in [**FvN**]. Assume, for instance, that A is a generator of a strongly continuous semigroup on X such that the abscissa of uniform boundedness of the resolvent $s_0(A)$ is negative. Define
$$rc_{s_0}(A, B, C) = \sup\{r \geq 0 : s_0(A + B\Delta C) < 0$$
$$\text{for all } \Delta \in \mathcal{L}(Y, U) \text{ with } \|\Delta\| \leq r\}.$$

THEOREM 5.29 ([**FvN**]).
$$rc_{s_0}(A, B, C) = \frac{1}{\sup_{s \in \mathbb{R}} \|C(A - is)^{-1}B\|}.$$

The following result is a generalization of Corollary 5.18.

COROLLARY 5.30 ([**FvN**]). *If A is the generator of an exponentially stable semigroup on X that is uniformly continuous for $t > 0$, then*
$$rc_{stab}(A, B, C) = \frac{1}{\sup_{s \in \mathbb{R}} \|C(A - is)^{-1}B\|}.$$

Note that the inequality "$\geq$" in Theorem 5.29 is proved similarly to the proof of (5.27) in Theorem 5.13. Indeed, if, for some r, there is an operator Δ such that
$$\|\Delta\| \leq r < \frac{1}{\sup_{s \in \mathbb{R}} \|C(A - is)^{-1}B\|} \leq \frac{1}{\sup_{\text{Re } \lambda \geq 0} \|C(A - \lambda)^{-1}B\|},$$
then the operators $I + C(A - \lambda)^{-1}B\Delta$ are invertible with their inverses bounded uniformly for Re $\lambda \geq 0$. The same is true for the operators $I + (A - \lambda)^{-1}B\Delta C$, see the proof of Theorem 5.13. Thus, the operators
$$A + B\Delta C - \lambda = (A - \lambda)[I + (A - \lambda)^{-1}B\Delta C]$$
are invertible with their inverses bounded uniformly for Re $\lambda \geq 0$. This shows "$\geq$" in Theorem 5.29. To prove "$\geq$" in Corollary 5.30, we note that $s_0(A) = \omega(A)$ provided that A generates a semigroup that is uniformly continuous for $t > 0$. If D is a bounded operator, then $A + D$ also generates a semigroup that is uniformly continuous for $t > 0$, see for example Nagel [**Na**]. Theorem 5.28 now implies the result.

Let $\mathcal{C}_n$ denote the set of operators of rank n, for each positive integer n, and $\mathcal{C}_\infty$ the set of compact operators. The following stability radii determine the robustness under the perturbation $\Delta \in \mathcal{C}_n$, for a positive integer n, or $n = \infty$:
$$rc_{stab}^{(n)}(A, B, C) := \sup\{r \geq 0 : \|\Delta\| \leq r \text{ and } \Delta \in \mathcal{C}_n \text{ implies } \omega(A + B\Delta C) < 0\}.$$

The following result is due to Arendt and Latushkin.

THEOREM 5.31. *Suppose that $\omega(A) < 0$. If n is a positive integer or $n = \infty$, then*
$$rc_{stab}^{(n)}(A, B, C) = \frac{1}{\sup_{s \in \mathbb{R}} \|C(A - is)^{-1}B\|}.$$

Indeed, the perturbation Δ constructed in the proof of the second inequality in Theorem 5.15 has rank one. On the other hand,
$$\omega(A + B\Delta C) = \max\{\omega_{\text{ess}}(A + B\Delta C), s(A + B\Delta C)\}.$$
Here, $\omega_{\text{ess}}(A)$ denotes the essential growth bound for a semigroup $\{e^{tA}\}_{t \geq 0}$; that is, $\omega_{\text{ess}}(A) = \log r_{\text{ess}}(e^A)$. Let us note that if Δ is a compact operator, then $\omega_{\text{ess}}(A + B\Delta C) = \omega_{\text{ess}}(A)$. Since the inequality $s(A + B\Delta C) < 0$ holds for each Δ with

$\|\Delta\| < 1/\sup_{s\in\mathbb{R}}\|C(A-is)^{-1}B\|$ (cf. the proof of Theorem 5.13), the result follows.

Persistence of dichotomy. Persistence of exponential dichotomies and hyperbolicity is important. Thus, almost every paper on dichotomy or hyperbolicity contains some information on persistence, see the bibliographical remarks for Chapter 3. Classical results on the persistence of a dichotomy are found in Coppel [**Co**], Daleckij and Krein [**DK**], Henry [**He**], and Massera and Schäffer [**MS2**] as well as in [**Co2**] and in the work by Palmer, see for example [**Pa**].

Bounded perturbations. Theorems 5.23 and 5.24 are taken from [**LMR**], see also [**CLRM**].

Note that if $\dot{x} = A(t)x(t)$ is well-posed and $t \mapsto B(t) \in \mathcal{L}(X)$ is continuous and bounded, then $\dot{x} = [A(t) + B(t)]x(t)$ is not, generally, well-posed, see the classical example given by Phillips [**Ph**] and the discussion on page 59. If the differential equation $\dot{x} = A(t)x(t)$ is well-posed, then its solutions are given by an evolution family $\{U(\theta,\tau)\}_{\theta\geq\tau}$ and that induces an evolution semigroup $\{E^t\}_{t\geq 0}$ on $L^p(\mathbb{R};X)$ with the generator Γ. Since $\mathcal{B}$ is a bounded operator, $\Gamma_1 = \Gamma + \mathcal{B}$ is a generator of a strongly continuous semigroup on $L^p(\mathbb{R};X)$. Moreover, using the characterization of the evolution semigroups from Theorem 3.28 one can check (see also Proposition 3.29) that Γ_1 is the generator of an *evolution* semigroup that corresponds to a certain strongly continuous evolution family $\{U_1(\theta,\tau)\}_{\theta\geq\tau}$. This evolution family satisfies (5.37). Generally, however, the evolution family $\{U_1(\theta,\tau)\}_{\theta\geq\tau}$ is not differentiable; if it is, then it solves the abstract Cauchy problem for $\dot{x} = [A(t) + B(t)]x(t)$. The proof of Theorem 5.24, in fact, shows that the strongly continuous evolution family $\{U_1(\theta,\tau)\}_{\theta\geq\tau}$ has a dichotomy provided $\{U(\theta,\tau)\}_{\theta\geq\tau}$ has a dichotomy and $\epsilon \geq \|B\|_\infty$ is sufficiently small.

Miyadera-type perturbations. In this subsection we follow Räbiger, Rhandi, Schnaubelt, and Voigt [**RRSV**] and Schnaubelt [**Sc**]. See also Chow and Leiva [**CLe4**] for the case of evolution families and [**CLe2**] for the case of linear skew-product flows.

CHAPTER 6

Linear Skew-Product Flows and Mather Evolution Semigroups

A linear skew-product flow is a dynamical system on a vector bundle such that each transformation is linear when restricted to a fiber of the bundle. For example, the tangent flow of a smooth flow on a manifold is a linear skew-product flow on the tangent bundle of the manifold.

Each linear skew-product flow induces a semigroup of operators, called the evolution or Mather semigroup, on the linear space of sections of the vector bundle. In this chapter we will give a characterization for the exponential dichotomy (hyperbolicity) of the linear skew-product flow in terms of the spectrum of the Mather semigroup and its generator defined on an appropriate space of sections. In particular, we will prove three main theorems about this evolution semigroup: the Spectral Projection Theorem, the Dichotomy Theorem, and the Spectral Mapping Theorem. A corollary of these results is an infinite dimensional version of the Sacker-Sell spectral theory.

In Section 6.1 we give basic definitions and examples of linear skew-product flows, cocycles, and their exponential dichotomy (hyperbolicity). In Section 6.2 the basic facts about the Mather evolution semigroup are presented; in particular, we will prove its strong continuity and describe its generator. Also, we will discuss a central idea, called "Mather localization", that is used to construct approximate eigenfunctions for the evolution operators that form the Mather semigroup. Roughly speaking, Mather localization allows us to produce approximate eigenfunctions for the evolution operators such that the eigenfunctions are localized along trajectories of the base flow. We will also generalize the idea of localization to prove the Spectral Mapping Theorems 6.30 and 6.37 for Mather semigroups in the case of an aperiodic base flow. In Section 6.3 we develop the infinite dimensional variant of the Sacker-Sell spectral theory for linear skew-product flows as a corollary of the Spectral Projection Theorem 6.38 and the Dichotomy Theorem 6.41.

6.1. Linear skew-product flows and dichotomy

We will state the basic definitions concerning cocycles, linear skew-product flows, and dichotomy.

6.1.1. Definition and examples of linear skew-product flows.
Before we give formal definitions, let us consider the prototypical example of a linear skew-product flow; namely, the skew-product flow associated with the solutions of a nonautonomous differential equation $\dot{x} = A(t)x$ on a Banach space X. For this case, consider the translation flow $\varphi^t : \theta \mapsto \theta + t$ on $\mathbb{R}$ and the trivial bundle $\mathbb{R} \times X$ over $\mathbb{R}$. A linear skew-product flow on $\mathbb{R} \times X$ is defined by $(\theta, x_0) \mapsto (\theta + t, x(t; \theta, x_0))$ where $t \mapsto x(t; \theta, x_0)$ is the solution of the differential equation with the initial condition

$x(\theta; \theta, x_0) = x_0$. If the solution is given by the evolution family $\{U(\theta, \tau)\}_{\theta \geq \tau}$, then the formula $\Phi^t(\theta) = U(\theta + t, \theta)$, for $t \geq 0$ and $\theta \in \mathbb{R}$, defines a cocycle $\{\Phi^t\}_{t \geq 0}$ over the flow $\{\varphi^t\}_{t \in \mathbb{R}}$.

To avoid unnecessary technical complications for the general case, we will define the notion of a cocycle and a linear skew-product flow in the setting of a trivial vector bundle. As it turns out, the theory that we will discuss is valid for general vector bundles, but the topology of nontrivial bundles plays no role in the analysis. In fact, the constructions of this chapter are local. They can always be carried out in a natural vector bundle chart.

Let (Θ, d) be a locally compact metric space and X a Banach space.

DEFINITION 6.1. A *continuous flow* on Θ is a one parameter group $\{\varphi^t\}_{t \in \mathbb{R}}$ of homeomorphisms of Θ such that the map $(t, \theta) \mapsto \varphi^t \theta$ is continuous. A *strongly continuous cocycle* $\{\Phi^t\}_{t \geq 0}$ over the continuous flow $\{\varphi^t\}_{t \in \mathbb{R}}$ is a family of bounded operators $\{\Phi^t(\theta) \in \mathcal{L}(X) : \theta \in \Theta, t \geq 0\}$, with the following properties: For each $x \in X$, the function $(\theta, t) \mapsto \Phi^t(\theta)x$ is continuous on $\Theta \times [0, \infty)$,

(a) $\Phi^{t+s}(\theta) = \Phi^t(\varphi^s \theta)\Phi^s(\theta)$ for $t \geq 0$ and $s \geq 0$, and
(b) $\Phi^0(\theta) = I$ for $\theta \in \Theta$.

If, in addition,

(c) there exist numbers $M > 0$ and $\omega > 0$ such that $\|\Phi^t(\theta)\|_{\mathcal{L}(X)} \leq M e^{\omega t}$ for all $t \in \mathbb{R}_+$ and $\theta \in \Theta$,

then the strongly continuous cocycle is called *exponentially bounded*. The *linear skew-product flow* (LSPF) associated with the cocycle $\{\Phi^t\}_{t \geq 0}$ is the dynamical system $\hat{\varphi}^t : \Theta \times X \to \Theta \times X$ defined by

(6.1) $$\hat{\varphi}^t : (\theta, x) \mapsto (\varphi^t \theta, \Phi^t(\theta)x), \qquad t \geq 0.$$

Let us note that the operators in a strongly continuous cocycle are not assumed to be invertible. For this reason, the cocycle is parameterized by $t \geq 0$, but *not* by $t \in \mathbb{R}$. Also, we note that each map $\hat{\varphi}^t$ is continuous if and only if the corresponding cocycle is strongly continuous. By the Uniform Boundedness Principle, if the base space Θ is compact, then a strongly continuous cocycle is exponentially bounded.

REMARK 6.2. If the metric space Θ is compact and the cocycle $\{\Phi^t\}_{t \geq 0}$ is continuous *in norm*; that is, the map $(t, \theta) \mapsto \Phi^t(\theta) \in \mathcal{L}(X)$ is continuous, then $\Phi^0(\theta) = I$ and $\|\Phi^t(\theta) - I\|_{\mathcal{L}(X)} \to 0$ as $t \to 0^+$. Hence, if $\delta > 0$ is sufficiently small, then the operator $\Phi^t(\theta)$ is invertible for each $t \in [0, \delta]$ and $\theta \in \Theta$. Also, if $t \geq 0$, then $t = \delta n + \rho$ for some $n \in \mathbb{N}$ and some $\rho \in [0, \delta)$. Using the cocycle property, each operator $\Phi^t(\theta) = \Phi^\delta(\varphi^{(n-1)\delta + \rho}\theta) \cdots \Phi^\delta(\varphi^\rho \theta)\Phi^\rho(\theta)$ is invertible. By defining $\Phi^{-t}(\theta) = [\Phi^t(\varphi^{-t}\theta)]^{-1}$, $t > 0$, the original family $\{\Phi^t(\theta)\}_{t \geq 0}$ is extended to $t \in \mathbb{R}$ while maintaining the cocycle property. In particular, if X is finite dimensional, then the strong continuity implies norm continuity; and therefore, if X is finite dimensional, then a strongly continuous cocycle can always be extended to $t \in \mathbb{R}$.
◇

The classic example of a cocycle appears as the solution operator for a variational equation. Indeed, let us assume that $\{\varphi^t\}_{t \in \mathbb{R}}$ is a continuous flow on the locally compact metric space Θ, and $\{A(\theta) : \theta \in \Theta\}$ is a family of (possibly unbounded) densely defined closed operators on the Banach space X. A strongly continuous cocycle $\{\Phi^t\}_{t \geq 0}$ is said to solve the (variational) equation

(6.2) $$\dot{x} = A(\varphi^t \theta)x, \quad \theta \in \Theta, \quad t \in \mathbb{R},$$

if, for every $\theta \in \Theta$, there exists a dense subset $Y_\theta \subseteq \mathcal{D}(A(\theta))$ such that, for every $x_\theta \in Y_\theta \subseteq \mathcal{D}(A(\theta))$, the function $t \mapsto x(t) := \Phi^t(\theta)x_\theta$ is differentiable for $t \geq 0$, the values $x(t) \in \mathcal{D}(A(\varphi^t\theta))$, and $t \mapsto x(t)$ satisfies the differential equation (6.2). More restrictive definition can be given if we require that $Y_\theta = \mathcal{D}(A(\theta))$ or even $Y_\theta = \mathcal{D}(A(\theta)) = \mathcal{D}$; that is, Y_θ is independent of θ, see the corresponding discussion after Definition 3.2.

Differential equations of type (6.2) arise from two basic sources that we will now describe.

First, consider a nonlinear differential equation $\dot{y} = F(y)$ on X where the function $F : X \to X$ is Fréchet differentiable. Suppose that this differential equation has a compact invariant set $\Theta \subset X$; that is, the solution $t \mapsto y(t;\theta)$ such that $y(0;\theta) = \theta$ has its values in Θ for all $t \in \mathbb{R}$ whenever the initial point $\theta \in \Theta$. In this case, the family of maps $\{\varphi^t : \theta \mapsto y(t;\theta) : t \in \mathbb{R}\}$, defines a flow on Θ. If $\theta \in \Theta$, then for a new initial condition x_0 in X, the difference $x(t) = y(t;x_0) - y(t;\theta)$ is such that

$$\dot{x}(t) = DF(y(t;\theta))x(t) + \eta(y,x), \quad |\eta(y,x)| = o(x), \quad |x| \to 0.$$

Thus, the differential equation (6.2) with $A(\theta) = DF(\theta)$, called the *variational equation*, determines the linearized flow of $\dot{y} = F(y)$. Note that, in an infinite dimensional setting, the operators $A(\theta)$ might be unbounded.

Second, define the *hull* of a continuous function $a : \mathbb{R} \to \mathcal{L}(X)$ to be the following set of operator-valued functions:

$$\text{Hull}(a) = \text{closure}\{a(\cdot + \tau) : \tau \in \mathbb{R}\}.$$

Under appropriate assumptions, the set $\Theta := \text{Hull}(a)$ may be a compact set of operator-valued functions on $\mathbb{R}$. For example, if $a : \mathbb{R} \to \mathcal{L}(\mathbb{R}^n)$ is almost-periodic and the closure is taken in the topology of uniform convergence on compact subsets of $\mathbb{R}$, then, by Bochner's Theorem, Θ is compact in the space of continuous matrix-valued functions. Let us define a flow on Θ by $\varphi^t(\theta)(s) = \theta(s+t)$ where $t \in \mathbb{R}$ and $s \in \mathbb{R}$. If we also set $A(\theta) = \theta(0) \in \mathcal{L}(X)$, then we obtain in (6.2) all differential equations of the form $\dot{x} = \theta(t)x$ where the function θ is in the hull of a.

As an illustration, let us consider the following "scalar" example of a linear skew-product flow. Later, see Example 6.18, we will continue the discussion of this example.

EXAMPLE 6.3. Suppose that $a : \mathbb{R} \to \mathbb{R}_+$ is a uniformly continuous function such that

(6.3) $$\beta := \lim_{t \to \infty} a(t) \geq \lim_{t \to -\infty} a(t) =: \alpha.$$

We will consider the nonautonomous differential equation $\dot{x} = a(t)x$, $t \in \mathbb{R}$, on a Banach space X. Endow the space $C(\mathbb{R};\mathbb{R})$ of continuous functions with the topology of uniform convergence on compact subsets of $\mathbb{R}$ and note that this space is metrizable with metric

(6.4) $$d(f,g) = \sum_{k=1}^{\infty} 2^{-k} \frac{d_k(f,g)}{1+d_k(f,g)}, \quad d_k(f,g) = \sup\{|f(t)-g(t)| : |t| \leq k\}.$$

Also, let $a_\tau(t) := a(t+\tau)$ and note that the hull of a is given by

(6.5) $$\Theta = \text{closure}\{a_\tau : \tau \in \mathbb{R}\} = \{a_\tau : \tau \in \mathbb{R}\} \cup \{a_{-\infty}(\cdot) \equiv \alpha\} \cup \{a_\infty(\cdot) \equiv \beta\}.$$

The formula
$$\Phi^t(\theta)x = \exp\Big(\int_0^t \theta(s)\,ds\Big)x, \quad x \in X, \quad t \geq 0,$$
defines a strongly continuous, exponentially bounded cocycle over the flow $\varphi^t : \Theta \mapsto \Theta$ given by $\varphi^t(\theta)(s) = \theta(t+s)$, $s \in \mathbb{R}$. The corresponding linear skew-product flow $\hat{\varphi}^t$ is defined by formula (6.1). ◇

The next four "examples" define other important classes of linear skew-product flows.

EXAMPLE 6.4. Suppose that $\Theta = \mathbb{R}$ and $A : \mathbb{R} \to \mathcal{L}(X)$ is a bounded, continuous function. The evolution family $\{U(\theta,\tau)\}_{(\theta,\tau)\in\mathbb{R}^2}$ for the differential equation

(6.6) $$\dot{x} = A(t)x, \quad t \in \mathbb{R},$$

is defined so that if $t \mapsto x(t)$ is a solution of the equation (6.6), then $x(\theta) = U(\theta,\tau)x(\tau)$ for all $(\theta,\tau) \in \mathbb{R}^2$. Let us consider the translation flow $\varphi^t : \theta \mapsto \theta + t$ on $\Theta = \mathbb{R}$ and, for each $\theta, t \in \mathbb{R}$, the corresponding operator defined by $\Phi^t(\theta) := U(\theta+t,\theta)$. Then $\{\Phi^t\}_{t\in\mathbb{R}}$ is a cocycle over $\{\varphi^t\}_{t\in\mathbb{R}}$. Moreover, we have that $A(\theta) = \frac{d}{dt}\Phi^t(\theta)\big|_{t=0}$ and the solution $t \mapsto x(t;x_0) = U(t,0)x_0$ with the initial condition $x(0;x_0) = x_0$ defines the linear skew-product flow $\hat{\varphi}^t : (\theta,x_0) \mapsto (\theta+t, x(\theta+t;x_0))$. ◇

EXAMPLE 6.5. Consider the translation flow $\varphi^t : \theta \mapsto \theta + t$, for $t \in \mathbb{R}$, on the space $\Theta = \mathbb{R}$, and let $\{U(\theta,\tau)\}_{\theta\geq\tau}$, $\theta,\tau \in \mathbb{R}$, be a strongly continuous, exponentially bounded evolution family on a Banach space X. As above, $\Phi^t(\theta) := U(\theta+t,\theta)$, $\theta \in \Theta$, $t \geq 0$, defines a cocycle $\{\Phi^t\}_{t\geq 0}$ over the translations $\{\varphi^t\}_{t\in\mathbb{R}}$. Suppose that $\{U(\theta,\tau)\}_{\theta\geq\tau}$ solves the abstract Cauchy problem

(ACP) $$\dot{x} = A(t)x, \quad x(\theta) = x_\theta \in \mathcal{D}(A(\theta)), \quad \theta \in \mathbb{R},$$

see Definition 3.2; that is, the function $t \mapsto x(\theta+t) = U(\theta+t,\theta)x_\theta$ is differentiable, the values $x(\theta+t) \in \mathcal{D}(A(\theta+t))$, $\theta \in \mathbb{R}$, $t \geq 0$, and the function satisfies (ACP). Then the cocycle $\{\Phi^t\}_{t\geq 0}$ is strongly differentiable with

$$\frac{d}{dt}\Phi^t(\theta)x = A(\varphi^t\theta)\Phi^t(\theta)x, \quad x \in \mathcal{D}(A(\theta)), \quad \theta \in \Theta = \mathbb{R},$$

and $\hat{\varphi}^t : \Theta \times X \to \Theta \times X$ defined by $(\theta,x_\theta) \mapsto (\theta+t, U(\theta+t,\theta)x_\theta)$ is a LSPF. ◇

EXAMPLE 6.6. Let $\{\varphi^t\}_{t\in\mathbb{R}}$ be a continuous flow on a compact metric space Θ, and $A : \Theta \to \mathcal{L}(X)$ a (norm) continuous function. If $\Phi^t(\theta)$, $t \in \mathbb{R}$, is the solution operator for the differential equation (6.2); that is, $x(t) = \Phi^t(\theta)x(0)$, then the cocycle $\{\Phi^t\}_{t\in\mathbb{R}}$ is (norm) continuous in θ. Moreover, this cocycle is *smooth* with

(6.7) $$A(\theta) = \frac{d}{dt}\Phi^t(\theta)\Big|_{t=0}, \quad \theta \in \Theta.$$

The corresponding LSPF $\{\hat{\varphi}^t\}_{t\in\mathbb{R}}$, as in Example 6.5, represents the solutions of the differential equation (6.2).

REMARK 6.7. There is no loss of generality if we assume that a norm continuous cocycle is smooth. Indeed, an arbitrary *norm* continuous cocycle $\{\Psi^t\}_{t\in\mathbb{R}}$ is cohomologous to a smooth cocycle $\{\Phi^t\}_{t\in\mathbb{R}}$. By definition, $\{\Psi^t\}_{t\in\mathbb{R}}$ is cohomologous to $\{\Phi^t\}_{t\in\mathbb{R}}$ if there is a continuous function $F : \Theta \to \mathcal{GL}(X)$ such that

$$\Phi^t(\theta) = F(\varphi^t\theta)\Psi^t(\theta)F^{-1}(\theta), \quad \theta \in \Theta, \quad t \in \mathbb{R}.$$

To construct F and $\{\Phi^t\}_{t\in\mathbb{R}}$, suppose that $r > 0$ and let $F(\theta) := \frac{1}{r}\int_0^r \Psi^s(\theta)\,ds$, see Johnson, Palmer, and Sell [**JPS**, p.8]. Since $(\theta, t) \mapsto \Psi^t(\theta) \in \mathcal{L}(X)$ is continuous and $\Psi^0(\theta) = I$, the operators $F(\theta)$ are invertible for sufficiently small $r > 0$ and all $\theta \in \Theta$. Moreover, for such an r, the cocycle

$$\Phi^t(\theta) := F(\varphi^t\theta)\Psi^t(\theta)F(\theta)^{-1} = \frac{1}{r}\int_t^{t+r} \Phi^s(\theta)\,ds\,F(\theta)^{-1}$$

is smooth. $\diamondsuit$

EXAMPLE 6.8. Let us suppose that each operator $A(\theta)$ in the family of differential equations (6.2) is of the special form $A(\theta) = A_0 + A_1(\theta)$ where A_0 is the generator of a strongly continuous semigroup $\{e^{tA_0}\}_{t\geq 0}$ on the Banach space X and $A_1(\cdot) \in C_b(\Theta; \mathcal{L}(X))$. If $A_1(\theta) = 0$, then the cocycle $\{\Phi^t(\theta)\}_{t\geq 0}$, defined by $\Phi^t(\theta) := e^{tA_0}$ is independent of $\theta \in \Theta$. For a general $A_1(\cdot)$, equation (6.2) may have classical; that is, differentiable solutions. Then, the corresponding family of solution operators $\{\Phi^t(\theta)\}_{t\geq 0}$ is a (strongly) differentiable cocycle. However, as we remarked on page 59, equation (6.2) may not be well-posed in the classical sense even though A_0 generates a strongly continuous semigroup. Therefore, we associate with (6.2) a mild integral equation and define a cocycle $\{\Phi^t\}_{t\geq 0}$ such that each function $t \mapsto \Phi^t(\theta)$ satisfies the "variation of constants" formula

(6.8) $$\Phi^t(\theta)x = e^{tA_0}x + \int_0^t e^{(t-\theta)A_0}A_1(\varphi^\tau\theta)\Phi^\tau(\theta)x\,d\tau$$

for $x \in X$ and $t \geq 0$. $\diamondsuit$

EXAMPLE 6.9. As a concrete example that satisfies the setting of Example 6.8 with the additional property that Θ is compact, consider the scalar parabolic partial differential equation

$$x_t = x_{ss} - a(t)x, \quad s \in (0,1), \quad t \in (0,1),$$

where $a \in C_b(\mathbb{R})$ is the function defined as follows: $a(t) = 0$ for $t \leq 0$, $a(t) = t$ for $0 < t < 1$, and $a(t) = 1$ for $t \geq 1$, see Chow and Leiva [**CLe**, Example 5.3]. This differential equation has a (weak) solution $x(t) = x(t;\cdot)$ in the Sobolev space $X = H^2(0,1) \cap H_0^1(0,1)$. For each $\tau \in \mathbb{R}$, let $a_\tau : \mathbb{R} \to \mathbb{R}$ be defined by $a_\tau(t) := a(t+\tau)$, and let Θ be the closure in $C_b(\mathbb{R}; X)$ of the set $\{a_\tau\}_{\tau\in\mathbb{R}}$. Also, for each $\theta \in \Theta$, note that $\theta : \mathbb{R} \to \mathbb{R}$ and define the operator $A(\theta)$ on X by $A(\theta)x = x_{ss} + \theta(0)x$. Also, define the flow $\{\varphi^t\}_{t\in\mathbb{R}}$ on Θ by $\varphi^t(\theta) = \theta_t$. Then, the differential equation (6.2) is the parabolic partial differential equation $x_t = x_{ss} + \theta(t)x$. By defining the operator $\Phi^t(\theta)$, $\theta \in \Theta$, as the solution operator of this differential equation, we obtain a cocycle $\{\Phi^t\}_{t\geq 0}$ that is a particular case of Example 6.8. $\diamondsuit$

The following example is a continuation of Example 6.3.

EXAMPLE 6.10. Consider the time dependent parabolic equation with Neumann boundary conditions:

(6.9) $$u_t = a(t)u_{ss}, \quad s \in (0,1), \quad u_s(t,0) = u_s(t,1) = 0,$$

where $a : \mathbb{R} \to \mathbb{R}_+$ satisfies condition (6.3). Consider the Hilbert space $X = L^2(0,1)$ and the operator $\Delta x := \frac{d^2}{ds^2}x$ — defined in the weak sense — with the domain $\mathcal{D}(\Delta) = \{x \in H^2(0,1) : x'(0) = x'(1) = 0\}$. Rewrite equation (6.9) as $\dot{x} = a(t)\Delta x$ and consider $\Theta := \text{closure}\{a_\tau : \tau \in \mathbb{R}\}$ so that Θ is the hull of the function a as defined in Example 6.3. For $\theta \in \Theta$, the equation $\dot{x} = \theta(t)\Delta x$ can be rewritten in the form of equation (6.2) by defining the function $\varphi^t : \Theta \mapsto \Theta$ by $(\varphi^t \theta)(s) = \theta(s+t)$ and $A(\theta) := \theta(0)\Delta$. If $T(t) = e^{t\Delta}$, $t \geq 0$, gives the corresponding semigroup on X, then the cocycle $\{\Phi^t\}_{t \in \mathbb{R}_+}$ is given by

$$(6.10) \qquad \Phi^t(\theta)x = T\Big(\int_0^t \theta(s)\,ds\Big)x, \quad x \in X, \quad t \geq 0, \quad \theta \in \Theta.$$

Note that the spectrum of Δ is given by $\sigma(\Delta) = \{\lambda_n = -(n-1)^2 \pi^2, n = 1, 2, \dots\}$ and the corresponding eigenfunctions are

$$x_n(s) = \sqrt{2}\cos(n-1)\pi s, \quad n = 2, 3, \dots, \quad x_1(s) = 1, \quad s \in (0,1).$$

If P_n denotes the orthogonal projection $P_n x = \langle x, x_n \rangle x_n$, then $\Phi^t(\theta)$ can be written as

$$(6.11) \qquad \Phi^t(\theta)x = \sum_{n=1}^{\infty} \exp\Big(\int_0^t \lambda_n \theta(s) ds\Big) P_n x,$$

which is, in fact, a direct sum of the "scalar" cocycles from Example 6.3. ◇

EXAMPLE 6.11. Consider the functional differential equation $\dot{x}(t) = a(t)x(t-1)$ on the Banach space $X = C([-1,0]; \mathbb{R})$ where $a \in C_b(\mathbb{R})$. Let $a_\tau(\cdot) = a(\cdot + \tau)$ and Θ be as in Example 6.9. If the flow $\{\varphi^t\}_{t \in \mathbb{R}}$ on the compact set Θ is defined by $\varphi^t \theta = \theta_t$, and $\Phi^t(\theta)$ is the solution operator for the equation $\dot{x}(t) = \theta(t)x(t-1)$, then formula (6.1) defines a linear skew-product flow, see Chow and Leiva [**CLe**, Ex. 5.2] for more details. ◇

The next example illustrates how a cocycle arises from the linearization of a nonlinear partial differential equation. We will briefly describe the situation for the linearized Navier-Stokes equation, see Sacker and Sell [**SS5**] for more details and further references.

EXAMPLE 6.12. Consider the Navier-Stokes equations

$$\frac{\partial \mathbf{v}}{\partial t} = \nu \Delta \mathbf{v} - \langle \mathbf{v}, \nabla \rangle \mathbf{v} - \operatorname{grad} \mathbf{p} + g, \quad \operatorname{div} \mathbf{v} = 0$$

on a bounded domain $\Omega \subset \mathbb{R}^2$ with zero boundary conditions. Here $\mathbf{v} : \Omega \to \mathbb{R}^2$ represents the velocity of an incompressible fluid, $\mathbf{p} : \Omega \to \mathbb{R}$ is the pressure, ν measures the viscosity of fluid, and $g : \mathbb{R} \times \Omega \to \mathbb{R}^2$ represents a time-dependent forcing term. Consider the orthogonal decomposition $L^2(\Omega; \mathbb{R}^2) = X \oplus H_\pi$ where X is the closure in $L^2(\Omega; \mathbb{R}^2)$ of the C^∞ divergence-free ($\nabla \cdot \mathbf{v} = 0$) vector fields with compact support in Ω and H_π is the closure in $L^2(\Omega; \mathbb{R}^2)$ of the gradients $\nabla \mathbf{p}$ of all $\mathbf{p} \in C^1(\Omega; \mathbb{R})$, see for example Constantin and Foias [**CF**] or Henry [**He**, Sect. 3.8]. Let $P : L^2(\Omega; \mathbb{R}^2) \to X$ be the corresponding orthogonal projection, and define $A = P\Delta$, $B(\mathbf{v}, \mathbf{u}) = -P\langle \mathbf{v}, \nabla \rangle \mathbf{u}$, and $f = Pg$. Thus, see for example Temam [**Te**], the Navier-Stokes equation can be rewritten as the following abstract equation on the Hilbert space X:

$$(6.12) \qquad \frac{d\mathbf{v}}{dt} = \nu A \mathbf{v} + B(\mathbf{v}, \mathbf{v}) + f, \quad \mathbf{v}(0) = \mathbf{v}_0.$$

The operator A with $\mathcal{D}(A) = X \cap H^2(\Omega; \mathbb{R}^2)$ is a negative operator, and therefore it generates an analytic semigroup on X, see for example [**Te**]. Also, we define $V = \mathcal{D}((-A)^{1/2})$.

Let us assume that the function F, defined by $F(t) := f(t, \cdot)$, $t \in \mathbb{R}$, is in $C_b(\mathbb{R}; X)$. Moreover, let us assume that the positive hull of F,
$$H^+(F) := \text{closure}_{C_b(\mathbb{R};X)}\{F_\tau := F(\cdot + \tau) : \tau \in \mathbb{R}_+\}$$
is a compact subset of $C_b(\mathbb{R}; X)$. Under this assumption, the omega-limit set $\omega(F) = \cap_{\tau \geq 0} H^+(F_\tau)$ is nonempty and compact. Moreover, there exists a global compact attractor $\Theta \subset \mathcal{D}(A) \times \omega(F)$ for the semiflow generated by the strong solutions of the abstract equation (6.12), see Raugel and Sell [**RSe**, Sect. 2.11–2.12] and the literature cited therein for the proof of this nontrivial fact. This attractor is invariant under the flow $\{\varphi^\tau\}_{\tau \in \mathbb{R}}$ defined by
$$\varphi^\tau : (\mathbf{v}, f) \mapsto \theta_\tau = (\mathbf{v}_\tau, f_\tau), \quad \tau \in \mathbb{R},$$
where $f_\tau(t, \cdot) = f(t + \tau, \cdot)$ and $\mathbf{v}_\tau(t, \cdot) = \mathbf{v}(t + \tau, \cdot)$ for the strong solution $\mathbf{v}(t, \cdot)$ of the equation (6.12).

If $\theta = (\mathbf{v}_0, f) \in \Theta$ and $\mathbf{v}(t) = \mathbf{v}(t; f; \mathbf{v}_0)$, $t \geq 0$, is the corresponding strong solution of equation (6.12), then (see, e.g., [**CF**])
$$\mathbf{v}(\cdot; f; \mathbf{v}_0) \in C([0, \infty); V) \cap L^\infty((0, \infty); V) \cap L^\infty_{\text{loc}}((0, \infty); \mathcal{D}(A)).$$
The linearized Navier-Stokes equation along the solution $\mathbf{v}$ is given by
$$(6.13) \qquad \frac{dx}{dt} = \nu A x + B(\mathbf{v}(t), x) + B(x, \mathbf{v}(t)), \quad x(0) = x_0 \in X.$$
Using the results in Sacker and Sell [**SS5**] and the references therein, if $x_0 \in V$, then there is a unique strong solution $x(t) = \Phi^t(\theta)x_0$ of the linearized equation (6.13) such that
$$x(\cdot) \in C([0, \infty), V) \cap L^\infty_{\text{loc}}((0, \infty), \mathcal{D}(A)), \quad x_t(\cdot) \in L^2_{\text{loc}}((0, \infty); X),$$
where $\Phi^t(\theta)$ is the solution operator for (6.13). Clearly, $\{\Phi^t\}_{t \geq 0}$ is a cocycle over the flow $\{\varphi^t\}_{t \in \mathbb{R}}$ on Θ. Furthermore, the mapping $\hat{\varphi}^t : (\theta, x_0) \mapsto (\varphi^t \theta, \Phi^t(\theta)x_0)$ is a continuous mapping of $\Theta \times V$ into $\Theta \times V$. $\diamondsuit$

So far, we have only considered examples of linear skew-product flows on trivial bundles. However, there are important linear skew-product flows on *nontrivial* bundles. Indeed, let $\{\varphi^t\}_{t \in \mathbb{R}}$ be a smooth flow on a compact smooth n-dimensional manifold Θ, and let $\Phi^t(\theta) = D\varphi^t(\theta)$, $t \in \mathbb{R}$, denote the differential of this flow. Note that $\Phi^t(\theta)$ maps the tangent space $\mathcal{T}_\theta \Theta$ at the point θ to the tangent space $\mathcal{T}_{\varphi^t \theta} \Theta$ at the point $\varphi^t \theta$. By the chain rule applied to the identity $\varphi^{t+s}(\theta) = \varphi^t(\varphi^s \theta)$, it follows that $\{\Phi^t\}_{t \in \mathbb{R}}$ is a cocycle. Here, the Banach space X is the model space for the fibers of the tangent bundle, and we must choose local coordinates to identify the family of differentials as maps on X. However, once this is done, we can write an associated LSPF $\{\hat{\varphi}^t\}_{t \in \mathbb{R}}$ on the tangent bundle $\mathcal{T}\Theta$ of Θ (locally) in a natural bundle chart by $\hat{\varphi}^t(\theta, x) = (\varphi^t \theta, D\varphi^t(\theta)x)$, $x \in X \cong \mathcal{T}_\theta \Theta$.

More generally, let $\mathcal{E} = \mathcal{E}(\Theta, p, X)$ be a vector bundle with a compact metric base space Θ, projection $p : \mathcal{E} \to \Theta$ and fibers $\mathcal{E}_\theta = p^{-1}\{\theta\}$ isomorphic to a Banach space X. Recall that a *linear skew-product flow* (linear bundle map) over a flow $\{\varphi^t\}_{t \in \mathbb{R}}$ in Θ is a dynamical system $\hat{\varphi}^t : \mathcal{E} \to \mathcal{E}$ such that the restriction $\Phi^t(\theta) := \hat{\varphi}^t|_{\mathcal{E}_\theta}$ is a linear map from $\mathcal{E}_\theta$ to $\mathcal{E}_{\varphi^t \theta}$. The linear skew-product flow $\{\hat{\varphi}^t\}_{t \geq 0}$ can be represented locally in a natural vector bundle chart by
$$\hat{\varphi}^t : (\theta, x_\theta) \mapsto (\varphi^t \theta, \Phi^t(\theta)x_\theta), \quad x_\theta \in E_\theta, \quad \theta \in \Theta,$$
where the operators $\Phi^t(\theta) : E_\theta \mapsto E_{\varphi^t \theta}$ give a linear cocycle.

6.1.2. Dichotomy and hyperbolicity of linear skew-product flows. In this subsection we will discuss the definitions of dichotomic and hyperbolic cocycles and linear skew-product flows in the setting of Definition 6.1.

Let us begin with the definition of "global" exponential dichotomy; that is, dichotomy on the entire base space Θ. For this definition, let us recall that if P is a projection, then the complementary projection is always denoted by $Q = I - P$, similarly for families of projections, e. g., $Q(\theta) = I - P(\theta)$, $Q_\tau = I - P_\tau$, etc.

DEFINITION 6.13. A linear skew-product flow $\{\hat{\varphi}^t\}_{t\geq 0}$ (or the corresponding cocycle $\{\Phi^t\}_{t\geq 0}$) has an *exponential dichotomy* on Θ over $\mathbb{R}$, respectively $\mathbb{Z}$, if there exist constants $\beta > 0$ and $M = M(\beta) > 0$ and a bounded, strongly continuous, projection-valued function $P : \Theta \to \mathcal{L}_s(X)$ such that, for all $\theta \in \Theta$ and $t \geq 0$ in $\mathbb{R}$, respectively $\mathbb{Z}$, the following conditions are satisfied:

(a) $P(\varphi^t\theta)\Phi^t(\theta) = \Phi^t(\theta)P(\theta)$.
(b) The restriction $\Phi_Q^t(\theta)$ of the operator $\Phi^t(\theta)$, defined by
$$\Phi_Q^t(\theta) := \Phi^t(\theta)Q(\theta) : \operatorname{Im} Q(\theta) \to \operatorname{Im} Q(\varphi^t\theta),$$
is invertible.
(c) $\|\Phi_P^t(\theta)\|_{\mathcal{L}(X)} \leq Me^{-\beta t}$ and $\|[\Phi_Q^t(\theta)]^{-1}\|_{\mathcal{L}(X)} \leq Me^{-\beta t}$ where $\Phi_P^t(\theta)$ is the restriction of the operator $\Phi^t(\theta)$ defined by
$$\Phi_P^t(\theta) := \Phi^t(\theta)P(\theta) : \operatorname{Im} P(\theta) \to \operatorname{Im} P(\varphi^t\theta).$$

The projection $P(\cdot)$ is called the *dichotomy projection*, the constants β and M are called the *dichotomy constants*, and $\beta_\Phi := \sup\{\beta > 0$ such that condition (c) holds for some $M = M(\beta)\}$ is called the *dichotomy bound*.

Suppose that the cocycle $\{\Phi^t\}_{t\geq 0}$ solves the differential equation (6.2). For each $\theta \in \Theta$, the nonautonomous differential equation $\dot{x} = A(\varphi^t\theta)x$ is solved by the evolution family $\{U_\theta(s,\tau)\}_{s\geq\tau}$ defined by $U_\theta(s,\theta) = \Phi^{s-\tau}(\varphi^\tau\theta)$. If the cocycle $\{\Phi^t\}_{t\geq 0}$ has an exponential dichotomy in the sense of Definition 6.13, then an exponential dichotomy in the sense of Definition 3.6 exists for each evolution family $\{U_\theta(s,\tau)\}_{s\geq\tau}$, the dichotomy projection is continuous in the parameter θ, and the dichotomy constants are uniform over $\theta \in \Theta$.

The next definition defines "pointwise" dichotomy; that is, dichotomy over the closure of a single orbit of the base flow.

DEFINITION 6.14. The linear skew-product flow $\{\hat{\varphi}\}_{t\geq 0}$ has an *exponential dichotomy over* $\mathbb{R}$, respectively $\mathbb{Z}$, *at the point* $\theta_0 \in \Theta$ if there is a family of projections $\{P_\tau \in \mathcal{L}(X) : \tau \in \mathbb{R}\}$, respectively $\{P_\tau \in \mathcal{L}(X) : \tau \in \mathbb{Z}\}$, and constants $\beta = \beta(\theta_0) > 0$ and $M = M(\beta) > 0$ such that the function $\tau \mapsto P_\tau$ is bounded, it is strongly continuous if $\tau \in \mathbb{R}$ and, for all τ and $t \geq 0$ in $\mathbb{R}$, respectively $\mathbb{Z}$, the following conditions are satisfied:

(a) $\Phi^t(\varphi^\tau\theta_0)P_\tau = P_{\tau+t}\Phi^t(\varphi^\tau\theta_0)$.
(b) The operator
$$\Phi_Q^t(\varphi^\tau\theta_0) := \Phi^t(\varphi^\tau\theta_0)Q_\tau : \operatorname{Im} Q_\tau \to \operatorname{Im} Q_{t+\tau}$$
is invertible.
(c) $\|\Phi_P^t(\varphi^\tau\theta_0)\|_{\mathcal{L}(X)} \leq Me^{-\beta t}$ and $\|[\Phi_Q^t(\varphi^\tau\theta_0)]^{-1}\|_{\mathcal{L}(X)} \leq Me^{-\beta t}$ where
$$\Phi_P^t(\varphi^\tau\theta_0) := \Phi^t(\varphi^\tau\theta_0)P_\tau : \operatorname{Im} P_\tau \to \operatorname{Im} P_{t+\tau}.$$

Clearly, the existence of an exponential dichotomy on Θ implies the existence of an exponential dichotomy at each $\theta_0 \in \Theta$. Let us note that if each operator $\Phi^t(\theta)$ is invertible, then $P_\tau = \Phi^\tau(\theta_0) P_0 [\Phi^\tau(\theta_0)]^{-1}$, and condition (c) in Definition 6.14 can be rewritten, for $P := P_0$, in the following form

$$\|\Phi^t(\theta_0) P [\Phi^\tau(\theta_0)]^{-1}\| \leq M e^{-\beta(t-\tau)}, \quad t \geq \tau,$$
$$\|\Phi^t(\theta_0) Q [\Phi^\tau(\theta_0)]^{-1}\| \leq M e^{-\beta(\tau-t)}, \quad t \leq \tau,$$

see for example Sacker and Sell [**SS3**].

Next, we give the definition of hyperbolicity over the base space Θ.

DEFINITION 6.15. The linear skew-product flow $\{\hat{\varphi}^t\}_{t \geq 0}$ is called *hyperbolic* on Θ if there exist constants $\beta > 0$ and $M = M(\beta) > 0$, and a projection-valued function $P \in C_b(\Theta; \mathcal{L}_s(X))$ such that, for all $\theta \in \Theta$ and $t \geq 0$, the following conditions are satisfied:

(i) $P(\varphi^t \theta) \Phi^t(\theta) = \Phi^t(\theta) P(\theta)$.
(ii)

$$|\Phi^t(\theta) x| \leq M e^{-\beta t} |x|, \quad x \in \operatorname{Im} P(\theta), \quad t > 0,$$
$$|\Phi^t(\theta) x| \geq M^{-1} e^{\beta t} |x|, \quad x \in \operatorname{Im} Q(\theta), \quad t > 0.$$

Note that if a linear skew-product flow has an exponential dichotomy, then it is hyperbolic. However, for infinite dimensional subspaces $\operatorname{Im} Q(\theta)$, the second inequality in condition (ii) of Definition 6.15 does not imply condition (b) of Definition 6.13 in general. Clearly, if $\{\Phi^t\}_{t \in \mathbb{R}}$ is a cocycle consisting of invertible operators, then the existence of an exponential dichotomy is equivalent to the hyperbolicity of the skew-product flow. For general cocycles, if $\dim \operatorname{Im} Q(\theta)$ is finite, then the second inequality in statement (ii) of Definition 6.15 implies statement (b) and the second inequality in statement (c) of Definition 6.13. If this is the case, then again the existence of an exponential dichotomy is equivalent to the hyperbolicity of the skew-product flow. We will see in Theorem 8.12 that if there is some number $T > 0$ such that the operator $\Phi^t(\theta)$ on X is compact for each $\theta \in \Theta$ and the function $\theta \mapsto \Phi^t(\theta)$ is continuous in norm whenever $t > T$, then $\dim \operatorname{Im} Q(\theta) < \infty$. Thus, for this class of cocycles, Definitions 6.13 and 6.15 are equivalent.

REMARK 6.16. A cocycle $\{\Phi^t\}_{t \in \mathbb{R}}$ consisting of invertible operators is called *weakly dichotomic* (quasi-Anosov, in another terminology) if $x = 0$ whenever $\theta \in \Theta$ and

$$C := \sup\{|\Phi^t(\theta) x| : t \in \mathbb{R}\} < \infty.$$

Of course, if a cocycle has an exponential dichotomy, then it is weakly dichotomic. Indeed, suppose that there is an exponential dichotomy and $\theta \in \Theta$. For each $x \in X$, we have that $x = P(\theta) x + Q(\theta) x$. But, by the second inequality in condition (ii) of Definition 6.15, we have that $Q(\theta) x = 0$ because $C < \infty$. Also, by the first inequality, $|P(\theta) x| = |\Phi^t(\varphi^{-t} \theta) \Phi^{-t}(\theta) P(\theta) x| \leq M e^{-\beta t} C$ for $t > 0$; and therefore, $P(\theta) x = 0$. $\diamondsuit$

For a strongly continuous, exponentially bounded cocycle $\{\Phi^t\}_{t \geq 0}$ and $\lambda \in \mathbb{R}$, we define the rescaled cocycle by $\Phi^t_\lambda(\theta) = e^{-\lambda t} \Phi^t(\theta)$ and the corresponding linear skew-product flow $\hat{\varphi}^t_\lambda : (\theta, x) \mapsto (\varphi^t \theta, e^{-\lambda t} \Phi^t(\theta) x)$. If the cocycle $\{\Phi^t\}_{t \geq 0}$ solves equation (6.2), then the cocycle $\{\Phi^t_\lambda\}_{t \geq 0}$ solves the equation $\dot{x} = [A(\varphi^t \theta) - \lambda] x$.

DEFINITION 6.17. The *dynamical (Sacker-Sell) spectrum* for the strongly continuous cocycle $\{\Phi^t\}_{t\geq 0}$, or the corresponding skew-product flow $\{\hat{\varphi}^t\}_{t\geq 0}$, is the set
$$\Sigma = \{\lambda \in \mathbb{R} : \{\hat{\varphi}^t_\lambda\}_{t\geq 0} \text{ does not have exponential dichotomy on } \Theta\}.$$

EXAMPLE 6.18. We will compute the dynamical spectrum for the "scalar" linear skew-product given in Example 6.3. In fact, we will show that $\Sigma = [\alpha, \beta]$ for the α and β defined in display (6.3). Fix $\lambda > \beta$, choose $\epsilon > 0$ such that $\beta + \epsilon < \lambda$, and, using the conditions in (6.3), select $\gamma \leq \delta$ such that $a(t) \leq \beta + \epsilon$ for $t \notin [\gamma, \delta]$. If $\tau \in (-\infty, \infty)$, then it is easy to check that
$$\int_\tau^{\tau+t} a(s) ds \leq \|a\|_\infty (\delta - \gamma) + t(\beta + \epsilon).$$
Thus, the cocycle $\{\Phi^t_\lambda\}_{t\geq 0}$ is uniformly exponentially stable; that is,
$$|\Phi^t_\lambda(\theta)x|_X = \Big|\exp\Big(\int_0^t a_\tau(s)\,ds - \lambda t\Big)x\Big| = \exp\Big(\int_\tau^{\tau+t} a(s)\,ds - \lambda t\Big)|x|$$
$$\leq M e^{t(\beta+\epsilon-\lambda)}$$
for $\theta = a_\tau$. If follows that $\lambda \notin \Sigma$, and similarly $\lambda \notin \Sigma$ for $\lambda < \alpha$. If $\lambda = \alpha$ or $\lambda = \beta$ and $x \in X$, then
$$\Phi^t_\alpha(a_{-\infty})x = \Phi^t_\beta(a_\infty)x = x, \quad t \in \mathbb{R}.$$
Thus, both α and β are in Σ by Remark 6.16. If $\lambda \in (\alpha, \beta)$ and $x \in X$, then, using the definition of $a_{\pm\infty}$ given in display (6.5), we have that
$$\lim_{t \to +\infty} |\Phi^t_\lambda(a_\infty)x| = \lim_{t \to +\infty} |e^{(\beta-\lambda)t}x| = \infty,$$
$$\lim_{t \to +\infty} |\Phi^t_\lambda(a_{-\infty})x| = \lim_{t \to +\infty} |e^{(\alpha-\lambda)t}x| = 0.$$
This contradicts the existence of a continuous function $P(\cdot)x$ on Θ such that $\lim_{t \to \infty}|\Phi^t(\theta)P(\theta)x| = 0$ for some x. Thus, there is no dichotomy for $\{\Phi^t_\lambda\}$, and therefore $\lambda \in \Sigma$. ◇

The next example gives a formula for the Sacker-Sell spectrum for the linear skew-product flow described in Example 6.10.

EXAMPLE 6.19. Consider the boundary value problem (6.9) and the cocycle (6.10) from Example 6.10. Using equation (6.11), we have
$$\Phi^t(\theta) = \sum_{n=1}^\infty \Phi^t_n(\theta), \quad \Phi^t_n(\theta)x = \exp\Big(\int_0^t \lambda_n \theta(s)\,ds\Big) P_n x, \quad \theta \in \Theta = \text{Hull}(a),$$
where $\{\lambda_n\}$ is the spectrum of the operator d^2/ds^2 and $\{P_n\}$ is the set of orthogonal projections on the corresponding eigenspaces. Note that each cocycle $\{\Phi^t_n\}_{t\geq 0}$ is generated by the equation $\dot{x} = \lambda_n a(t)x$ exactly as described in Example 6.3. By the result proved in Example 6.18, its Sacker-Sell spectrum is given by $\Sigma_n = [\lambda_n \alpha, \lambda_n \beta]$. The spectrum of $\{\Phi^t\}_{t\geq 0}$, the cocycle represented as the direct sum of the family of cocycles $\{\{\Phi^t_n\}_{t\geq 0} : n > 0\}$, is given by
$$\Sigma = \bigcup_{n=1}^\infty \Sigma_n = \bigcup_{n=1}^\infty [-(n-1)^2 \pi^2 \alpha, -(n-1)^2 \pi^2 \beta].$$
◇

6.2. The Mather semigroup

In this section we begin the study of the evolution semigroup $\{E^t\}_{t\geq 0}$ induced by a cocycle over a flow on a locally compact metric space Θ. In Subsection 6.2.1 we will prove the strong continuity of the evolution semigroup on the space of continuous functions and give examples to show how to compute its generator. In Subsection 6.2.2 we restrict our attention to the single evolution operator $E := E^1$. Under the assumption that the underlying map of Θ is aperiodic, we will prove that the spectrum of E is rotationally invariant. This proof involves a construction that we call "Mather localization" (cf. Mather [**Ma**]). This procedure allows us to construct approximate eigenfunctions for the operator E supported in sufficiently long tubes formed by the trajectories of sufficiently small balls centered at aperiodic points in Θ. In the finite dimensional case, we will show how to use a similar idea to obtain an important fact that is known in the literature as Mañe's Lemma (Lemma 6.29). In Subsection 6.2.3 we will develop further the localization concept and then use the analysis to prove the Spectral Mapping Theorem — in the case of aperiodic base flows — for evolution semigroups on the space of continuous functions. Finally, in Subsection 6.2.4, we will prove the Spectral Mapping Theorem for evolution semigroups on L^p-spaces.

Let us suppose that Θ is a locally compact metric space. The space $C_0(\Theta; X)$, informally referred to as the space of continuous X-valued functions that vanish at infinity, is defined to be the set of all continuous functions $f : \Theta \to X$ such that for each $\epsilon > 0$ there exists a compact subset $K_\epsilon \subset \Theta$ with $|f(\theta)| < \epsilon$ for all $\theta \in \Theta \setminus K_\epsilon$. The space $C_0(\Theta; X)$ is equipped with the supremum norm. If Θ is a *compact* metric space, then, with a slight abuse of notation, we define $C_0(\Theta; X) = C_b(\Theta; X) = C(\Theta; X)$ where $C(\Theta; X)$ is the space of continuous X-valued functions on Θ.

Let $\{\varphi^t\}_{t\geq 0}$ be a continuous flow on a locally compact metric space Θ. For a cocycle $\{\Phi^t\}_{t\geq 0}$ over $\{\varphi^t\}_{t\in\mathbb{R}}$, the corresponding *evolution (Mather) semigroup* $\{E^t\}_{t\geq 0}$ is defined on $C_0(\Theta; X)$ by the rule

$$(E^t f)(\theta) = \Phi^t(\varphi^{-t}\theta)f(\varphi^{-t}\theta), \quad \theta \in \Theta, \quad t \geq 0, \quad f \in C_0(\Theta; X).$$

This is the evolution semigroup that we will study.

6.2.1. Strong continuity for evolution semigroups. In this subsection we show that our assumptions on the base flow and the cocycle are exactly the requirements to obtain a *strongly continuous* evolution semigroup on $C_0(\Theta; X)$.

THEOREM 6.20. *The semigroup $\{E^t\}_{t\geq 0}$ is strongly continuous on $C_0(\Theta; X)$ if and only if $\{\Phi^t\}_{t\geq 0}$ is an exponentially bounded, strongly continuous cocycle.*

PROOF. Let us suppose that the cocycle $\{\Phi^t\}_{t\geq 0}$ is strongly continuous and exponentially bounded. Then, in particular, there is a constant $C > 0$ such that $\sup\{\|\Phi^t(\theta)\|_{\mathcal{L}(X)} : \theta \in \Theta, t \in [0,1]\} \leq C$.

Fix $f \in C_0(\Theta; X)$. We will prove the following proposition: For each $\epsilon > 0$, there is some $\delta > 0$ such that if $t \in [0, \delta]$, then

$$\|E^t f - f\|_\infty = \sup_{\theta \in \Theta} |\Phi^t(\varphi^{-t}\theta)f(\varphi^{-t}\theta) - f(\theta)| = \sup_{\theta \in \Theta} |\Phi^t(\theta)f(\theta) - f(\varphi^t\theta)|$$

(6.14) $$\leq \epsilon \max\{C + 3, 2C + 1\}.$$

First, we claim that for each $\epsilon > 0$ there is some $\delta_\epsilon \in (0, 1)$ such that if $|t| \leq \delta$ and $\theta \in \Theta$, then $|f(\varphi^t\theta) - f(\theta)| < \epsilon$. To prove the claim, let us suppose that for some $\epsilon > 0$ there exists a sequence of points $\{\theta_n\}_{n=1}^{\infty}$ in Θ and a sequence $\{t_n\}_{n=1}^{\infty} \subset \mathbb{R}$ such that $\lim_{n \to \infty} t_n = 0$ and

(6.15) $$|f(\varphi^{t_n}\theta_n) - f(\theta_n)| \geq \epsilon.$$

Define $\eta_n := \varphi^{t_n}\theta_n$. There are two possibilities:

(a) For each compact set $K \subset \Theta$, there is some positive integer N such that for all $n \geq N$ the points η_n and θ_n do not belong to K.
(b) There exists a compact set $K \subset \Theta$ such that if N is a positive integer, then there is some $n \geq N$ such that at least one of the points θ_n or η_n belongs to K.

If statement (a) is true, then set $K = K_{\epsilon/4}$; where K_ϵ is chosen such that $|f(\theta)| < \epsilon$ for $\theta \in \Theta \setminus K_\epsilon$. Since $\theta_n, \eta_n \in \Theta \setminus K_{\epsilon/4}$, it follows immediately that $|f(\eta_n) - f(\theta_n)| \leq 2(\epsilon/4)$, in contradiction to the inequality (6.15).

If statement (b) is true, then, by choosing an appropriate subsequence, we can assume that there is a point $\theta_* \in K$ such that $\lim_{n \to \infty} \theta_n = \theta_*$. Since the function $(t, \theta) \mapsto \varphi^t \theta$ is continuous, $\lim_{n \to \infty} \eta_n = \lim_{n \to \infty} \varphi^{t_n}\theta_n = \theta_*$. But then, since f is continuous, $\lim_{n \to \infty} |f(\eta_n) - f(\theta_n)| = 0$, in contradiction to inequality (6.15). Similarly, if $\lim_{n \to \infty} \eta_n = \theta_* \in K$, then $\lim_{n \to \infty} \theta_n = \lim_{n \to \infty} \varphi^{-t_n}\eta_n = \theta_*$, and this proves the claim.

Fix $\epsilon > 0$. Since $f(K_\epsilon)$ is compact, there is a number $k < \infty$ and points $\theta_i \in K_\epsilon$, for $i = 1, 2, \ldots k$, such that if $\theta \in K_\epsilon$, then there is some θ_i for which $|f(\theta) - f(\theta_i)| < \epsilon$. Since the function $(t, \theta) \mapsto \Phi^t(\theta)f(\theta_i)$ is continuous on $\mathbb{R}_+ \times \Theta$, it is uniformly continuous on the compact subset $[0, 1] \times K_\epsilon$ for each $i = 1, 2, \ldots k$. For $\delta_\epsilon \in (0, 1)$ defined as in the claim stated above, fix $\delta \in (0, \delta_\epsilon)$ such that if $t \in [0, \delta]$ and $\theta \in K_\epsilon$, then $|\Phi^t(\theta)f(\theta_i) - f(\theta_i)| < \epsilon$.

For each $\theta \in \Theta$ and $t \in [0, \delta]$, we will estimate $|\Phi^t(\theta)f(\theta) - f(\varphi^t\theta)|$. If $\theta \in \Theta \setminus K_\epsilon$, then, by the definition of K_ϵ, the estimate $\|\Phi^t(\theta)\| \leq C$, and the claim proved above, we have that

$$|\Phi^t(\theta)f(\theta) - f(\varphi^t\theta)| \leq |\Phi^t(\theta)f(\theta) - f(\theta)| + |f(\theta) - f(\varphi^t\theta)| \leq 2C\epsilon + \epsilon.$$

If $\theta \in K_\epsilon$ and $t \in [0, \delta]$, then, by using the set of points $\{\theta_i : i = 1, \ldots, k\}$, it follows that

$$|\Phi^t(\theta)f(\theta) - f(\varphi^t\theta)| \leq |\Phi^t(\theta)[f(\theta) - f(\theta_i)]| + |\Phi^t(\theta)f(\theta_i) - f(\theta_i)|$$
$$+ |f(\theta_i) - f(\theta)| + |f(\theta) - f(\varphi^t\theta)| \leq C\epsilon + \epsilon + \epsilon + \epsilon.$$

Therefore, inequality (6.14) holds and $\{E^t\}_{t \geq 0}$ is a strongly continuous semigroup, as required.

Conversely, if $\{E^t\}_{t \geq 0}$ is a strongly continuous semigroup on $C_0(\Theta; X)$, then there are constants $M \geq 1$ and $\omega > 0$ such that $\|E^t\|_{\mathcal{L}(C_0(\Theta;X))} \leq Me^{\omega t}$ for all $t \geq 0$. Fix a point $\theta_0 \in \Theta$ and a vector $x \in X$. Let $f \in C_0(\Theta; X)$ be a function such that $f(\theta_0) = x$ and $|x| = \|f\|_\infty$. We have that

$$|\Phi^t(\theta_0)x| \leq \sup_{\theta \in \Theta} |\Phi^t(\theta)f(\theta)| = \sup_{\theta \in \Theta} |\Phi^t(\varphi^{-t}\theta)f(\varphi^{-t}\theta)| = \|E^t f\|_\infty \leq Me^{\omega t}|x|;$$

and therefore, the cocycle $\{\Phi^t\}_{t \geq 0}$ is exponentially bounded.

Fix $t_0 \in \mathbb{R}$ and $\epsilon > 0$. For $\theta \in \Theta$ and $t \geq 0$ we have that

(6.16) $$|\Phi^t(\theta)x - \Phi^{t_0}(\theta_0)x| \leq |\Phi^t(\theta)x - \Phi^t(\theta_0)x| + |\Phi^t(\theta_0)x - \Phi^{t_0}(\varphi^{t-t_0}\theta_0)x|$$
$$+ |\Phi^{t_0}(\varphi^{t-t_0}\theta_0)x - \Phi^{t_0}(\theta_0)x|.$$

Let D be a compact set in Θ such that θ_0 is in the interior of D, and choose $\alpha : \Theta \to [0,1]$ with compact support such that $\alpha(\theta) = 1$ for all $\theta \in D$. Also, define $f \in C_0(\Theta; X)$ by $f(\theta) = \alpha(\theta)x$ and note that if $\theta \in D$ and $t \geq 0$, then $\Phi^t(\theta)x = (E^t f)(\varphi^t \theta)$.

First consider the middle term on the right-hand side of (6.16) and choose $\delta_1 > 0$ such that $\varphi^{t-t_0}\theta_0 \in D$ for all $|t - t_0| < \delta_1$. Also, choose $\delta_2 \in (0, \delta_1)$ such that $\|E^t f - E^{t_0} f\|_\infty \leq \epsilon/3$ whenever $|t - t_0| < \delta_2$. Then, for each t such that $|t - t_0| < \delta_2$, we have

$$|\Phi^t(\theta_0)x - \Phi^{t_0}(\varphi^{t-t_0}\theta_0)x| = |(E^t f)(\varphi^t \theta_0) - (E^{t_0} f)(\varphi^{t_0}(\varphi^{t-t_0}\theta_0))|$$
$$= |(E^t f)(\varphi^t \theta_0) - (E^{t_0} f)(\varphi^t \theta_0)| \leq \|E^t f - E^{t_0} f\|_\infty < \epsilon/3.$$

Since $E^{t_0} f$ is a continuous function with a compact support, there exists $\delta' > 0$ such that $|(E^{t_0} f)(\theta_1) - (E^{t_0} f)(\theta_2)| < \epsilon/3$ whenever $d(\theta_1, \theta_2) < \delta'$. Also, since the function $t \mapsto \varphi^t \theta_0$ is continuous, there exists $\delta_3 \in (0, \delta_2]$ such that $d(\varphi^t \theta_0, \varphi^{t_0} \theta_0) < \delta'$ whenever $|t - t_0| < \delta_3$. Hence, if $|t - t_0| < \delta_3$, then

$$|\Phi^{t_0}(\varphi^{t-t_0}\theta_0)x - \Phi^{t_0}(\theta_0)x| = |E^{t_0} f(\varphi^{t_0}(\varphi^{t-t_0}\theta_0)) - E^{t_0} f(\varphi^{t_0}\theta_0)|$$
$$= |E^{t_0} f(\varphi^t \theta_0) - E^{t_0} f(\varphi^{t_0}\theta_0)| < \epsilon/3.$$

Let $B(\theta, r)$ denote the ball in Θ centered at θ with radius r. Choose $\delta'' > 0$ such that $B(\theta_0, \delta'') \subset D$, and $|E^t f(\varphi^t \theta) - E^t f(\varphi^t \theta_0)| < \epsilon/3$ whenever $\theta \in B(\theta_0, \delta'')$. The number $\delta'' = \delta''(t)$ seems to depend on t. But, the function $t \mapsto E^t f(\varphi^t \cdot) \in C_0(\Theta; X)$ is uniformly continuous on the compact interval $[t_0 - \delta_3, t_0 + \delta_3]$; and therefore, δ'' may be chosen independent of t. Hence, if $|t - t_0| < \delta_3$ and $\theta \in B(\theta_0, \delta'')$, then, by using the inequality (6.16), we have that $|\Phi^t(\theta)x - \Phi^{t_0}(\theta_0)x| < \epsilon$. $\square$

Let Γ denote the generator of the strongly continuous semigroup $\{E^t\}_{t \geq 0}$ on $C_0(\Theta; X)$. We will calculate Γ for several examples.

EXAMPLE 6.21. For the norm continuous compact setting as in equation (6.7), define the operator $\mathbf{d}$ on $C_0(\Theta; X)$, with maximal domain, given by

$$(\mathbf{d}f)(\theta) = \frac{d}{dt}(f \circ \varphi^t)(\theta)\Big|_{t=0}.$$

The generator Γ of the semigroup $(E^t f)(\theta) = \Phi^t(\varphi^{-t}\theta)f(\varphi^{-t}\theta)$ with $\frac{d}{dt}\Phi^t(\theta)\big|_{t=0} = A(\theta)$ is given as follows:

$$(\Gamma f)(\theta) = -(\mathbf{d}f)(\theta) + A(\theta)f(\theta), \quad \mathcal{D}(\Gamma) = \mathcal{D}(\mathbf{d}).$$

$\diamondsuit$

EXAMPLE 6.22. Recall Example 6.8, and suppose that $\Phi^t(\theta) = e^{tA_0}$ is the solution operator for the differential equation (6.2) where $A(\theta) \equiv A_0$, $\theta \in \Theta$, is the generator of a strongly continuous semigroup $\{e^{tA_0}\}_{t \geq 0}$ on X. In the tensor product $C_0(\Theta; X) = C_0(\Theta) \hat{\otimes} X$, the evolution semigroup $\{E^t\}_{t \geq 0}$ is represented in the form $E^t = V^t \otimes e^{tA_0}$ where $V^t f(\theta) = f(\varphi^{-t}\theta)$. Then, the generator Γ_0 of $\{E^t\}_{t \geq 0}$ is the closure of the operator $\Gamma_0' f = -\mathbf{d}f + A_0 f$ with the domain

$$\mathcal{D}(\Gamma_0') = \{f \in C_0(\Theta, X) : f \in \mathcal{D}(\mathbf{d}), \ f : \Theta \to \mathcal{D}(A_0), \ -\mathbf{d}f + A_0 f \in C_0(\Theta; X)\},$$

see for example Nagel [**Na**, p.23]. $\diamondsuit$

Suppose, in the setting of Example 6.8, that

(6.17) $$A(\theta) = A_0 + A_1(\theta), \quad A_1(\cdot) \in C_b(\Theta; \mathcal{L}_s(X)).$$

Recall that $A_1(\cdot)$ defines the multiplication operator $\mathcal{A}_1 \in \mathcal{L}(C_0(\Theta; X))$ by the formula $(\mathcal{A}_1 f)(\theta) = A_1(\theta) f(\theta)$, $\theta \in \Theta$, see page 22. Then, for Γ_0 as in Example 6.22, we have the following proposition.

PROPOSITION 6.23. *If the strongly continuous cocycle $\{\Phi^t\}_{t \geq 0}$ solves the differential equation (6.2) with $A(\cdot)$ as in equation (6.17), then the generator Γ of the evolution semigroup $\{E^t\}_{t \geq 0}$ is given by $\mathcal{D}(\Gamma) = \mathcal{D}(\Gamma_0)$ and*

(6.18) $$(\Gamma f)(\theta) = -(\mathbf{d}f)(\theta) + A_0 f(\theta) + A_1(\theta) f(\theta), \quad f \in \mathcal{D}(\Gamma).$$

PROOF. Since the cocycle $\{\Phi^t\}_{t \geq 0}$ solves the differential equation (6.2) with $A(\cdot)$ as in equation (6.17), this cocycle also defines mild solutions; that is, $t \mapsto \Phi^t(\theta)$ satisfies equation (6.8). Since Γ_0 generates a strongly continuous semigroup and $\mathcal{A}_1 \in \mathcal{L}(C_0(\Theta; X))$ is a bounded perturbation, the operator $\Gamma = \Gamma_0 + \mathcal{A}_1$ also generates a strongly continuous semigroup $\{S^t\}_{t \geq 0}$ that is the unique solution of the integral equation

(6.19) $$S^t f = e^{t\Gamma_0} f + \int_0^t e^{(t-\tau)\Gamma_0} \mathcal{A}_1 S^\tau f \, d\tau, \quad f \in C_0(\Theta; X), \quad t \geq 0,$$

see for example Pazy [**Pz**, p.11]. We will show that $S^t = E^t$, where $\{E^t\}_{t \geq 0}$ is the evolution semigroup generated by $\{\Phi^t\}_{t \geq 0}$. Indeed, for each $f \in C_0(\Theta, X)$, define

$$g = \int_0^t e^{(t-\tau)\Gamma_0} \mathcal{A}_1 E^\tau f \, d\tau$$

and note that, by equation (6.8),

$$g(\theta) = \int_0^t e^{(t-\tau)A_0} A_1(\varphi^\tau(\varphi^{-t}\theta)) \Phi^\tau(\varphi^{-t}\theta) f(\varphi^{-t}\theta) \, dt$$
$$= \Phi^t(\varphi^{-t}\theta) f(\varphi^{-t}\theta) - e^{tA_0} f(\varphi^{-t}\theta) = (E^t f)(\theta) - (e^{t\Gamma_0} f)(\theta).$$

Therefore, $\{E^t\}_{t \geq 0}$ satisfies (6.19) and so $E^t = S^t$. The generator $\Gamma = \Gamma_0 + \mathcal{A}_1$ of this semigroup is given by (6.18). □

EXAMPLE 6.24. In the situation described in Examples 6.4–6.5, the evolution semigroup $(E^t f)(\theta) = \Phi^t(\varphi^{-t}\theta) f(\varphi^{-t}\theta)$ is exactly the evolution semigroup

$$(E^t f)(\theta) = U(\theta, \theta - t) f(\theta - t)$$

studied in Chapters 3–4; its generator $\Gamma = -\frac{d}{dt} + \mathcal{A}$ is described in Theorem 3.12. ◇

EXAMPLE 6.25. Let Θ be a smooth compact manifold, $\mathbf{v}$ a smooth vectorfield (a section of the tangent bundle $\mathcal{T}\Theta$), and $\{\varphi^t\}_{t \in \mathbb{R}}$ the flow on Θ generated by $\mathbf{v}$. Consider the differential $\Phi^t(\theta) = D\varphi^t(\theta)$ and the corresponding evolution semigroup $\{E^t\}_{t \in \mathbb{R}}$ defined on the space $C(\mathcal{T}\Theta)$ of continuous sections of $\mathcal{T}\Theta$ by $E^t f(\theta) = D\varphi^t(\varphi^{-t}\theta) f(\varphi^{-t}\theta)$. Each operator E^t is a "push-forward operator" as defined in differential geometry, see for example Abraham, Marsden, and Ratiu [**AMR**]. The corresponding generator Γ, the Lie derivative in the direction of $\mathbf{v}$, is given by the formula $\Gamma f = -\langle \mathbf{v}, \nabla \rangle f + \langle f, \nabla \rangle \mathbf{v}$. ◇

6.2.2. Mather localization and Mañe's Lemma.

In this subsection we will explain an approach to the description of the structure of the approximate eigenfunctions for evolution operators. In particular, we will show that if $\{E^t\}_{t\geq 0}$ is an evolution semigroup over an aperiodic base flow $\{\varphi^t\}_{t\in\mathbb{R}}$, then, for each $t > 0$, the spectrum $\sigma(E^t)$ is rotationally invariant. This result uses an important idea introduced by Mather [**Ma**] that concerns the "localization property" for the approximate eigenfunctions of E^t. We will also use a similar idea to prove Mañe's Lemma 6.29 for the case where the Banach space X is finite dimensional.

We start with the definition of aperiodic flows and maps.

DEFINITION 6.26. A point $\theta \in \Theta$ is called a *periodic point* for a continuous flow $\{\varphi^t\}_{t\in\mathbb{R}}$ on a locally compact metric space Θ if there is a number $T > 0$ such that $\varphi^T\theta = \theta$. A point $\theta \in \Theta$ is called *aperiodic* if no such T exists. A point $\theta \in \Theta$ is called a *periodic point* for a homeomorphism φ of Θ if there is a positive integer T such that $\varphi^T\theta = \theta$; otherwise θ is called *aperiodic*. For a flow $\{\varphi^t\}_{t\in\mathbb{R}}$ (respectively, homeomorphism φ) we define the of *prime period function* $p : \Theta \to \mathbb{R}_+ \cup \{\infty\}$ (respectively, $p : \Theta \to \mathbb{Z}_+ \cup \{\infty\}$) by setting $p(\theta) = \infty$ if θ is aperiodic, and $p(\theta) = \inf\{T > 0 : \varphi^t\theta = \theta\}$ if θ is periodic. Also, we define

$$\mathcal{B}(\Theta) := \{\theta \in \Theta : p \text{ is bounded in some neighborhood of } \theta\}.$$

The flow $\{\varphi^t\}_{t\in\mathbb{R}}$ or the homeomorphism φ on Θ is called *aperiodic* if $\mathcal{B}(\Theta) = \emptyset$.

We remark that φ is aperiodic in the sense of Definition 6.26 provided that the set of aperiodic points is dense in Θ.

Recall that the *approximate point spectrum* $\sigma_{\mathrm{ap}}(T)$ of an operator T on a Banach space Y is the set of all approximate eigenvalues $\lambda \in \mathbb{C}$; that is, $\lambda \in \sigma_{\mathrm{ap}}(T)$ if, for each $\epsilon > 0$, there exists $y \in Y$, called an *approximate eigenvector*, such that $|(\lambda - T)y| < \epsilon|y|$.

Let us consider a flow $\{\varphi^t\}_{t\in\mathbb{R}}$ on a locally compact metric space Θ, a cocycle $\{\Phi^t\}_{t\geq 0}$ over $\{\varphi^t\}_{t\in\mathbb{R}}$, and the homeomorphism $\varphi := \varphi^1$ and operator $\Phi(\theta) := \Phi^1(\theta)$. Also, define the evolution operator E on the space $C_0(\Theta, X)$ by $(Ef)(\theta) = \Phi(\varphi^{-1}\theta)f(\varphi^{-1}\theta)$.

One important property of the approximate eigenvalues $\lambda \in \sigma_{\mathrm{ap}}(E)$ is that their corresponding approximate eigenfunctions can be localized along the trajectories of an aperiodic map φ. More precisely, suppose that the base map φ is aperiodic and $\lambda \in \sigma_{\mathrm{ap}}(E)$. Then for each positive integer N, there is a point $\theta_N \in \Theta$, a neighborhood B of this point, and a nonzero function $g \in C_0(\Theta; X)$ with $\operatorname{supp} g \subset \bigcup_{k=0}^{2N} \varphi^k(B)$ such that

$$\|\lambda g - Eg\|_\infty = O(1/N)\|g\|_\infty \quad \text{as} \quad N \to \infty.$$

We will use the localization method to show that if $\lambda \in \sigma_{\mathrm{ap}}(E)$, then $\lambda z \in \sigma_{\mathrm{ap}}(E)$ as long as $|z| = 1$. In other words, we will show that the approximate point spectrum is rotationally invariant. Also, under the assumption that X is finite dimensional, we will prove that if $1 \in \sigma_{\mathrm{ap}}(E)$, then there exists a point $\theta_0 \in \Theta$ and a nonzero vector $x_0 \in X$ such that $\sup_{t\in\mathbb{R}} |\Phi^t(\theta_0)x_0| < \infty$. In the next section, an "integral" variant of the localization method is used to prove the Spectral Mapping Theorem 6.30 for the evolution semigroup $\{E^t\}_{t\geq 0}$.

In what follows, we will only be interested in nonzero points λ in the approximate point spectrum. By rescaling, it suffices to consider only $\lambda = 1$. Indeed, under the assumption that $\lambda \neq 0$ we have that $\lambda \in \sigma_{\mathrm{ap}}(E)$ if and only if $1 \in \sigma_{\mathrm{ap}}(\lambda^{-1}E)$.

Thus, we can replace the operator E by $\lambda^{-1}E$ and consider the approximate eigenvalue $\lambda = 1$ for the operator $\lambda^{-1}E$.

PROPOSITION 6.27. *Suppose that T is a bounded operator on a Banach space Y and $N \in \mathbb{Z}_+$. If $1 \in \sigma_{ap}(T)$, then there exists $y \in Y$ such that $|y| = 1$,*

(a) $|T^N y - y|_Y \leq 1/8$, *and*

(b) $|T^j y|_Y \leq 2$ *for each $j = 0, 1, \ldots, 2N$.*

If T is invertible, then there is some $y \in Y$ such that $|y| = 1$ and

(6.20) $$|T^j y| \leq 2 \quad \text{for} \quad |j| \leq N.$$

PROOF. Set $c = \sum_{j=0}^{2N} \|T\|_{\mathcal{L}(Y)}^j$ and note that $c \geq 1$. By using the identity

$$T^j - I = (T^{j-1} + \ldots + T + I)(T - I)$$

that holds for all integers $j \geq 1$, it follows that if $y \in Y$ and $j = 0, 1, \ldots, 2N$, then

(6.21) $$|(T^j - I)y| \leq c|(T - I)y|.$$

Also, since $1 \in \sigma_{ap}(T)$, there exists $y \in Y$ such that $|y| = 1$ and $|(T-I)y| \leq 1/(8c)$. Hence, if we set $j = N$ in (6.21), then we obtain inequality (a). If $j = 0, 1, \ldots, 2N$, then, by inequality (6.21), we have that

$$|T^j y| \leq |(T^j - I)y| + |y| \leq c|(T-I)y| + 1 \leq 2,$$

as required for inequality (b). A similar argument proves inequality (6.20). □

LEMMA 6.28. *Suppose that φ is aperiodic on Θ. If $1 \in \sigma_{ap}(E)$, then $\mathbb{T} \subseteq \sigma_{ap}(E)$. In other words, $\sigma(E)$ is rotationally invariant.*

PROOF. For each $N \in \mathbb{Z}_+$, apply Proposition 6.27 to $Y = C_0(\Theta; X)$ and $T = E$ to obtain a function $f \in C_0(\Theta, X)$ with $\|f\|_\infty = 1$ such that

(a) $\|E^N f - f\|_\infty \leq 1/8$ and (b) $\|E^j f\|_\infty \leq 2$, for $j = 0, \ldots, 2N$.

Since φ is aperiodic on Θ and $\|f\|_\infty = 1$, there is point θ_0 such that $|f(\theta_0)| \geq 7/8$ and its prime period satisfies $p(\theta_0) \geq 2N + 1$. Set $\theta_N = \varphi^{-N}\theta_0$ and note that since $p(\theta_0) \geq 2N + 1$, there is an open neighborhood B of θ_N such that the sets $\varphi^j(B)$ are pairwise disjoint for $j = 0, \ldots, 2N$. Let $\alpha : \Theta \to [0,1]$ be a continuous bump-function that is supported in B such that $\|\alpha\|_\infty = 1$ and $\alpha(\theta_N) = 1$, and note that $\mathrm{supp}(\alpha \circ \varphi^{-j}) \subset \varphi^j(B)$ for each $j = 0, \ldots, 2N$.

Let $\gamma : \mathbb{Z} \to [0,1]$ be a bump-function on the integers such that $\mathrm{supp}\,\gamma \subset (0, 2N)$, its maximum value is given by $\gamma(N) = 1$, and $|\gamma(j-1) - \gamma(j)| \leq c/N$ for $j = 1, \ldots 2N+1$ and a constant $c > 0$ that is independent of N. For instance, the function γ defined by $\gamma(j) = (N - |j - N|)/N$ for $j = 0, \ldots 2N$, and $\gamma(j) = 0$ otherwise, satisfies these requirements.

Fix $z \in \mathbb{C}$ with $|z| = 1$ and define $g \in C_0(\Theta; X)$ by

(6.22) $$g(\theta) = \sum_{j=0}^{2N} z^{N-j} \gamma(j)(E^j \alpha f)(\theta).$$

Since $E^j \alpha f = \alpha \circ \varphi^{-j} E^j f$, for $j = 0, \ldots, 2N$, the summands in the definition of g have disjoint supports. In fact, g is supported in the disjoint union $\bigcup_{j=0}^{2N} \varphi^j(B)$ such that if $\theta \in \varphi^j(B)$, then

$$g(\theta) = z^{N-j} \gamma(j)(E^j \alpha f)(\theta).$$

Since $\gamma(N) = 1$ and $\alpha \circ \varphi^{-N}(\theta_0) = 1$, we have that $g(\theta_0) = (E^N \alpha f)(\theta_0) = (E^N f)(\theta_0)$. Hence, by inequality (a),

$$|g(\theta_0)| = |E^N f(\theta_0)| \geq |f(\theta_0)| - \|E^N f - f\|_\infty \geq 3/4;$$

and therefore, $\|g\|_\infty \geq 3/4$.

Because $\gamma(0) = \gamma(2N) = 0$, it follows that

$$zg(\theta) - (Eg)(\theta) = \sum_{j=0}^{2N} z^{N-j+1}\gamma(j)(E^j \alpha f)(\theta) - \sum_{j=0}^{2N} z^{N-j}\gamma(j)(E^{j+1}\alpha f)(\theta)$$

$$= z^{N+1}\gamma(0)(\alpha f)(\theta) + \sum_{j=1}^{2N} z^{N-j+1}[\gamma(j) - \gamma(j-1)](E^j \alpha f)(\theta)$$

$$- z^{-N}\gamma(2N)(E^{2N+1}\alpha f)(\theta)$$

$$= \sum_{j=1}^{2N} z^{N-j+1}[\gamma(j) - \gamma(j-1)](E^j \alpha f)(\theta).$$

Using the fact that the summands in the last sum have disjoint supports, the choice of the bump functions γ and α, the estimate (b), and the equality $|z| = 1$, we have the estimate

$$\|zg - Eg\|_\infty \leq \max_{1 \leq j \leq 2N} \left(|\gamma(j) - \gamma(j-1)| \cdot \|\alpha \circ \varphi^{-j}\|_\infty \|E^j f\|_\infty\right)$$

$$\leq \frac{2c}{N} \leq \frac{8c}{3N}\|g\|_\infty.$$

Finally, since $\|g\|_\infty \neq 0$ and N can be chosen as large as we like independent of the positive constant c, we have that $z \in \sigma_{\mathrm{ap}}(E)$. □

The following finite dimensional result is known as Mañe's Lemma. It holds without the assumption that φ is aperiodic.

LEMMA 6.29. *Suppose that* $\dim X < \infty$ *and* Θ *is compact. If* $1 \in \sigma_{ap}(E)$ *on* $C(\Theta; X)$, *then there exist a point* $\theta_0 \in \Theta$ *and a vector* $x_0 \in X$ *such that* $|x_0| = 1$ *and*

(6.23) $$\sup\{|\Phi^t(\theta_0)x_0| : t \in \mathbb{R}\} < \infty.$$

PROOF. Assume that $1 \in \sigma_{\mathrm{ap}}(E)$. For each $N \in \mathbb{N}$, use Proposition 6.27 and the estimate (6.20) to find $f_N \in C(\Theta; X)$ such that $\|f_N\|_\infty = 1$ and $\|E^j f_N\|_\infty \leq 2$ for $|j| \leq N$. Choose θ_N such that $|f_N(\theta_N)| = \|f_N\|_\infty = 1$ and define $x_N = f_N(\theta_N)$. Then, we have the estimates

$$|\Phi^j(\theta_N)x_N| \leq \sup_{\theta \in \Theta} |\Phi^j(\theta)f_N(\theta)| = \|E^j f_N\|_\infty \leq 2$$

for $|j| \leq N$. Since Θ and the unit sphere in X are compact, there is no loss of generality if we assume that $(\theta_N, x_N) \to (\theta_0, x_0)$ for some $\theta_0 \in \Theta$ and $|x_0| = 1$. We claim that the inequality (6.23) holds for this θ_0 and x_0. Indeed, for each fixed $j \in \mathbb{Z}$, we have that $|\Phi^j(\theta_0)x_0| = \lim_{N \to \infty} |\Phi^j(\theta_N)x_N| \leq 2$. The desired inequality (6.23) follows from the estimate

$$\sup_{t \in \mathbb{R}} |\Phi^t(\theta_0)x_0| \leq \sup_{j \in \mathbb{Z}} \sup_{\tau \in [0,1]} \|\Phi^\tau(\varphi^j \theta_0)\| \cdot |\Phi^j(\theta_0)x_0|.$$

□

The point θ_0, respectively the vector x_0, as in display (6.23) is called a *Mañe point*, respectively a *Mañe vector*. In Section 7.2 we will give an infinite dimensional generalization of these notions. However, for $\dim X < \infty$, we will also prove, see Corollary 7.30, a converse for the Mañe Lemma 6.29: If a Mañe point and a Mañe vector exist, then $\sigma_{\mathrm{ap}}(E) \cap \mathbb{T} \neq \emptyset$. A similar statement, see Corollary 7.29, holds for the infinite dimensional case.

6.2.3. The Spectral Mapping Theorem. In this subsection we prove the Spectral Mapping Theorem for the evolution semigroup $\{E^t\}_{t \geq 0}$ on $C_0(\Theta; X)$. As shown in the previous subsection, see formula (6.22), the approximate eigenfunctions for E can be "localized" along trajectories. That is, starting with an aperiodic point θ_0 where the norm of the value $f(\theta_0)$ of an eigenfunction f is "big", we can "spread" this value to a small neighborhood of θ_0 using a bump-function α, and then "propagate" the new function αf using the operator E. A similar construction will be used here to create an approximate eigenfunction for the generator Γ from an approximate eigenfunction f for E, see formula (6.34) below. The main difference in the construction is that the sum in equation (6.22) will be replaced by an integral in the definition (6.34). In this subsection we assume that $\{\varphi^t\}_{t \in \mathbb{R}}$ is a continuous *aperiodic* flow on a locally compact metric space Θ.

THEOREM 6.30 (Spectral Mapping Theorem). *Let Γ be the generator of the evolution semigroup $\{E^t\}_{t \geq 0}$ given by $(E^t f)(\theta) = \Phi^t(\varphi^{-t}\theta) f(\varphi^{-t}\theta)$ on $C_0(\Theta; X)$ where $\{\Phi^t\}_{t \geq 0}$ is a strongly continuous, exponentially bounded cocycle over the flow $\{\varphi^t\}_{t \geq 0}$. If the flow $\{\varphi^t\}_{t \in \mathbb{R}}$ is aperiodic, then*

$$\sigma(E^t) \setminus \{0\} = e^{t\sigma(\Gamma)}, \quad t > 0. \tag{6.24}$$

Moreover, the spectrum $\sigma(\Gamma)$ is invariant with respect to translations along the imaginary axis, and the spectrum $\sigma(E^t)$, for $t > 0$, is invariant with respect to rotations centered at origin.

PROOF. By the Spectral Inclusion Theorem 2.6, we have $e^{t\sigma(\Gamma)} \subset \sigma(E^t) \setminus \{0\}$, $t \geq 0$. By the Spectral Mapping Theorems for the point and residual spectra (see Theorems 2.8 and Theorem 2.9), to show the equality (6.24) it suffices to prove that $\sigma_{\mathrm{ap}}(E^t) \setminus \{0\} \subset e^{t\sigma_{\mathrm{ap}}(\Gamma)}$. Moreover, by rescaling, it suffices to show that if $1 \in \sigma_{\mathrm{ap}}(E)$ where $E = E^1$, then $0 \in \sigma_{\mathrm{ap}}(\Gamma)$. In fact, we will show that $\sigma_{\mathrm{ap}}(\Gamma)$ contains the entire imaginary axis whenever $1 \in \sigma_{\mathrm{ap}}(E)$. Since $\sigma_{\mathrm{ap}}(\Gamma)$ contains the boundary of $\sigma(\Gamma)$, all assertions of Theorem 6.30 follow from this result.

Let $1 \in \sigma_{\mathrm{ap}}(E)$ and $\xi \in \mathbb{R}$. By Proposition 6.27 with $Y = C_0(\Theta; X)$ and $T = E$, if N is a positive integer, then there is a function $f \in C_0(\Theta; X)$ with the following properties:

$$\|f\|_\infty = 1, \tag{6.25}$$

$$\|E^N f - f\|_\infty \leq 1/8, \tag{6.26}$$

$$\|E^k f\|_\infty \leq 2, \quad \text{for} \quad k = 0, \ldots, 2N. \tag{6.27}$$

For this f, fix $s \in (0, 1)$ such that

$$\|E^{t+N} f - E^N f\|_\infty \leq 1/16 \tag{6.28}$$

and

$$|e^{-i\xi t} - 1| \leq 1/32, \tag{6.29}$$

6.2. THE MATHER SEMIGROUP

whenever $|t| \leq s$. Also, choose a smooth bump-function $\gamma : \mathbb{R} \to [0,1]$ such that

(6.30) $\qquad\qquad\qquad \gamma(t) = 0 \quad \text{for} \quad t \notin (0, 2N),$

(6.31) $\qquad\qquad\qquad |\gamma'(t)| \leq 2/N \quad \text{for} \quad t \in \mathbb{R}, \text{ and}$

(6.32) $\qquad\qquad\qquad \gamma(t) = 1 \quad \text{for} \quad |t - N| \leq s.$

By the aperiodicity of the flow $\{\varphi^t\}_{t \in \mathbb{R}}$, there is some $\theta_0 \in \Theta$ such that its prime period satisfies $p(\theta_0) \geq 5N + 1$ and

(6.33) $$|f(\theta_0)| \geq \frac{7}{8}\|f\|_\infty = \frac{7}{8}.$$

We will also use the point $\theta_N := \varphi^{-N}\theta_0$.

The remainder of the proof will be given in a series of lemmas that formalize the following strategy. The idea is to choose an open neighborhood B of the point θ_N and another bump-function α that is supported in B. The precise properties of B and α are given in Lemma 6.31. Then, for $\xi \in \mathbb{R}$, we will define $g \in C_0(\Theta; X)$ by

(6.34) $$g(\theta) = \int_{-\infty}^{\infty} e^{-i\xi t}\gamma(t)(E^t\alpha f)(\theta)\, dt.$$

Here, in view of the condition (6.30), the integrand is supported in the interval $[0, 2N]$; and therefore, the indefinite integral converges. Also, we have that

$$\Gamma g = i\xi g - \int_{-\infty}^{\infty} e^{-i\xi t}\gamma'(t)(E^t\alpha f)\, dt.$$

Indeed,

$$\Gamma g = \frac{d}{d\tau}\bigg|_{\tau=0} E^\tau g = \frac{d}{d\tau}\int_{-\infty}^{\infty} e^{-i\xi t}\gamma(t)(E^{t+\tau}\alpha f)\, dt\bigg|_{\tau=0}$$

$$= \frac{d}{d\tau}\int_{-\infty}^{\infty} e^{-i\xi(t-\tau)}\gamma(t-\tau)(E^t\alpha f)\, dt\bigg|_{\tau=0}$$

$$= \int_{-\infty}^{\infty} [i\xi e^{-i\xi t}\gamma(t)(E^t\alpha f) - e^{-i\xi t}\gamma'(t)(E^t\alpha f)]\, dt.$$

Using the specific choice of the neighborhood B and the bump-function α, we will show that there are constants $C_1 > 0$ and $C_2 > 0$, independent of the choice of N, such that $\|g\|_\infty \geq C_1 s$ and $\|\Gamma g - i\xi g\|_\infty < C_2 s/N$ where s is as in the conditions (6.28)–(6.29). As a result, for each positive integer N, there is a corresponding function $g \in C_0(\Theta; X)$ such that

$$\|\Gamma g - i\xi g\|_\infty < \frac{C_2}{C_1 N}\|g\|_\infty.$$

Hence, $i\xi \in \sigma_{\mathrm{ap}}(\Gamma)$, as required.

The next lemma states the existence of a neighborhood B and bump-function α with the properties that are needed to complete the proof. We will use the following notation: If $\theta \in \Theta$ and B is an open subset of Θ, then

(6.35) $\qquad\qquad R(\theta) = \{t \in \mathbb{R} : |t| \leq N \text{ and } \varphi^t\theta \in B\}.$

Also, the Lebesgue measure on $\mathbb{R}$ will be denoted by mes.

LEMMA 6.31. *Suppose N is a positive integer, $s \in (0,1)$, and $\theta_* \in \Theta$ is a point whose prime period is $p(\theta_*) \geq 5N + 1$. Then, there is a neighborhood B of θ_* and a continuous bump-function $\alpha : \Theta \to [0,1]$ such that $\operatorname{supp}\alpha \subset B$,*

(a) $\alpha(\varphi^t \theta_*) = 1$ for $|t| \leq s/4$,
(b) $\alpha(\varphi^t \theta_*) = 0$ for $s \leq |t| \leq 2N$, and
(c) $mes(R(\theta)) \leq 2s$ for all $\theta \in \Theta$.

PROOF. The proof is very easy for the important special case where $\Theta = \mathbb{R}$ and $\varphi^t \theta = \theta + t$. Indeed, for $\theta_* \in \mathbb{R}$, let $B := (\theta_* - s, \theta_* + s)$ and define $\alpha : \mathbb{R} \to [0, 1]$ such that $\operatorname{supp} \alpha \subset B$ and $\alpha(\theta) = 1$ for $\theta \in (\theta_* - s/4, \theta_* + s/4)$.

For the general case of a flow on a locally compact metric space Θ, the proof is based on the following two claims. Recall that $B(\theta, r)$ denotes the ball in Θ centered at θ with radius r.

Claim 1. There is some $\epsilon_* > 0$ such that if $\delta < \epsilon_*$ and $\theta \in B(\theta_*, \delta)$, then $\varphi^t \theta \notin B(\theta_*, \delta)$ for all t that satisfy the inequality $s/2 \leq |t| \leq 5N$.

Proof of Claim 1. Suppose, to the contrary, that there is a sequence $\{\delta_n\}_{n=1}^\infty$ of positive numbers that converges to zero and, for each $n \geq 1$, some $\theta_n \in B(\theta_*, \delta_n)$ and t_n with $s/2 \leq |t_n| \leq 5N$ such that $\varphi^{t_n} \theta_n \in B(\theta_*, \delta_n)$. Then, $\lim_{n \to \infty} \theta_n = \theta_*$. By compactness, we can assume that $\lim_{n \to \infty} t_n = t_*$ for some t_* with $s/2 \leq |t_*| \leq 5N$. By the continuity of the map $(\theta, t) \mapsto \varphi^t \theta$, from $\Theta \times \mathbb{R}$ to Θ, it follows that $\lim_{n \to \infty} \varphi^{t_n} \theta_n \to \varphi^{t_*} \theta_*$. On the other hand, because $\varphi^{t_n} \theta_n \in B(\theta_*, \delta_n)$, we must have $\lim_{n \to \infty} \varphi^{t_n} \theta_n = \theta_*$ and $\varphi^{t_*} \theta_* = \theta_*$, in contradiction to the assumption that $p(\theta_*) \geq 2N + 1$. This proves Claim 1.

With ϵ_* as in Claim 1, pick $\delta > 0$ such that $\delta < \epsilon_*$ and define $B' := B(\theta_*, \delta)$. Also, choose an open neighborhood B'' of θ_* such that $\overline{B''} \subset B'$ and define the sets

(6.36) $$B := \bigcup_{|t| \leq s/4} \varphi^t(B'), \quad C := \bigcup_{|t| \leq s/4} \varphi^t(B'').$$

Since $\overline{C} \subset B$, there exists a continuous $\alpha : \Theta \to [0, 1]$ such that $\alpha(\theta) = 1$ for $\theta \in C$, and $\alpha(\theta) = 0$ for $\theta \notin B$. Since $\theta_* \in B''$, it follows that $\{\varphi^t \theta_* : |t| \leq s/4\} \subset C$ and so $\alpha(\varphi^t \theta_*) = 1$ whenever $|t| \leq s/4$. This proves statement (a).

To prove statement (b), we will use the next claim.

Claim 2. If $\theta \in B$ and $s \leq |t| \leq 2N$, then $\varphi^t \theta \notin B$.

Proof of Claim 2. Suppose, to the contrary, that there is some $\theta \in B$ and $t_* \in \mathbb{R}$ such that $s \leq |t_*| \leq 2N$ and $\varphi^{t_*} \theta \in B$. By the definition (6.36) of B, there are two points θ_1 and θ_2 in B' such that $\theta = \varphi^{t_1} \theta_1$ and $\varphi^{t_*} \theta = \varphi^{t_2} \theta_2$ where $\max\{|t_1|, |t_2|\} \leq s/4$. Hence, we have

(6.37) $$\varphi^{t_* + t_1 - t_2} \theta_1 = \theta_2 \in B'.$$

But, $|t_* + t_1 - t_2| \leq 2N + s/2 \leq 5N$ and $|t_* + t_1 - t_2| \geq |t_*| - |t_1| - |t_2| \geq s/2$, so (6.37) contradicts the result of Claim 1. This proves Claim 2.

Since, in particular, $\theta_* \in B$ and $\alpha(\theta) = 0$ for $\theta \notin B$, part (b) of the lemma follows from Claim 2.

To prove statement (c), fix $\theta_1 \in \Theta$ and consider the segment of its orbit given by $\mathcal{O}(\theta_1) := \{\varphi^t \theta_1 : |t| \leq N\}$. Clearly, $mes(R(\theta_1)) = 0$ if $\mathcal{O}(\theta_1) \cap B = \emptyset$. So, let us consider the case where $\mathcal{O}(\theta_1) \cap B \neq \emptyset$. Note that if $\theta \in \mathcal{O}(\theta_1) \cap B$, then $\theta = \varphi^{t_1} \theta_1$ for some t_1 with $|t_1| \leq N$. Moreover, if $t_* \in R(\theta_1)$, then $|t_* - t_1| \leq 2N$ and

$$\varphi^{t_*} \theta_1 = \varphi^{t_* - t_1} \theta \in B.$$

Since $\theta \in B$, Claim 2 shows that the last equation can only hold provided that $|t_* - t_1| < s$. Because t_* was chosen arbitrarily in $R(\theta_1)$, it follows that $R(\theta_1) \subseteq (t_1 - s, t_1 + s)$; and therefore, $mes(R(\theta_1)) \leq 2s$. This proves Lemma 6.31 □

Returning to the proof of the Spectral Mapping Theorem 6.30, let us select B and α as in Lemma 6.31, with $\theta_* = \theta_N$ and s satisfying (6.28)–(6.29), and then define g by equation (6.34). We will show that $\|\Gamma g - i\xi g\| = O(1/N)\|g\|$ as $N \to \infty$. Denote $C := \max\{\|E^t\|_{\mathcal{L}(C_0(\Theta;X))} : 0 \leq t \leq 1\}$.

Claim A. $\|\Gamma g - i\xi g\|_\infty \leq 8Cs/N$.

Proof of Claim A. First use the inequalities (6.30) and (6.31) to obtain the estimate

$$\|\Gamma g - i\xi g\|_\infty = \left\| -\int_0^{2N} e^{-i\xi t}\gamma'(t)\alpha(\varphi^{-t}\cdot)E^t f(\cdot)\,dt \right\|_\infty$$

(6.38)
$$\leq \frac{2}{N}\max_{0\leq t\leq 2N}\|E^t f\| \cdot \max_{\theta\in\Theta}\int_0^{2N}\alpha(\varphi^{-t}\theta)\,dt.$$

By the bound (6.27), we have that

(6.39) $$\max_{0\leq t\leq 2N}\|E^t f\|_\infty \leq \max_{0\leq k\leq 2N}\max_{0\leq \tau\leq 1}\|E^\tau E^k f\|_\infty \leq 2C.$$

Also, using the change of variables $t \mapsto -t + N$, it follows that

$$\max_{\theta\in\Theta}\int_0^{2N}\alpha(\varphi^{-t}\theta)\,dt = \max_{\theta\in\Theta}\int_{-N}^{N}\alpha(\varphi^t(\varphi^{-N}\theta))\,dt = \max_{\theta\in\Theta}\int_{-N}^{N}\alpha(\varphi^t\theta)\,dt.$$

Now, for each fixed $\theta \in \Theta$, the statement in Lemma 6.31(c) implies the set $R(\theta) = \{t \in \mathbb{R} : |t| \leq N, \varphi^t(\theta) \in B\}$ has measure $\mathrm{mes}(R(\theta)) \leq 2s$. Hence, if $|t| \leq N$, then $\alpha(\varphi^t\theta) \leq 1$ for $t \in R(\theta)$, and $\alpha(\varphi^t\theta) = 0$ for $t \notin R(\theta)$. Hence,

$$\max_{\theta\in\Theta}\int_{-N}^{N}\alpha(\varphi^t\theta)\,dt \leq \max_{\theta\in\Theta}\int_{R(\theta)}dt \leq 2s.$$

By inserting this result and the estimate (6.39) into the inequality (6.38), Claim A follows immediately.

Claim B. $\|g\|_\infty \geq s/128$.

Proof of Claim B. Recall that $\theta_N = \varphi^{-N}\theta_0$. Using the condition (6.30) on γ and the change of variable $t \mapsto t + N$, note that

$$g(\theta_0) = \int_0^{2N} e^{-i\xi t}\gamma(t)\alpha(\varphi^{-t}\theta_0)(E^t f)(\theta_0)\,dt$$
$$= \int_{-N}^{N} e^{-i\xi(t+N)}\gamma(t+N)\alpha(\varphi^{-t}\theta_N)(E^{t+N}f)(\theta_0)\,dt.$$

Also, by Lemma 6.31(b) and the "flat" choice of γ specified in equation (6.32), it follows that

$$g(\theta_0) = \int_{-s}^{s} e^{-i\xi(t+N)}\gamma(t+N)\alpha(\varphi^{-t}\theta_N)(E^{t+N}f)(\theta_0)\,dt$$
$$= e^{-i\xi N}\int_{-s}^{s} e^{-i\xi t}\alpha(\varphi^{-t}\theta_N)(E^{t+N}f)(\theta_0)\,dt = e^{-i\xi N}(I_1 + I_2 + I_3)$$

where

$$I_1 := f(\theta_0)\int_{-s}^{s} e^{-i\xi t}\alpha(\varphi^{-t}\theta_N)\,dt, \quad I_2 := \left[(E^N - I)f\right](\theta_0)\int_{-s}^{s} e^{-i\xi t}\alpha(\varphi^{-t}\theta_N)\,dt,$$
$$I_3 := \int_{-s}^{s} e^{-i\xi t}\alpha(\varphi^{-t}\theta_N)[(E^{t+N} - E^N)f](\theta_0)\,dt.$$

To estimate $|I_1|_X$ from below, recall that $|f(\theta_0)|$ is "big" as in (6.33), while $e^{-i\xi t}$ is "almost" equal to 1 for $|t| \leq s$, as in (6.29). Also, $0 \leq \alpha(\theta) \leq 1$; and, by Lemma 6.31(a), $\alpha(\varphi^{-t}\theta_N) = 1$ for $|t| \leq s/4$. Therefore, we have

$$|I_1|_X = |f(\theta_0)|_X \left| \int_{-s}^{s} e^{-i\xi t} \alpha(\varphi^{-t}\theta_N)\, dt \right|$$

$$\geq \frac{7}{8}\left(\left| \int_{-s}^{s} \alpha(\varphi^{-t}\theta_N)\, dt \right| - \left| \int_{-s}^{s} (e^{-i\xi t} - 1)\alpha(\varphi^{-t}\theta_N)\, dt \right| \right)$$

$$\geq \frac{7}{8}\left(\int_{-s/4}^{s/4} \alpha(\varphi^{-t}\theta_N)\, dt - \int_{-s}^{s} |e^{-i\xi t} - 1|\, dt \right) \geq \frac{7}{8}\left(\frac{s}{2} - \frac{1}{32}\cdot 2s \right) = \frac{49}{128}s.$$

To estimate $|I_2|_X$ and $|I_3|_X$ from above, note that $\|(E^N - I)f\|_\infty$ is "small" as in (6.26) while $\|(E^{t+N} - E^N)f\|_\infty$ for $|t| \leq s$ is "small" as in (6.28). Thus, because $\alpha(\theta) \leq 1$, we have

$$|I_2|_X \leq \|(E^N - I)f\|_\infty \left| \int_{-s}^{s} e^{-i\xi t}\alpha(\varphi^{-t}\theta_N)\, dt \right| \leq \frac{1}{8}\cdot 2s = \frac{s}{4},$$

$$|I_3|_X \leq \int_{-s}^{s} |e^{-i\xi t}\alpha(\varphi^{-t}\theta_0)| \cdot \|(E^{t+N} - E^N)f\|_\infty\, dt \leq 2s \cdot \frac{1}{16} = \frac{s}{8}.$$

As a result,

$$\|g\|_\infty \geq |g(\theta_0)|_X = |I_1 + I_2 + I_3|_X \geq |I_1| - |I_2| - |I_3| \geq s/128,$$

and Claim B is proved.

To finish the proof of the Spectral Mapping Theorem 6.30, combine Claim A and Claim B to obtain the estimate

$$\|\Gamma g - i\xi g\|_\infty \leq \frac{8C}{N} \cdot s \leq \frac{8C}{N} \cdot 128\|g\|_\infty.$$

Since the integer $N \geq 2$ is arbitrary and $\|g\|_\infty \neq 0$, it follows that $i\xi \in \sigma_{\mathrm{ap}}(\Gamma)$, as required. $\square$

We stress the fact that the Spectral Mapping Theorem 6.30 does not hold without the assumption of the aperiodicity of the flow $\{\varphi^t\}_{t\in\mathbb{R}}$, see Example 6.32. However, in case the flow is not aperiodic, we will formulate and prove a replacement for the Spectral Mapping Theorem called the Annular Hull Theorem, see Theorem 7.25.

EXAMPLE 6.32. Let $\Theta = \mathbb{R}/2\pi\mathbb{Z} = [0, 2\pi)$ and $\varphi^t\theta = \theta + t(\mathrm{mod}\, 2\pi)$. Note that $\varphi^{2\pi}\theta = \theta$. Thus, every point in Θ is periodic. Define $X = \mathbb{R}^1$ and $\Phi^t(\theta) \equiv 1$. We have the corresponding evolution group $(E^t f)(\theta) = f(\theta - t(\mathrm{mod}\, 2\pi))$ with generator $(\Gamma f)(\theta) = -\frac{d}{d\theta}f(\theta)$ on $C(\Theta; X) = C_{\mathrm{per}}([0, 2\pi])$. If $t \in \mathbb{R}$ is incommensurate with 2π, then the Spectral Mapping Theorem 6.30 for the group $\{E^t\}_{t\in\mathbb{R}}$ fails; that is, $e^{t\sigma(\Gamma)} \neq \sigma(E^t)$. Indeed, as we will show, $\sigma(\Gamma) = \sigma(-d/d\theta) = \{ik : k \in \mathbb{Z}\}$. However, because $t/2\pi$ is irrational and $\sigma(E^t)$ is closed, it follows from the Spectral Mapping Theorem 2.8 for the point spectrum that

$$\sigma(E^t) = \overline{\{e^{ikt} : k \in \mathbb{Z}\}} \neq e^{t\sigma(\Gamma)}.$$

To see that $\sigma(\Gamma) = \{ik : k \in \mathbb{Z}\}$, define $\varepsilon_\lambda(\theta) := e^{-\lambda\theta}$, for $\theta \in [0, 2\pi]$ and $\lambda \in \mathbb{C}$, and note that if $\lambda = ik$, then $\varepsilon_\lambda \in C_{\mathrm{per}}(0, 2\pi)$ and $(\lambda I - \Gamma)\varepsilon_\lambda = 0$. On the other hand, if $\lambda \neq ik$, then define

$$(R_\lambda f)(\theta) = \frac{1}{e^{2\pi\lambda} - 1} \int_0^{2\pi} e^{\lambda s} f(s+\theta)\, ds = \frac{1}{e^{2\pi\lambda} - 1} \int_\theta^{2\pi+\theta} e^{\lambda(s-\theta)} f(s)\, ds.$$

Since $\frac{d}{d\theta}(R_\lambda f)(\theta) = f(\theta) - \lambda R_\lambda f(\theta)$, it follows that $R_\lambda = (\lambda - \Gamma)^{-1}$. $\diamond$

6.2.4. The Mather semigroup on L^p-spaces. In this subsection we will study the evolution semigroup $\{E^t\}_{t \geq 0}$, $E^t = e^{t\Gamma}$, defined on the space $L^p = L^p(\Theta, \mu; X)$, $1 \leq p < \infty$, by

(6.40) $$(E^t f)(\theta) = \left(\frac{d\mu \circ \varphi^{-t}}{d\mu}(\theta)\right)^{\frac{1}{p}} \Phi^t(\varphi^{-t}\theta) f(\varphi^{-t}\theta).$$

Here $\{\Phi^t\}_{t \geq 0}$ is a strongly continuous, exponentially bounded cocycle over the flow $\{\varphi^t\}_{t \in \mathbb{R}}$. In this subsection we assume that Θ is a locally compact metric space, μ is a σ-finite regular Borel measure on Θ that is positive on open sets and *quasi-invariant* with respect to the flow $\{\varphi^t\}_{t \in \mathbb{R}}$; that is, the Radon-Nikodým derivative $d\mu \circ \varphi^t/d\mu$ belongs to $L^\infty(\Theta)$ and is uniformly bounded for $t \in \mathbb{R}$. By Lemma 6.33 below, $\{E^t\}_{t \geq 0}$ is a strongly continuous semigroup on $L^p(\Theta, \mu; X)$. In addition, let us suppose that the set of aperiodic points (see Definition 6.26) of the flow $\{\varphi^t\}_{t \in \mathbb{R}}$ has full μ-measure in Θ. In particular, by this assumption, the aperiodic points are dense in Θ. Under these assumptions we will prove the Spectral Mapping Theorem 6.30 for the evolution semigroup (6.40) on L^p.

By standard arguments involving rescaling, the Spectral Inclusion Theorem, and the Spectral Mapping Theorem for the residual spectrum, it suffices to prove the following proposition: *If $1 \in \sigma_{ap}(E)$, then $i\mathbb{R} \subseteq \sigma_{ap}(\Gamma)$.* The essential element of our proof is an explicit construction of an approximate eigenfunction of the generator Γ. This eigenfunction is "localized" in a long and thin tube given by a "flow-box" formed by trajectories of φ^t. To achieve this localization we use *weighted shift operators* $\pi_\theta(E)$ acting on the sequence space $\ell^p = \ell^p(\mathbb{Z}, X)$ by

(6.41) $$\pi_\theta(E)v = (\Phi(\varphi^{k-1}\theta)x_{k-1})_{k \in \mathbb{Z}}, \quad v = (x_k)_{k \in \mathbb{Z}} \in \ell^p.$$

The appearance of the family of operators $\{\pi_\theta(E) : \theta \in \Theta\}$ is related to a certain algebraic structure that will be discussed in Section 7.1. In fact, π_θ defines a representation of a certain *weighted translation algebra* of operators of the type (6.40) in ℓ^p.

As a preparatory result, we prove in Proposition 6.36 that the uniform injectivity of $I - E$ on $L^p(\Theta, \mu; X)$ is equivalent to the uniform injectivity of each operator $I - \pi_\theta(E)$ on ℓ^p and the existence of a positive uniform lower bound over $\theta \in \Theta$ for the family of these operators. This result is of independent interest, cf. Theorem 7.4. However, we will use it to construct an approximate eigenfunction for Γ by choosing a point θ_0 where $I - \pi_{\theta_0}(E)$ "almost" loses its uniform injectivity, and then by spreading the corresponding approximate eigenvector $v = (x_k)_{k \in \mathbb{Z}} \in \ell^p$ for $\pi_{\theta_0}(E)$ in a tube around the orbit through θ_0.

We remark that, for a given flow $\{\varphi^t\}_{t \in \mathbb{R}}$, there exists a regular Borel measure μ with $\operatorname{supp} \mu = \Theta$ such that $\mu \circ \varphi^t$ is absolutely continuous with respect to μ, see for example Antonevich [**An2**, §9]. For brevity, denote the Radon-Nikodým derivative by

$$J^t(\theta) = \frac{d\mu \circ \varphi^t}{d\mu}(\theta)$$

and note that we have

(6.42) $$J^{t+s}(\theta) = J^t(\varphi^s\theta)J^s(\theta), \qquad \left(J^t(\theta)\right)^{-1} = J^{-t}(\varphi^t\theta) > 0$$

for $t, s \in \mathbb{R}$ and almost every $\theta \in \Theta$.

LEMMA 6.33. *Under the assumptions in this subsection, the family of operators $\{E^t\}_{t\geq 0}$, defined in display (6.40), is a strongly continuous semigroup on L^p.*

PROOF. Clearly, $\{E^t\}_{t\geq 0}$ is a semigroup on L^p. Define a family of isometries $\{V^t\}_{t\in\mathbb{R}}$ on L^p by $(V^t f)(\theta) = J^{-t}(\theta)^{\frac{1}{p}} f(\varphi^{-t}\theta)$. Since

$$\|E^t f - f\| \leq M e^{\omega t} \|V^t f - f\| + \|(\Phi^t(\varphi^{-t}\cdot) - I)f\|$$

for all $f \in L^p$, it suffices to show the strong continuity of the family $\{V^t\}_{t\in\mathbb{R}}$. We claim that

(6.43) if $t_n \to 0$, then there is a subsequence t_{n_k} such that $J^{t_{n_k}}(\theta) \to 1$ for a.e. θ.

Otherwise, there is a subsequence $t_k \to 0$ and a compact set W of positive measure so that $|J^{t_k}(\theta) - 1| \geq \delta > 0$ for $\theta \in W$. We may assume that $J^{t_k}(\theta) \geq 1$. Hence, $\mu(\varphi^{t_k}W) - \mu(W) \geq \delta\mu(W)$. However, $\mu(\varphi^{t_k}W) \to \mu(W)$ by the Dominated Convergence Theorem, and therefore the claim (6.43) is true.

If V^t is not strongly continuous, then there is a continuous function f with compact support and a sequence $t_n \to 0$ so that $\|V^{t_n} f - f\| \geq \delta > 0$. By (6.43), this result leads to a contradiction. □

We will need two technical lemmas.

LEMMA 6.34. *If $\theta_0 \in \Theta$ is aperiodic and $n \in \mathbb{N}$, then there exists an open set $\mathcal{U}$ with compact closure, a set Σ such that $\theta_0 \in \Sigma \subseteq \mathcal{U}$, and a continuous function $t : \mathcal{U} \to (-n, n)$ such that $\sigma = \sigma(\theta) := \varphi^{t(\theta)}\theta \in \Sigma$.*

The set $\mathcal{U}$ in the lemma is called a *flow-box* of length n with *cross-section* Σ at θ_0. Lemma 6.34 is proved for example in the book of Bhatia and Szegö [**BS**, Thm. IV.2.11].

The next result is a simple consequence of the Halmos-Rokhlin Lemma.

LEMMA 6.35. *Suppose that $g_1, \ldots, g_m$ are nonnegative functions in the space $L^1(\Theta, \mu; \mathbb{R})$. If $\epsilon > 0$ and $n \in \mathbb{N}$, then there is a measurable set $F \subseteq \Theta$ such that the sets $\varphi^k F$ and $\varphi^l F$ are pairwise disjoint for $k, l \in \{-n, \ldots, n\}$, and moreover*

$$\int_{\Theta \setminus U} g_j(\theta) \, d\mu(\theta) \leq \epsilon$$

for $j = 1, \ldots, m$ and $U := \bigcup_{|k|\leq n} \varphi^k F$.

PROOF. Fix $\epsilon > 0$ and $n \in \mathbb{N}$. There is a set $\Theta_0 \subseteq \Theta$ of finite measure so that $\int_{\Theta \setminus \Theta_0} g_j(\theta) \, d\mu(\theta) \leq \epsilon/2$ for $j = 1, \ldots, m$. Choose a positive function $\alpha \in L^1(\Theta, \mu; \mathbb{R}) \cap L^\infty(\Theta, \mu; \mathbb{R})$ with $\alpha = 1$ on Θ_0. Also, set $\nu := \alpha\mu$ and note that $g_j \in L^1(\Theta, \nu; \mathbb{R})$. Since the set of aperiodic points of the flow $\{\varphi^t\}_{t\in\mathbb{R}}$ has full measure, the Halmos-Rokhlin Lemma (see Jones and Krengel [**JK**, Thm. 1.11]) applied to the map φ^1 on the finite measure space (Θ, ν) yields the following result: For each $\delta > 0$, there is a measurable subset F_δ of Θ such that the sets $\varphi^k F_\delta$

and $\varphi^l F_\delta$ are pairwise disjoint for $k, l \in \{-n, \ldots, n\}$, and $\nu(\Theta \setminus U_\delta) \leq \delta$ where $U_\delta := \bigcup_{|k| \leq n} \varphi^k F_\delta$. Hence, we have that

$$\int_{\Theta \setminus U_\delta} g_j(\theta)\, d\mu(\theta) \leq \int_{\Theta_0 \cap (\Theta \setminus U_\delta)} \alpha(\theta) g_j(\theta)\, d\mu(\theta) + \int_{\Theta \setminus \Theta_0} g_j(\theta)\, d\mu(\theta)$$

$$\leq \int_{\Theta \setminus U_\delta} g_j(\theta)\, d\nu(\theta) + \frac{\epsilon}{2} \leq \epsilon$$

for $j = 1, \ldots, m$ and $\delta > 0$ sufficiently small. The desired result follows by setting $F := F_\delta$. $\square$

Consider the operator $E := E^1$ as defined in display (6.40) and the family of operators $\{\pi_\theta(E) : \theta \in \Theta\}$ as defined in (6.41).

PROPOSITION 6.36. *The following statements are equivalent:*
(a) *There is a constant $c > 0$ such that $\|(I - E)f\|_{L^p} \geq c\|f\|_{L^p}$ for all $f \in L^p$.*
(b) *There is a constant $c > 0$ such that $\|(I - \pi_\theta(E))v\|_{\ell^p} \geq c\|v\|_{\ell^p}$ for all $\theta \in \Theta$ and $v \in \ell^p$.*

PROOF. **(a)$\Rightarrow$(b).** Consider an aperiodic point $\theta_0 \in \Theta$ and a finitely supported sequence $v = (x_k) \in \ell^p$ with $x_k = 0$ for $|k| > n$. Fix $\epsilon > 0$. There is an open set U with finite measure such that $\theta_0 \in U$, the sets $\varphi^k U$ and $\varphi^l U$ are pairwise disjoint for $k, l \in \{-n, \ldots, n+1\}$, and

(6.44) $$|\Phi(\varphi^k \theta) x_k - \Phi(\varphi^k \theta_0) x_k| \leq \epsilon$$

for $|k| \leq n$ and $\theta \in U$. Define $f : \Theta \to X$ by $f(\theta) := \left(J^{-k}(\theta)\right)^{\frac{1}{p}} x_k$ if $\theta \in \varphi^k U$ and $|k| \leq n$, and $f(\theta) = 0$ otherwise. Clearly, $f \in L^p$ and

(6.45) $$\|f\|_{L^p}^p = \sum_{k=-n}^{n} |x_k|^p \int_{\varphi^k U} J^{-k}(\theta)\, d\mu(\theta) = \mu(U) \|v\|_{\ell^p}^p.$$

On the other hand, using the properties in display (6.42), especially the fact that

$$J^{-1}(\varphi^k(\theta)) = (J^1(\varphi^{k-1}(\theta)))^{-1}$$

and the inequality (6.44), we compute

$$\|(I - E)f\|_{L^p}^p = \int_{\bigcup_{k=-n}^{n+1} \varphi^k U} \left| f(\theta) - \left(J^{-1}(\theta)\right)^{\frac{1}{p}} \Phi(\varphi^{-1}\theta) f(\varphi^{-1}\theta) \right|^p d\mu(\theta)$$

$$= \int_U \sum_{k=-n}^{n+1} \left| \left(J^k(\theta)\right)^{-\frac{1}{p}} x_k - \left(J^1(\varphi^{k-1}\theta) J^{k-1}(\theta)\right)^{-\frac{1}{p}} \Phi(\varphi^{k-1}\theta) x_{k-1} \right|^p J^k(\theta)\, d\mu(\theta)$$

$$= \int_U \sum_{k=-n}^{n+1} |x_k - \Phi(\varphi^{k-1}\theta) x_{k-1}|^p\, d\mu(\theta)$$

$$\leq \int_U \sum_{k=-n}^{n+1} \left(|x_k - \Phi(\varphi^{k-1}\theta_0)x_{k-1}| + |\Phi(\varphi^{k-1}\theta_0)x_{k-1} - \Phi(\varphi^{k-1}\theta)x_{k-1}|\right)^p d\mu(\theta)$$

$$\leq \int_U \sum_{k=-n}^{n+1} \left(|x_k - \Phi(\varphi^{k-1}\theta_0)x_{k-1}| + \epsilon\right)^p d\mu(\theta)$$

$$= \mu(U) \sum_{k=-n}^{n+1} \left(|x_k - \Phi(\varphi^{k-1}\theta_0)x_{k-1}| + \epsilon\right)^p.$$

This estimate, together with equation (6.45) and assumption (a), implies the inequalities

$$c^p \|v\|_{\ell^p}^p = \frac{c^p}{\mu(U)} \|f\|_{L^p}^p \leq \frac{1}{\mu(U)} \|(I - E)f\|_{L^p}^p$$

$$\leq \sum_{k=-n}^{n+1} \left(|x_k - \Phi(\varphi^{k-1}\theta_0)x_{k-1}| + \epsilon\right)^p.$$

Passing to the limit as $\epsilon \to 0$, we derive the inequality $\|(I - \pi_\theta(E))v\|_{\ell^p} \geq c\|v\|_{\ell^p}$. Thus we have proved statement (b) for the sets of finitely supported $v \in \ell^p$ and aperiodic $\theta \in \Theta$. The desired result follows from the density of these sets.

(b)$\Rightarrow$(a). Fix $f \in L^p$ and $\epsilon > 0$. By Lemma 6.35, there is an integer n and a measurable set $F \subseteq \Theta$ such that the sets in the family $\{\varphi^k F : |k| \leq n\}$ are pairwise disjoint,

(6.46) $$\|f\|_{L^p}^p \leq \epsilon + \int_{\bigcup_{|k|\leq n} \varphi^k F} |f(\theta)|^p \, d\mu(\theta),$$

and

(6.47) $$\epsilon \geq \int_{\varphi^{n+1} F} |Ef(\theta)|^p \, d\mu(\theta).$$

For almost every $\theta \in F$, we define the sequence $v(\theta) = (x_k(\theta))_{k\in\mathbb{Z}} \in \ell^p$ by $x_k(\theta) := \left(J^k(\theta)\right)^{\frac{1}{p}} f(\varphi^k \theta)$ if $|k| \leq n$, and $x_k(\theta) = 0$ otherwise. Then by the properties (6.42), the assumption (b), and the estimates (6.46)–(6.47), we have that

$$\|f - Ef\|_{L^p}^p$$
$$\geq \int_{\bigcup_{|k|\leq n} \varphi^k F} \left|f(\theta) - \left(J^{-1}(\theta)\right)^{\frac{1}{p}} \Phi(\varphi^{-1}\theta) f(\varphi^{-1}\theta)\right|^p d\mu(\theta)$$
$$= \int_F \sum_{k=-n}^{n} \left|f(\varphi^k \theta) - \left(J^{-1}(\varphi^k\theta)\right)^{\frac{1}{p}} \Phi(\varphi^{k-1}\theta) f(\varphi^{k-1}\theta)\right|^p J^k(\theta) \, d\mu(\theta)$$
$$= \int_F \sum_{k=-n}^{n} |x_k(\theta) - \Phi(\varphi^{k-1}\theta) x_{k-1}(\theta)|^p d\mu(\theta)$$
$$= \int_F \sum_{k=-n}^{n+1} |x_k(\theta) - \Phi(\varphi^{k-1}\theta) x_{k-1}(\theta)|^p d\mu(\theta) - \int_F |\Phi(\varphi^n \theta) x_n(\theta)|^p d\mu(\theta)$$
$$= \int_F \|(I - \pi_\theta(E))v(\theta)\|_{\ell^p}^p \, d\mu(\theta) - \int_{\varphi^{n+1} F} |Ef(\theta)|^p \, d\mu(\theta)$$
$$\geq -\epsilon + c^p \int_F \|v(\theta)\|_{\ell^p}^p \, d\mu(\theta) = -\epsilon + c^p \int_{\bigcup_{|k|\leq n} \varphi^k F} |f(\theta)|^p d\mu(\theta)$$
$$\geq -(c^p + 1)\epsilon + c^p \|f\|_{L^p}^p.$$

Statement (b) follows from the last inequality because $\epsilon > 0$ is arbitrary. □

THEOREM 6.37 (Spectral Mapping Theorem). *Under the assumptions made in this subsection, the spectral mapping theorem holds; that is, $e^{t\sigma(\Gamma)} = \sigma(E^t) \setminus \{0\}$ for $t > 0$. Also, $\sigma(E^t)$, $t > 0$, is invariant with respect to rotations centered at the origin, and $\sigma(\Gamma)$ is invariant with respect to vertical translations.*

6.2. THE MATHER SEMIGROUP

PROOF. **Step 1.** Assume $1 \in \sigma_{\mathrm{ap}}(E)$. By Proposition 6.36, if $n \in \mathbb{N}$, then there is a vector $v^n \in \ell^p$ and a point $\theta^n \in \Theta$ such that

(6.48) $$\|(I - \pi_{\theta^n}(E))v^n\|_{\ell^p} < n^{-1}\|v^n\|_{\ell^p}.$$

Moreover, since the aperiodic points are dense in Θ, let us choose θ^n to be aperiodic.

Fix $n \geq 2$, define $v = (x_k) := v^n$, and, for each $k \in \mathbb{Z}$, let $\theta_k := \varphi^k \theta^n$. Since $\pi_{\theta_0}(E)v \in \ell^p$, there is some positive integer $N \geq n$ such that

(6.49) $$\frac{1}{2}\sum_{k=-\infty}^{\infty}|\Phi(\theta_k)x_k|^p \leq \sum_{k=-N}^{N-1}|\Phi(\theta_k)x_k|^p.$$

By Lemma 6.34, there is a flow-box $\mathcal{U}$ of length $2N$ with a cross-section Σ at θ_0 such that $\mu(\mathcal{U}) < \infty$,

(6.50) $$|\Phi(\varphi^{k-1}\sigma)x_{k-1} - x_k| \leq 2|\Phi(\theta_{k-1})x_{k-1} - x_k|,$$

(6.51) $$\frac{1}{2}|\Phi(\theta_{k-1})x_{k-1}| \leq |\Phi(\varphi^{k-1}\sigma)x_{k-1}| \leq 2|\Phi(\theta_{k-1})x_{k-1}|$$

as long as $\sigma \in \Sigma$ and $|k| \leq 2N$. Finally, using the inequality (6.48), it follows that

(6.52) $$\sum_{k=-\infty}^{\infty}|\Phi(\theta_{k-1})x_{k-1} - x_k|^p < 2^p n^{-p}\sum_{k=-\infty}^{\infty}|\Phi(\theta_{k-1})x_{k-1}|^p.$$

We will construct an approximate eigenfunction g of Γ. The idea is essentially the same as in the proof of Lemma 2.29. Set $\mathcal{U}_0 = \{\theta = \varphi^t\sigma : \sigma \in \Sigma, 0 < t \leq 1\}$ and $\mathcal{U}_1 = \{\varphi^t\sigma : \sigma \in \Sigma, \frac{3}{4} < t \leq 1\}$. Choose a smooth function $\alpha : [0,1] \to [0,1]$ such that $\alpha = 0$ on $[0, \frac{1}{4}]$, $\alpha = 1$ on $[\frac{3}{4}, 1]$, and $\|\alpha'\|_\infty \leq 3$. Choose a smooth function $\gamma : \mathbb{R} \to [0,1]$ such that $\gamma = 1$ on $[-N, N]$, $\mathrm{supp}\,\gamma \subseteq (-2N, 2N)$, and $\|\gamma'\|_\infty \leq \frac{2}{N}$. For $\theta \in \mathcal{U}$, we set $\theta = \varphi^{k+\tau}\sigma$ where $\sigma \in \Sigma$ and $t = k + \tau \in (-2N, 2N)$ with $k = -2N, \ldots, 2N-1$ and $0 \leq \tau < 1$. Also, for each $\xi \in \mathbb{R}$, we define a function $g \in L^p$ as follows:

$$g(\theta) = \gamma(t)e^{-i\xi t}\left(J^{-t}(\theta)\right)^{\frac{1}{p}}\Phi^\tau(\varphi^k\sigma)[(1-\alpha(\tau))\Phi(\varphi^{k-1}\sigma)x_{k-1} + \alpha(\tau)x_k]$$

for $\theta \in \mathcal{U}$ and $g(\theta) = 0$ for $\theta \notin \mathcal{U}$.

Step 2. We will show that $g \in \mathcal{D}(\Gamma)$ and we will compute Γg. Consider $\theta \in \Theta$ and $h > 0$ such that $\varphi^{-h}\theta = \varphi^{k+\tau-h}\sigma \in \mathcal{U}$. First, assume that $\tau > 0$. If $h < \tau$ and $t = k + \tau$, then

$$(E^h g)(\theta) = \left(J^{-h}(\theta)\right)^{\frac{1}{p}}\Phi^h(\varphi^{-h}\theta)g(\varphi^{-h}\theta)$$
$$= \gamma(t-h)e^{i\xi(h-t)}\left(J^{-t+h}(\varphi^{-h}\theta)J^{-h}(\theta)\right)^{\frac{1}{p}}\Phi^h(\varphi^{k+\tau-h}\sigma)\Phi^{\tau-h}(\varphi^k\sigma)$$
$$\cdot[(1-\alpha(\tau-h))\Phi(\varphi^{k-1}\sigma)x_{k-1} + \alpha(\tau-h)x_k]$$
$$= \gamma(t-h)e^{i\xi h}e^{-i\xi t}\left(J^{-t}(\theta)\right)^{\frac{1}{p}}\Phi^\tau(\varphi^k\sigma)$$
$$\cdot[(1-\alpha(\tau-h))\Phi(\varphi^{k-1}\sigma)x_{k-1} + \alpha(\tau-h)x_k].$$

If $\tau = 0$, then, for sufficiently small $h > 0$, we have that

$$E^h g(\theta) = \gamma(t-h)e^{i\xi(h-t)}\left(J^{-t}(\theta)\right)^{\frac{1}{p}}\Phi^h(\varphi^{k-h}\sigma)\Phi^{1-h}(\varphi^{k-1}\sigma)x_{k-1}$$
$$= \gamma(t-h)e^{i\xi h}e^{-i\xi t}\left(J^{-t}(\theta)\right)^{\frac{1}{p}}\Phi(\varphi^{k-1}\sigma)x_{k-1}.$$

Using these results, we have the following propositions: if h is sufficiently small and $\theta \notin \mathcal{U}$, then $E^h g(\theta) = g(\theta) = 0$; if $\theta \in \mathcal{U}$ and $\tau = 0$, then

$$\lim_{h \searrow 0} h^{-1}(E^h g(\theta) - g(\theta))$$
$$= \lim_{h \searrow 0} h^{-1}(\gamma(t-h)e^{i\xi h} - \gamma(t))e^{-i\xi t} \left(J^{-t}(\theta)\right)^{\frac{1}{p}} \Phi(\varphi^{k-1}\sigma)x_{k-1}$$
$$= i\xi g(\theta) - \gamma'(t)e^{-i\xi t} \left(J^{-t}(\theta)\right)^{\frac{1}{p}} \Phi(\varphi^{k-1}\sigma)x_{k-1};$$

and if $\theta \in \mathcal{U}$ and $\tau \neq 0$, then

$$\lim_{h \searrow 0} h^{-1}(E^h g(\theta) - g(\theta))$$
$$= e^{-i\xi t} \left(J^{-t}(\theta)\right)^{\frac{1}{p}} \Phi^\tau(\varphi^k \sigma) \lim_{h \searrow 0} \left\{ \gamma(t-h) \left[\left(h^{-1}(e^{i\xi h} - 1) - h^{-1}(e^{i\xi h}\right.\right.\right.$$
$$\cdot \alpha(\tau - h) - \alpha(\tau))) \Phi(\varphi^{k-1}\sigma)x_{k-1} + h^{-1}(e^{i\xi h}\alpha(\tau - h) - \alpha(\tau))x_k \bigr]$$
$$+ h^{-1}(\gamma(t-h) - \gamma(t))\left[(1 - \alpha(\tau))\Phi(\varphi^{k-1}\sigma)x_{k-1} + \alpha(\tau)x_k\right] \Big\}$$
$$= i\xi g(\theta) + e^{-i\xi t} \left(J^{-t}(\theta)\right)^{\frac{1}{p}} \Phi^\tau(\varphi^k \sigma) \left[\gamma(t)\alpha'(\tau)(\Phi(\varphi^{k-1}\sigma)x_{k-1} - x_k)\right.$$
$$\left. - \gamma'(t)((1 - \alpha(\tau))\Phi(\varphi^{k-1}\sigma)x_{k-1} + \alpha(\tau)x_k)\right].$$

It follows that $g \in \mathcal{D}(\Gamma)$. Moreover,

$$\Gamma g(\theta) - i\xi g(\theta) = e^{-i\xi t} \left(J^{-t}(\theta)\right)^{\frac{1}{p}} \Phi^\tau(\varphi^k \sigma) \left[(\gamma(t)\alpha'(\tau) + \gamma'(t)\alpha(\tau))\right.$$
$$\left. \cdot (\Phi(\varphi^{k-1}\sigma)x_{k-1} - x_k) - \gamma'(t)\Phi(\varphi^{k-1}\sigma)x_{k-1}\right]$$

for $\theta \in \mathcal{U}$, and $\Gamma g(\theta) - i\xi g(\theta) = 0$ for $\theta \notin \mathcal{U}$.

Step 3. We will prove the following proposition: There is a constant $\tilde{C} > 0$, which is independent of n and g, such that

$$\|\Gamma g - i\xi g\|_{L^p} < \frac{\tilde{C}}{n} \|g\|_{L^p}.$$

Define $M := \sup\{\|\Phi^\tau(\theta)\| : 0 \leq \tau \leq 1, \theta \in \Theta\}$ and $C := \sup\{J^t(\theta) : t \in \mathbb{R}, \theta \in \Theta\}$. By our assumptions, M and C are finite, and $\mu(\mathcal{U}_1) \geq \mu(\mathcal{U}_0)/(4C)$. Hence, by the inequalities (6.50) and (6.51), if $\theta = \varphi^{k+\tau}\sigma \in \mathcal{U}$, then there is a constant $C_1 > 0$, which is independent of k, n, and g, such that

$$|\Gamma g(\theta) - i\xi g(\theta)|^p \leq CM^p \left(8|\Phi(\theta_{k-1})x_{k-1} - x_k| + \frac{4}{N}|\Phi(\theta_{k-1})x_{k-1}|\right)^p$$
$$\leq C_1 \left(|\Phi(\theta_{k-1})x_{k-1} - x_k| + \frac{1}{n}|\Phi(\theta_{k-1})x_{k-1}|\right)^p.$$

Consequently, we have that

$$\|\Gamma g - i\xi g\|_{L^p} = \left(\int_{\bigcup_{k=-2N}^{2N-1} \varphi^k \mathcal{U}_0} |\Gamma g(\theta) - i\xi g(\theta)|^p \, d\mu(\theta)\right)^{\frac{1}{p}}$$
$$\leq C_1^{\frac{1}{p}} \left[\sum_{k=-\infty}^{\infty} \mu(\varphi^k \mathcal{U}_0) \left(|\Phi(\theta_{k-1})x_{k-1} - x_k| + \frac{1}{n}|\Phi(\theta_{k-1})x_{k-1}|\right)^p\right]^{\frac{1}{p}}$$
$$(6.53) \qquad < (CC_1\mu(\mathcal{U}_0))^{\frac{1}{p}} \frac{3}{n} \left(\sum_{k=-\infty}^{\infty} |\Phi(\theta_k)x_k|^p\right)^{\frac{1}{p}}$$

where, in the last step, we have used the inequality (6.52). On the other hand, if $\theta \in \varphi^k \mathcal{U}_1$ and $k = -N, \ldots, N-1$, then
$$g(\theta) = e^{-i\xi t} \left(J^{-t}(\theta)\right)^{\frac{1}{p}} \Phi^\tau(\varphi^k \sigma) x_k.$$
Moreover, for $\theta \in \Theta$, $0 \leq \tau \leq 1$, and $x \in X$, we have
$$|\Phi(\theta)x| \leq \|\Phi^{1-\tau}(\varphi^\tau \theta)\| \, |\Phi^\tau(\theta)x| \leq M|\Phi^\tau(\theta)x|.$$
Thus, the properties of the Radon-Nikodým derivative in (6.42) and the inequality (6.51) imply
$$|g(\theta)|^p = \left|(J^{-t}(\theta))^{\frac{1}{p}} \Phi^\tau(\varphi^k \sigma) x_k\right|^p \geq C^{-1} M^{-p} |\Phi(\varphi^k \sigma) x_k|^p$$
$$\geq C^{-1}(2M)^{-p} |\Phi(\theta_k) x_k|^p$$
for $\theta \in \varphi^k \mathcal{U}_1$ and $k = -N, \ldots, N-1$. Using the inequality (6.49), it follows that
$$\|g\|_{L^p}^p \geq \int_{\bigcup_{k=-N}^{N-1} \varphi^k \mathcal{U}_1} |g(\theta)|^p \, d\mu(\theta) \geq \frac{1}{C(2M)^p} \sum_{k=-N}^{N-1} \mu(\varphi^k \mathcal{U}_1) |\Phi(\theta_k) x_k|^p$$
(6.54)
$$\geq \frac{\mu(\mathcal{U}_0)}{8C^3 (2M)^p} \sum_{k=-\infty}^{\infty} |\Phi(\theta_k) x_k|^p.$$

By combining the inequalities (6.53) and (6.54), the theorem is established. $\square$

6.3. Sacker-Sell spectral theory

In this section we will characterize the exponential dichotomy of a strongly continuous cocycle $\{\Phi^t\}_{t\geq 0}$ or linear skew-product flow $\{\hat{\varphi}^t\}_{t\geq 0}$ over a flow $\{\varphi^t\}_{t\in\mathbb{R}}$ on a locally compact metric space Θ in terms of the hyperbolicity of the corresponding evolution semigroup $\{E^t\}_{t\geq 0}$ on $C_0(\Theta; X)$ (Dichotomy Theorem 6.41). The main tool used here is the Spectral Projection Theorem 6.38 (see Theorem 3.14 for the case $\Theta = \mathbb{R}$). This theorem states that the Riesz spectral projection for the hyperbolic operator E^t, $t > 0$, is a multiplication operator whose multiplier is a bounded, strongly continuous, projection valued function on Θ. We will use these results to give a spectral theory for the linear skew-product flow $\{\hat{\varphi}^t\}_{t\geq 0}$. In particular, we will describe its Sacker-Sell spectrum Σ and the corresponding invariant spectral subbundles. In this section, the flow $\{\varphi^t\}_{t\in\mathbb{R}}$ is *not* assumed to be aperiodic.

Since a semigroup $\{E^t\}_{t\geq 0}$ is hyperbolic if and only if E^1 is hyperbolic, and since for each $t > 0$ the Riesz projection for the operator E^t that corresponds to the set $\sigma(E^t) \cap \mathbb{D}$ is the same as the Riesz projection for E^1 (see Lemma 2.15), it suffices to restrict attention to the case $t = 1$. Let us define $E := E^1$, $\Phi(\theta) := \Phi^1(\theta)$, $\hat{\varphi} := \hat{\varphi}^1$, and $\varphi := \varphi^1$.

6.3.1. The Spectral Projection Theorem. Let $\{\Phi^t\}_{t\geq 0}$ be a strongly continuous, exponentially bounded cocycle over a continuous flow $\{\varphi^t\}_{t\geq 0}$ on a locally compact metric space Θ.

THEOREM 6.38 (Spectral Projection Theorem). *If $\sigma(E^t) \cap \mathbb{T} = \emptyset$, $t > 0$, then there exists a bounded, strongly continuous, projection-valued function $P : \Theta \to \mathcal{L}_s(X)$ such that the Riesz projection $\mathcal{P}$ that corresponds to the operator E^t and the spectral set $\sigma(E^t) \cap \mathbb{D}$ is given by the formula $(\mathcal{P}f)(\theta) = P(\theta)f(\theta)$.*

PROOF. We will consider the case $t = 1$, and use, as usual, the complementary projection $\mathcal{Q} = I - \mathcal{P}$ and the notations E_P and E_Q for the restrictions of E on $\operatorname{Im} \mathcal{P}$ and $\operatorname{Im} \mathcal{Q}$, respectively. Also, for each $\chi \in C_b(\Theta; \mathbb{R})$, we will also use χ to denote the multiplication operator on $C_0(\Theta; X)$ given by $(\chi f)(\theta) = \chi(\theta) f(\theta)$.

LEMMA 6.39. *If $\chi : \Theta \to \mathbb{R}$ is bounded and continuous, then $\chi \mathcal{P} = \mathcal{P} \chi$.*

PROOF. Since $E\chi = (\chi \circ \varphi^{-1}) E$, the proof repeats word-for-word the proof of Lemma 3.15. □

Fix $\theta_0 \in \Theta$ and $x \in X$. Also, choose $f \in C_0(\Theta; X)$ such that $f(\theta_0) = x$. Note that $\mathcal{P} f \in C_0(\Theta; X)$, and define

$$P(\theta_0) x = (\mathcal{P} f)(\theta_0). \tag{6.55}$$

Step 1. $P(\theta_0)$ is well-defined. To see this, suppose that $f_i \in C_0(\Theta; X)$ and $f_i(\theta_0) = x$ for $i = 1, 2$. We will show that $(\mathcal{P} f_1)(\theta_0) = (\mathcal{P} f_2)(\theta_0)$. Let $B(\theta, r)$ denote the open ball in Θ centered at $\theta \in \Theta$ with radius $r > 0$. Define $f := f_1 - f_2$. Also, for each integer $n \geq 2$, use Urysohn's Lemma to obtain a function $\chi_n \in C_b(\Theta; \mathbb{R})$ such that

$$\chi_n(\theta) = 1 \text{ for } \theta \in \overline{B(\theta_0, 1/(n+1))}, \text{ and } \chi_n(\theta) = 0 \text{ for } \theta \in \Theta \setminus B(\theta_0, 1/n). \tag{6.56}$$

Since the function f is continuous, $f(\theta_0) = 0$, and

$$\|\chi_n f\|_\infty = \sup_{\theta \in \Theta} |\chi_n(\theta) f(\theta)| \leq \sup_{\theta \in B(\theta_0, 1/n)} |f(\theta)|,$$

we have that $\lim_{n \to \infty} \|\chi_n f\|_\infty = 0$. But, by using Lemma 6.39, we have the equalities

$$(\mathcal{P} f)(\theta_0) = \chi_n(\theta_0)(\mathcal{P} f)(\theta_0) = (\mathcal{P} \chi_n f)(\theta_0);$$

and therefore,

$$|(\mathcal{P} f)(\theta_0)| \leq \|\mathcal{P} \chi_n f\|_\infty \leq \|\mathcal{P}\|_{\mathcal{L}(C_0(\Theta; X))} \|\chi_n f\|_\infty.$$

By passing to the limit as $n \to \infty$, it follows that $|(\mathcal{P} f)(\theta_0)| = 0$.

Step 2. $P(\theta_0)$ is a linear projection. Clearly, $P(\theta_0)$ is linear. To see that it is a projection, suppose that $g = \mathcal{P} f$ and $f(\theta_0) = x$, and compute

$$P^2(\theta_0) x = P(\theta_0)[(\mathcal{P} f)(\theta_0)] = P(\theta_0) g(\theta_0) = (\mathcal{P} g)(\theta_0)$$
$$= (\mathcal{P}^2 f)(\theta_0) = (\mathcal{P} f)(\theta_0) = P(\theta_0) x.$$

Also, if $f \in C_0(\Theta; X)$ with $\|f\|_\infty = |f(\theta_0)| = |x|$, then

$$|P(\theta_0) x|_X = |(\mathcal{P} f)(\theta_0)| \leq \|\mathcal{P} f\|_\infty \leq \|\mathcal{P}\|_{\mathcal{L}(C_0(\Theta; X))} \|f\|_\infty = \|\mathcal{P}\| \cdot |x|.$$

Hence, $\sup_{\theta \in \Theta} \|P(\theta)\|_{\mathcal{L}(X)} \leq \|\mathcal{P}\|_{\mathcal{L}(C_0(\Theta; X))} < \infty$, and therefore $P(\theta_0)$ is bounded.

Step 3. If $x \in X$, then the function given by $\theta \mapsto P(\theta) x$ is continuous. To prove this fact, fix $\theta_0 \in \Theta$ and $x \in X$. Also, consider χ_1 and χ_2 as defined in display (6.56) and define $f(\theta) := \chi_2(\theta) x$. In view of equation (6.55) and the fact that $f(\theta) \equiv x$ for $\theta \in B(\theta_0, 1/3)$, we have that $P(\theta) x = (\mathcal{P} f)(\theta)$ for $\theta \in B(\theta_0, 1/3)$. Also, $\chi_1 \chi_2 = \chi_2$, $\chi_1 f = f$, and, by Lemma 6.39, $\mathcal{P} f = \mathcal{P} \chi_1 f = \chi_1 \mathcal{P} f$. Thus, the continuous function $\mathcal{P} f$ is supported in $\overline{B(\theta_0, 1)}$. Therefore, $\mathcal{P} f$ is uniformly continuous. Moreover, if $\epsilon > 0$, then there is an integer $N \geq 3$ such that

$$|P(\theta) x - P(\theta_0) x| = |(\mathcal{P} f)(\theta) - (\mathcal{P} f)(\theta_0)| \leq \epsilon$$

whenever $n > N$ and $\theta \in B(\theta_0, 1/n)$. In other words, $\theta \mapsto P(\theta) x$ is continuous at $\theta = \theta_0$. □

For each $\lambda \in \mathbb{R}$, define $\mathbb{T}_\lambda := \{z \in \mathbb{C} : |z| = e^\lambda\}$ and note that $\mathbb{T} = \mathbb{T}_0$. By rescaling, we obtain the following corollary from Theorem 6.38.

COROLLARY 6.40. *Suppose that $t > 0$. If $\sigma(E^t) \cap \mathbb{T}_\lambda = \emptyset$, then the Riesz projection $\mathcal{P}_\lambda$ that corresponds to the operator E^t and the spectral set $\sigma(E^t) \cap \{|z| < e^\lambda\}$ is of the form $(\mathcal{P}_\lambda f)(\theta) = P_\lambda(\theta) f(\theta)$ for some $P_\lambda \in C_b(\Theta; \mathcal{L}_s(X))$.*

6.3.2. The Dichotomy Theorem. In this subsection we will prove the Dichotomy Theorem for a continuous linear skew-product flow $\hat{\varphi}^t : \Theta \times X \to \Theta \times X$.

THEOREM 6.41 (Dichotomy Theorem). *Let $\{\Phi^t\}_{t \geq 0}$ be a strongly continuous, exponentially bounded cocycle over a continuous flow $\{\varphi^t\}_{t \geq 0}$ on a locally compact metric space Θ, and let*

$$(E^t f)(\theta) = \Phi^t(\varphi^{-t}\theta) f(\varphi^{-t}\theta), \quad t \geq 0,$$

be the corresponding evolution semigroup on $C_0(\Theta; X)$. The following statements are equivalent:

(1) *The linear skew-product flow $\hat{\varphi}^t : (\theta, x) \mapsto (\varphi^t \theta; \Phi^t(\theta)x)$ has exponential dichotomy on Θ.*
(2) *The semigroup $\{E^t\}_{t \geq 0}$ is hyperbolic.*

Moreover, the Riesz projection $\mathcal{P}$ that corresponds to the operator E^1 and the spectral set $\sigma(E^1) \cap \mathbb{D}$ is related to the dichotomy projection $P : \Theta \to \mathcal{L}_s(X)$ that satisfies Definition 6.13 by the formula $(\mathcal{P}f)(\theta) = P(\theta) f(\theta)$ where $\theta \in \Theta$ and $f \in C_0(\Theta; X)$.

PROOF. (1) $\Rightarrow$ (2). Assume that $\{\Phi^t\}_{t \geq 0}$ has exponential dichotomy with projections $P(\cdot)$ and define $\mathcal{P}$ on $C_0(\Theta; X)$ by $(\mathcal{P}f)(\theta) = P(\theta) f(\theta)$. Note that $\mathcal{P} E^t = E^t \mathcal{P}$ by condition (a) in Definition 6.13. The inequality $\|E_\mathcal{P}^t\|_{\mathcal{L}(C_0(\Theta;X))} \leq M e^{-\beta t}$ follows from the first inequality in condition (c) of the definition. Also, the operator E_Q^t on $\operatorname{Im} \mathcal{Q}$ is invertible with inverse $(R^t f)(\theta) = [\Phi_Q^t(\theta)]^{-1} f(\varphi^t \theta)$. Again, by (c), $\|[E_Q^t]^{-1}\|_{\mathcal{L}(C_0(\Theta;X))} \leq M e^{-\beta t}$ for $t \geq 0$. Hence, $\{E^t\}_{t \geq 0}$ is hyperbolic.

(2) $\Rightarrow$ (1). Assume that $\{E^t\}_{t \geq 0}$ is hyperbolic and let $\mathcal{P}$ be the Riesz projection for E^1. The Spectral Projection Theorem 6.38 shows that $(\mathcal{P}f)(\theta) = P(\theta) f(\theta)$ for some bounded, strongly continuous, projection valued function $P : \Theta \to \mathcal{L}_s(X)$. Since $\mathcal{P}$ is the Riesz projection for E^1, the operator E_Q^t is invertible on $\operatorname{Im} \mathcal{Q}$ and there exist constants $\beta > 0$ and $M > 0$ such that

(6.57) $$\|E_P^t\|_{\mathcal{L}(C_0(\Theta;X))} \leq M e^{-\beta t}, \quad t \geq 0,$$
(6.58) $$\|E_Q^t \mathcal{Q} f\|_\infty \geq M^{-1} e^{\beta t} \|\mathcal{Q} f\|_\infty, \quad t \geq 0.$$

Clearly, condition (a) in Definition 6.13 holds. To prove condition (c), consider $\theta \in \Theta$ and $x \in X$, and choose a function $f \in C_0(\Theta; X)$ such that $f(\theta) = x$ and $\|f\|_\infty = |x|$. Using the estimate (6.57), we obtain the inequality

$$|\Phi_P^t(\theta) x| \leq \|\Phi_P^t(\cdot) f(\cdot)\|_\infty = \|E_P^t f\|_\infty \leq M e^{-\beta t} \|f\|_\infty = M e^{-\beta t} |x|,$$

and therefore the first inequality in condition (c) holds.

To prove condition (b), fix $\theta_0 \in \Theta$ and $x \in \operatorname{Im} Q(\theta_0)$. For $n \in \mathbb{N}$, let $f_n(\theta) = \chi_n(\theta) Q(\theta) x$ where χ_n is defined in (6.56). Note that $f_n(\theta_0) = x$ and $f \in \operatorname{Im} \mathcal{Q}$. Using the estimate (6.58) we obtain the inequality

$$M^{-1} e^{\beta t} |x| = M^{-1} e^{\beta t} |f_n(\theta_0)| \leq M^{-1} e^{\beta t} \|\mathcal{Q} f_n\|_\infty \leq \|E_Q^t f_n\|_\infty$$
$$= \sup_{\theta \in \Theta} |\Phi_Q^t(\theta) f_n(\theta)| = \sup_{\theta \in \Theta} |\chi_n(\theta) \Phi_Q^t(\theta) x| \leq \sup_{\theta \in B(\theta_0, 1/n)} |\Phi_Q^t(\theta) x|.$$

Since the function $\theta \mapsto \Phi_Q^t(\theta)x$ is continuous at $\theta = \theta_0$, the last supremum approaches $|\Phi_Q^t(\theta_0)x|$ as $n \to \infty$. Thus, for all $\theta \in \Theta$, we have

(6.59) $$|\Phi_Q^t(\theta)x| \geq M^{-1}e^{\beta t}|x|.$$

In particular, the operator $\Phi_Q^t(\theta) : \operatorname{Im} Q(\theta) \to \operatorname{Im} Q(\varphi^t\theta)$ is injective. To see that it is surjective, choose $y \in \operatorname{Im} Q(\varphi^t\theta)$ and a continuous, compactly supported function $\alpha : \Theta \mapsto [0, 1]$ such that $\alpha(\varphi^t\theta) = 1$. Then, define $f := \alpha(\cdot)Q(\cdot)y$ and note that $f \in \operatorname{Im} \mathcal{Q}$. Since E_Q^t is invertible on $\operatorname{Im} \mathcal{Q}$, there exists $g \in \operatorname{Im} \mathcal{Q}$ such that $E^t \mathcal{Q} g = f$. Hence, for $x := g(\theta) \in \operatorname{Im} Q(\theta)$, this gives the equality

$$\Phi_Q^t(\theta)x = \Phi_Q^t(\theta)g(\theta) = f(\varphi^t\theta) = \alpha(\varphi^t\theta)Q(\varphi^t\theta)y = y.$$

Thus, condition (b) in Definition 6.13 holds. Combining this fact with the inequality (6.59), the second inequality in condition (c) of this definition is verified. □

COROLLARY 6.42. *The dichotomy projection is uniquely determined by the conditions (a)–(c) of Definition 6.13.*

Theorem 6.41 combined with the argument used in part (1) ⇒ (2) of its proof can be used to prove the following corollary.

COROLLARY 6.43. *The existence of an exponential dichotomy for the cocycle $\{\Phi^t\}_{t\geq 0}$ with continuous time $t \in \mathbb{R}_+$ is equivalent to the existence of an exponential dichotomy for the cocycle $\{\Phi^n\}_{n\in\mathbb{N}}$ with discrete time $n \in \mathbb{N}$.*

Using the results of this section, it is easy to derive the robustness of exponential dichotomy.

COROLLARY 6.44. *Let $\{\Phi^t\}_{t\geq 0}$ be an exponentially bounded, strongly continuous cocycle and $\{E^t\}_{t\geq 0}$ the corresponding evolution semigroup. If $\{\Phi^t\}_{t\geq 0}$ has an exponential dichotomy and $\{\Psi^t\}_{t\geq 0}$ is an exponentially bounded, strongly continuous cocycle such that*

$$\sup_{\theta \in \Theta} \|\Phi^t(\theta) - \Psi^t(\theta)\|_{\mathcal{L}(X)} < \inf_{\lambda \in \mathbb{T}} \|(\lambda - E^t)^{-1}\|^{-1}$$

for some $t > 0$, then $\{\Psi^t\}_{t\geq 0}$ has an exponential dichotomy.

PROOF. Let $\{F^t\}_{t\geq 0}$ denote the evolution semigroup induced by $\{\Psi^t\}_{t\geq 0}$ and note that

$$\|E^t - F^t\|_{\mathcal{L}(C_0(\Theta;X))} = \sup_{\theta \in \Theta} \|\Phi^t(\theta) - \Psi^t(\theta)\|_{\mathcal{L}(X)} < \inf_{\lambda \in \mathbb{T}} \|(\lambda - E^t)^{-1}\|^{-1}.$$

For $\lambda \in \mathbb{T}$, use the identity

$$\lambda - F^t = \left(I - (F^t - E^t)(\lambda - E^t)^{-1}\right)(\lambda - E^t)$$

and the last inequality to show that the operator $\lambda - F^t$ is invertible. It follows that $\{F^t\}_{t\geq 0}$ is hyperbolic. Hence, $\{\Psi^t\}_{t\geq 0}$ has an exponential dichotomy. □

6.3.3. Spectral theory. We derive several simple corollaries of Theorem 6.41 to give an account of the Sacker-Sell spectral theory for a strongly continuous, exponentially bounded cocycle $\{\Phi^t\}_{t\geq 0}$ on an infinite dimensional Banach space X. Recall Definition 6.17 and the notation Σ for the Sacker-Sell spectrum of the linear skew-product flow $\{\hat{\varphi}^t\}_{t\geq 0}$.

COROLLARY 6.45. *If $\{\hat{\varphi}^t\}_{t\geq 0}$ is a continuous linear skew-product flow over a continuous flow $\{\varphi^t\}_{t\in\mathbb{R}}$, then*

(6.60) $$\Sigma = \ln|\sigma(E^1) \setminus \{0\}|.$$

If, in addition, $\{\varphi^t\}_{t\in\mathbb{R}}$ is aperiodic, then $\Sigma = \sigma(\Gamma) \cap \mathbb{R}$.

PROOF. By definition, $\lambda \notin \Sigma$ if and only if the rescaled cocycle $\{e^{-\lambda t}\Phi^t\}_{t\geq 0}$ has an exponential dichotomy. In view of Theorem 6.41, this condition is equivalent to the statement $\sigma(e^{-\lambda}E^1) \cap \mathbb{T} = \emptyset$, or $|z| \neq e^\lambda$ for all $z \in \sigma(E^1)$. The second statement of the theorem follows from the Spectral Mapping Theorem 6.30. □

By Corollary 6.45, Σ is a closed set. Moreover, Σ is bounded provided that each operator $\Phi^t(\theta)$ is invertible.

Define the "main" Lyapunov exponent $\omega(\hat{\varphi}^t)$ for the linear skew-product flow $\{\hat{\varphi}^t\}_{t\geq 0}$ as follows:

$$\omega(\hat{\varphi}^t) = \inf\{\omega \in \mathbb{R} : \text{ there exists } M = M(\omega) > 0 \text{ such that}$$
$$\sup_{\theta \in \Theta}\|\Phi^t(\theta)\|_{\mathcal{L}(X)} \leq Me^{\omega t} \text{ for all } t \geq 0\}.$$

By equation (6.60), $\omega(\hat{\varphi}^t)$ coincides with the growth bound $\omega(\Gamma)$ for the semigroup $\{E^t\}_{t\geq 0}$. Let us also define the *general* Lyapunov exponent $\omega_g(\hat{\varphi}^t)$ by

$$\omega_g(\hat{\varphi}^t) = \inf\{\omega \in \mathbb{R} : \text{ there exists } M = M(\omega) > 0 \text{ such that,}$$
(6.61) $\quad$ for all $\theta \in \Theta$ and $x \in X$, $|\Phi^t(\theta)x| \leq Me^{\omega(t-\tau)}|\Phi^\tau(\theta)x|$ for $t \geq \tau\}.$

To prove the equality $\omega(\hat{\varphi}^t) = \omega_g(\hat{\varphi}^t)$, let us note that if $\omega_1 > \omega(\hat{\varphi}^t)$, then there is some $M = M(\omega_1)$ such that, for all $t \geq \tau$ and all $(\theta, x) \in \Theta \times X$, the following inequality holds:

$$|\Phi^t(\theta)x| = |\Phi^{t-\tau+\tau}(\theta)x| = |\Phi^{t-\tau}(\varphi^\tau\theta)\Phi^\tau(\theta)x|$$
$$\leq \sup_\theta \|\Phi^{t-\tau}(\varphi^\tau\theta)\|_{\mathcal{L}(X)}|\Phi^\tau(\theta)x| \leq Me^{\omega_1(t-\tau)}|\Phi^\tau(\theta)x|.$$

Thus, $\omega_1 \geq \omega_g(\hat{\varphi}^t)$ and $\omega(\hat{\varphi}^t) \geq \omega_g(\hat{\varphi}^t)$. On the other hand, if $\omega_2 > \omega_g(\hat{\varphi}^t)$ and we take $\tau = 0$ in (6.61), then $\omega_2 \geq \omega(\hat{\varphi}^t)$; and therefore, $\omega_g(\hat{\varphi}^t) \geq \omega(\hat{\varphi}^t)$. Now, using the general formula, $\omega(\Gamma) = t^{-1}\log r(E^t)$ where $r(\cdot)$ is the spectral radius (see (2.6)), we conclude that

(6.62) $$\omega(\hat{\varphi}^t) = \omega_g(\hat{\varphi}^t) = t_0^{-1}\log r(E^{t_0}) = \log r(E)$$

for all $t_0 > 0$.

Let us use the Riesz projections for E^1 to construct the $\hat{\varphi}^t$-invariant spectral subbundles in $\Theta \times X$ that correspond to open-closed components in Σ. For this, fix $\lambda < \lambda'$, two numbers that belong to different connected components of $\mathbb{R} \setminus \Sigma$, and consider the following Riesz projections for E on $C_0(\Theta; X)$:

$$\mathcal{P}_\lambda = \frac{1}{2\pi i}\int_{\mathbb{T}_\lambda} R(z; E)\,dz, \quad \mathcal{P}_\lambda^{\lambda'} = \mathcal{P}_{\lambda'} - \mathcal{P}_\lambda, \quad \mathcal{P}^{\lambda'} = I - \mathcal{P}_{\lambda'}$$

where $\mathbb{T}_\lambda := \{z \in \mathbb{C} : |z| = e^\lambda\}$. These projections correspond to the subsets of $\sigma(E)$ in $\{z \in \mathbb{C} : |z| < e^\lambda\}$, $\{z \in \mathbb{C} : e^\lambda < |z| < e^{\lambda'}\}$, and $\{z \in \mathbb{C} : |z| > e^{\lambda'}\}$, respectively. It is important to note that these projections do not depend on the choice of λ and λ' in the corresponding connected components of $\mathbb{R} \setminus \Sigma$. Also, we have that $\mathcal{P}_\lambda \mathcal{P}_{\lambda'} = \mathcal{P}_{\lambda'} \mathcal{P}_\lambda$. Note that $\operatorname{Im}\mathcal{P}_{\lambda'} \cap \operatorname{Im}\mathcal{P}^\lambda$ is a direct complement of $\operatorname{Im}\mathcal{P}_\lambda$ to $\operatorname{Im}\mathcal{P}_{\lambda'}$, and $\mathcal{P}_\lambda^{\lambda'}$ projects $C_0(\Theta; X)$ on $\operatorname{Im}\mathcal{P}_{\lambda'} \cap \operatorname{Im}\mathcal{P}^\lambda$ parallel to $\operatorname{Im}\mathcal{P}_\lambda \dotplus \operatorname{Im}\mathcal{P}^{\lambda'}$.

By the Spectral Projection Theorem 6.38 (see Corollary 6.40), the projections $\mathcal{P}_\lambda$, $\mathcal{P}_\lambda^{\lambda'}$, $\mathcal{P}^{\lambda'}$ are multiplication operators corresponding to the projection-valued functions P_λ, $P_\lambda^{\lambda'}$, $P^{\lambda'} \in C_b(\Theta; \mathcal{L}_s(X))$. Let

(6.63) $\quad \mathbb{S}_\theta^\lambda := \operatorname{Im} P_\lambda(\theta), \quad \mathbb{E}_\theta^{\lambda\lambda'} := \operatorname{Im} P_\lambda^{\lambda'}(\theta), \quad \mathbb{U}_\theta^{\lambda'} := \operatorname{Im} P^{\lambda'}(\theta), \quad \theta \in \Theta,$

and note that we have obtained a $\hat{\varphi}^t$-invariant direct sum decomposition

(6.64) $\qquad\qquad \Theta \times X = \mathbb{S}^\lambda \dotplus \mathbb{E}^{\lambda\lambda'} \dotplus \mathbb{U}^{\lambda'}, \quad X = \mathbb{S}_\theta^\lambda \dotplus \mathbb{E}_\theta^{\lambda\lambda'} \dotplus \mathbb{U}_\theta^{\lambda'}$

that is continuous with respect to $\theta \in \Theta$ and such that the direct summands, called *spectral subbundles*, have the following properties:

$$\Sigma(\hat{\varphi}^t|\mathbb{S}^\lambda) = \Sigma \cap [-\infty, \lambda), \quad \Sigma(\hat{\varphi}^t|\mathbb{E}^{\lambda\lambda'}) = \Sigma \cap (\lambda, \lambda'), \quad \Sigma(\hat{\varphi}^t|\mathbb{U}^{\lambda'}) = \Sigma \cap (\lambda', \infty).$$

REMARK 6.46. Suppose that Θ is connected. Since the functions $P_\lambda(\cdot)$, $P_\lambda^{\lambda'}(\cdot)$, and $P^{\lambda'}(\cdot)$ are continuous, the dimensions of $\mathbb{S}_\theta^\lambda$, $\mathbb{E}_\theta^{\lambda\lambda'}$, and $\mathbb{U}_\theta^{\lambda'}$ do not depend on $\theta \in \Theta$. Note that $\mathbb{E}_\theta^{\lambda\lambda'} = \mathbb{S}_\theta^{\lambda'} \cap \mathbb{U}_\theta^\lambda$. Moreover, since λ and λ' belong to different connected components of $\mathbb{R} \setminus \Sigma$, we have $\dim \mathbb{E}_\theta^{\lambda\lambda'} \geq 1$. As a result, the number of maximal segments $[r^-, r^+]$ with $r^- \leq r^+$ whose union is Σ, as well as the number of corresponding spectral subbundles, does not exceed $\dim X$. In particular, if $\dim X < \infty$ and the function $\theta \mapsto \Phi^t(\theta) \in \mathcal{L}(X)$ is continuous, then each operator $\Phi^t(\theta)$ is invertible and there is some $N \leq \dim X$ such that $\Sigma = \cup_{k=1}^N [r_k^-, r_k^+]$. If numbers $\lambda_0, \lambda_1, \ldots, \lambda_N$ are chosen such that

$$\lambda_0 > r_1^+ \geq r_1^- > \lambda_1 > \ldots > r_N^+ \geq r_N^- > \lambda_N > -\infty,$$

then

$$X = \sum_{k=1}^N \mathbb{E}_\theta^{\lambda_k \lambda_{k-1}}, \quad \theta \in \Theta.$$

$\diamondsuit$

Going back to the infinite dimensional situation, we will describe the spectral subbundles in display (6.64) and the spectral subspaces of E in terms of Lyapunov numbers. Here, we will give the definitions for the case of discrete time $n \in \mathbb{Z}_+$.

DEFINITION 6.47. For $\theta \in \Theta$ and $x \in X \setminus \{0\}$, define

$$\lambda_s^+(\theta, x) = \limsup_{n \to \infty} n^{-1} \log |\Phi^n(\theta)x|, \quad \lambda_i^+(\theta, x) = \liminf_{n \to \infty} n^{-1} \log |\Phi^n(\theta)x|.$$

Also, if there exists a sequence $\{x_n\}_{n=0}^\infty \subset X$ such that

(6.65) $\qquad\qquad x = x_0, \quad \text{and} \quad x_{n-1} = \Phi^1(\varphi^{-n}\theta) x_n \quad \text{for } n \geq 1,$

then define

(6.66) $\quad \lambda_i^-(\theta, x) = -\limsup_{n \to \infty} n^{-1} \log |x_n|, \quad \lambda_s^-(\theta, x) = -\liminf_{n \to \infty} n^{-1} \log |x_n|.$

If there is no sequence $\{x_n\}_{n=0}^\infty$ as in display (6.65), then set $\lambda_s^-(\theta, x) := \lambda_i^-(\theta, x) = -\infty$. The numbers $\lambda_s^\pm(\theta, x)$ and $\lambda_i^\pm(\theta, x)$ are called *Lyapunov numbers*.

We stress the fact that $\lambda_s^-(\theta, x)$ and $\lambda_i^-(\theta, x)$, as defined in display (6.66), may depend on the choice of the sequence $\{x_n\}_{n=0}^\infty$. Thus, when this dependence is important, we will write $\lambda_{s,i}^-(\theta, x) = \lambda_{i,s}^-(\theta, \{x_n\}_{n=0}^\infty)$. If each operator $\Phi^n(\theta)$ in the cocycle $\{\Phi^n\}_{n\in\mathbb{Z}}$ is invertible, then $x_n = [\Phi^n(\varphi^{-n}\theta)]^{-1}x = \Phi^{-n}(\theta)x$ for each $n \in \mathbb{N}$. In this case, we have the equalities

$$\lambda_i^-(\theta,x) = -\limsup_{n\to\infty} n^{-1}\log|\Phi^{-n}(\theta)x| = \liminf_{n\to\infty}(-n)^{-1}\log|\Phi^{-n}(\theta)x|$$
$$= \liminf_{k\to-\infty} k^{-1}\log|\Phi^k(\theta)x|$$

and

$$\lambda_s^-(\theta,x) = -\liminf_{n\to\infty} n^{-1}\log|\Phi^{-n}(\theta)x| = \limsup_{n\to\infty}(-n)^{-1}\log|\Phi^{-n}(\theta)x|$$
$$= \limsup_{k\to-\infty} k^{-1}\log|\Phi^k(\theta)x|.$$

It follows that $\lambda_s^+(\theta,x) \geq \lambda_i^+(\theta,x)$ and $\lambda_s^-(\theta,x) \geq \lambda_i^-(\theta,x)$.

THEOREM 6.48. *If $\lambda \in \mathbb{R}\setminus\Sigma$ and $x \neq 0$, then*
  (i) $x \in \mathbb{S}_\theta^\lambda$ *implies* $\lambda_s^+(\theta,x), \lambda_s^-(\{x_n\}_{n=0}^\infty) \in \Sigma \cap [-\infty,\lambda)$ *for each* $\{x_n\}_{n=0}^\infty$ *as in (6.65);*
  (ii) $\lambda_i^+(\theta,x) < \lambda$ *implies* $x \in \mathbb{S}_\theta^\lambda$;
  (iii) $x \in \mathbb{U}_\theta^\lambda$ *implies* $\lambda_i^+(\theta,x), \lambda_i^-(\theta,\{x_n\}_{n=0}^\infty) \in \Sigma \cap (\lambda,\infty)$ *for some* $\{x_n\}_{n=0}^\infty$ *as in (6.65); and*
  (iv) $\lambda_s^-(\theta,\{x_n\}_{n=0}^\infty) > \lambda$ *for some* $\{x_n\}_{n=0}^\infty$ *as in (6.65) implies* $x \in \mathbb{U}_\theta^\lambda$.

PROOF. For $\lambda \in \mathbb{R}\setminus\Sigma$, the linear skew-product flow

$$\hat{\varphi}_\lambda^t : \Theta \times X \to \Theta \times X : (\theta,x) \mapsto (\varphi^t\theta, e^{-\lambda t}\Phi^t(\theta)x)$$

has an exponential dichotomy. Thus, for the decomposition $X = \mathbb{S}_\theta^\lambda \dotplus \mathbb{U}_\theta^\lambda$, there are numbers $\beta > 0$ and $M > 0$ such that

(6.67) $\qquad |e^{-\lambda t}\Phi^t(\theta)x| \leq Me^{-\beta t}|x|, \quad x \in \mathbb{S}_\theta^\lambda,$

(6.68) $\qquad |e^{-\lambda t}\Phi^t(\theta)x| \geq M^{-1}e^{\beta t}|x|, \quad x \in \mathbb{U}_\theta^\lambda, \quad t > 0, \quad \theta \in \Theta.$

To prove statement (i), fix $x \in \mathbb{S}_\theta^\lambda$. By inequality (6.67), we have the inequality $\lambda_s^+(\theta,x) < \lambda$. To see that $\lambda_s^-(\theta,x) < \lambda$, let $\{x_n\}_{n=0}^\infty$ be a sequence as in display (6.65); that is, $x = x_0$ and $\Phi^n(\varphi^{-n}\theta)x_n = x_0$. We will show that $x_n \in \mathbb{S}_{\varphi^{-n}\theta}^\lambda$ for each positive integer n. To see this, decompose x_n as $x_n = x_n^s + x_n^u$ with $x_n^s \in \mathbb{S}_{\varphi^{-n}\theta}^\lambda$ and $x_n^u \in \mathbb{U}_{\varphi^{-n}\theta}^\lambda$. Since $\mathbb{S}^\lambda$ and $\mathbb{U}^\lambda$ are $\hat{\varphi}^t$-invariant, and $\text{Ker}(\Phi^n(\theta)|\mathbb{U}_\theta^\lambda) = \{0\}$ by (b) in the Definition 6.13 of exponential dichotomy, if $x_n^u \neq 0$, then $x_0^u = \Phi^n(\theta)x_n^u \neq 0$, in contradiction to the fact that $x = x_0 \in \mathbb{S}_\theta^\lambda$. Moreover, since $x_n \in \mathbb{S}_{\varphi^{-n}\theta}^\lambda$, it follows from inequality (6.67) that

$$e^{-\lambda n}|x| = |e^{-\lambda n}\Phi^n(\varphi^{-n}\theta)x_n| \leq Me^{-\beta n}|x_n|;$$

and therefore, $\liminf_{n\to\infty} n^{-1}\log|x_n| \geq -\lambda + \beta$. Thus we have $\lambda_s^-(\theta,x) < \lambda$ as required in statement (i).

To prove statement (ii), decompose x as $x = x^s + x^u$ with $x^s \in \mathbb{S}_\theta^\lambda$ and $x^u \in \mathbb{U}_\theta^\lambda$. Using inequalities (6.67) and (6.68), note that

$$|x^u| \leq Me^{-\beta n}|e^{-\lambda n}\Phi^n(\theta)x^u| \leq Me^{-\beta n}\{|e^{-\lambda n}\Phi^n(\theta)x| + |e^{-\lambda n}\Phi^n(\theta)x^s|\}$$
$$\leq Me^{-\beta n}\{|e^{-\lambda n}\Phi^n(\theta)x| + Me^{-\beta n}|x^s|\}.$$

Select a sequence $\{n_k\}_{k=1}^\infty \subset \mathbb{N}$ such that $\lambda_i^+(\theta,x) = \lim_{k\to\infty} n_k^{-1}\log|\Phi^{n_k}(\theta)x|$. Also, note that if $\epsilon > 0$, then there is some $K > 0$ such that

$$e^{-\lambda n_k}|\Phi^{n_k}(\theta)x| \leq e^{n_k(\lambda_i^+ + \epsilon)}e^{-\lambda n_k}$$

whenever $k \geq K$. Thus, if we take $\epsilon < \beta + \lambda - \lambda_i^+$, then

$$|x^u| \leq \lim_{k \to \infty} Me^{-\beta n_k}\{e^{n_k(\lambda_i^+ + \epsilon - \lambda)} + Me^{-\beta n_k}|x^s|\} = 0,$$

and $x = x^s \in \mathbb{S}_\theta^\lambda$.

To prove statement (iii), fix $x \in \mathbb{U}_\theta^\lambda$, and note that $\lambda_i^+(\theta, x) > \lambda$ by inequality (6.68). Since the restriction $\Phi^n(\varphi^{-n}\theta)|\mathbb{U}_{\varphi^{-n}\theta}^\lambda$ is an invertible operator from $\mathbb{U}_{\varphi^{-n}\theta}^\lambda$ to $\mathbb{U}_\theta^\lambda$ by condition (b) in the Definition 6.13, there exists a sequence given by $x_n := [\Phi^n(\varphi^{-n}\theta)|\mathbb{U}_{\varphi^{-n}\theta}^\lambda]^{-1}x$ as in display (6.65). Note that $x_n \in \mathbb{U}_{\varphi^{-n}\theta}^\lambda$ and $x = \Phi^n(\varphi^{-n}\theta)x_n$. Then use (6.68) to conclude that

$$|x_n| \leq Me^{-\beta n}|e^{-\lambda n}\Phi^n(\varphi^{-n}\theta)x_n| = Me^{-\beta n}e^{-\lambda n}|x|.$$

This proves $\lambda_i^-(\theta, x) > \lambda$, as required.

To prove statement (iv), fix a sequence $\{x_n\}_{n=0}^\infty$ as in display (6.65) such that $\Phi^n(\varphi^{-n}\theta)x_n = x$ and $\lambda_s^-(\theta, \{x_n\}_{n=0}^\infty) > \lambda$, and decompose x_n as $x_n = x_n^s + x_n^u$. By the $\hat{\varphi}^t$-invariance, $x = \Phi^n(\varphi^{-n}\theta)x_n^s + \Phi^n(\varphi^{-n}\theta)x_n^u$, and using inequalities (6.67) and (6.68), it follows that

$$e^{-\lambda n}|x^s| = e^{-\lambda n}|\Phi^n(\varphi^{-n}\theta)x_n^s| \leq Me^{-\beta n}|x_n^s| \leq Me^{-\beta n}\{|x_n| + |x_n^u|\}$$
$$\leq Me^{(-\lambda-\beta)n}\{e^{\lambda n}|x_n| + Me^{-\beta n}|x^u|\}.$$

Now choose a sequence $\{n_k\}_{k=1}^\infty \subset \mathbb{N}$ such that $\lambda_s^-(\theta, x) = -\lim_{k \to \infty} n_k^{-1}\log|x_{n_k}|$. If $\epsilon > 0$ is sufficiently small and k is sufficiently large, then

$$|x^s| \leq Me^{-\beta n_k}\left\{e^{(\lambda - \lambda_s^- + \epsilon)n_k} + Me^{-\beta n_k}|x^u|\right\}.$$

Thus, $x \in \mathbb{U}_\theta^\lambda$, as required.

To finish the proof of the inclusions $\lambda_s^\pm(\theta, x), \lambda_i^\pm(\theta, x) \in \Sigma$, let us check that there is no point $\lambda \in \mathbb{R} \setminus \Sigma$ between $\lambda_i^+(\theta, x) \leq \lambda_s^+(\theta, x)$. Suppose that $\lambda_i^+(\theta, x) < \lambda < \lambda_s^+(\theta, x)$ for some $\lambda \in \mathbb{R} \setminus \Sigma$. Because $\lambda_i^+(\theta, x) < \lambda$, it follows that $x \in \mathbb{S}_\theta^\lambda$ by statement (ii), and, in turn, this implies $\lambda_s^+(\theta, x) < \lambda$ by statement (i). But $\lambda_i^+(\theta, x) = \lambda_s^+(\theta, x) \in \mathbb{R} \setminus \Sigma$ is also impossible. Indeed, suppose that $\lambda_i^+(\theta, x) = \lambda_s^+(\theta, x) \in \mathbb{R} \setminus \Sigma$ and choose $\lambda' < \lambda_i^+(\theta, x) = \lambda_s^+(\theta, x) < \lambda''$ where λ' and λ'' belong to the same open component of $\mathbb{R} \setminus \Sigma$ as $\lambda_i^+(\theta, x)$ and $\lambda_s^+(\theta, x)$. We have $\mathbb{S}_\theta^{\lambda'} = \mathbb{S}_\theta^{\lambda''}$ and $\mathbb{U}_\theta^{\lambda'} = \mathbb{U}_\theta^{\lambda''}$ because the spectral projections for the operator E do not depend upon the choice of the circle $\mathbb{T}_\lambda$ for λ within an open component of $\mathbb{R} \setminus \Sigma$. But, by statement (ii), $\lambda_i^+(\theta, x) < \lambda''$ implies $x \in \mathbb{S}_\theta^{\lambda''} = \mathbb{S}_\theta^{\lambda'}$, which by statement (i), implies $\lambda_s^+(\theta, x) < \lambda'$, a contradiction. Similarly, $\lambda_i^-(\theta, x), \lambda_s^-(\theta, x) \in \Sigma$. □

COROLLARY 6.49. *If $\theta \in \Theta$, then the subspace $\mathbb{E}_\theta^{\lambda\lambda'}$ is the closed linear span of the nonzero vectors x such that $\lambda_i^\pm(\theta, x)$ and $\lambda_s^\pm(\theta, x)$ are in $\Sigma \cap (\lambda, \lambda')$.*

PROOF. The corollary follows from Theorem 6.48 and the formula $\mathbb{E}_\theta^{\lambda\lambda'} = \mathbb{S}_\theta^{\lambda'} \cap \mathbb{U}_\theta^\lambda$. □

By the corollary, each component $[r^-, r^+]$ of the set Σ has a corresponding spectral subbundle with the fibers

$$\mathbb{E}_\theta^{\lambda\lambda'} = \bigcap_{\lambda, \lambda'}(\mathbb{S}_\theta^{\lambda'} \cap \mathbb{U}_\theta^\lambda), \quad \lambda < r^- \leq r^+ < \lambda', \quad \lambda, \lambda' \in \mathbb{R} \setminus \Sigma,$$

that consist of the closed linear span of the nonzero vectors x such that $\lambda_i^\pm(\theta,x)$, $\lambda_s^\pm(\theta,x) \in [r^-, r^+]$.

The Sacker-Sell spectrum Σ is a union of, generally, infinitely many spectral segments $[r^-, r^+]$, corresponding to the spectral subbundles $\mathbb{S}^\lambda$, $\mathbb{E}^{\lambda\lambda'}$, and $\mathbb{U}^{\lambda'}$ as in display (6.64). Consider $\mathbb{E}^{\lambda\lambda'}$ and define its *general upper Lyapunov exponent* $\omega_g^+ = \omega_g^+(P_\lambda^{\lambda'})$ as follows:

$$\omega_g^+ = \inf\{\omega \in \mathbb{R} : \exists M = M(\omega) \text{ such that } |\Phi^n(\theta)x| \le Me^{\omega(n-k)}|\Phi^k(\theta)x|$$
$$\text{for all } \theta \in \Theta,\ x \in \mathbb{E}_\theta^{\lambda\lambda'},\ k, n \in \mathbb{N},\ 0 \le k \le n\}.$$

The *general lower Lyapunov exponent* $\omega^- = \omega^-(P_\lambda^{\lambda'})$ is defined as

$$\omega_g^- = \sup\{\omega \in \mathbb{R} : \exists M = M(\omega) \text{ such that } |\Phi^n(\theta)x| \ge Me^{\omega(n-k)}|\Phi^k(\theta)x|$$
$$\text{for all } \theta \in \Theta,\ x \in \mathbb{E}_\theta^{\lambda\lambda'},\ k, n \in \mathbb{N},\ 0 \le k \le n\}.$$

Clearly, $\omega_g^\pm = r^\pm$. Also, for the spectral radius $r(\cdot)$ of the restriction $E|\operatorname{Im}\mathcal{P}_\lambda^{\lambda'}$ as in (6.62) we have

$$r_+ = \omega_g^+ = \log r(E|\operatorname{Im}\mathcal{P}_\lambda^{\lambda'}), \quad r_- = \omega_g^- = -\log r([E|\operatorname{Im}\mathcal{P}_\lambda^{\lambda'}]^{-1}).$$

In other words, the spectrum of the Lyapunov numbers $\lambda_s^\pm$ and $\lambda_i^\pm$ is split by the segments $[r_-, r_+]$ determined by the general Lyapunov exponents or the spectral radii of the restrictions of E on its spectral subspaces corresponding to annular components of its spectrum.

Our point of view, that the spectral subbundles are "fiberwise" images of the Riesz projections for E, while Σ is given by $\log|\sigma(E) \setminus \{0\}|$, is convenient for answering many questions. We will illustrate this for a question concerning the normal and tangential spectrum.

Let F be an $\{\hat\varphi^t\}$-invariant subbundle in $\Theta \times X$; that is, a bundle with the base Θ and fibers $F_\theta = \operatorname{Im} R(\theta)$ where $R : \Theta \to \mathcal{L}_s(X)$ is a bounded, continuous, projection-valued function such that $\Phi^t(\theta)R(\theta) = R(\varphi^t\theta)\Phi^t(\theta)$ for $t \ge 0$. Consider F', a direct complement to F relative to $\Theta \times X$. In other words, $F'_\theta = \operatorname{Im} R'(\theta)$ where $R' : \Theta \to \mathcal{L}_s(X)$ is some bounded, continuous, projection-valued function such that $\operatorname{Ker} R'(\theta) = \operatorname{Im} R(\theta)$. We stress the point that F' is not assumed to be $\hat\varphi^t$-invariant.

Define $\hat\varphi_N^t$ to be the restriction of $\hat\varphi^t$ on F'. Locally, this skew-product flow is given by

$$\hat\varphi_N^t : (\theta, x) \mapsto (\varphi^t\theta, R'(\varphi^t\theta)\Phi^t(\theta)x), \quad \theta \in \Theta, \quad x \in \operatorname{Im} R'(\theta).$$

Also, define the tangential and normal spectra to be, respectively,

$$\Sigma_T := \Sigma(\hat\varphi^t|F), \qquad \Sigma_N := \Sigma(\hat\varphi_N^t).$$

To clarify this terminology, consider, for the moment, a smooth flow $\{\varphi^t\}_{t \in \mathbb{R}}$ on a smooth manifold Θ and let $\Phi^t(\theta) = D\varphi^t(\theta)$ denote the differential of φ^t. Let $F = F_T$ be the subbundle of the tangent bundle $\mathcal{T}\Theta$ that consists of the vectors tangent to the trajectories of the flow $\{\varphi^t\}_{t \in \mathbb{R}}$. Clearly, F is invariant under $\hat\varphi^t$. Let $F' = F_N$ be the subbundle in $\mathcal{T}\Theta$ of the vectors that are normal to the trajectories. Then Σ_T and Σ_N are the tangential and normal spectra.

Returning to the general situation, we have the following result.

THEOREM 6.50. *(a) The normal spectrum Σ_N does not depend on the choice of the direct complement F'.*

(b) $\Sigma \subset \Sigma_N \cup \Sigma_T$.

(c) $\Sigma = \Sigma_N \cup \Sigma_T$ provided that either $\Sigma_N \cap \Sigma_T = \emptyset$ or F has a $\hat{\varphi}^t$-invariant direct complement.

PROOF. Since F is $\hat{\varphi}^t$-invariant, the matrix of the operator E in the decomposition $C_0(\Theta; X) = \operatorname{Im}\mathcal{R} + \operatorname{Im}(I - \mathcal{R})$ is upper triangular where, as usual, $(\mathcal{R}f)(\theta) = R(\theta)f(\theta)$. By formula (6.60) in Corollary 6.45, we have that
$$\Sigma = \log|\sigma(E) \setminus \{0\}|, \quad \Sigma_T = \log|\sigma(E|\operatorname{Im}\mathcal{R}) \setminus \{0\}|, \quad \Sigma_N = \log|\sigma(\mathcal{R}'E\mathcal{R}') \setminus \{0\}|.$$
Also, the equality $\operatorname{Ker}\mathcal{R}' = \operatorname{Im}\mathcal{R}$ implies that $\mathcal{R}'(I - \mathcal{R}) = \mathcal{R}'$ and $(I - \mathcal{R})\mathcal{R}' = I - \mathcal{R}$. Thus, the operators
$$\mathcal{R}'(I - \mathcal{R}) : \operatorname{Im}(I - \mathcal{R}) \to \operatorname{Im}\mathcal{R}' \quad \text{and} \quad (I - \mathcal{R})\mathcal{R}' : \operatorname{Im}\mathcal{R}' \to \operatorname{Im}(I - \mathcal{R})$$
are mutually inverse operators. Therefore, by the identity
$$(I - \mathcal{R})\mathcal{R}'(\mathcal{R}'E\mathcal{R}')\mathcal{R}'(I - \mathcal{R}) = (I - \mathcal{R})E(I - \mathcal{R})$$
we have that $\sigma(\mathcal{R}'E\mathcal{R}') = \sigma((I-\mathcal{R})E(I-\mathcal{R}))$. This gives statement (a). Statements (b) and (c) follow immediately from the triangular representation of E. □

We conclude this subsection with a remark that relates the subbundles $\mathbb{S}^\lambda$, $\mathbb{E}^{\lambda\lambda'}$, and $\mathbb{U}^{\lambda'}$ defined in display (6.2) by means of the spectral projection for the evolution operator E and subbundles $\mathcal{S}^\lambda$, $\mathcal{U}^{\lambda'}$ defined in the classical Sacker-Sell theory (see for example Sacker and Sell [**SS3**] and also [**SS5**] as well as Chow and Leiva [**CLe2, CLe3**]). Assume that operators $\Phi^t(\theta)$ in the cocycle $\{\Phi^t\}_{t\in\mathbb{R}}$ are invertible. For $\lambda < \lambda'$, two numbers belonging to different components of $\mathbb{R} \setminus \Sigma$, let $\mathcal{E}^{\lambda\lambda'} = \mathcal{S}^{\lambda'} \cap \mathcal{U}^\lambda$ where $\mathcal{S}^{\lambda'}$ and $\mathcal{U}^\lambda$ are the bundles with the fibers
$$\mathcal{S}_\theta^{\lambda'} = \{x : |e^{-\lambda' n}\Phi^n(\theta)x| \to 0 \text{ as } n \to \infty\},$$
$$\mathcal{U}_\theta^\lambda = \{x : |e^{\lambda n}\Phi^{-n}(\theta)x| \to 0 \text{ as } n \to \infty\}.$$
We claim that $\mathbb{E}_\theta^{\lambda\lambda'} = \mathcal{S}_\theta^{\lambda'} \cap \mathcal{U}_\theta^\lambda$. Indeed, to show the equality $\mathcal{S}_\theta^{\lambda'} = \mathbb{S}_\theta^{\lambda'}$, note that $0 \neq x \in \mathcal{S}_\theta^{\lambda'}$ is equivalent to the inequality $\lambda_s^+(\theta, x) < \lambda'$. By Theorem 6.48, it follows that $x \in \mathbb{S}_\theta^{\lambda'}$. Similarly, $\mathcal{U}_\theta^\lambda = \mathbb{U}_\theta^\lambda$.

6.4. Bibliography and remarks

Linear skew-product flows and dichotomy. General references on skew-product flows are Arnold [**A**], Bronshtein [**Br**], Bronshtein and Kopanskii [**BKo**], Hale [**Ha2**], Hale and Verduyn Lunel [**HL**], Hirsch, Pugh, Shub [**HPS**], Johnson, Palmer, and Sell [**JPS**], Katok and Hasselblatt [**KH**], and Sacker and Sell [**SS5**]. This topic is very broad and ranges from random dynamical systems to smooth maps on manifolds, see the bibliography compiled by Sell [**Se2**]. Our exposition is related to the classical papers by Sacker and Sell [**Sa, Sa2, SS, SS2, SS3, SS4**], Selgrade [**Sg**], and Sell [**Se**]. There is an excellent review of the work done from this viewpoint in Johnson, Palmer, and Sell [**JPS**], and infinite dimensional generalizations in [**SS5**]. Magalhães [**Mg**], and Chow and Leiva [**CLe, CLe2, CLe3**] have recently obtained additional results for the infinite dimensional case. In addition, we mention that applications of skew-product flows are common. As important examples, we mention the recent work by Mischaikow, Smith, and Thieme [**MST**] and Shen and Yi [**SY, SY2, Yi**].

Definition and examples of linear skew-product flows. For the case where Θ is compact and X is finite dimensional, see [**JPS**] for the relations between the variational equation (6.2) and cocycles. In the Russian literature, the term "linear extension" is used for linear skew-product flow, see Bronshtein and Kopanskii [**Br**, **BKo**], Latushkin and Stepin [**LSt2**], and Mitropolskij, Samojlenko, and Kulik [**Sm**, **MSK**]. Also, we note that our setting is more general than in [**CLe2, CLe3, SS5**].

Examples 6.3 and 6.19 are taken from Leiva [**Le**]. Similar calculations, see [**Le**], work for more general boundary value problems

$$u_t = a(t)u_{ss} + b(t)u, \quad u_x(t,0) = u_s(t,1), \quad s \in (0,1),$$

where $a(\cdot)$ and $b(\cdot)$ are continuous functions with

$$\alpha := \lim_{t \to -\infty} a(t) \leq \lim_{t \to \infty} a(t) =: \beta, \quad \gamma := \lim_{t \to -\infty} b(t) \leq \lim_{t \to \infty} b(t) =: \delta.$$

Here, the Sacker-Sell spectrum is the set

$$\Sigma = \bigcup_{n=1}^{\infty} [-(n-1)^2 \pi^2 \alpha + \gamma, -(n-1)^2 \pi^2 \beta + \delta].$$

If a and b are assumed to be ω-periodic, then [**Le**]

$$\Sigma = \left\{ \frac{-(n-1)^2 \pi^2}{\omega} \int_0^\omega a(s)ds + \frac{1}{\omega} \int_0^\omega b(s)ds : n = 1, 2, \ldots \right\}.$$

Similar formulas hold for the case of Dirichlet boundary conditions. There are further generalizations in [**Le2**].

The idea of associating a linear skew-product flow, over the translations of $\mathbb{R}$, to the nonautonomous equation (6.6) as in Examples 6.4 and 6.5, is used, for example, by Chow and Leiva [**CLe, CLe2**]. Also, the cocycles described in Example 6.8 are studied in [**CLe2**]. Example 6.9 is taken from [**CLe**, Example 5.3].

Among other things, a paper by Magalhães [**Mg2**] contains very interesting examples of the calculation of the spectra for the linear skew-product flows described in Example 6.11. For instance in Example 6.11, if we define $a(t) = -2\sin^2 \pi t$ for $t \in [2n, 2n+1]$ where $n \in \mathbb{Z}$, and $a(t) = 0$ otherwise, then $\Sigma = \emptyset$. If on the other hand, $a(t) = 0$ for $t \leq 0$, $a(t) = -t$ for $t \in (0,1)$, and $a(t) = -1$ for $t \geq 1$, then $\Sigma = (-\infty, 0]$. The paper [**Mg2**] also contains examples of a cocycle whose spectrum equals the union of an unbounded interval and a compact interval, as well as an example where Σ consists of infinitely many nontrivial compact intervals.

In Example 6.12 we follow Sacker and Sell [**SS5**]; see Constantin and Foias [**CF**], Henry [**He**], and Temam [**Te**] for further references on Navier-Stokes equations.

Dichotomy and hyperbolicity of linear skew-product flows. General references on dichotomy are Coppel [**Co**], Daleckij and Krein [**DK**], Henry [**He**], and Massera and Schäffer [**MS2**]. Our definitions are essentially the same as in Johnson, Palmer, and Sell [**JPS**], Sacker and Sell [**SS3, SS5**] and Chow and Leiva [**CLe2, CLe3**]. Condition (*b*) for the infinite dimensional case of Definitions 6.13 and 6.14 is quite natural; it appears already in Henry [**He**]. For a modern account of the theory of finite dimensional hyperbolic (Anosov) systems see Katok and Hasselblatt [**KH**] and Pesin [**Ps**]. Hyperbolicity of linear skew-product flows and more general ("affine") extensions, in the spirit of this book, is studied by Antonevich in [**An2, An3**].

Weakly dichotomic (in other terminologies, quasi-hyperbolic or quasi-Anosov) cocycles, see Remark 6.16, were studied by many authors, see for example Churchill, Franks and Selgrade [**CFS**], Franks and Robinson [**FRo**], Sacker and Sell [**SS**], and Selgrade [**Sg**]. We note that one of the starting points in [**SS2, SS3**] is to see under which conditions weak dichotomy implies dichotomy (see also Problem 10 in Palis and Pugh [**PP**] and related work by Mañe [**Mn2**]).

Using negative continuations, weakly dichotomic systems can be defined also for cocycles with noninvertible operators $\Phi^t(\theta)$ as follows: We say that a point $(\theta,x) \in \Theta \times X$ has *a negative continuation* ϕ if there exists a continuous function $\phi : (-\infty, 0] \to X$ such that: (1) $\phi(t)$ lies in the fiber over $\varphi^t(\theta)$; (2) $\phi(0) = x$; and (3) $\Phi^t(\varphi^s\theta)\phi(s) = \phi(s+t)$ for $s \le 0$ and $0 \le t \le -s$. Thus, the existence of a unique negative continuation allows one to extend $\{\Phi^t(\theta)x\}_{t \ge 0}$ to negative values of t. Let $\mathcal{B}^+ = \{(\theta,x) \in \Theta \times X : \sup_{t \ge 0} |\Phi^t(\theta)x| < \infty\}$ and $\mathcal{B}^- = \{(\theta,x) \in \Theta \times X :$ there exists a negative continuation ϕ of (θ,x) such that $\sup_{t \le 0} |\phi(t)| < \infty\}$, and define $\mathcal{B} = \mathcal{B}^+ \cap \mathcal{B}^-$. We say that $\{\Phi^t\}_{t \ge 0}$ is *weakly dichotomic* provided that $\mathcal{B} = \Theta \times \{0\}$, see Sacker and Sell [**SS5**].

Definition 6.17 of the dynamical spectrum is due to Sacker and Sell, see [**SS2**] and [**SS3**], and also Selgrade [**Sg**]. The similar notion of Bohl spectrum in Definition 3.9 is taken from Daleckij and Krien [**DK**]. Examples 6.18 and 6.19 are taken from Leiva [**Le**].

Mather semigroup. As far as we know, Mather [**Ma**] was the first to realize that the "push-forward" operator $E : f \mapsto \hat\varphi \circ f \circ \varphi^{-1}$ on $C(\mathcal{T}\Theta)$ is hyperbolic if and only if $\{\hat\varphi^t\}_{t \in \mathbb{R}}$ is Anosov (cf. Dichotomy Theorem 6.41). Mañe proved in [**Mn2**] that $1 \notin \sigma_{\mathrm{ap}}(E)$ if and only if the linear skew-product flow $\{\hat\varphi^t\}_{t \in \mathbb{R}}$ generated by the differential is quasi-Anosov. Also, the evolution operator E was systematically used by Hirsch, Pugh and Shub in [**HPS**]. Semigroups of Mather operators are studied by Johnson [**Jo, Jo2**] and Chicone and Swanson [**CS, CS2, CS3**], see also papers by Swanson [**Sn, Sn2**] and Otsuki [**Ot**]. An important contribution was made by Antonevich [**An2, An3**], see also a review on weighted composition operators in Latushkin and Stepin [**LSt2**]. The papers [**LSt2, LSt3**] contain a systematic study of Mather evolution semigroups for the case of norm continuous cocycles in the case where X is a Hilbert space. Rau [**Ra, Ra3**] studied this semigroup for the case of strongly continuous cocycles. Our exposition follows Latushkin, Montgomery-Smith, Randolph and Schnaubelt [**LMR, LS**].

Strong continuity for evolution semigroups. Theorem 6.20 is taken from the paper [**LMR**, Thm. 2.1] and [**Ra3**, Prop.3]. See Chapter 8 for the spectral theory of the semigroup of push-forward operators on the space of divergence-free sections of $\mathcal{T}\Theta$. Proposition 6.23 is taken from [**LMR**], see also Theorem 5.1 in Chow and Leiva [**CLe**] concerning Example 6.8.

Mather localization and Mañe's Lemma. Lemma 6.28 is essentially contained in the paper by Mather [**Ma**]. See also the excellent exposition of this result in Proposition 6.22 by Bronshtein [**Br**], and Section 2.7 on Mather theory in the review [**Ps**] by Pesin. Mather's spectrum in [**Ps**] is defined as $\ln|\sigma(E) \setminus \{0\}|$; it is exactly the Sacker-Sell dynamical spectrum by Corollary 6.45. R. de la Llave [**dL2**] recently used Mather localization to study the spectrum of E on $C^1(\mathcal{T}\Theta)$, see Section 8.3 below. We remark that a localization method for scalar evolution operators was independently developed by Arendt and Greiner in [**AG**].

Mañe's Lemma 6.29 is a part of the following deep theorem of Mañe [**Mn2**].

THEOREM 6.51 (Mañe's Theorem). *For a C^r-diffeomorphism φ on a closed C^∞ manifold Θ with $\Phi(\theta) = D\varphi(\theta)$, the following statements are equivalent:*

(a) *$\hat\varphi$ is weakly dichotomic (quasi-Anosov).*

(b) φ satisfies Axiom A ([**KH**]) and the following transversality condition:
$$\mathcal{T}_\theta W^s(\theta) \cap \mathcal{T}_\theta W^u(\theta) = \{0\}$$
for each $\theta \in \Theta$ where W^s denotes the stable manifold and W^u the unstable manifold.

(c) There exists a closed C^∞-manifold Θ_1, a C^∞-embedding $i : \Theta \to \Theta_1$ and C^r-diffeomorphism ψ of Θ_1 such that $\psi \circ i = i \circ \varphi$ and $\hat{\psi}$ has dichotomy (is hyperbolic) on $i(\Theta)$.

(d) $1 \notin \sigma_{ap}(E)$ on $C(\mathcal{T}\Theta)$.

Note that (d) implies that the aperiodic points of φ are dense ([**Mn2**, p.366]). The converse to (a)$\Rightarrow$(d), recorded above as Lemma 6.29, was originally used by Chicone and Swanson in [**CS2, CS3**] to construct the approximate eigenfunctions for Γ in order to prove the Spectral Mapping Theorem for the evolution semigroup $\{E^t\}_{t \geq 0}$.

The Spectral Mapping Theorem. The proof of Theorem 6.30 is taken from the paper by Latushkin, Montgomery-Smith, and Randolph [**LMR**], see Chapter 8 for a similar proof of its variant for the case of the space of divergence free vector fields.

The Mather semigroup on L^p-spaces. The content of this subsection is taken from Latushkin and Schnaubelt [**LS2**]. The Mather semigroup on the spaces $L^2(\Theta, \mu; \mathcal{H})$ is studied by Latushkin and Stepin [**LSt2, LSt3**]. In [**LSt2**], the Gearhart Spectral Mapping Theorem 2.10 was applied to prove the Spectral Mapping Theorem for $\{E^t\}_{t \geq 0}$. See also [**LSt2**] for more references on weighted composition operators on L^p-spaces. Using the Halmos-Rokhlin Lemma, Kitover [**Ki**], cf. Lemma 6.35, proved the rotational symmetry of $\sigma(E)$ on L^p. Proposition 6.36 is parallel to Theorem 4.9 and Lemma 4.11. The Spectral Mapping Theorem 6.37 under more restrictive assumptions was proved in Chicone and Swanson [**CS3**] and Latushkin and Stepin [**LSt2**]. Note that for $\dim X < \infty$, see [**CS3**], one has $\sigma_{\mathrm{ap}}(E) = \sigma(E)$ and $\sigma(E; C(\Theta; X)) = \sigma(E; L^2(\Theta; X))$.

Sacker-Sell spectral theory. For $\dim X < \infty$, the spectral theory of linear skew-product flows was constructed by Sacker [**Sa, Sa2**], Sacker and Sell [**SS2, SS3**], and Selgrade [**Sg**], see also the paper by Johnson, Palmer, and Sell [**JPS**]. For the infinite dimensional case see the paper [**SS5**] where certain additional conditions (on compactness of the operators $\Phi^t(\theta)$) were imposed. Magalhães in [**Mg**] developed the spectral theory of infinite dimensional linear skew-product flows and gave applications to the theory of invariant manifolds for functional differential equations. The general Banach space case is analyzed by Chow and Leiva [**CLe, CLe2, CLe3**] along the lines of Sacker and Sell [**SS3, SS5**].

In this section, we use a completely different approach that goes back to [**An3, LSt2, Ra3**]: Starting with spectral information on E, we derive the existence of spectral subbundles for $\{\hat{\varphi}^t\}_{t \geq 0}$.

The Spectral Projection Theorem. Theorem 6.38 was proved by Rau [**Ra3**, Lemma 7], see Antonevich, Latushkin, Räbiger, Schnaubelt, and Stepin [**An2, LM, LMR, LSt2, RS**] for earlier and different versions. Also, cf. Theorem 3.14. The proof is taken from [**LS**].

The Dichotomy Theorem. The proof of Theorem 6.41 is a modification of the proof in Rau [**Ra3**, Thms.10,12] and is taken from the paper of Latushkin and Schnaubelt [**LS**]; see also [**An2, LM, LMR, LSt2, RS**], and especially Hirsch, Pugh, and Shub [**HPS**, Thm.2.5], for earlier versions of this result. Corollary 6.42

corresponds to the uniqueness of the dichotomy projection mentioned in Bart, Gohberg, and Kaashoek [**BrGK**, Prop.2.1]. Corollary 6.43, of course, was known in the framework of Sacker-Sell theory, see [**SS2, SS3**]. Corollary 6.44 is the simplest robustness result, we stress the fact that its proof is quite short. See similar results in [**CLe2**, Thm.3.2] and more advanced versions in [**CLe4**].

Spectral theory. Most of the content of this subsection is taken from Latushkin and Stepin [**LSt2, LSt3**]. Corollary 6.45 for $\dim X < \infty$ was known already in Chicone and Swanson [**CS3**] and Johnson [**Jo2**]; we stress again the fact that in these papers the Sacker-Sell theory was *used*, while here we *derive* it from the spectral theory for the operator E. A variety of different Lyapunov exponents are studied in Bylov, Vinograd, Grobman, and Nemytski [**BVGN**].

Remark 6.46 summarizes the core content of the finite dimensional Sacker-Sell Spectral Theory [**SS3**]. Of course, the Sacker-Sell theory of linear skew-product flows gives much more. As an example, we give a variant of the following Compatibility Theorem from [**SS5**]. Assume that $T \geq 0$, $\{\Phi^t(\theta) : t > T\}$ is a family of *compact* operators, and the cocycle $\{\Phi^t\}_{t \geq 0}$ is weakly dichotomic. Let $\mathcal{S}_\theta$ and $\mathcal{U}_\theta$ denote the stable and unstable fibers in X over $\theta \in \Theta$.

THEOREM 6.52 ([**SS5**]). *If either* $\mathrm{codim}\,\mathcal{S}_\theta$ *assumes the same value on all minimal sets in* Θ *or* $\dim \mathcal{U}_\theta$ *assumes the same value on all minimal sets in* Θ, *then* $\{\Phi^t\}_{t \geq 0}$ *has an exponential dichotomy on* Θ.

This result follows from the Alternative Theorem (see [**SS5**, Thm.E]) that describes dichotomies over Morse sets in Θ.

Theorem 6.48 and Corollary 6.49 were proved in [**SS3**] for $\dim X < \infty$, see [**CLe2**] for an infinite dimensional version. Our proofs are taken from [**LSt2, LSt3**].

Many examples of skew-product flows with infinitely many spectral components in Σ are given in Magalhães [**Mg2**].

Theorem 6.50 for the case where $\dim X < \infty$ is a central result of the paper [**SS4**] by Sacker and Sell. This paper has an example where inclusion in statement (b) is strict; it is also proved that statement (c) holds provided φ is chain-recurrent. See, for example, Bronshtein [**Br**] and Colonius and Kliemann [**CK2, CK3**] for definitions and more information about linear skew-product flows over chain-recurrent base flows and various spectra for linear skew-product flows.

CHAPTER 7

Characterizations of Dichotomy for Linear Skew-Product Flows

In this chapter we will continue our study of the existence of exponential dichotomies for linear skew-product flows by using the associated Mather evolution semigroups. The relationship between "global" and "trajectorial" (pointwise) dichotomies is discussed in Section 7.1 where we also develop an algebraic technique that parallels Section 4.1. The main result in this section is the Discrete Dichotomy Theorem 7.9. As a byproduct, we will give an infinite dimensional version of the Sacker-Sell Perturbation Theorem: it states that exponential dichotomies on compact invariant sets of the base flow persist under small perturbations of the compact invariant sets. In Section 7.2 we return to the question of how the spectrum of an evolution (Mather) operator is related to the spectrum of the generator of the Mather semigroup. In particular, we will prove the Annular Hull Theorem 7.25. This result replaces the Spectral Mapping Theorem 6.30 for the general case where the base flow is not required to be aperiodic. In Section 7.3 "mild" variational equations are discussed, cf. Sections 4.2 and 4.3, and a Perron-type characterization of dichotomy in terms of the existence of bounded mild solutions of the variational equation (Theorem 7.34) is proved. We will also discuss Green's functions for linear skew-product flows. Sections 7.4 and 7.5 contain some results for Hilbert spaces. Section 7.4 is a brief exposition of a general C^*-algebraic technique that underlies the "discrete dichotomies" argument in Section 7.1, and in Section 7.5 there are some characterizations of dichotomy for linear skew-product flows using quadratic Lyapunov functions.

7.1. Pointwise dichotomies

In this section we will consider a continuous linear skew-product flow $\{\hat{\varphi}^t\}_{t \geq 0}$ and relate the existence of an exponential dichotomy for $\{\hat{\varphi}^t\}_{t \geq 0}$ on Θ (see Definition 6.13) to its "pointwise" dichotomies; that is, its dichotomies on the closure of orbits of $\{\varphi^t\}_{t \in \mathbb{R}}$ (see Definition 6.14). By the Dichotomy Theorem 6.41, the existence of a "global" dichotomy (on the entire set Θ) is equivalent to the hyperbolicity of the "global" evolution semigroup $\{E^t\}_{t \geq 0}$ on $C_0(\Theta; X)$. As we will see, the existence of a "pointwise" dichotomy on the closure of the orbit through a point $\theta \in \Theta$ is equivalent to the hyperbolicity of the "discrete" evolution operator $\pi_\theta(E)$, $E := E^1$, acting on the sequence space $c_0(\mathbb{Z}; X)$ by the rule

(7.1) $$\pi_\theta(E) : (x_n)_{n \in \mathbb{Z}} \mapsto (\Phi(\varphi^{n-1}\theta)x_{n-1})_{n \in \mathbb{Z}},$$

see (6.41). We will relate the hyperbolicity of the operator E on $C_0(\Theta; X)$ to the hyperbolicity of the set of operators $\{\pi_\theta(E) : \theta \in \Theta\}$ on $c_0(\mathbb{Z}; X)$. In addition, we will introduce a corresponding set of evolution semigroups $\{\{\Pi_\theta^t\}_{t \geq 0} : \theta \in \Theta\}$

on $C_0(\mathbb{R}; X)$, and relate their hyperbolicity to the existence of a dichotomy for the skew-product flow $\{\hat{\varphi}^t\}_{t\geq 0}$.

The results of this section can be viewed as a continuation of the theory developed in Subsection 4.1.2. However, in this section we work with the space $C_0(\Theta; X)$ while in Subsection 4.1.2 we presented the results obtained for L^p-spaces. We also mention that we will *use* the Spectral Projection Theorem 6.38, while in Subsection 4.1.2 we gave a proof of the corresponding Spectral Projection Theorem 3.14.

In Subsection 7.1.1 we will develop some algebraic machinery that will be used in the proofs to follow. The main result of the section, the Discrete Dichotomy Theorem 7.9, is proved in Subsection 7.1.2 where we also derive some of its consequences for the set of semigroups $\{\{\Pi_\theta^t\}_{t\geq 0} : \theta \in \Theta\}$, see Theorem 7.12. In Subsection 7.1.3 we use the Discrete Dichotomy Theorem 7.9 to prove that the dynamical spectrum $\Sigma(\Theta_0)$ is semicontinuous as a function of the compact φ^t-invariant set $\Theta_0 \subset \Theta$ (Theorem 7.14).

7.1.1. Weighted translation algebras. Suppose that $\{\varphi^t\}_{t\in\mathbb{R}}$ is a continuous flow on a locally compact metric space Θ, and define the corresponding translation group $\{V^t\}_{t\in\mathbb{R}}$ by $(V^t f)(\theta) = f(\varphi^{-t}\theta)$ for $t \in \mathbb{R}$ on the space $C_0(\Theta; X)$. Also, let $\{\Phi^t\}_{t\geq 0}$ be an exponentially bounded, strongly continuous cocycle over $\{\varphi^t\}_{t\in\mathbb{R}}$ and define $a_t(\theta) := \Phi^t(\varphi^{-t}\theta)$ for $t \geq 0$. Finally, for notational convenience, let us define $V := V^1$, $E := E^1$, $\varphi := \varphi^1$, and $\Phi := \Phi^1$.

We denote by $\tilde{\mathfrak{A}} = C_b(\Theta; \mathcal{L}_s(X))$ the algebra of bounded, strongly continuous, operator valued functions $a = a(\cdot)$ on Θ with point-wise multiplication, and supremum norm given by

$$\|a\|_{\tilde{\mathfrak{A}}} = \|a\|_\infty = \sup_{\theta \in \Theta} \|a(\theta)\|_{\mathcal{L}(X)}.$$

Clearly, $\tilde{\mathfrak{A}}$ is isometrically isomorphic to the subalgebra $\mathfrak{A}$ of $\mathcal{L}(C_0(\Theta; X))$ that consists of the multiplication operators of the form $(af)(\theta) = a(\theta)f(\theta)$ where $a : \Theta \to \mathcal{L}(X)$ is a bounded, strongly continuous function and $f \in C_0(\Theta; X)$.

Let $\hat{\mathfrak{B}}$ denote the set of all sequences $(a_k)_{k\in\mathbb{Z}}$ with entries $a_k \in \tilde{\mathfrak{A}}$ such that

$$(\|a_k\|_{\tilde{\mathfrak{A}}})_{k\in\mathbb{Z}} \in \ell^1(\mathbb{Z}),$$

and note that $\hat{\mathfrak{B}}$ is a Banach space when equipped with the norm

$$\|(a_k)_{k\in\mathbb{Z}}\|_1 := \sum_{k=-\infty}^{\infty} \|a_k\|_{\tilde{\mathfrak{A}}} = \sum_{k=-\infty}^{\infty} \sup_{\theta \in \Theta} \|a_k(\theta)\|_{\mathcal{L}(X)}.$$

We introduce a product on $\hat{\mathfrak{B}}$ that makes it a Banach algebras as follows: For $\hat{b}' = (a'_n)_{n\in\mathbb{Z}}$ and $\hat{b}'' = (a''_n)_{n\in\mathbb{Z}}$ in $\hat{\mathfrak{B}}$, let us define the product $\hat{b} := \hat{b}' * \hat{b}''$ of $\hat{b}'$ and $\hat{b}''$ as the "convolution" $\hat{b} = (a_n)_{n\in\mathbb{Z}}$ where $a_n(\theta) = \sum_{k\in\mathbb{Z}} a'_k(\theta) a''_{n-k}(\varphi^{-k}\theta)$ for $\theta \in \Theta$ and $n \in \mathbb{Z}$. To see that $\hat{b} = \hat{b}' * \hat{b}''$ belongs to $\hat{\mathfrak{B}}$, we notice that $\|a_n(\theta)\|_{\mathcal{L}(X)} \leq \|\hat{b}'\|_1 \|\hat{b}''\|_1$, the function $\theta \mapsto a'_k(\theta) a''_{n-k}(\varphi^{-k}\theta)x$ is continuous for each $x \in X$ and $k, n \in \mathbb{Z}$, and

$$|(a_n(\theta_1) - a_n(\theta_2))x| \leq \sum_{|k|\leq N} |(a'_k(\theta_1) a''_{n-k}(\varphi^{-k}\theta_1) - a'_k(\theta_2) a''_{n-k}(\varphi^{-k}\theta_2))x|$$

$$+ 2|x| \|\hat{b}''\|_1 \sum_{|k|>N} \|a'_k\|_\infty$$

for $\theta_1, \theta_2 \in \Theta$ and $N \in \mathbb{N}$. Therefore, $a_n \in \tilde{\mathfrak{A}}$. By the estimate

$$\|\hat{b}\|_1 \leq \sum_n \sum_k \|a'_k\|_\infty \|a''_{n-k} \circ \varphi^{-k}\|_\infty = \sum_k \sum_n \|a'_k\|_\infty \|a''_{n-k}\|_\infty \leq \|\hat{b}'\|_1 \|\hat{b}''\|_1,$$

we have that $\hat{b} \in \hat{\mathfrak{B}}$. Therefore, $(\hat{\mathfrak{B}}, *, \|\cdot\|_1)$ is a Banach algebra. The unit element of $\hat{\mathfrak{B}}$ is the sequence $\hat{e} = e \otimes (\delta_{n,0})_{n \in \mathbb{Z}}$ where $e(\theta) = I_X$ for $\theta \in \Theta$, and $\delta_{n,k}$ is the Kronecker delta. In other words, $\hat{e} = (e_n)_{n \in \mathbb{Z}}$ where $e_0(\theta) = I$, and $e_n(\theta) = 0$ for $n \neq 0, \theta \in \Theta$.

To relate the algebra $\hat{\mathfrak{B}}$ to operators on $C_0(\Theta; X)$, we define the (algebra) homomorphism

$$(7.2) \qquad \rho: \hat{\mathfrak{B}} \to \mathcal{L}(C_0(\Theta; X)) : \hat{b} = (a_k)_{k \in \mathbb{Z}} \mapsto \sum_{k=-\infty}^{\infty} a_k V^k.$$

Clearly, $\rho(\hat{e}) = I$ and $\|\rho(\hat{b})\|_{\mathcal{L}(C_0(\Theta; X))} \leq \|\hat{b}\|_1$.

We stress the fact that ρ is not always injective. As an example, suppose that φ is the identity map on Θ. Then, $V = I$ and each $a \in \mathfrak{A} \subset \mathcal{L}(C_0(\Theta; X))$ can be represented as $\rho(\hat{b})$ in infinitely many different ways; for example,

$$a = \tfrac{1}{2}a + \tfrac{1}{2}aV = \tfrac{1}{3}a + \tfrac{2}{3}aV.$$

However, we will show that ρ is injective whenever φ is aperiodic in the sense of Definition 6.26. Indeed, let $\mathfrak{B}$ denote the subalgebra of bounded operators on $C_0(\Theta; X)$ of the form

$$(7.3) \qquad b = \sum_{k=-\infty}^{\infty} a_k V^k \text{ where } a_k \in \mathfrak{A} \text{ and } \|b\|_1 := \sum_{k=-\infty}^{\infty} \|a_k\|_{\mathfrak{A}} < \infty.$$

PROPOSITION 7.1. *If φ is aperiodic on Θ, then the representation of $b \in \mathfrak{B}$ as in (7.3) is unique and $(\mathfrak{B}, \|\cdot\|_1)$ is a Banach algebra.*

PROOF. The uniqueness of the representation is a consequence of the the following observation: If $b_N = \sum_{k=-N}^{N} a_k V^k$ in $\mathcal{L}(C_0(\Theta; X))$ is a *polynomial*, then

$$(7.4) \qquad \|b_N\|_{\mathcal{L}(C_0(\Theta;X))} \geq \|a_k\|_{\mathcal{L}(C_0(\Theta;X))}, \quad |k| \leq N.$$

To prove (7.4), note that, by replacing b_N with $b_N V^{-k}$, it suffices to consider only the case $k = 0$. Fix $\epsilon > 0$. For $a \in \mathfrak{A}$, we have that

$$\|a\|_{\mathfrak{A}} = \|a\|_{\mathcal{L}(C_0(\Theta;X))} = \|a\|_\infty = \sup_{\theta \in \Theta} \sup_{|x|=1} |a(\theta)x|.$$

Hence, by the assumption, there is a point $\theta_0 \in \Theta$ and a vector $x \in X$ such that $|x| = 1$, the prime period of θ_0 satisfies $p(\theta_0) \geq 2N + 1$, and

$$|a_0(\theta_0)x| \geq \|a_0\|_{\mathfrak{A}} - \epsilon.$$

Choose $\delta > 0$ such that, for the ball $B = B(\theta_0, \delta)$, the sets in the family $\{\varphi^k(B) : |k| \leq N\}$ are pairwise disjoint. Also, choose a continuous bump-function $\alpha : \Theta \to [0, 1]$ such that $\alpha(\theta_0) = 1$ and $\alpha(\theta) = 0$ for $\theta \notin B$, and define $f \in C_0(\Theta; X)$ by $f(\theta) = \alpha(\varphi^{-k}\theta)x$ whenever $\theta \in \varphi^k(B)$ and $|k| \leq N$, and $f(\theta) = 0$ otherwise. Then, we have that $\|f\|_\infty = 1$ and

$$\|b_N\|_{\mathcal{L}(C_0(\Theta;X))} \geq \|b_N f\|_\infty = \max_{\theta \in \Theta} |\sum_{k=-N}^{N} a_k(\theta)\alpha(\varphi^{-k}\theta)x|$$

$$= \max_\theta \max_{|k| \leq N} |a_k(\theta)\alpha(\varphi^{-k}\theta)x| \geq |a_0(\theta_0)x| \geq \|a_0\|_{\mathcal{L}(C_0(\Theta;X))} - \epsilon.$$

Thus, inequality (7.4) is proved.

To see that $(\mathfrak{B}, \|\cdot\|_1)$ is complete, consider a $\|\cdot\|_1$-Cauchy sequence $\{b^{(n)}\}_{n=1}^\infty \subset \mathfrak{B}$ where $b^{(n)} = \sum a_k^{(n)} V^k$. Since, for every $m, n \in \mathbb{Z}$, we have

$$\left|\sum_{k\in\mathbb{Z}} \|a_k^{(m)}\|_{\mathfrak{A}} - \sum_{k\in\mathbb{Z}} \|a_k^{(n)}\|_{\mathfrak{A}}\right| \leq \|b^{(m)} - b^{(n)}\|_1,$$

there exists $a_k := \|\cdot\|_1\text{-lim}_{m\to\infty} a_k^{(m)}$ for each $k \in \mathbb{Z}$. The sequence $(\|a_k^{(m)}\|_{\mathfrak{A}})_{k\in\mathbb{Z}} \in \ell^1(\mathbb{Z})$ converges in $\ell^1(\mathbb{Z})$ to $(\|a_k\|_{\mathfrak{A}})_{k\in\mathbb{Z}} \in \ell^1(\mathbb{Z})$ as $m \to \infty$. Hence, $b = \sum a_k V^k$ is an element of $\mathfrak{B}$. For every $\epsilon > 0$ and for sufficiently large integers m and n, the inequality

$$\epsilon \geq \|b^{(m)} - b^{(n)}\|_1 = \sum_{k\in\mathbb{Z}} \|a_k^{(m)} - a_k^{(n)}\|_{\mathfrak{A}} \geq \sum_{k=-N}^{N} \|a_k^{(m)} - a_k^{(n)}\|_{\mathfrak{A}}$$

holds for every $N \in \mathbb{N}$. By first passing to the limit as $n \to \infty$, and then the limit as $N \to \infty$, we have that $\lim_{m\to\infty} \|b^{(m)} - b\|_1 = 0$. $\square$

Therefore, if φ is aperiodic, then for each $\hat{b} = (a_k)_{k\in\mathbb{Z}}$ in $\hat{\mathfrak{B}}$ there is a corresponding operator $b = \rho(\hat{b})$ given by $b = \sum_{k=-\infty}^{\infty} a_k V^k$. This operator is unique by Proposition 7.1.

Let us return to the general case where φ is not necessarily aperiodic. Recall, see page 88, that $\mathfrak{D} \subset \mathcal{L}(c_0(\mathbb{Z}; X))$ is the algebra of all diagonal operators $d = \text{diag}(d^{(n)})_{n\in\mathbb{Z}}$ that are bounded on $c_0(\mathbb{Z}; X)$, let $S : (x_n)_{n\in\mathbb{Z}} \to (x_{n-1})_{n\in\mathbb{Z}}$ denote the shift operator on $c_0(\mathbb{Z}; X)$, and define

$$\mathfrak{C} := \{D = \sum_{k=-\infty}^{\infty} d_k S^k \in \mathcal{L}(c_0(\mathbb{Z}; X)) : d_k \in \mathfrak{D} \text{ and}$$

$$\|D\|_{\mathfrak{C}} := \sum_{k=-\infty}^{\infty} \|d_k\|_{\mathcal{L}(c_0(\mathbb{Z}; X))} < \infty\}.$$

REMARK 7.2. The representation of $D \in \mathfrak{C}$ as $D = \sum d_k S^k$ is unique. Indeed, for $m_0, n_0 \in \mathbb{Z}$ and $x \in X$ with $|x| = 1$, we have

$$\|D\|_{\mathcal{L}(c_0(\mathbb{Z};X))} \geq \|D(x \otimes \delta_{n,n_0})\|_{c_0(\mathbb{Z};X)}$$
$$= \left\|\sum_k \text{diag}(d_k^{(n)})_{n\in\mathbb{Z}}(x \otimes \delta_{n-k,n_0})_{n\in\mathbb{Z}}\right\|_{c_0(\mathbb{Z};X)}$$
$$\geq \left|\sum_k d_k^{(m_0)} x\, \delta_{m_0-k,n_0}\right| = |d_{m_0-n_0}^{(m_0)} x|.$$

$\diamond$

We will use the family of maps defined by

(7.5) $\qquad \hat{\pi}_\theta : \hat{\mathfrak{B}} \to \mathcal{L}(c_0(\mathbb{Z}; X)) : \hat{b} = (a_k)_{k\in\mathbb{Z}} \mapsto \sum_{k=-\infty}^{\infty} \text{diag}\left(a_k(\varphi^n \theta)\right)_{n\in\mathbb{Z}} S^k$

for $\theta \in \Theta$. Observe that $\hat{\pi}_\theta : \hat{\mathfrak{B}} \to \mathfrak{C}$ and

$$\|\hat{\pi}_\theta(\hat{b})\|_{\mathfrak{C}} = \sum_{k=-\infty}^{\infty} \|\text{diag}(a_k(\varphi^n \theta))_{n\in\mathbb{Z}}\|_{\mathcal{L}(c_0(\mathbb{Z};X))}$$
$$\leq \sum_{k=-\infty}^{\infty} \sup_{n\in\mathbb{Z}} \sup_{\theta\in\Theta} \|a_k(\varphi^n \theta)\|_{\mathcal{L}(X)} = \sum_{k=-\infty}^{\infty} \|a_k\|_\infty = \|\hat{b}\|_1.$$

Also, for $\hat{b} = \hat{b}' * \hat{b}''$, we have that

$$\hat{\pi}_\theta(\hat{b}) = \sum_{k=-\infty}^{\infty} \mathrm{diag}\Big(\sum_{l=-\infty}^{\infty} a'_l(\varphi^n\theta) a''_{k-l}(\varphi^{n-l}\theta) \Big)_{n\in\mathbb{Z}} S^k$$

$$= \sum_{l=-\infty}^{\infty} \Big[\mathrm{diag}(a'_l(\varphi^n\theta))_{n\in\mathbb{Z}} S^l S^{-l} \sum_{k=-\infty}^{\infty} \mathrm{diag}(a''_{k-l}(\varphi^{n-l}\theta))_{n\in\mathbb{Z}} S^k \Big]$$

$$= \sum_{l=-\infty}^{\infty} \Big[\mathrm{diag}(a'_l(\varphi^n\theta))_{n\in\mathbb{Z}} S^l \sum_{k=-\infty}^{\infty} \mathrm{diag}(a''_{k-l}(\varphi^n\theta))_{n\in\mathbb{Z}} S^{k-l} \Big]$$

$$= \hat{\pi}_\theta(\hat{b}') \cdot \hat{\pi}_\theta(\hat{b}'');$$

and therefore, $\hat{\pi}_\theta : \hat{\mathfrak{B}} \to \mathfrak{C}$ is a continuous homomorphism for each $\theta \in \Theta$.

PROPOSITION 7.3. $\bigcap_{\theta\in\Theta} \ker \hat{\pi}_\theta = \{0\}$.

PROOF. Assume that $\hat{\pi}_\theta(\hat{b}) = 0$ for all $\theta \in \Theta$ and some $\hat{b} = (a_n)_{n\in\mathbb{Z}} \in \hat{\mathfrak{B}}$; or, in other words,

$$0 = \hat{\pi}_\theta(\hat{b}) = \sum_k \mathrm{diag}(a_k(\varphi^n\theta))_{n\in\mathbb{Z}} S^k \in \mathfrak{C}$$

for all $\theta \in \Theta$. Since the representation $D = \sum_k \mathrm{diag}(d_k^{(n)}) S^k$ for $D \in \mathfrak{C}$ is unique, we conclude that $a_k(\varphi^n\theta) = 0$ for each $k, n \in \mathbb{Z}$ and $\theta \in \Theta$. Thus, $\hat{b} = 0$ in $\hat{\mathfrak{B}}$. □

For $a \in \mathfrak{A}$, let us define the operator $b = I - aV \in \mathcal{L}(C_0(\Theta; X))$ and the associated sequence $\hat{b} = (a_k)_{k\in\mathbb{Z}} \in \hat{\mathfrak{B}}$ by $a_0 = e$, $a_1 = -a$, and $a_k = 0$ otherwise. Then, by the definition in display (7.2), we have that $\rho(\hat{b}) = I - aV = b$. Also, for each such $b = I - aV$ and each $\theta \in \Theta$, let us introduce the operator $\pi_\theta(b) \in \mathcal{L}(c_0(\mathbb{Z}; X))$ by the rule

(7.6) $$\pi_\theta(b) = \pi_\theta(I - aV) := I - \mathrm{diag}(a(\varphi^n\theta))_{n\in\mathbb{Z}} S.$$

These operators are related to the evolution semigroup $\{E^t\}_{t\geq 0}$ on $C_0(\Theta; X)$ that is induced by the cocycle $\{\Phi^t\}_{t\geq 0}$. In fact, the operator $I - E$ is of the form $b = I - E = I - aV$ with $a(\theta) = \Phi(\varphi^{-1}\theta)$ for $\theta \in \Theta$. The next theorem gives a sufficient condition for the invertibility of the operator $I - E$ in terms of the family of operators $\{I - \pi_\theta(E) : \theta \in \Theta\}$.

THEOREM 7.4. Suppose that $b = I - aV$ for $a \in \mathfrak{A}$ and $\pi_\theta(b)$ is invertible in $\mathcal{L}(c_0(\mathbb{Z}; X))$ for each $\theta \in \Theta$. If there exists a constant $B > 0$ such that

(7.7) $$\|[\pi_\theta(b)]^{-1}\|_{\mathcal{L}(c_0(\mathbb{Z};X))} \leq B$$

uniformly for $\theta \in \Theta$, then b^{-1} exists in $\mathcal{L}(C_0(\Theta; X))$.

PROOF. Let $\hat{b} = (a_k)_{k\in\mathbb{Z}} \in \hat{\mathfrak{B}}$ be given by $a_0 = I$, $a_1 = -a$, and $a_k = 0$ for $k \neq 0, 1$. By the definitions in displays (7.5) and (7.6), if $\theta \in \Theta$, then

(7.8) $$\hat{\pi}_\theta(\hat{b}) = \pi_\theta(b).$$

Also, $\rho(\hat{b}) = b$ by the definition in (7.2). Thus, if $\theta \in \Theta$ and we set

$$d_0(\theta) = I, \quad d_1(\theta) = \mathrm{diag}(a(\varphi^n\theta))_{n\in\mathbb{Z}}, \quad \text{and} \quad d_k(\theta) = 0, \ k \neq 1, 2,$$

then

$$\hat{\pi}_\theta(\hat{b}) = \sum_{k=-\infty}^{\infty} d_k(\theta) S^k \in \mathfrak{C}.$$

Next, we will show that the inverse of $\hat{\pi}_\theta(\hat{b})$ also belongs to $\mathfrak{C}$. The proof of this fact is essentially the same as the proof of Lemma 4.1.

PROPOSITION 7.5. *Under the assumptions of Theorem 7.4, the operator*
$$[\hat{\pi}_\theta(\hat{b})]^{-1} = [\pi_\theta(b)]^{-1}$$
belongs to $\mathfrak{C}$ *for each* $\theta \in \Theta$; *that is,*

(7.9) $$[\hat{\pi}_\theta(\hat{b})]^{-1} = \sum_{k=-\infty}^{\infty} C_k(\theta) S^k$$

where

(7.10) $$C_k(\theta) = \operatorname{diag}(C_k^{(n)}(\theta))_{n\in\mathbb{Z}}$$

and

(7.11) $$\sum_{k=-\infty}^{\infty} \sup_{\theta\in\Theta} \|C_k(\theta)\|_{\mathcal{L}(c_0(\mathbb{Z};X))} < \infty.$$

Proof of Proposition 7.5. Set $D_\theta = \hat{\pi}_\theta(\hat{b}) = I - \operatorname{diag}(a(\varphi^n\theta))_{n\in\mathbb{Z}} S$. Assumption (7.7) and identity (7.8) yield

(7.12) $$\sup_{\theta\in\Theta} \|D_\theta^{-1}\| \leq B.$$

For each $\gamma > 0$, define the corresponding operator
$$D_\theta(\gamma) = I - \operatorname{diag}(e^{\gamma(|n|-|n-1|)} a(\varphi^n\theta))_{n\in\mathbb{Z}} S$$
on $c_0(\mathbb{Z}; X)$ and note that $D_\theta(\gamma) = J_\gamma^{-1} D_\theta J_\gamma$ for $J_\gamma = \operatorname{diag}(e^{-\gamma|n|})_{n\in\mathbb{Z}}$. Clearly,
$$\|D_\theta - D_\theta(\gamma)\|_{\mathcal{L}(c_0(\mathbb{Z};X))} \leq \max\{|1-e^\gamma|, |1-e^{-\gamma}|\} \|a\|_{\mathfrak{A}};$$
and therefore,
$$\lim_{\gamma\to 0} \|D_\theta - D_\theta(\gamma)\|_{\mathcal{L}(c_0(\mathbb{Z};X))} = 0.$$
Choose γ_0 such that
$$q := B \max\{|1 - e^{\pm\gamma_0}|\} \|a\|_{\mathfrak{A}} < 1.$$
Then, $\|D_\theta - D_\theta(\gamma)\| \cdot \|D_\theta^{-1}\| \leq q$ for all $\theta \in \Theta$ and $0 \leq \gamma \leq \gamma_0$. Moreover, since D_θ is invertible in $\mathcal{L}(c_0(\mathbb{Z}; X))$, we conclude that $D_\theta(\gamma)$ is invertible and
$$\|[D_\theta(\gamma)]^{-1}\| \leq R := \frac{B}{1-q}$$
for all $0 \leq \gamma \leq \gamma_0$ and $\theta \in \Theta$.

Consider D_θ^{-1} as an operator matrix $D_\theta^{-1} = [C_{kj}(\theta)]_{k,j\in\mathbb{Z}}$ where the operators $C_{kj}(\theta) \in \mathcal{L}(X)$ are defined for $x \in X$ by
$$C_{kj}(\theta) x := (D_\theta^{-1}(\bar{x}_j))_k \quad \text{for} \quad \bar{x}_j = (\delta_{jn} x)_{n\in\mathbb{Z}},$$
and use the identity $D_\theta(\gamma)^{-1} = J_\gamma^{-1} D_\theta^{-1} J_\gamma$ to obtain the equation $D_\theta(\gamma)^{-1} = [e^{\gamma(|k|-|j|)} C_{kj}(\theta)]_{k,j\in\mathbb{Z}}$. Then, by the inequality
$$R \geq \|D_\theta(\gamma)^{-1}\|_{\mathcal{L}(c_0(\mathbb{Z};X))} \geq e^{\gamma(|k|-|j|)} \|C_{kj}(\theta)\|_{\mathcal{L}(X)}$$
for $k, j \in \mathbb{Z}$, we have the estimate
$$\|C_{k,0}(\theta)\|_{\mathcal{L}(X)} \leq R e^{-\gamma|k|} \quad \text{for} \quad \theta \in \Theta \text{ and } k \in \mathbb{Z}.$$

Observe that if $j, k, m \in \mathbb{Z}$, then

$$(7.13) \qquad D_{\varphi^m \theta} = S^{-m} D_\theta S^m \quad \text{and} \quad C_{kj}(\varphi^m \theta) = C_{k+m, j+m}(\theta).$$

In particular, using the equation $C_{k+m,m}(\theta) = C_{k,0}(\varphi^m \theta)$, it follows that

$$(7.14) \qquad \|C_{k+m,m}(\theta)\|_{\mathcal{L}(X)} \le R e^{-\gamma |k|} \quad \text{for} \quad k, m \in \mathbb{Z} \text{ and } \theta \in \Theta.$$

Thus, we can express D_θ^{-1} as

$$D_\theta^{-1} = \sum_k C_k(\theta) S^k = \sum_k \operatorname{diag}(C_k^{(n)}(\theta))_{n \in \mathbb{Z}} S^k$$

for $C_k^{(n)}(\theta) = C_{n, n-k}(\theta)$, and we have the estimate

$$\sum_{k \in \mathbb{Z}} \sup_{\theta \in \Theta} \|C_k(\theta)\|_{\mathcal{L}(c_0(\mathbb{Z}; X))} = \sum_{k \in \mathbb{Z}} \sup_{\theta \in \Theta} \sup_{n \in \mathbb{Z}} \|C_{n,n-k}(\theta)\|_{\mathcal{L}(X)} \le \sum_{k \in \mathbb{Z}} R e^{-\gamma |k|} < \infty.$$

Thus, Proposition 7.5 is proved. $\square$

To conclude the proof of Theorem 7.4, we will use the following claim.

Claim. The function $C_k^{(0)} : \Theta \to \mathcal{L}_s(X)$, where $C_k^{(0)}$ is as in display (7.10), is continuous, bounded, and such that

$$\sum_{k \in \mathbb{Z}} \sup_{\theta \in \Theta} \|C_k^{(0)}(\theta)\|_{\mathcal{L}(X)} < \infty.$$

Let us finish the proof of the theorem; the Claim will be proved below. First, notice that the identities (7.13) imply the equalities

$$(7.15) \qquad C_k^{(n)}(\theta) = C_{n, n-k}(\theta) = C_{0, -k}(\varphi^n \theta) = C_k^{(0)}(\varphi^n \theta)$$

for $k, n \in \mathbb{Z}$ and $\theta \in \Theta$. Moreover, the sequence $\hat{d} = (C_k^{(0)})_{k \in \mathbb{Z}}$ belongs to $\hat{\mathfrak{B}}$ by the Claim. Thus, we can apply the homomorphism $\rho : \hat{\mathfrak{B}} \to \mathcal{L}(C_0(\Theta; X))$ to $\hat{d}$. Let $d = \rho(\hat{d})$. We will show that $db = bd = I$.

Consider the sequence $\hat{r} = \hat{d} * \hat{b} - \hat{e} \in \hat{\mathfrak{B}}$. Fix $\theta \in \Theta$ and apply $\hat{\pi}_\theta : \hat{\mathfrak{B}} \to \mathfrak{C}$. Using (7.5), (7.15), and Proposition 7.5, we compute

$$\hat{\pi}_\theta(\hat{r}) = \hat{\pi}_\theta(\hat{d}) \cdot \hat{\pi}_\theta(\hat{b}) - I = \left[\sum_k \operatorname{diag}(C_k^{(0)}(\varphi^n \theta))_{n \in \mathbb{Z}} S^k \right] \cdot \hat{\pi}_\theta(\hat{b}) - I$$

$$= \left[\sum_k \operatorname{diag}(C_k^{(n)}(\theta))_{n \in \mathbb{Z}} S^k \right] \cdot \hat{\pi}_\theta(\hat{b}) - I = \left[\sum_k C_k(\theta) S^k \right] \cdot \hat{\pi}_\theta(\hat{b}) - I = 0.$$

Since $\theta \in \Theta$ is arbitrary, Proposition 7.3 implies $\hat{r} = 0$ in $\hat{\mathfrak{B}}$. As a consequence, we have that

$$0 = \rho(\hat{r}) = \rho(\hat{d} * \hat{b} - \hat{e}) = \rho(\hat{d}) \cdot \rho(\hat{b}) - I = d \cdot b - I.$$

Similarly, we obtain the equality $0 = bd - I$. This completes the proof of the theorem.

Proof of Claim. Suppose that $k \in \mathbb{Z}$, $x \in X$, and $\theta_0 \in \Theta$; and define $\bar{x} = (x_n) \in c_0(\mathbb{Z}; X)$ by $x_n = x$ if $n = -k$, and $x_n = 0$ if $n \ne -k$. Then, for each $\theta \in \Theta$, we have that

$$|[C_k^{(0)}(\theta) - C_k^{(0)}(\theta_0)]x| = \Big|\sum_{l=-\infty}^{\infty}[C_l^{(0)}(\theta) - C_l^{(0)}(\theta_0)]x_{-l}\Big|$$

$$\leq \sup_{n\in\mathbb{Z}}\Big|\sum_l [C_l^{(n)}(\theta) - C_l^{(n)}(\theta_0)]x_{n-l}\Big|$$

$$= \Big\|\big(\sum_l \operatorname{diag}(C_l^{(n)}(\theta) - C_l^{(n)}(\theta_0))_{n\in\mathbb{Z}} S^l\big)\bar{x}\Big\|_{c_0(\mathbb{Z};X)}$$

$$= \Big\|\big([\hat{\pi}_\theta(\hat{b})]^{-1} - [\hat{\pi}_{\theta_0}(\hat{b})]^{-1}\big)\bar{x}\Big\|_{c_0(\mathbb{Z};X)}$$

(7.16)
$$\leq \Big\|[\hat{\pi}_\theta(\hat{b})]^{-1}\Big\|_{\mathcal{L}(c_0(\mathbb{Z};X))} \cdot \Big\|[\hat{\pi}_{\theta_0}(\hat{b}) - \hat{\pi}_\theta(\hat{b})]\bar{y}\Big\|_{c_0(\mathbb{Z};X)}$$

where $\bar{y} := [\hat{\pi}_{\theta_0}(\hat{b})]^{-1}\bar{x} \in c_0(\mathbb{Z};X)$. It is easy to see that the function from Θ to $\mathcal{L}_s(c_0(\mathbb{Z};X))$ defined by $\theta \mapsto \pi_\theta(b)$ is *strongly* continuous. Thus, the inequalities (7.16) and (7.12) imply the strong continuity of the function $\theta \mapsto C_k^{(0)}(\theta) \in \mathcal{L}_s(X)$. Finally, using (7.15) and (7.11), we conclude that

$$\sum_{k\in\mathbb{Z}}\sup_{\theta\in\Theta}\|C_k^{(0)}(\theta)\|_{\mathcal{L}(X)} = \sum_{k\in\mathbb{Z}}\sup_{\theta\in\Theta}\sup_{n\in\mathbb{Z}}\|C_k^{(0)}(\varphi^n\theta)\|_{\mathcal{L}(X)}$$

$$= \sum_{k\in\mathbb{Z}}\sup_{\theta\in\Theta}\|C_k(\theta)\|_{\mathcal{L}(c_0(\mathbb{Z};X))} < \infty.$$

$\square$

REMARK 7.6. If φ is aperiodic, then $\rho : \hat{\mathfrak{B}} \to \mathfrak{B}$ is an isomorphism and $\pi_\theta = \hat{\pi}_\theta \circ \rho^{-1}$. The proof of Theorem 7.4 shows that, in this case, if $b = I - aV$, then b^{-1} belongs to $\mathfrak{B}$, cf. Theorem 4.9. $\diamond$

In the course of the proof of Theorem 7.4 we showed that, for $b = I - aV$, the element $\hat{b}$ is invertible in $\hat{\mathfrak{B}}$ under condition (7.7). Moreover, the inverse of $\hat{b}$ is given by $\hat{d} = (C_k^{(0)})_{k\in\mathbb{Z}}$. Further on, where we study cocycles that are eventually norm-continuous in θ, we will need the following refinement of this fact. Let $\tilde{\mathfrak{A}}_{\text{norm}}$ be the closed subalgebra $C_b(\Theta; \mathcal{L}(X))$ in $\tilde{\mathfrak{A}}$ of all *norm* continuous, bounded, operator valued functions $a : \Theta \to \mathcal{L}(X)$. Also, let $\hat{\mathfrak{B}}_{\text{norm}}$ denote the corresponding closed subalgebra of $\hat{\mathfrak{B}}$; that is, $\hat{\mathfrak{B}}_{\text{norm}} = \{(a_k)_{k\in\mathbb{Z}} \in \hat{\mathfrak{B}} : a_k \in \tilde{\mathfrak{A}}_{\text{norm}}\}$.

COROLLARY 7.7. *Suppose that $a \in C_b(\Theta; \mathcal{L}(X))$; that is, a is norm continuous. If the conditions of Theorem 7.4 hold, then $\hat{d} = \hat{b}^{-1}$ belongs to $\hat{\mathfrak{B}}_{\text{norm}}$.*

PROOF. We have to check that the function $\theta \mapsto C_k^{(0)}(\theta) \in \mathcal{L}(X)$ is norm continuous (cf. the claim in the proof of Theorem 7.4). Fix $k \in \mathbb{Z}$ and recall the estimate (7.16)

$$|[C_k^{(0)}(\theta) - C_k^{(0)}(\theta_0)]x| \leq B \,\|[\hat{\pi}_{\theta_0}(\hat{b}) - \hat{\pi}_\theta(\hat{b})]\bar{y}\|_{c_0(\mathbb{Z};X)}$$

where $x \in X$ and the constant B is given in display (7.7). Also, using the identities (7.15), note that

$$\bar{y} = [\hat{\pi}_{\theta_0}(\hat{b})]^{-1}\bar{x} = \sum_{l=-\infty}^{\infty}\operatorname{diag}(C_l^{(n)}(\theta_0))_{n\in\mathbb{Z}} S^l \bar{x}$$

$$= (C_{n-k}^{(n)}(\theta_0)x)_{n\in\mathbb{Z}} = (C_{n,k}(\theta_0)x)_{n\in\mathbb{Z}}$$

where $\bar{x} = x \otimes (\delta_{n,-k})$. The inequality (7.14) yields $\|C_{n,k}(\theta_0)\| \leq Re^{-\gamma|n-k|}$ for $n \in \mathbb{Z}$. Fix $\varepsilon > 0$ and take $N = N_\varepsilon$ such that $Re^{-\gamma|n-k|} < \varepsilon$ for $|n| \geq N$. Using the continuity of the function $a : \Theta \to \mathcal{L}(X)$, choose $\delta > 0$ such that
$$\|a(\varphi^n\theta) - a(\varphi^n\theta_0)\|_{\mathcal{L}(X)} < \varepsilon \quad \text{whenever} \quad |n| < N \text{ and } d(\theta, \theta_0) < \delta.$$
Hence, if $d(\theta, \theta_0) < \delta$, then
$$\|[\hat{\pi}_{\theta_0}(\hat{b}) - \hat{\pi}_\theta(\hat{b})]\bar{y}\|_{c_0(\mathbb{Z};X)} \leq \max\{\sup_{|n|<N} |[a(\varphi^n\theta) - a(\varphi^n\theta_0)]C_{n,k}(\theta_0)x|,$$
$$\sup_{|n|\geq N} |[a(\varphi^n\theta) - a(\varphi^n\theta_0)]C_{n,k}(\theta_0)x|\}$$
$$\leq \varepsilon \max\{R, 2\sup_{\theta\in\Theta} \|a(\theta)\|_{\mathcal{L}(X)}\} |x|.$$
$\square$

7.1.2. The Discrete Dichotomy Theorem. The algebraic method developed in the previous section will be used to prove a result that relates several concepts; namely, global exponential dichotomy as defined in Definition 6.13, exponential dichotomies at points in the base space as defined in Definition 6.14, the hyperbolicity of the operator given by $(Ef)(\theta) = \Phi(\varphi^{-1}\theta)f(\varphi^{-1}\theta)$ on $C_0(\Theta; X)$, and the hyperbolicity of the operator $\pi_\theta(E) = \text{diag}(\Phi(\varphi^{n-1}\theta))_{n\in\mathbb{Z}} S$ on $c_0(\mathbb{Z}; X)$.

Let us observe that
$$(7.17) \qquad \pi_\theta(E)^k = \pi_\theta(E^k) = \text{diag}(\Phi^k(\varphi^{n-k}\theta))_{n\in\mathbb{Z}} S^k$$
for $k \in \mathbb{N}$ and $\theta \in \Theta$. Moreover, the operator $\pi_\theta(E)$ is also defined on $\ell^\infty(\mathbb{Z}, X)$, the space of bounded sequences endowed with the supremum norm.

PROPOSITION 7.8. *If $z \in \mathbb{T}$ and $\theta \in \Theta$, then $\|I - \pi_\theta(E)\|_\bullet = \|z - \pi_\theta(E)\|_\bullet$ on $c_0(\mathbb{Z}; X)$ or $\ell^\infty(\mathbb{Z}, X)$, and the spectrum $\sigma(\pi_\theta(E))$ in $c_0(\mathbb{Z}; X)$ or $\ell^\infty(\mathbb{Z}, X)$ is invariant with respect to rotations centered at the origin. Moreover, if $\sigma(\pi_\theta(E)) \cap \mathbb{T} = \emptyset$, then $\|(I - \pi_\theta(E))^{-1}\| = \|(z - \pi_\theta(E))^{-1}\|$.*

PROOF. Define $L_\xi = \text{diag}(e^{-in\xi})_{n\in\mathbb{Z}}$ on $c_0(\mathbb{Z}; X)$ or $\ell^\infty(\mathbb{Z}, X)$ for $\xi \in \mathbb{R}$, and note that L_ξ is an invertible isometry on both spaces. The desired result is a direct consequence of the following identity:
$$L_\xi [z - \pi_\theta(E)] L_\xi^{-1} = z - e^{-i\xi} \text{diag}(\Phi(\varphi^{n-1}\theta))_{n\in\mathbb{Z}} S = e^{-i\xi}(ze^{i\xi} - \pi_\theta(E)).$$
$\square$

By Corollary 6.43, the existence of exponential dichotomies for continuous and discrete time cocycles are equivalent. Thus, with no loss of generality, let us consider the case of discrete time.

THEOREM 7.9 (Discrete Dichotomy Theorem). *Let $\varphi : \Theta \to \Theta$ be a homeomorphism on a locally compact metric space Θ and $\{\Phi^t\}_{t\in\mathbb{N}}$, with $\Phi^1 = \Phi \in C_b(\Theta; \mathcal{L}_s(X))$, a discrete time cocycle over $\{\varphi^t\}_{t\in\mathbb{Z}}$. The following statements are equivalent.*

(a) *The LSPF $\{\hat{\varphi}^t\}_{t\in\mathbb{N}}$ has an exponential dichotomy on Θ over $\mathbb{Z}$.*
(b) *The LSPF $\{\hat{\varphi}^t\}_{t\in\mathbb{N}}$ has an exponential dichotomy over $\mathbb{Z}$ at each point $\theta \in \Theta$ with dichotomy constants $\beta(\theta) \geq \beta > 0$ and $M(\beta(\theta)) \leq M$.*
(c) *The operator $\pi_\theta(E)$ is hyperbolic on $c_0(\mathbb{Z}; X)$ for each $\theta \in \Theta$ and*
$$\sup_{z\in\mathbb{T}} \sup_{\theta\in\Theta} \|[z - \pi_\theta(E)]^{-1}\|_{\mathcal{L}(c_0(\mathbb{Z};X))} < \infty.$$

(d) *The operator E is hyperbolic on $C_0(\Theta; X)$.*

If (a)–(d) hold, let $P(\cdot)$ denote the dichotomy projection for $\{\hat\varphi^t\}_{t\in\mathbb{N}}$ on Θ, let $\{P_n : n \in \mathbb{Z}\}$ denote the family of the dichotomy projections for $\{\hat\varphi^t\}_{t\in\mathbb{N}}$ at $\theta \in \Theta$, let $p_\theta \in \mathcal{L}(c_0(\mathbb{Z}; X))$, for $\theta \in \Theta$, be the Riesz projection for $\pi_\theta(E)$ on $c_0(\mathbb{Z}; X)$ that corresponds to $\sigma(\pi_\theta(E)) \cap \mathbb{D}$, and let $\mathcal{P}$ be the Riesz projection for E on $C_0(\Theta; X)$ that corresponds to $\sigma(E) \cap \mathbb{D}$. Then,

$$\mathcal{P} = P(\cdot), \quad P_n = P(\varphi^n \theta), \quad \text{and} \quad p_\theta = \operatorname{diag}(P_n)_{n\in\mathbb{Z}}.$$

PROOF. (a) $\Rightarrow$ (b). Suppose that $\{\hat\varphi^t\}_{t\in\mathbb{N}}$ has an exponential dichotomy on Θ for discrete time with dichotomy constants β and M and dichotomy projection $P(\cdot) \in C_b(\Theta; \mathcal{L}_s(X))$ as in Definition 6.13. For $\theta \in \Theta$, the projections defined by $P_n := P(\varphi^n \theta)$, for $n \in \mathbb{Z}$, clearly satisfy Definition 6.14 with constants $\beta(\theta) = \beta$ and $M(\beta(\theta)) = M$.

(b) $\Rightarrow$ (c). Fix $\theta \in \Theta$. Let $\{P_n : n \in \mathbb{Z}\}$ denote the family of dichotomy projections for $\{\hat\varphi^t\}_{t\in\mathbb{N}}$ at θ as in Definition 6.14, and define $p_\theta = \operatorname{diag}(P_n)_{n\in\mathbb{Z}}$ on $c_0(\mathbb{Z}; X)$. By condition (a) in Definition 6.14, we have $p_\theta \pi_\theta(E) = \pi_\theta(E) p_\theta$. The identity (7.17) and condition (b) of Definition 6.14 imply that $(\pi_\theta(E) p_\theta)^k = \operatorname{diag}(\Phi^k(\varphi^{n-k}\theta) P_{n-k})_{n\in\mathbb{Z}} S^k$ and

$$(\pi_\theta(E) q_\theta)^{-k} = \operatorname{diag}([\Phi_Q^k(\varphi^n\theta) Q_n]^{-1})_{n\in\mathbb{Z}} S^{-k} = \operatorname{diag}(\Phi_Q^{-k}(\varphi^{n+k}\theta) Q_{n+k})_{n\in\mathbb{Z}} S^{-k}$$

for $k \in \mathbb{N}$. Thus, the estimates in condition (c) of Definition 6.14 yield the inequalities

$$\|(\pi_\theta(E) p_\theta)^k\|_{\mathcal{L}(c_0(\mathbb{Z};X))} \leq M e^{-\beta k} \quad \text{and} \quad \|(\pi_\theta(E) q_\theta)^{-k}\|_{\mathcal{L}(c_0(\mathbb{Z};X))} \leq M e^{-\beta k}.$$

As a result, $\pi_\theta(E)$ is hyperbolic and $(z, \theta) \mapsto \|[z - \pi_\theta(E)]^{-1}\|_{\mathcal{L}(c_0(\mathbb{Z};X))}$ is bounded.

(c) $\Rightarrow$ (d). For each $z \in \mathbb{T}$, apply Theorem 7.4 to $b = I - aV$ where $a(\theta) = z^{-1}\Phi(\varphi^{-1}\theta)$ and $(Vf)(\theta) = f(\varphi^{-1}\theta)$ for $\theta \in \Theta$.

The implication (d) $\Rightarrow$ (a) and the identity $\mathcal{P} = P(\cdot)$ are proved in the Dichotomy Theorem 6.41. $\square$

The equivalence (b)$\Leftrightarrow$(c) can be replaced by the following stronger result.

COROLLARY 7.10. *The LSPF $\{\hat\varphi^t\}_{t\geq 0}$ has an exponential dichotomy at $\theta \in \Theta$ over $\mathbb{Z}$ if and only if $\pi_\theta(E)$ is hyperbolic on $c_0(\mathbb{Z}; X)$.*

PROOF. Using the proof of the implication (b)$\Rightarrow$(c) of the Discrete Dichotomy Theorem 7.9, it suffices to prove that if $\pi_\theta(E)$ is hyperbolic, then $\{\hat\varphi^t\}_{t\geq 0}$ has an exponential dichotomy at $\theta \in \Theta$.

Let us suppose that $\pi_\theta(E)$ is hyperbolic. As in Commutation Lemma 3.15, we see that $p_\theta \chi = \chi p_\theta$ for every $\chi = (\chi_n) \in \ell^\infty(\mathbb{Z})$. By arguments similar to those used in the proof of the Spectral Projection Theorem 6.38, it follows that $p_{\theta_0} = \operatorname{diag}(P_n)_{n\in\mathbb{Z}}$ for a family of uniformly bounded projections $\{P_n \in \mathcal{L}(X) : n \in \mathbb{Z}\}$. Also, from the identity $p_\theta \pi_\theta(E) = \pi_\theta(E) p_\theta$, we have statement (a) in Definition 6.14. Further, by the identities in display (7.17) and the surjectivity of $\pi_\theta(E)^k$ on $\operatorname{Im} q_\theta$, the function $\Phi^k(\varphi^{n-k}\theta) : \operatorname{Im} Q_{n-k} \to \operatorname{Im} Q_n$ is surjective. Finally, the estimates in condition (c) of Definition 6.14 follow as in the proof of Theorem 6.41. $\square$

For each $\theta \in \Theta$, let us define the semigroup $\{\Pi_\theta^t\}_{t\geq 0}$ on $C_0(\mathbb{R}; X)$, called an *evolution semigroups along the trajectory through* $\theta \in \Theta$, by

$$(\Pi_\theta^t h)(s) := \Phi^t(\varphi^{s-t}\theta) h(s - t), \quad s \in \mathbb{R}, \ t \geq 0, \ h \in C_0(\mathbb{R}; X).$$

Pointwise dichotomies of linear skew-product flows can be characterized using these semigroups.

Since $\{\Phi^t\}_{t\geq 0}$ is an exponentially bounded, strongly continuous cocycle, the evolution family $\{U_\theta(s,\tau)\}_{s\geq\tau}$ defined by $U_\theta(s,\tau) := \Phi^{s-\tau}(\varphi^\tau\theta)$ is exponentially bounded and strongly continuous. In particular, $(\Pi_\theta^t h)(s) = U_\theta(s, s-t)h(s-t)$ and $\{\Pi_\theta^t\}_{t\geq 0}$ is a Howland evolution semigroup on $C_0(\mathbb{R}; X)$, see Chapter 3 for the definitions. By the Dichotomy Theorem 3.17, the evolution family $\{U_\theta(s,\tau)\}_{s\geq\tau}$ has an exponential dichotomy on $\mathbb{R}$ if and only if $\{\Pi_\theta^t\}_{t\geq 0}$ is hyperbolic or, equivalently, if and only if the generator Γ_θ of the semigroup $\{\Pi_\theta^t\}_{t\geq 0}$ is invertible. We summarize this information in the following lemma.

LEMMA 7.11. *The spectrum $\sigma(\Pi_\theta^t)$ is invariant under rotations centered at the origin of the complex plane, $\sigma(\Gamma_\theta)$ is invariant under translations along $i\mathbb{R}$, and the spectral mapping property holds; that is,*

$$\sigma(\Pi_\theta^t)\setminus\{0\} = e^{t\sigma(\Gamma_\theta)}, \quad t > 0.$$

Further, if $\sigma(\Pi_\theta^1) \cap \mathbb{T} = \emptyset$, then the Riesz projection $\mathcal{P}_\theta$ that corresponds to the operator Π_θ^1 and the spectral set $\sigma(\Pi_\theta^1) \cap \mathbb{D}$ is a multiplication operator; that is, $(\mathcal{P}_\theta f)(s) = P_\theta(s)f(s)$ for some bounded, continuous, projection-valued function $P_\theta : \mathbb{R} \to \mathcal{L}_s(X)$.

We remark that the existence of an exponential dichotomy for $\{U_\theta(s,\tau)\}_{s\geq\tau}$ is equivalent to the existence of an exponential dichotomy for $\{\Phi^t\}_{t\geq 0}$ at θ over $\mathbb{R}$. The next result concerning the exponential dichotomy for $\{\Phi^t\}_{t\geq 0}$ on Θ now follows from the Dichotomy Theorem 6.41.

THEOREM 7.12. *Suppose that $\{\Phi^t\}_{t\geq 0}$ is an exponentially bounded, strongly continuous cocycle over a continuous flow $\{\varphi^t\}_{t\in\mathbb{R}}$ on Θ. The following statements are equivalent.*

(i) *The cocycle $\{\Phi^t\}_{t\geq 0}$ has an exponential dichotomy on Θ.*
(ii) *$\sigma(\Pi_\theta^1) \cap \mathbb{T} = \emptyset$ for each $\theta \in \Theta$ and*

(7.18) $$\sup_{z\in\mathbb{T}}\sup_{\theta\in\Theta} \|[z - \Pi_\theta^1]^{-1}\|_{\mathcal{L}(C_0(\mathbb{R};X))} < \infty.$$

(iii) *$\sigma(\Gamma_\theta) \cap i\mathbb{R} = \emptyset$ for each $\theta \in \Theta$ and*

(7.19) $$\sup_{\xi\in\mathbb{R}}\sup_{\theta\in\Theta} \|[i\xi - \Gamma_\theta]^{-1}\|_{\mathcal{L}(C_0(\mathbb{R};X))} < \infty.$$

Moreover, the dichotomy projection $P(\cdot)$ on Θ for the cocycle $\{\Phi^t\}_{t\geq 0}$ and the Riesz projection $\mathcal{P}_\theta = P_\theta(\cdot)$ for Π_θ^1 are related as follows: $P_\theta(s) = P(\varphi^s\theta)$, $s \in \mathbb{R}$, $\theta \in \Theta$.

PROOF. To prove (i) $\Leftrightarrow$ (ii), let us introduce operators on the space

$$C_0(\mathbb{R} \times \Theta, X) = C_0(\Theta, C_0(\mathbb{R}, X)) = C_0(\mathbb{R}, C_0(\Theta, X))$$

that are defined by

$$(\Pi g)(s,\theta) = \Phi(\varphi^{s-1}\theta)g(s-1,\theta), \quad (\hat{E}g)(s,\theta) = \Phi(\varphi^{-1}\theta)g(s-1,\varphi^{-1}\theta),$$
$$(Jg)(s,\theta) = g(s,\varphi^s\theta), \quad (J^{-1}g)(s,\theta) = g(s,\varphi^{-s}\theta)$$

for $s \in \mathbb{R}$, $\theta \in \Theta$, and $g \in C_0(\mathbb{R} \times \Theta, X)$. These operators satisfy the identity

(7.20) $$J^{-1}\Pi J = \hat{E}.$$

Indeed, for $f(s,\theta) = g(s, \varphi^s \theta)$, we have that
$$(\Pi f)(s, \theta) = \Phi(\varphi^{s-1}\theta) g(s-1, \varphi^{s-1}\theta)$$
and
$$(J^{-1}\Pi J g)(s,\theta) = \Phi(\varphi^{s-1}(\varphi^{-s}\theta)) g(s-1, \varphi^{s-1}(\varphi^{-s}\theta)) = (\hat{E}g)(s,\theta).$$

Next, note that if $F: \Theta \to C_0(\mathbb{R}, X)$, then the operator Π acts as a multiplication operator with multiplier Π_θ^1: $(\Pi F)(\theta) = \Pi_\theta^1 F(\theta)$. Hence, for $z \in \mathbb{T}$, the operator $z - \Pi$ of multiplication by $z - \Pi_\theta^1$ is invertible on $C_0(\Theta, C_0(\mathbb{R}, X))$ if and only if $z - \Pi_\theta^1$ is invertible on $C_0(\mathbb{R}, X)$ for each $\theta \in \Theta$ and $\|(z - \Pi_\theta^1)^{-1}\| \le C$ for some $C > 0$, all $z \in \mathbb{T}$, and $\theta \in \Theta$. In other words, the equality $\sigma(\Pi) \cap \mathbb{T} = \emptyset$ is equivalent to statement (ii). Moreover, the Riesz projection $\mathbf{P}_\Pi$ that corresponds to the operator Π and the spectral set $\sigma(\Pi) \cap \mathbb{D}$ acts by the formula $(\mathbf{P}_\Pi h)(s, \theta) = P_\theta(s) h(s, \theta)$ where $\mathcal{P}_\theta = P_\theta(\cdot)$ is the Riesz projection that corresponds to Π_θ^1 and the spectral set $\sigma(\Pi_\theta^1) \cap \mathbb{D}$ as described in Lemma 7.11.

By Theorem 6.41, statement (i) is equivalent to the equality $\sigma(E) \cap \mathbb{T} = \emptyset$ on $C_0(\Theta; X)$. Also, by the equivalence (1)$\Leftrightarrow$(2) of Theorem 2.39, we conclude that $\sigma(E) \cap \mathbb{T} = \emptyset$ on $C_0(\Theta; X)$ if and only if $\sigma(\hat{E}) \cap \mathbb{T} = \emptyset$ on $C_0(\mathbb{R}, C_0(\Theta; X))$. Indeed, for $f: \mathbb{R} \to C_0(\Theta, X)$, the operator $\hat{E}$ acts as $(\hat{E}f)(s) = Ef(s-1)$, exactly the case considered in Theorem 2.39. Hence, statement (i) is equivalent to the equality $\sigma(\hat{E}) \cap \mathbb{T} = \emptyset$. Equation (7.20) implies $\sigma(\Pi) = \sigma(\hat{E})$, and hence (i) $\Leftrightarrow$ (ii). Also, the statement concerning the Riesz and dichotomy projections holds.

To prove the equivalence (iii) $\Leftrightarrow$ (ii), observe that $\sigma(\Pi_\theta^1) \cap \mathbb{T} = \emptyset$ if and only if $\sigma(\Gamma_\theta) \cap \mathbb{T} = \emptyset$ by Lemma 7.11.

To prove that (7.18) implies (7.19), we first remark that
$$\|(i\xi - \Gamma_\theta)^{-1}\| = \|\Gamma_\theta^{-1}\| \quad \text{for each} \quad \xi \in \mathbb{R} \quad \text{and} \quad \theta \in \Theta,$$
see display (3.8). For each $\theta \in \Theta$, consider the Green's operator $\mathbb{G}_\theta$ defined as in display (4.25) for the evolution family $\{U_\theta(s,\tau)\}_{s \ge \tau}$. By Theorem 4.25 we have $\Gamma_\theta^{-1} = -\mathbb{G}_\theta$. Since statement (ii) in Theorem 7.12 implies $\{\Phi^t\}_{t \ge 0}$ has an exponential dichotomy on $\{\Phi^t\}_{t \ge 0}$ with the dichotomy projection $P(\cdot)$, formula (4.25), applied to the exponentially dichotomic evolution family $\{U_\theta(s,\tau)\}_{s \ge \tau}$ and the dichotomy projection $P_\theta(\cdot) = P(\varphi^{(\cdot)}\theta)$, yields the identity

$$(\mathbb{G}_\theta h)(s) = \int_{-\infty}^{s} U_{\theta, P}(s, \tau) h(\tau)\, d\tau - \int_{s}^{\infty} [U_{\theta, Q}(\tau, s)]^{-1} h(\tau)\, d\tau$$
$$= \int_{-\infty}^{s} \Phi_P^{s-\tau}(\varphi^\tau \theta) h(\tau)\, d\tau - \int_{s}^{\infty} [\Phi_Q^{\tau-s}(\varphi^s \theta)]^{-1} h(\tau)\, d\tau$$

for each $h \in C_0(\mathbb{R}; X)$ and $s \in \mathbb{R}$. Hence, $\|\Gamma_\theta^{-1}\| \le 2M/\beta$ where M and β are the dichotomy constants for the cocycle $\{\Phi^t\}_{t \ge 0}$, and the bound (7.19) follows.

To prove that (7.19) implies (7.18), we will use Baskakov's Theorem 4.37. First, observe that the operator $\pi_\theta(\Pi_\theta^1)$ defined on $c_0(\mathbb{Z}; X)$ by formula (4.36) for the evolution operator Π_θ^1 *is equal to* the operator $\pi_\theta(E)$ defined in display (7.1) for the evolution operator E. Indeed,
$$\pi_\theta(\Pi_\theta^1): (x_n)_{n \in \mathbb{Z}} \mapsto (U_\theta(n, n-1) x_{n-1})_{n \in \mathbb{Z}} = (\Phi(\varphi^{n-1}\theta) x_{n-1})_{n \in \mathbb{Z}}, \quad \theta \in \Theta.$$

Using (7.19) and formula (4.37) in the Baskakov Theorem 4.37 for the operator Γ_θ, we conclude that

$$\sup_{\theta \in \Theta} \|[I - \pi_\theta(E)]^{-1}\| = \sup_{\theta \in \Theta} \|[I - \pi_0(\Pi_\theta^1)]^{-1}\| \leq 1 + C \sup_{\theta \in \Theta} \|\Gamma_\theta^{-1}\| < \infty.$$

This gives (c) in the Discrete Dichotomy Theorem 7.9, and (7.18) follows. $\square$

7.1.3. A Perturbation Theorem. We will generalize the Sacker-Sell Perturbation Theorem (Theorem 6 in Sacker and Sell [**SS3**]) to an infinite dimensional setting, and thereby establish the semicontinuity of the dynamical spectrum as a function of φ^t-invariant compact subsets of Θ. To this end, let us assume that Θ is a compact metric space and the cocycle $\{\Phi^t\}_{t \geq 0}$ is *eventually norm continuous in θ*; that is, for some $t_0 > 0$, the map from Θ to $\mathcal{L}(X)$ given by $\theta \mapsto \Phi^{t_0}(\theta)$ is continuous in the operator norm, see Nagel [**Na**, p. 38]. By rescaling the time, it suffices to suppose that $t_0 = 1$ and consider the case of discrete time $t \in \mathbb{N}$. Thus, we will consider a homeomorphism φ of Θ and the cocycle $\{\Phi^t\}_{t \in \mathbb{N}}$ over $\{\varphi^t\}_{t \in \mathbb{Z}}$ where $\Phi(\cdot)$ is norm-continuous. As usual, we define $\Phi := \Phi^1$ and $E := E^1$.

The results of Subsection 7.1.1 are used to obtain the following variant of the Spectral Projection Theorem 6.38.

PROPOSITION 7.13. *Suppose that the function from Θ to $\mathcal{L}(X)$ given by $\theta \mapsto \Phi^1(\theta)$ is norm continuous. If $\sigma(E) \cap \mathbb{T} = \emptyset$, then $\{\Phi^n\}_{n \in \mathbb{N}}$ has an exponential dichotomy with a norm continuous dichotomy projection $P(\cdot)$.*

PROOF. Assume that E is hyperbolic. By the Discrete Dichotomy Theorem 7.9, the Riesz projection p_θ for $\pi_\theta(E) = \pi_\theta(aV)$ is given by

$$p_\theta = \text{diag}(P(\varphi^n \theta))_{n \in \mathbb{Z}}.$$

Notice that $a = \Phi^1 \circ \varphi^{-1} \in \tilde{\mathfrak{A}}_{\text{norm}}$, see the notations on page 212. Define $\hat{a} = (a_n)_{n \in \mathbb{Z}} \in \hat{\mathfrak{B}}_{\text{norm}}$ by $a_1 = a$ and $a_n = 0$ otherwise. Also, let $\hat{b}_z = \hat{e} - z^{-1}\hat{a}$ for $z \in \mathbb{T}$.

By equation (7.8), we have that $\hat{\pi}_\theta(\hat{b}_z) = \pi_\theta(I - z^{-1}aV) = I - z^{-1}\pi_\theta(E)$; and, by the Discrete Dichotomy Theorem 7.9,

$$\|(I - z^{-1}\pi_\theta(E))^{-1}\|_{\mathcal{L}(c_0(\mathbb{Z};X))} \leq C \quad \text{for all} \quad \theta \in \Theta \text{ and } z \in \mathbb{T}.$$

Also, by an application of Corollary 7.7 to $b_z = I - z^{-1}aV$, the element $z\hat{e} - \hat{a} \in \hat{\mathfrak{B}}_{\text{norm}}$ is invertible in $\hat{\mathfrak{B}}_{\text{norm}}$ for each $z \in \mathbb{T}$.

Define the idempotent $\hat{p}$ in the algebra $\hat{\mathfrak{B}}_{\text{norm}}$ by

$$\hat{p} = \frac{1}{2\pi i} \int_{\mathbb{T}} (z\hat{e} - \hat{a})^{-1} dz.$$

Since $\hat{\pi}_\theta : \hat{\mathfrak{B}} \to \hat{\mathfrak{C}}$ is a continuous homomorphism for each $\theta \in \Theta$, we obtain the identity

$$\hat{\pi}_\theta(\hat{p}) = \frac{1}{2\pi i} \int_{\mathbb{T}} \hat{\pi}_\theta([z\hat{e} - \hat{a}]^{-1}) \, dz = \frac{1}{2\pi i} \int_{\mathbb{T}} [z - \pi_\theta(aV)]^{-1} \, dz = p_\theta$$

for each $\theta \in \Theta$. On the order hand, let $\hat{p} = (P_k)_{k \in \mathbb{Z}}$ for $P_k \in \tilde{\mathfrak{A}}_{\text{norm}}$. Then, by the definition in display (7.5) of $\hat{\pi}_\theta$, we have

$$0 = \hat{\pi}_\theta(\hat{p}) - p_\theta = \pi_\theta(P_0) - p_\theta + \sum_{k \neq 0} \pi_\theta(P_k) S^k$$

where $\pi_\theta(P_k) = \mathrm{diag}(P_k(\varphi^n\theta))_{n\in\mathbb{Z}}$. Since the representation of $\hat{\pi}_\theta(\hat{p}) - p_\theta \in \mathfrak{C}$ as a series in powers of S is unique (Remark 7.2), we conclude that $\pi_\theta(P_0) = p_\theta$ and $\pi_\theta(P_k) = 0$ for $k \neq 0$ and each $\theta \in \Theta$. So Proposition 7.3 yields $P_k = 0$ for $k \neq 0$; hence, $P = P_0 \in \hat{\mathfrak{A}}_{\mathrm{norm}}$ (use the sequence $(\delta_{n0}P_k)_n \in \hat{\mathfrak{B}}$). $\square$

Let $\mathcal{K}$ be the set of all φ-invariant compact subsets of Θ and let d denote the metric on Θ. For $\Theta_0 \in \mathcal{K}$, some fixed element of $\mathcal{K}$, let us define the function $\mathrm{dist}_{\Theta_0} : \mathcal{K} \to \mathbb{R}$ by

$$\mathrm{dist}_{\Theta_0}(\Theta_1) = \sup_{\theta_1 \in \Theta_1} \inf_{\theta_0 \in \Theta_0} d(\theta_0, \theta_1).$$

Further, let us denote by $\Sigma(\Theta_1) = \Sigma(\{\hat{\varphi}^t | \Theta_1 \times X\})$ the Sacker-Sell dynamical spectrum (see Definition 6.17) of the linear skew-product flow $\{\hat{\varphi}^t\}_{t\in\mathbb{N}}$ restricted to $\Theta_1 \times X$. The next result establishes the upper semicontinuity of the function $\Theta_1 \mapsto \Sigma(\Theta_1)$ on $\mathcal{K}$.

THEOREM 7.14 (Perturbation Theorem). *Suppose that $\{\Phi^t\}_{t\in\mathbb{N}}$ is a discrete time cocycle over a homeomorphism φ on a compact metric space Θ generated by $\Phi \in C_b(\Theta; \mathcal{L}(X))$. If the corresponding LSPF $\{\hat{\varphi}^t\}_{t\in\mathbb{N}}$ has an exponential dichotomy on $\Theta_0 \in \mathcal{K}$, then there exists $\delta_* > 0$ such that $\{\hat{\varphi}^t\}_{t\in\mathbb{N}}$ has an exponential dichotomy on each $\Theta_1 \in \mathcal{K}$ whenever $\delta \in (0, \delta_*)$ and*

(7.21) $$\mathrm{dist}_{\Theta_0}(\Theta_1) \leq \delta.$$

The proof of Theorem 7.14 uses only elementary calculations with Neumann series and the Discrete Dichotomy Theorem 7.9. We will sketch the argument before presenting the formal proof. For $\varepsilon > 0$, a sequence $\bar{\theta} = \{\theta_n\}_{n\in\mathbb{Z}} \subset \Theta$ is called an ε-pseudo-orbit for φ if $d(\varphi\theta_n, \theta_{n+1}) \leq \varepsilon$ for all $n \in \mathbb{Z}$. In analogy with the family of operators $\{\pi_\theta(E) : \theta \in \Theta\}$ (see display (7.1)), let us define, for the ε-pseudo-orbit $\bar{\theta} = \{\theta_n\}_{n\in\mathbb{Z}} \subset \Theta_0$, the operator $T_{\bar{\theta}} : (x_n)_{n\in\mathbb{Z}} \mapsto (\Phi(\theta_{n-1})x_{n-1})_{n\in\mathbb{Z}}$ on $c_0(\mathbb{Z}; X)$.

Under the assumption that the skew-product flow $\{\hat{\varphi}^t\}_{t\in\mathbb{N}}$ has an exponential dichotomy on Θ_0, we will show that if ε is sufficiently small, then each operator $I - T_{\bar{\theta}}$ is invertible on $c_0(\mathbb{Z}, X)$ and there is a finite uniform bound $\sup\{\|(I - T_{\bar{\theta}})^{-1}\| : \bar{\theta}\}$ where the supremum is taken over the set of all ε-pseudo-orbits in Θ_0. Then, we will fix some $\theta \in \Theta_1$ and choose an ε-pseudo-orbit $\bar{\theta} = \{\theta_n\} \subset \Theta_0$ such that $d(\theta_n, \varphi^n\theta)$ is small for each $n \in \mathbb{Z}$. The continuity of $\Phi(\cdot)$ is used to show that, for this choice of $\{\theta_n\}$, the norm $\|\pi_\theta(E) - T_{\bar{\theta}}\|_{\mathcal{L}(c_0(\mathbb{Z};X))}$ is small for sufficiently small $\varepsilon > 0$. Since $I - T_{\bar{\theta}}$ is invertible for $\bar{\theta} \subset \Theta_0$, we conclude that $I - \pi_\theta(E)$ is invertible and $\|(I - \pi_\theta(E))^{-1}\| \leq C$ for some $C > 0$ and all $\theta \in \Theta_1$. Hence, $\{\hat{\varphi}^t\}_{t\in\mathbb{N}}$ has an exponential dichotomy on Θ_1 by the Discrete Dichotomy Theorem 7.9.

PROOF. Assume that $\{\hat{\varphi}^t\}_{t\in\mathbb{N}}$ has an exponential dichotomy on $\Theta_0 \in \mathcal{K}$ with dichotomy constants M and β and dichotomy projection $P(\cdot)$. Set

$$c = \max_{\theta \in \Theta_0}\{\|P(\theta)\|, 1\}$$

and choose $\gamma > 0$ such that

(7.22) $$\gamma \leq \frac{1}{16c}.$$

7.1. POINTWISE DICHOTOMIES

Let us fix $N \in \mathbb{N}$ such that $Me^{-\beta N} \leq \gamma$, and define $\psi := \varphi^N$ and $\Psi(\theta) := \Phi^N(\theta)$. Since $\hat{\varphi}^N$ has an exponential dichotomy on Θ_0, if $\theta \in \Theta_0$, then

(7.23) $$\Psi(\theta)P(\theta) = P(\psi\theta)\Psi(\theta),$$
(7.24) $$\Psi_Q(\theta) : \operatorname{Im} Q(\theta) \to \operatorname{Im} Q(\psi\theta) \text{ is invertible,}$$
(7.25) $$\|\Psi_P(\theta)\| \leq \gamma,$$
(7.26) $$\|\Psi_Q^{-1}(\theta)\| \leq \gamma.$$

Recall that $P(\cdot)$ and $\Psi(\cdot)$ are norm-continuous and Θ_0 and Θ are compact. Fix $\varepsilon_* > 0$ such that, for every $\varepsilon \in (0, \varepsilon_*)$, the following inequalities hold:

(7.27) $$\sup\nolimits_{d(\theta,\eta)\leq\varepsilon,\,\theta,\eta\in\Theta} \|\Psi(\theta) - \Psi(\eta)\| \leq 3/16,$$
(7.28) $$\sup\nolimits_{d(\theta,\eta)\leq\varepsilon,\,\theta,\eta\in\Theta_0} \|P(\theta) - P(\eta)\| \leq \frac{1}{2(1+2c)},$$
(7.29) $$c\sup\nolimits_{d(\theta,\eta)\leq\varepsilon,\,\theta,\eta\in\Theta_0} \|P(\theta) - P(\eta)\| \cdot \max\nolimits_{\theta\in\Theta}\|\Psi(\theta)\| \leq 3/8.$$

Also, for a sequence $\bar{\theta} = \{\theta_n\}_{n\in\mathbb{Z}} \subset \Theta_0$, let us define four operators on $c_0(\mathbb{Z}; X)$ by

$$T_{\bar{\theta}} : (x_n)_{n\in\mathbb{Z}} \mapsto (\Psi(\theta_{n-1})x_{n-1})_{n\in\mathbb{Z}}, \qquad P_{\bar{\theta}} = \operatorname{diag}\left(P(\theta_n)\right)_{n\in\mathbb{Z}},$$

$$Q_{\bar{\theta}} = I - P_{\bar{\theta}}, \qquad \hat{p}_{\bar{\theta}} = \operatorname{diag}\left(P(\psi\theta_{n-1})\right)_{n\in\mathbb{Z}}.$$

Then, using the identity (7.23), we have that

(7.30) $$T_{\bar{\theta}} P_{\bar{\theta}} = \hat{p}_{\bar{\theta}} T_{\bar{\theta}} \quad \text{and} \quad T_{\bar{\theta}} Q_{\bar{\theta}} = (I - \hat{p}_{\bar{\theta}}) T_{\bar{\theta}}.$$

The following lemma shows that if an exponential dichotomy exists; that is, the properties (7.23)–(7.26) hold, then the operator $I - T_{\bar{\theta}}$ is invertible whenever $\bar{\theta}$ is an ε-pseudo-orbit for ψ in Θ_0 and ε is given by (7.27)–(7.29).

LEMMA 7.15. *If $\bar{\theta} = \{\theta_n\} \subset \Theta_0$ is an ε-pseudo-orbit for ψ and ε satisfies (7.27)–(7.29), then $I - T_{\bar{\theta}}$ is invertible on $c_0(\mathbb{Z}, X)$ and*

(7.31) $$\|(I - T_{\bar{\theta}})^{-1}\| \leq 8/3.$$

Let us finish the proof of Theorem 7.14; the proof of Lemma 7.15 will be given below. Fix ε satisfying (7.27)–(7.29) and $\delta_* \leq \varepsilon/4$ such that $d(\psi\theta, \psi\eta) \leq \varepsilon/2$ whenever $d(\theta, \eta) < 2\delta_*$. Let $\delta < \delta_*$ and let $\Theta_1 \in \mathcal{K}$ satisfy the inequality (7.21). Also, for $\theta \in \Theta_1$, define the operator

$$T_\theta := \pi_\theta(E^N) : (x_n)_{n\in\mathbb{Z}} \mapsto (\Psi(\psi^{n-1}\theta)x_{n-1})_{n\in\mathbb{Z}}.$$

For $\theta \in \Theta_1$ and each $n \in \mathbb{Z}$, use the inequality (7.21) to find $\theta_n \in \Theta_0$ such that $d(\theta_n, \psi^n\theta) < 2\delta$. The sequence $\bar{\theta} = \{\theta_n\} \subset \Theta_0$ is an ε-pseudo-orbit for ψ because

$$d(\theta_n, \psi\theta_{n-1}) \leq d(\theta_n, \psi^n\theta) + d(\psi(\psi^{n-1}\theta), \psi(\theta_{n-1})) \leq 2\delta_* + \varepsilon/2 \leq \varepsilon.$$

Thus, Lemma 7.15 can be applied to the operator $I - T_{\bar{\theta}}$ to obtain the inequality $\|(I - T_{\bar{\theta}})^{-1}\| \leq 8/3$. Also, by the inequality (7.27), we have that

(7.32) $$\|T_\theta - T_{\bar{\theta}}\| = \sup_{n\in\mathbb{Z}}\|\Psi(\psi^n\theta) - \Psi(\theta_n)\| \leq \sup_{d(\vartheta,\eta)\leq\varepsilon}\|\Psi(\vartheta) - \Psi(\eta)\| \leq 3/16.$$

In view of this result and the inequalities (7.31) and (7.32), it follows that the operator

$$I - T_\theta = I - T_{\bar{\theta}} - (T_\theta - T_{\bar{\theta}}) = (I - T_{\bar{\theta}})\left[I - (I - T_{\bar{\theta}})^{-1}(T_\theta - T_{\bar{\theta}})\right]$$

is invertible. Moreover, by the usual estimate for the Neumann series, we have that $\|(I - T_\theta)^{-1}\| \leq 16/3$.

As a consequence of the Discrete Dichotomy Theorem 7.9, E^N is hyperbolic on $C_0(\Theta_1; X)$. Hence, E is hyperbolic on $C_0(\Theta_1; X)$ and, by the Dichotomy Theorem 6.41, $\{\hat{\varphi}^t\}_{t \in \mathbb{N}}$ has an exponential dichotomy on Θ_1. □

Proof of Lemma 7.15. In the decomposition $c_0(\mathbb{Z}, X) = \operatorname{Im} P_{\bar\theta} \oplus \operatorname{Im} Q_{\bar\theta}$, we represent $T_{\bar\theta}$ as $T_{\bar\theta} = A + B$ where

$$(7.33) \qquad A = \begin{bmatrix} P_{\bar\theta} T_{\bar\theta} P_{\bar\theta} & 0 \\ 0 & Q_{\bar\theta} T_{\bar\theta} Q_{\bar\theta} \end{bmatrix} \quad \text{and} \quad B = \begin{bmatrix} 0 & P_{\bar\theta} T_{\bar\theta} Q_{\bar\theta} \\ Q_{\bar\theta} T_{\bar\theta} P_{\bar\theta} & 0 \end{bmatrix}.$$

The proof of the lemma is completed in the following four steps.

Step 1. We will show that

$$(7.34) \qquad \max\{\|P_{\bar\theta} T_{\bar\theta} P_{\bar\theta}\|, \|[Q_{\bar\theta} T_{\bar\theta} Q_{\bar\theta}]^{-1}\|\} \le 1/8.$$

First, the inequalities (7.22) and (7.25) imply

$$\|P_{\bar\theta} T_{\bar\theta} P_{\bar\theta}\| = \sup_{n \in \mathbb{Z}} \|P(\theta_n) \Psi(\theta_{n-1}) P(\theta_{n-1})\| \le c\gamma \le 1/16.$$

Further, by the identities (7.30), we have

$$(7.35) \qquad Q_{\bar\theta} T_{\bar\theta} Q_{\bar\theta} = Q_{\bar\theta}(I - \hat{p}_{\bar\theta}) \cdot (I - \hat{p}_{\bar\theta}) T_{\bar\theta} Q_{\bar\theta}.$$

The operator $(I - \hat{p}_{\bar\theta}) T_{\bar\theta} Q_{\bar\theta} : \operatorname{Im} Q_{\bar\theta} \to \operatorname{Im}(I - \hat{p}_{\bar\theta})$ is invertible by (7.24) and (7.26), and

$$(7.36) \qquad \|[(I - \hat{p}_{\bar\theta}) T_{\bar\theta} Q_{\bar\theta}]^{-1}\| = \sup_{n \in \mathbb{Z}} \|\Psi_Q^{-1}(\theta_n)\| \le \gamma.$$

To prove that the operator $Q_{\bar\theta}(I - \hat{p}_{\bar\theta}) : \operatorname{Im}(I - \hat{p}_{\bar\theta}) \to \operatorname{Im} Q_{\bar\theta}$ is invertible, let us consider the operator

$$R := I - (P_{\bar\theta} - Q_{\bar\theta})(P_{\bar\theta} - \hat{p}_{\bar\theta}) = Q_{\bar\theta}(I - \hat{p}_{\bar\theta}) + P_{\bar\theta} \hat{p}_{\bar\theta}$$

on $c_0(\mathbb{Z}, X)$ and its decomposition

$$R : \operatorname{Im} \hat{p}_{\bar\theta} \oplus \operatorname{Im}(I - \hat{p}_{\bar\theta}) \to \operatorname{Im} P_{\bar\theta} \oplus \operatorname{Im} Q_{\bar\theta};$$

$$R = \begin{bmatrix} P_{\bar\theta} \\ Q_{\bar\theta} \end{bmatrix} R \, [\hat{p}_{\bar\theta}, I - \hat{p}_{\bar\theta}] = \begin{bmatrix} P_{\bar\theta} \hat{p}_{\bar\theta} & 0 \\ 0 & Q_{\bar\theta}(I - \hat{p}_{\bar\theta}) \end{bmatrix}.$$

Since $\bar\theta = \{\theta_n\}$ is an ε-pseudo-orbit for ψ, the inequality (7.28) yields

$$\|(P_{\bar\theta} - Q_{\bar\theta})(P_{\bar\theta} - \hat{p}_{\bar\theta})\| \le (1 + 2c)\|P_{\bar\theta} - \hat{p}_{\bar\theta}\| = (1 + 2c) \sup_{n \in \mathbb{Z}} \|P(\theta_n) - P(\psi \theta_{n-1})\| \le 1/2.$$

It follows that R is invertible. Therefore, $Q_{\bar\theta}(I - \hat{p}_{\bar\theta})$ is invertible with $\|[Q_{\bar\theta}(I - \hat{p}_{\bar\theta})]^{-1}\| \le \|R^{-1}\| \le 2$. This estimate together with (7.35), (7.36), and (7.22) gives

$$\|[Q_{\bar\theta} T_{\bar\theta} Q_{\bar\theta}]^{-1}\| \le 2\gamma \le 1/8.$$

Step 2. We will show that $I - A$ is invertible and

$$(7.37) \qquad \|(I - A)^{-1}\| \le 4/3.$$

The inequality (7.34) implies that the operators

$$P_{\bar\theta} - P_{\bar\theta} T_{\bar\theta} P_{\bar\theta} \quad \text{and} \quad Q_{\bar\theta} - Q_{\bar\theta} T_{\bar\theta} Q_{\bar\theta} = -Q_{\bar\theta} T_{\bar\theta} Q_{\bar\theta} \left(Q_{\bar\theta} - (Q_{\bar\theta} T_{\bar\theta} Q_{\bar\theta})^{-1}\right)$$

are invertible on $\operatorname{Im} P_{\bar\theta}$ and $\operatorname{Im} Q_{\bar\theta}$, respectively. Using this fact, we have the equality

$$(I - A)^{-1} = (P_{\bar\theta} - P_{\bar\theta} T_{\bar\theta} P_{\bar\theta})^{-1} \oplus (Q_{\bar\theta} - Q_{\bar\theta} T_{\bar\theta} Q_{\bar\theta})^{-1}$$

$$= \sum_{k=0}^{\infty} (P_{\bar\theta} T_{\bar\theta} P_{\bar\theta})^k \ominus \sum_{k=1}^{\infty} (Q_{\bar\theta} T_{\bar\theta} Q_{\bar\theta})^{-k},$$

and, as a result, $\|(I - A)^{-1}\| \leq 4/3$.

Step 3. We will estimate $\|B\| = \max\{\|P_{\bar\theta} T_{\bar\theta} Q_{\bar\theta}\|, \|Q_{\bar\theta} T_{\bar\theta} P_{\bar\theta}\|\}$. By the identities (7.30),
$$P_{\bar\theta} T_{\bar\theta} Q_{\bar\theta} = P_{\bar\theta}(P_{\bar\theta} - \hat{p}_{\bar\theta}) T_{\bar\theta} \quad \text{and} \quad Q_{\bar\theta} T_{\bar\theta} P_{\bar\theta} = (\hat{p}_{\bar\theta} - P_{\bar\theta}) \hat{p}_{\bar\theta} T_{\bar\theta}.$$
We also have $\|P_{\bar\theta}\| \leq c$, $\|\hat{p}_{\bar\theta}\| \leq c$,
$$\|\hat{p}_{\bar\theta} - P_{\bar\theta}\| \leq \sup_{d(\theta,\eta) \leq \varepsilon} \|P(\theta) - P(\eta)\|,$$
and $\|T_{\bar\theta}\| \leq \sup_{\theta \in \Theta} \|\Psi(\theta)\|$. Thus, by the estimate (7.29), we obtain the inequality
$$(7.38) \qquad \|B\| \leq c \sup_{d(\theta,\eta) \leq \varepsilon} \|P(\theta) - P(\eta)\| \cdot \sup_{\theta \in \Theta} \|\Psi(\theta)\| \leq 3/8\,.$$

Step 4. The estimates in (7.37) and (7.38) are used to finish the proof. Indeed, we have that $\|(I - A)^{-1} B\| \leq \frac{1}{2}$ and
$$\|(I - T_{\bar\theta})^{-1}\| = \|[(I - A)(I - (I - A)^{-1} B]^{-1}\|$$
$$\leq \|(I - A)^{-1}\| \|[I - (I - A)^{-1} B]^{-1}\| \leq 8/3. \qquad \square$$

We conclude this section with a typical application of Theorem 7.14. For a variational equation, this theorem proves the roughness of dichotomy with respect to small perturbation in the usual metric on the space $C(\mathbb{R}; \mathcal{L}(X))$ with the topology of uniform convergence on compact sets, see (6.4). This is to be compared with the roughness of dichotomy with respect to the norm on $C_b(\mathbb{R}; \mathcal{L}(X))$, see Sacker and Sell [**SS3**, §5].

Consider the variational equation
$$(7.39) \qquad \dot{x} = [A + B(\varphi^t \theta)] x$$
where A is a generator of a strongly continuous semigroup on X, and B is a bounded function from Θ to $\mathcal{L}(X)$ such that the function from Θ to $C(\mathbb{R}; \mathcal{L}(X))$ defined by
$$(7.40) \qquad \theta \mapsto B(\varphi^{(\cdot)} \theta)$$
is continuous; that is,
$$\sup_{t \in K} \|B(\varphi^t \theta_1) - B(\varphi^t \theta_2)\|_{\mathcal{L}(X)} \to 0 \quad \text{as} \quad \theta_1 \to \theta_2$$
for each compact set $K \subset \mathbb{R}$. Assume that the equation (7.39) has an exponential dichotomy at a point $\theta_0 \in \Theta$. By the Perturbation Theorem 7.14, there is an $\varepsilon > 0$ such that the differential equation (7.39) has an exponential dichotomy for each $\theta \in \Theta$ with the property that the orbits $\mathcal{O}(\theta)$ and $\mathcal{O}(\theta_0)$ satisfy $\text{dist}_{\overline{\mathcal{O}(\theta_0)}}(\overline{\mathcal{O}(\theta)}) < \varepsilon$. Indeed, the cocycle $\{\Phi^t\}_{t \geq 0}$ for (7.39) satisfies the identity
$$\Phi^t(\theta) x = e^{tA} x + \int_0^t e^{(t-\tau)A} B(\varphi^\tau \theta) \Phi^\tau(\theta) x \, d\tau, \quad x \in X,$$
see Chow and Leiva [**CLe2**, Thm. 5.1] and Example 6.8. Using Gronwall's inequality, the boundedness of B, and condition (7.40), it is easy to check that the function from Θ to $\mathcal{L}(X)$ defined by $\theta \mapsto \Phi^1(\theta)$ is norm continuous.

7.2. The Annular Hull Theorem

In this section we will characterize the existence of an exponential dichotomy for the cocycle $\{\Phi^t\}_{t\geq 0}$ using the spectrum of the generator Γ of the corresponding evolution semigroup $\{E^t\}_{t\geq 0}$ on the space $\mathcal{F} = C_0(\Theta; X)$. The main result here is the Annular Hull Theorem. We will also give an alternative proof of the Spectral Mapping Theorem for $\{E^t\}_{t\geq 0}$. In particular, under the assumption that $\sigma_{\mathrm{ap}}(E) \cap \mathbb{T} \neq \emptyset$, we will give an explicit construction of the approximate eigenfunctions for Γ. These approximate eigenfunctions are supported in flow boxes selected to contain *Mañe sequences* for $\{\Phi^t\}_{t\in\mathbb{N}}$. Throughout this section, the cocycle $\{\Phi^t\}_{t\geq 0}$ is assumed to be exponentially bounded and strongly continuous over a continuous base flow $\{\varphi^t\}_{t\in\mathbb{R}}$ on a locally compact metric space Θ.

7.2.1. Mañe sequences. We will define and study an infinite dimensional generalization of the notion of a Mañe point, a concept discussed in Subsection 6.2.2.

DEFINITION 7.16. A sequence $\{\theta_m\}_{m=1}^\infty \subset \Theta$ is called a *Mañe sequence* if there is a sequence of vectors $\{x_m\}_{m=1}^\infty$ in X, together with positive constants C and c, such that, for each positive integer m,

$$|\Phi^m(\theta_m)x_m| \geq c \tag{7.41}$$

and, for $k = 0, 1, \ldots, 2m$,

$$|\Phi^k(\theta_m)x_m| \leq C. \tag{7.42}$$

REMARK 7.17. The concept of Mañe sequence is related to the concept of Mañe point and Mañe vector, see also Lemma 6.29, Corollary 7.28, Corollary 7.29, and Corollary 7.30. Indeed, let us suppose that $\{\Phi^n\}_{n\in\mathbb{Z}}$ is a cocycle with *invertible* values and $\theta_0 \in \Theta$ is a Mañe point with $x_0 \in X$ a corresponding Mañe vector; that is, $x_0 \neq 0$ and

$$C := \sup\{|\Phi^n(\theta_0)x_0| : n \in \mathbb{Z}\} < \infty,$$

see Subsection 6.2.2. We claim that the points $\theta_m = \varphi^{-m}\theta_0$ form a Mañe sequence with corresponding Mañe vectors $x_m = \Phi^{-m}(\theta_0)x_0$. To see this fact, simply note that

$$|\Phi^m(\theta_m)x_m| = |\Phi^m(\varphi^{-m}\theta_0)\Phi^{-m}(\theta_0)x_0| = |x_0| =: c > 0$$

and, for $k \in \mathbb{Z}$,

$$|\Phi^k(\theta_m)x_m| = |\Phi^k(\varphi^{-m}\theta_0)\Phi^{-m}(\theta_0)x_0| = |\Phi^{k-m}(\theta_0)x_0| \leq C.$$

Thus, Definition 7.16 extends the notion of a Mañe point to cocycles consisting of non-invertible operators. ◇

Returning to the general case of cocycles consisting of possibly noninvertible operators, we have the following proposition, cf. Remark 6.16.

PROPOSITION 7.18. *If $\{\Phi^t\}_{t\in\mathbb{Z}_+}$ has an exponential dichotomy on Θ, then there are no Mañe sequences in Θ.*

PROOF. Suppose that $\{\theta_m\}_{m=1}^\infty$ is a Mañe sequence in Θ and $\{x_m\}_{m=1}^\infty \subset X$ is the corresponding sequence of Mañe vectors. Decompose x_m as $x_m = x_m^s + x_m^u$ where $x_m^s \in \mathrm{Im}\, P(\theta_m)$ and $x_m^u \in \mathrm{Im}\, Q(\theta_m)$. Let $y_m = \Phi^m(\theta_m)x_m$, $y_m^s =$

$\Phi^m(\theta_m)x_m^s$, and $y_m^u = \Phi^m(\theta_m)x_m^u$. Using the dichotomy and the inequality (7.42) with $k = 2m$ and $k = m$, we obtain the estimates

$$|y_m^u| \le Me^{-\beta m}|\Phi_Q^m(\varphi^m\theta_m)y_m^u| \le Me^{-\beta m}\{|\Phi^{2m}(\theta_m)x_m| + |\Phi_P^m(\varphi^m\theta_m)y_m^s|\}$$
$$\le Me^{-\beta m}\{C + Me^{-\beta m}(|y_m| + |y_m^u|)\} \le Me^{-\beta m}\{C + Me^{-\beta m}(C + |y_m^u|)\}.$$

Thus, for sufficiently large m, we have

$$|y_m^u| \le Me^{-\beta m}\frac{C + MCe^{-\beta m}}{1 - M^2 e^{-2\beta m}};$$

and therefore, $\lim_{m\to\infty}|y_m^u| = 0$. This fact, together with the inequalities (7.41) and (7.42) for $k = m$, implies $|y_m^s| \in [c/2, 2C]$ for sufficiently large m. Since $\{\Phi^t\}_{t\ge 0}$ has an exponential dichotomy, we conclude that

$$\frac{c}{2} \le |y_m^s| = |\Phi^m(\theta_m)x_m^s| \le Me^{-\beta m}|x_m^s|,$$
$$2C \ge |y_m^u| = |\Phi^m(\theta_m)x_m^u| \ge M^{-1}e^{\beta m}|x_m^u|.$$

Hence, $|x_m^s| \to \infty$ and $|x_m^u| \to 0$ as $m \to \infty$. This contradicts the existence of the bound (7.42) for $k = 0$. $\square$

We now give sufficient conditions for the existence of a Mañe sequence, using the following refinement of the implication (c)$\Rightarrow$(d) of the Discrete Dichotomy Theorem 7.9, cf. Proposition 6.36. Recall that the operator $\pi_\theta(E)$ acts on the spaces $c_0(\mathbb{Z}; X)$ or $\ell^\infty(\mathbb{Z}; X)$ by the formula

$$\pi_\theta(E) = \operatorname{diag}(\Phi(\varphi^{n-1}\theta))_{n\in\mathbb{Z}}S : (x_n)_{n\in\mathbb{Z}} \mapsto (\Phi(\varphi^{n-1}\theta)x_{n-1})_{n\in\mathbb{Z}}.$$

LEMMA 7.19. *If there is a constant $c > 0$ such that*

$$\|I - \pi_\theta(E)\|_{\bullet,\ell^\infty(\mathbb{Z};X)} \ge c \quad \text{for all} \ \theta \in \Theta,$$

then $\|z - E\|_{\bullet,C_0(\Theta;X)} \ge c$ for all $z \in \mathbb{T}$.

PROOF. If $f \in C_0(\Theta; X)$ and $z \in \mathbb{T}$, then the sequence $\bar{x}_\theta := (f(\varphi^k\theta))_{k\in\mathbb{Z}}$ is bounded for each $\theta \in \Theta$. By the hypothesis of the lemma and Proposition 7.8, we have

$\|zf - Ef\|_{C_0(\Theta;X)}$
$= \sup_{\theta\in\Theta}|zf(\theta) - \Phi^1(\varphi^{-1}\theta)f(\varphi^{-1}\theta)| = \sup_{\theta\in\Theta}\sup_{k\in\mathbb{Z}}|zf(\varphi^k\theta) - \Phi^1(\varphi^{k-1}\theta)f(\varphi^{k-1}\theta)|$
$= \sup_{\theta\in\Theta}\|[z - \pi_\theta(E)]\bar{x}_\theta\|_{\ell^\infty(\mathbb{Z};X)} \ge c\sup_{\theta\in\Theta}\|\bar{x}_\theta\|_{\ell^\infty(\mathbb{Z};X)} = c\|f\|_{C_0(\Theta;X)}.$

$\square$

PROPOSITION 7.20. *If either (a) $\inf_{\theta\in\Theta}\|I - \pi_\theta(E)\|_{\bullet,\ell^\infty} = 0$ or (b) $\sigma_{ap}(E) \cap \mathbb{T} \ne \emptyset$, then there is a Mañe sequence in Θ.*

PROOF. In view of Lemma 7.19, it suffices to consider statement (a). Since $\{\Phi^t\}_{t\ge 0}$ is exponentially bounded, $N := \sup_\theta\|\pi_\theta(E)\|$ is finite. Fix $m \in \mathbb{N}$ and let $\varepsilon := (4\sum_{k=0}^{2m} N^k)^{-1}$. By the assumption, there exists $\sigma_m \in \Theta$ and $\bar{y}^{(m)} = (y_l^{(m)})_{l\in\mathbb{Z}} \in \ell^\infty(\mathbb{Z}; X)$ such that $\|\bar{y}^{(m)}\|_\infty = 1$ and $\|[I - \pi_{\sigma_m}(E)]\bar{y}^{(m)}\|_\infty < \varepsilon$. Hence, by Proposition 6.27,

$$\tfrac{1}{2} < \|[\pi_{\sigma_m}(E)]^k \bar{y}^{(m)}\|_{\ell^\infty(\mathbb{Z};X)} < 2 \quad \text{for} \quad k = 0, 1, \ldots 2m.$$

For a fixed $k \in \{0, 1, \ldots, 2m\}$, we have
$$\|[\pi_{\sigma_m}(E)]^k \bar{y}^{(m)}\|_{l^\infty(\mathbb{Z};X)} = \sup_{l \in \mathbb{Z}} |\Phi^k(\varphi^{l-k}\sigma_m) y^{(m)}_{l-k}| = \sup_{l \in \mathbb{Z}} |\Phi^k(\varphi^{l-m}\sigma_m) y^{(m)}_{l-m}|.$$

Select l such that $\frac{1}{2} \leq |\Phi^m(\varphi^{l-m}\sigma_m) y^{(m)}_{l-m}|$ and note that $|\Phi^k(\varphi^{l-m}\sigma_m) y^{(m)}_{l-m}| \leq 2$ holds for $k = 0, 1, \ldots 2m$. Therefore, $\theta_m := \varphi^{l-m}\sigma_m$ defines a Mañe sequence with the corresponding sequence of Mañe vectors vectors defined by $x_m := y^{(m)}_{l-m}$. □

7.2.2. The Annular Hull and Spectral Mapping Theorems. In this subsection we will prove the Annular Hull Theorem and the Spectral Mapping Theorem 7.25. The proof of the Spectral Mapping Theorem given here is different from the proof given in Subsection 6.2.3. Recall that in the proof of Theorem 6.30, we used the generalization (6.34) of the Mather localization formula (6.22); here we will make use of Mañe sequences. In the course of the proof we will give another set of explicit formulas for the approximate eigenfunctions for Γ.

Let us recall, see Lemma 6.34, that a *flow box* at $\theta_0 \in \Theta$ of *length* τ is a subset of Θ of the form $\mathcal{U} = \{\theta = \varphi^t \sigma : |t| < \tau, \sigma \in \Sigma\}$ where $\Sigma \ni \theta_0$ is a *cross section* through θ_0. Also, let $\mathcal{O}(\theta) := \{\varphi^t \theta : t \in \mathbb{R}\}$ denote the orbit through $\theta \in \Theta$ and let $p : \theta \to \mathbb{R}_+ \cup \{\infty\}$ be the prime period function for the flow $\{\varphi^t\}_{t \in \mathbb{R}}$ on Θ; that is, $p(\theta) = \inf\{T > 0 : \varphi^T \theta = \theta\}$. We will use the following sets of φ-periodic points in Θ, see Definition 6.26:

(7.43)
$$\mathcal{B}(\Theta) := \{\theta \in \Theta : p \text{ is bounded in a neighborhood of } \theta\},$$
$$\mathcal{BC}(\Theta) := \{\theta \in \mathcal{B}(\Theta) : p \text{ is continuous at } \theta\}.$$

The first result is a variant of Lemma 6.34, see Bhatia and Szegö [**BS**, Thm. IV.2.11].

LEMMA 7.21. *Suppose that $\{\varphi^t\}_{t \in \mathbb{R}}$ is a continuous flow on a locally compact metric space Θ and p is the corresponding prime period function. If $\theta_0 \in \Theta$ and τ are such that $0 < \tau < p(\theta_0)/4$, then there exists an open flow box $\mathcal{U} = \{\theta = \varphi^t \sigma : |t| < \tau, \sigma \in \Sigma\}$ at $\theta_0 \in \Sigma$ such that $\theta \mapsto t = t(\theta)$ is a continuous map from $\mathcal{U}$ to $(-\tau, \tau)$.*

LEMMA 7.22. *If $\{\varphi^t\}_{t \in \mathbb{R}}$ is a continuous flow on the locally compact metric space Θ and p is the prime period function, then*

(a) *p is lower semicontinuous; in particular, the set $\{\theta : p(\theta) \leq d\}$ is closed for each $d \geq 0$;*
(b) *the points of continuity of p are dense in Θ; and*
(c) *for $\theta_0 \in \mathcal{BC}(\Theta)$ with $p(\theta_0) > 0$, there exists a relatively compact, open set $\mathcal{U}$ and a set $\Sigma \ni \theta_0$ such that $\mathcal{O}(\theta_0) \subset \mathcal{U} \subset \mathcal{BC}(\Theta)$, while for $\theta \in \mathcal{U}$ there is a unique $t = t(\theta) \in [0, p(\theta))$ with $\varphi^{-t}\theta \in \Sigma$, and the map $\mathcal{U} \ni \theta \mapsto t(\theta)$ is continuous.*

PROOF. Statement (a) is proved in Arendt and Greiner [**AG**, p.314]. Statement (b) follows from (a) and the results in Engelking [**En**, pp.87]. To prove statement (c), we use Lemma 7.21 to find a relatively compact, open flow box $\mathcal{U}_0$ at θ_0 of length $\tau \in (0, p(\theta_0)/4)$ with cross section Σ_0 such that $\theta \mapsto t(\theta)$ is continuous, the function p, restricted to $\mathcal{U}_0$, takes values in the interval $(3p(\theta_0)/4, 5p(\theta_0)/4)$, and

(7.44) $$\mathcal{O}(\theta_0) \cap \Sigma_0 = \{\theta_0\}.$$

Let Σ_1 be a relatively open subset of Σ_0 containing θ_0 such that $\overline{\Sigma}_1 \subset \mathcal{U}_0$. We claim that $\mathcal{O}(\theta_0)$ has an open neighborhood $\mathcal{U} \subset \mathcal{U}_0$ that satisfies the following condition:

(7.45) $\quad$ for each $\theta \in \mathcal{U}$, the set $\mathcal{O}(\theta) \cap \Sigma_1$ contains exactly one element.

Suppose that no such neighborhood exists. Then there exists a sequence of points $\{\theta_n\}_{n=0}^\infty$ such that $\lim_{n \to \infty} \theta_n = \varphi^t \theta_0$ for some $0 \leq t < p(\theta_0)$ and such that the set $\mathcal{O}(\theta_n) \cap \Sigma_1$ is either empty or contains more than one element. But, since $\lim_{n \to \infty} \varphi^{-t} \theta_n = \theta_0$, we obtain $\varphi^{-t} \theta_n \in \mathcal{U}_0$. Hence, there are some points $\sigma_n \in \mathcal{O}(\theta_n) \cap \Sigma_0$ for large n. Moreover, $t(\varphi^{-t} \theta_n) \to t(\theta_0) = 0$ so that $\sigma_n \to \theta_0$. In particular, $\sigma_n \in \Sigma_1$ for large n. Using the continuity of p at θ_0 we conclude that there must exist $\sigma'_n = \varphi^{s_n} \sigma_n \in \Sigma_1$ for some s_n with

$$0 < \tau \leq s_n \leq p(\sigma_n) - \tau < p(\theta_0) - \frac{\tau}{2}$$

and n sufficiently large. Taking an appropriate subsequence, we may assume that $s_n \to s$ so that

$$\lim_{n \to \infty} \sigma'_n = \varphi^s \theta_0 \in \overline{\Sigma}_1 \subset \mathcal{U}_0 \,.$$

Again by the continuity of $t(\cdot)$, we find that $\varphi^s \theta_0 \in \Sigma_0$. By the equality (7.44), it follows that $\varphi^s \theta_0 = \theta_0$. But this contradicts the inequality $0 < s < p(\theta_0)$. Hence, there exists a neighborhood $\mathcal{U}$ of $\mathcal{O}(\theta_0)$ satisfying statement (7.45).

It remains to show the continuity of $p(\cdot)$ and $t(\cdot)$ on $\mathcal{U}$. Let $\theta_n \to \theta$ in $\mathcal{U}$. Suppose that the sequence $\{p(\theta_n)\}_{n=1}^\infty$ does not converge to $p(\theta)$. But, some subsequence, again denoted by $\{p(\theta_n)\}_{n=1}^\infty$, converges to some number p_0. Then, $\lim_{n \to \infty} \theta_n = \lim_{n \to \infty} \varphi^{p(\theta_n)} \theta_n = \varphi^{p_0} \theta$. Hence, $p_0 = kp(\theta)$ for some $k \neq 1$. This is impossible since $p(\theta)$ and p_0 belong to $[3p(\theta_0)/4, 5p(\theta_0)/4]$. Moreover, $\lim_{n \to \infty} \varphi^r \theta_n = \varphi^r \theta$ for some r; and therefore, $\varphi^r \theta_n \in \mathcal{U}_0$ for sufficiently large n. This proves the continuity of $\theta \mapsto t(\theta)$ on $\mathcal{U}$. $\square$

Recall that Γ denotes the infinitesimal generator of the evolution semigroup $\{E^t\}_{t \geq 0}$ on $\mathcal{F} = C_0(\Theta; X)$ defined by $(E^t f)(\theta) = \Phi^t(\varphi^{-t} \theta) f(\varphi^{-t} \theta)$. Our next goal is to construct the approximate eigenfunctions for Γ corresponding to a Mañe sequence. Let us consider first the case where the periods of the points in the Mañe sequence are unbounded.

LEMMA 7.23. *If $\{\theta_m\}_{m=1}^\infty$ is a Mañe sequence such that $\theta_m \notin \mathcal{B}(\Theta)$ or $p(\theta_m) \to \infty$, then, for each $\xi \in \mathbb{R}$, there is a sequence of functions $f_n \in \mathcal{D}(\Gamma)$ with $\|f_n\|_\infty = 1$ such that $\lim_{n \to \infty} \|(i\xi - \Gamma) f_n\|_\infty = 0$.*

PROOF. Without loss of generality, we will assume that $p(\theta_m) \to \infty$ as $m \to \infty$. Define $n := \min\{m, \left[\frac{1}{5} p(\theta_m)\right]\}$ where $[\cdot]$ is the greatest integer function, and note that $n = n(m) \to \infty$ as $m \to \infty$. Also, set $\sigma_0 = \varphi^m(\theta_m)$ and $x = \Phi^{m-n}(\theta_m) x_m$. By the Definition 7.16 of Mañe sequences, there are positive constants C and c such that $c \leq |\Phi^m(\theta_m) x_m|$ and $|\Phi^k(\theta_m) x_m| \leq C$ for $k = 0, 1, \ldots, 2m$. Thus, we have that

(7.46) $\quad 0 < c \leq |\Phi^n(\varphi^{-n} \sigma_0) x|$ $\quad$ and $\quad$ $|\Phi^k(\varphi^{-n} \sigma_0) x| \leq C$ $\quad$ for $\quad k = 0, 1, \ldots, 2n$.

Using Lemma 7.21, there is an open flow box $\mathcal{U}$ of length n whose cross-section Σ contains σ_0 and such that

(7.47) $\qquad |\Phi^t(\varphi^{-n} \sigma) x - \Phi^t(\varphi^{-n} \sigma_0) x| \leq 1$

whenever $0 \leq t \leq 2n$ and $\sigma \in \Sigma$. Let us choose two bump-functions: a compactly supported function $\beta \in C(\mathcal{U})$ such that $0 \leq \beta \leq 1$ and $\beta(\sigma_0) = 1$, and a function $\gamma \in C^1(\mathbb{R})$ with $0 \leq \gamma \leq 1$, $\mathrm{supp}\,\gamma \subset (0, 2n)$, $\gamma(n) = 1$, and $\|\gamma'\|_\infty \leq 2/n$.

For each $\xi \in \mathbb{R}$, let

(7.48) $$f_n(\theta) := e^{-i\xi t} \beta(\sigma) \gamma(t) \Phi^t(\varphi^{-n}\sigma) x$$

if $\theta = \varphi^{t-n}\sigma \in \mathcal{U}$, $t \in (0, 2n)$, and $\sigma \in \Sigma$; and $f_n(\theta) := 0$ if $\theta \notin \mathcal{U}$. Clearly, $f_n \in \mathcal{F}$ and $\|f_n\|_\infty \geq |f_n(\sigma_0)| \geq c > 0$ by (7.46). For $h > 0$ so small that $\gamma = 0$ on $[2n - h, 2n]$, we have that

$$(E^h f_n)(\theta) = e^{-i\xi(t-h)} \beta(\sigma) \gamma(t - h) \Phi^t(\varphi^{-n}\sigma) x$$

for $\theta = \varphi^{t-n}\sigma \in \mathcal{U}$ and $(E^h f_n)(\theta) = 0$ for $\theta \notin \mathcal{U}$. Thus, $f_n \in \mathcal{D}(\Gamma)$ and

(7.49) $$(\Gamma f_n)(\theta) - i\xi f_n(\theta) = -\gamma'(t) \beta(\sigma) e^{-i\xi t} \Phi^t(\varphi^{-n}\sigma) x$$

for $\theta = \varphi^{t-n}\sigma$ with $t \in (0, 2n)$ and $\sigma \in \Sigma$, while $(\Gamma f_n)(\theta) = 0$ for $\theta \notin \mathcal{U}$. Let $N := \sup\{\|\Phi^\tau(\theta)\| : \tau \in [0,1], \theta \in \Theta\}$ and use the inequalities (7.47) and (7.46) to show that

$$\|\Gamma f_n - i\xi f_n\|_\infty \leq \frac{2N}{n} \max_{k=0,\ldots,2n} |\Phi^k(\varphi^{-n}\sigma) x| \leq \frac{2N(C+1)}{n}.$$

$\square$

We will use the next lemma in case the Mañe sequence consists of periodic points with uniformly bounded periods.

LEMMA 7.24. *Suppose that there is a sequence $\{\theta_m\}_{m=1}^\infty \subset \mathcal{BC}(\Theta)$ with $0 < d_0 \leq p(\theta_m) \leq d_1$ and a sequence $\{x_m\}_{m=1}^\infty \subset X$ with $|x_m| = 1$ such that*

$$\lim_{m \to \infty} |[z - \Phi^{p(\theta_m)}(\theta_m)] x_m| = 0$$

for some $z \in \mathbb{T}$. Then, there is some $\xi \in \mathbb{R}$ and a sequence $\{f_n\}_{n=1}^\infty \subset \mathcal{D}(\Gamma)$ such that $\|f_n\|_\infty = 1$ and $\lim_{n \to \infty} \|(i\xi - \Gamma)f_n\|_\infty = 0$.

PROOF. We can assume that $z = e^{i\eta} = \exp(i\frac{\eta}{p(\theta_n)} p(\theta_n))$ and $p(\theta_n) \to p_0 \in [d_0, d_1]$. Set $\xi = \eta/p_0$. By Proposition 7.22, there exist, for each n, a corresponding open tubular neighborhood $\mathcal{U}_n$ of θ_n, with section $\Sigma_n \ni \theta_n$, such that the function p is continuous on $\mathcal{U}_n$, takes values in $[d_0/2, 2d_1]$, and

(7.50) $$\lim_{n \to \infty} |x_n - e^{-i\xi p(\sigma)} \Phi^{p(\sigma)}(\sigma) x_n| = 0$$

for $\sigma \in \Sigma_n$. Choose a continuous compactly supported bump-function $\beta : \mathcal{U}_n \to [0, 1]$ with $\beta(\theta_n) = 1$. Also, choose a function $\alpha \in C^1[0, 1]$ such that $0 \leq \alpha \leq 1$, $\alpha = 0$ on $[0, 1/3]$, $\alpha = 1$ on $[2/3, 1]$, and $\|\alpha'\|_\infty \leq 4$. Then, define

(7.51) $$f_n(\theta) = e^{-i\xi t} \beta(\sigma) \Phi^t(\sigma)[\alpha(t/p(\sigma)) x_n + (1 - \alpha(t/p(\sigma))) e^{-i\xi p(\sigma)} \Phi^{p(\sigma)}(\sigma) x_n]$$

for $\theta = \varphi^t \sigma \in \mathcal{U}_n$ where $\sigma \in \Sigma_n$ and $0 \leq t < p(\sigma)$, and $f_n(\theta) = 0$ for $\theta \notin \mathcal{U}_n$, cf. the proof of Lemma 2.29. Clearly, $f_n \in C_0(\Theta; X)$ and

$$\|f_n\|_\infty \geq |f_n(\theta_n)| = |\Phi^{p(\theta_n)}(\theta_n) x_n| \geq 1/2$$

for large n. Moreover, for $0 < h < d_0/6$, we compute

$(E^h f_n)(\theta) =$
$$\begin{cases} e^{-i\xi(t-h)} \beta(\sigma) \Phi^t(\sigma)[\alpha(\frac{t-h}{p(\sigma)}) x_n + (1 - \alpha(\frac{t-h}{p(\sigma)})) e^{-i\xi p(\sigma)} \Phi^{p(\sigma)}(\sigma) x_n], & t \geq h, \\ e^{-i\xi(t-h)} \beta(\sigma) \Phi^t(\sigma) e^{-i\xi p(\sigma)} \Phi^{p(\sigma)}(\sigma) x_n, & t < h, \end{cases}$$

for $\theta = \varphi^t \sigma \in \mathcal{U}_n$ and $(E^h f_n)(\theta) = 0$ otherwise. (If $\theta \in \mathcal{U}_n$ and $t - h < 0$, then we write $\varphi^{-h}\theta = \varphi^{p(\theta)+t-h}\sigma$.) Thus, $f_n \in \mathcal{D}(\Gamma)$ and

(7.52)
$$\Gamma f_n(\theta) - i\xi f_n(\theta) = -e^{-i\xi t}\beta(\sigma)\frac{1}{p(\sigma)}\alpha'(\tfrac{t}{p(\sigma)})\Phi^t(\sigma)[x_n - e^{-i\xi p(\sigma)}\Phi^{p(\sigma)}(\sigma)x_n]$$

for $\theta = \varphi^t \sigma \in \mathcal{U}_n$, and $\Gamma f_n(\theta) - i\xi f_n(\theta) = 0$ otherwise. If $N := \sup\{\|\Phi^\tau(\theta)\| : \theta \in \Theta, 0 \le \tau \le 2d_1\}$, then

$$\|(\Gamma - i\xi)f_n\|_\infty \le \frac{8N}{d_0}|x_n - e^{-i\xi p(\sigma)}\Phi^{p(\sigma)}(\sigma)x_n|;$$

and therefore, using the limit in (7.50), we have that $\lim_{n\to\infty}\|(\Gamma - i\xi)f_n\|_\infty = 0$, as required. $\square$

We are now ready to prove the Annular Hull Theorem and to give an alternative proof of the Spectral Mapping Theorem for the evolution semigroup $(E^t f)(\theta) = \Phi^t(\varphi^{-t}\theta)f(\varphi^{-t}\theta)$ defined on the space $\mathcal{F} = C_0(\Theta; X)$ with the generator Γ. The Spectral Mapping Theorem states that if $t \ge 0$, then

(7.53)
$$\sigma(E^t)\setminus\{0\} = \exp(t\sigma(\Gamma)).$$

Recall that, in general, this result does not hold if the base flow $\{\varphi^t\}_{t\in\mathbb{R}}$ is not aperiodic, see Example 6.32. The Annular Hull Theorem states that if $t \ge 0$, then

$$\exp(t\sigma(\Gamma)) \subseteq \sigma(E^t)\setminus\{0\} \subseteq \mathcal{H}(\exp(t\sigma(\Gamma)))$$

where $\mathcal{H}(S)$ denotes the annular hull of the set $S \subset \mathbb{C}$; that is, the union of all circles centered at the origin that intersect S. Equivalently, the Annular Hull Theorem states that if $t \ge 0$, then

(7.54)
$$\mathbb{T}\cdot(\sigma(E^t)\setminus\{0\}) = \mathbb{T}\cdot\exp(t\sigma(\Gamma)).$$

Observe that the Annular Hull Theorem is not generally valid if the base flow has fixed points. Indeed, for the one-point set $\Theta = \{\theta\}$ and the flow $\varphi^t(\theta) \equiv \theta$, if we define a cocycle by $\Phi^t(\theta) = e^{tA}$, where $\{e^{tA}\}_{t\ge 0}$ is a strongly continuous semigroup on X that does not satisfy the Annular Hull Theorem (see Chapter 2 for examples), then the identity (7.54) does not hold for the corresponding evolution semigroup. Recall definition (7.43) of the sets $\mathcal{B}(\Theta)$ and $\mathcal{BC}(\Theta)$.

THEOREM 7.25 (Spectral Mapping/Annular Hull Theorem). *Let $\{E^t\}_{t\ge 0}$ denote the evolution semigroup on $\mathcal{F} = C_0(\Theta; X)$ that is induced by an exponentially bounded, strongly continuous cocycle $\{\Phi^t\}_{t\ge 0}$ over a continuous flow $\{\varphi^t\}_{t\in\mathbb{R}}$ on a locally compact metric space Θ, let Γ be the generator of $\{E^t\}_{t\ge 0}$, and let p be the prime period function for $\{\varphi^t\}$.*

(a) *If $\mathcal{B}(\Theta) = \emptyset$, then $\sigma(\Gamma)$ is invariant with respect to vertical translations, $\sigma(E^t)$, $t > 0$, is invariant with respect to rotations centered at zero, and the spectral mapping property (7.53) holds.*
(b) *If $p(\theta) \ge d_0 > 0$ for all $\theta \in \mathcal{B}(\Theta)$, then the annular hull property (7.54) holds.*

PROOF. By the Spectral Inclusion Theorem 2.6, the spectral mapping property for the residual spectrum, and a standard rescaling, statement (b) holds provided that

$$\sigma_{\text{ap}}(E) \cap \mathbb{T} \ne \emptyset \quad \text{implies} \quad \sigma(\Gamma) \cap i\mathbb{R} \ne \emptyset$$

for $E := E^1$. In view of this fact, let $z \in \sigma_{\mathrm{ap}}(E) \cap \mathbb{T}$; and note that, by Proposition 7.20, there exists a Mañé sequence $\{\theta_m\}_{m=1}^\infty$ in Θ. We have to consider two cases.

Aperiodic Case: There is a subsequence, again denoted by $\{\theta_m\}_{m=1}^\infty$, of the Mañé sequence with either $\theta_m \notin \mathcal{B}(\Theta)$ or $p(\theta_m) \to \infty$ as $m \to \infty$.
Periodic Case: $\theta_m \in \mathcal{B}(\Theta)$ and $p(\theta_m) < d_1$ for some positive constant d_1.

In the aperiodic case, Lemma 7.23 implies $i\mathbb{R} \subseteq \sigma_{\mathrm{ap}}(\Gamma)$, and this inclusion implies the equality (7.53). The other implications in statement (a) follow from the Spectral Inclusion Theorem 2.6 and the fact that the boundary of $\sigma(\Gamma)$ is contained in $\sigma_{\mathrm{ap}}(\Gamma)$.

In the periodic case, using Lemma 7.22, we can and will assume that $\theta_m \in \Omega$ where

$$(7.55) \qquad \Omega := \{\theta \in \mathcal{BC}(\Theta) : d_0 \leq p(\theta) \leq d_1\}.$$

The set Ω is φ^t-invariant and locally compact by Lemma 7.22. Let us define a new flow $\{\psi^t\}_{t\in\mathbb{R}}$ on Ω by $\psi^t(\theta) = \varphi^{tp(\theta)}(\theta)$ and a new cocycle $\{\Psi^t\}_{t\geq 0}$ over $\{\psi^t\}_{t\in\mathbb{R}}$ by $\Psi^t(\theta) = \Phi^{tp(\theta)}(\theta)$. Clearly, $\{\psi^t\}_{t\in\mathbb{R}}$ is continuous, and $\{\Psi^t\}_{t\geq 0}$ is strongly continuous and exponentially bounded on Ω. Hence, there is a corresponding evolution semigroup $\{E_\Psi^t\}_{t\geq 0}$ on $C_0(\Omega; X)$. Let $E_\Psi = E_\Psi^1$. We stress the fact that $(E_\Psi f)(\theta) = \Phi^{p(\theta)}(\theta) f(\theta)$ for $f \in C_0(\Omega; X)$ and $\theta \in \Omega$.

LEMMA 7.26. *Let Ω be as defined in display (7.55). If $\sigma(\Phi^{p(\theta)}(\theta)) \cap \mathbb{T} = \emptyset$ for each $\theta \in \Omega$ and*

$$\sup_{\theta \in \Omega} \|[z - \Phi^{p(\theta)}(\theta)]^{-1}\|_{\mathcal{L}(X)} < \infty$$

for each $z \in \mathbb{T}$, then $\{\Phi^t\}_{t\geq 0}$ has an exponential dichotomy on Ω.

PROOF. By the assumptions in the lemma, we have that $\sigma(E_\Psi) \cap \mathbb{T} = \emptyset$. Moreover,

$$(z - E_\Psi)^{-1} f(\theta) = [z - \Phi^{p(\theta)}(\theta)]^{-1} f(\theta)$$

for $\theta \in \Omega$, $z \in \mathbb{T}$, and $f \in C_0(\Omega; X)$. By an application of the Dichotomy Theorem 6.41 to E_Ψ, the cocycle $\{\Psi^t\}_{t\geq 0}$ has an exponential dichotomy on Ω. If β and M are the dichotomy constants and P is the dichotomy projection for $\{\Psi^t\}_{t\geq 0}$, then the cocycle $\{\Phi^t\}_{t\geq 0}$ has an exponential dichotomy on Ω with projection $P(\cdot)$, and constants β/d_1 and M. □

Since Ω contains a Mañé sequence, the cocycle $\{\Phi^t\}_{t\geq 0}$ does not have an exponential dichotomy on Ω by Proposition 7.18. Therefore, the assumptions in Lemma 7.26 cannot be satisfied. In other words, either

(1) $\|z - \Phi^{p(\theta_n)}(\theta_n)\|_{\bullet,X} \to 0$ for some $z \in \mathbb{T}$ and $\theta_n \in \Omega$; or
(2) there exist $z \in \mathbb{T}$, $\theta_0 \in \Omega$, $y \in X$, and $\delta > 0$ such that

$$|zx - \Phi^{p(\theta_0)}(\theta_0)x - y| \geq \delta$$

for all $x \in X$.

In case (1), Lemma 7.24 implies $\sigma_{\mathrm{ap}}(\Gamma) \cap i\mathbb{R} \neq \emptyset$. In case (2), choose a function $g \in C_0(\Theta; X)$ such that $g(\theta_0) = y$. Then, for all $f \in C_0(\Theta; X)$, we have

$$\|zf - E^{p(\theta_0)} f - g\|_\infty \geq |zf(\theta_0) - \Phi^{p(\theta_0)} f(\theta_0) - y| \geq \delta;$$

that is, $z \in \sigma_r(E^{p(\theta_0)}) \cap \mathbb{T}$. Hence, $\sigma_r(\Gamma) \cap i\mathbb{R} \neq \emptyset$ by the spectral mapping property for the residual spectrum. This finishes the proof of statement (b) in Theorem 7.25. □

Combining the above result with the Dichotomy Theorem 6.41, we obtain the following corollary.

COROLLARY 7.27. *Let $\{\Phi^t\}_{t \geq 0}$ be an exponentially bounded, strongly continuous cocycle over a continuous flow $\{\varphi^t\}_{t \in \mathbb{R}}$ on a locally compact metric space Θ. Let Γ be the generator of the induced evolution semigroup on $C_0(\Theta; X)$. Suppose that $p(\theta) \geq d_0 > 0$ for all $\theta \in \mathcal{B}(\Theta)$. Then, the cocycle has an exponential dichotomy on Θ if and only if $i\mathbb{R} \subset \rho(\Gamma)$. Moreover, if $\mathcal{B}(\Theta) = \emptyset$, then the cocycle $\{\Phi^t\}_{t \geq 0}$ has an exponential dichotomy on Θ if and only if $\rho(\Gamma) \cap i\mathbb{R} \neq \emptyset$.*

By Proposition 7.20, we see that if $\sigma_{\mathrm{ap}}(E) \cap \mathbb{T} \neq \emptyset$, then there is a Mañe sequence. We will state some converse results.

COROLLARY 7.28. *Suppose that $\sigma_r(E) \cap \mathbb{T} = \emptyset$. A Mañe sequence exists if and only if $\sigma_{ap}(E) \cap \mathbb{T} \neq \emptyset$.*

PROOF. If $\sigma_{\mathrm{ap}}(E) \cap \mathbb{T} \neq \emptyset$, then a Mañe sequence exists by Proposition 7.20. If a Mañe sequence exists, then $\sigma(E) \cap \mathbb{T} \neq \emptyset$ by Proposition 7.18 and Dichotomy Theorem 6.41. Hence, the assumption $\sigma_r(E) \cap \mathbb{T} = \emptyset$ implies the desired result. □

COROLLARY 7.29. *Suppose that $\{\theta_m\}_{m=1}^\infty$ is a Mañe sequence.*
(a) *If the aperiodic case holds for $\{\theta_m\}_{m=1}^\infty$; that is, there is a subsequence such that $\theta_m \notin \mathcal{B}(\Theta)$ or $p(\theta_m) \to \infty$ as $m \to \infty$, then $\mathbb{T} \subset \sigma_{ap}(E)$.*
(b) *If the periodic case holds for $\{\theta_m\}_{m=1}^\infty$; that is, $\theta_m \in \Omega = \{\theta \in BC(\Theta) : d_0 \leq p(\theta) \leq d_1\}$ for some constant d_1, and if $\sigma_r(\Phi^{p(\theta)}(\theta)) \cap \mathbb{T} = \emptyset$ for all $\theta \in \Omega$, then $\sigma_{ap}(E) \cap \mathbb{T} \neq \emptyset$.*

PROOF. In the aperiodic case (a), we have $i\mathbb{R} \subset \sigma_{\mathrm{ap}}(\Gamma)$ for the generator Γ of the evolution semigroup $\{E^t\}_{t \geq 0}$ on $C_0(\Theta; X)$ by Lemma 7.23. Then, the Spectral Inclusion Theorem 2.6 yields $\mathbb{T} \subset \sigma_{\mathrm{ap}}(E)$.

In the periodic case (b), the cocycle $\{\Phi^t\}_{t \geq 0}$ does not have an exponential dichotomy on Ω by Proposition 7.18. Therefore, the assumptions of Lemma 7.26 do not hold. Since, by our assumption, $\sigma_r(\Phi^{p(\sigma)}(\sigma)) \cap \mathbb{T} = \emptyset$ for all $\theta \in \Omega$, we conclude that $\|z - \Phi^{p(\theta_n)}(\theta_n)\|_{\bullet, X} \to 0$ for some $z \in \mathbb{T}$ and $\theta_n \in \Omega$. Lemma 7.24 implies $i\xi \in \sigma_{\mathrm{ap}}(\Gamma)$ for some $\xi \in \mathbb{R}$, and hence $\sigma_{\mathrm{ap}}(E) \cap \mathbb{T} \neq \emptyset$ by the Spectral Inclusion Theorem. □

Next, we have the following converse of the Mañe Lemma 6.29.

COROLLARY 7.30. *Suppose that Θ is compact and $\dim X < \infty$. If θ_0 is a Mañe point and x_0 a corresponding Mañe vector; that is, $C := \sup\{|\Phi^t(\theta_0)x_0| : t \in \mathbb{R}\} < \infty$, then $\sigma_{ap}(E) \cap \mathbb{T} \neq \emptyset$.*

PROOF. Let $\theta_m = \varphi^{-m}\theta_0$. Then, see Remark 7.17, $\{\theta_m\}_{m \in \mathbb{Z}}$ is a Mañe sequence.

If $\theta_0 \notin \mathcal{B}(\Theta)$, then $\theta_m \notin \mathcal{B}(\Theta)$ and $\mathbb{T} \subset \sigma_{\mathrm{ap}}(E)$ by Lemma 7.23 (or Corollary 7.29(a)). If $\theta_0 \in \mathcal{B}(\Theta)$ and $p(\theta_0) > 0$, then $\sigma(\Phi^{p(\theta_0)}(\theta_0)) \cap \mathbb{T} \neq \emptyset$. Moreover, recall that $\dim X < \infty$ implies $\sigma(\Phi^{p(\theta_0)}(\theta_0)) = \sigma_p(\Phi^{p(\theta_0)}(\theta_0))$. Therefore, the hypotheses of Lemma 7.24 hold. Thus, by the Spectral Inclusion Theorem, we have that $\sigma_{\mathrm{ap}}(E) \cap \mathbb{T} \neq \emptyset$.

The only remaining case is $\theta_0 \in \mathcal{B}(\Theta)$ and $p(\theta_0) = 0$. Since $\varphi^t \theta_0 = \theta_0$, the infinitesimal generator of the group of operators $\{\Phi^t(\theta_0)\}_{t \in \mathbb{R}}$ is some matrix $A \in \mathcal{L}(X)$. Because $C < \infty$, the group $\{e^{tA}\}_{t \in \mathbb{R}}$ is not hyperbolic; hence, $Ax = i\xi x$ for some $\xi \in \mathbb{R}$ and $x \in X$ with $|x| = 1$. Choose $U \ni \theta_0$ with $p(\theta) \leq d_1$ for all $\theta \in U$ and define $\delta(\theta) = \sup_{0 \leq \tau \leq d_1} d(\varphi^\tau \theta, \theta_0)$ for $\theta \in U$. The function $\delta(\cdot)$ is continuous and $\delta(\theta_0) = 0$. Note that $\delta(\varphi^t \theta) = \delta(\theta)$, for $t \in \mathbb{R}$, and fix $\epsilon > 0$. Using the equation $\Phi(\theta_0)x = e^{i\xi}x$, choose a neighborhood $V \subset U$ of θ_0 such that $\sup\{|\Phi(\theta)x - e^{i\xi}x| : \theta \in V\} < \epsilon$. Find $c > 0$ so small that the set $W := \{\theta \in U : \delta(\theta) \leq c\}$ is contained in V. Construct a bump-function α such that $\alpha(\theta) = 1 - c^{-1}\delta(\theta)$ for $\theta \in W$, and $\alpha(\theta) = 0$ for $\theta \notin W$. Note that $\alpha(\varphi^t \theta) = \alpha(\theta)$, for $t \in \mathbb{R}$, and let $f(\theta) = \alpha(\theta)x$, for $\theta \in \Theta$. Then, by the identity $(Ef)(\theta) = \Phi(\varphi^{-1}\theta)\alpha(\theta)x$, it follows that

$$\|Ef - e^{i\xi}f\|_\infty = \sup_{\theta \in \Theta} |\Phi(\varphi^{-1}\theta)\alpha(\theta)x - e^{i\xi}\alpha(\theta)x|$$
$$= \sup_{\theta \in \Theta} |(\Phi(\theta)x - e^{i\xi}x)\alpha(\theta)| \leq \epsilon \|\alpha\|_\infty \leq 2\epsilon.$$

□

7.3. Dichotomy, mild solutions, and Green's function

In this section we discuss two related topics: a characterization of dichotomy in terms of the existence and uniqueness of solutions of mild inhomogeneous variational equations and the existence of a Green's function for linear skew-product flows. In particular, Subsection 7.3.1 is a continuation of Section 4.3 while Subsection 7.3.2 is a continuation of Section 4.2. Our main tool, as usual, is the Mather evolution semigroup $\{E^t\}_{t \geq 0}$ acting on $C_0(\Theta; X)$. However, while the space $C_b(\Theta; X)$ of *bounded* continuous functions is of special interest in this section, the semigroup $\{E^t\}_{t \geq 0}$ is *not* a strongly continuous semigroup on this space. This fact results in some complications that will be handled by using the family of discrete operators $\{\pi_\theta(E) : \theta \in \Theta\}$ and the explicit formulas for the approximate eigenfunctions for Γ given in the previous Subsection 7.2.2.

7.3.1. Theorems of Perron type. Let $\{\Phi^t\}_{t \geq 0}$ be a strongly continuous, exponentially bounded cocycle over a continuous flow $\{\varphi^t\}_{t \in \mathbb{R}}$ on a locally compact metric space Θ, and let $\{\hat{\varphi}^t\}_{t \geq 0}$ be the corresponding linear skew-product flow. We will relate the existence of a dichotomy for the linear skew-product flow $\{\hat{\varphi}^t\}_{t \geq 0}$ on $\Theta \times X$ to the existence and uniqueness of the solution u of the mild form of the inhomogeneous equation $\frac{d}{dt}(u(\varphi^t\theta)) = (A(\varphi^t\theta) - \lambda)u(\varphi^t\theta) + g(\varphi^t\theta)$. Note that $x(t) = u(\varphi^t\theta)$ gives a solution of the inhomogeneous variational equation $\dot{x} = (A(\varphi^t\theta) - \lambda)x + g(\varphi^t\theta)$. Let us recall the notation $\Phi^t_\lambda(\theta) = e^{-\lambda t}\Phi^t(\theta)$ for $\lambda \in \mathbb{C}$, $t \geq 0$, and $\theta \in \Theta$; and consider the mild inhomogeneous integral equation

$$(7.56) \qquad u(\varphi^t\theta) = \Phi^t_\lambda(\theta)u(\theta) + \int_0^t \Phi^{t-\tau}_\lambda(\varphi^\tau\theta)g(\varphi^\tau\theta)\,d\tau, \qquad t \geq 0, \quad \theta \in \Theta,$$

on $C_0(\Theta; X)$ or the space $C_b(\Theta; X)$ of bounded continuous functions $f : \Theta \to X$ endowed with the supremum norm.

DEFINITION 7.31. Let $\mathcal{F} = C_0(\Theta; X)$ or $C_b(\Theta; X)$ and $\lambda \in \mathbb{C}$. We say that *condition* $(M_\lambda, \mathcal{F})$ *holds* if for each $g \in \mathcal{F}$ there exists a unique $u \in \mathcal{F}$ satisfying the equation (7.56).

7.3. DICHOTOMY, MILD SOLUTIONS, AND GREEN'S FUNCTION

If the condition $(M_\lambda, \mathcal{F})$ holds for some $\lambda \in \mathbb{C}$, then we can define a linear operator $R_\lambda : g \mapsto u$ on $\mathcal{F}$ that recovers the unique solution u of equation (7.56) for a given g. Moreover, this mapping is continuous.

LEMMA 7.32. *If condition $(M_\lambda, \mathcal{F})$ holds for some $\lambda \in \mathbb{C}$ and $\mathcal{F}$ is one of the spaces $C_0(\Theta; X)$ or $C_b(\Theta; X)$, then R_λ is a bounded operator on $\mathcal{F}$.*

PROOF. By the Closed Graph Theorem, it suffices to show that the operator R_λ is closed. Suppose that $\{g_n\}_{n=1}^\infty$ is a convergent sequence in $\mathcal{F}$ with limit g. Define $u_n := R_\lambda g_n$, and suppose that the sequence $\{u_n\}_{n=1}^\infty$ converges to u in $\mathcal{F}$. Since equation (7.56) holds for u_n and g_n, we have

$$u(\theta) = \lim_{n\to\infty} u_n(\theta) = \lim_{n\to\infty} \left[\Phi_\lambda^t(\varphi^{-t}\theta) u_n(\varphi^{-t}\theta) + \int_0^t \Phi_\lambda^{t-\tau}(\varphi^{\tau-t}\theta) g_n(\varphi^{\tau-t}\theta)\, d\tau \right]$$

$$= \Phi_\lambda^t(\varphi^{-t}\theta) u(\varphi^{-t}\theta) + \int_0^t \Phi_\lambda^{t-\tau}(\varphi^{\tau-t}\theta) g(\varphi^{\tau-t}\theta)\, d\tau.$$

In other words, $u = R_\lambda g$, as required. $\square$

The next proposition states that the existence of an exponential dichotomy for the cocycle $\{\Phi^t\}_{t\geq 0}$ implies the condition $(M_\lambda, \mathcal{F})$. This result does not require the cocycle to be exponentially bounded.

PROPOSITION 7.33. *Suppose that $\{\Phi^t\}_{t\geq 0}$ is a strongly continuous cocycle over a continuous flow $\{\varphi^t\}_{t\in\mathbb{R}}$ on a locally compact metric space Θ. If $\{\Phi^t\}_{t\geq 0}$ has an exponential dichotomy on Θ, then condition $(M_\lambda, \mathcal{F})$ holds for $\mathcal{F} = C_0(\Theta; X)$ or $C_b(\Theta; X)$ and for all $\lambda = i\xi \in i\mathbb{R}$.*

PROOF. If $\{\Phi^t\}_{t\geq 0}$ has an exponential dichotomy, then so does $\{\Phi_\lambda^t\}_{t\geq 0}$ with $\lambda = i\xi \in i\mathbb{R}$. Hence, it is enough to prove the theorem for $\lambda = 0$.

Let M and β be the dichotomy constants and $P(\cdot)$ the dichotomy projection as in Definition 6.13. Also, define Green's function $G(\theta, t)$ by

$$G(\theta, t) = \begin{cases} \Phi_P^t(\theta) = \Phi^t(\theta) P(\theta), & t \geq 0,\ \theta \in \Theta, \\ -\Phi_Q^t(\theta) = -[\Phi_Q^{-t}(\varphi^t\theta) Q(\varphi^t\theta)]^{-1}, & t < 0,\ \theta \in \Theta. \end{cases}$$

Clearly, $\|G(\theta, t)\| \leq M e^{-\beta|t|}$ and the function $(\theta, t) \mapsto G(\theta, t)$ is strongly continuous on the set $\Theta \times (\mathbb{R} \setminus \{0\})$. Moreover, Green's operator $\mathbb{G}$, defined by

$$(\mathbb{G}f)(\theta) = \int_{-\infty}^\infty G(\varphi^{-\tau}\theta, \tau) f(\varphi^{-\tau}\theta)\, d\tau$$

for $f \in \mathcal{F}$ and $\theta \in \Theta$, is a bounded operator on $C_0(\Theta; X)$ and $C_b(\Theta; X)$. In case Θ is compact, we recall the notation $C_0(\Theta; X) = C_b(\Theta; X) = C(\Theta; X)$. Hence, $\mathbb{G}$ maps $C_0(\Theta; X)$ into $C_0(\Theta; X)$. To see that the same is true for a general locally compact metric space Θ, choose a function $f \in C_0(\Theta; X)$ whose support is in a compact subset K of Θ, and let $\{\theta_n\}_{n=1}^\infty$ be a sequence in the complement of K that "tends to ∞". Define

$$t_n := \sup\{t \geq 0 : \varphi^\tau(\theta_n) \notin K \text{ for all } \tau \in [-t, t]\}.$$

If $t_n < T < \infty$, then, for each positive integer n, there exists $\tau_n \in [-T, T]$ such that $\varphi^{\tau_n}\theta_n \in K$. In other words, $\theta_n \in \bigcup_{|\tau|\leq T} \varphi^\tau(K)$, in contradiction. As a result, $t_n \to \infty$ and, in view of the estimate $\|G(\theta, \tau)\| \leq M e^{-\beta|\tau|}$, it follows that $\lim_{n\to\infty} (\mathbb{G}f)(\theta_n) = 0$.

Let $u = \mathbb{G}g$ for $g \in \mathcal{F}$ and compute

$$u(\varphi^t \theta) - \Phi^t(\theta)u(\theta)$$

$$= \int_0^\infty \Phi^\tau(\varphi^{t-\tau}\theta)P(\varphi^{t-\tau}\theta)g(\varphi^{t-\tau}\theta)\,d\tau - \Phi^t(\theta)\int_0^\infty \Phi^\tau(\varphi^{-\tau}\theta)P(\varphi^{-\tau}\theta)g(\varphi^{-\tau}\theta)\,d\tau$$

$$- \int_{-\infty}^0 \Phi_Q^\tau(\varphi^{t-\tau}\theta)Q(\varphi^{t-\tau}\theta)g(\varphi^{t-\tau}\theta)\,d\tau$$

$$+ \Phi^t(\theta)\int_{-\infty}^0 \Phi_Q^\tau(\varphi^{-\tau}\theta)Q(\varphi^{-\tau}\theta)g(\varphi^{-\tau}\theta)\,d\tau$$

$$= \int_{-\infty}^t \Phi^{t-\tau}(\varphi^\tau\theta)P(\varphi^\tau\theta)g(\varphi^\tau\theta)\,d\tau - \int_{-\infty}^0 \Phi^{t-\tau}(\varphi^\tau\theta)P(\varphi^\tau\theta)g(\varphi^\tau\theta)\,d\tau$$

$$- \int_t^\infty \Phi_Q^{t-\tau}(\varphi^\tau\theta)Q(\varphi^\tau\theta)g(\varphi^\tau\theta)\,d\tau + \int_0^\infty \Phi_Q^{t-\tau}(\varphi^\tau\theta)Q(\varphi^\tau\theta)g(\varphi^\tau\theta)\,d\tau$$

$$= \int_0^t \Phi^{t-\tau}(\varphi^\tau\theta)P(\varphi^\tau\theta)g(\varphi^\tau\theta)\,d\tau + \int_0^t \Phi^{t-\tau}(\varphi^\tau\theta)Q(\varphi^\tau\theta)g(\varphi^\tau\theta)\,d\tau$$

$$= \int_0^t \Phi^{t-\tau}(\varphi^\tau\theta)g(\varphi^\tau\theta)\,d\tau$$

for $t \geq 0$ and $\theta \in \Theta$. It follows that $u := \mathbb{G}g$ satisfies the equation (7.56) with $\lambda = 0$.

To show the uniqueness, take $g = 0$ and let $u \in \mathcal{F}$ satisfy $u(\varphi^t \theta) = \Phi^t(\theta)u(\theta)$ for $\theta \in \Theta$ and $t \geq 0$. Since $\{\Phi^t\}$ has an exponential dichotomy, we have

$$P(\theta)u(\theta) = \Phi_P^t(\varphi^{-t}\theta)P(\varphi^{-t}\theta)u(\varphi^{-t}\theta) \text{ and } Q(\theta)u(\theta) = [\Phi_Q^t(\theta)]^{-1}Q(\varphi^t\theta)u(\varphi^t\theta)$$

for $t \geq 0$ and $\theta \in \Theta$. Finally, using the estimates

$$|P(\theta)u(\theta)| \leq Me^{-\beta t}\|u\|_\infty, \qquad |Q(\theta)u(\theta)| \leq Me^{-\beta t}\|u\|_\infty,$$

it follows that $u = 0$. $\square$

Let us consider the converse of Proposition 7.33. In case $\mathcal{F} = C_0(\Theta; X)$ the following result is an easy consequence of Corollary 7.27.

THEOREM 7.34. *Suppose that $\{\Phi^t\}_{t \geq 0}$ is an exponentially bounded, strongly continuous cocycle over a continuous flow $\{\varphi^t\}_{t \in \mathbb{R}}$ on a locally compact metric space Θ.*
   (a) *If $p(\theta) \geq d_0 > 0$ for all $\theta \in \mathcal{B}(\Theta)$ and the condition $(M_\lambda, C_0(\Theta; X))$ holds for all $\lambda = i\xi \in i\mathbb{R}$, then $\{\Phi^t\}$ has an exponential dichotomy on Θ.*
   (b) *If $\mathcal{B}(\Theta) = \emptyset$ and the condition $(M_0, C_0(\Theta; X))$ holds, then $\{\Phi^t\}$ has an exponential dichotomy on Θ.*

PROOF. In view of Corollary 7.27, it suffices to show that $i\mathbb{R} \subset \rho(\Gamma)$ in case (a), and $0 \in \rho(\Gamma)$ in case (b). Fix $g \in C_0(\Theta; X)$. By assumption, the function $u := R_\lambda g$ is in $C_0(\Theta; X)$. Also, by applying the operator E^h to both sides of the equation (7.56), we obtain the equation

$$e^{-\lambda h}(E^h u)(\theta) = \Phi_\lambda^h(\varphi^{-h}\theta)u(\varphi^{-h}\theta) = u(\theta) - \int_0^h \Phi_\lambda^{h-\tau}(\varphi^{\tau-h}\theta)g(\varphi^{\tau-h}\theta)\,d\tau.$$

Therefore,
$$h^{-1}[e^{-\lambda h}(E^h u)(\theta) - u(\theta)] = -h^{-1}\int_0^h \Phi_\lambda^\tau(\varphi^{-\tau}\theta)g(\varphi^{-\tau}\theta)\,d\tau,$$
and we conclude that $u \in \mathcal{D}(\Gamma)$ and $(\lambda - \Gamma)u = g$. In other words, $\lambda - \Gamma$ is surjective for all $\lambda \in i\mathbb{R}$ in case (a) and for $\lambda = 0$ in case (b).

If $(\Gamma - \lambda)u = 0$, then $e^{-\lambda t}E^t u = u$ or $u(\varphi^t \theta) = \Phi_\lambda^t(\theta)u(\theta)$ for $\theta \in \Theta$ and $t \geq 0$. Hence, the uniqueness part of the condition $(M_\lambda, C_0(\Theta; X))$ implies the injectivity of $\lambda - \Gamma$ for all $\lambda \in i\mathbb{R}$ in case (a) and for $\lambda = 0$ in case (b). $\square$

The case where $\mathcal{F} = C_b(\Theta; X)$ cannot be reduced as above to the results of the previous sections since the evolution semigroup is not strongly continuous on $C_b(\Theta; X)$. To remedy this problem, we will use the discrete operators $\pi_\theta(E)$. We will need the next two preliminary lemmas. The proof of the first one is exactly the same as the proof of Lemma 2.37.

LEMMA 7.35. *Suppose that $\theta \in \Theta$ is a fixed point of φ^1. Then $\pi_\theta(E)$ is hyperbolic on $c_0(\mathbb{Z}; X)$ if and only if $\Phi^1(\theta)$ is hyperbolic on X. Moreover, if $1 \notin \sigma_{ap}(\pi_\theta(E))$, then $\sigma_{ap}(\Phi^1(\theta)) \cap \mathbb{T} = \emptyset$.*

LEMMA 7.36. *If the condition $(M_\lambda, \mathcal{F})$ holds for $\mathcal{F} = C_0(\Theta; X)$ or $C_b(\Theta; X)$ and some $\lambda \in \mathbb{C}$, then, for each periodic point $\theta_0 \in \Theta$ with $p(\theta_0) > 0$, the operator $e^{\lambda p(\theta_0)} - \Phi^{p(\theta_0)}(\theta_0)$ is surjective in X.*

PROOF. Fix $y \in X$. Choose a function $\alpha \in C([0, p(\theta_0)]; \mathbb{R})$ such that $\alpha(0) = 0$, $\alpha(p(\theta_0)) = 1/p(\theta_0)$, and $\int_0^{p(\theta_0)}\alpha(\tau)\,d\tau = 1$; and define $g : \mathcal{O}(\theta_0) \to X$ by
$$g(\varphi^\tau \theta_0) = \alpha(\tau)\Phi_\lambda^\tau(\theta_0)y + \left[\tfrac{1}{p(\theta_0)} - \alpha(\tau)\right]\Phi_\lambda^{p(\theta_0)+\tau}(\theta_0)y.$$
The function g is continuous and it can be extended to a continuous function on Θ so that the extension, again denoted by g, is a function $g \in \mathcal{F}$ with compact support. Let $u = R_\lambda g$ and observe that
$$\Phi_\lambda^{p(\theta_0)-\tau}(\varphi^\tau \theta_0)\Phi_\lambda^{p(\theta_0)+\tau}(\theta_0) = \Phi^{2p(\theta_0)}(\theta_0).$$
Using this fact, equation (7.56), and the equation $\int_0^{p(\theta_0)}[\tfrac{1}{p(\theta_0)} - \alpha(\tau)]d\tau = 0$, we obtain the identities
$$u(\varphi^{p(\theta_0)}\theta_0) = u(\theta_0) = \Phi_\lambda^{p(\theta_0)}(\theta_0)u(\theta_0) + \Phi_\lambda^{p(\theta_0)}(\theta_0)y.$$
In particular, for $x = u(\theta_0)$, we have
$$x - e^{-\lambda p(\theta_0)}\Phi^{p(\theta_0)}(\theta_0)x = e^{-\lambda p(\theta_0)}\Phi^{p(\theta_0)}(\theta_0)y;$$
and therefore, $[e^{\lambda p(\theta_0)} - \Phi^{p(\theta_0)}(\theta_0)](e^{-\lambda p(\theta_0)}(x+y)) = y$. $\square$

The following result together with Proposition 7.33 gives a characterization of the existence of an exponential dichotomy for $\{\Phi^t\}_{t \geq 0}$ in terms of the existence and uniqueness of mild solutions of the variational equation.

THEOREM 7.37. *Let $\{\Phi^t\}_{t \geq 0}$ be an exponentially bounded, strongly continuous cocycle over a continuous flow $\{\varphi^t\}_{t \in \mathbb{R}}$ on a locally compact metric space Θ.*
  (a) *If $p(\theta) \geq d_0 > 0$ for all $\theta \in \mathcal{B}(\Theta)$ and condition $(M_\lambda, C_b(\Theta, X))$ holds for all $\lambda = i\xi \in i\mathbb{R}$, then $\{\Phi^t\}_{t \geq 0}$ has an exponential dichotomy on Θ.*
  (b) *If $\mathcal{B}(\Theta) = \emptyset$ and condition $(M_0, C_b(\Theta; X))$ holds, then $\{\Phi^t\}_{t \geq 0}$ has an exponential dichotomy on Θ.*

PROOF. We will show that the operator $I - \pi_\theta(E)$ is invertible on the space $c_0(\mathbb{Z}; X)$ and

(7.57) $$\|(I - \pi_\theta(E))^{-1}\|_{\mathcal{L}(c_0(\mathbb{Z};X))} \leq C$$

for all $\theta \in \Theta$. The theorem then follows from the Discrete Dichotomy Theorem 7.9 and Proposition 7.8.

Step I. We will prove that

(7.58) $$\inf_{\theta \in \Theta} \|I - \pi_\theta(E)\|_{\bullet, c_0(\mathbb{Z};X)} > 0.$$

Let us suppose that the inequality (7.58) is not valid; that is,

(7.59) $$\inf_{\theta \in \Theta} \|I - \pi_\theta(E)\|_{\bullet, c_0(\mathbb{Z};X)} = 0,$$

and derive a contradiction. Recall that, by Proposition 7.20, there exists a Mañe sequence $\{\theta_m\}_{m=1}^\infty$ in Θ. We will consider aperiodic, respectively periodic, base flows separately in paragraph (1), respectively (2), below.

(1) Suppose that there is a subsequence such that $p(\theta_m) \to \infty$ or $\theta_m \notin \mathcal{B}(\Theta)$. By an application of Lemma 7.23, there is a sequence of functions $\{f_n\}_{n=1}^\infty$ in $\mathcal{D}(\Gamma)$ such that $\|f_n\| = 1$ and $\lim_{n \to \infty} \Gamma f_n = 0$ where Γ denotes the generator of the corresponding evolution semigroup on $C_0(\Theta; X)$. Recall that each function f_n is given explicitly as in display (7.48) and a formula for Γf_n is given in display (7.49). Using the integral equation (7.56), it is easy to verify that $R_0 \Gamma f_n = -f_n$. But, this contradicts the continuity of the operator R_0, see Lemma 7.32.

(2) By Lemma 7.22, it remains only to consider the case where the sequence $\{\theta_m\}_{m=1}^\infty$ is in the set $\Omega = \{\theta \in \mathcal{BC}(\Theta) : d_0 \leq p(\theta) \leq d_1\}$ for some $d_1 > 0$. We claim that there is a constant c such that

(7.60) $$\|z - \Phi^{p(\theta)}(\theta)\|_{\bullet, X} \geq c > 0$$

for $\theta \in \Omega$ and $z \in \mathbb{T}$. The combination of this fact, the assumptions in part (a), and Lemma 7.36, can be used to verify the hypotheses of Lemma 7.26. But the conclusion of Lemma 7.26 contradicts the equality (7.59) by virtue of the Discrete Dichotomy Theorem 7.9. This would prove the desired result.

We will now prove the claim (7.60). If $\inf\{\|z - \Phi^{p(\theta)}(\theta)\|_{\bullet, X} : \theta \in \Omega\} = 0$ for some $z \in \mathbb{T}$, then the conclusion of Lemma 7.24 holds; that is, there is a sequence of functions $\{f_n\}_{n=1}^\infty$ in $\mathcal{D}(\Gamma)$ given by the formula (7.51) such that, for each positive integer n, we have $\|f_n\|_\infty \geq 1/2$, and $\lim_{n \to \infty} \|(\Gamma - i\xi)f_n\|_\infty = 0$ for some $\xi \in \mathbb{R}$, see formula (7.52). Set $g_n = \Gamma f_n - i\xi f_n$ and $\lambda = i\xi$. In the next lemma, we will show that $f_n = -R_\lambda g_n$. But this contradicts the fact that R_λ is a bounded operator on $C_b(\Theta; X)$. Here, the fact that R_λ is bounded follows from the condition $(M_\lambda, C_b(\Theta; X))$ and Lemma 7.32.

LEMMA 7.38. *If f_n is the function defined in display (7.51) and $g_n := (\Gamma - i\xi)f_n$, then $f_n = -R_\lambda g_n$; that is, for $t \geq 0$ and $\theta \in \Theta$, we have that*

(7.61) $$f_n(\varphi^t \theta) - \Phi^t_{i\xi}(\theta) f_n(\theta) = -\int_0^t \Phi^{t-\tau}_{i\xi}(\varphi^t \theta) g_n(\varphi^\tau \theta) \, d\tau.$$

PROOF. Recall that the functions f_n and g_n are supported in the set $\mathcal{U}_n = \{\varphi^s \sigma : 0 \leq s < p(\sigma), \sigma \in \Sigma\}$, see the formulas in displays (7.51) and (7.52), and fix $\theta = \varphi^s \sigma \in \mathcal{U}_n$ together with some $t \geq 0$.

If $s + t < p(\sigma)$, then, in view of the definition (7.51),

$$f_n(\varphi^t\theta) - \Phi^t_{i\xi}(\theta)f_n(\theta) = f_n(\varphi^{t+s}\sigma) - \Phi^t_{i\xi}(\varphi^s\sigma)f_n(\varphi^s\sigma)$$
$$= \beta(\sigma)\Phi^{t+s}_{i\xi}(\sigma)\left[\alpha(\tfrac{t+s}{p(\sigma)})\,(x_n - \Phi^{p(\sigma)}_{i\xi}x_n) + \Phi^{p(\sigma)}_{i\xi}(\sigma)x_n\right]$$
$$- \beta(\sigma)\Phi^{t+s}_{i\xi}(\sigma)\left[\alpha(\tfrac{s}{p(\sigma)})\,(x_n - \Phi^{p(\sigma)}_{i\xi}x_n) + \Phi^{p(\sigma)}_{i\xi}(\sigma)x_n\right]$$
$$= \beta(\sigma)\Phi^{t+s}_{i\xi}(\sigma)\left[\alpha(\tfrac{t+s}{p(\sigma)}) - \alpha(\tfrac{s}{p(\sigma)})\right](x_n - \Phi^{p(\sigma)}_{i\xi}x_n).$$

On the other hand, using formula (7.52), we have that

$$-\int_0^t \Phi^{t-\tau}_{i\xi}(\varphi^\tau\theta)g_n(\varphi^\tau\theta)\,d\tau$$
$$= \beta(\sigma)\int_0^t \Phi^{t-\tau}_{i\xi}(\varphi^{\tau+s}\sigma)\Phi^{\tau+s}_{i\xi}(\sigma)[x_n - \Phi^{p(\sigma)}_{i\xi}(\sigma)x_n]\frac{1}{p(\sigma)}\alpha'(\tfrac{\tau+s}{p(\sigma)})\,d\tau$$
$$= \beta(\sigma)\Phi^{t+s}_{i\xi}(\sigma)(x_n - \Phi^{p(\sigma)}_{i\xi}(\sigma)x_n)\int_0^t \alpha'(\tfrac{\tau+s}{p(\sigma)})\frac{d\tau}{p(\sigma)},$$

and therefore we have the identity (7.61), as required.

If $s + t \in [kp(\sigma), (k+1)p(\sigma))$ for some positive integer k, then $\varphi^t\theta = \varphi^{t+s}\sigma = \varphi^{t+s-kp(\sigma)}\sigma$ where $t+s-kp(\sigma) \in [0, p(\sigma))$, and the left hand side of equation (7.61) can be written as

$$\beta(\sigma)\,\Phi^{t+s-kp(\sigma)}_{i\xi}(\sigma)\,\alpha(\tfrac{t+s-kp(\sigma)}{p(\sigma)})\,[x_n - \Phi^{p(\sigma)}_{i\xi}(\sigma)x_n] + \beta(\sigma)\Phi^{t+s-(k-1)p(\sigma)}_{i\xi}(\sigma)\,x_n$$
$$-\beta(\sigma)\,\Phi^{t+s+p(\sigma)}_{i\xi}(\sigma)x_n - \beta(\sigma)\,\Phi^{t+s}_{i\xi}(\sigma)\,\alpha(\tfrac{s}{p(\sigma)})\,[x_n - \Phi^{p(\sigma)}_{i\xi}(\sigma)x_n].$$

Let us split the integral on the right hand side of equation (7.61) into the sum of three integrals and treat each of them separately. By the choice of the function α, we have that $\alpha(1) = 1$. This fact is used to compute the first summand:

$$-\int_0^{p(\sigma)-s} \Phi^{t-\tau}_{i\xi}(\varphi^\tau\theta)g_n(\varphi^\tau\theta)\,d\tau$$
$$= \beta(\sigma)\int_0^{p(\sigma)-s} \frac{1}{p(\sigma)}\alpha'(\tfrac{\tau+s}{p(\sigma)})\Phi^{t-\tau}_{i\xi}(\varphi^{s+\tau}\sigma)\Phi^{\tau+s}_{i\xi}(\sigma)[x_n - \Phi^{p(\sigma)}_{i\xi}(\sigma)x_n]\,d\tau$$
$$= \beta(\sigma)\left(1 - \alpha(\tfrac{s}{p(\sigma)})\right)\Phi^{t+s}_{i\xi}(\sigma)[x_n - \Phi^{p(\sigma)}_{i\xi}(\sigma)x_n].$$

Next, we use the equalities $\alpha(0) = 0$ and $\alpha(1) = 1$, and the cocycle property to obtain

$$-\int_{p(\sigma)-s}^{kp(\sigma)-s} \Phi^{t-\tau}_{i\xi}(\varphi^\tau\theta)g_n(\varphi^\tau\theta)\,d\tau$$
$$= \beta(\sigma)\sum_{\ell=1}^{k-1}\int_{\ell p(\sigma)-s}^{(\ell+1)p(\sigma)-s} \frac{1}{p(\sigma)}\,\alpha'(\tfrac{s+\tau-\ell p(\sigma)}{p(\sigma)})\,\Phi^{t-\tau}_{i\xi}(\varphi^{s+\tau-\ell p(\sigma)}\sigma)\Phi^{s+\tau-\ell p(\sigma)}_{i\xi}(\sigma)$$
$$\cdot [x_n - \Phi^{p(\sigma)}_{i\xi}(\sigma)x_n]\,d\tau$$
$$= \beta(\sigma)[\Phi^{t+s-(k-1)p(\sigma)}_{i\xi}(\sigma)x_n - \Phi^{t+s}_{i\xi}(\sigma)x_n].$$

Finally, we have that

$$-\int_{kp(\sigma)-s}^{t} \Phi_{i\xi}^{t-\tau}(\varphi^\tau\theta) g_n(\varphi^\tau\theta)\, d\tau$$

$$= \beta(\sigma) \int_{kp(\sigma)-s}^{t} \frac{1}{p(\sigma)} \alpha'\left(\frac{\tau+s-kp(\sigma)}{p(\sigma)}\right) \Phi_{i\xi}^{t-\tau}(\varphi^{\tau+s-kp(\sigma)}\sigma) \Phi_{i\xi}^{\tau+s-kp(\sigma)}(\sigma)$$

$$[x_n - \Phi_{i\xi}^{p(\sigma)}(\sigma)x_n]\, d\tau$$

$$= \beta(\sigma)\, \Phi_{i\xi}^{t+s-kp(\sigma)}(\sigma)\, \alpha\left(\frac{t+s-kp(\sigma)}{p(\sigma)}\right) [x_n - \Phi_{i\xi}^{p(\sigma)}(\sigma)x_n].$$

Lemma 7.38 follows by combining these identities. □

As a result, the inequality (7.58) is verified.

Step II. We will prove that the operator $I - \pi_{\theta_0}(E)$ is invertible on $c_0(\mathbb{Z}; X)$ for each $\theta_0 \in \Theta$. This fact together with the inequality (7.58) implies the inequality (7.57) and hence the theorem. We will consider three cases.

(1) Assume that $p(\theta_0) = \infty$. For each $\gamma \geq 0$, define the corresponding operator

$$D(\gamma) : (x_k)_{k\in\mathbb{Z}} \mapsto (x_k - e^{-\gamma(|k|-|k-1|)} \Phi^1(\varphi^{k-1}\theta_0) x_{k-1})_{k\in\mathbb{Z}}$$

on $c_0(\mathbb{Z}; X)$, and note that

(7.62) $$\lim_{\gamma \to 0} D(\gamma) = D(0) = I - \pi_{\theta_0}(E) \quad \text{in } \mathcal{L}(c_0(\mathbb{Z}; X)).$$

This limit, together with inequality (7.58), implies that there are positive constants c and γ_0 such that

(7.63) $$\|D(\gamma)\|_{\bullet, c_0(\mathbb{Z};X)} \geq c$$

for all $0 \leq \gamma \leq \gamma_0$.

Let $\bar{y} = (y_k)_{k\in\mathbb{Z}} \in c_0(\mathbb{Z}; X)$ be a sequence such that $y_k = 0$ for $|k| \geq n = n(\bar{y})$. Also, choose a function $\alpha \in C(\mathbb{R})$ such that $\operatorname{supp}\alpha \subset (0,1)$ and $\int_\mathbb{R} \alpha(\tau)\, d\tau = 1$. For $m = n+1, n+2, \cdots$ and $0 \leq \gamma \leq \gamma_0$, define

$$\tilde{g}_m(\varphi^t\theta_0) = \alpha(t-k+1)\, \Phi^{t-k+1}(\varphi^{k-1}\theta_0)\, e^{\gamma|k-1|}\, y_{k-1}$$

for $t \in [k-1, k)$ and $k = -m+1, \ldots, m$. We can extend $\tilde{g}_m$ to a function $g_m \in C(\Theta; X)$ with compact support such that

$$\|g_m\|_\infty \leq \|\alpha\|_\infty \sup_{0 \leq \tau \leq 1,\, \theta \in \Theta} \|\Phi^\tau(\theta)\|\, e^{\gamma_0 n}\, \|\bar{y}\|_\infty.$$

Condition $(M_0, C_b(\Theta; X))$ yields the function $u_m = R_0 g_m \in C_b(\Theta; X)$; that is,

$$u_m(\varphi^k\theta_0) = \Phi^1(\varphi^{k-1}\theta_0) u_m(\varphi^{k-1}\theta_0) + \int_0^1 \Phi^{1-\tau}(\varphi^{\tau+k-1}\theta_0) g(\varphi^{\tau+k-1}\theta_0)\, d\tau$$

$$= \Phi^1(\varphi^{k-1}\theta_0) u_m(\varphi^{k-1}\theta_0) + \Phi^1(\varphi^{k-1}\theta_0) e^{\gamma|k-1|} y_{k-1}$$

for $k = -m+1, \ldots, m$. If $v_m := (u_m(\varphi^k\theta_0) + e^{\gamma|k|} y_k)_{k\in\mathbb{Z}} \in \ell^\infty(\mathbb{Z}; X)$ and

$$r_m := (r_{mk})_{k\in\mathbb{Z}} = (I - \pi_{\theta_0}(E))v_m - (e^{\gamma|k|} y_k)_{k\in\mathbb{Z}}$$

for $m > n$, then $(r_{mk})_{k\in\mathbb{Z}} \in \ell^\infty(\mathbb{Z}; X)$ and $r_{mk} = 0$ for $|k| \leq m$. Since $y_k = 0$ for $|k| \geq n$, we obtain

$$\|r_m\|_\infty \leq \|I - \pi_{\theta_0}(E)\|_{\mathcal{L}(\ell^\infty(\mathbb{Z};X))} \|u_m\|_\infty \leq \|I - \pi_{\theta_0}(E)\|\, \|R_0\|\, \|g_m\|_\infty \leq C$$

where C depends on $\bar{y}$ but not on m and $\gamma \in [0, \gamma_0]$. Moreover, we have that $(e^{-\gamma|k|} v_{mk})_{k \in \mathbb{Z}} \in c_0(\mathbb{Z}; X)$ and

$$\|D(\gamma)(e^{-\gamma|k|} v_{mk})_{k \in \mathbb{Z}} - \bar{y}\|_\infty = \|(e^{-\gamma|k|}[(I - \pi_{\theta_0}(E))v_m]_k)_{k \in \mathbb{Z}} - \bar{y}\|_\infty$$
$$= \sup_{k \in \mathbb{Z}} |e^{-\gamma|k|} r_{mk}| \le C\, e^{-\gamma m}$$

for $m > n$. As a consequence, the operator $D(\gamma)$ has dense range on $c_0(\mathbb{Z}; X)$ and is invertible by inequality (7.63). Therefore, in view of the limit (7.62), the operator $I - \pi_{\theta_0}(E)$ is invertible on $c_0(\mathbb{Z}; X)$.

(2i) Suppose that $p(\theta_0) = 1$. By Lemma 7.35 and inequality (7.58), we have that $\sigma_{\mathrm{ap}}(\Phi^1(\theta_0)) \cap \mathbb{T} = \emptyset$. Also, by Lemma 7.36, the operator $e^\lambda - \Phi^1(\theta_0)$ is surjective for each $\lambda = i\xi \in i\mathbb{R}$ in case (a) and for $\lambda = 0$ in case (b) of Theorem 7.37. Therefore, $e^\lambda \in \rho(\Phi^1(\theta_0))$ and the boundary of $\sigma(\Phi^1(\theta_0))$ does not intersect $\mathbb{T}$. As a result, $\Phi^1(\theta_0)$ is hyperbolic and, by Lemma 7.35, the operator $I - \pi_{\theta_0}(E)$ is invertible on $c_0(\mathbb{Z}; X)$.

(2ii) Suppose that $p(\theta_0) = T \in (0, \infty)$. In this case, define the continuous flow $\psi^t(\theta) = \varphi^{tT}(\theta)$ and the exponentially bounded, strongly continuous cocycle $\Psi^t(\theta) = \Phi^{tT}(\theta)$ over the flow $\{\psi^t\}_{t \in \mathbb{R}}$ on Θ. This cocycle satisfies condition $(M_{\lambda T}, C_b(\Theta; X))$. Also, for the flow $\{\psi^t\}_{t \in \mathbb{R}}$, we have that $p_\psi(\theta_0) = 1$. By an application of parts (I) and (II.2i) to the closed ψ^t-invariant set $\mathcal{O}(\theta_0)$ and the cocycle $\{\Psi^t\}_{t \ge 0}$, it follows that each operator $I - \pi_\theta(E_\Psi)$ is invertible on $c_0(\mathbb{Z}; X)$ and $\|(I - \pi_\theta(E_\Psi))^{-1}\| \le C$ for $\theta \in \mathcal{O}(\theta_0)$. Therefore, $\{\Psi^t\}_{t \ge 0}$ has an exponential dichotomy on $\mathcal{O}(\theta_0)$ by the Discrete Dichotomy Theorem 7.9. It follows that, $\{\Phi^t\}_{t \ge 0}$ has an exponential dichotomy on $\mathcal{O}(\theta_0)$, and this fact implies the invertibility of $I - \pi_{\theta_0}(E)$.

(3) Suppose that $p(\theta_0) = 0$. In this case, $\Phi^t(\theta_0) = e^{tA}$ give a strongly continuous semigroup on X. For $y \in X$, choose $g \in C_b(\Theta; X)$ with $g(\theta_0) = y$. By condition $(M_\lambda, \mathcal{F})$ (for all $\lambda = i\xi \in i\mathbb{R}$ in case (a) and for $\lambda = 0$ in case (b) of Theorem 7.37), there is a function $f \in C_b(\Theta; X)$ such that $x := f(\theta_0)$ satisfies

$$e^{\lambda t} x - e^{tA} x = \int_0^t e^{(t-\tau)A} e^{\lambda \tau} y\, d\tau.$$

Consequently, $x \in \mathcal{D}(A)$ and $(\lambda - A)x = y$; that is, $\lambda \notin \sigma_r(A)$. Also, by Lemma 7.35 and the inequality (7.58), it follows that $\sigma_{\mathrm{ap}}(e^A) \cap \mathbb{T} = \emptyset$. Hence, $\sigma_{\mathrm{ap}}(A) \cap i\mathbb{R} = \emptyset$ by the Spectral Inclusion Theorem. In particular, $\lambda \notin \sigma(A)$ and the boundary of $\sigma(A)$ does not intersect $i\mathbb{R}$. So we infer that $\sigma(A) \cap i\mathbb{R} = \emptyset$ and, by the Spectral Mapping Theorem for the residual spectrum, $\sigma_r(e^A) \cap \mathbb{T} = \emptyset$. As a result, $\Phi^1(\theta_0)$ is hyperbolic. By Lemma 7.35, $1 \in \rho(\pi_{\theta_0}(E))$. □

Let us combine the results of this subsection into the following characterization.

COROLLARY 7.39. *Let $\{\Phi^t\}_{t \ge 0}$ be an exponentially bounded, strongly continuous cocycle over a continuous flow $\{\varphi^t\}_{t \in \mathbb{R}}$ on a locally compact metric space Θ and let $\mathcal{F} = C_b(\Theta, X)$ or $C_0(\Theta, X)$.*

(a) *If $p(\theta) \ge d_0 > 0$ for all $\theta \in \mathcal{B}(\Theta)$, then condition $(M_\lambda, \mathcal{F})$ holds for each $\lambda = i\xi \in i\mathbb{R}$ if and only if $\{\Phi^t\}_{t \ge 0}$ has an exponential dichotomy on Θ.*
(b) *If $\mathcal{B}(\Theta) = \emptyset$, then condition $(M_0, \mathcal{F})$ holds if and only if $\{\Phi^t\}_{t \ge 0}$ has an exponential dichotomy on Θ.*

7.3.2. Green's functions.

In this subsection we will prove that the existence of a dichotomy for a linear skew-product flow is equivalent to the existence of a unique Green's function. This will give a generalization of the results of Section 4.2 for the case of linear skew-product flows.

As usual, the term *splitting projection* will refer to a projection-valued function $P(\cdot) \in C_b(\Theta, \mathcal{L}_s(X))$ such that statements (a) and (b) in the Definition 6.13 of dichotomy hold; that is, the identity $P(\varphi^t \theta)\Phi^t(\theta) = \Phi^t(\theta)P(\theta)$ is satisfied and the restriction $\Phi_Q^t(\theta)$ of $\Phi^t(\theta)$ on $\operatorname{Im} Q(\theta)$ is invertible as an operator from $\operatorname{Im} Q(\theta)$ to $\operatorname{Im} Q(\varphi^t \theta)$ for $\theta \in \Theta$ and $t \geq 0$.

If a splitting projection P is given, then we have a projection $\mathcal{P}$ on $C_0(\Theta; X)$ defined by $(\mathcal{P}f)(\theta) = P(\theta)f(\theta)$. This projection satisfies $\mathcal{P}E^t = E^t \mathcal{P}$ where $\{E^t\}_{t \geq 0}$ is the evolution semigroup. Note that by restricting each operator E^t to $\operatorname{Im} \mathcal{Q}$, $\mathcal{Q} = I - \mathcal{P}$, we define an evolution semigroup $\{E_\mathcal{Q}^t\}_{t \geq 0}$ induced the flow $\{\varphi^t\}_{t \in \mathbb{R}}$ and the restricted cocycle $\{\Phi_Q^t\}_{t \geq 0}$. Since each operator $\Phi_Q^t(\theta)$, for $t > 0$, is invertible, each operator $E_\mathcal{Q}^t$, for $t > 0$, is invertible as well. Therefore, the semigroup $\{E_\mathcal{Q}^t\}_{t \geq 0}$ can be extended to a strongly continuous evolution *group* $\{E_\mathcal{Q}^t\}_{t \in \mathbb{R}}$ induced by the flow $\{\varphi^t\}_{t \in \mathbb{R}}$ and the cocycle $\{\Phi_Q^t\}_{t \in \mathbb{R}}$ where $\Phi_Q^t(\theta) := -[\Phi_Q^{-t}(\varphi^t \theta)Q(\varphi^t \theta)]^{-1}$ for $t < 0$. Using Theorem 6.20 we conclude, in particular, that the cocycle $\{\Phi_Q^t\}_{t \in \mathbb{R}}$ is exponentially bounded.

Suppose that a splitting projection P is given. The *Green's function* $G_P : \Theta \times (\mathbb{R}\setminus\{0\}) \to \mathcal{L}_s(X)$ is defined by

$$G_P(\theta, t) = \begin{cases} \Phi_P^t(\theta), & t > 0; \\ -\Phi_Q^t(\theta) := -[\Phi_Q^{-t}(\varphi^t \theta)Q(\varphi^t \theta)]^{-1}, & t < 0. \end{cases}$$

As we have seen in the proof of Proposition 7.33, if $\{\Phi^t\}_{t \geq 0}$ has an exponential dichotomy with the dichotomy projection P, then the *Green's operator*

$$(\mathbb{G}_P f)(\theta) = \int_{-\infty}^{\infty} G(\varphi^{-\tau}\theta, \tau) f(\varphi^{-\tau}\theta) d\tau, \quad \theta \in \Theta,$$

is a bounded operator on the spaces $C_b(\Theta; X)$ and $C_0(\Theta; X)$. We will concentrate on the converse result for the case of the space $C_0(\Theta; X)$.

THEOREM 7.40. *Let $\{\Phi^t\}_{t \geq 0}$ be an exponentially bounded, strongly continuous cocycle over a continuous flow $\{\varphi^t\}_{t \in \mathbb{R}}$ on a locally compact metric space Θ, and let p be the prime period function for $\{\varphi^t\}_{t \in \mathbb{R}}$.*

(a) If $\mathcal{B}(\Theta) = \emptyset$, then $\{\Phi^t\}_{t \geq 0}$ has an exponential dichotomy if and only if the Green's operator $\mathbb{G}_P$ is bounded on $C_0(\Theta; X)$.

(b) If $p(\theta) \geq d_0 > 0$ for all $\theta \in \mathcal{B}(\Theta)$, then $\{\Phi^t\}_{t \geq 0}$ has an exponential dichotomy on Θ if and only if, for each $\xi \in \mathbb{R}$, the Green's operator

$$(\mathbb{G}_{P,\xi} f)(\theta) = \int_{-\infty}^{\infty} e^{-i\xi\tau} G_P(\varphi^{-\tau}\theta, \tau) f(\varphi^{-\tau}\theta) d\tau, \quad \theta \in \Theta,$$

is bounded on $C_0(\Theta; X)$.

PROOF. The "only if" parts follow, as in the proof of Proposition 7.33, from the exponential estimates given in part (c) of Definition 6.13.

By Corollary 7.27 of the Annular Hull/Spectral Mapping Theorem 7.25, for the "if" part it is enough to prove that Γ^{-1} (respectively, $(\Gamma - i\xi)^{-1}$, for $\xi \in \mathbb{R}$) is bounded on $C_0(\Theta; X)$ in case (a) (respectively, in case (b)). We will concentrate on case (b).

Consider the rescaled evolution semigroup $\{T^t\}_{t\geq 0}$ where $T^t = e^{-i\xi t}E^t$, $\xi \in \mathbb{R}$, $t \geq 0$, on $C_0(\Theta; X)$ and define the operator $\tilde{G}$ on $C_0(\Theta; X)$, see (4.24), as follows:
$$\tilde{G}f = \int_0^\infty T_Q^{-t}f\,dt - \int_0^\infty T_P^t f.$$
Here $P(\cdot)$ is the splitting projection that was used to define the Green's function $G_P(\theta, t)$, $T_P^t := e^{-i\xi t}E_P^t$, and $T_Q^t := e^{-i\xi t}E_Q^t$. By Proposition 4.21, the generator $A = \Gamma - i\xi$ of the semigroup $\{T^t\}_{t\geq 0}$ has bounded inverse provided that $\tilde{G}$ is bounded on $C_0(\Theta; X)$. However,

$$(\tilde{G}f)(\theta) = \int_{-\infty}^0 e^{-i\xi t}\Phi_Q^t(\varphi^{-t}\theta)f(\varphi^{-t}\theta)\,dt$$
$$- \int_0^\infty e^{-i\xi t}\Phi_P^t(\varphi^{-t}\theta)f(\varphi^{-t}\theta)\,dt = -(\mathbb{G}_{P,\xi}f)(\theta);$$

and therefore, $\tilde{G}$ is bounded since $\mathbb{G}_{P,\xi}$ is bounded by the assumption. □

The next proposition states a simple sufficient condition for the operator $\mathbb{G}_{P,\xi}$ to be bounded on $C_0(\Theta; X)$.

PROPOSITION 7.41. *If*
$$\int_{-\infty}^\infty \sup_{\theta \in \Theta} \|G_P(\theta, \tau)\|_{\mathcal{L}(X)}\,d\tau < \infty$$
and $\xi \in \mathbb{R}$, then $\mathbb{G}_{P,\xi}$ is bounded on $C_0(\Theta; X)$.

PROOF. Clearly, $\mathbb{G}_{P,\xi}$ is bounded on $C_b(\Theta; X)$. Hence, it suffices to check that $\mathbb{G}_{P,\xi}$ maps $C_0(\Theta; X)$ into $C_0(\Theta; X)$. For each compactly supported $f \in C_0(\Theta; X)$, define θ_n and $t_n \to \infty$ as in the proof of Proposition 7.33 and note that
$$|\mathbb{G}_{P,\xi}f(\theta_n)| \leq \|f\|_{C_0(\Theta;X)} \int_{|\tau|\geq t_n} \sup_{\theta \in \Theta}\|G_P(\theta,\tau)\|_{\mathcal{L}(X)}\,d\tau.$$
By our assumption, the right hand side of this inequality converges to zero as n tends to infinity. The proposition follows from this fact. □

However, we can replace the assumption in Proposition 7.41 by a milder hypothesis.

THEOREM 7.42. *An exponentially bounded, strongly continuous cocycle $\{\Phi^t\}_{t\geq 0}$ has an exponential dichotomy on Θ if and only if there exists a unique Green's function G_P such that*

(7.64) $$K := \sup_{\theta \in \Theta} \int_{-\infty}^\infty \|G_P(\varphi^{-\tau}\theta, \tau)\|_{\mathcal{L}(X)}\,d\tau < \infty.$$

PROOF. The "only if" part follows from the exponential estimates in part (c) of Definition 6.13. To prove the "if" part, let us note that, by the implication (2)⇒(1) of Dichotomy Theorem 6.41, it suffices to check the hyperbolicity of the evolution semigroup $\{E^t\}_{t\geq 0}$ on $C_0(\Theta; X)$. For the projection $P(\cdot)$ as in the definition of the Green's function, note that $\mathcal{P} = P(\cdot)$ is a splitting projection for the semigroup $\{E^t\}$ on $C_0(\Theta; X)$. Consider the Green's operator $\mathbb{G}_\mathcal{P}$ on $C_b(\mathbb{R}; C_0(\Theta; X))$ corresponding to $\{E^t\}_{t\geq 0}$ that is defined, as indicated in Subsection 4.2.1, by

$$(\mathbb{G}_\mathcal{P}F)(\tau) = \int_{-\infty}^\infty G_\mathcal{P}(\tau - s)F(s)\,ds, \quad F(\tau) = f(\tau, \cdot) \in C_0(\Theta; X), \quad \tau \in \mathbb{R}.$$

Here, $G_\mathcal{P}$ is the Green's function for the semigroup $\{E^t\}_{t\geq 0}$; that is, $G_\mathcal{P}(\tau) = E_\mathcal{P}^\tau$ for $\tau > 0$, and $G_\mathcal{P}(\tau) = -E_\mathcal{Q}^\tau$ for $\tau < 0$. As usual, $E_\mathcal{P}^\tau$ and $E_\mathcal{Q}^\tau$ are the restrictions of E^τ on $\operatorname{Im}\mathcal{P}$ and $\operatorname{Im}\mathcal{Q}$, respectively: $(E_\mathcal{P}^\tau h)(\theta) = \Phi_\mathcal{P}^\tau(\varphi^{-\tau}\theta)h(\varphi^{-\tau}\theta)$ and

$$(E_\mathcal{Q}^\tau h)(\theta) = \Phi_\mathcal{Q}^\tau(\varphi^{-\tau}\theta)h(\varphi^{-\tau}\theta), \quad \theta \in \Theta, \quad h \in C_0(\Theta; X).$$

We claim that $\mathbb{G}_\mathcal{P}$ is a bounded operator on $C_b(\mathbb{R}; C_0(\Theta; X))$. Since

$$(\mathbb{G}_\mathcal{P} F)(\tau) = \int_{-\infty}^\infty G_\mathcal{P}(s) F(\tau - s)\, ds$$

$$= \int_0^\infty E_\mathcal{P}^s F(\tau - s)\, ds - \int_{-\infty}^0 E_\mathcal{Q}^s F(\tau - s)\, ds$$

for $F(\tau) = f(\tau, \cdot)$, if we use the definition of E^s, then we have that

(7.65)
$$(\mathbb{G}_\mathcal{P} f)(\tau, \theta) = \int_0^\infty \Phi_\mathcal{P}^s(\varphi^{-s}\theta) f(\tau - s, \varphi^{-s}\theta)\, ds - \int_{-\infty}^0 \Phi_\mathcal{Q}^s(\varphi^{-s}\theta) f(\tau - s, \varphi^{-s}\theta)\, ds.$$

Then, using the hypothesis (7.64),

$$\|\mathbb{G}_\mathcal{P} F\|_{C_b(\mathbb{R}; C_0(\Theta; X))} = \sup_{\tau \in \mathbb{R}} \sup_{\theta \in \Theta} |\mathbb{G}_\mathcal{P} f(\tau, \theta)| \leq K \|F\|_{C_b(\mathbb{R}; C_0(\Theta; X))},$$

as claimed.

Now apply Proposition 4.29 for the (autonomous) evolution family $\{U(t, \tau)\}_{t \geq \tau}$ given by $U(t, \tau) = E^{t-\tau}$ on the space $C_0(\Theta; X)$ instead of X. By Proposition 4.29, if $H \in C_b(\mathbb{R}; C_0(\Theta; X))$, then there exists a solution $F \in C_b(\mathbb{R}; C_0(\Theta; X))$ for the equation

$$F(t) = E^{t-\tau} F(\tau) + \int_\tau^t E^{t-s} H(s)\, ds, \quad t \geq \tau.$$

Let us show that this solution is unique. Indeed, take $H = 0$ and suppose there exists $F \in C_b(\mathbb{R}; C_0(\Theta; X))$ such that $F(t) = E^{t-\tau} F(\tau)$ holds for $t \geq \tau$. Using the definition of E^t, if $F(t) = f(t, \cdot)$, then we have $f(t, \theta) = \Phi^{t-\tau}(\varphi^{-(t-\tau)}\theta) f(\tau, \varphi^{-(t-\tau)}\theta)$ for $t \geq \tau$ and $\theta \in \Theta$. From this fact and the identity (7.65), we have that

$$(\mathbb{G}_\mathcal{P} f)(\tau, \theta) = \int_0^\infty \left[P(\theta) f(\tau, \theta) - Q(\theta) f(\tau, \theta) \right] ds.$$

But this is possible only if $P(\theta) f(\tau, \theta) = Q(\theta) f(\tau, \theta)$. Thus, $f(\tau, \theta) = 0$, or $F = 0$. As a result, we have proved that Condition (M) holds for $\{E^{t-\tau}\}_{t \geq \tau}$ on $C_b(\mathbb{R}; C_0(\Theta; X))$. By Theorem 4.28, it follows that $\{E^{t-\tau}\}_{t \geq \tau}$ has an exponential dichotomy; that is, the evolution semigroup $\{E^t\}_{t \geq 0}$ is hyperbolic on $C_0(\Theta; X)$. □

7.4. Isomorphism Theorems

Let us restrict attention to the case where the space X is replaced by a Hilbert space. In Subsection 7.4.1 we will first discuss a certain algebraic approach related to the discrete operators introduced in Subsection 7.1.1 above. Next, we will show in Subsection 7.4.2 that this approach is in fact connected to a general theory of cross-products of C^*-algebras by groups of automorphisms. In this section the evolution operator E is considered on the space $L^2(\Theta, \mu; \mathcal{H})$ where $\mathcal{H}$ is a Hilbert space, see also Subsection 6.2.4.

7.4.1. Evolution semigroups and Isomorphism Theorems for C^* algebras.
In this subsection we show that there is a general algebraic structure present in the background of our previous discussions on translation algebras, discrete operators and pointwise dichotomies for linear skew-product flows. This structure is related to a canonical construction of a C^*-algebra $\mathfrak{A} \otimes G$ called *cross-product*. The algebra $\mathfrak{A} \otimes G$ is related to a given C^*-algebra $\mathfrak{A}$ and a group G acting on $\mathfrak{A}$. Suppose that $\mathfrak{A}$ is the algebra of multiplication operators on $L^2(\Theta, \mu; \mathcal{H})$ and G acts on $\mathfrak{A}$ by means of the set of translation operators $\{V_g : g \in G\}$ induced by the family of homeomorphisms $\{\varphi_g : g \in G\}$ on Θ. As we will see, if appropriate conditions are met, then the C^*-algebra $\mathfrak{A} \otimes G$ is *isomorphic* to the smallest C^*-subalgebra $C^*(\mathfrak{A}, V)$ in $\mathcal{L}(L^2(\Theta, \mu; \mathcal{H}))$ generated by the multiplication operators in $\mathfrak{A}$ and $\{V_g : g \in G\}$. We start with an isomorphism theorem of this type for the group $G = \mathbb{Z}$.

Let Θ be a compact metric space and $\varphi : \Theta \to \Theta$ a homeomorphism with a dense set of aperiodic points. Also, let μ be a Borel probability measure on Θ that is positive on open subsets and quasi-invariant with respect to φ; that is, for each Borel subset $\Theta' \subset \Theta$ we have $\mu(\Theta') = 0$ if and only if $\mu(\varphi^{-1}(\Theta')) = 0$. As we already mentioned in Subsection 6.2.4, there always exists a quasi-invariant measure μ that is positive on open subsets of Θ, see for example Antonevich [**An2**, §9].

Let $\mathcal{H}$ be a Hilbert space, $\Phi : \Theta \to \mathcal{L}(\mathcal{H})$ a *norm* continuous operator-valued function, and E the corresponding evolution operator on $L^2(\Theta; \mu; \mathcal{H})$ defined as follows (cf. (6.40)):

$$(7.66) \qquad (Ef)(\theta) = \left[\frac{d\mu\varphi^{-1}(\theta)}{d\mu}\right]^{\frac{1}{2}} \Phi(\varphi^{-1}\theta) f(\varphi^{-1}\theta).$$

Here $d\mu\varphi^{-1}/d\mu \in L^\infty(\Theta)$ is the Radon-Nikodým derivative of the measure

$$\mu\varphi^{-1}(\Theta') = \mu[\varphi^{-1}(\Theta')]$$

with respect to the measure μ. The Radon-Nikodým derivative appears in the formula (7.66) to make the operator

$$(Vf)(\theta) = \left[\frac{d\mu\varphi^{-1}(\theta)}{d\mu}\right]^{1/2} f(\varphi^{-1}\theta), \quad \theta \in \Theta, \quad f \in L^2(\Theta; \mu; \mathcal{H}),$$

unitary on $L^2(\Theta; \mu; \mathcal{H})$. Let us also note that with $a(\theta) = \Phi(\varphi^{-1}\theta)$ and $a \in C(\Theta; \mathcal{L}(\mathcal{H}))$, we have the identity $E = aV$.

Let $\mathfrak{A}$ be the C^*-algebra of multiplication operators a on $L^2(\Theta, \mu; \mathcal{H})$ with norm continuous operator-valued multipliers, $(af)(\theta) = a(\theta)f(\theta)$, and let $C^*(\mathfrak{A}; V)$ be the smallest C^*-algebra of operators on $L^2(\Theta, \mu; \mathcal{H})$ generated by V, V^*, and all operators $a \in \mathfrak{A}$. Note that we have $V^{-1} = V^*$ where

$$(V^* f)(\theta) = \left[\frac{d\mu\varphi(\theta)}{d\mu}\right]^{\frac{1}{2}} f(\varphi\theta).$$

The homeomorphism $\varphi : \Theta \to \Theta$ defines an action

$$(7.67) \qquad \bar{\varphi} : \mathfrak{A} \to \mathfrak{A} : a \mapsto VaV^*, \quad (\bar{\varphi}a)(\theta) = a(\varphi^{-1}\theta),$$

of the group $\mathbb{Z}$ on $\mathfrak{A}$.

There exists a canonical construction of a new C^*-algebra associated with the given C^*-algebra $\mathfrak{A}$ and the action of the group $\mathbb{Z}$ on $\mathfrak{A}$, see for example Antonevich and Lebedev [**AL2**], Bratteli and Robinson [**BR**], or Pedersen [**Pe**]. This new C^*-algebra $\mathfrak{A} \otimes_{\bar\varphi} \mathbb{Z}$ is called the *cross-product of $\mathfrak{A}$ and the group action*.

Let us recall the definition of the cross-product $\mathfrak{A} \otimes_{\bar\varphi} \mathbb{Z}$. We start with the set $\mathfrak{B}^0$ of formal finite sums $b = \sum_k a_k \otimes \bar\varphi^k$, where $a_k \in \mathfrak{A}$, endowed with product defined as the "convolution"

$$(7.68) \qquad (a_k \otimes \bar\varphi^k) \cdot (a_m \otimes \bar\varphi^m) := a_k \bar\varphi^k(a_m) \otimes \bar\varphi^{m+k},$$

and the involution $[a_k \otimes \bar\varphi^k]^* = [\bar\varphi^{-k}(a)]^* \otimes \bar\varphi^{-k}$. Then, the cross-product $\mathfrak{A} \otimes_{\bar\varphi} \mathbb{Z}$ is defined as the closure of $\mathfrak{B}^0$ in the norm given by $\sup\{\|\pi(b)\|\}$, where π ranges over the set of all representations of $\mathfrak{B}^0$ that are continuous with respect to the norm $\|b\|_1 = \sum_k \|a_k\|_{\mathfrak{A}}$.

There is a natural correspondence $a \otimes \bar\varphi \mapsto aV$ of an element $a \otimes \bar\varphi \in \mathfrak{A} \otimes_{\bar\varphi} \mathbb{Z}$ to the operator $aV \in \mathcal{L}(L^2(\Theta, \mu; \mathcal{H}))$ that can be extended by linearity to the set $\mathfrak{B}^0$. The following *isomorphism theorem* shows that if appropriate conditions are met, then the algebra $\mathfrak{A} \otimes_{\bar\varphi} \mathbb{Z}$ is isomorphic to the subalgebra $C^*(\mathfrak{A}, V)$ of $\mathcal{L}(L^2(\Theta, \mu; \mathcal{H}))$.

THEOREM 7.43 (Antonevich-Lebedev). *If the set of aperiodic points of φ is dense, then the correspondence $a \otimes \bar\varphi \mapsto aV$ can be extended to an isomorphism of the C^*-algebras $\mathfrak{A} \otimes_{\bar\varphi} \mathbb{Z}$ and $C^*(\mathfrak{A}; V)$.*

For the proof of this isomorphism theorem see Antonevich and Lebedev [**An3, AL2**]. One of the crucial steps in the proof uses the *Neumann map* defined on $\mathfrak{B}^0$ by the rule $\sum_k a_k \otimes \bar\varphi^k \mapsto a_0 \in \mathcal{L}(L^2(\Theta, \mu; \mathcal{H}))$. In fact, the Neumann map was used in the proof of Proposition 7.1 in a slightly different situation.

The Isomorphism Theorem 7.43 has many consequences, see [**An3, AL2**]. We briefly discuss its applications to rotational invariance of the spectrum of evolution operators, to the "inverse-closedness" of translation algebras, and to discrete operators.

First, we remark that the evolution operator $E = aV$ with $a(\theta) = \Phi(\varphi^{-1}\theta)$ belongs to the C^*-algebra $C^*(\mathfrak{A}, V)$. The Isomorphism Theorem 7.43 immediately gives the following result, which is a variant of Lemma 6.28 above.

COROLLARY 7.44. *The spectrum of $E = aV$ is invariant with respect to rotations centered at the origin of the complex plane.*

PROOF. If $|z| = 1$, then the automorphism of $\mathfrak{A}$ given by

$$\bar\varphi_z : a \mapsto (zV)a(z^{-1}V^*)$$

coincides with $\bar\varphi$. Therefore, $\mathfrak{A} \otimes_{\bar\varphi_z} \mathbb{Z} = \mathfrak{A} \otimes_{\bar\varphi} \mathbb{Z}$. By the Isomorphism Theorem 7.43, the C^*-algebras $C^*(\mathfrak{A}, V)$ and $C^*(\mathfrak{A}; zV)$ are isomorphic. Recall that $C^*(\mathfrak{A}, V)$ is a C^*-subalgebra of $\mathcal{L}(L^2(\Theta, \mu; \mathcal{H}))$. Hence, the operator $\lambda - aV$ is invertible on $L^2(\Theta, \mu; \mathcal{H})$ (that is, it is invertible in $\mathcal{L}(L^2(\Theta, \mu; \mathcal{H}))$) if and only if $\lambda - aV \in C^*(\mathfrak{A}; V)$ is invertible in $C^*(\mathfrak{A}; V)$, see for example Proposition 2.2.7 in Bratteli and Robinson [**BR**]. Therefore, the operator $\lambda - zaV$ is invertible on $L^2(\Theta, \mu, \mathcal{H})$ if and only if $\lambda - aV$ is invertible on $L^2(\Theta, \mu, \mathcal{H})$. $\square$

Next, we discuss the invertibility of elements of weighted translation algebras, cf. Subsection 4.1.2 and Theorem 7.4. Let us consider the subalgebra $\mathfrak{B}$ of bounded operators on $L^2(\Theta, \mu; \mathcal{H})$ given by the set of all operators of the form

$$b = \sum_{k=-\infty}^{\infty} a_k V^k \text{ where } a_k \in \mathfrak{A} \text{ and } \|b\|_1 := \sum_{k=-\infty}^{\infty} \|a_k\|_{\mathfrak{A}} < \infty.$$

Since φ is aperiodic, $(\mathfrak{B}, \|\cdot\|_1)$ is a well-defined Banach algebra (see Proposition 7.1). The next theorem is proved in Antonevich and Lebedev [**An3, AL2**] as a corollary of Theorem 7.43.

THEOREM 7.45. *The algebra $\mathfrak{B}$ is inverse-closed in $\mathcal{L}(L^2(\Theta, \mu; \mathcal{H}))$; that is, if an operator $b \in \mathfrak{B}$ has inverse $b^{-1} \in \mathcal{L}(L^2(\Theta, \mu; \mathcal{H}))$, then $b^{-1} \in \mathfrak{B}$.*

This fact implies the conclusion of the Spectral Projection Theorem 6.38.

COROLLARY 7.46. *If E is hyperbolic on $L^2(\Theta, \mu; \mathcal{H})$, then the Riesz projection $\mathcal{P}$ that corresponds to E and the spectral set $\sigma(E) \cap \mathbb{D}$ is an element of $\mathfrak{A}$; that is, $(\mathcal{P}f)(\theta) = P(\theta)f(\theta)$ for some norm-continuous projection-valued function $P: \Theta \to \mathcal{L}(\mathcal{H})$.*

PROOF. If we use the Commutation Lemma 6.39, then $V^k \chi = \chi \circ \varphi^{-k} V^k$ for every $\chi \in C(\Theta; \mathbb{R})$ and $k \in \mathbb{Z}$. The required statement now follows from Theorem 7.45 exactly as in the proof of Proposition 4.18. □

Corollary 7.46 implies the following L^2-version of the Dichotomy Theorem 6.41.

COROLLARY 7.47. *The linear skew-product flow $\{\hat{\varphi}^t\}_{t \geq 0}$ has an exponential dichotomy on Θ if and only if the operator E is hyperbolic on $(L^2(\Theta, \mu; \mathcal{H})$.*

Finally, we turn our attention to the family of discrete operators $\{\pi_\theta(E) : \theta \in \Theta\}$, see Section 7.1. For each $\theta \in \Theta$ and $a \in \mathfrak{A}$ we define the maps

$$\pi_\theta : a \mapsto \pi_\theta(a) = \mathrm{diag}(a(\varphi^n \theta))_{n \in \mathbb{Z}}, \qquad \pi_\theta : V \mapsto S,$$

where S is the shift operator and the operators $\pi_\theta(a)$ and $\pi_\theta(V)$ act on the sequence space $\ell^2(\mathbb{Z}; \mathcal{H})$. Of course, by linearity, each map π_θ can be extended to the set of all finite sums $\sum_k a_k V^k \in C^*(\mathfrak{A}; V)$. In addition, π_θ can be extended to $C^*(\mathfrak{A}, V)$ to get a representation

$$\pi_\theta : C^*(\mathfrak{A}, V) \mapsto \mathcal{L}(\ell^2(\mathbb{Z}; \mathcal{H}))$$

of the C^*-algebra $C^*(\mathfrak{A}; V)$ in the Hilbert space $\ell^2(\mathbb{Z}; \mathcal{H})$. Moreover, the Isomorphism Theorem 7.43 can be used to prove the following important fact: The set of representations $\{\pi_\theta\}_{\theta \in \Theta}$ is *sufficient* in the sense that the invertibility of $b \in C^*(\mathfrak{A}; V)$ is equivalent to the invertibility of $\pi_\theta(b)$ in $\ell^2(\mathbb{Z}; \mathcal{H})$ for all $\theta \in \Theta$. This result is due to A. Antonevich, Karlovich, and Lebedev (see [**An3, AL2, Ka**]); it has the following corollary.

COROLLARY 7.48. *Suppose that φ is a homeomorphism of a compact metric space Θ with a dense set of aperiodic points, and $\Phi: \Theta \to \mathcal{L}(\mathcal{H})$ is norm continuous. The evolution operator E is hyperbolic on $L^2(\Theta, \mu; \mathcal{H})$ if and only if, for each $\theta \in \Theta$, the operator $\pi_\theta(E)$ is hyperbolic on $\ell^2(\mathbb{Z}; \mathcal{H})$. Moreover, $\{\hat{\varphi}^t\}_{t \geq 0}$ has an exponential dichotomy on Θ if and only if $\{\hat{\varphi}^t\}_{t \geq 0}$ has an exponential dichotomy at each point $\theta_0 \in \Theta$.*

We stress the fact that Corollary 7.48 *does not require* the uniform boundedness of the pointwise dichotomy constants $\beta(\theta)$ and $M(\beta(\theta))$, cf. (b) in the Discrete Dichotomy Theorem 7.9.

7.4.2. General Isomorphism Theorems.

In this subsection we outline a general context wherein the action of the group $\mathbb{Z}$ on $\mathfrak{A}$ described in the previous Subsection 7.4.1 is a special case, see for example Antonevich and Lebedev [**AL2**], Bratteli and Robinson [**BR**], or Pedersen [**Pe**].

Let $\mathfrak{A}$ be a unitary C^*-subalgebra of the algebra $\mathcal{L}(\mathcal{H})$ of bounded linear operators on a Hilbert space $\mathcal{H}$. Let G be a discrete group and $V : g \mapsto V_g$ a unitary representation of G in $\mathcal{H}$ with the additional property that $V_g \mathfrak{A} V_g^* = \mathfrak{A}$, for all $g \in G$. The triple $(\mathfrak{A}, V, G)$ is sometimes called a non-commutative (or C^*-) dynamical system, see Bratteli and Robinson [**BR**]. Let us consider the C^*-algebra $C^*(\mathfrak{A}, V, G)$ generated by all operators $a \in \mathfrak{A}$ and V_g, for $g \in G$. Also, consider another C^*-dynamical system $(\mathfrak{A}', V', G)$ with the associated C^*-algebra $C^*(\mathfrak{A}', V', G)$ such that there is an isomorphism $i : \mathfrak{A} \to \mathfrak{A}'$ with the property that $i(V_g a V_g^*) = V_g' i(a) V_g'^*$ for all $a \in \mathfrak{A}$ and $g \in G$. We are interested in conditions on G, V, and $\mathfrak{A}$ under which the mappings $a \mapsto i(a)$, $V_g \mapsto V_g'$ can be extended to an isomorphism of the C^*-algebras $C^*(\mathfrak{A}, V, G)$ and $C^*(\mathfrak{A}', V', G)$.

To formulate the corresponding isomorphism theorem, let us recall first that a group G is called *amenable* if it has an invariant mean χ; that is, a functional $\chi : 2^G \to \mathbb{R}$ such that $\chi(G') = \chi(gG')$ for every $G' \subset G$ and $g \in G$. All commutative groups, finite groups, and groups with a polynomial word-growth rate are amenable. Also, amenable groups satisfy all propositions of the usual "discrete time" ergodic theory; for example, the Rokhlin-Halmos Lemma, the Individual Ergodic Theorem, etc.

Second, let us define the notion of a topologically free action of G on $\mathfrak{A}$. We refer to Bratteli and Robinson [**BR**] for the notions needed to give this definition. Let Z be a central C^*-subalgebra of $\mathfrak{A}$ with unity such that $V_g Z V_g^* = Z$, and let Θ be the compact set of maximal ideals of Z. Note that for each $g \in G$ the map $\bar{g} : a \to V_g a V_g$ defines an automorphism of Z. With the usual identifications, let $\theta \in \Theta$ be a homomorphism from Z to $\mathbb{C}$ with kernel θ. Then, each $g \in G$ induces a homeomorphism $\varphi_g : \Theta \to \Theta$ by the rule $(\varphi_g \theta)(a) = \theta(\bar{g}(a))$. Consider a pure state $\nu \in P_\mathfrak{A}$; that is, an extreme point of positive homomorphisms of $\mathfrak{A}$ to $\mathbb{C}$, and define $\theta_\nu \in \Theta$ by $\theta_\nu = \text{Ker } \nu \cap Z$. We say that a group G acts *topologically freely* on $\mathfrak{A}$ if for every $g \in G$ (different from unity) and every open set $\mathcal{U} \subset P_\mathfrak{A}$, there exists a $\nu \in \mathcal{U}$ such that $\varphi_g(\theta_\nu) \neq \theta_\nu$.

We are ready to formulate the following isomorphism theorem proved in Antonevich and Lebedev [**AL2**].

THEOREM 7.49 (Isomorphism Theorem). *If (a) G is amenable and (b) G acts on $\mathfrak{A}$ topologically freely, then the mappings*

$$a \mapsto i(a), \quad a \in \mathfrak{A}, \quad V_g \mapsto V_g', \quad g \in G,$$

can be extended to an isomorphism of the algebras $C^*(\mathfrak{A}, V, G)$ and $C^*(\mathfrak{A}', V', G)$.

We will give several examples where the Isomorphism Theorem 7.49 is applicable.

EXAMPLE 7.50. If V is a unitary representation of G such that $V_g \mathfrak{A} V_g^* = \mathfrak{A}$, then there is an action of G on $\mathfrak{A}$ defined as $\bar{g} : a \mapsto V_g a V_g^*$ (for example, see (7.67)). Consider $\mathfrak{B}^0$, the set of formal finite sums $b = \sum_g a_g \otimes \bar{g}$, $a_g \in \mathfrak{A}$, with the product and involution defined as follows

$$(a_g \otimes \bar{g}) \cdot (a_h \otimes \bar{h}) = a_g \cdot \bar{g}(a_h) \otimes \overline{gh}, \quad (a_g \otimes \bar{g})^* = \overline{g^{-1}}(a_g) \otimes \overline{g^{-1}},$$

cf. (7.68). Then, $\mathfrak{B}^0$ is a (nonclosed) $*$-algebra. Define the norm $\|b\| = \sup_\pi \|\pi(b)\|$ on $\mathfrak{B}^0$ where the supremum is taken over all representations of $\mathfrak{B}^0$ that are continuous with respect to the norm $\|b\|_1 = \sum \|a_g\|$ on $\mathfrak{B}^0$. The completion of $\mathfrak{B}^0$ with respect to this norm is called the *cross product of* $\mathfrak{A}$ *and* G; we will denote it by $\mathfrak{A} \otimes_V G$. The Isomorphism Theorem 7.49 states that if G is amenable and acts on $\mathfrak{A}$ topologically freely, then the mappings

$$a \mapsto a \otimes \bar{1}, \quad V_g \mapsto e \otimes \bar{g},$$

can be extended to an isomorphism of $C^*(\mathfrak{A}; V; G)$ and $\mathfrak{A} \otimes_V G$. Here 1 and e are unit elements in G and $\mathfrak{A}$, respectively. $\diamond$

EXAMPLE 7.51. Suppose that $\mathfrak{A}$ is a C^*-algebra of operators acting on the Hilbert space $\mathfrak{H}$ and let $\tilde{\mathfrak{H}} := \ell^2(G; \mathfrak{H})$. Each $a \in \mathfrak{A}$ generates an operator $\tilde{a}$ on $\tilde{\mathfrak{H}}$ by the formula

(7.69) $$\tilde{a} : (x_h)_{h \in G} \mapsto (\bar{h}(a) x_h)_{h \in G}.$$

For example, let $\mathfrak{H} = L^2(\Theta, \mu; \mathcal{H})$, $G = \mathbb{Z}$, and $(V_k f)(\theta) = f \circ \varphi^{-k}$, for $k \in \mathbb{Z}$, where $\varphi : \Theta \to \Theta$ is a homeomorphism that leaves μ invariant. Let $\mathfrak{A}$ be the algebra of multiplication operators with norm continuous multipliers $a(\cdot) \in C(\Theta; \mathcal{L}(\mathcal{H}))$. Then $\bar{k}(a) = a \circ \varphi^{-k}$, for $k \in \mathbb{Z}$, and, on the space $\tilde{\mathfrak{H}} := \ell^2(\mathbb{Z}; L^2(\Theta, \mu; \mathcal{H}))$, we have

$$\tilde{a} : (f_k(\cdot))_{k \in \mathbb{Z}} \mapsto (a \circ \varphi^{-k}(\cdot) f_k(\cdot))_{k \in \mathbb{Z}}, \quad f_k(\cdot) \in L^2(\Theta, \mu; \mathcal{H}),$$

or, equivalently, $\tilde{a}(\theta) = \text{diag}(a \circ \varphi^{-k}(\theta))_{k \in \mathbb{Z}}$, $\theta \in \Theta$.

Returning to the definition in display (7.69), let us note that the set of operators $\{\tilde{a} : a \in \mathfrak{A}\}$ forms a C^*-algebra $\tilde{\mathfrak{A}}$ that is isomorphic to $\mathfrak{A}$. Let us define a unitary representation $\tilde{V}$ of G in $\tilde{\mathfrak{H}}$ by

$$\tilde{V}_g : (x_h)_{h \in G} \mapsto (x_{g^{-1}h})_{h \in G}.$$

In particular, for our example,

$$\tilde{V}_n = S^n : (f_k(\cdot))_{k \in \mathbb{Z}} \mapsto (f_{k-n}(\cdot))_{k \in \mathbb{Z}}$$

where S is the shift operator on $\ell^2(\mathbb{Z}; L^2(\Theta, \mu; \mathcal{H}))$.

The algebra $C^*(\tilde{\mathfrak{A}}, \tilde{V}, G)$ is called the *regular representation* of the C^*-dynamical system $(\mathfrak{A}, V, G)$ and is denoted by $R^*(\mathfrak{A}, V, G)$. By the Isomorphism Theorem 7.49, $C^*(\mathfrak{A}, V, G)$ is isomorphic to the regular representation $R^*(\mathfrak{A}, V, G)$ provided that G is an amenable group which acts on $\mathfrak{A}$ topologically freely. $\diamond$

EXAMPLE 7.52. Let G be a discrete group, Id its trivial action on $\mathfrak{A} = \mathbb{C}$; that is, the action $a \mapsto \bar{g}(a) = a$. The cross-product $\mathbb{C} \otimes_{Id} G$ is called the *group C^*-algebra* and is denoted by $C^*(G)$. The regular representation $R^*(\mathbb{C}, Id, G)$ is just the subalgebra in $\mathcal{L}(\ell^2(G; \mathbb{R}))$ generated by the shift operators $V_g : (x_h)_{h \in G} \mapsto (x_{g^{-1}h})_{h \in G}$. The group C^*-algebra $C^*(G)$ is isomorphic to $R^*(\mathbb{C} Id, G)$ if and only if G is amenable (see Pedersen [**Pe**, p.243]). $\diamond$

Note that because $C^*(G) \simeq R^*(\mathbb{C}, Id, G)$ if and only if G is amenable, condition (a) in the Isomorphism Theorem 7.49 is necessary. Condition (b) in the Isomorphism Theorem 7.49 for the situation of the previous subsection; that is, $\mathfrak{H} = L^2(\Theta, \mu; \mathcal{H})$, $G = \mathbb{Z}$, $(V_k f)(\theta) = f(\varphi^{-k}\theta)$, etc., is equivalent to the density of the set of aperiodic points of the map $\varphi : \Theta \to \Theta$.

We mention a few corollaries of the Isomorphism Theorem 7.49; see Antonevich and Lebedev [**AL2**] for more results of this type.

1. If there are no compact operators in $\mathfrak{A}$, then the set of invertible operators and the set of Fredholm operators in $C^*(\mathfrak{A}, V, G)$ coincide.

2. Let $G = \mathbb{Z}^n$ and $g_1, \ldots, g_n$ be the generators of G. Also, assume that $g \in G$ is of the form $g = g_1^{m_1} \cdot \ldots \cdot g_n^{m_n}$, and for this g, let $z_g = (m_1, \ldots, m_n) \in \mathbb{Z}^n$. Moreover, for $\lambda = (\lambda_1, \ldots, \lambda_n) \in \mathbb{C}^n$, let $\lambda^{z_g} = (\lambda_1^{m_1}, \ldots, \lambda_n^{m_n})$. The Isomorphism Theorem 7.49 implies that the set of those $\lambda \in \mathbb{C}^n$, for which an element $\sum_{g \in G} \lambda^{z_g} a_g V_g \in C^*(\mathfrak{A}, V, G)$ is not invertible, is invariant under all transformations $\lambda \mapsto \omega \lambda$, for $\omega \in \mathbb{T}^n$. Note that, for the case $n = 1$, this gives Corollary 7.44.

7.5. Dichotomy and quadratic Lyapunov function

In this section we will show that the existence of a dichotomy for a linear skew-product flow is equivalent to the existence of a quadratic Lyapunov function. This section parallels Section 4.4 for the situation of linear skew-product flows. In fact, we will apply the results of Section 4.4 to the Mather evolution semigroup and related evolution semigroups along trajectories of the flow. In Subsection 7.5.1 we will describe the quadratic Lyapunov function defined on all of Θ and give a construction of the Lyapunov function "along trajectories" using connections between global and pointwise dichotomies. In Subsection 7.5.2 we will discuss connections of our techniques to Riccati and Hamiltonian equations, and to geodesic flows.

7.5.1. Global and trajectorial Lyapunov functions.
Let $\mathcal{H}$ be a Hilbert space with the scalar product $\langle \cdot, \cdot \rangle$, let $\{\Phi^t\}_{t \geq 0}$, $\Phi^t(\theta) \in \mathcal{L}(\mathcal{H})$, be a cocycle over a flow $\{\varphi^t\}_{t \in \mathbb{R}}$ on Θ. A global quadratic Lyapunov function l, cf. Section 4.4, is determined by an operator-function $W(\cdot)$ on Θ with self-adjoint values such that

$$l : t \mapsto \langle \Phi^t(\theta) x, W(\varphi^t \theta) \Phi^t(\theta) x \rangle_{\mathcal{H}}$$

decreases monotonically, uniformly for $\theta \in \Theta$ and $x \in \mathcal{H}$. Trajectorial Lyapunov functions are defined similarly for a fixed θ by an appropriate operator-function $w_\theta(\cdot)$ defined on $\mathbb{R}$. In this subsection we will describe these concepts in more detail and relate them to global and pointwise dichotomies.

7.5.1.1. *The setting.* To avoid additional technical difficulties, we will assume in this section that the flow $\{\varphi^t\}_{t \in \mathbb{R}}$ has a dense set of aperiodic points and the cocycle $\{\Phi^t\}_{t \in \mathbb{R}}$ is norm-continuous in θ and smooth in t, see page 217 and Remark 6.7. For convenience, the assumptions for this section are summarized as follows:

(1) Θ is a compact metric space, $\{\varphi^t\}_{t \in \mathbb{R}}$ is a continuous flow on Θ, and μ is a Borel measure on Θ which is quasi-invariant with respect to $\{\varphi^t\}_{t \in \mathbb{R}}$ and positive on open subsets of Θ. Also, see Subsection 6.2.4, the set of aperiodic points has full measure in Θ. In particular, the aperiodic points are dense in Θ.

(2) $X = \mathcal{H}$ is a separable Hilbert space and the cocycle $\{\Phi^t\}_{t \in \mathbb{R}}$ over $\{\varphi^t\}_{t \in \mathbb{R}}$ is norm-continuous and smooth; that is, the map from $\Theta \times \mathbb{R}$ to $\mathcal{L}(\mathcal{H})$ defined by $(\theta, t) \mapsto \Phi^t(\theta)$ is continuous (in norm!) and there exists a continuous function $A : \Theta \to \mathcal{L}(\mathcal{H})$ such that $A(\theta) = \frac{d}{dt} \Phi^t(\theta) \big|_{t=0}$ in $\mathcal{L}(\mathcal{H})$.

(3) The evolution group $\{E^t\}_{t \in \mathbb{R}}$ acts on the space $L^2 = L^2(\Theta, \mu; \mathcal{H})$ by the rule

$$(E^t f)(\theta) = \left(\frac{d\mu \circ \varphi^{-t}(\theta)}{d\mu} \right)^{\frac{1}{2}} \Phi^t(\varphi^{-t}\theta) f(\varphi^{-t}\theta), \quad \theta \in \Theta, \quad t \in \mathbb{R}, \quad f \in L^2.$$

Let us note that the generator Γ of the evolution semigroup is computed by the formula $\Gamma = -\mathbf{d} + \mathcal{A}$, where $\mathbf{d}$ is the generator of the group $\{V^t\}_{t \in \mathbb{R}}$ defined by
$$(V^t f)(\theta) = (d\mu \circ \varphi^{-t}(\theta)/d\mu)^{\frac{1}{2}} f(\varphi^{-t}\theta)$$
and $(\mathcal{A}f)(\theta) = A(\theta)f(\theta)$, $\theta \in \Theta$, $f \in L^2$, see Example 6.21. Also, for $\Gamma^* = \mathbf{d} + \mathcal{A}^*$, we have that $\mathcal{D}(\Gamma) = \mathcal{D}(\mathbf{d}) = \mathcal{D}(\Gamma^*)$.

Under these assumptions, the dichotomy (or hyperbolicity) of $\{\Phi^t\}_{t \in \mathbb{R}}$ is equivalent (Corollary 7.47 and Dichotomy Theorem 6.41) to the hyperbolicity of the group $\{E^t\}_{t \in \mathbb{R}}$ or, equivalently, (see Theorem 6.37) to the requirement that $\Gamma^{-1} \in \mathcal{L}(L^2)$ or $\sigma(\Gamma) \cap i\mathbb{R} = \emptyset$. Also, let us define, for each $\theta \in \Theta$, the corresponding "discrete" operator on $\ell^2(\mathbb{Z}; \mathcal{H})$ by

(7.70) $\quad \pi_\theta(E) : (x_n)_{n \in \mathbb{Z}} \mapsto (\Phi(\varphi^{n-1}\theta)x_{n-1})_{n \in \mathbb{Z}}, \quad (x_n)_{n \in \mathbb{Z}} \in \ell^2(\mathbb{Z}; \mathcal{H})$.

Then, by Corollary 7.48, the existence of an exponential dichotomy for $\{\Phi^t\}_{t \in \mathbb{R}}$ is equivalent to the hyperbolicity of the operator $\pi_\theta(E)$ on $\ell^2(\mathbb{Z}; \mathcal{H})$ for each $\theta \in \Theta$. Moreover, to apply Theorem 7.12, let us define, for each $\theta \in \Theta$, the group $\{\Pi_\theta^t\}_{t \in \mathbb{R}}$ on $L^2(\mathbb{R}; \mathcal{H})$ by

(7.71) $\quad (\Pi_\theta^t h)(s) = \Phi^t(\varphi^{s-t}\theta)h(s-t), \quad s \in \mathbb{R}, \quad t \geq 0, \quad h \in L^2(\mathbb{R}; \mathcal{H})$,

and let Γ_θ denote its generator. Then, the existence of an exponential dichotomy for $\{\Phi^t\}_{t \in \mathbb{R}}$ is equivalent to the hyperbolicity of $\{\Pi_\theta^t\}_{t \in \mathbb{R}}$ for all $\theta \in \Theta$ and, equivalently, to the invertibility of Γ_θ or $\sigma(\Gamma_\theta) \cap i\mathbb{R} = \emptyset$ on $L^2(\mathbb{R}; \mathcal{H})$ for all $\theta \in \Theta$. Let us also note that if h is smooth, then $(\Gamma_\theta h)(s) = -\frac{d}{ds}h(s) + A(\varphi^s\theta)h(s)$ and $\mathcal{D}(\Gamma_\theta) = \mathcal{D}(d/ds)$ on $L^2(\mathbb{R}; \mathcal{H})$. Also, $\mathcal{D}(\Gamma_\theta^*) = \mathcal{D}(d/ds)$ where $(\Gamma_\theta^* h)(s) = \frac{d}{ds}h(s) + A^*(\varphi^s\theta)h(s)$.

In this section, the set of invertible operators in $\mathcal{L}(\mathcal{H})$ will be denoted by $\mathcal{GL}(\mathcal{H})$ and the set of self-adjoint operators will be denoted by $\mathcal{SL}(\mathcal{H})$. Also, let $C_\varphi^1(\Theta; Y)$ be the set of functions $f : \Theta \to Y$ that are smooth in the direction of the flow $\{\varphi^t\}$; that is, $f \in C_\varphi^1(\Theta; Y)$ if there exists a function $h \in C(\Theta; Y)$ such that $h(\theta) = \dot{f}(\theta)$ where "·" denotes the derivative along the flow: $\dot{f}(\theta) = \frac{d}{dt}f(\varphi^t\theta)\big|_{t=0} = (\mathbf{d}f)(\theta)$.

7.5.1.2. *The global Lyapunov function.* A review of Section 4.4 is recommended before reading this section. In particular, we recall the notation $W << 0$; it means that $\langle Wx, x \rangle \leq -\delta|x|^2$ for some $\delta > 0$ and all $x \in \mathcal{H}$.

Our first result is a direct application of the Dichotomy Theorem and Corollary 4.43 to the evolution group $\{E^t\}_{t \in \mathbb{R}}$, $E^t = e^{t\Gamma}$.

THEOREM 7.53. *With the assumptions listed above, the cocycle $\{\Phi^t\}_{t \in \mathbb{R}}$ has an exponential dichotomy on Θ if and only if there is an invertible operator $W \in \mathcal{SL}(L^2(\Theta, \mu; \mathcal{H}))$ such that the operator $\hat{W}$, defined on $L^2(\Theta, \mu; \mathcal{H})$ by*

(7.72) $\quad \hat{W}f := \frac{d}{dt}(E^{t*}WE^t f)\Big|_{t=0}$,

is bounded and uniformly negative ($\hat{W} << 0$). If $\{\Phi^t\}_{t \in \mathbb{R}}$ has an exponential dichotomy, then W can be chosen as an operator of multiplication whose multiplier is a function $W = W(\cdot) \in C_\varphi^1(\Theta; \mathcal{SL}(\mathcal{H}) \cap \mathcal{GL}(\mathcal{H}))$ with the following properties:

(i) *If $H(\cdot) \in C(\Theta; \mathcal{SL}(\mathcal{H}))$ is such that there exists $\delta > 0$ with $\langle H(\theta)x, x \rangle \geq \delta|x|^2$ for all $x \in \mathcal{H}$ and all $\theta \in \Theta$, and $H_{PQ} := \mathcal{Q}^*H\mathcal{Q} + \mathcal{P}^*H\mathcal{P}$ for the spectral projections $\mathcal{P}$ and $\mathcal{Q} = I - \mathcal{P}$ corresponding to $\{E^t\}$, then W is given by the formula*

(7.73) $\quad W = \int_0^\infty \{E^{t*}\mathcal{P}H\mathcal{P}E^t - E^{-t*}\mathcal{Q}^*H\mathcal{Q}E^{-t}\} dt.$

(ii) *The operator W in (7.73) satisfies the equation*
$$Re\,(W\Gamma) = -\frac{1}{2}H_{PQ}.$$

(iii) *With the choice of W as in (7.73), the following inequalities hold:*
$$Re\,\langle \Gamma f, Wf\rangle_{L^2} \le -\delta\|f\|^2, \quad f \in \mathcal{D}(\Gamma),$$
$$Re\,\langle \Gamma^* f, W^{-1}f\rangle_{L^2} \le -\delta\|f\|^2, \quad f \in \mathcal{D}(\Gamma^*).$$

(iv) *The subspaces* $\mathrm{Im}\,\mathcal{P}$ *and* $\mathrm{Im}\,\mathcal{Q}$ *are W-orthogonal; that is, $\langle Wf, g\rangle = 0$ for all $f \in \mathrm{Im}\,\mathcal{P}$ and $g \in \mathrm{Im}\,\mathcal{Q}$.*

PROOF. **"If part"**. If W exists, then for each $f \in \mathcal{D}(\Gamma)$ such that $Wf \in \mathcal{D}(\Gamma^*)$ we have the equation $\hat{W}f = (\Gamma^*W + W\Gamma)f$. We will show first that if $\hat{W}$ is bounded on L^2, then $W: \mathcal{D}(\Gamma) \to \mathcal{D}(\Gamma^*)$. Indeed, for $f \in \mathcal{D}(\Gamma)$, the derivative
$$\left.\frac{d}{dt}E^{t*}Wf\right|_{t=0} = \left.\frac{d}{dt}(E^{t*}WE^tE^{-t}f)\right|_{t=0} = \hat{W}f - W\Gamma f$$
exists in L^2 because $\hat{W}f \in L^2$. Moreover, since $\hat{W} << 0$, there is some $\delta > 0$ such that
$$Re\,\langle \Gamma f, Wf\rangle = 1/2\langle (W\Gamma + \Gamma^*W)f, f\rangle = 1/2\langle \hat{W}f, f\rangle \le -1/2\delta\|f\|^2$$
whenever $f \in \mathcal{D}(\Gamma)$. Therefore, Γ is W-dissipative.

To apply Corollary 4.43, we will show that Γ^* is W^{-1}-dissipative. First, we claim that the operator on L^2 given by
$$\check{W}^{-1}f := \left.\frac{d}{dt}(E^t W^{-1}E^{t*}f)\right|_{t=0}$$
is well-defined. To this end, consider the identity
$$t^{-1}[E^t W^{-1}E^{t*} - W^{-1}]f = W^{-1}E^{-t*}t^{-1}[E^{t*}WE^t - W]W^{-1}E^{t*}f,$$
and note that if $f \in L^2$, then $\lim_{t\to 0} W^{-1}E^{-t*}f = W^{-1}f$, $\lim_{t\to 0}t^{-1}[E^{t*}WE^t - W]f = \hat{W}f$, and $\lim_{t\to 0}W^{-1}E^{t*}f = W^{-1}f$. Thus, the operator
$$\check{W}^{-1}f = W^{-1}\hat{W}W^{-1}f$$
is defined, bounded, and uniformly negative on L^2.

As above, $W^{-1}: \mathcal{D}(\Gamma^*) \to \mathcal{D}(\Gamma)$ provided that $\check{W}^{-1}$ is bounded on L^2. Also as above, if $f \in \mathcal{D}(\Gamma^*)$ such that $W^{-1}f \in \mathcal{D}(\Gamma)$, then $\check{W}^{-1}f = (W^{-1}\Gamma^* + \Gamma W^{-1})f$. Therefore, Γ^* is W^{-1}-dissipative.

By Corollary 4.43, the group $\{E^t\}_{t\in\mathbb{R}}$ is hyperbolic on L^2, and therefore $\{\Phi^t\}_{t\in\mathbb{R}}$ has an exponential dichotomy.

"Only if part". Since $\{\Phi^t\}_{t\in\mathbb{R}}$ has an exponential dichotomy, $\{E^t\}_{t\in\mathbb{R}}$ is hyperbolic on L^2. Choose $H(\cdot) >> 0$ as in statement (i). Also, recall that $\mathcal{P}$ and $\mathcal{Q}$ are multiplication operators, see Dichotomy Theorem 6.41. If W is defined as in equation (7.73), then, by Theorem 4.40 and Remark 4.42, statement (ii) holds and, for each $f \in L^2$,
$$\hat{W}f = \left.\frac{d}{dt}(E^{t*}WE^tf)\right|_{t=0} = -H_{PQ}f, \quad -H_{PQ} << 0.$$
Now, $W: \mathcal{D}(\Gamma) \to \mathcal{D}(\Gamma^*)$, the operator $\check{W}^{-1}$ is defined, and $W^{-1}: \mathcal{D}(\Gamma) \to \mathcal{D}(\Gamma^*)$. Hence, Γ is W-dissipative and Γ^* is W^{-1}-dissipative. An application of Theorem 4.40 completes the proof. $\square$

REMARK 7.54. Let $W = W(\cdot)$ denote the operator of multiplication by
$$W(\cdot) \in C^1_\varphi(\Theta; \mathcal{SL}(\mathcal{H}) \cap \mathcal{GL}(\mathcal{H})).$$
Because $W(\cdot)$ is smooth and invertible, and $\mathcal{D}(\Gamma^*) = \mathcal{D}(\Gamma) = \mathcal{D}(\mathbf{d})$, it follows that
$$W : \mathcal{D}(\Gamma) \to \mathcal{D}(\Gamma^*) \quad \text{and} \quad W^{-1} : \mathcal{D}(\Gamma^*) \to \mathcal{D}(\Gamma).$$
Since $V^{t*}WV^t = W \circ \varphi^t$, the operator $\hat{W}$, defined in (7.72), is given by the formula
$$\hat{W} = \dot{W} + \mathcal{A}^*W + W\mathcal{A}, \quad \dot{W} = \frac{d}{dt}W \circ \varphi^t\Big|_{t=0}, \quad A(\cdot) = \frac{d}{dt}\Phi^t(\cdot)\Big|_{t=0}.$$
◇

COROLLARY 7.55. *The cocycle $\{\Phi^t\}_{t\in\mathbb{R}}$ has an exponential dichotomy on Θ if and only if there exists a function $W = W(\cdot) \in C^1_\varphi(\Theta; \mathcal{SL}(\mathcal{H}) \cap \mathcal{GL}(\mathcal{H}))$ such that the function $\hat{W}(\theta) := \dot{W}(\theta) + A^*(\theta)W(\theta) + W(\theta)A(\theta)$ satisfies the following property: There is constant $\delta > 0$ such that*
$$\langle \hat{W}(\theta)x, x\rangle_\mathcal{H} \leq -\delta|x|^2$$
for all $\theta \in \Theta$ and $x \in \mathcal{H}$. Such a W satisfies properties (i)–(iv) listed in Theorem 7.53.

Corollary 4.45 gives the following result.

COROLLARY 7.56. *The hyperbolicity of E^t, for $t \neq 0$, is equivalent to the existence of a continuous function $W : \Theta \to \mathcal{GL}(\mathcal{H}) \cap \mathcal{SL}(\mathcal{H})$ such that, in the Hilbert space $\mathcal{H}$, we have*

(7.74) $$\Phi^{t*}(\theta)W(\varphi^t\theta)\Phi^t(\theta) << W(\theta),$$
(7.75) $$\Phi^t(\theta)W^{-1}(\theta)\Phi^{t*}(\theta) << W^{-1}(\varphi^t\theta)$$

uniformly with respect to $\theta \in \Theta$.

Note that the condition (7.75) is redundant if $\dim \mathcal{H} < \infty$ (see Proposition 7.57 and Lewowicz [**Lw**]). On the other hand, if $\mathcal{H}$ is infinite dimensional, then Example I.7.3 from Daleckij and Krein [**DK**] shows that condition (7.74) does not imply the hyperbolicity of $\{\Phi^t\}_{t\in\mathbb{R}}$, even for the case when Θ is a singleton. In effect, condition (7.74) gives the left invertibility of each operator $z - E^t$ for $z \in \mathbb{T}$, while condition (7.75) gives its right invertibility.

PROPOSITION 7.57. *Suppose that $\dim \mathcal{H} < \infty$. The hyperbolicity of E^t, for $t \neq 0$, is equivalent to the existence of a continuous function $W : \Theta \to \mathcal{GL}(\mathcal{H}) \cap \mathcal{SL}(\mathcal{H})$ such that condition (7.74) holds uniformly with respect to $\theta \in \Theta$.*

PROOF. By Corollary 4.45, it suffices to show that (7.74) implies the hyperbolicity of $\{E^t\}_{t\in\mathbb{R}}$. In turn, this fact is implied by the hyperbolicity of all operators in the family $\{\pi_\theta(E) : \theta \in \Theta\}$ on $\ell^2(\mathbb{Z}; \mathcal{H})$ defined in (7.70). In view of these facts, let us fix $\theta \in \Theta$ and, for convenience, define $\pi := \pi_\theta(E)$. We will prove that π is hyperbolic.

Consider the operator $w = \text{diag}\left(W(\varphi^n\theta)\right)_{n\in\mathbb{Z}}$ on $\ell^2(\mathbb{Z}; \mathcal{H})$. Using the fact that $\pi^* : (x_n) \mapsto (\Phi(\varphi^n\theta)x_{n+1})$ and the definition $\Phi(\theta) = \Phi^1(\theta)$, we have that
$$\pi^*w\pi = \text{diag}\left(\Phi^*(\varphi^n\theta)W(\varphi^{n+1}\theta)\Phi(\varphi^n\theta)\right)_{n\in\mathbb{Z}}.$$

Also, by condition (7.74), $\pi^*w\pi << \pi$ in $\ell^2(\mathbb{Z}; \mathcal{H})$; that is, π is a uniform w-contraction. Now, Corollary I.7.1 and Lemma I.7.3 from Daleckij and Krein [**DK**] imply that $z - \pi$ is left invertible for all $z \in \mathbb{T}$. Moreover, by these results from [**DK**], to show the two-sided invertibility of $z - \pi$ *for all* $z \in \mathbb{T}$ it suffices to prove that $z - \pi$ has a right inverse for *some* z, say, for $z = 1$.

We will prove that $\mathrm{Ker}(I - \pi^*) = \{0\}$. Note that $(x_n)_{n \in \mathbb{Z}} \in \mathrm{Ker}(I - \pi^*)$ if and only if $x_0 = \Phi^{n*}(\theta)x_n$, $n \in \mathbb{Z}$, or $x_n = [\Phi^{n*}(\theta)]^{-1}x_0 = \Phi^{-n*}(\varphi^n\theta)x_0$. Thus, the proof will be completed as soon as we infer from the limits

$$|\Phi^{-n*}(\varphi^n\theta)x_0| \to 0 \quad \text{as} \quad n \to \infty \quad \text{and} \quad n \to -\infty$$

the fact that $x_0 = 0$. To this end, consider the following subsets of $\mathcal{H}$:

$$S = \{x : \lim_{n \to \infty} |\Phi^n(\theta)x| = 0\}, \qquad S^* = \{x : \lim_{n \to \infty} |\Phi^{-n*}(\varphi^n\theta)| = 0\}$$
$$U = \{x : \lim_{n \to -\infty} |\Phi^n(\theta)x| = 0\}, \quad U^* = \{x : \lim_{n \to -\infty} |\Phi^{-n*}(\varphi^n\theta)| = 0\}.$$

The inequality

$$|\langle x, y \rangle| = |\langle x, [\Phi^n(\theta)]^{-1}\Phi^n(\theta)y \rangle| = |\langle \Phi^{-n*}(\varphi^n\theta)x, \Phi^n(\theta)y \rangle|$$
$$\leq |\Phi^{-n*}(\varphi^n\theta)x| \cdot |\Phi^n(\theta)y|$$

shows that $S^* \subset S^\perp$ and $U^* \subset U^\perp$. Thus, to prove the required equality $S^* \cap U^* = \{0\}$, we must show that $S + U = \mathcal{H}$.

Apply condition (7.74) at $\varphi^n\theta$ for $n \in \mathbb{Z}$. There is some $\delta > 0$ such that if $n > m$ and $x \in \mathcal{H}$, then

$$(7.76) \quad \langle W(\varphi^n\theta)\Phi^n(\theta)x, \Phi^n(\theta)x \rangle \leq \langle W(\varphi^m\theta)\Phi^m(\theta)x, \Phi^m(\theta)x \rangle - \delta \sum_{j=m}^{n-1} |\Phi^j(\theta)x|^2.$$

In particular, the function

$$g(\theta, n, x) := \langle W(\varphi^n\theta)\Phi^n(\theta)x, \Phi^n(\theta)x \rangle, \quad x \in \mathcal{H}, \quad \theta \in \Theta,$$

decreases monotonically with respect to $n \in \mathbb{Z}$. Since $\sup\{\|W(\varphi^n\theta)\| : n \in \mathbb{Z}\} < \infty$ and $W(\theta)$ is invertible, the following equalities follow from inequality (7.76):

$$S = \{x \in \mathcal{H} : g(\theta, n, x) \geq 0, n \geq 0\}, \quad U = \{x \in \mathcal{H} : g(\theta, n, x) \leq 0, n \leq 0\}.$$

In particular, S and U are closed subspaces with $S \cap U = \{0\}$.

In order to prove that $S + U = \mathcal{H}$, let us denote by $\mathcal{H}_n^+$ and $\mathcal{H}_n^-$ the spectral subspaces of the self-adjoint and invertible operator $W(\varphi^n\theta)$ that corresponds to the positive and negative part of its spectrum. In particular, $\langle W(\varphi^n\theta)x, x \rangle > 0$ (resp., < 0) for $x \in \mathcal{H}_n^+$ (resp., for $x \in \mathcal{H}_n^-$). Since $W(\cdot)$ is continuous, $\dim \mathcal{H}_n^+$ and $\dim \mathcal{H}_n^-$ do not depend on $n \in \mathbb{Z}$. Since $\dim \mathcal{H} < \infty$, there exist sequences $\{n_j^+\} \subset \mathbb{Z}_+$ and $\{n_j^-\} \subset \mathbb{Z}_-$ such that the following limit subspaces exist in $\mathcal{H}$:

$$\mathcal{H}_{-\infty} := \lim_{j \to \infty} [\Phi^{n_j^-}(\theta)]^{-1} \mathcal{H}_{n_j^-}^+, \quad \mathcal{H}_{+\infty} := \lim_{j \to \infty} [\Phi^{n_j^+}(\theta)]^{-1} \mathcal{H}_{n_j^+}^-.$$

Let $x = \lim_{j \to \infty} [\Phi^{n_j^+}(\theta)]^{-1} x_j \in \mathcal{H}_{+\infty}$, where $x_j \in \mathcal{H}_{n_j^+}^-$. Since

$$\langle W(\varphi^{n_j^+}\theta)x_j, x_j \rangle = g(\varphi^{n_j^+}\theta, 0, x_j) < 0,$$

the monotonicity of g implies the inequality $g(\varphi^{n_j^+}\theta, n - n_j^+, x_j) < 0$ for $n < n_j^+$. From the cocycle property we have

$$g(\theta, n, x) = \lim_{j \to \infty} g(\varphi^{n_j^+}\theta, n - n_j^+, x_j) \leq 0,$$

and therefore $\mathcal{H}_{+\infty} \subset U$. Similarly, $\mathcal{H}_{-\infty} \subset S$. Hence, we have
$$\dim S + \dim U \geq \dim \mathcal{H}_{+\infty} + \dim \mathcal{H}_{-\infty} \geq \dim \mathcal{H}_n^- + \dim \mathcal{H}_n^+ = \dim \mathcal{H}.$$
□

7.5.1.3. *Trajectorial Lyapunov functions.* In this subsection we show that it is possible to replace the condition of uniformity with respect to $\theta \in \Theta$ in Theorem 7.53 and Corollaries 7.55 and 7.56 by conditions that hold pointwise. The idea is to apply Theorems 4.38 and 4.40 for each group $\{\Pi_\theta^t\}_{t\in\mathbb{R}}$, see (7.71).

Recall, that if $\mathcal{P} \in \mathcal{L}(L^2(\Theta, \mu; \mathcal{H}))$ is the splitting (Riesz) projection for the hyperbolic group $\{E^t\}_{t\in\mathbb{R}}$, then $(\mathcal{P}f)(\theta) = P(\theta)f(\theta)$ for some $P(\cdot) \in C(\Theta; \mathcal{L}(\mathcal{H}))$, see Proposition 7.13. Moreover, the splitting (Riesz) projection for the hyperbolic group $\{\Pi_\theta^t\}_{t\in\mathbb{R}}$, $\theta \in \Theta$, on $L^2(\mathbb{R}; \mathcal{H})$ is a multiplication operator with multiplier $p: \mathbb{R} \to \mathcal{L}(\mathcal{H})$ given by $p(s) = P(\varphi^s \theta)$ for $s \in \mathbb{R}$, see Theorem 7.12. By assumption (2) of Subsection 7.5.1.1, there exists $A(\theta) = \frac{d}{dt}\Phi^t(\theta)\big|_{t=0}$ in $\mathcal{L}(\mathcal{H})$ and $A(\cdot)$ is continuous. Fix $\theta \in \Theta$, and define $a(s) := A(\varphi^s \theta)$ for $s \in \mathbb{R}$. Given a function $w: \mathbb{R} \to \mathcal{GL}(\mathcal{H})$, let w^{-1} denote the function $s \mapsto [w(s)]^{-1}$, and let $C_b^1(\mathbb{R}; Y)$ denote the space of bounded, continuously differentiable functions from $\mathbb{R}$ to Y.

THEOREM 7.58. *For a fixed $\theta \in \Theta$, the group $\{\Pi_\theta^t\}_{t\in\mathbb{R}}$ defined in display (7.71) is hyperbolic on $L^2(\mathbb{R}; \mathcal{H})$ if and only if there exists a function w, with*
$$w^{\pm 1} \in C_b^1(\mathbb{R}, \mathcal{SL}(\mathcal{H}) \cap \mathcal{GL}(\mathcal{H})),$$
such that the function $\hat{w}$, defined by

(7.77)
$$\hat{w}(s) = \frac{dw}{ds} + a^*(s)w(s) + w(s)a(s), \quad s \in \mathbb{R},$$

is negative on $\mathcal{H}$ uniformly with respect to $s \in \mathbb{R}$; that is, there is a $\delta > 0$ such that $\langle \hat{w}(s)x, x \rangle_\mathcal{H} \leq -\delta |x|^2$ for all $x \in \mathcal{H}$ and $s \in \mathbb{R}$. Moreover, if $\{\Pi_\theta^t\}_{t\in\mathbb{R}}$ is hyperbolic on $L^2(\mathbb{R}; \mathcal{H})$ and $p = p(\cdot)$, $q = I - p$, are the Riesz projections for Π_θ^1, then $\operatorname{Im} p(s)$ is $w(s)$-positive and $\operatorname{Im} q(s)$ is $w(s)$-negative in $\mathcal{H}$ uniformly for all $s \in \mathbb{R}$. Also, for each $h \in C_b(\mathbb{R}; \mathcal{SL}(\mathcal{H}))$ that is uniformly positive with respect to $s \in \mathbb{R}$, the function
$$w := \int_0^\infty \{\Pi_\theta^{\tau*} p^* h p \Pi_\theta^\tau - \Pi_\theta^{-\tau*} q^* h q \Pi_\theta^{-\tau}\} d\tau$$
is the solution of the equation

(7.78)
$$\frac{dw}{ds} + a^*(s)w(s) + w(s)a(s) = -h_{pq}(s)$$

where $h_{pq}(s) := q^*(s)h(s)q(s) + p^*(s)h(s)p(s)$. For this w, the subspaces $\operatorname{Im} p(s)$ and $\operatorname{Im} q(s)$ are $w(s)$-orthogonal in $\mathcal{H}$ for all $s \in \mathbb{R}$.

PROOF. The proof is similar to the proof of Theorem 7.53. Recall that the action of the generator Γ_θ of the group $\{\Pi_\theta^t\}_{t\in\mathbb{R}}$ on a smooth function h is given by $(\Gamma_\theta h)(s) = -\frac{dh}{ds} + a(s)h(s)$, and $\mathcal{D}(\Gamma_\theta) = \mathcal{D}(\Gamma_\theta^*) = \mathcal{D}(d/ds)$ on $L^2(\mathbb{R}; \mathcal{H})$. If $w: \mathbb{R} \to \mathcal{SL}(\mathcal{H}) \cap \mathcal{GL}(\mathcal{H})$ satisfies the conditions of the theorem, then the operator
$$\hat{w}h := \frac{d}{dt}(\Pi_\theta^{t*} w \Pi_\theta^t h)|_{t=0}$$
is defined and bounded on $L^2(\mathbb{R}; \mathcal{H})$. As in the proof of Theorem 7.53, the operator Γ_θ is w-dissipative and Γ_θ^* is w^{-1}-dissipative for the operator w defined by $(wh)(s) = w(s)h(s)$. □

Combining Theorem 7.58 with the results listed in Subsection 7.5.1.1, we obtain the following characterization for the existence of an exponential dichotomy for $\{\Phi^t\}_{t\in\mathbb{R}}$ on Θ.

COROLLARY 7.59. *The cocycle $\{\Phi^t\}_{t\in\mathbb{R}}$ has an exponential dichotomy on Θ if and only if, for each $\theta \in \Theta$, there exists a function $w = w_\theta$ with the properties listed in Theorem 7.58.*

Under additional assumptions, it is possible to avoid the requirement to check the condition $\hat{w}_\theta \ll 0$ for all θ. This is the content of the next corollary.

COROLLARY 7.60. *If the orbit of $\{\varphi^t\}_{t\in\mathbb{R}}$ through the point $\theta_0 \in \Theta$ is dense in Θ, then $\{\Phi^t\}_{t\in\mathbb{R}}$ has an exponential dichotomy on Θ if and only if there exists w, with $w^{\pm 1} \in C_b^1(\mathbb{R}; \mathcal{SL}(\mathcal{H}) \cap \mathcal{GL}(\mathcal{H}))$, such that $\hat{w} \ll 0$ for $\hat{w} := \hat{w}_{\theta_0}$ as in equation (7.77).*

PROOF. By Theorem 7.58, the operator $\Pi^1_{\theta_0}$ is hyperbolic on $L^2(\mathbb{R}; \mathcal{H})$ provided that $\hat{w}_{\theta_0} \ll 0$. Choose $\theta \in \Theta$ and use the hypothesis to construct a convergent sequence, with elements of the form $\theta_n = \varphi^{t_n}\theta_0$, that has limit θ. Since the map $(\theta, t) \mapsto \Phi^t(\theta)$ from $\Theta \times \mathbb{R}$ to $\mathcal{L}(\mathcal{H})$ is continuous, $\Pi^1_{\theta_n}$ converges to Π^1_θ, and $\Pi^{1*}_{\theta_n}$ converges to Π^{1*}_θ strongly on $L^2(\mathbb{R}; \mathcal{H})$ as $n \to \infty$. Let V^t denote the translation operator $V^t h(s) = h(s-t)$ on $L^2(\mathbb{R}; \mathcal{H})$ and note that $\Pi^1_{\theta_0} = V^{t_n}\Pi^1_{\theta_n}V^{-t_n}$. Because $\Pi^1_{\theta_0}$ is hyperbolic, we have that

$$\|(\Pi^1_{\theta_n} - z)^{-1}\| < \infty, \qquad \|(\Pi^{1*}_{\theta_n} - z)^{-1}\| < \infty$$

uniformly with respect to n and $z \in \mathbb{T}$. By Lemma III.1.1 in Gohberg and Fel'dman [**GF**], the operator Π^1_θ is hyperbolic. Since θ is arbitrary in Θ, the dichotomy of $\{\Phi^t\}_{t\in\mathbb{R}}$ follows as indicated in Subsection 7.5.1.1. □

EXAMPLE 7.61. Consider the Schrödinger operator on $L^2(\mathbb{R})$ of the form

$$L_{\theta_0} y = y'' + q(\varphi^t \theta_0)$$

where θ_0 is a point whose φ^t-orbit is dense in Θ and where the potential $s \mapsto q(\varphi^s \theta_0)$ is defined by a continuous function $q : \Theta \to \mathbb{R}$. As is well-known, see for example Johnson [**Jo3**] or Johnson, Palmer, and Sell [**JPS**], every Schrödinger operator with an almost periodic potential can be written in this form.

To describe $\sigma(L_{\theta_0})$, rewrite the second order differential equation $L_\theta y - \lambda y = 0$ as an equivalent first order system

$$\frac{dx}{dt} = A_\lambda(\varphi^t \theta) x(t), \quad A_\lambda(\theta) = \begin{pmatrix} 0 & 1 \\ \lambda - q(\theta) & 0 \end{pmatrix}, \quad x = \begin{pmatrix} y \\ y' \end{pmatrix}, \quad t \in \mathbb{R}.$$

Let $\Phi_\lambda^t(\theta)$ be the solution operator for this equation; that is, $x(t) = \Phi_\lambda^t(\theta) x(0)$. As shown in Johnson [**Jo4**], $\lambda \in \rho(L_{\theta_0})$ if and only if the cocycle $\{\Phi_\lambda^t\}_{t\in\mathbb{R}}$ has an exponential dichotomy on Θ. Corollary 7.60 now implies that $\lambda \in \rho(L_{\theta_0})$ if and only if there exists $w = w_{\theta_0}$, with $w^{\pm 1} \in C_b^1(\mathbb{R}; \mathcal{SL}(\mathbb{R}^2) \cap \mathcal{GL}(\mathbb{R}^2))$, such that the function $\hat{w}_\lambda$ given by

$$\hat{w}_\lambda(s) = \frac{dw}{ds} + a_\lambda^*(s) w(s) + w(s) a_\lambda(s), \quad a_\lambda(s) = A_\lambda(\varphi^s \theta_0),$$

is uniformly negative on $\mathcal{H}$ for all $s \in \mathbb{R}$. ◇

REMARK 7.62. The results above can be easily modified to apply to the "stable" case; that is, the case where $P = I$. In particular, Theorems 7.53 and 7.58 and Corollaries 7.55–7.60 hold when the word "dichotomic" is replaced by the word "stable" and the conclusions are modified to require the operator W (or $W(\theta)$, or $w(s)$) to be *positive*. ◇

7.5.2. Applications. In Subsection 7.5.1.3 we will describe how the trajectorial quadratic Lyapunov functions are related to the existence of solutions to Riccati and Hamiltonian equations, and in Subsection 7.5.2.2 we will show how our techniques can be applied to the geodesic flow: a very important example in the theory of hyperbolic dynamical systems.

7.5.2.1. Hamiltonian and Riccati equations. Corollary 7.59 and Theorem 7.58 describe the existence of an exponential dichotomy for $\{\Phi^t\}_{t\in\mathbb{R}}$ in terms of the existence of bounded solutions for the family of nonhomogeneous linear Riccati equations (7.78) parameterized by $\theta \in \Theta$. The existence of exponential dichotomies for $\{\Phi^t\}_{t\in\mathbb{R}}$ is determined in the next corollary in terms of the existence of bounded solutions for the family of homogeneous, but nonlinear, Riccati equations on $\mathbb{R}$ given by

$$(7.79) \qquad w' + wa + a^*w + wbw = 0, \quad a(s) = A(\varphi^s\theta), \quad \theta \in \Theta, \ s \in \mathbb{R},$$

where $b : \mathbb{R} \to \mathcal{SL}(\mathcal{H})$ is a bounded function. We will show that if $\{\Phi^t\}_{t\in\mathbb{R}}$ has an exponential dichotomy, then for each θ there exists a uniformly positive $b = b_\theta$ such that the differential equation (7.79) has a bounded solution. Moreover, b_θ can be calculated via the spectral projection for the operator Π_θ^1. Conversely, if for each $\theta \in \Theta$ there is a $b_\theta \gg 0$ such that the equation (7.79) has a bounded solution, then $\{\Phi^t\}_{t\in\mathbb{R}}$ has an exponential dichotomy.

COROLLARY 7.63. *The cocycle $\{\Phi^t\}_{t\in\mathbb{R}}$ has an exponential dichotomy if and only if for each $\theta \in \Theta$ there exists a function $b = b_\theta \in C_b^1(\mathbb{R}; \mathcal{SL}(\mathcal{H}))$ that is uniformly positive with respect to $s \in \mathbb{R}$ such that the Riccati equation (7.79) has a solution $w = w_\theta$ with $w^{\pm 1} \in C_b^1(\mathbb{R}; \mathcal{SL}(\mathcal{H}) \cap \mathcal{GL}(\mathcal{H}))$.*

PROOF. Assume that the cocycle $\{\Phi^t\}_{t\in\mathbb{R}}$ has an exponential dichotomy. By Corollary 7.59, for each point $\theta \in \Theta$, there is a $w = w_\theta$, with $w^{\pm 1} \in C_b^1(\mathbb{R}; \mathcal{SL}(\mathcal{H}) \cap \mathcal{GL}(\mathcal{H}))$, such that the function $\hat{w}$, computed as in formula (7.77), is uniformly negative. If for this w and $\hat{w}$ we define $b := -w^{-1}\hat{w}w^{-1}$, then w is a solution for the Riccati equation (7.79) with this $b \gg 0$.

Conversely, suppose that, for each $\theta \in \Theta$, there is some $b = b_\theta \in C_b^1(\mathbb{R}; \mathcal{SL}(\mathcal{H}))$ with $b \gg 0$ such that equation (7.79) has a solution w with $w^{\pm 1} \in C_b^1(\mathbb{R}; \mathcal{SL}(\mathcal{H}) \cap \mathcal{GL}(\mathcal{H}))$. Compute $\hat{w}$ for this w by formula (7.77). Then $\hat{w} = -wbw \ll 0$ since $b \gg 0$, and $\{\Phi^t\}_{t\in\mathbb{R}}$ has dichotomy by Corollary 7.59. □

It is well-known that the Hamiltonian equation

$$(7.80) \qquad \mathcal{J}y' + \mathcal{A}y = 0, \quad \mathcal{J} = \begin{bmatrix} 0 & -I \\ I & 0 \end{bmatrix}, \quad \mathcal{A} = \begin{bmatrix} 0 & -a^* \\ -a & -b \end{bmatrix}$$

is related to the Riccati equation (7.79). This allows another formulation of the dichotomy result.

COROLLARY 7.64. *The cocycle $\{\Phi^t\}_{t\in\mathbb{R}}$ has an exponential dichotomy if and only if, for each $\theta \in \Theta$, there exists a function $b = b_\theta \in C_b^1(\mathbb{R}; \mathcal{SL}(\mathcal{H}))$ that is uniformly positive with respect to $s \in \mathbb{R}$ such that the Hamiltonian equation (7.80)* has a solution $y : \mathbb{R} \to \mathcal{L}(\mathcal{H}) \dot{+} \mathcal{L}(\mathcal{H})$ which can be represented as $y = \begin{pmatrix} u \\ v \end{pmatrix}$ with $u, v \in C^1(\mathbb{R}; \mathcal{GL}(\mathcal{H}))$ and $uv^{-1}, vu^{-1} \in C_b^1(\mathbb{R}; \mathcal{SL}(\mathcal{H})) \cap \mathcal{GL}(\mathcal{H}))$.

PROOF. The Hamiltonian equation (7.80) is given as the system

(7.81) $$v' + a^*v = 0, \quad u' = au + bv.$$

If for some b there is a solution y of the differential equation (7.80) with the properties stated in the corollary, then define $w = vu^{-1}$. A direct computation using the system (7.81) shows that w is a solution of the equation (7.79). The existence of an exponential dichotomy for $\{\Phi^t\}_{t \in \mathbb{R}}$ now follows from Corollary 7.64.

Conversely, by Corollary 7.59, if $\{\Phi^t\}_{t \in \mathbb{R}}$ has an exponential dichotomy, then, for each $\theta \in \Theta$, there exists a function w, such that $w^{\pm 1} \in C_b^1(\mathbb{R}; \mathcal{SL}(\mathcal{H}) \cap \mathcal{GL}(\mathcal{H}))$ and $\hat{w}$, computed by formula (7.77), is uniformly negative. Set

(7.82) $$b = -w^{-1}\hat{w}w^{-1} = -w^{-1}w'w^{-1} - aw^{-1} - w^{-1}a^*,$$

and consider system (7.81) with this b.

Recall that $\Phi^s(\theta)$ is the solution operator for the equation $dx/ds = A(\varphi^s\theta)x$, and therefore

$$\frac{d}{ds}\Phi^s(\theta) = A(\varphi^s\theta)\Phi^s(\theta), \quad \frac{d}{ds}\Phi^{s*}(\theta) = \Phi^{s*}(\theta)A^*(\varphi^s\theta),$$

$$\frac{d}{ds}[\Phi^{s*}(\theta)]^{-1} = -A^*(\varphi^s\theta)[\Phi^{s*}(\theta)]^{-1}.$$

Define $v(s) = [\Phi^{s*}(\theta)]^{-1}$ and $u(s) = [w(s)]^{-1}v(s)$, and note that this pair of functions is a solution of system (7.81). □

7.5.2.2. *Geodesic flows.* Let M denote a two dimensional compact oriented Riemannian manifold with Riemann metric d, and let $\mathcal{T}_m M$ denote the tangent plane at $m \in M$. The Riemannian metric d assigns a scalar product in each tangent plane such that the assignment varies smoothly with the base point. If $\delta : (0, 1) \to M$ is a smooth function — its image is a curve in M — then we let $\dot{\delta}(t) \in \mathcal{T}_{\delta(t)}M$ denote the tangent vector to δ at $\delta(t)$. The set

$$\Theta := \{(m, x) : m \in M, x \in \mathcal{T}_m M, d_m(x, x) = 1\}$$

is a three dimensional compact manifold called the *unit tangent bundle* of M. If $\nu : (0, 1) \to \Theta$ is a curve in Θ with $\nu(0) = \theta$, then, in local coordinates, $\nu(t) = (\gamma(t), x(t))$ with $x(t) \in \mathcal{T}_{\gamma(t)}M$. Thus, for example, the tangent to ν at $t = 0$ is the vector $\nu \in \mathcal{T}_\theta\Theta$ given by $\nu = \dot{\nu}(0) = (\dot{\gamma}(0), \dot{x}(0))$. Here $\dot{\gamma}(0)$ is a two dimensional vector tangent to Θ while $\dot{x}(0)$ is a one dimensional vector tangent to the unit circle.

We will use several standard facts about geodesics on M. If $(m, x) \in \Theta$, then there is a unique geodesic $\gamma : \mathbb{R} \to M$ with initial value (m, x) that remains in Θ for all time; that is, $\gamma(0) = m$, $\dot{\gamma}(0) = x$, and $(\gamma(t), \dot{\gamma}(t)) \in \Theta$ for each $t \in \mathbb{R}$. The geodesic flow $\{\varphi^t\}_{t \in \mathbb{R}}$ on Θ is defined as follows: If $\theta = (m, x) \in \Theta$ and γ is the geodesic starting at m with tangent vector x, then $\varphi^t : \theta = (m, x) \mapsto (\gamma(t), \dot{\gamma}(t))$. This flow has an invariant Borel measure μ that is absolutely continuous with respect to the Lebesgue measure on Θ. The generator $\mathbf{v}$ of the geodesic flow is a vector field on Θ called the geodesic vector field; it is defined by $\mathbf{v}(\theta) = \frac{d}{dt}\varphi^t\theta\big|_{t=0}$. Moreover, there is a decomposition of the tangent bundle $\mathcal{T}\Theta$ of Θ as a direct sum $\mathcal{T}^v\Theta \dot{+} [\mathbf{v}]$ where $[\mathbf{v}]$ is the line bundle generated by $\mathbf{v}$ and the complementary bundle

$\mathcal{T}^{\mathbf{v}}\Theta$, with two dimensional fibers, is invariant with respect to the linear skew-product flow $\{\hat{\varphi}^t\}_{t\in\mathbb{R}}$ generated by the differential $\Phi^t(\theta)$ of φ^t. We will use the same symbols $\hat{\varphi}^t$ and Φ^t to denote the restrictions $\hat{\varphi}^t = \hat{\varphi}^t|\mathcal{T}^{\mathbf{v}}\Theta$ and $\Phi^t(\theta) = \Phi^t(\theta)|\mathcal{T}_\theta^{\mathbf{v}}\Theta$. With this notation, the geodesic flow $\{\varphi^t\}_{t\in\mathbb{R}}$ on M is called *Anosov* if the linear skew-product flow $\{\hat{\varphi}^t\}_{t\in\mathbb{R}}$ on $\mathcal{T}^{\mathbf{v}}\Theta$ has an exponential dichotomy.

Let $\kappa : M \to \mathbb{R}$ be the scalar curvature function and let us, for notational convenience, consider κ as a function on Θ given by $\kappa(\theta) = \kappa(m, x) = \kappa(m)$. Also, consider the operator $(\Gamma f)(\theta) = -\mathbf{d}f(\theta) + A(\theta)f(\theta)$ on $L^2(\Theta, \mu; \mathbb{R}^2)$ where

$$\mathbf{d}f(\theta) = \frac{d}{dt}f(\varphi^t\theta)\Big|_{t=0}, \quad A(\theta) := \begin{bmatrix} 0 & -1 \\ \kappa(\theta) & 0 \end{bmatrix}. \tag{7.83}$$

PROPOSITION 7.65. *The geodesic flow on Θ is Anosov if and only if the evolution group $\{E^t\}_{t\in\mathbb{R}}$, $E^t = e^{t\Gamma}$, is hyperbolic on $L^2(\Theta, \mu; \mathbb{R}^2)$.*

PROOF. An operator in the evolution group acts on a function $f \in L^2(\Theta, \mu; \mathbb{R}^2)$ by $(E^t f)(\theta) = \Phi^t(\varphi^{-t}\theta)f(\varphi^{-t}\theta)$. Since Φ^t is the differential of φ^t, the generator of the evolution group is the operator of Lie differentiation in the direction of the geodesic vector field denoted $L_{\mathbf{v}}$.

The geodesic vector field $\mathbf{v}$ on Θ has no zeros. Moreover, since M is two dimensional and oriented, there are two additional nonvanishing vector fields on Θ, which we denote by $\mathbf{y}$ and $\mathbf{z}$, such that $\mathbf{v}(\theta)$, $\mathbf{y}(\theta)$, and $\mathbf{z}(\theta)$ are linearly independent for each $\theta \in \Theta$. Indeed, using the Riemannian structure, the orientation, and the fact that M is two dimensional, there is a well-defined notion of rotation through a right angle given by the map $(m, x) \mapsto (m, x^\perp)$ such that $d_m(x, x^\perp) = 0$ and $d_m(x^\perp, x^\perp) = d_m(x, x)$. The orthogonal geodesic vector field $\mathbf{y}$ is the generator of the flow $(m, x) \mapsto \varphi^t(m, x^\perp)$. The vector field $\mathbf{z}$ is the generator of the fiber rotation flow on Θ given by $(m, x) \mapsto (m, R_m(t)x)$ where $R_m(t)x := e^{it}x$ is unit speed rotation of the unit circle in the positive sense given by the orientation in $\mathcal{T}_m M$. The following Lie derivatives are computed by Green [**Gr2**]:

$$L_{\mathbf{v}}\mathbf{z} = -\mathbf{y}, \quad L_{\mathbf{v}}\mathbf{y} = \kappa\mathbf{z}.$$

Identify $L^2(\Theta, \mu; \mathbb{R}^2)$ with $\mathcal{F} = \{f_1\mathbf{y} + f_2\mathbf{z} : f_1, f_2 \in L^2(\Theta, \mu; \mathbb{R})\}$ and use the above relations to obtain the equation

$$L_{\mathbf{v}}(f_1\mathbf{y} + f_2\mathbf{z}) = (-\mathbf{d}f_1 - f_2)\mathbf{y} + (-\mathbf{d}f_2 + \kappa f_1)\mathbf{z}.$$

It follows that the operator Γ, as defined in display (7.83), is the closed extension of the operator given by $L_{\mathbf{v}}$. Using this fact, the proposition is an immediate consequence of the Dichotomy Theorem 6.41. □

In the next result we apply the dichotomy conditions obtained in Corollary 7.59 by exploiting the specific form of A as given in display (7.83).

COROLLARY 7.66. *The geodesic flow on a compact oriented two dimensional Riemannian manifold with unit tangent bundle Θ is Anosov if and only if, for each $\theta \in \Theta$, there exist functions $h_1, h_2, h_3 : \mathbb{R} \to \mathbb{R}$ that satisfy the following conditions:*

(1) *h_1, h_2, h_3 are bounded on $\mathbb{R}$.*
(2) *$h_1 < 0$ and $h_1 h_3 - |h_2|^2 > 0$ uniformly on $\mathbb{R}$.*

(3) *The system of ordinary differential equations*

(7.84)
$$w'_1 + 2\kappa(\varphi^s\theta)\operatorname{Re}(w_2(s)) = h_1(s),$$
$$w'_2 + \kappa(\varphi^s\theta)w_3(s) - w_1(s) = h_2(s),$$
$$w'_3 - 2\operatorname{Re}(w_2(s)) = h_3(s)$$

has a bounded solution (w_1, w_2, w_3) *such that* $\left|w_1 w_3 - |w_2|^2\right| > 0$ *uniformly on* $\mathbb{R}$.

PROOF. By Proposition 7.65 and Corollary 7.59 the geodesic flow is Anosov if and only if for each $\theta \in \Theta$ there is a bounded self-adjoint solution $w = w_\theta$ of the matrix equation $\hat{w} := w' + a^*w + wa = h$ for some uniformly negative matrix-function $h \in C_b^1(\mathbb{R}; \mathcal{SL}(\mathbb{R}^2))$. Here $a(s) = A(\varphi^s\theta)$ where A is defined via the curvature in display (7.83). If the matrix-functions w and h are defined by

$$w = \begin{bmatrix} w_1 & w_2 \\ \bar{w}_2 & w_3 \end{bmatrix}, \quad w_1(s), \ w_3(s) \in \mathbb{R}, \quad w_2(s) \in \mathbb{C},$$

$$h = \begin{bmatrix} h_1 & h_2 \\ \bar{h}_2 & h_3 \end{bmatrix}, \quad h_1(s), \ h_3(s) \in \mathbb{R}, \quad h_2(s) \in \mathbb{C},$$

then system (7.84) is exactly the equation $\hat{w} = h$. The condition $h \ll 0$ is equivalent to statement (2). Also, by the inequality $|w_1 w_3 - |w_2|^2| > 0$, we have that $w^{-1} \in C_b^1(\mathbb{R}; \mathcal{SL}(\mathbb{R}^2) \cap \mathcal{GL}(\mathbb{R}^2))$. □

EXAMPLE 7.67. Suppose the curvature κ is constant. By the substitution $\tilde{w}_1 = w_1 + \kappa w_3$, the first equation in system (7.84) is reduced to $\tilde{w}'_1 = h_1 + \kappa h_3$. If $\kappa \geq 0$, then, for any choice of $h_1 < h_1^0 < 0$, and $h_3 < h_3^0 < 0$, this equation has no bounded solution. Hence, if the flow is Anosov, we must have $\kappa < 0$. ◇

EXAMPLE 7.68. Suppose κ is not necessarily constant, but $\kappa(\theta) \leq \kappa_0 < 0$. If $h_1 := 2\kappa < 0$, $h_2 := 0$, and $h_3 := -2$; then $h_1 h_3 - |h_2|^2 = -4\kappa > 0$ and $(w_1, w_2, w_3) := (0, 1, 0)$ is a bounded solution of system (7.84). Hence, the geodesic flow is Anosov. In this case, note that the "Lyapunov" metric defined by W, $W(\theta) = W_\theta(0)$, is exactly the Vilms metric defined in [**Vi**]. ◇

7.6. Bibliography and remarks

Pointwise dichotomies. The content of this section is taken, mainly, from Latushkin and Schnaubelt [**LS**], see also Latushkin, Montgomery-Smith, Randolph, and Stepin [**LMR, LR, LSt2**]. The idea to apply algebraic techniques to study dichotomy of linear extensions (skew-product flows) goes back to Antonevich [**An2**], see also Section 7.4. His paper [**An2**] is concerned with the finite dimensional case (see also [**An3**]); the infinite dimensional Hilbert space case for *norm* continuous $\Phi(\cdot)$ is treated by Antonevich and Lebedev [**AL2**]. The norm-continuity of $\Phi(\cdot)$ and the Hilbert space setting are essential for their results since they use C^*-algebraic techniques. For a *strongly* continuous function $\theta \mapsto \Phi(\theta)$, the operator-valued function $\theta \mapsto \Phi^*(\theta)$ is generally not strongly continuous. Thus, C^*-algebra methods are not applicable in the strongly continuous setting.

Weighted translation algebras. The results in this subsection are due to Latushkin and Schnaubelt [**LS**], see also [**LMR, LR, LSt2**] for earlier versions. They can be viewed as the development of a Banach algebra variant of the approach taken by Antonevich and Lebedev [**An2, An3, AL2**], see also Hadwin and

Hoover [**HH, HH2, HH3**]. However, we derive only those results from Banach algebras that are needed to prove the Discrete Dichotomy Theorem 7.9. The operation "*" on $\ell^1(\mathbb{Z};\mathfrak{A})$ is a "skewed" convolution, see the definition of $\mathfrak{A} \otimes_V G$ in Example 7.50 or Section 2.7.1 in Bratteli and Robinson [**BR**]. Proposition 7.1 is taken from Latushkin, Montgomery-Smith, and Randolph [**LMR**]. Its analogue in C^*-algebra theory (see, e.g. Lemma 7.1 in Antonevich [**An3**]) ensures that the Neumann map $b_N \mapsto a_0$ is a continuous map on $\mathfrak{B}$. Theorem 7.4, taken from [**LS**], is parallel to Theorem 4.9, but we conclude less: The statement that $b^{-1} \in \mathfrak{B}$ has no meaning without the assumption that φ is aperiodic because there is no corresponding algebra $\mathfrak{B}$. The idea of the proof of Proposition 7.5 goes back to Kurbatov [**Ku**]. Note (see Proposition 7.8) that the condition (7.7) implies the hyperbolicity of $\pi_\theta(aV)$, which, in turn, implies the *hyperbolicity* of aV. However, the converse of Theorem 7.4, cf. Lemma 4.11, is not valid without the assumption that φ is aperiodic since $\sigma(aV)$ is not generally rotationally invariant. Corollary 7.7 is needed to prove $P(\cdot) \in \mathfrak{A}_{\text{norm}}$ in Proposition 7.13. A similar result is contained in the original proof by Antonevich ([**An2**, Thm. 2]) of the C^*-algebraic result.

The Discrete Dichotomy Theorem. The Discrete Dichotomy Theorem 7.9 is proved in [**LS**], see also [**LMR, LR, LSt2**] for earlier versions. The main part of the Discrete Dichotomy Theorem 7.9 is the implication (b) $\Rightarrow$ (a) and this is, in a sense, the main step in Sacker and Sell [**SS2, SS3**], see [**SS2**, Thm. 5,1] and [**SS3**, Lemma 2]. Indeed, see [**SS3**, p. 327], the Sacker-Sell spectrum Σ was originally defined as the set of all $\lambda \in \mathbb{R}$ such that $\{\Phi_\lambda^t\}$ does not have dichotomy at least at one point $\theta \in \Theta$. For $\dim X < \infty$ and compact Θ, Antonevich [**An2**, Prop. 8] uses C^*-algebraic methods to show that even if β and M depend on $\theta \in \Theta$, the implication (b)$\Rightarrow$(a) is still true (cf. Corollary 7.48).

For the infinite dimensional case, the results of the type (b) $\Rightarrow$ (a) as in the Discrete Dichotomy Theorem 7.9 are proved in Chow and Leiva [**CLe2**, Thm. 4.1] using an approach that goes back to Henry [**He**, Sect. 7.6]; however, some additional assumptions are imposed.

A variant of Corollary 7.10 was proved in Latushkin and Stepin [**LSt2**, Thm. 3.2], see also Antonevich [**An2**]. Evolution semigroups along trajectories, denoted here by $\{\Pi_\theta^t\}_{t \geq 0}$, are introduced in [**LSt2**], see also [**Lt2**]. Theorem 7.12, for the case of aperiodic φ, appeared in Latushkin, Montgomery-Smith and Randolph [**LMR**]; our exposition follows [**LS**].

A Perturbation Theorem. The idea of the proof of Proposition 7.13, taken from [**LS**], goes back to Antonevich [**An2**], see also [**An3**, Thm. 9.2] and Latushkin and Stepin [**LSt2**, Thm.2.4]. The main point here is the term-by-term integration of a series that gives the resolvent operator for $\hat{a}$ or aV, cf. the proof of Proposition 4.18.

The finite dimensional Perturbation Theorem 7.14 is due to Sacker and Sell [**SS3**, Thm.6]. It gives the persistence of dichotomy for $\dot{x} = A(t)x$ provided that the perturbation of the Hull (A) is "small" (see pp. 344–346 in [**SS3**]). The idea to use ϵ-pseudo-orbits, employed in the proof, is due to A. Stepin, see [**LSt4**].

The Annular Hull Theorem. The content of this section is taken from Latushkin and Schnaubelt [**LS**], see Chicone and Swanson [**CS2**] and [**CS3**], Johnson [**Jo2**], and Latushkin and Stepin [**LSt2**] for the earlier versions.

Mañe sequences. See the remarks on Mañe points in Chapter 6. Mañe sequences were introduced in [**LS**]; they are related to the set $\mathcal{B} = \{(\theta, x) : \sup_{t \in \mathbb{R}} |\Phi^t(\theta)x| < \infty\}$ discussed in the Bibliography and Remarks to Chapter 6.

The Annular Hull and the Spectral Mapping Theorems. Theorem 7.25 is proved in [**LS**], see also [**LMR, LR, LSt2**] for earlier versions. The difference between the cases $\theta_0 \in \mathcal{B}(\Theta)$ and $\theta_0 \notin \mathcal{B}(\Theta)$ for the Mañe point θ_0 was discussed for the finite dimensional case by Chicone and Swanson [**CS, CS2, CS3**]. The constructions of flow-boxes in Lemmas 7.21–7.22, taken from [**LS**], are needed to cover the case where Θ is not a smooth manifold, cf. [**CS, CS2, CS3**]. Formula (7.48) for the approximate eigenfunction for Γ in Lemma 7.23 is taken from [**LS**], the idea goes back to Chicone and Swanson [**CS2**] for flows and to Mañé [**Mn2**] for diffeomorphisms, cf. the proofs of the spectral mapping theorems by Räbiger and Schnaubelt [**RS2, Sc**]. Formula (7.51) in Lemma 7.24 is also taken from [**LS**], this approach goes back to [**CS, CS2**] and uses the same idea of interpolation as the proof of Lemma 2.29.

The Spectral Mapping/Annular Hull Theorem 7.25 holds for compact Θ such that $\dim X < \infty$ without the restriction $p(\theta) \geq d_0 > 0$. A variant of the "change-of-time" trick used in the proof of Lemma 7.26 can be found in Latushkin and Stepin [**LSt2**, Thm. 1.11]. Corollary 7.27 implies $\Sigma = |\sigma(\Gamma)|$. For Corollaries 7.28–7.30 we remark that $\dim X < \infty$ implies $\sigma(E) = \sigma_{\mathrm{ap}}(E)$, see Chicone and Swanson [**CS3**]. The proof of Corollary 7.30 is taken from [**CS2**].

Dichotomy, mild solutions and Green's functions. The results of this section are adapted from Latushkin and Schnaubelt [**LS**], see also Montgomery-Smith, Randolph, and Stepin [**Lt2, LMR, LRS, LSt2**] for earlier versions.

Theorems of Perron type. Theorems 7.34 and 7.37 are proved in [**LS**]. The "uniqueness" part of the condition $(M_0, C_b(\Theta; X))$ is related to weak dichotomy, see Remark 6.16 and comments in Section 6.4. Indeed, suppose that the equation $u(\varphi^t \theta) = \Phi^t(\theta) u(\theta)$ implies $u = 0$ whenever $u \in C_b(\Theta; X)$. Then, $\{\Phi^t\}_{t \geq 0}$ has a weak dichotomy. In case $\Phi^t(\theta)$ is uniformly α-contracting for each $t \geq 0$ and $\theta \in \Theta$, the following sufficient condition for the existence of an exponential dichotomy is given by Sacker and Sell [**SS5**, Thm. B] (cf. Theorem 6.52 above): A weakly dichotomic cocycle $\{\Phi^t\}_{t \geq 0}$ has an exponential dichotomy on a compact base space Θ provided that the stable sets $\mathcal{S}_\theta = \{x \in X : |\Phi^t(\theta) x| \to 0 \text{ as } t \to \infty\}$ have a fixed finite codimension. If this condition on the codimension is not satisfied, then the behavior of $\{\hat{\varphi}^t\}_{t \geq 0}$ is more complicated, see for example the Alternative Theorem E in [**SS5**]. From this point of view, the Perron-type Theorem 7.37 gives yet another sufficient condition for dichotomy.

The adjective "Perron-type" is related to the classical paper by Perron [**Pn**], see also Section 4.3. Proposition 7.33 is classical; it is just the Lyapunov-Perron formula, cf. Sacker and Sell [**SS5**, p. 27].

The simplicity of the proof of Theorem 7.34 on the space $C_0(\Theta; X)$ (versus $C_b(\Theta; X)$) demonstrates an advantage of using evolution semigroups. In particular, recall that $\{E^t\}_{t \geq 0}$ is *not* strongly continuous on $C_b(\Theta; X)$. Thus, for the proof of Theorem 7.37, it is enough to check that the Green's operator maps $C_0(\Theta; X)$ to $C_0(\Theta; X)$. To do so, we have used some heavy machinery; for example, in the proof of Lemma 7.38. Perhaps there is a simpler argument.

Green's functions. Theorems 7.40 and 7.42, proved in [**LS**], see also [**Lt2, LMR, LRS, LSt2**], are infinite dimensional generalizations of results that can be found, for example, in Mitropolskij, Samojlenko, and Kulik [**MSK, Sm**], see also numerous references therein and an excellent exposition in Bronshtein and Kopanskii [**BKo**]. Of course, our proofs are completely different, see Latushkin

[**Lt2**] for more information. The main focus in [**Sm**] is on conditions for the existence of invariant tori. Recall that an *invariant torus*, in this context, is the image of a function $v : \Theta \to X$ where v is a solution of the inhomogeneous equation $\dot{v} = A(\varphi^t \theta) v + f(\varphi^t \theta)$ for some $f : \Theta \to X$. In other words, an invariant torus is given by a solution of the operator equation $\Gamma v = f$ where $\Gamma = -\mathbf{d} + A(\cdot)$ acts from $C^1_\varphi(\Theta; X)$ to $C(\Theta; X)$ (we assume here that Θ is compact). The Green's function G_P gives the (right) inverse operator $\mathbb{G}_P$ for Γ.

Isomorphisms Theorems. This section is just a brief outline of an extensive theory, see Antonevich and Lebedev [**An3, AL2**], the review by Karlovich, Kravchenko, and Litvinchuk [**KKL**], and Latushkin and Stepin [**LSt2**] for further references. See also, the important papers [**HH, HH2, HH3**] by Hadwin and Hoover.

Evolution semigroups and Isomorphism Theorems for C^* algebras. The C^*-algebra approach was the main point in Latushkin and Stepin [**LSt2**]. We stress the fact that, in Corollary 7.48, the implication "pointwise dichotomy implies global dichotomy" does not require uniform constants β and M in the Definition 6.14, in contrast to the requirements for the implication (b) $\Rightarrow$ (a) in the Discrete Dichotomy Theorem 7.9. The reason is given by a nontrivial fact from C^*-algebra theory (see Lemma 10.1 in [**An3**]): The set of representations π_θ is sufficient. As mentioned previously, the norm-continuity of $\Phi(\cdot)$ and the fact that X is a Hilbert space are very important here. On the other hand, Sacker-Sell theory provides the same conclusion (see [**SS2**] and Lemma 2 in [**SS3**]): The implication (b)$\Rightarrow$(a) in the Discrete Dichotomy Theorem 7.9 is valid, if Θ is compact and $\dim X < \infty$, without assuming that the pointwise dichotomy constants $\beta(\theta)$ and $M(\theta)$ are uniformly bounded in θ.

General Isomorphism Theorems. General references for this topic are Arveson and Josephson [**AJ**], Bratteli and Robinson [**BR**], and Pedersen [**Pe**]. Our exposition outlines only a fraction of a theory presented in Antonevich and Lebedev [**An3, AL2**].

Quadratic Lyapunov functions. This section follows the presentation in Chicone and Latushkin [**ChLt**]. See Daleckij and Krein [**DK**] for the generalized Lyapunov Theorem in the case $\Theta = \mathbb{R}$, and Samojlenko [**Sm**] for the case of linear skew-product flows in finite dimensional setting.

The setting. The conditions on aperiodicity of φ can be relaxed, see Chicone and Latushkin [**ChLt**].

The global Lyapunov function. Theorem 7.53 and Corollary 7.55 are taken from [**ChLt**]. They generalize finite dimensional results from Samojlenko [**Sm**] with completely different proofs. Corollary 7.56 generalizes a result by Lewowicz [**Lw**]. He considers the case of a cocycle generated by the differential of a diffeomorphism of a finite dimensional smooth manifold and proves that the existence of $W : \Theta \to \mathcal{GL}(\mathbb{R}^n)$ satisfying condition (7.74) is equivalent to the hyperbolicity of the diffeomorphism. Some of his arguments are used in the proof of Proposition 7.57.

Trajectorial Lyapunov functions. Theorem 7.58 is proved in [**ChLt**]. Example 7.61 is presented in more detail in Latushkin and Stepin [**LSt2**].

Hamiltonian and Riccati equations. General information on connections to Hamiltonian and Riccati equations can be found in the literature on control theory; for example, in Curtain and Pritchard [**CP**] and Zabczyk [**Za2**].

Geodesic flows. Corollary 7.66 is proved in [**ChLt**]. For more information, see Chicone and Swanson [**CS**], Chicone [**Ch**], the references therein, and also Beem, Chicone, and Ehrlich [**BCE, CE**].

CHAPTER 8

Evolution Operators and Exact Lyapunov Exponents

One of the main themes of this chapter is the use of exact Oseledets-Lyapunov exponents and the Multiplicative Ergodic Theorem to study spectral properties of evolution operators. In Section 8.1 we relate the spectrum and spectral subspaces of the evolution operator and the dynamical spectrum and spectral subbundles of the corresponding linear skew-product flow to the Oseledets-Lyapunov spectrum and measurable subbundles as they appear in the Multiplicative Ergodic Theorem. The main result here is a formula that expresses the spectral radius of the evolution operator (or the largest Lyapunov exponent of the cocycle) in terms of the exact Oseledets-Lyapunov exponents. An application is given in Section 8.2 where we will study the kinematic dynamo operator for an ideally conducting fluid. In this case, the corresponding evolution semigroup acts on the space of divergence free sections of the tangent bundle. We will prove the Spectral Mapping Theorem and Annular Hull Theorem for this semigroup, and give a full description of the spectrum in terms of the exact Lyapunov exponents. Finally, in Section 8.3 we give a brief account of the theory of the matrix-coefficient Ruelle transfer operator, which is a generalization of the evolution operator for the case where the underlying map is not injective.

8.1. Oseledets' Theorem and linear skew-product flows

In this section we describe the dynamical Sacker-Sell spectrum and continuous spectral subbundles for a linear skew-product flow in terms of the Oseledets-Lyapunov exponents and measurable subbundles given in the Multiplicative Ergodic Theorem.

Let us consider a connected *compact* metric space Θ, a homeomorphism $\varphi : \Theta \to \Theta$, a Hilbert space $X = \mathcal{H}$, and a *norm* continuous function $\Phi : \Theta \to \mathcal{L}(\mathcal{H})$ given by $\theta \mapsto \Phi(\theta)$ such that $\{\Phi^n\}_{n \in \mathbb{Z}_+}$ is a discrete time cocycle over $\{\varphi\}_{n \in \mathbb{Z}}$ with corresponding discrete time linear skew-product flow $\{\hat{\varphi}^n\}_{n \in \mathbb{Z}_+}$. By Proposition 6.43, the existence of an exponential dichotomy for a continuous time skew-product flow is equivalent to the existence of an exponential dichotomy for a corresponding discrete time skew-product flow. Thus, there is no loss of generality in the technically convenient restriction to discrete time. We will make the additional assumption that each operator $\Phi(\theta)$, for $\theta \in \Theta$, is a compact operator on $\mathcal{H}$ or an invertible operator expressible as the sum of a unitary and a compact operator. Under either of these assumptions, the Multiplicative Ergodic Theorem applies to the cocycle $\{\Phi^n\}_{n \in \mathbb{Z}_+}$.

8.1.1. The Multiplicative Ergodic Theorem. In this subsection we formulate the Multiplicative Ergodic Theorem for the infinite dimensional cocycle

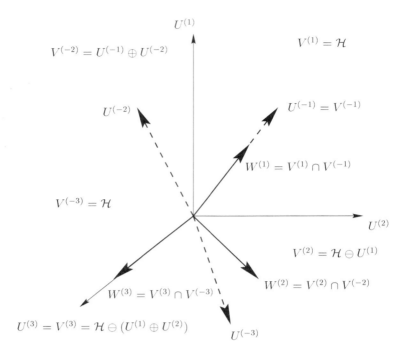

FIGURE 1. A schematic diagram for the Multiplicative Ergodic Theorem

$\{\Phi^n\}_{n\in\mathbb{Z}_+}$. In addition to the cocycle $\{\Phi^n\}_{n\in\mathbb{Z}_+}$ where
$$\Phi^n(\theta) = \Phi(\varphi^{n-1}\theta)\cdots\Phi(\theta),$$
let us consider the adjoint operators $\Phi^*(\theta)$ and the cocycle given by
$$\tilde{\Phi}^n(\theta) = \Phi^*(\varphi^{-n}\theta)\cdots\Phi^*(\varphi^{-1}\theta).$$
Also, we will consider the linear skew-product flow $\{\hat{\varphi}^n\}_{n\in\mathbb{Z}_+}$ on $\Theta \times \mathcal{H}$ given by $\hat{\varphi}^n : (\theta, x) \mapsto (\varphi^n\theta, \Phi^n(\theta)x)$, and the family of evolution operators $\{E^n\}_{n\in\mathbb{Z}_+}$ defined by $(E^n f)(\theta) = \Phi^n(\varphi^{-n}\theta)f(\varphi^{-n}\theta)$ on the space $C(\Theta; \mathcal{H})$ of continuous functions from Θ to $\mathcal{H}$ with the supremum norm. As usual, we will abbreviate E^1 by E.

Let $\mathrm{Erg} = \mathrm{Erg}(\varphi) = \mathrm{Erg}(\Theta;\varphi)$ denote the set of Borel probability measures that are ergodic with respect to the homeomorphism φ on Θ. The Oseledets' Multiplicative Ergodic Theorem (MET) describes the asymptotic behavior of the operators in the family $\{\Phi^n(\theta)\}_{n\in\mathbb{Z}_+}$ for almost all $\theta \in \Theta$ with respect to each measure $\nu \in \mathrm{Erg}$. We will use an infinite dimensional version of this theorem, which is due to Ruelle [**Ru3**]. To state it, let $\mathcal{C}(\mathcal{H})$ denote the set of compact operators on $\mathcal{H}$.

THEOREM 8.1 (MET-1). *Suppose that $\nu \in \mathrm{Erg}$. If $\Phi(\theta) \in \mathcal{C}(\mathcal{H})$ for each $\theta \in \Theta$, then there is a φ-invariant Borel set $\Theta_\nu \subset \Theta$ such that $\nu(\Theta_\nu) = 1$. For each point $\theta \in \Theta_\nu$, the following limits exist in $\mathcal{L}(\mathcal{H})$:*

(8.1) $\quad \Lambda_\theta = \lim_{n\to\infty} [\Phi^n(\theta)^*\Phi^n(\theta)]^{\frac{1}{2n}}, \quad \tilde{\Lambda}_\theta = \lim_{n\to\infty} \left[\tilde{\Phi}^n(\theta)^*\tilde{\Phi}^n(\theta)\right]^{\frac{1}{2n}},$

and $\sigma(\Lambda_\theta) = \sigma(\tilde{\Lambda}_\theta)$.

The set Θ_ν in Theorem 8.1 is called an *Oseledets set*. We stress the fact that the limits in (8.1) are *exact*; that is, lim sup and lim inf of the corresponding sequences coincide.

The limit operators Λ_θ and $\tilde{\Lambda}_\theta$ are nonnegative compact operators on $\mathcal{H}$. Thus, their eigenvalues are nonnegative, the origin is the only possible accumulation point of these eigenvalues, and each positive eigenvalue has a finite dimensional eigenspace. Let $e^{\lambda^1} > e^{\lambda^2} > \ldots > e^{\lambda^s} > \ldots$ be the sequence of different eigenvalues of the operators Λ_θ and $\tilde{\Lambda}_\theta$. We set $s = \infty$ if the sequence is infinite. By convention, if $s = \infty$ and we write, for example, $j = 1, 2, \ldots, s$, then we mean that j takes each positive integer as a value. For each eigenvalue e^{λ^j}, let $U_\theta^{(j)}$ denote the eigenspace for Λ_θ, and $U_\theta^{(-j)}$ the eigenspace for $\tilde{\Lambda}_\theta$ that corresponds to the eigenvalue e^{λ^j}. Moreover, for $j = 1, 2, \ldots, s$, define

$$V_\theta^{(0)} = \{0\}, \quad V_\theta^{(1)} = \mathcal{H}, \quad V_\theta^{(j+1)} = \mathcal{H} \ominus (U_\theta^{(1)} + \cdots + U_\theta^{(j)}),$$
$$V_\theta^{(-j)} = U_\theta^{(-1)} + \cdots + U_\theta^{(-j)}, \quad W_\theta^j = V_\theta^{(j)} \cap V_\theta^{(-j)},$$
$$V_\theta^{(s+1)} = \operatorname{Ker} \Lambda_\theta.$$

Thus, we obtain two filtrations:

$$V_\theta^{(s+1)} = \operatorname{Ker} \Lambda_\theta \subset \cdots \subset V_\theta^{(j)} \subset \cdots \subset V_\theta^{(1)} = \mathcal{H},$$
$$V_\theta^{(0)} = \{0\} \subset V_\theta^{(-1)} \subset \cdots \subset V_\theta^{(-j)} \subset \cdots \subset V_\theta^{(-s-1)} = \mathcal{H}.$$

Also, W_θ^j contains those vectors from $U_\theta^{(-1)} + \cdots + U_\theta^{(-j)}$ that are orthogonal to $U_\theta^{(1)} + \cdots + U_\theta^{(j)}$. See Figure 1 for a schematic diagram of the positions of these spaces for the case where $\dim \mathcal{H} = 3$.

THEOREM 8.2 (MET-2). *Assume the hypotheses of Theorem 8.1. The subspaces U_θ, V_θ, and W_θ depend measurably on $\theta \in \Theta_\nu$. For each $\theta \in \Theta_\nu$, we have $\Phi(\theta)V_\theta^{(j)} \subset V_{\varphi\theta}^{(j)}$ and, for each $j = 1, 2, \ldots, s$,*

(8.2) $$\mathcal{H} = V_\theta^{(j+1)} \dotplus V_\theta^{(-j)}, \quad V_\theta^{(-j)} = W_\theta^j \dotplus \cdots \dotplus W_\theta^1,$$

(8.3) $$V_\theta^{(j)} = V_\theta^{(j+1)} \dotplus W_\theta^j.$$

Though the definitions of the subspaces V and W as well as the formulas (8.2)-(8.3) appear to be complicated, their dynamical characterization is transparent. Indeed, the norm of a vector x in $V^{(1)} = \mathcal{H}$ and *not* in $V^{(2)}$ grows under the action of the cocycle like $e^{n\lambda^1}$, the norm of a vector in $V^{(2)}$ and *not* in $V^{(3)}$ grows like $e^{n\lambda^2}$, and so forth.

THEOREM 8.3 (MET-3). *Under the hypotheses of Theorem 8.1, if $\theta \in \Theta_\nu$ and $x \in X$, then there is an* exact *Oseledets-Lyapunov exponent*

(8.4) $$\lambda(\theta, x) = \lim_{n \to \infty} n^{-1} \log |\Phi^n(\theta)x|.$$

Moreover, $\lambda(\theta, x) \in \{\lambda^1, \lambda^2, \ldots\}$ such that

$$\lambda(\theta, x) = \lambda^j \quad \text{if and only if} \quad x \in V_\theta^{(j)} \setminus V_\theta^{(j+1)}, \quad j = 1, 2, \ldots, s,$$

and

$$\lambda(\theta, x) = -\infty \quad \text{if and only if} \quad x \in V_\theta^{(s+1)}.$$

In particular, the largest Oseledets-Lyapunov exponent $\lambda_\nu^1 = \lambda^1$ can be computed as follows:

(8.5) $$\lambda_\nu^1 = \lim_{n \to \infty} n^{-1} \log \|\Phi^n(\theta)\|_{\mathcal{L}(\mathcal{H})}, \quad \theta \in \Theta_\nu.$$

The formulation of the Multiplicative Ergodic Theorem in the finite dimensional case is simpler than in the infinite dimensional case.

THEOREM 8.4 (MET-FD). *If* $\dim \mathcal{H} = \ell < \infty$ *and* $\Phi(\theta)$ *is invertible for all* $\theta \in \Theta$, *then, for some integer* $s \leq \ell$, *there are* s *different Lyapunov exponents* $\lambda^1 > \ldots > \lambda^s$ *and a corresponding* $\{\hat{\varphi}^n\}_{n \in \mathbb{Z}_+}$-*invariant decomposition*

$$\mathcal{H} = W_\theta^1 + \cdots + W_\theta^s, \quad \theta \in \Theta,$$

such that the W_θ^j *are characterized by the equalities*

(8.6) $$\lambda^j = \lim_{n \to \pm\infty} n^{-1} \log |\Phi^n(\theta) x|, \quad x \in W_\theta^j, \quad j = 1, \ldots, s,$$

where $\Phi^{-n}(\theta) = \Phi^{-1}(\varphi^{-n}\theta) \cdots \Phi^{-1}(\varphi^{-1}\theta)$.

REMARK 8.5. *The Multiplicative Ergodic Theorem 8.3 (MET-3) is valid for endomorphisms* $\varphi: \Theta \to \Theta$, *see for example Arnold* [**A**] *or Walters* [**Wa2**]. ◇

REMARK 8.6. Since $\Lambda_\theta \in \mathcal{C}(\mathcal{H})$, the *multiplicities*

$$m_j = \dim W_\theta^j \quad \text{and} \quad \dim V_\theta^{(-j)} = \sum_{i=1}^j m_i, \quad j = 1, 2, \ldots, s,$$

are finite. If $s = \infty$, then $-\infty$ is the only possible accumulation point of the sequence $\{\lambda^j\}_{j=1}^\infty$. Due to the ergodicity of $\nu \in \text{Erg}$, the quantities λ^j, W_θ^j, and s do not depend on $\theta \in \Theta_\nu$. However, all objects mentioned in the Multiplicative Ergodic Theorem, generally, *do depend* on $\nu \in \text{Erg}$. Therefore, we will sometimes write λ_ν^j, $W_{\theta,\nu}^j$, s_ν, etc. Also, for an ℓ-dimensional cocycle, we will often use the notation $\ell_\nu' = s_\nu$, $\ell_\nu' \leq \ell$. ◇

The *Oseledets-Lyapunov spectrum* of the cocycle $\{\Phi^n\}_{n \geq 0}$ is defined to be

$$\Sigma_L = \{\lambda_\nu^j : j = 1, 2, \ldots, s_\nu, \quad \nu \in \text{Erg}(\Theta, \varphi)\}.$$

REMARK 8.7. Assume that $\dim \mathcal{H} = \infty$ and, for each $\theta \in \Theta$, the operator $\Phi(\theta)$ is invertible and given as a sum of a unitary and a compact operator. Then, as in Theorem 8.1 (MET-1), the limit Λ_θ exists for each $\theta \in \Theta_\nu$. In this case Λ_θ is invertible and $\Lambda_\theta - I$ is a compact operator. For $\lambda \in \sigma(\log \Lambda_\theta)$, let $V_\theta^{(\lambda)}$ denote the spectral subspace of Λ_θ that corresponds to the part of $\sigma(\log \Lambda_\theta)$ in the interval $(-\infty, \lambda]$. Then, as in Theorem 8.3 (MET-3), exact limits (8.4) exist for all $x \in \mathcal{H}$ and

(8.7) $$\lambda(\theta, x) = \lambda \quad \text{if and only if} \quad x \in V_\theta^{(\lambda)} \setminus \bigcup_{\lambda' < \lambda} V_\theta^{(\lambda')}.$$

Moreover, the only possible accumulation point of the sequence $\lambda^1 > \ldots > \lambda^s > \ldots > 0 > \lambda^{(-s')} > \ldots > \lambda^{(-1)}$ of the eigenvalues of $\log \Lambda_\theta$ is $0 \in \sigma(\log \Lambda_\theta)$. Also, the eigenspace corresponding to each nonzero $\lambda^{(j)}$ is finite dimensional. ◇

One of the advantages of the finite dimensional formula (8.6) is the possibility to compute the limit as $n \to -\infty$. This is because the family $\{\Phi^n(\theta)\}_{n \in \mathbb{Z}}$ in Theorem 8.4 consists of invertible operators. For the general situation, where $\dim \mathcal{H} = \infty$

and $\Phi(\theta) \in \mathcal{C}(\mathcal{H})$, this is not the case. However, nonzero vectors from $V_\theta^{(-j)}$ do have "backward trajectories". To be more precise, the following proposition holds, see Ruelle [**Ru3**, p. 261].

PROPOSITION 8.8. *For each $\theta \in \Theta_\nu$ and each nonzero vector $x = x_0 \in V_\theta^{(-j)}$, $j = 1, 2, \ldots, s$, there exists a unique sequence $\{x_n\}_{n=1}^\infty$ such that $\Phi(\varphi^{-n}\theta)x_n = x_{n-1}$ for all positive integers n, and*

$$\lim_{n \to +\infty} n^{-1} \log |x_n| \leq -\lambda^j$$

where the limit is exact.

8.1.2. Spectral and measurable subbundles. In this subsection we will relate the measurable subbundles W_θ^j, as in the Multiplicative Ergodic Theorem, to the continuous spectral subbundles $\mathbb{E}_\theta^{\lambda\lambda'}$ that appear in the Sacker-Sell spectral theory, see Section 6.3. Also, we will describe the dynamical spectrum as a set. A review of Subsection 6.3.3 is recommended before reading this section.

Let us assume that $\Phi : \Theta \to \mathcal{C}(\mathcal{H})$ is norm continuous, and $\{\Phi^n\}_{n \in \mathbb{Z}_+}$ has an exponential dichotomy with positive dichotomy constants M and β, and a corresponding *norm* continuous dichotomy projection $P(\cdot)$. Recall that $(\mathcal{P}f)(\theta) = P(\theta)f(\theta)$ defines the Riesz projection on $C(\Theta; \mathcal{H})$ for the hyperbolic evolution operator given by $(Ef)(\theta) = \Phi(\varphi^{-1}\theta)f(\varphi^{-1}\theta)$. Also, the dichotomy projection $P(\cdot)$ defines a *continuous* splitting $\Theta \times \mathcal{H} = \mathbb{S} + \mathbb{U}$ where $\mathbb{S}_\theta = \operatorname{Im} P(\theta)$, $\mathbb{U}_\theta = \operatorname{Im} Q(\theta)$, and $Q(\theta) = I - P(\theta)$, see the notations (6.63)–(6.64). Here, $\mathbb{S}_\theta$ is the stable fiber and $\mathbb{U}_\theta$ the unstable fiber over $\theta \in \Theta$. By Definition 6.13 of exponential dichotomy, there exist positive constants β and M such that if n is a nonnegative integer, then

$$\begin{aligned} |\Phi^n(\theta)x| &\leq Me^{-\beta n}, \quad x \in \mathbb{S}_\theta = \operatorname{Im} P(\theta), \\ |\Phi^n(\theta)x| &\geq M^{-1}e^{\beta n}, \quad 0 \neq x \in \mathbb{U}_\theta = \operatorname{Im} Q(\theta), \quad \theta \in \Theta. \end{aligned} \tag{8.8}$$

REMARK 8.9. The estimates in display (8.8) imply the Oseledets-Lyapunov spectrum Σ_L is bounded away from zero for every cocycle with an exponential dichotomy. Moreover, if $\theta \in \Theta_\nu$ for $\nu \in \operatorname{Erg}(\Theta, \varphi)$, and $x \neq 0$, then

$$(8.9) \quad \lambda(\theta, x) \leq -\beta < 0 \iff x \in \mathbb{S}_\theta \quad \text{and} \quad \lambda(\theta, x) \geq \beta > 0 \iff x \in \mathbb{U}_\theta.$$

Taking into account Remark 8.6, if $\nu \in \operatorname{Erg}$, then there exists some $j_\nu < \infty$ so that $\lambda_\nu^{(j_\nu+1)} < 0 < \lambda_\nu^{j_\nu}$. Here, set $j_\nu = 0$ if $\lambda_\nu^1 < 0$. $\diamond$

The next proposition shows that if the cocycle $\{\Phi^n\}_{n \in \mathbb{Z}_+}$ has an exponential dichotomy, then the measurable subbundles $V_{\theta,\nu}$, $\theta \in \Theta_\nu$, that satisfy the identities (8.2)–(8.3) in the Multiplicative Ergodic Theorem 8.2 (MET-2) and correspond to the negative and positive Oseledets-Lyapunov exponents, respectively, are in fact the "traces" on Θ_ν of the continuous subbundles $\mathbb{S}_\theta = \operatorname{Im} P(\theta)$ and $\mathbb{U}_\theta = \operatorname{Im} Q(\theta)$ for $\theta \in \Theta$.

PROPOSITION 8.10. *If $\{\Phi^n\}_{n \in \mathbb{Z}_+}$ has an exponential dichotomy, then, for each $\nu \in \operatorname{Erg}(\Theta, \varphi)$ and $\theta \in \Theta_\nu$,*

$$(8.10) \qquad V_{\theta,\nu}^{(j_\nu+1)} = \operatorname{Im} P(\theta) \quad \text{and} \quad \sum_{j=1}^{j_\nu} W_{\theta,\nu}^j = \operatorname{Im} Q(\theta).$$

PROOF. The first equality is obvious. Indeed, if $x \in V_\theta^{(j_\nu+1)}$, then, by Theorem 8.3 (MET-3), $\lambda(\theta, x) \leq \lambda_\nu^{(j_\nu+1)} < 0$ and $x \in \mathbb{S}_\theta$ by (8.9). Conversely, for $x \in \mathbb{S}_\theta$, equation (8.9) and Theorem 8.3 (MET-3) imply $x \in V_\theta^{(j_\nu+1)}$.

Using the first equality in (8.10) and formula (8.2) in Theorem 8.2 (MET-2), let us note that both subspaces $\mathbb{U}_\theta$ and $V_{\theta,\nu}^{(-j_\nu)} = \sum_{j=1}^{j_\nu} W_{\theta,\nu}^j$ are direct complements of the subspace $\mathbb{S}_\theta = V_{\theta,\nu}^{(j_\nu+1)}$ in $\mathcal{H}$. Therefore, to prove the second equality in (8.10), it suffices to show the inclusion $V_{\theta,\nu}^{(-j_\nu)} \subset \mathbb{U}_\theta$. For this, fix a nonzero vector $x = x_0 \in V_{\theta,\nu}^{(-j_\nu)}$. By Proposition 8.8, there is a sequence $\{x_n\}_{n=1}^\infty$ such that $\Phi(\varphi^{-n}\theta)x_n = x_{n-1}$ and

$$\lim_{n \to \infty} n^{-1} \log |x_n| \leq -\lambda_\nu^{(j_\nu)} < 0.$$

Decompose each $x_n = x_n^s + x_n^u$ so that $x_n^s \in \mathbb{S}_{\varphi^{-n}\theta}$ and $x_n^u \in \mathbb{U}_{\varphi^{-n}\theta}$. Since $\mathbb{S}$ and $\mathbb{U}$ are $\{\hat{\varphi}^n\}_{n \in \mathbb{Z}_+}$-invariant, the estimates (8.8) give

$$|x_0^s| = |\Phi^n(\varphi^{-n}\theta)x_n^s| \leq Me^{-\beta n}|x_n^s|,$$
$$|x_0^u| = |\Phi^n(\varphi^{-n}\theta)x_n^u| \geq M^{-1}e^{\beta n}|x_n^u|.$$

Therefore, if $\epsilon > 0$ is sufficiently small and $M_1 > 0$ is sufficiently large, by Proposition 8.8, we have that

$$|x_0^s| \leq Me^{-\beta n}|x_n - x_n^u| \leq Me^{-\beta n}\left[M_1 e^{(\epsilon-\lambda_\nu^{(j_\nu)})n} + Me^{-\beta n}|x_0^u|\right].$$

Passing to the limit as $n \to \infty$, we have the required equality $x_0^s = 0$. □

REMARK 8.11. Recall that Θ is connected. Since $\theta \mapsto P(\theta) \in \mathcal{L}(\mathcal{H})$ is continuous, the quantities $\dim \mathbb{S}_\theta$ and $\dim \mathbb{U}_\theta$ do not depend on $\theta \in \Theta$. Proposition 8.10 and Remark 8.6 imply that $\dim \mathbb{U}_\theta = \sum_{j=1}^{j_\nu} m_\nu^{(j)}$ is finite. Also, the last sum does not depend on the choice of $\nu \in \mathrm{Erg}(\Theta; \varphi)$. ◇

For numbers $\gamma < \gamma'$ that belong to different connected components of the complement $\mathbb{R} \setminus \Sigma$ of the Sacker-Sell spectrum $\Sigma = \Sigma(\varphi)$, see displays (6.63)-(6.64), we will use the notations $\mathbb{S}^\gamma$, $E^{\gamma\gamma'}$, and $\mathbb{U}^{\gamma'}$ to denote the $\{\hat{\varphi}^n\}_{n \in \mathbb{Z}_+}$-invariant complementary subbundles of $\Theta \times \mathcal{H}$ that correspond, respectively, to the parts of the spectrum $\Sigma \cap (-\infty, \gamma)$, $\Sigma \cap (\gamma, \gamma')$, and $\Sigma \cap (\gamma', \infty)$.

THEOREM 8.12. *Suppose that the function* $\Phi : \Theta \to \mathcal{C}(\mathcal{H})$ *is norm continuous.*

(1) *The Sacker-Sell spectrum* $\Sigma(\hat{\varphi})$ *consists of finitely many or countably many spectral segments:*

$$\Sigma = \cup_{k=1}^N [r_k^-, r_k^+], \quad N \leq \dim \mathcal{H}.$$

(2) *If, in addition,* $\gamma_0, \gamma_1, \ldots$ *are chosen as follows:*

(8.11) $$\gamma_0 > r_1^+ \geq r_1^- > \gamma_1 > r_2^+ \geq \ldots \geq -\infty,$$

then the fibers of the spectral subbundles that correspond to the bounded parts $[r_k^-, r_k^+]$ *of* Σ *are finite dimensional; that is,*

$$\dim E_\theta^{\gamma_k \gamma_{k-1}} < \infty, \quad k = 1, 2, \ldots, N-1.$$

(3) *If* $\dim \mathcal{H} = \infty$, *then* $-\infty \in \Sigma$.
(4) *If* $N = \infty$, *then the sequences* $e^{r_k^\pm}$ *converge to zero as* $k \to \infty$.

PROOF. By Remark 8.11, if $\gamma \in \mathbb{R} \setminus \Sigma$, then the subspace $\mathbb{U}_\theta^\gamma$ is finite dimensional. Since $\dim E_\theta^{\gamma_k \gamma_{k-1}} \geq 1$ (see Remark 6.46), this implies statements (1) and (2). Because $\Sigma = \log |\sigma(E) \setminus \{0\}|$, see (6.60), we have that $-\infty \in \Sigma$ if and only if $0 \in \sigma(E)$. Note that $-\infty \in \Sigma$ whenever $\dim \mathcal{H} = \infty$. Indeed, $E = aV$ is not invertible. This fact follows because the operator defined by $(Vf)(\theta) = f(\varphi^{-1}\theta)$ is invertible on $C(\Theta; \mathcal{H})$, but each operator defined by $a(\theta) = \Phi(\varphi^{-1}\theta)$ is compact, hence noninvertible on an infinite dimensional space. This gives statement (3). To complete the proof, let us suppose that there is some $\omega \in \mathbb{R}$ such that infinitely many spectral segments $[r_k^-, r_k^+]$ lie to the right of ω. Take $\nu \in \mathrm{Erg}$, $\theta \in \Theta_\nu$ and choose $x \in E_\theta^{\gamma_k \gamma_{k-1}}$. By Theorem 6.48, the exact Oseledets-Lyapunov exponent $\lambda^k = \lambda(\theta, x)$ belongs to $[r_k^-, r_k^+]$. This is a contradiction: By Remark 8.6, zero is the only possible accumulation point of the sequence $\{e^{\lambda^k}\}_{n=1}^\infty$. □

The next theorem is a generalization of Proposition 8.10. It states a relationship between the continuous spectral subbundles $\mathbb{S}$ and $\mathbb{E}$, and the measurable subbundles V and W that appear in the Multiplicative Ergodic Theorem 8.2 (MET-2).

THEOREM 8.13. *Suppose that* $\Phi : \Theta \to \mathcal{C}(\mathcal{H})$ *is norm continuous and fix numbers* $\gamma_0, \gamma_1, \ldots$ *as in display* (8.11). *If* $\nu \in \mathrm{Erg}$, *then there exist indices*

$$j_1 = 1 < j_2 < \ldots < j_{N+1} = s_\nu + 1, \quad N \leq \infty,$$

for which

$$(8.12) \qquad \lambda_\nu^{(j_k)} < \gamma_{k-1} < \lambda_\nu^{(j_k - 1)}, \quad k = 1, 2, \ldots, N.$$

The corresponding continuous and measurable spectral subbundles coincide on Θ_ν; *that is,*

$$(8.13) \qquad \mathbb{S}_\theta^{\gamma_{k-1}} = V_{\theta,\nu}^{(j_k)} \quad \text{for} \quad k = 1, 2, \ldots, N, \quad \theta \in \Theta_\nu,$$

$$(8.14) \qquad \mathbb{E}_\theta^{\gamma_k \gamma_{k-1}} = \sum_{j=j_k}^{j_{k+1}-1} W_{\theta,\nu}^j, \quad k = 1, 2, \ldots, N-1, \quad \theta \in \Theta_\nu.$$

The corresponding Oseledets-Lyapunov exponents satisfy

$$(8.15) \qquad \lambda_\nu^j \in [r_k^-, r_k^+], \quad j_k \leq j < j_{k+1}, \quad k = 1, 2, \ldots, N.$$

PROOF. Since $\gamma_{k-1} \in \mathbb{R} \setminus \Sigma$, the cocycle $\{e^{-n\gamma_{k-1}}\Phi^n\}_{n \in \mathbb{Z}_+}$ has an exponential dichotomy. By Remark 8.9, the Oseledets-Lyapunov spectrum $\Sigma_L - \gamma_{k-1}$ of this cocycle is bounded away from zero. So, there are natural numbers $j_1, j_2, \ldots$ as in display (8.12). Recall the notations in displays (6.63)–(6.64) and note that $P_{\gamma_{k-1}}(\cdot)$ is the dichotomy projection for the cocycle $\{e^{-n\gamma_{k-1}}\Phi^n\}_{n \in \mathbb{Z}_+}$. The equalities in display (8.13) now follow from Proposition 8.10. By statement (8.3) in the Multiplicative Ergodic Theorem 8.2 (MET-2) and Proposition 8.10, we have that

$$V_{\theta,\nu}^{(j_k)} = V_{\theta,\nu}^{(j_{k+1})} \dotplus \sum_{j=j_k}^{j_{k+1}-1} W_{\theta,\nu}^j, \quad \mathbb{S}_\theta^{\gamma_{k-1}} = V_{\theta,\nu}^{(j_k)}, \quad \text{and} \quad \mathbb{U}_\theta^{\gamma_k} = \sum_{j=1}^{j_{k+1}-1} W_{\theta,\nu}^j.$$

Since $\mathcal{H} = \mathbb{S}_\theta^{\gamma_k} \dotplus E_\theta^{\gamma_k \gamma_{k-1}} \dotplus \mathbb{U}_\theta^{\gamma_{k-1}}$, this yields statement (8.14). Finally, statement (8.15) follows from Corollary 6.49. □

We conclude this section by giving a description of Σ, or $\sigma(E)$, for the situation where $\Phi : \Theta \to \mathcal{L}(\mathcal{H})$ is norm continuous, $\dim \mathcal{H} = \infty$, and for each $\theta \in \Theta$ the *invertible* operator $\Phi(\theta)$ is a sum of a unitary and a compact operator. For this setting, the Multiplicative Ergodic Theorem is described in Remark 8.7.

THEOREM 8.14. *Suppose that* $\dim \mathcal{H} = \infty$, *the function* $\Phi : \Theta \to \mathcal{L}(\mathcal{H})$ *is norm continuous,* $\Phi(\theta)$ *is invertible for each* $\theta \in \Theta$, *and* $\Phi(\theta) = U(\theta) + K(\theta)$ *where* $U(\theta)$ *is a unitary operator on* $\mathcal{H}$ *and* $K(\theta) \in \mathcal{C}(\mathcal{H})$.

(1) *The Sacker-Sell spectrum* $\Sigma(\hat{\varphi})$ *consists of finitely many or countably many spectral segments:*
$$\Sigma = \cup_{k=-N'}^{N} [r_k^-, r_k^+], \quad N, N' \leq \infty.$$

(2) *If* $\gamma_0, \gamma_{\pm 1}, \gamma_{\pm 2}, \ldots$ *are such that*
$$\gamma_{-1} < r_{-1}^- \leq r_{-1}^+ < \gamma_{-2} < r_{-2}^- \leq \ldots \leq 0 \leq \ldots r_2^- \leq r_2^+ < \gamma_1 < r_1^- \leq r_1^+ < \gamma_0,$$
then the fiber of the spectral subbundle $\mathbb{E}^{\gamma_k \gamma_{k-1}}$ *that correspond to the bounded part* $[r_k^-, r_+^+]$ *of* Σ *is finite dimensional; that is,* $\dim \mathbb{E}_\theta^{\gamma_k \gamma_{k-1}} < \infty$ *for each* $\theta \in \Theta$ *and* $k \in (-N', N) \cap \mathbb{Z}$.

(3) *The number 0 is in* Σ.

(4) *If* $r_N^+ > 0$, *respectively* $r_N^- < 0$, *then* $N < \infty$, *respectively* $N' < \infty$.

(5) *If* $N = \infty$, *respectively* $N' = \infty$, *then*
$$\lim_{k \to \infty} r_k^\pm = 0, \quad \text{respectively} \quad \lim_{k \to -\infty} r_k^\pm = 0.$$

PROOF. Choose $\gamma \in \mathbb{R} \setminus \Sigma$ such that $\gamma > 0$ and, if possible, fix a number $\gamma' > \gamma$ that is in a connected component of $\mathbb{R} \setminus \Sigma$ not containing γ. Let $\nu \in \text{Erg}$ and $\theta \in \Theta_\nu$. Also, let $V_\theta^{(\lambda)}$ (see Remark 8.7) be the spectral subspace for Λ_θ corresponding to the part of $\sigma(\log \Lambda_\theta)$ that lies in $(-\infty, \lambda]$ for some $\lambda \in \sigma(\log \Lambda_\theta)$.

We claim that (see notations (6.63)-(6.64))

(8.16) $$\mathbb{S}_\theta^\gamma = \bigcup_\lambda V_{\theta,\nu}^{(\lambda)}, \quad \lambda \in \sigma(\log \Lambda_\theta) \cap (-\infty, \gamma).$$

Indeed, by Theorem 6.48, $x \in \mathbb{S}_\theta^\gamma$ if and only if $\lambda_s^+ = \lambda_i^+ = \lambda(\theta, x) < \gamma$. This fact, together with (8.7), gives statement (8.16). Recall that the eigenspaces of $\log \Lambda_\theta$ that correspond to nonzero λ are finite dimensional. Also, there are finitely many $\lambda > \gamma > 0$ in $\sigma(\log \Lambda_\theta)$. Since the maps $\theta \mapsto \mathbb{S}_\theta^\gamma$, $\theta \mapsto \mathbb{E}_\theta^{\gamma\gamma'}$, and $\theta \mapsto \mathbb{U}_\theta^{\gamma'}$ are continuous, the dimensions of $\mathbb{S}_\theta^\gamma$, $\mathbb{E}_\theta^{\gamma\gamma'}$, and $\mathbb{U}_\theta^{\gamma'}$ do not depend on $\theta \in \Theta$. Moreover, since $\dim \mathbb{E}^{\gamma\gamma'} \geq 1$, we conclude that only finitely many spectral segments lie to the right of $\gamma > 0$ and each $\dim \mathbb{E}_\theta^{\gamma\gamma'}$ is finite. For the operator E^{-1}, or the cocycle given by $\Phi^{-n}(\theta) = \Phi^{-1}(\varphi^{-n}\theta) \cdots \Phi^{-1}(\varphi^{-1}\theta)$, similar arguments show that at most finitely many of the intervals $[r_k^-, r_k^+]$ lie to the left of $\gamma' < 0$ for $\gamma' \in \mathbb{R} \setminus \Sigma$. Thus, statements (1) and (2) are proved. For statement (3), it suffices to recall that the existence of an exponential dichotomy for $\{\hat{\varphi}^n\}_{n \in \mathbb{Z}}$ implies $0 \notin \Sigma_L$, in contradiction to Remark 8.7. Also, by Remark 8.7, the only possible accumulation point of the sequence $\{\lambda_\nu^j\}_{j \in \mathbb{Z}}$ is zero. Statements (4) and (5) follow from this fact as in the proof of statement (4) in Theorem 8.12. □

8.1.3. Calculation of the spectrum. In this subsection, under the assumption that $\Phi : \Theta \to \mathcal{C}(\mathcal{H})$ is compact valued, we will propose an inductive procedure that can be used to describe the spectrum $\Sigma = \cup_{k=1}^N [r_k^-, r_k^+]$, $N \leq \infty$, or $\sigma(E)$, in terms of the set of Oseledets-Lyapunov exponents λ_ν^j, $j = 1, 2, \ldots, s_\nu$ and measurable subbundles $W_{\theta,\nu}^j$, $V_{\theta,\nu}^j$, for $\theta \in \Theta_\nu$ and $\nu \in \text{Erg}(\Theta; \varphi)$.

Recall that the rightmost point r_1^+ of $\Sigma(\hat{\varphi})$ coincides with $\log r(E)$, the log of the spectral radius $r(E)$ of the evolution operator E on $C(\Theta; \mathcal{H})$ defined by $(Ef)(\theta) = \Phi(\varphi^{-1}\theta)f(\varphi^{-1}\theta)$. The main technical difficulties in the proof of the following result are related to the fact that $\mathcal{H}$ might be infinite dimensional or $\Phi(\theta)$ might be noninvertible.

THEOREM 8.15. *If $\mathcal{H}$ is separable and $\Phi : \Theta \to \mathcal{C}(\mathcal{H})$ is norm continuous, then*
$$\log r(E) = \sup\{\lambda_\nu^1 : \nu \in \mathrm{Erg}(\Theta; \varphi)\}$$
where $r(E)$ is the spectral radius of E and λ_ν^1 is the largest Oseledets-Lyapunov exponent of the cocycle $\{\Phi^n\}_{n \geq 0}$ with respect to the measure ν.

PROOF. Note that $(E^n f)(\theta) = \Phi^n(\varphi^{-n}\theta)f(\varphi^{-n}\theta)$ and
$$(8.17) \qquad r(E) = \lim_{n \to \infty} \|E^n\|_{\mathcal{L}(C(\Theta; \mathcal{H}))}^{1/n} = \lim_{n \to \infty} \max_{\theta \in \Theta} \|\Phi^n(\theta)\|_{\mathcal{L}(\mathcal{H})}^{1/n}.$$

Using formula (8.5) for the largest Oseledets-Lyapunov exponent λ_ν^1, it follows that if $\nu \in \mathrm{Erg}(\Theta; \varphi)$, then there is a point $\theta \in \Theta_\nu$ such that
$$\lambda_\nu^1 = \lim_{n \to \infty} n^{-1} \log \|\Phi^n(\theta)\|_{\mathcal{L}(\mathcal{H})} \leq \lim_{n \to \infty} n^{-1} \log \max_{\theta \in \Theta} \|\Phi^n(\theta)\|_{\mathcal{L}(\mathcal{H})} = \log r(E).$$

To complete the proof, we will show that if $\epsilon > 0$, then there exists a measure $\nu = \nu(\epsilon) \in \mathrm{Erg}(\Theta; \varphi)$ such that $\lambda_\nu^1 \geq \log r(E) - \epsilon$. The strategy of the proof is to construct a measure μ on the "projective bundle" $\Theta \times B$ where B is the unit ball in $\mathcal{H}$, and then to obtain ν as the projection of μ on Θ.

Let S and B denote, respectively, the unit sphere and the unit ball in $\mathcal{H}$ equipped with the weak topology. Also consider the sets $\Theta_B := \Theta \times B$ and $\Theta_S := \Theta \times S$ with the product topology. Let us define the subset $\Theta_B^0 \subset \Theta_B$ by
$$\Theta_B^0 = \{(\theta, x) \in \Theta_B : \Phi(\theta)x \neq 0\}$$
and note that if $\dim \mathcal{H} < \infty$ and $\Phi(\theta)$ is invertible for each $\theta \in \Theta$, then $\Theta_B^0 = \Theta_B$. However, in the general case, $\Theta_B^0 \subsetneq \Theta_B$. Indeed, if $\dim \mathcal{H} = \infty$, then the range $\mathrm{Im}\,\Phi(\theta)$ of the compact operator $\Phi(\theta)$ is not closed and $\inf\{|\Phi(\theta)x| : x \in S\} = 0$.

Consider the scalar-valued function
$$\tilde{\Phi}(\theta, x) = \log |\Phi(\theta)x|, \quad (\theta, x) \in \Theta_B^0,$$
and recall that each compact operator $\Phi(\theta)$ maps weakly convergent nets to strongly convergent nets. Also, the function $\theta \mapsto \Phi(\theta) \in \mathcal{L}(\mathcal{H})$ is continuous. Hence, if $(\theta_n, x_n) \to (\theta_0, x_0)$ in Θ_B, then
$$|\Phi(\theta_n)x_n - \Phi(\theta_0)x_0| \leq |(\Phi(\theta_n) - \Phi(\theta_0))x_n| + |\Phi(\theta_0)(x_n - x_0)| \to 0,$$
and we conclude that $\tilde{\Phi}$ is continuous on Θ_B^0.

Define the "projectivization" $\tilde{\varphi} : \Theta_B \to \Theta_B$ of the homeomorphism $\varphi : \Theta \to \Theta$ as follows:
$$\tilde{\varphi}(\theta, x) = \left(\varphi\theta, \frac{\Phi(\theta)x}{|\Phi(\theta)x|}\right) \quad \text{for} \quad (\theta, x) \in \Theta_B^0, \quad \text{and}$$
$$\tilde{\varphi}(\theta, x) = (\varphi\theta, 0) \quad \text{for} \quad (\theta, x) \in \Theta_B \setminus \Theta_B^0.$$

Note that $\tilde{\varphi}$ is a Borel endomorphism of Θ_B since $\tilde{\varphi}$ is the composition of the map $(\theta, x) \mapsto (\varphi\theta, \Phi(\theta)x)$ and the map given by
$$(\theta, x) \mapsto \left(\theta, \frac{x}{|x|}\right) \quad \text{for} \quad x \neq 0, \quad \text{and} \quad (\theta, 0) \mapsto (\theta, 0).$$

Also, the image $\tilde{\varphi}(\Theta_B) = \tilde{\varphi}(\Theta_B^0) \cap (\Theta \times \{0\})$ is a Borel subset of Θ_B. Indeed,
$$\tilde{\varphi}(\Theta_B^0) = \Theta_S \cap \{(\theta, x) : \theta \in \Theta, x \in \operatorname{Im} \Phi(\theta)\}$$
while $\operatorname{Im} \Phi(\theta) = \cup_{k=0}^{\infty} \operatorname{Im}(\Phi(\theta)|\{x : |x| \leq k\})$. But the operator $\Phi(\theta)$ on the Hilbert space $\mathcal{H}$ maps each closed ball, given by $\{x : |x| \leq k\}$, into a strongly closed set, see for example Halmos [**Hm**, p. 90].

Let us observe that if $\tilde{\varphi}^j(\theta, x) \in \Theta_B^0$ for $j = 0, \ldots, n-1$, then
$$\tag{8.18} \tilde{\varphi}^n(\theta, x) = (\varphi^n \theta, \Phi^n(\theta)x/|\Phi^n(\theta)x|).$$

Since each $\Phi^n(\theta)$ is a compact operator and Θ is compact, there is a point $(\theta_n, x_n) \in \Theta_S$ such that
$$\tag{8.19} |\Phi^n(\theta_n)x_n| = \max_{\theta \in \Theta} \|\Phi^n(\theta)\|_{\mathcal{L}(\mathcal{H})}.$$

Without loss of generality, we can assume that $r(E) > 0$. Thus, using formula (8.17), we have
$$\tilde{\varphi}^j(\theta_n, x_n) \in \Theta_B^0, \quad j = 0, \ldots n-1.$$

The desired measure $\nu = \nu(\epsilon)$, with $\lambda_\nu^1 > \log r(E) - \epsilon$, will be defined below as the projection onto Θ of a Borel probability measure μ on Θ_B which we will now construct. For each $n \in \mathbb{N}$, define a Borel probability measure μ_n on Θ_B by
$$\tag{8.20} \mu_n(g) = n^{-1} \sum_{j=0}^{n-1} g(\tilde{\varphi}^j(\theta_n, x_n)), \quad g \in C(\Theta_B, \mathbb{R}).$$

Since Θ_B is compact, the unit ball in the adjoint space $C^*(\Theta_B, \mathbb{R})$ is compact, and we can choose a subsequence $\{\mu_{n_k}\}_{k=1}^{\infty}$ that converges to some Borel measure μ on Θ_B; that is,
$$\tag{8.21} \mu(g) = \lim_{k \to \infty} \mu_{n_k}(g) \quad \text{for each} \quad g \in C(\Theta_B; \mathbb{R}).$$

It is important to note that $\tilde{\Phi}$ does *not* belong to $C(\Theta_B; \mathbb{R})$. However, the values $\mu_n(\tilde{\Phi})$ are defined by equation (8.20). Also, by the choice of (θ_n, x_n) in display (8.19), formula (8.18) and formula (8.17) for the spectral radius, it follows that
$$\tag{8.22} \log r(E) = \lim_{k \to \infty} \mu_{n_k}(\tilde{\Phi}).$$

The next proposition shows that even though $\tilde{\Phi} \notin C(\Theta_B; \mathbb{R})$, the function $\tilde{\Phi}$ is μ-integrable. Thus, the limit in (8.22) gives $\mu(\tilde{\Phi})$.

PROPOSITION 8.16. *With the notation defined above, the following statements are true:*
 (a) $\mu(\Theta_B^0) = 1$.
 (b) *The function* $\tilde{\Phi}$ *is μ-integrable on* Θ_B.
 (c) $\mu(\tilde{\Phi}) \geq \log r(E)$.

PROOF. For each positive integer N, define the set
$$F_N = \{(\theta, x) : \tilde{\Phi}(\theta, x) \leq -N\} \supset \Theta_B \setminus \Theta_B^0$$
and the continuous function $\tilde{\Phi}_N$ on Θ_B as follows: $\tilde{\Phi}_N(\theta, x) = \tilde{\Phi}(\theta, x)$ for $(\theta, x) \in \Theta_B \setminus F_N$, and $\tilde{\Phi}_N(\theta, x) = -N$ for $(\theta, x) \in F_N$.

Let $M := \max_{\theta \in \Theta} \log \|\Phi(\theta)\|_{\mathcal{L}(\mathcal{H})}$ and note that

(8.23) $$\tilde{\Phi}(\theta, x) \leq \tilde{\Phi}_N(\theta, x) \leq M, \quad (\theta, x) \in \Theta_B,$$

for each N.

From the first inequality in (8.23) we see that $\mu_{n_k}(\tilde{\Phi}) \leq \mu_{n_k}(\tilde{\Phi}_N)$. Also, we have $\tilde{\Phi}_N \in C(\Theta_B; \mathbb{R})$ for each fixed N. Hence, by the definition (8.21) of μ, we have that $\lim_{k \to \infty} \mu_{n_k}(\tilde{\Phi}_N) = \mu(\tilde{\Phi}_N)$. On the other hand, using formula (8.22), it follows that

(8.24) $$\mu(\tilde{\Phi}_N) = \lim_{k \to \infty} \mu_{n_k}(\tilde{\Phi}_N) \geq \lim_{k \to \infty} \mu_{n_k}(\tilde{\Phi}) = \log r(E).$$

From the second inequality in display (8.23) we have the inequality $\mu(\tilde{\Phi}_N) \leq -N\mu(F_N) + M$; and, using (8.24), we conclude that $\lim_{N \to \infty} \mu(F_N) = 0$. Thus, since $F_N \supset \Theta_B \setminus \Theta_B^0$, we have that $\mu(\Theta_B \setminus \Theta_B^0) = 0$. This proves statement (a).

To prove statement (b), let us observe that $\tilde{\Phi}_N \searrow \tilde{\Phi}$ pointwise in Θ_B^0; that is, the convergence is monotone and holds μ-everywhere. Therefore, by the Monotone Convergence Theorem, the function $\tilde{\Phi}$ is μ-integrable, and, in addition,

$$\mu(\tilde{\Phi}) = \lim_{N \to \infty} \mu(\tilde{\Phi}_N) \geq \log r(E)$$

where the last inequality is from display (8.24). This proves statement (c), and the proof of Proposition 8.16 is complete. $\square$

Our next objective is to show that the pointwise Birkhoff-Khinchine Ergodic Theorem can be applied to the function $\tilde{\Phi}$, measure μ, and the Borel endomorphism $\tilde{\varphi}$.

PROPOSITION 8.17. *The measure μ on Θ_B is $\tilde{\varphi}$-invariant.*

PROOF. It suffices to prove that $\mu(h \circ \tilde{\varphi}) = \mu(h)$ for all $h \in C(\Theta_B; \mathbb{R})$. Let $C_b(\Theta_B^0; \mathbb{R})$ denote the set of all continuous functions on Θ_B^0 that are bounded on Θ_B. Since $\tilde{\varphi}$ is continuous on Θ_B^0, we have $h \circ \tilde{\varphi} \in C_b(\Theta_B^0; \mathbb{R})$ for every $h \in C(\Theta_B; \mathbb{R})$.

Fix $g \in C(\Theta_B^0; \mathbb{R})$. We claim that $\mu(g) = \lim_{k \to \infty} \mu_{n_k}(g)$. To see this, fix $\delta > 0$ and choose closed subsets $\Theta_B^1 \subset \Theta_B^2 \subset \Theta_B$ such that

$$\Theta_B \setminus \Theta_B^0 \subset \Theta_B^1, \quad \Theta_B^1 \cap (\Theta_B \setminus \Theta_B^2) = \emptyset, \quad \text{and} \quad \mu(\Theta_B^2) < \delta.$$

Since $\mathcal{H}$ is separable, $\Theta_B = \Theta \times \{x : |x| \leq 1\}$ is metrizable. Choose a function $g_1 \in C(\Theta_B; \mathbb{R})$ such that $g_1(\theta, x) = g(\theta, x)$ for $(\theta, x) \in \Theta_B \setminus \Theta_B^1$ and $\|g_1\|_\infty \leq 2\|g\|_\infty$. Note that $m := 3\|g\|_\infty \geq \sup\{|g(\theta, x) - g_1(\theta, x)| : (\theta, x) \in \Theta_B^1\}$, and choose a nonnegative $g_2 \in C(\Theta_B; \mathbb{R})$ with $\|g_2\|_\infty \leq m$ such that $g_2(\theta, x) = 0$ for $(\theta, x) \notin \Theta_B^2$, and $g_2(\theta, x) = m$ for $(\theta, x) \in \Theta_B^1$. With these definitions, we have that

$$\begin{aligned}|\mu(g) - \mu_{n_k}(g)| &\leq |\mu(g - g_1)| + |\mu(g_1) - \mu_{n_k}(g_1)| + \mu_{n_k}(|g - g_1|) \\ &\leq m\mu(\Theta_B^1) + |\mu(g_1) - \mu_{n_k}(g_1)| + \mu_{n_k}(g_2) \\ &= m\mu(\Theta_B^1) + |\mu(g_1) - \mu_{n_k}(g_1)| + |\mu_{n_k}(g_2) - \mu(g_2)| + m\mu(\Theta_B^2).\end{aligned}$$

Since $g_i \in C(\Theta_B; \mathbb{R})$, it follows that $\lim_{k \to \infty} \mu_{n_k}(g_i) = \mu(g_i)$ for $i = 1, 2$. This proves the claim.

Since $h \in C(\Theta_B; \mathbb{R})$ implies $h \circ \tilde{\varphi} \in C_b(\Theta_B^0; \mathbb{R})$, the desired $\tilde{\varphi}$-invariance of μ on Θ_B is proved as follows:

$$\mu(h \circ \tilde{\varphi}) - \mu(h) = \lim_{k \to \infty} [\mu_{n_k}(h \circ \tilde{\varphi}) - \mu_{n_k}(h)]$$

$$= \lim_{k \to \infty} \Big[\frac{1}{n_k} \sum_{j=0}^{n_k-1} h \circ \tilde{\varphi}^{j+1}(\theta_{n_k}, x_{n_k}) - \frac{1}{n_k} \sum_{j=0}^{n_k-1} h \circ \tilde{\varphi}^j(\theta_{n_k}, x_{n_k})\Big]$$

$$= \lim_{k \to \infty} \frac{1}{n_k}[h \circ \tilde{\varphi}^{h_k}(\theta_{n_k}, x_{n_k}) - h(\theta_{n_k}, x_{n_k})] = 0.$$

$\square$

As we have remarked above, $\tilde{\varphi}(\Theta_B)$ is a Borel set. By Proposition 8.17, we conclude that $\mu(\tilde{\varphi}(\Theta_B)) = \mu(\Theta_B) = 1$. Thus, $\tilde{\varphi}$ is a $(\mu - \mathrm{mod}\, 0)$-endomorphism of Θ_B. By statement (b) of Proposition 8.16, $\tilde{\Phi}$ is μ-integrable. Therefore, by the Birkhoff-Khinchine Ergodic Theorem, there is a subset $\Theta_B(\mu)$ in Θ_B of full μ-measure such that, for each point $(\theta, x) \in \Theta(\mu)$, the following limit exists:

$$(8.25) \qquad \bar{\Phi}(\theta, x) := \lim_{n \to \infty} n^{-1} \sum_{j=0}^{n-1} \tilde{\Phi}(\tilde{\varphi}^j(\theta, x)), \quad \text{and} \quad \mu(\bar{\Phi}) = \mu(\tilde{\Phi}).$$

Moreover, by the Birkhoff-Khinchine Ergodic Theorem, for each given $\epsilon > 0$, the set

$$\Theta_B^\epsilon(\mu) := \{(\theta, x) \in \Theta_B(\mu) : \bar{\Phi}(\theta, x) \geq \mu(\tilde{\Phi}) - \epsilon\}$$

has positive measure: $\mu(\Theta_B^\epsilon(\mu)) > 0$.

To construct the measure ν on Θ, let $\pi_1(\theta, x) = \theta$, $\pi_2(\theta, x) = x$, and let ν and $\tilde{\nu}$ be the projections of μ on Θ and B, respectively; that is, $\nu = \mu \circ \pi_1^{-1}$ and $\tilde{\nu} = \mu \circ \pi_2^{-1}$. By Fubini's Theorem,

$$\mu(\Theta_B^\epsilon(\mu)) = \int_{\pi_2(\Theta_B^\epsilon(\mu))} \nu(\{\theta \in \Theta : (\theta, x) \in \Theta_B^\epsilon(\mu)\})\, d\tilde{\nu}(x).$$

Since $\mu(\Theta_B^\epsilon(\mu)) > 0$, there is a vector $\bar{x} \in \mathcal{H}$ such that $|\bar{x}| = 1$ and the set $\Theta_\mu^\epsilon := \{\theta \in \Theta : (\theta, \bar{x}) \in \Theta_B^\epsilon(\mu)\}$ has positive ν-measure. Passing, if necessary, to a decomposition of Θ into ergodic components, let us assume, without loss of generality, that ν is ergodic. Apply the Multiplicative Ergodic Theorem for ν to obtain a full ν-measure subset $\Theta_\nu \subset \Theta$ such that, for each point $\theta \in \Theta_\nu$, the limit (8.5) exists. It follows that if $\bar{\theta} \in \Theta_\nu \cap \Theta_\mu^\epsilon$, then

$$\lambda_\nu^1 = \lim_{n \to \infty} n^{-1} \log \|\Phi^n(\bar{\theta})\|_{\mathcal{L}(\mathcal{H})}.$$

Recall that $\mu(\bar{\Phi}) = \mu(\tilde{\Phi}) \geq \log r(E)$ by (8.25) and statement (c) of Proposition 8.16. Since $(\bar{\theta}, \bar{x}) \in \Theta_B^\epsilon(\mu)$, by the definition of $\Theta_B^\epsilon(\mu)$ and (8.25), we have the inequalities

$$\log r(E) - \epsilon \leq \mu(\tilde{\Phi}) - \epsilon \leq \bar{\Phi}(\bar{\theta}, \bar{x}) = \lim_{n \to \infty} n^{-1} \sum_{j=0}^{n-1} \tilde{\Phi}(\tilde{\varphi}^j(\bar{\theta}, \bar{x})).$$

Using the definition (8.18), the last limit is equal to

$$\lim_{n \to \infty} n^{-1} \log |\Phi^n(\bar{\theta})\bar{x}| \leq \lim_{n \to \infty} n^{-1} \log \|\Phi^n(\bar{\theta})\|_{\mathcal{L}(\mathcal{H})} = \lambda_\nu^1.$$

This completes the proof of Theorem 8.15. $\square$

REMARK 8.18. If $\dim \mathcal{H} < \infty$ and $\Phi(\theta)$ is invertible for each $\theta \in \Theta$, then $\tilde{\Phi} \in C(\Theta_B; \mathbb{R})$; and, as a result, $\log r(E) = \lambda_\nu^1$ for $\nu = \mu \circ \pi_1^{-1}$ where ν is the projection of μ on Θ. Thus, the supremum in Theorem 8.15 is attained. $\diamond$

Let us specialize to the one dimensional case.

COROLLARY 8.19. *If $\Phi : \Theta \to \mathbb{R}$ is a continuous scalar valued function, then*

$$\log r(E) = \sup \left\{ \int_\Theta \log \Phi \, d\nu : \nu \in Erg\,(\Theta, \varphi) \right\}$$

where $r(E)$ is the spectral radius of E on $C(\Theta)$.

Returning to the general infinite dimensional case, let us consider the numbers $\gamma_0, \gamma_1, \ldots$ as indicated in Theorem 8.12, and the set of indices $\{j_k = j_{k,\nu} : k = 1, 2, \ldots, N\}$ as in Theorem 8.13. The operator E on $C(\Theta; \mathcal{H})$ has a diagonal decomposition that corresponds to the decomposition $\Theta \times \mathcal{H} = \mathbb{S}^{\gamma_{k-1}} \oplus \mathbb{E}^{\gamma_k \gamma_{k-1}}$, see also (6.63)–(6.64). Let $C(\mathbb{E}^{\gamma_k \gamma_{k-1}})$ be the space of continuous sections of the subbundle $\mathbb{E}^{\gamma_k \gamma_{k-1}}$. Using Theorem 8.15 for each restriction E_k of E on $C(\mathbb{E}^{\gamma_k \gamma_{k-1}})$ with $r(E_k) = e^{r_k^+}$ and $r([E_k]^{-1}) = e^{-r_k^-}$, where $\Sigma = \cup_{k=1}^N [r_k^-, r_k^+]$ as in Theorem 8.12, we obtain the following result.

COROLLARY 8.20. *For $k = 1, 2, \ldots, N$,*

$$r_k^- = \inf\{\lambda_\nu^j : j_k \leq j < j_{k+1}, \nu \in \mathrm{Erg}\} = \inf\{\lambda_\nu^j \in [r_k^-, r_k^-] : \nu \in \mathrm{Erg}\},$$
$$r_k^+ = \sup\{\lambda_\nu^j : j_k \leq j < j_{k+1}, \nu \in \mathrm{Erg}\} = \sup\{\lambda_\nu^j \in [r_k^-, r_k^+] : \nu \in \mathrm{Erg}\}.$$

Next, we will describe an inductive procedure for computing the spectrum Σ, or $\sigma(E)$, in terms of the objects given in the Multiplicative Ergodic Theorem. This procedure consists of sequentially "splitting off" the unstable spectral subbundles in display (6.64) that are continuous extensions to all of Θ of measurable subbundles on Θ_ν with fibers

$$V_{\theta,\nu}^{(-j)} = \sum_{i=1}^j W_{\theta,\nu}^i, \quad \theta \in \Theta_\nu, \quad \nu \in \mathrm{Erg}\,.$$

The procedure consists of five steps.

Step 1. Compute $r_1^+ = \log r(E)$ using the largest Oseledets-Lyapunov exponents of the cocycle according to Theorem 8.15:

$$r_1^+ = \sup\{\lambda_\nu^1 : \nu \in \mathrm{Erg}\}.$$

Step 2. For each $\nu \in \mathrm{Erg}$, compute the sums $m_\nu(k) = m_\nu^1 + \ldots + m_\nu^k$ where $m_\nu^j = \dim W_\nu^j$ are the multiplicities for $k = 1, 2, \ldots$. Also, fix the minimal $k = k(\nu)$ so that $m = m_\nu(k(\nu))$ is independent of $\nu \in \mathrm{Erg}$.

Step 3. For $\nu \in \mathrm{Erg}$, consider the subbundles from the Multiplicative Ergodic Theorem 8.2 (MET-2) (see (8.2)–(8.3)) with the base Θ_ν, the m-dimensional fiber

$$V_{\theta,\nu}^{(-k(\nu))} = W_{\theta,\nu}^1 + \ldots + W_{\theta,\nu}^{k(\nu)},$$

and the complementary fiber $V_\theta^{k(\nu)+1}$, and "check" the existence of their extensions from Θ_ν up to $\hat{\varphi}^n$-invariant continuous subbundles of $\Theta \times X$. If there is no such extension, return to Step 2 and increase m. We stress the point that in the generality of our setting it does not seem possible to find an effective way to complete this step.

Note that the existence of a continuous extension from Θ_ν to Θ means the following: There exist $\hat{\varphi}$-invariant subbundles $\mathbb{U}$ and $\mathbb{S}$ of $\Theta \times X$ over the entire set Θ such that $\mathbb{U}_\theta \dotplus \mathbb{S}_\theta = \mathcal{H}$ for all $\theta \in \Theta$, the maps $\theta \mapsto \mathbb{U}_\theta$, $\theta \mapsto \mathbb{S}_\theta$ are continuous, and
$$\mathbb{U}_\theta = W^1_{\theta,\nu} + \ldots + W^{k(\nu)}_{\theta,\nu}, \quad \mathbb{S}_\theta = V^{k(\nu)+1}_{\theta,\nu} \quad \text{for all} \quad \theta \in \Theta_\nu.$$
Also, if this continuous extension from Θ_ν to Θ exists for some measure $\nu \in \text{Erg}$, then, since m is independent of ν, it exists for all $\nu \in \text{Erg}$.

Step 4. Compute $r_1^- = \inf\{\lambda_\nu^{k(\nu)} : \nu \in \text{Erg}\}$ and $r_2^+ = \sup\{\lambda_\nu^{k(\nu)+1} : \nu \in \text{Erg}\}$. If $r_1^- \leq r_2^+$, then go back to Step 2 and increase m. Otherwise, fix γ_1 such that $r_2^+ < \gamma_1 < r_1^-$.

Step 5. Construct the cocycle using $\Phi(\theta)|\mathbb{S}_\theta$, for $\theta \in \Theta$, and repeat the procedure starting with Step 2.

8.2. The kinematic dynamo operator

In this section we will study the evolution semigroup $\{E^t\}_{t \in \mathbb{R}}$ defined by the formula $(E^t f)(\theta) = \Phi^t(\varphi^{-t}\theta) f(\varphi^{-t}\theta)$ on the space of *divergence free* continuous sections of the tangent bundle of a smooth compact finite dimensional manifold Θ with $\dim \Theta \geq 3$. In Subsection 8.2.1 we will prove the Spectral Mapping Theorem and the Annular Hull Theorem for this group. In Subsection 8.2.2 we give a description of the spectrum and annular hull of the spectrum of this group and its generator in terms of the exact Oseledets-Lyapunov exponents. If the flow $\{\varphi^t\}_{t \in \mathbb{R}}$ that generates the group $\{E^t\}_{t \in \mathbb{R}}$ is viewed as the flow of an incompressible Eulerian conducting fluid, and the cocycle $\{\Phi^t\}_{t \in \mathbb{R}}$ is the differential of $\{\varphi^t\}_{t \in \mathbb{R}}$, then the generator Γ of the evolution (push-forward) group can be interpreted to be the kinematic dynamo operator for the ideally conducting fluid. The kinematic dynamo operator governs the evolution of the induction of the magnetic field by the flow. The connection of our results to the "fast dynamo" problem for magnetic induction is described in Subsection 8.2.3.

8.2.1. The Spectral Mapping and Annular Hull Theorems.
The two main results of this subsection, the Spectral Mapping Theorem 8.22 and the Annular Hull Theorem 8.23 for the evolution semigroup on the space of divergence free vector fields are formulated in Subsection 8.2.1.1. Also, we will state the main technical result, Theorem 8.26, and show how to derive Theorems 8.22 and 8.23 from Theorem 8.26. In Subsection 8.2.1.2, we collect some preliminary "topological" results needed for the proof of Theorem 8.26. The proof itself is divided in two parts which are considered in Subsections 8.2.1.3 and 8.2.1.4, respectively.

8.2.1.1. *Setting and main results.* Let Θ be a smooth compact Riemannian manifold with $\dim \Theta \geq 3$, and $\{\varphi^t\}_{t \in \mathbb{R}}$ a flow on Θ induced by a vector field $\mathbf{v}$ with zero divergence: $\text{div } \mathbf{v} = 0$. Later on we will take $\mathbf{v}$ to be a steady-state solution of the Euler equation. Let $\{\Phi^t\}_{t \in \mathbb{R}}$ be a cocycle over $\{\varphi^t\}_{t \in \mathbb{R}}$ on the tangent bundle $\mathcal{T}\Theta$; that is, for $t, s \in \mathbb{R}$ and $\theta \in \Theta$, we require that $\Phi^t(\theta) : \mathcal{T}_\theta \Theta \to \mathcal{T}_{\varphi^t \theta} \Theta$ for $t \in \mathbb{R}$, $\Phi^{t+s}(\theta) = \Phi^t(\varphi^s \theta)\Phi^s(\theta)$, and $\Phi^0 = I$.

In the space $C(\Theta; \mathcal{T}\Theta)$ of continuous vector fields (continuous sections of $\mathcal{T}\Theta$), consider a closed subspace $C_{ND} = C_{ND}(\Theta; \mathcal{T}\Theta)$ of divergence free vector fields. In fact, there are at least two choices for the space of continuous divergence free vector fields, depending on whether the divergence is understood in the classical sense or in the sense of distributions. These spaces are defined, respectively, as follows:

(8.26) $C_{ND}^0(\Theta; \mathcal{T}\Theta) := \text{closure } \{f \in C^\infty(\Theta; \mathcal{T}\Theta) : \text{div } f = 0\},$

$C_{ND}^1(\Theta; \mathcal{T}\Theta) := \{f \in C(\Theta; \mathcal{T}\Theta) :$

(8.27) $\qquad \int_\Theta \langle f, \text{grad } g \rangle \, d\mu = 0 \quad \text{for all} \quad g \in C^\infty(\Theta; \mathbb{R})\}.$

The closure in (8.26) is taken with respect to the supremum norm while the scalar product $\langle \cdot, \cdot \rangle$ and gradient in (8.27) are taken with respect to a Riemannian metric and associated volume on Θ. We note that the space $C_{ND}^1(\Theta; \mathcal{T}\Theta)$ is a closed subspace of $C_{ND}^0(\Theta; \mathcal{T}\Theta)$ which, in turn, is a closed subspace of $C(\Theta; \mathcal{T}\Theta)$:

$$C_{ND}^1(\Theta; \mathcal{T}\Theta) \subset C_{ND}^0(\Theta; \mathcal{T}\Theta) \subset C(\Theta; \mathcal{T}\Theta).$$

Throughout this section the space $C_{ND} = C_{ND}(\Theta; \mathcal{T}\Theta)$ may be taken to be either $C_{ND}^0(\Theta; \mathcal{T}\Theta)$ or $C_{ND}^1(\Theta; \mathcal{T}\Theta)$.

Let us consider the evolution group $\{E^t\}_{t \in \mathbb{R}}$, defined as usual on $C(\Theta; \mathcal{T}\Theta)$ by $(E^t f)(\theta) = \Phi^t(\varphi^{-t}\theta)f(\varphi^{-t}\theta)$, and let Γ be its infinitesimal generator. *Throughout this section we assume that $C_{ND}(\Theta; \mathcal{T}\Theta)$ is E^t-invariant.*

EXAMPLE 8.21. Let $\mathbf{v} \in C(\Theta; \mathcal{T}\Theta)$ be a divergence free vector field, $\varphi^t : \Theta \to \Theta$ the flow tangent to $\mathbf{v}$, and $\Phi^t(\theta) = D\varphi^t(\theta)$ the differential of this flow. Then, the generator Γ of the evolution semigroup $(E^t f)(\theta) = D\varphi^t(\varphi^{-t}\theta)f(\varphi^{-t}\theta)$ is just the Lie derivative in the direction of $\mathbf{v}$; that is, if f is a smooth vector field, then the operator Γ acts as follows:

$$\Gamma : f \mapsto -\langle \mathbf{v}, \nabla \rangle f + \langle f, \nabla \rangle \mathbf{v}.$$

Clearly, E^t preserves $C_{ND}(\Theta; \mathcal{T}\Theta)$ whenever $\mathbf{v}$ has zero divergence. $\diamond$

Recall the definition (see Definition 6.26) of the prime period function $p : \Theta \to \mathbb{R}_+ \cup \{\infty\}$. It is given by $p(\theta) = \inf\{T > 0 : \varphi^T \theta = \theta\}$ for each periodic point $\theta \in \Theta$, and by $p(\theta) = \infty$ in case θ is aperiodic. Also, recall the set

$\mathcal{B}(\Theta) = \{\theta \in \Theta : \text{ there exists a constant } \bar{p} \in \mathbb{R}_+ \text{ and a}$

$\qquad \text{neighborhood } U \ni \theta \text{ such that } p(\eta) \leq \bar{p} \text{ for all } \eta \in U\},$

see (7.43). We will call the flow $\{\varphi^t\}_{t \in \mathbb{R}}$ *aperiodic* if $\mathcal{B}(\Theta) = \emptyset$. Clearly, if the set of aperiodic trajectories of $\{\varphi^t\}_{t \in \mathbb{R}}$ is dense in Θ, then $\mathcal{B}(\Theta) = \emptyset$.

Our objective in this subsection is to prove the following theorems relating the spectra of E^t and Γ.

THEOREM 8.22 (Spectral Mapping Theorem). *Let $\{\varphi^t\}_{t \in \mathbb{R}}$ denote the flow that is generated by the divergence free vector field $\mathbf{v}$ on Θ, let $\{\Phi^t\}_{t \in \mathbb{R}}$ be a cocycle over $\{\varphi^t\}_{t \in \mathbb{R}}$, let $\{E^t\}_{t \in \mathbb{R}}$ denote the corresponding evolution group on the space $C_{ND}(\Theta; \mathcal{T}\Theta)$ of continuous divergence free vector fields, and let Γ denote the infinitesimal generator of $\{E^t\}_{t \in \mathbb{R}}$. If the flow $\{\varphi^t\}_{t \in \mathbb{R}}$ is aperiodic on Θ, then*

$$\sigma(E^t; C_{ND}) = \exp(t\sigma(\Gamma; C_{ND})), \quad t \neq 0.$$

Moreover, $\sigma(E^t; C_{ND})$ is invariant with respect to rotations centered at the origin, and $\sigma(\Gamma; C_{ND})$ is invariant with respect to translations along the imaginary axis.

Example 8.34 below (cf. Example 6.32) shows that neither of the statements of the Spectral Mapping Theorem hold in general without the assumption that the flow is aperiodic. The following Annular Hull Theorem does not require the base flow to be aperiodic. To state it, recall that the annular hull of a set $S \subset \mathbb{C}$, denoted by $\mathcal{AH}(S)$, is the union of all circles centered at the origin that intersect the set S.

THEOREM 8.23 (Annular Hull Theorem). *If $\{\varphi^t\}_{t\in\mathbb{R}}$ is the flow on Θ generated by the divergence free vector field $\mathbf{v}$, $\{\Phi^t\}_{t\in\mathbb{R}}$ is a cocycle over the flow, and $\{E^t\}_{t\in\mathbb{R}}$, $E^t = e^{t\Gamma}$, is the corresponding evolution group on $C_{ND}(\Theta;\mathcal{T}\Theta)$, then*

$$\exp(t\sigma(\Gamma;C_{ND})) \subset \sigma(e^{t\Gamma};C_{ND}) \subset \mathcal{AH}(\exp(t\sigma(\Gamma;C_{ND}))), \quad t \neq 0.$$

Recall, see Subsection 2.1.3, that the growth and spectral bounds for a semigroup $\{e^{t\Gamma}\}_{t\geq 0}$ are, respectively,

$$\omega(\Gamma) := \lim_{t\to\infty} t^{-1} \log \|e^{t\Gamma}\| \quad \text{and} \quad s(\Gamma) := \sup\{\operatorname{Re} z : z \in \sigma(\Gamma)\}.$$

Theorem 8.23 implies that $\omega(\Gamma) = s(\Gamma)$.

COROLLARY 8.24. *If $\{\varphi^t\}_{t\in\mathbb{R}}$ is the flow on Θ generated by the divergence free vector field $\mathbf{v}$, then the spectral and growth bounds for the group $\{E^t\}_{t\in\mathbb{R}}$, $E^t = e^{t\Gamma}$, on $C_{ND}(\Theta;\mathcal{T}\Theta)$ are equal: $s(\Gamma) = \omega(\Gamma)$.*

The growth bounds $\omega(\Gamma)$ and $-\omega(-\Gamma)$ are just the logarithms of the spectral radii of the operators E and E^{-1}, respectively (see Subsection 2.1.3). Thus, $\omega(\Gamma)$ and $\omega(-\Gamma)$ can be computed, using Theorem 8.15 and Remark 8.6, via the exact Oseledets-Lyapunov exponents $\lambda_\nu^1 > \ldots > \lambda_\nu^{s_\nu}$, $s_\nu \leq \dim \Theta$, of the cocycle $\{\Phi^n\}_{n\in\mathbb{N}}$ with respect to all φ^t-ergodic measures on Θ.

COROLLARY 8.25. *Under the assumptions of the Annular Hull Theorem,*

$$\omega(\Gamma) = s(\Gamma) = \sup\{\lambda_\nu^1 : \nu \in \operatorname{Erg}(\Theta;\varphi)\},$$
$$-\omega(-\Gamma) = -s(-\Gamma) = \inf\{\lambda_\nu^{s_\nu} : \nu \in \operatorname{Erg}(\Theta;\varphi)\}.$$

Note that the supremum and infimum in Corollary 8.25 are both attained (see Remark 8.18). That is, there exist measures ν_+ and ν_-, points θ_+ and θ_-, vectors x_+ and x_-, and Oseledets-Lyapunov exponents $\lambda_{\nu_+}(\theta_+,x_+)$ and $\lambda_{\nu_-}(\theta_-,x_-)$ such that $\log r_+ = \lambda_{\nu_+}(\theta_+,x_-)$ and $\log r_- = \lambda_{\nu_-}(\theta_-,x_-)$ where r_+ denotes the spectral radius $r(E) = \sup\{|z| : z \in \sigma(E;C_{ND})\}$ on $C_{ND}(\Theta,\mathcal{T}\Theta)$ and r_- denotes $(r(E^{-1}))^{-1} = \inf\{|z| : z \in \sigma(E;C_{ND})\}$.

We will see in the next subsection that the set $\mathcal{AH}(e^{t\Gamma};C_{ND})$ is, in fact, exactly one annulus. Therefore, these formulas give a full description of this set.

To begin the proof of Theorems 8.22 and 8.23, let us note that, by the standard rescaling, Spectral Inclusion and the Spectral Mapping Theorems for the point and residual spectra, Theorems 8.22 and 8.23 follow from the next result where, as usual, we let $E := E^1$.

THEOREM 8.26. *Under the hypotheses of the Annular Hull Theorem, if $1 \in \sigma_{ap}(E,C(\Theta;\mathcal{T}\Theta))$, then*

$$\sigma_{ap}(\Gamma;C_{ND}(\Theta;\mathcal{T}\Theta)) \cap i\mathbb{R} \neq \emptyset.$$

In addition, if the flow $\{\varphi^t\}_{t\in\mathbb{R}}$ is aperiodic, then

$$\sigma_{ap}(\Gamma;C_{ND}(\Theta;\mathcal{T}\Theta)) \supset i\mathbb{R}.$$

Let us show that Theorems 8.22 and 8.23 follow from Theorem 8.26.

PROOF. First, note that
$$\exp(t\sigma(\Gamma; C_{ND})) \subset \sigma(e^{t\Gamma}; C_{ND})$$
by the Spectral Inclusion Theorem 2.6. Also, by our assumptions, $C_{ND}(\Theta; \mathcal{T}\Theta)$ is a closed E^t-invariant subspace of $C(\Theta; \mathcal{T}\Theta)$. Therefore,
$$\sigma_{ap}(E, C_{ND}(\Theta; \mathcal{T}\Theta)) \subset \sigma_{ap}(E, C(\Theta; \mathcal{T}\Theta)).$$

Fix $z = |z|e^{i\arg z} \in \sigma_{ap}(E, C_{ND}(\Theta; \mathcal{T}\Theta))$, and note that $z = e^{t\lambda}$ for $\lambda = \log|z| + i\arg z$. Also, define a new cocycle by $\tilde{\Phi}^t(\theta) = e^{-t\lambda}\Phi^t(\theta)$ and let $\tilde{\Gamma}$ denote the generator of the corresponding evolution group $\{\tilde{E}^t\}_{t \in \mathbb{R}}$. With these definitions, if $z \in \sigma_{ap}(E; C_{ND}(\Theta; \mathcal{T}\Theta))$, then $1 \in \sigma_{ap}(\tilde{E}; C_{ND})$; and therefore, $1 \in \sigma_{ap}(\tilde{E}; C(\Theta; \mathcal{T}\Theta))$.

By an application of Theorem 8.26 to $\tilde{E}$, if the flow $\{\varphi^t\}_{t \in \mathbb{R}}$ is aperiodic, then $\sigma_{ap}(\tilde{\Gamma}; C_{ND}(\Theta, \mathcal{T}\Theta)) \supseteq i\mathbb{R}$. Since $\sigma_{ap}(\Gamma) = \lambda + \sigma_{ap}(\tilde{\Gamma})$, it follows that $\lambda \in \sigma_{ap}(\Gamma; C_{ND}(\Theta, \mathcal{T}\Theta))$, as required in Theorem 8.22.

For an arbitrary flow $\{\varphi^t\}_{t \in \mathbb{R}}$, Theorem 8.26 gives the existence of a real number ξ such that $i\xi \in \sigma_{ap}(\tilde{\Gamma}; C_{ND}(\Theta; \mathcal{T}\Theta))$. Thus, $\lambda + i\xi \in \sigma_{ap}(\Gamma; C_{ND}(\Theta; \mathcal{T}\Theta))$ and
$$z = e^{t\lambda} = e^{t(\lambda + i\xi)}e^{-ti\xi} \in \mathcal{AH}(\exp(t\sigma_{ap}(\Gamma; C_{ND}(\Theta; \mathcal{T}\Theta)))),$$
as required in Theorem 8.23. □

Let us now begin the proof of Theorem 8.26. Recall that, by the Mañe Lemma 6.29, if $1 \in \sigma_{ap}(E; C(\Theta; \mathcal{T}\Theta))$, then there is a point $\bar{\theta} \in \Theta$ (the Mañe point) and a vector $\bar{x} \in \mathcal{T}_{\bar\theta}\Theta$ with $|\bar{x}| = 1$ (the Mañe vector) such that
$$\sup_{t \in \mathbb{R}} |\Phi^t(\bar{\theta})\bar{x}| < \infty. \tag{8.28}$$

Also, recall that the prime period function p is lower semicontinuous.

Suppose that $\bar\theta$ is a Mañe point as in display (8.28). There are two cases to consider.

Case 1. *Uniformly bounded periods.* There is a constant $\bar{p} \in \mathbb{R}_+$ and a neighborhood $U \ni \bar\theta$ such that $p(\eta) \leq \bar{p}$ for all $\eta \in U$.

Case 2. *Unbounded periods.* For every $\bar{p} \in \mathbb{R}_+$ and every neighborhood $U \ni \bar\theta$, there is an $\eta \in U$ such that $p(\eta) \geq \bar{p}$.

These cases will be considered, respectively, in Subsection 8.2.1.3 and Subsection 8.2.1.4 below. Our strategy is as follows. Using the Mañe point $\bar\theta$ and the Mañe vector $\bar{x}$ as in (8.28), we will choose an appropriate point $\theta_0 \in \Theta$ near $\bar\theta$ and a vector $x_0 \in \mathcal{T}_{\theta_0}\Theta$. Then, we define a divergence free vector field f supported in a sufficiently small neighborhood D of θ_0 that has "constant value x_0" in a smaller neighborhood $B \subset D$ of θ_0. Using f, we will construct a divergence free approximate eigenfunction $g \in C_{ND}(\Theta; \mathcal{T}\Theta)$ for Γ by integrating the function $t \mapsto E^t f(\theta)$ along sufficiently long segments of the trajectories of $\{\varphi^t\}_{t \in \mathbb{R}}$ through the points $\theta \in D$ (cf. formula (6.34)). Finally, we will estimate $\|\Gamma g - i\xi g\|_\infty$ from above and $\|g\|_\infty$ from below to show that g is indeed an approximate eigenfunction for Γ, cf. the proof of the Spectral Mapping Theorem 6.30.

8.2.1.2. *Preliminary results.* The following lemma is used to construct $B \subset D$ and f as indicated above.

LEMMA 8.27. *Suppose that $\theta_0 \in \Theta$ and $x_0 \in \mathcal{T}_{\theta_0}\Theta$. If numbers $\epsilon > 0$ and $\delta > 0$ are given, then there are neighborhoods B and D of θ_0 such that $B \subset D$ and* diam $D < \epsilon$ with the following additional property: There exists a smooth bump-function $\alpha : \Theta \to [0,1]$ with $\alpha(\theta) = 1$ for $\theta \in B$, and $\alpha(\theta) = 0$ for $\theta \notin D$ together with a continuously differentiable vector field supported in $D \setminus B$ such that

(1) the vector field $f(\theta) = \alpha(\theta)x_0 + f_0(\theta)$ is divergence free and has value x_0 in B, and
(2) $\|f_0\|_\infty \leq \delta$.

PROOF. The geometrical idea of the proof is the following: Choose positive numbers $a < b < \epsilon$. There is a thin and long ellipsoid B centered at θ_0 whose longest axis, directed along x_0, has length $b/2$ while all its other axes have length $a/2$. The ellipsoid B is contained in a longer ellipsoid D whose two longest axes, which are directed, respectively, along x_0 and some direction perpendicular to x_0, both have length b, while all of its other axes have length a.

For a function α as in the statement of the lemma, one can compensate for the nonzero divergence of $\alpha(\cdot)x_0$ by taking an appropriate vector field f_0 whose norm satisfies the estimate $\|f_0\|_\infty \leq C \cdot a/b$ for some constant $C > 0$ that does not depend on a and b. To see this, imagine a fluid flowing out from the top of a thin vertical pipe B, then slowly recirculating to the bottom of B through the much wider pipe D. The lemma is proved by taking a/b sufficiently small. To make this argument rigorous, a lengthy proof seems to be required. It starts with the following simple proposition. Let $n := \dim \Theta$.

PROPOSITION 8.28. *If $\theta_0 \in \Theta$, then there is a coordinate chart at θ_0 with coordinate functions $(\theta_1, \ldots, \theta_n)$, such that the local representation of the volume element in Θ is just the usual volume $d\theta_1 \wedge \ldots \wedge d\theta_n$. Moreover, if $x_0 \in \mathcal{T}_{\theta_0}\Theta$, then the coordinates can be chosen so that the local representative of x_0 is $|x_0|\frac{\partial}{\partial \theta_2}$.*

PROOF. Let $\eta_1, \ldots, \eta_n$ denote local coordinates at θ_0. Clearly, there is a non-vanishing density function $\rho : \mathbb{R}^n \to \mathbb{R}$ such that the volume element is given by

$$\rho(\eta_1, \ldots, \eta_n) d\eta_1 \wedge \ldots \wedge d\eta_n.$$

Let us seek new coordinates in the form

$$\eta_1 = \chi(\theta_1, \ldots, \theta_n), \quad \eta_2 = \theta_2, \ldots, \eta_n = \theta_n$$

where the volume element is given by

$$\rho(\chi(\theta_1, \ldots, \theta_n), \theta_2, \ldots, \theta_n) \frac{\partial \chi}{\partial \theta_1}(\theta_1, \ldots, \theta_n) d\theta_1 \wedge \ldots \wedge d\theta_n.$$

There is a smooth function χ defined in a neighborhood of the origin in $\mathbb{R}^n$, such that

$$\frac{\partial \chi}{\partial \theta_1}(0, \ldots, 0) \neq 0, \quad \frac{\partial \chi}{\partial \theta_1}(\theta_1, \ldots, \theta_n) = \rho(\chi(\theta_1, \ldots, \theta_n), \theta_2, \ldots, \theta_n)^{-1}.$$

The first condition together with the Implicit Function Theorem implies the change of coordinates is invertible; the second condition ensures that the volume element in the new coordinates has the desired form.

For the second statement of the proposition, note that the volume element is invariant under rigid rotations of Euclidean space. □

A coordinate chart, as in the Proposition 8.28, is called *adapted* to the volume on Θ and the vector x_0. We make the following observation: If $f = (f_1, \ldots, f_n)$ is a vector field defined in an adapted coordinate chart, then

$$\text{div } f = \sum_{i=1}^{n} \frac{\partial f_i}{\partial \theta_i}. \tag{8.29}$$

Returning to the proof of Lemma 8.27, consider $\theta_0 \in \Theta$ and $x_0 \in \mathcal{T}_{\theta_0}\Theta$, and let $(\theta_1, \ldots, \theta_n)$ denote local coordinates at θ_0 adapted to the volume on Θ and the vector x_0. In particular, this means that x_0 is parallel to the direction vector $\partial/\partial\theta_2$.

For an adapted coordinate system defined in a coordinate ball around θ_0 of diameter $\epsilon > 0$ and for $a, b \in \mathbb{R}$ such that $0 < a < b < \epsilon/8$, let

$$D_{a,b} := \{(\theta_1, \ldots, \theta_n) : |\theta_j| \leq 4b, j = 1, 2, |\theta_j| \leq a, j = 3, \ldots, n\},$$

$$B_{a,b} := \{(\theta_1, \ldots, \theta_n) : |\theta_j| \leq \frac{a}{2}, j = 1, 3, \ldots, n, |\theta_2| \leq \frac{b}{2}\}.$$

Note that the closure of $B_{a,b}$ is contained in $D_{a,b}$. We say there is an (a,b) *divergence free extension* of the vector x_0 at θ_0 if there is a smooth bump-function $\alpha : \mathbb{R}^n \to [0,1]$ with $\alpha(\theta) = 1$ for $\theta \in B_{a,b}$ and $\alpha(\theta) = 0$ for $\theta \notin D_{a,b}$, and a continuously differentiable vector field f_0 with support in $D_{a,b}$ such that

(i) the vector field $f(\theta) := \alpha(\theta_0)|x_0|\frac{\partial}{\partial\theta_2} + f_0(\theta)$ is divergence free and has value $|x_0|\frac{\partial}{\partial\theta_2}$ in $B_{a,b}$, and

(ii) there is a number $C > 0$ independent of a, b such that $\|f_0\|_\infty \leq Ca/b$.

PROPOSITION 8.29. *Every tangent vector x_0 on Θ has an (a,b) divergence free extension.*

PROOF. Let $x_0 \in \mathcal{T}_{\theta_0}\Theta$. We will first prove the lemma for the case $n = 2$. To construct a vector field $f = W$ with the required properties in the (θ_1, θ_2) coordinate plane, consider the curves given by

$$\frac{\theta_1^4}{a^4} + \frac{\theta_2^4}{b^4} = 1, \quad \frac{\theta_1^4}{a^4} + \frac{\theta_2^4}{b^4} = 2.$$

Let $\rho : \mathbb{R} \to [0,1]$ denote a smooth function such that $\rho(t) = 1$ for $t \leq 1$, $\rho(t) = 0$ for $t \geq 2$, and $|\rho'(t)| \leq 3$, $t \in \mathbb{R}$. Also, define the sets

$$R = \{(\theta_1, \theta_2) : |\theta_1| \leq 2^{1/4}a, |\theta_2| \leq 2^{1/4}b\},$$

$$S^+ = \{(\theta_1, \theta_2) : \frac{(\theta_1 - 2^{1/4}a)^4}{a^4} + \frac{\theta_2^4}{b^4} \leq 2, \theta_1 \geq 2^{1/4}a\},$$

$$S^- = \{(\theta_1, \theta_2) : \frac{(\theta_1 + 2^{1/4}a)^4}{a^4} + \frac{\theta_2^4}{b^4} \leq 2, \theta_1 \leq -2^{1/4}a\},$$

and the functions $\nu : \mathbb{R}^2 \to \mathbb{R}$, $\mu : \mathbb{R}^2 \to \mathbb{R}$, and $\eta^\pm : \mathbb{R} \to \mathbb{R}$ by

$$\nu(\theta_1, \theta_2) = \rho\left(\frac{\theta_1^4}{a^4} + \frac{\theta_2^4}{b^4}\right),$$

$$\mu(\theta_1, \theta_2) = -\frac{4\theta_2^3 |x_0|}{b^4} \int_0^{\theta_1} \rho'\left(\frac{s^4}{a^4} + \frac{\theta_2^4}{b^4}\right) ds,$$

$$\eta^+(\tau) = -\frac{4|x_0|}{b^4} \int_0^{2^{1/4}a} \rho'\left(\frac{s^4}{a^4} + \frac{\tau}{b^4}\right) ds,$$

$$\eta^-(\tau) = \frac{4|x_0|}{b^4} \int_{-2^{1/4}a}^0 \rho'\left(\frac{s^4}{a^4} + \frac{\tau}{b^4}\right) ds,$$

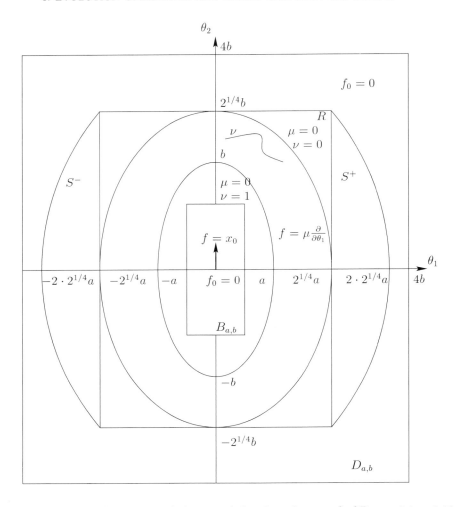

FIGURE 2. The configuration of the sets defined in the proof of Proposition 8.29

The vector field f_0 is defined in R by $\mu(\theta_1, \theta_2)\partial/\partial\theta_1$, in S^+ by

$$\eta^+((\theta_1 - 2^{1/4}a)^4 + \theta_2^4)\theta_2^3 \frac{\partial}{\partial \theta_1} - \eta^+((\theta_1 - 2^{1/4}a)^4 + \theta_2^4)(\theta_1 - 2^{1/4}a)^3 \frac{\partial}{\partial \theta_2},$$

in S^- by

$$\eta^-((\theta_1 + 2^{1/4}a)^4 + \theta_2^4)\theta_2^3 \frac{\partial}{\partial \theta_1} - \eta^-((\theta_1 + 2^{1/4}a)^4 + \theta_2^4)(\theta_1 + 2^{1/4}a)^3 \frac{\partial}{\partial \theta_2},$$

and f_0 is defined to vanish on the complement of $R \cup S^+ \cup S^-$.

We will complete the proof for $n = 2$ by showing that the vector field $f = W$ where

$$W(\theta_1, \theta_2) := \nu(\theta_1, \theta_2)|x_0|\frac{\partial}{\partial \theta_2} + f_0(\theta_1, \theta_2)$$

is the required extension of x_0.

A direct computation using formula (8.29) shows $\operatorname{div} W = 0$ in the coordinate chart. Also, using the definition of ρ, let us note that the support of W is in $D_{a,b}$. To show W is C^1, just observe that f_0 and each of its first partial derivatives is continuous on the lines $\theta_1 = \pm 2^{1/4} a$ and on the boundary of $R \cup S^+ \cup S^-$. (We remark that additional smoothness can be obtained, if desired, by using the function $\theta_1^k / a^k + \theta_2^k / b^k$ with k a sufficiently large positive integer in place of the choice $k = 4$ used here.)

To obtain the required norm bound, let $G(\theta_1, \ldots, \theta_n)$ denote the matrix of the components g_{ij} of the Riemannian metric in the adapted coordinates. Then, using the usual estimate for the integral in the definition of μ, we estimate the norm of f_0 in R by

$$\|G\| \sup |\mu(\theta_1, \theta_2)| \leq \|G\|(4|x_0|2^{3/4} b^3 / b^4) 3(2^{1/4} a) \leq C_1 a/b$$

where the constant C_1 does not depend on a or b. The norm of f_0 in $S^\pm$ is estimated in a similar manner. For example, in S^+, the upper bound is

$$\|G\| \sup \left| \eta^+ ((\theta_1 - 2^{1/4} a)^4 + \theta_2^4) \right| ((2^{1/4} b)^6 + (2^{1/4} a)^6)^{1/2} \leq C_2 a/b.$$

To prove the proposition for the case $n \geq 3$, we will show how to extend the vector field W defined above to a vector field f on $\mathbb{R}^n$ with the required properties. To do this, let $\chi : \mathbb{R} \to [0, 1]$ denote a smooth bump-function such that $\chi(t) = 1$ for $|t| \leq \dfrac{a}{2}$ and $\chi(t) = 0$ for $|t| \geq a$, and define

$$\Psi(\theta_3, \ldots, \theta_n) = \prod_{j=3}^n \chi(\theta_j).$$

The required vector field f is given by

$$f(\theta_1, \ldots, \theta_n) = \Psi(\theta_3, \ldots, \theta_n) W(\theta_1, \theta_2).$$

The fact that the new vector field f is continuously differentiable, agrees with $|x_0| \partial/\partial \theta_2$ in $B_{a,b}$, and is supported in $D_{a,b}$ is clear. Also, since the range of Ψ is the unit interval, the norm bound on ΨW is the same as the norm bound on W. To complete the proof, we must show that $\operatorname{div} f = 0$. But, since the vector field f has nonzero components only in the first two coordinate directions,

$$\operatorname{div} f(\theta_1, \ldots, \theta_n) = \Psi(\theta_3, \ldots, \theta_n) \operatorname{div} W(\theta_1, \theta_2) = 0,$$

as required. This completes the proof of Proposition 8.29. $\square$

To finish the proof of Lemma 8.27, take a/b so small that $Ca/b < \delta$, see statement (ii) of the definition of divergence free extension given on page 279. Then, $B = B_{a,b}$ and $D = D_{a,b}$, together with α and f_0 from statement (i) of the definition, give the desired objects in Lemma 8.27. $\square$

To carry out the estimates for the approximate eigenfunction g mentioned above and needed for the proof of Theorem 8.26, we must control the sojourn time of a trajectory segment that passes through the sets D and B constructed in Lemma 8.27. To do this, let us first recall the definition of $R(\theta)$ in (6.35) and Lemma 6.31. Then, for an open set U containing $\theta \in \Theta$ and $s > 0$, define

(8.30) $\quad R_{s,U}(\theta) := \{ t \in \mathbb{R} : |t| \leq s, \varphi^t \theta \in U \}, \quad M_{s,U}(\theta) := \operatorname{mes} R_{s,U}(\theta).$

LEMMA 8.30. *Suppose that $n = \dim \Theta \geq 3$. There exists a constant $K > 0$ with the following property: If $\theta_0 \in \Theta$ has prime period $p(\theta_0) > 0$, the open set V contains θ_0, and $0 < s < p(\theta_0)/8$, then there are open sets $B \subset D$ with all properties listed in Lemma 8.27 such that*

(8.31) $$M_{s,D}(\theta) \leq K M_{s,B}(\theta_0)$$

for every $\theta \in \Theta$.

PROOF. We start with the following proposition.

PROPOSITION 8.31. *For every $\theta_0 \in \Theta$ and $s \in (0, p(\theta_0)/8)$, there exists $\epsilon > 0$ such that if U is a neighborhood of θ_0 with diam $U < \epsilon$, $\eta \in U$, and $s < |t| < 2s$, then $\varphi^t \eta \notin U$.*

PROOF. Suppose that the proposition is false. Choose $U_k \ni \theta_0$ with diam $U_k \to 0$ and $\eta_k \in U_k$ together with t_k such that $s < |t_k| < 2s$ but $\varphi^{t_k} \eta_k \in U_k$. By compactness, we may assume $t_k \to t_*$ for some t_* such that $s \leq |t_*| \leq 2s$. By continuity, $\varphi^{t_k} \eta_k \to \varphi^{t_*} \theta_0$ since $t_k \to t_*$ and $\eta_k \to \theta_0$. Also $\varphi^{t_k} \eta_k \to \theta_0$ since $\varphi^{t_k} \eta_k \in U_k$ and diam $U_k \to 0$. Hence, $\varphi^{t_*} \theta_0 = \theta_0$. But $|t_*| > 0$ and $|t_*| \leq p(\theta_0)/4$ with $p(\theta_0)$ being the *prime* period, in contradiction. □

Returning to the proof of Lemma 8.30, fix $\epsilon > 0$ as in Proposition 8.31 and choose $U \ni \theta_0$ with diam $U < \epsilon$. We claim that

(8.32) $$R_{s,U}(\eta) = R_{2s,U}(\eta) \quad \text{for each} \quad \eta \in U.$$

Clearly $R_{s,U}(\eta) \subseteq R_{2s,U}(\eta)$. On the other hand, if $t \in R_{2s,U}(\eta)$, then $|t| \leq s$ since $\varphi^t \eta \notin U$ for $|t| \in (s, 2s)$ by Proposition 8.31. Thus, we have that $t \in R_{s,U}(\eta)$.

Let us suppose that there exists a constant $K > 0$ such that if $B \subset D \subset U$ (B and D are as in Lemma 8.27 and dim $U < \epsilon$), then the following statement holds:

(8.33) $$M_{s,D}(\eta) \leq K M_{s,B}(\theta_0) \quad \text{for all} \quad \eta \in D.$$

We claim that this implies the required inequality (8.31) *for all $\theta \in \Theta$*. Indeed, using equation (8.32) for the case $U = D$, it follows that

(8.34) $$M_{s,D}(\eta) = M_{2s,D}(\eta) \quad \text{for each} \quad \eta \in D.$$

If $\theta \in \Theta$ is such that $D \cap \{\varphi^t \theta : |t| \leq s\} = \emptyset$, then $M_{s,D}(\theta) = 0$ and (8.31) does hold for this θ. For $\theta \in \Theta$ such that $D \cap \{\varphi^t \theta : |t| \leq s\} \neq \emptyset$, fix $\eta \in D \cap \{\varphi^t \theta : |t| \leq s\}$. In other words, fix t_1 with $|t_1| \leq s$ such that $\eta = \varphi^{t_1} \theta$. Let us note that

(8.35) $$R_{s,D}(\theta) \subset t_1 + R_{2s,D}(\eta).$$

Indeed, if $t \in R_{s,D}(\theta)$, then D contains $\varphi^t \theta = \varphi^{t-t_1} \eta$ and $|t| \leq s$. Thus, $\tau = t - t_1 \in R_{2s,D}(\eta)$ because $\varphi^\tau \eta \in D$ and $|\tau| \leq 2s$. By (8.34), (8.35), and (8.33), we have

$$M_{s,D}(\theta) = \text{mes} R_{s,D}(\theta) \leq \text{mes}\{t_1 + R_{2s,D}(\eta)\} = M_{2s,D}(\eta)$$
$$\leq \max_{\eta \in D} M_{2s,D}(\eta) = \max_{\eta \in D} M_{s,D}(\eta) \leq K M_{s,B}(\theta_0).$$

Therefore, the inequality (8.31) holds for all $\theta \in \Theta$.

To complete the proof of Lemma 8.30, we will determine a constant K and construct $B \subset D$ with the properties listed in Lemma 8.27 such that the statement in display (8.33) holds. This will require several steps.

Step 1. We will work in an adapted coordinate system at θ_0 with coordinate functions $(\theta_1, \ldots, \theta_n)$, and use it to determine the required sets B and D for appropriate a and b of the form

$$B := \left\{ \theta : \frac{\theta_1^4}{(a/2)^4} + \frac{\theta_2^4}{(b/2)^4} + \sum_{j=3}^{4} \frac{\theta_j^4}{(a/2)^4} \leq 1 \right\},$$

$$D := \left\{ \theta : \frac{\theta_1^4}{(4nb)^4} + \frac{\theta_2^4}{(4nb)^4} + \sum_{j=3}^{4} \frac{\theta_j^4}{(na)^4} \leq 1 \right\}.$$

If $a < b$, then, clearly, $B \subset B_{a,b}$ and $D_{a,b} \subset D$. We will show that there is a constant K such that, for some choice of a, b, and a/b all sufficiently small, the statement in display (8.33) is valid for the corresponding set D.

For the following useful auxiliary constructions, let us recall that the flow $\{\varphi^t\}_{t \in \mathbb{R}}$ is generated by the divergence free vector field $\mathbf{v}$.

For each $\delta > 0$, let S_δ denote a cross-section for $\mathbf{v}$ at the origin of the coordinate system; that is, S_δ contains θ_0 and has Riemannian diameter such that diam $S_\delta < \delta$. Also, define $\Sigma_\delta := \{\varphi^t \sigma : \sigma \in S_\delta, |t| \leq \delta\}$. Consider the local representation of $\mathbf{v}$ given by $\sum_{i=1}^{n} v_i(\theta) \partial/\partial \theta_i$. By a rigid rotation, if necessary, we will arrange the adapted coordinates so that $v_1(0) = 0$. Also, we define V by

$$V(\theta) := \sum_{i=3}^{n} v_i^4(\theta), \quad \theta \in \Sigma_\delta,$$

and the number

$$M_\delta := \max \left\{ \max_{\theta \in \Sigma_\delta} |v_1(\theta)|, \max_{\theta \in \Sigma_\delta} \left(\sum_{i=3}^{n} v_i^4(\theta) \right)^{1/4} \right\}.$$

If $V(\theta_0) = 0$, then $\lim_{\delta \to 0} M_\delta = 0$. In case $V(\theta_0) = 0$ and $M_\delta \neq 0$ for every δ, we will only consider a and b such that

(8.36) $$a = b \cdot (M_\delta)^{1/2}.$$

Of course, even under the restriction just imposed, a, b, and a/b can each be chosen arbitrarily small. If $V(\theta_0) \neq 0$ or if $V(\theta_0) = 0$ and $M_\delta \equiv 0$ for all sufficiently small δ, then we ignore this restriction.

Step 2. For each δ, Proposition 8.31 and the definition of Σ_δ together imply that there is an open ball $A_\delta \subset \Sigma_\delta$ at the origin such that, for each $\theta \in A_\delta$ and for each time t with $|t| > \delta$, the point $\varphi^t \theta$ is not in Σ_δ. If $\delta > 0$ is given, choose a and b as required in (8.36) and so small that D is in A_δ. If $\theta \in D$, let θ' denote the point on ∂D, the boundary of D, where the segment of the trajectory $\{\varphi^t \theta : |t| \leq N\}$ first enters D, and let $\theta'' = \varphi^{t_D} \theta'$ denote the point of ∂D where the segment of the trajectory $\{\varphi^t \theta : |t| \leq N\}$ last exits D. Clearly, $M_{N,D}(\theta) \leq t_D$.

We use the Mean Value Theorem for integrals on the j-th component of the vector field $\mathbf{v}$ to obtain a point $\xi^j \in \Sigma_\delta$ such that

$$\theta''_j = \theta'_j + \int_0^{t_D} v_j(\varphi^t \theta) \, dt = \theta'_j + t_D v_j(\xi^j).$$

For each $j = 1, \ldots, n$, let $v_j^* = v_j(\xi^j)$. Also, as an abbreviation, define $\alpha_j = 4nb$ for $j = 1, 2$, and $\alpha_j = na$ for $j = 3, \ldots, n$.

Since θ' and θ'' both belong to ∂D, we have

$$\sum_{i=1}^n \left(\frac{\theta'_i}{\alpha_i}\right)^4 = 1 \quad \text{and} \quad \sum_{i=1}^n \left(\frac{\theta'_i + t_D v^*_i}{\alpha_i}\right)^4 = 1.$$

Using a standard inequality for the norm $\|(\gamma_i)\| = \left(\sum_i |\gamma_i|^4\right)^{1/4}$, we find that

$$1 = \sum_{i=1}^n \left(\frac{\theta'_i + t_D v^*_i}{\alpha_i}\right)^4 \geq \left(\sum_{i=1}^n \left(\frac{t_D v^*_i}{\alpha_i}\right)^4\right)^{1/4} - \left(\sum_{i=1}^n \left(\frac{\theta'_i}{\alpha_i}\right)^4\right)^{1/4}$$

$$= \left(\sum_{i=1}^n \left(\frac{t_D v^*_i}{\alpha_i}\right)^4\right)^{1/4} - 1.$$

This computation yields the estimate

$$(8.37) \qquad t_D^4 \sum_{i=1}^n \left(\frac{v^*_i}{\alpha_i}\right)^4 \leq 2^4.$$

Step 3. Consider the time t_B when the segment of the trajectory $\{\varphi^t \theta_0 : |t| \leq N\}$ first leaves B. Clearly, $M_{N,B}(\theta_0) \geq t_B$.

Recall that in our local coordinates, θ_0 resides at the origin and define $\tilde\theta = \varphi^{t_B}(\theta_0)$. As in Step 2, by the the Mean Value Theorem, there is some $\eta^j \in \Sigma_\delta$ such that

$$\tilde\theta_j = \int_0^{t_B} v_j(\varphi^t(\theta_0))\, dt = t_B v_j(\eta^j).$$

Define $\tilde v_j := v_j(\eta^j)$, the numbers $\beta_j = a/2$ for $j = 1, 3, \ldots, n$ and $\beta_2 = b/2$. Since $\tilde\theta \in \partial B$, we have

$$(8.38) \qquad t_B^4 \sum_{j=1}^n \left(\frac{\tilde v_j}{\beta_j}\right)^4 = 1.$$

Step 4. In accordance with the previous notation, we define

$$\tilde V := \sum_{i=3}^n \tilde v_i^4, \quad V^* := \sum_{i=3}^n (v^*_i)^4,$$

and use (8.37)–(8.38) to obtain the estimate $(t_D/t_B)^4 \leq 2^{16} n^4 \cdot d$ where

$$d := \frac{\tilde v_1^4 + \left(\frac{a}{b} \tilde v_2\right)^4 + \tilde V}{\left(\frac{a}{b} v^*_1\right)^4 + \left(\frac{a}{b} v^*_2\right)^4 + 4^4 V^*}.$$

We will show that for all sufficiently small $\delta > 0$, there are some choices of a and b such that $d < 2$. There are several cases.

Case 1. If $V(\theta_0) \neq 0$, then $\lim_{\delta \to 0} d = 4^{-4}$.

Case 2. If $V(\theta_0) = 0$ and $M_\delta \equiv 0$ for all sufficiently small δ, then, since $v_2(\theta_0) \neq 0$, we have $\lim_{\delta \to 0} d = 1$.

Case 3. Suppose $V(\theta_0) = 0$ and $M_\delta \neq 0$. Also, note that we still have $\lim_{\delta \to 0} M_\delta = 0$. Moreover, by the restriction we imposed in (8.36), we have that

$$ba^{-1}\tilde v_1^{1/2} \leq 1, \quad ba^{-1}\tilde V^{1/8} \leq 1.$$

Therefore, in this case,
$$d = \frac{(ba^{-1}\tilde{v}_1)^4 + \tilde{v}_2^4 + (ba^{-1})^4 \tilde{V}}{(v_1^*)^4 + (v_2^*)^4 + 4^4(ba^{-1})^4 V^*} \leq \frac{\tilde{v}_1^2 + \tilde{v}_2^4 + \tilde{V}^{1/2}}{(v_2^*)^4}.$$

Passing to the limit as $\delta \to 0$, the last expression converges to 1, and the proof of Lemma 8.30 is complete. $\square$

We are ready to prove Theorem 8.26; that is, using a Mañe point $\bar{\theta}$ and Mañe vector $\bar{x}$ as in (8.28), we will construct the approximate eigenfunction for Γ. Recall, see page 277, that there are two cases to consider: uniformly bounded periods and unbounded periods.

8.2.1.3. *Case 1. Uniformly bounded periods.* Consider the Mañe point $\bar{\theta}$ as in (8.28) and let the open set $U \ni \bar{\theta}$ be as in Case 1, see page 277. In particular, $\bar{\theta}$ is periodic. Using the lower semicontinuity of the prime period function p, we can and will choose U sufficiently small so that we have one of the following alternatives:

Subcase 1.1: $p(\bar{\theta}) = 0$.
Subcase 1.2: $p(\theta) > 0$ for all $\theta \in U$.

For Subcase 1.1, we have $\varphi^t \bar{\theta} = \bar{\theta}$ for all $t \in \mathbb{R}$. Since $\{\Phi^t\}_{t \in \mathbb{R}}$ is a cocycle, $\{\Phi^t(\bar{\theta})\}_{t \in \mathbb{R}}$ is a group. Let $A(\bar{\theta})$ denote the matrix that generates this group: $\Phi^t(\bar{\theta}) = e^{tA(\bar{\theta})}$, $t \in \mathbb{R}$, and note that the bound (8.28) implies $\sigma(e^{tA(\bar{\theta})}) \cap \mathbb{T} \neq \emptyset$, for $t \neq 0$. By the spectral mapping theorem for matrices, there exists $\xi \in \mathbb{R}$ so that $i\xi \in \sigma(A(\bar{\theta}))$. For a notational convenience, define $\theta_0 = \bar{\theta}$. Choose $x_0 \in \mathcal{T}_{\theta_0}\theta$ with $|x_0| = 1$ so that $A(\theta_0)x_0 = i\xi x_0$, and note that $\Phi^t(\theta_0)x_0 = e^{i\xi t}x_0$ for $t \in \mathbb{R}$. We will construct a sequence of approximate eigenfunctions for Γ that corresponds to the approximate eigenvalue $i\xi$.

Fix $N \in \mathbb{N}$ and take a smooth function $\gamma : \mathbb{R} \to [0,1]$ with $\operatorname{supp}\gamma \subset [-N,N]$ such that $|\gamma'(t)| \leq 2/N$ for all $t \in \mathbb{R}$, and $\gamma(t) \geq 1/4$ for $|t| \leq N/2$. Use the continuity of $\Phi(\cdot)$ to find an open set $V \ni \theta_0$ with $V \subset U$ where U is the neighborhood mentioned in Case 1, see page 277, so that

(8.39) $$\sup\{\|\Phi^t(\varphi^{-t}\theta) - \Phi^t(\theta_0)\| : \theta \in V, |t| \leq 3N\} \leq 1.$$

Recall that, in accordance with Case 1 on page 277, $p(\eta) \leq \bar{p}$ for each $\eta \in U$. Also, use the continuity of φ to find $\epsilon > 0$ so small that, for every open set $D \ni \theta_0$ with $\operatorname{diam} D < \epsilon$, we have

$$D \subset U \quad \text{and} \quad \cup_{0 \leq t \leq \bar{p}} \varphi^t(D) \subset V.$$

Fix a positive number δ such that $\delta < 1/\sup\{\|E^t\|_{\mathcal{L}(C(\Theta;\mathcal{T}\Theta))} : |t| \leq 3N\}$. Apply Lemma 8.27 for this ϵ and δ to find $B \subset D$ and $f = \alpha(\cdot)x_0 + f_0 \in C_{ND}(\Theta;\mathcal{T}\Theta)$ where $\alpha(\theta) \leq 1$ for $\theta \in \Theta$, and $\alpha(\theta_0) = 1$, as specified in this lemma.

Define a function g by

(8.40) $$g(\theta) = \int_{-\infty}^{\infty} \gamma(t) e^{-i\xi t} (E^t f)(\theta) \, dt, \quad \theta \in \Theta.$$

Since, by our assumptions, the evolution group $\{E^t\}_{t \in \mathbb{R}}$ preserves $C_{ND}(\Theta;\mathcal{T}\Theta)$, we clearly have that $g \in C_{ND}(\Theta;\mathcal{T}\Theta)$. Since $\operatorname{supp} f \subset D$, we have $\operatorname{supp} g \subset V$. Also, since $\operatorname{supp} \gamma \subset [-N,N]$, integration in (8.40) is unchanged if restricted to $|t| \leq N$. Because $\Gamma g = \frac{d}{dt} E^t g|_{t=0}$, it follows directly that

(8.41) $$\Gamma g(\theta) - i\xi g(\theta) = -\int_{-\infty}^{\infty} \gamma'(t) e^{-i\xi t} (E^t f)(\theta) \, dt.$$

Our choice of x_0 and γ allows us to estimate $\|g\|_\infty$ from below as follows:

(8.42)
$$|g(\theta_0)| = \left|\int_{-\infty}^{\infty} \gamma(t) e^{-i\xi t} \Phi^t(\theta_0) x_0 \, dt\right| = |x_0| \int_{-\infty}^{\infty} \gamma(t) \, dt \geq \int_{-N/2}^{N/2} \frac{1}{4} \, dt = \frac{N}{4}.$$

If $\theta \in V$, the norm of the integral in the right hand side of (8.41) is bounded above by $I(\theta) + J(\theta)$ where

$$I(\theta) := \int_{-\infty}^{\infty} |\gamma'(t)| \|\Phi^t(\varphi^{-t}\theta) x_0| \alpha(\varphi^{-t}\theta) \, dt,$$

$$J(\theta) := \int_{-\infty}^{\infty} |\gamma'(t)| \|E^t f_0\|_\infty \alpha(\varphi^{-t}\theta) \, dt.$$

Using the choice of x_0 and (8.39), we have the estimates

$$I(\theta) \leq \int_{-\infty}^{\infty} |\gamma'(t)| |\Phi^t(\theta_0) x_0| \alpha(\varphi^{-t}\theta) \, dt$$
$$+ \int_{-\infty}^{\infty} |\gamma'(t)| |[\Phi^t(\varphi^{-t}\theta) - \Phi^t(\theta_0)] x_0| \alpha(\varphi^{-t}\theta) \, dt \leq 2 \int_{-\infty}^{\infty} |\gamma'(t)| \, dt.$$

Also, the inequality (2) in Lemma 8.27 and our choice of δ imply that $J(\theta) \leq \int_{-\infty}^{\infty} |\gamma'(t)| \, dt$. In view of these estimates and (8.42), let us use the inequality $|\gamma'(t)| \leq 2/N$ to obtain

$$\|\Gamma_g - i\xi g\|_\infty \leq 3 \int_{-\infty}^{\infty} |\gamma'(t)| \, dt \leq 48 \|g\|_\infty / N.$$

Thus, $i\xi \in \sigma_{ap}(\Gamma; C_{ND}(\Theta; \mathcal{T}\Theta))$, and the proof of Theorem 8.26 is complete for Subcase 1.1 on page 285.

Consider Subcase 1.2. The neighborhood U does not contain rest points. By the semicontinuity of the prime period function p we can and will assume that there is a number $p_0 > 0$ such that $p(\theta) \geq p_0$ for all $\theta \in U$.

We note that $\sigma(\Phi^{p(\bar\theta)}(\bar\theta)) \cap \mathbb{T} \neq \emptyset$. Indeed, if $\varphi^{p(\bar\theta)}(\bar\theta) = \bar\theta$, then $\Phi^{kp(\bar\theta)}(\bar\theta) = [\Phi^{p(\bar\theta)}(\bar\theta)]^k$ for $k \in \mathbb{Z}$. However, by the definition (8.28) of the Mañe point and vector, the sequence $\{|[\Phi^{p(\bar\theta)}(\bar\theta)]^k \bar x|\}_{k\in\mathbb{Z}}$ is bounded. Therefore, the matrix $\Phi^{p(\bar\theta)}(\bar\theta)$ can not be hyperbolic. Thus, using the notation $\theta_0 := \bar\theta$, there exist some $x \in \mathcal{T}_{\theta_0}\Theta$ and $\xi \in \mathbb{R}$ such that

(8.43) $$\Phi^{p(\theta_0)}(\theta_0) x = e^{i\xi} x, \quad |x| = 1.$$

As a result, starting from the assumption that $1 \in \sigma_{ap}(E; C_{ND}(\Theta; \mathcal{T}\Theta))$, we have found a periodic point θ_0 with period $p(\theta_0) > 0$, a neighborhood U of θ_0 consisting entirely of periodic points whose periods are uniformly bounded and separated from zero, as well as a number $\xi \in \mathbb{R}$ and a vector x as in (8.43). To complete the proof of Theorem 8.26 for Subcase 1.2, we will show that $\lambda := i\xi/p(\theta_0)$ is an approximate eigenvalue for Γ.

Let $p := p(\theta_0)$ and let $\mu_1 > 0$ be a constant such that

(8.44) $$p(\theta) \leq \mu_1 p, \quad \theta \in U.$$

Fix a natural number $N > \mu_1 p + 1$. In what follows, we use the letter μ (resp., ν) with subscripts to denote "big" (resp., "small") constants that do not depend

on N. Select (small) constants $\nu_i < 1$, $i = 0, \ldots, 5$. The required values of these constants will be determined later. Define
$$\mu_2 := 2\sup\{\|\Phi^t(\theta_0)\| : |t| \leq \mu_1 p\} \quad \text{and} \quad R := N + \mu_1 p.$$
Also, fix $s > 0$ so that $s < \min(p/8, 1/4)$ and so that the following estimates hold:

(8.45) $$\max_{|t| \leq s} \left| \exp\left(-\frac{i\xi t}{p}\right) - 1 \right| \leq \nu_4,$$

(8.46) $$\sup_{|t| \leq s} |\Phi^t(\varphi^{-t}\theta_0)x - x| \cdot \sup_{|n| \leq 3N/p} \|[\Phi^p(\theta_0)]^n\| \leq \nu_2.$$

The following fact is similar to Proposition 8.31.

PROPOSITION 8.32. *There exists $\epsilon > 0$ such that, for every open set $D \ni \theta_0$ with $\operatorname{diam} D \leq \epsilon$ and every $\theta \in D$, the following implication holds:*

If $\varphi^t \theta \in D$ and $|t| \leq R$, then $t \in \cup_{n \in \mathbb{Z}} (-s + np, s + np)$.

PROOF. Suppose not. Choose $D_k \ni \theta_0$ with $\operatorname{diam} D_k \to 0$ and $\theta_k \in D_k$. Also, choose $t_k \in \mathbb{R}$ and $|t_k| \leq R$ so that $\varphi^{t_k}\theta_k \in D_k$, but $t_k \notin \cup_{n \in \mathbb{Z}}(-s+np, s+np)$. By compactness, we may assume $t_k \to t_*$, $|t_*| \leq R$. Also, by continuity, $\varphi^{t_k}\theta_k \to \varphi^{t_*}\theta_0$; and therefore, $\varphi^{t_*}\theta_0 = \theta_0$. Since $t_* = np + \tau$ for some τ with $s \leq \tau \leq p - s$ and $n \in \mathbb{Z}$, it follows that $\varphi^{\tau}\theta_0 = \theta_0$. But $p = p(\theta_0)$ is the prime period, and we have a contradiction. □

Fix $\epsilon > 0$ as in Proposition 8.32. Choose a neighborhood $V \ni \theta_0$ with $\operatorname{diam} V < \epsilon$ so small that the following inequalities hold:

(8.47) $$\sup\{\|\Phi^t(\theta)\| : |t| \leq \mu_1 p, \theta \in V\} \leq \mu_2,$$

(8.48) $$\sup_{\theta \in V} \sup_{|t| \leq 3N} \|\Phi^t(\varphi^{-t}\theta) - \Phi^t(\varphi^{-t}\theta_0)\| \leq \nu_1.$$

Recall that $\Phi^0(\theta) = I$, for $\theta \in \Theta$ and fix $\delta > 0$ such that

(8.49) $$\delta < \nu_3,$$

(8.50) $$\delta \max_{|t| \leq s} \|\Phi^t(\varphi^{-t}\theta_0)\| \cdot \sup_{|n| \leq 3N/p} \|[\Phi^p(\theta_0)]^n\| \leq \nu_5.$$

Let $B \subset D$ and $D \subset V$ be the open sets from Lemma 8.30 with the prescribed choices of θ_0, s, V, ϵ, and δ; and let $m := M_{s,B}(\theta_0)$, see (8.30). Finally, choose a smooth function $\gamma : \mathbb{R} \to [0, 1]$ with $\operatorname{supp} \gamma \subset [-N, N]$ so that $|\gamma'(t)| \leq 2/N$, and

(8.51) $$\gamma(t) \geq 1/4 \quad \text{for} \quad |t| \leq N/2.$$

We are ready to define g, the approximate eigenfunction for Γ.

For $f = \alpha(\cdot)x + f_0$ as in Lemma 8.27, define g as follows:

(8.52) $$g(\theta) = \int_{-\infty}^{\infty} \gamma(t) e^{-\frac{i\xi t}{p}} (E^t f)(\theta) \, dt, \quad \theta \in \Theta.$$

Since f is divergence free and E^t preserves $C_{ND}(\Theta, \mathcal{T}\Theta)$ by the assumption, the vector field g is divergence free. We claim that there is a constant $\mu > 0$, independent of N, such that

(8.53) $$\left\| \Gamma g - \frac{i\xi}{p} g \right\|_{\infty} \leq \frac{\mu}{N} \|g\|_{\infty}.$$

The inequality (8.53) guarantees that $\sigma_{ap}(\Gamma) \cap i\mathbb{R} \neq \emptyset$. Thus, this result would complete the proof of Theorem 8.26 for Subcase 1.2 on page 285. The proof of (8.53) is accomplished in two steps.

Step 1. The Upper Estimate. We will show that there is a constant $\mu > 0$ such that

(8.54) $$\left\|\Gamma g - \frac{i\xi}{p}g\right\|_\infty \leq \mu \cdot m$$

for $m = M_{s,B}(\theta_0)$. A direct calculation yields

$$(\Gamma g)(\theta) - \frac{i\xi}{p}g(\theta) = -\int_{-\infty}^\infty \gamma'(t) e^{-\frac{i\xi t}{p}} (E^t f)(\theta)\, dt.$$

Recall that $f(\theta) = \alpha(\theta)x + f_0(\theta)$, $\theta \in \Theta$ and define

$$I(\theta) = \int_{-\infty}^\infty |\gamma'(t)| \|\Phi^t(\varphi^{-t}\theta)x| \alpha(\varphi^{-t}\theta)\, dt,$$

$$J(\theta) = \int_{-\infty}^\infty |\gamma'(t)| \|\Phi^t(\varphi^{-t}\theta) f_0(\varphi^{-t}\theta)|\, dt.$$

To prove (8.54), it suffices to show there is a constant $\mu > 0$ such that, for all $\theta \in \Theta$,

(8.55) $$I(\theta) \leq \mu \cdot m, \quad J(\theta) \leq \mu \cdot m.$$

For $\theta \notin \cup_{\eta \in D} \cup_{0 \leq t \leq p(\eta)} \varphi^t \eta$, we have that $\alpha(\varphi^{-t}\theta) = 0$ and $f_0(\varphi^{-t}\theta) = 0$. Therefore, $I(\theta) = J(\theta) = 0$. So, fix $\theta \in \cup_{\eta \in D} \cup_{0 \leq t \leq p(\eta)} \varphi^t \eta$ and choose $\eta \in D$ such that $\theta = \varphi^\tau \eta$. In accordance with (8.44), let us assume that $0 \leq \tau \leq \mu_1 p$.

To estimate $I(\theta)$ from above, we have the inequality

$$I(\theta) \leq \int_{-\infty}^\infty |\gamma'(t+\tau)| \|\Phi^{t+\tau}(\varphi^{-t}\eta)x| \alpha(\varphi^{-t}\eta)\, dt.$$

Since $\tau \leq \mu_1 p$ and $\operatorname{supp} \gamma' \subset [-N, N]$, the integration is unchanged if restricted to $|t| \leq R := \mu_1 p + N \leq 3N$. Also, $\eta \in D \subset V$. Using the inequalities (8.47) and (8.48), we obtain the following estimate:

$$I(\theta) \leq \int_{-\infty}^\infty |\gamma'(t+\tau)| \|\Phi^\tau(\eta)\| (|\Phi^t(\varphi^{-t}\theta_0)x|$$
$$+ |\Phi^t(\varphi^{-t}\eta)x - \Phi^t(\varphi^{-t}\theta_0)x|) \alpha(\varphi^{-t}\eta)\, dt$$
$$\leq \mu_2 (I_1(\eta) + \nu_1 I_2(\eta))$$

where

$$I_1(\eta) := \int_{-\infty}^\infty |\gamma'(t+\tau)| \|\Phi^t(\varphi^{-t}\theta_0)x| \alpha(\varphi^{-t}\eta)\, dt,$$

$$I_2(\eta) := \int_{-\infty}^\infty |\gamma'(t+\tau)| \alpha(\varphi^{-t}\eta)\, dt.$$

Let us use Proposition 8.32 to estimate $I_1(\eta)$ and $I_2(\eta)$. Since $\eta \in D$ and $\operatorname{supp} \alpha \subset D$, the integration in $I_1(\eta)$ and $I_2(\eta)$ is unchanged if restricted to $t \in \cup_{n \in \mathbb{Z}} (-s + np, s + np)$. Also, $\Phi^{t+np}(\varphi^{-t}\theta_0) = [\Phi^p(\theta_0)]^n \Phi^t(\varphi^{-t}\theta_0)$; and, as a result,

(8.56) $$I_1(\eta) = \sum_{n \in \mathbb{Z}} \int_{|t| \leq s} |\gamma'(t + np + \tau)| \|[\Phi^p(\theta_0)]^n \Phi^t(\varphi^{-t}\theta_0)x| \cdot \alpha(\varphi^{-t-np}\eta)\, dt.$$

We will use the inclusion $\operatorname{supp} \gamma' \subset [-N, N]$ to prove that the integrals in (8.56) are all equal to zero if $|n| > 2N/p$. Recall the inequalities $N \geq \mu_1 p + 1$ and $s < 1$. Since

$|t| \leq s$ and $|\tau| \leq \mu_1 p$ in (8.56), if $|n| > 2N/p$, then $|t + \tau + np| \geq |n|p - s - \mu_1 p \geq |n|p - N > N$. Therefore, $I_1(\eta)$ does not exceed

$$\sum_{|n| \leq \frac{2N}{p}} \int_{|t| \leq s} |\gamma'(t + np + \tau)|(|[\Phi^p(\theta_0)]^n x|$$

$$+ |[\Phi^p(\theta_0)]^n [\Phi^t(\varphi^{-t}\theta_0)x - x]|)\alpha(\varphi^{-t-np}\eta)\, dt.$$

We use (8.43) and (8.46) and the last expression to obtain the estimate

$$(8.57) \quad I_1(\eta) \leq (1 + \nu_2) \sum_{|n| \leq 2N/p} \int_{|t| \leq s} |\gamma'(t + np + \tau)|\alpha(\varphi^{-t-np}\eta)\, dt.$$

Similarly, for $I_2(\eta)$, we have

$$(8.58) \quad I_2(\eta) \leq \sum_{|n| \leq 2N/p} \int_{|t| \leq s} |\gamma'(t + np + \tau)|\alpha(\varphi^{-t-np}\eta)\, dt.$$

Since $\operatorname{supp}\alpha \subset D$, the integration in (8.57)–(8.58) can be restricted to $t \in R_{s,D}(\varphi^{-np}\eta)$ for each $n \in \mathbb{Z}$, see (8.30). Recall that, if $t \in \mathbb{R}$ and $\theta \in \Theta$, then $|\gamma'(t)| \leq 2/N$ and $\alpha(\theta) \leq 1$. As a result, there is a constant $\mu' > 0$ such that

$$I(\theta) \leq \frac{\mu'}{N} \sum_{|n| \leq 2N/p} M_{s,D}(\varphi^{-np}\eta).$$

Finally, we use estimate (8.31) from Lemma 8.30 to obtain the first inequality in (8.55); namely, the inequality

$$(8.59) \quad I(\theta) \leq \frac{\mu'}{N} \left[\frac{2N}{p}\right] \cdot K M_{s,B}(\theta_0) \leq \mu m$$

where the square brackets denote the integer part of the enclosed expression.

For the proof of the second inequality in (8.55), we arrive, as in (8.56) above, at the following estimate:

$$J(\theta) \leq \mu_2 [J_1(\eta) + \nu_1 J_2(\eta)]$$

where

$$J_1(\eta) := \int_{-\infty}^{\infty} |\gamma'(t + \tau)||\Phi^t(\varphi^{-t}\theta_0) f_0(\varphi^{-t}\eta)|\, dt,$$

$$J_2(\eta) := \int_{-\infty}^{\infty} |\gamma'(t + \tau)||f_0(\varphi^{-t}\eta)|\, dt.$$

Let us use the estimate $\|f_0\| \leq \delta$ in statement (2) of Lemma 8.27. Also, since $\operatorname{supp}\gamma' \subset [-N, N]$, and in view of Proposition 8.32, the integration in $J_1(\eta)$ and $J_2(\eta)$ can be restricted, as for $I_1(\eta)$ and $I_2(\eta)$ above, to $t \in \cup_{|n| \leq 2N/p}(-s + np, s + np)$. Since $\operatorname{supp} f_0 \subset D$, as in (8.56)–(8.58) above, we use the estimates (8.49) and (8.50) to obtain

$$J_1(\eta) \leq \nu_5 \sum_{|n| \leq 2N/p} \int_{|t| \leq s} |\gamma'(t + np + \tau)|\chi_D(\varphi^{-t-np}\eta)\, dt,$$

$$J_2(\eta) \leq \nu_3 \sum_{|n| \leq 2N/p} \int_{|t| \leq s} |\gamma'(t + np + \tau)|\chi_D(\varphi^{-t-np}\eta)\, dt$$

where $\chi_D(\cdot)$ is the characteristics function of D. Using (8.31) in Lemma 8.30, as in (8.55), there are positive constants μ' and μ'' such that

(8.60) $$J_1(\eta) \leq \mu' m, \quad J_2(\eta) \leq \mu'' m.$$

Therefore, we obtain the second inequality in (8.55), and (8.54) is proved.

Step 2. The Lower Estimate. We will prove that there is a constant $\nu > 0$ such that $\|g\|_\infty \geq |g(\theta_0)| \geq \nu Nm$. This, together with the inequality (8.54), gives the desired estimate (8.53).

Let us define
$$I(\theta_0) := \left| \int_{-\infty}^{\infty} \gamma(t) e^{-\frac{i\xi t}{p}} \Phi^t(\varphi^{-t}\theta_0) x \alpha(\varphi^{-t}\theta_0) \, dt \right|,$$
$$J(\theta_0) := \left| \int_{-\infty}^{\infty} \gamma(t) e^{-\frac{i\xi t}{p}} \Phi^t(\varphi^{-t}\theta_0) f_0(\varphi^{-t}\theta_0) \, dt \right|,$$

and note that $|g(\theta_0)| \geq I(\theta_0) - J(\theta_0)$. Recall that $\operatorname{supp} \gamma \subset [-N, N]$ and $0 \leq \gamma(t) < 1$. We can use arguments similar to those employed to prove the inequalities (8.60) and (8.56)–(8.59). Indeed, the estimate $|\gamma'(t)| \leq 2/N$ can be replaced by $\gamma(t) \leq 1$. Also, an estimate based on $\|f_0\| \leq \delta$ and (8.50) can be used. As a result, there is a constant $\mu > 0$ such that

$$J(\theta_0) \leq \sum_{|n| \leq 2N/p} \int_{|t| \leq s} |\gamma(t+np)| \, \|[\Phi^p(\theta_0)]^n\| \, |\Phi^t(\varphi^{-t}\theta_0) f_0(\varphi^{-t}\theta_0)| \, dt$$

(8.61) $$\leq \nu_5 \sum_{|n| \leq 2N/p} \int_{|t| \leq s} |\gamma(t+np)| \chi_D(\varphi^{-t}\theta_0) \, dt \leq \mu \nu_5 Nm.$$

Let us use equation (8.43) to estimate $I(\theta_0)$ from below as follows:
$$I(\theta_0) \geq \left| \sum_{n \in \mathbb{Z}} \int_{|t| \leq s} \gamma(t+np) e^{-\frac{i\xi t}{p}} [e^{-i\xi} \Phi^p(\theta_0)]^n \Phi^t(\varphi^{-t}\theta_0) x \alpha(\varphi^{-t}\theta_0) \, dt \right|$$
$$\geq I_1(\theta_0) - I_2(\theta_0) - I_3(\theta_0)$$

where this time we define
$$I_1(\theta_0) := \left| \sum_{n \in \mathbb{Z}} \int_{|t| \leq s} \gamma(t+np) x \alpha(\varphi^{-t}\theta_0) \, dt \right|,$$
$$I_2(\theta_0) := \left| \sum_{n \in \mathbb{Z}} \int_{|t| \leq s} \gamma(t+np) \left(e^{-\frac{i\xi t}{p}} - 1 \right) x \alpha(\varphi^{-t}\theta_0) \, dt \right|,$$
$$I_3(\theta_0) := \left| \sum_{n \in \mathbb{Z}} \int_{|t| \leq s} \gamma(t+np) \left[e^{-i\xi} \Phi^p(\theta_0) \right]^n \left[\Phi^t(\varphi^{-t}\theta_0) x - x \right] \alpha(\varphi^{-t}\theta_0) \, dt \right|.$$

We will estimate $I_1(\theta_0)$ from below. By inequality (8.51), $\gamma(\tau) \geq 1/4$ provided that $|\tau| \leq N/2$. If $|n| \leq (N-1)/2p$, then $|t+np| \leq s + \frac{N-1}{2p} \cdot p \leq N/2$. Hence, $\gamma(t+np) \geq 1/4$ for those n and $|t| \leq s$. Also, $|x| = 1$; and therefore,

$$I_1(\theta_0) = \sum_{n \in \mathbb{Z}} \int_{|t| \leq s} \gamma(t+np) \alpha(\varphi^{-t}\theta_0) \, dt \geq \frac{1}{4} \sum_{|n| \leq \frac{N-1}{2p}} \int_{|t| \leq s} \alpha(\varphi^{-t}\theta_0) \, dt.$$

Since $\alpha(\varphi^{-t}\theta_0) = 1$ for $t \in R_{s,B}(\theta_0)$, see (8.30), we have that
$$I_1(\theta_0) \geq \frac{1}{4} \sum_{|n| \leq \frac{N-1}{2p}} \int_{R_{s,B}(\theta_0)} dt \geq \frac{1}{4} m \cdot \operatorname{card} \left\{ n : |n| \leq \frac{N-1}{2p} \right\} \geq Nm\nu_0$$

for some $\nu_0 > 0$.

As in the inequality (8.61), using also inequality (8.45), there is some $\mu' > 0$ such that
$$I_2(\theta_0) \leq \nu_4 \sum_{n \in \mathbb{Z}} \int_{|t| \leq s} \gamma(t+np) \alpha(\varphi^{-t}\theta_0) \, dt \leq \nu_4 \mu' Nm.$$

Similarly, using (8.46), we have that

(8.62) $$I_3(\theta_0) \leq \nu_2 \sum_{n \in \mathbb{Z}} \int_{|t| \leq s} \gamma(t+np) \alpha(\varphi^{-t}\theta_0) \, dt \leq \nu_2 \mu' Nm.$$

Combining the estimates (8.61)–(8.62), we obtain the lower bound
$$|g(\theta_0)| \geq I(\theta_0) - J(\theta_0) \geq mN(\nu_0 - \nu_5 \mu - \nu_2 \mu' - \nu_4 \mu') \geq \nu N_m$$
for the constant $\nu := \nu_0 - \nu_5 \mu - \nu_2 \mu' - \nu_4 \mu' > 0$, provided that ν_2, ν_4, and ν_5 are sufficiently small. Finally, let us use inequality (8.54) to obtain the desired estimate (8.53). This completes the proof of Theorem 8.26 for Case 1 on page 277.

8.2.1.4. *Case 2. Unbounded periods.* In this subsection let us assume that, for the Mañe point $\bar{\theta}$ in display (8.28), Case 2 on page 277 holds. Note that Case 2 holds provided that the flow $\{\varphi^t\}_{t \in \mathbb{R}}$ is aperiodic; that is, $\mathcal{B}(\Theta) = \emptyset$. Our objective is the proof of the following implication: $1 \in \sigma_{ap}(E, C(\Theta, \mathcal{T}\Theta))$ implies $i\mathbb{R} \subseteq \sigma_{ap}(\Gamma; C_{ND}(\Theta; \mathcal{T}\Theta))$. This will finish the proof of Theorem 8.26.

PROPOSITION 8.33. *Suppose that, for the Mañe point, Case 2 on page 277 holds. For each $N \in \mathbb{N}$ there is a point θ_0, a vector $x \in \mathcal{T}_{\theta_0}\Theta$ with $|x| \geq 1/2$, and an open set $U \ni \theta_0$ such that*

(a) $\sup\{|\Phi^t(\theta)x| : \theta \in U, |t| \leq 8N\} < \mu$ *for some constant $\mu > 0$ that is independent of N, and*

(b) $p(\theta) \geq 9N$ *for all $\theta \in U$.*

PROOF. Use (8.28) to define $\mu = 2\sup\{|\Phi^t(\bar{\theta})\bar{x}| : t \in \mathbb{R}\}$. Fix $N \in \mathbb{N}$. In the notations of Case 2, let $\bar{p} = 9N$ and set $x = \bar{x}$. Use the continuity of $\Phi(\cdot)$ to find $U' \ni \bar{\theta}$ such that statement (a) holds. Since we are in Case 2, for this U' and $\bar{p}$ there is a point $\theta_0 \in U'$ with $p(\theta_0) \geq 10N$. The prime period function p is lower semicontinuous. Hence, we can find a smaller neighborhood $U \subset U'$ such that statement (b) holds. □

To finish the proof of Theorem 8.26, fix $\xi \in \mathbb{R}$ and $N \in \mathbb{N}$, and choose θ_0, U, and x as indicated in Proposition 8.33. We will construct $g \in C_{ND}(\Theta, \mathcal{T}\Theta)$ such that

(8.63) $$\|\Gamma g - i\xi g\|_\infty \leq \frac{\mu}{N} \|g\|_\infty$$

for some constant μ that is independent of N.

Choose "small" constants $\nu_1, \ldots, \nu$ and "big" constants $\mu_1, \ldots, \mu$; the required values of these constants will be specified later. Pick a number $s > 0$ such that, for $|t| \leq s$, we have the estimates

(8.64) $$|\Phi^t(\varphi^{-t}\theta_0)x - x| \leq \nu_2 \quad \text{and} \quad |e^{-i\xi t} - 1| \leq \nu_4,$$

cf. (8.45)–(8.46). Select a smooth function $\gamma : \mathbb{R} \to [0,1]$ with supp $\gamma \subset [-N, N]$ such that $\|\gamma'\|_\infty \leq 2/N$, and $\gamma(t) = 1$ for $|t| \leq s$. Also, choose $\delta \leq \nu_3$ such that (cf.

(8.49)–(8.50))

(8.65)
$$\delta \max\{\|E^t\| : |t| \leq 2N\} \leq \nu_5.$$

Arguing as in the proof of Proposition 8.32, we use statement (b) in Proposition 8.33 to find an $\epsilon > 0$ such that, for every $V \ni \theta_0$ with diam $V < \epsilon$, the segment of the trajectory $\{\varphi^t \eta : s \leq |t| \leq N\}$ does not intersect V whenever $\eta \in V$. Choose $V \ni \theta_0$ with diam $V < \epsilon$ so that $V \subset U$ and

(8.66)
$$\sup_{\eta \in V} \sup_{|t| \leq 3N} \|\Phi^t(\varphi^{-t}\eta) - \Phi^t(\varphi^{-t}\theta_0)\| \leq \nu_1,$$

cf. inequality (8.48).

Apply Lemma 8.30 with $s = N$ and the prescribed choice of V. Choose $B \subset D$ where $D \subset V$ as in Lemma 8.27 with the ϵ and δ selected above. For the vector field $f = \alpha(\cdot)x + f_0$ given by Lemma 8.27, we define (cf. (8.52)) the a function g by

(8.67)
$$g(\theta) = \int_{-\infty}^{\infty} e^{-i\xi t} \gamma(t)(E^t f)(\theta)\, dt, \quad \theta \in \Theta.$$

To prove the estimate (8.63), we will show that

(8.68) $\quad |(\Gamma g)(\theta) - i\xi g(\theta)| = \left| \int_{-\infty}^{\infty} e^{-i\xi t} \gamma'(t)(E^t f)(\theta)\, dt \right| \leq \dfrac{\mu}{N} M_{N,B}(\theta_0), \quad \theta \in \Theta,$

for some $\mu > 0$, and

(8.69)
$$\|g\|_\infty \geq |g(\theta_0)| \geq \nu M_{N,B}(\theta_0)$$

for some $\nu > 0$ independent of N, see (8.30) for the definition of $M_{N,B}(\theta_0)$.

Step 1. The Upper Estimate. To prove the estimate in (8.68), we note first that the integration in (8.67) is unchanged if restricted to $t \in R_{N,D}(\theta)$. Also, $g(\theta) = 0$ provided that $\theta \notin \cup_{|t| \leq 2N} \varphi^t(D)$. Besides, for s as above, the segment of the trajectory $\{\varphi^t \theta : |t| \leq 2N\}$, for $\theta \in \Theta$, spends no more than time $2s$ in D. As a result, the left hand side of (8.68) is estimated from above by

$$\left| \int_{-\infty}^{\infty} e^{-i\xi t} \gamma'(t) [\alpha(\varphi^{-t}\theta) \Phi^t(\varphi^{-t}\theta)x + (E^t f_0)(\theta)]\, dt \right| \leq I(\theta) + J(\theta)$$

where, using $\|\gamma'\|_\infty \leq 2/N$,

$$I(\theta) := \frac{2}{N} \int_{R_{N,D}(\theta)} \alpha(\varphi^{-t}\theta) |\Phi^t(\varphi^{-t}\theta)x|\, dt,$$

$$J(\theta) := \frac{2}{N} \int_{R_{N,D}(\theta)} \|E^t\| \|f_0\|\, dt.$$

Using statement (a) in Proposition 8.33 and supp $\alpha \subset D \subset V \subset U$, we have

$$I(\theta) \leq \frac{2}{N} \mu M_{N,D}(\theta) \leq \frac{2}{N} \mu K M_{N,B}(\theta_0)$$

where Lemma 8.30 is also used to obtain the last inequality. Recall that $\|f_0\|_\infty \leq \delta$ by statement (2) in Lemma 8.27. Now the inequality (8.65) and Lemma 8.30 imply

$$J(\theta) \leq \frac{2}{N} \nu_5 M_{N,D}(\theta) \leq \frac{2}{N} \nu_5 K M_{N,B}(\theta_0).$$

This completes the proof of the upper estimate (8.68).

Step 2. The Lower Estimate. To prove the estimate (8.69), note that $f = \alpha(\cdot)x + f_0$ gives $\|g\|_\infty \geq |g(\theta_0)| \geq I_1 - I_2 - J$ where

$$I_1 := \left| \int_{-\infty}^{\infty} e^{-i\xi t} \gamma(t) \alpha(\varphi^{-t}\theta_0) x \, dt \right|,$$

$$I_2 := \left| \int_{-\infty}^{\infty} e^{-i\xi t} \gamma(t) \alpha(\varphi^{-t}\theta_0) [\Phi^t(\varphi^{-t}\theta_0)x - x] \, dt \right|,$$

$$J := \left| \int_{-\infty}^{\infty} e^{-i\xi t} \gamma(t) (E^t f_0)(\theta_0) \, dt \right|.$$

Again, since supp $\gamma \subset [-N, N]$, supp $\alpha \subset D$, and supp $f_0 \subset D$, each integral is equal to its restriction to $R_{N,D}(\theta_0)$.

First, we estimate I_1 from below:

$$I_1 = |x| \left| \int_{R_{N,D}(\theta_0)} e^{-i\xi t} \gamma(t) \alpha(\varphi^{-t}\theta_0) \, dt \right| \geq I_{11} - I_{12},$$

where

$$I_{11} := \left| \int_{R_{N,D}(\theta_0)} \gamma(t) \alpha(\varphi^{-t}\theta_0) \, dt \right| \cdot |x|,$$

$$I_{12} := \left| \int_{R_{N,D}(\theta_0)} (e^{-i\xi t} - 1) \gamma(t) \alpha(\varphi^{-t}\theta_0) \, dt \right| \cdot |x|.$$

Recall that the segment of the trajectory $\{\varphi^t \theta_0 : s \leq |t| \leq N\}$ does not intersect $D \subset V$. Thus, because supp $\alpha \subset D$, we have $\alpha(\varphi^{-t}\theta_0) = 0$ for $s \leq |t| \leq N$. On the other hand, γ is chosen such that $\gamma(t) = 1$ for $|t| \leq s$. Since $|x| \geq 1/2$ in Proposition 8.33, and $\alpha(\eta) = 1$ for $\eta \in B$ in Lemma 8.27, we have

$$I_{11} = \frac{1}{2} \int_{R_{N,D}(\theta_0)} \alpha(\varphi^{-t}\theta_0) \, dt \geq \frac{1}{2} \int_{R_{N,B}(\theta_0)} \alpha(\varphi^{-t}\theta_0) \, dt = \frac{1}{2} M_{N,B}(\theta_0).$$

Using the second inequality in (8.64), and Lemma 8.30, we see the inequality $I_{12} \leq \nu_4 K M_{N,B}(\theta_0)$.

Next, we estimate I_2 from above. Again, the integral in I_2 is unchanged if restricted to $t \in \{|t| \leq s\} \cap R_{N,D}(\theta_0)$. Therefore, we can use the first inequality in (8.64) and Lemma 8.30 to conclude that $I_2 \leq \nu_2 K M_{N,B}(\theta_0)$.

Finally, to estimate J from above, we use $\|f_0\|_\infty \leq \delta$ from (2) in Lemma 8.27 and (8.65) to get $J \leq \nu_5 K M_{N,B}(\theta_0)$.

Inequality (8.69) now follows with the appropriate choice of $\nu_1, \ldots, \nu_5$. This completes the proof of Theorem 8.26 and, as a result, the proofs of Theorems 8.22–8.23.

EXAMPLE 8.34. We will show that none of the statements of Theorem 8.23 hold without the assumption that the flow $\{\varphi^t\}_{t \in \mathbb{R}}$ is aperiodic.

Specifically, we will construct an Eulerian vector field **v** so that the Spectral Mapping Theorem for the corresponding group $\{E^t\}_{t \in \mathbb{R}}$ does not hold. For this, consider the three dimensional torus $\mathbb{T}^3$, viewed as $\mathbb{R}^3 / \mathbb{Z}^3$ with coordinates $(\theta_1, \theta_2, \theta_3)$ in $\mathbb{R}^3$ modulo 2π, and recall that Euler's equations are

(8.70) $$\mathbf{v}_t + \langle \mathbf{v} \cdot \nabla \rangle \mathbf{v} = -\frac{1}{\rho} \nabla \mathbf{p}, \quad \text{div } \mathbf{v} = 0$$

where ρ is the fluid density and **p** is the pressure. The constant vector field **v** given by $\mathbf{v}(\theta_1, \theta_2, \theta_3) = (1, 0, 0)$ together with a constant pressure provides a (steady state) solution for (8.70). Define Γ to be the Lie derivative in the direction of **v**. We will show that the Spectral Mapping Theorem, considered on the space C_{ND} of divergence free vector fields on $\mathbb{T}^3$, does not hold for the group $\{E^t\}_{t\in\mathbb{R}}$ generated by Γ. Moreover, $\sigma(E^t)$ and $\sigma(\Gamma)$ are not invariant with respect to arbitrary rotations and vertical translations.

The tangent bundle of $\mathbb{T}^3$ is trivial, so we consider the elements of the Banach space C_{ND} of divergence free sections as maps $f : \mathbb{T}^3 \to \mathbb{C}^3$. With this representation, we have

$$(\Gamma f)(\theta_1, \theta_2, \theta_3) = \frac{\partial}{\partial \theta_1} f(\theta_1, \theta_2, \theta_3), \quad (E^t f)(\theta) = f(\theta_1 + t, \theta_2, \theta_3).$$

The spectrum of Γ is $\{ik : k \in \mathbb{Z}\}$. Indeed, the function $F(\theta_1, \theta_2, \theta_3) := (0, e^{ik\theta_1}, 0)$ is a divergence free eigenfunction of Γ with eigenvalue ik, and, on the other hand, if λ is not of the form ik, then $\Gamma - \lambda$ is invertible with its inverse given by

$$(\Gamma - \lambda)^{-1} g(\theta_1, \theta_2, \theta_3) = \frac{e^{2\pi\lambda}}{1 - e^{2\pi\lambda}} \int_0^{2\pi} e^{-\lambda s} g(s + \theta_1, \theta_2, \theta_3)\, ds.$$

Thus, the property $\sigma(\Gamma) = i\xi + \sigma(\Gamma)$ does not hold for all $\xi \in \mathbb{R}$.

Let t denote a real number that is incommensurate with 2π. Then $\sigma(E^t) = \mathbb{T}$. This follows from the Spectral Inclusion Theorem 2.6 and the fact that the spectrum of the bounded operator E^t is closed. Since $\exp t\sigma(\Gamma)$ is not the whole circle, the Spectral Mapping Theorem fails. It is also easy to give a direct verification that $\sigma(E^t) = \mathbb{T}$. For this, choose a point $\lambda = e^{i\xi} \in \mathbb{T}$. If $\epsilon > 0$, then there is an integer k so that $|e^{ikt} - e^{i\xi}| < \epsilon$. But, for F defined above,

$$|(E^t F)(\theta_1, \theta_2, \theta_3) - e^{i\xi} F(\theta_1, \theta_2, \theta_3)| = |e^{ikt} - e^{i\xi}| < \epsilon;$$

and therefore, $e^{i\xi} \in \sigma_{ap}(E^t)$.

If $t/2\pi$ is a rational number p/q, $p \in \mathbb{Z}$, $q \in \mathbb{N}$, then $\sigma(E) = \{z \in \mathbb{C} : z^q = 1\}$. Hence, this spectrum is not invariant with respect to all rotations centered at origin. However, let us note that the Annular Hull Theorem is verified directly: $\mathcal{AH}(\sigma(E^t)) = \mathbb{T} = \mathcal{AH}(\exp t\sigma(\Gamma))$. $\diamond$

8.2.2. The spectrum. We will give a description of the annular hull, denoted $\mathcal{AH}(\sigma(E^t; C_{ND}))$, of the spectrum of the operator E^t defined on the space C_{ND} of divergence free vector fields. Recall that $\sigma(E^t; C_{ND})$ is rotationally invariant by Theorem 8.22 provided that the flow $\{\varphi^t\}_{t\in\mathbb{R}}$ is aperiodic. Thus, in this case, the annular hull

$$\mathcal{AH}(\sigma(E^t; C_{ND})) := \mathbb{T} \cdot \sigma(E^t; C_{ND}), \quad t \neq 0,$$

is equal to $\sigma(E^t; C_{ND})$. Also, by Theorem 8.23,

$$\sigma(\Gamma; C_{ND}) + i\mathbb{R} = \log|\mathcal{AH}(\sigma(E^t; C_{ND}))| + i\mathbb{R}, \quad t \neq 0;$$

and, for an aperiodic flow, $\sigma(\Gamma; C_{ND}) + i\mathbb{R} = \sigma(\Gamma; C_{ND})$ by Theorem 8.22.

We will compare the spectrum $\sigma(E^t; C_{ND})$ to the spectrum $\sigma(E^t; C)$ of E^t on the space of continuous vector fields (with no divergence free condition). For $E := E^1$, let $\mathcal{AH}_C$ denote the smallest annulus $\{z \in \mathbb{C} : r_- \leq |z| \leq r_+\}$ that contains $\sigma(E, C)$. Here, $r_+ = \exp \omega(\Gamma)$ and $r_- = \exp(-\omega(-\Gamma))$, see Corollary 8.25 for the formulas expressing $r_\pm$ in terms of the Oseledets-Lyapunov exponents. Of course, in the general case, $\mathcal{AH}(\sigma(E; C))$ consists of *several* disjoint annuli. For instance, if the flow $\{\varphi^t\}_{t\in\mathbb{R}}$ is Anosov, then $\mathbb{T} \cap \sigma(E; C) \neq \emptyset$. But, there is always an annulus of positive measure in $\rho(E;C)\cup\mathbb{T}$ that contains $\mathbb{T}$. In other words, the set $\mathcal{AH}_C \setminus \mathcal{AH}(\sigma(E;C))$ might contain spectral gaps.

An interesting phenomenon was discovered by de la Llave [**dL**]. He found that, in passing from the space C to C_{ND}, a drastic change occurs in the spectrum: the spectral gaps, if any, are filled by the circles passing through the points of the residual spectrum of E on C_{ND}. As we will see, this can be explained by the fact that the Riesz projections corresponding to components of $\mathcal{AH}(\sigma(E;C_{ND}))$ are operators of multiplication by continuous operator-valued functions exactly as in the Spectral Projection Theorem 6.38. These multiplication operators must preserve the space C_{ND} of divergence free vector fields. But, this is only possible if the projections are trivial. More precisely, we have the following theorem.

THEOREM 8.35. *If $\{\varphi^t\}_{t\in\mathbb{R}}$ is the flow of a smooth divergence free vector field on Θ and $\{E^t\}_{t\in\mathbb{R}}$ is an evolution group that preserves C_{ND}, then*

(8.71) $$\mathcal{AH}(\sigma(E;C_{ND})) = \mathcal{AH}_C.$$

Moreover, if $\{\varphi^t\}_{t\in\mathbb{R}}$ is aperiodic, then $\sigma(E;C_{ND}) = \mathcal{AH}_C$.

We remark that the second assertion follows from the first. Also, by Theorem 8.22, it follows that $\sigma_{ap}(E;C_{ND})$ is rotationally invariant provided that $\{\varphi^t\}_{t\in\mathbb{R}}$ is aperiodic.

PROOF. Since C_{ND} is a closed subspace of C, we have that

$$\sigma_{ap}(E, C_{ND}) \subset \sigma_{ap}(E;C)$$

and, obviously, $\mathcal{AH}(\sigma_{ap}(E;C_{ND})) \subset \mathcal{AH}(\sigma_{ap}(E;C))$. We will show that, in fact,

(8.72) $$\mathcal{AH}(\sigma_{ap}(E;C_{ND})) = \mathcal{AH}(\sigma_{ap}(E;C)).$$

Indeed, if $\mathbb{T}^{(r)} := \{z \in \mathbb{C} : |z| = r\} \subset \mathcal{AH}(\sigma_{ap}(E;C))$, then there is a point $z := re^{i\alpha} \in \sigma_{ap}(E;C)$. Consider the group $\{\tilde{E}^t\}_{t\in\mathbb{R}}$, $\tilde{E}^t = r^{-t}e^{-i\alpha t}E^t$, generated by $\Gamma_{\alpha,r} := -\log r - i\alpha + \Gamma$. Since $1 \in \sigma_{ap}(\tilde{E};C)$, Theorem 8.26 applied to $\{\tilde{E}^t\}_{t\in\mathbb{R}}$ implies there is some $\xi \in \mathbb{R}$ such that

$$i\xi \in \sigma_{ap}(\Gamma_{\alpha,r};C_{ND}) = -\log r - i\alpha + \sigma_{ap}(\Gamma;C_{ND}).$$

By the Spectral Inclusion Theorem 2.6 on C_{ND}, we have $re^{i(\xi+\alpha)} \in \sigma_{ap}(E;C_{ND})$; and therefore, $\mathbb{T}^{(r)} \subset \mathcal{AH}(\sigma_{ap}(E;C_{ND}))$. This proves (8.72).

Since the boundary of $\sigma(E;C_{ND})$ belongs to $\sigma_{ap}(E;C_{ND})$, the inclusion

$$\mathcal{AH}(\sigma(E;C_{ND})) \subset \mathcal{AH}_C$$

in (8.71) follows directly from (8.72).

To prove the inverse inclusion, suppose that $\mathbb{T}^{(r)} \subset \mathcal{AH}_C$ but $\sigma(E;C_{ND}) \cap \mathbb{T}^{(r)} = \emptyset$. By rescaling, we can and will assume that $r = 1$; that is, $r_- < 1 < r_+$ for the boundaries $r_\pm$ of $\mathcal{AH}_C$, but $\sigma(E;C_{ND}) \cap \mathbb{T} = \emptyset$.

Let $\mathcal{P}_{ND}$ denote the Riesz projection for the operator E on C_{ND} corresponding to the part of $\sigma(E;C_{ND})$ that lies inside of the unit disc $\mathbb{D}$. We claim that both $\sigma(E;C_{ND}) \cap \mathbb{D}$ and $\sigma(E;C_{ND}) \cap (\mathbb{C}\setminus\overline{\mathbb{D}})$ are nonempty. Indeed, since $r_- < 1 < r_+$, the set $\mathcal{AH}(\sigma_{ap}(E;C))$ contains a circle with radius greater than one and a circle with radius less than one. By (8.72), $\mathcal{AH}(\sigma_{ap}(E;C_{ND}))$ also contains these circles. This proves the claim.

Choose r_1 and r_2 such that $r_- < r_1 < 1 < r_2 < r_+$ and the annulus $\{z : r_1 \leq |z| \leq r_2\}$ is contained in the resolvent set of E on C_{ND}. There exist positive constants C_1 and C_2 such that

(8.73) $\qquad \operatorname{Im} \mathcal{P}_{ND} = \{f \in C_{ND} : \|E^n f\|_\infty \leq C_1 r_1^n \|f\|_\infty, n \in \mathbb{N}\},$

(8.74) $\qquad \operatorname{Im} \mathcal{Q}_{ND} = \{f \in C_{ND} : \|E^{-n} f\|_\infty \leq C_2 r_2^{-n} \|f\|_\infty, n \in \mathbb{N}\}$

where, as usual, we denote $\mathcal{Q}_{ND} = I - \mathcal{P}_{ND}$.

We will construct a projection $\mathcal{P}$ on $C = C(\Theta; \mathcal{T}\Theta)$ that commutes with E and has the following two properties:

$$\sigma(E|\operatorname{Im}\mathcal{P}; C) \subseteq \mathbb{D}, \quad \sigma([E|\operatorname{Im}\mathcal{Q}]^{-1}; C) \subseteq \mathbb{D}.$$

In other words, the operator E is hyperbolic on $C = C(\Theta; \mathcal{T}\Theta)$ and $\mathcal{P}$ is the Riesz projection for E in $C(\Theta; \mathcal{T}\Theta)$. By the Spectral Projection Theorem 6.38, the projection $\mathcal{P}$ has a form $(\mathcal{P}f)(\theta) = P(\theta)f(\theta)$ where $P : \Theta \to \operatorname{proj}(\mathcal{T}_\theta\Theta)$ is a continuous projection-valued function.

Note that $\mathcal{P} = \mathcal{P}_{ND}$ on $C_{ND}(\Theta; \mathcal{T}\Theta)$. Hence, $\mathcal{P}$ maps C_{ND} into itself. We claim that $\mathcal{P}$ is either the identity or the zero operator, in contradiction to the fact that E is a hyperbolic operator on C_{ND} with nonempty spectral components in both $\mathbb{D}$ and $\mathbb{C} \setminus \mathbb{D}$.

To prove the claim, consider local (adapted) coordinates $(\theta_1, \ldots, \theta_n)$ so that the divergence operator is given as in formula (8.29). The projection $\mathcal{P}$ is represented by a matrix-valued function $P(\cdot)$ with components $P_{ij}(\theta)$. For each divergence free vector field f, we then have

(8.75) $\qquad 0 = \operatorname{div}(\mathcal{P}f) = \sum_{ij} \frac{\partial P_{ij}}{\partial \theta_i} f_j + P_{ij} \frac{\partial f_j}{\partial \theta_i}.$

For each point θ in the coordinate chart and each index pair i, j with $i \neq j$, there is a divergence free vector field f such that $f(\theta) = 0$ and $\partial f_j(\theta)/\partial \theta_i = \delta_{ij}$ where δ_{ij} is Kronecker's delta. With this choice of f, equation (8.75) implies $P_{ij} = 0$ for $i \neq j$. Since $\mathcal{P}(\mathcal{P} - I) = 0$, each diagonal element $P_{ii}(\theta)$ is either zero or one. Since $\mathcal{P}$ preserves all divergence free vector fields, it is easy to see that all diagonal elements must be equal and $P(\theta)$ is as required in the coordinate chart. The desired result follows by the continuity of $P(\cdot)$.

Thus, to finish the proof of Theorem 8.35 it remains to construct the required projection $\mathcal{P}$ on the space $C(\Theta; \mathcal{T}\Theta)$. This is done in five steps.

Step 1. Introduce "step-functions" in $C(\Theta; \mathcal{T}\Theta)$.

Since Θ is compact, there is a partition of unity $\{\rho_k\}_{k=1}^N$ with $N < \infty$; that is, for each integer $1 \leq k \leq N$, the function $\rho_k : \Theta \to [0, 1]$ is continuous and, for each $\theta \in \Theta$,

(8.76) $\qquad \sum_{k=1}^N \rho_k(\theta) = 1,$

(8.77) $\qquad \operatorname{supp} \rho_k \setminus (\cup_{\ell \neq k} \operatorname{supp} \rho_\ell) \neq \emptyset.$

In particular, there is some $\theta_k \in \operatorname{supp} \rho_k \setminus (\cap_{\ell \neq k} \operatorname{supp} \rho_\ell)$ such that $\rho_k(\theta_k) = 1$.

For each set of vectors $x_1, \ldots, x_N$, the vector field $g(\theta) = \sum_{k=1}^N \rho_k(\theta) x_k$ is continuous and $\|g\|_C = \sup_k |x_k|$. Indeed, if $|x_{k'}| = \sup_k |x_k|$, then, by the statement (8.77), there is some θ_0 such that $\rho_{k'}(\theta_0) = 1$ for some $\theta_0 \in \operatorname{supp} \rho_{k'} \setminus \cup_{\ell \neq k'} \operatorname{supp} \rho_\ell$;

and therefore,
$$\|g\|_\infty \geq |g(\theta_0)| = |\sum_k \rho_k(\theta_0)x_k| = |\rho_{k'}(\theta_0)x_{k'}| = |x_{k'}|.$$

On the other hand, using (8.76),
$$\|g\|_\infty = \max_{\theta \in \Theta} |\sum_k \rho_k(\theta)x_k| \leq \max_{\theta \in \Theta} \sum_k \rho_k(\theta)|x_k| \leq \sup_k |x_k|.$$

It is also easy to see that the set $\mathfrak{G}$ of all such "step-functions" of the form $g = \sum_k \rho_k x_k$ is dense in $C(\Theta; \mathcal{T}\Theta)$.

Step 2. Define $\mathcal{P}$ for $g \in \mathfrak{G}$. The main tool here is the following simplified version of Lemma 8.27: If $\theta_0 \in \Theta$ and $x_0 \in \mathcal{T}_{\theta_0}\Theta$, then for each sufficiently small neighborhood $B \subset D$ of θ_0 there exists a vector field $f \in C_{ND}(\Theta; \mathcal{T}\Theta)$ with supp $f \subset D$ such that $f(\theta) = x_0$ for $\theta \in B$.

Assuming that the partition of unity $\{\rho_k\}_{k=1}^N$ is fine enough, there is, for each k, a section $f_k \in C_{ND}(\Theta; \mathcal{T}\Theta)$ such that $f_k(\theta) = x_k$ for $\theta \in \text{supp}\,\rho_k$ and $\|f_k\|_C = |x_k|$. If we use these sections, then, for each $\theta \in \Theta$,

$$(8.78) \qquad g(\theta) = \sum_k \rho_k(\theta)f_k(\theta), \quad f_k \in C_{ND}(\Theta; \mathcal{T}\Theta).$$

Using the Riesz projection $\mathcal{P}_{ND}$ for E on $C_{ND}(\Theta; \mathcal{T}\Theta)$ and $\mathcal{Q}_{ND} = I - \mathcal{P}_{ND}$, define $\mathcal{P}g$ and $\mathcal{Q}g$ for $g \in \mathfrak{G}$ as follows:

$$(8.79) \qquad \mathcal{P}g(\theta) = \sum_k \rho_k(\theta)(\mathcal{P}_{ND}f_k)(\theta), \quad \mathcal{Q}g(\theta) = \sum_k \rho_k(\theta)(\mathcal{Q}_{ND}f_k)(\theta).$$

Formally, the decomposition $g = \mathcal{P}g + \mathcal{Q}g$ depends upon the choice of f_k. However, we will show that, in fact, the definition (8.79) does not depend on this choice and $\mathcal{P}$ is a bounded linear operator on $\mathfrak{G}$. Once this is proved, the unique bounded linear extension of $\mathcal{P}$ to $C(\Theta; \mathcal{T}\Theta)$ is the desired projection.

Step 3. The sets
$$\mathcal{F}_+ := \{f \in C_{ND}(\Theta; \mathcal{T}\Theta) : \lim_{n\to\infty} \|E^n f\|_\infty = 0\},$$
$$\mathcal{F}_- := \{f \in C_{ND}(\Theta; \mathcal{T}\Theta) : \lim_{n\to\infty} \|E^{-n} f\|_\infty = 0\}$$

are such that $\mathcal{F}_+ \cap \mathcal{F}_- = \{0\}$. To prove this, we remark first that the $\mathcal{F}_\pm$ are linear (not necessarily closed) subspaces in $C(\Theta; \mathcal{T}\Theta)$. Next, if $f \in \mathcal{F}_+ \cap \mathcal{F}_-$, then

$$\lim_{n\to\pm\infty} \|E^n f\|_\infty = \lim_{n\to\pm\infty} \max_{\theta \in \Theta} |\Phi^n(\theta)f(\theta)| = 0.$$

If we suppose that $f \neq 0$, then there is some $\theta_0 \in \Theta$ such that $f(\theta_0) \neq 0$ and $\sup\{|\Phi^n(\theta_0)f(\theta_0)| : n \in \mathbb{Z}\} < \infty$. Thus, θ_0 is a Mañe point and $x_0 = f(\theta_0)/|f(\theta_0)|$ is a Mañe vector, see (6.23). By Corollary 7.30, we have that $\sigma_{ap}(E; C(\Theta; \mathcal{T}\Theta)) \cap \mathbb{T} \neq \emptyset$ and, by formula (8.72), $\sigma_{ap}(E; C_{ND}(\Theta; \mathcal{T}\Theta)) \cap \mathbb{T} \neq \emptyset$, in contradiction to the hyperbolicity of E on $C_{ND}(\Theta; \mathcal{T}\Theta)$.

Step 4. The projections $\mathcal{P}$ and $\mathcal{Q}$ are well-defined on $\mathfrak{G}$.

Suppose that $g = \sum_k \rho_k x_k \in \mathfrak{G}$ and, for each k, the section f_k is chosen as in (8.78). Also, define $g_+ := \mathcal{P}g$ and $g_- := \mathcal{Q}g$. We will show that $g_\pm \in \mathcal{F}_\pm$.

Indeed, using (8.73), if $\mathcal{P}_{ND}f_k \in \operatorname{Im}\mathcal{P}_{ND}$, then

$$\|E^n g_+\| = \|\sum_k \rho_k \circ \varphi^{-n} E^n \mathcal{P}_{ND} f_k\|_\infty$$
$$\leq \sup_k \|E^n \mathcal{P}_{ND} f_k\|_\infty \max_\theta \sum_k \rho_k \circ \varphi^{-n}(\theta)$$
$$\leq C_1 r_1^n \sup_k \|\mathcal{P}_{ND} f_k\|_\infty \leq C_1 r_1^n \|\mathcal{P}_{ND}\| \sup_k \|f_k\|_\infty$$
$$= C_1 r_1^n \|\mathcal{P}_{ND}\| \sup_k |x_k| = C_1 r_1^n \|\mathcal{P}_{ND}\| \cdot \|g\|_\infty.$$

In particular, $\lim_{n\to\infty} \|E^n g_+\| = 0$ and $g_+ \in \mathcal{F}_+$. Similarly, using (8.74), we have

$$\|E^{-n} g_-\| = \|\sum_k \rho_k \circ \varphi^n E^{-n} \mathcal{Q}_{ND} f_k\|_\infty$$
$$\leq \sup_k \|E^{-n} \mathcal{Q}_{ND} f_k\|_\infty \max_\theta \sum_k \rho_k \circ \varphi^n(\theta)$$
$$\leq C_2 r_2^{-n} \|\mathcal{Q}_{ND}\| \sup_k \|f_k\|_\infty = C_2 r_2^{-n} \|\mathcal{Q}_{ND}\| \cdot \|g\|_\infty.$$

It follows that $\lim_{n\to\infty} \|E^{-n} g_-\| = 0$, and therefore, $g_- \in \mathcal{F}_-$.

By Step 3, we have $\mathcal{F}_+ \cap \mathcal{F}_- = 0$. Hence, the functions $g_\pm$, in the decomposition $g = g_+ + g_-$ with $g_\pm \in \mathcal{F}_\pm$, are uniquely defined. In particular, the definitions of $\mathcal{P}g$ and $\mathcal{Q}g$ in (8.79) do not depend on the choice of f_k.

Step 5. Extend $\mathcal{P}$ and $\mathcal{Q}$ from $\mathfrak{G}$ to $C(\Theta; \mathcal{T}\Theta)$.

Using the calculations in Step 4, if $C_1' := C_1 \|\mathcal{P}_{ND}\|$, $C_2' := C_2 \|\mathcal{Q}_{ND}\|$, and $n \in \mathbb{N}$, then

$$\|E^n g_+\|_\infty \leq C_1' r_1^n \|g\|_\infty, \quad \|E^{-n} g_-\|_\infty \leq C_2' r_2^{-n} \|g\|_\infty.$$

These inequalities, for $n = 0$, show that $\mathcal{P}$ and $\mathcal{Q}$ are bounded on $\mathfrak{G}$. To complete the proof, we will demonstrate that these operators are linear on $\mathfrak{G}$. Indeed, for

$$g_1 = \sum_{i=1}^{i_0} \rho_i^1 x_i^1 \quad \text{and} \quad g_2 = \sum_{j=1}^{j_0} \rho_j^2 x_j^2,$$

there are functions $f_i^1, f_j^2 \in C_{ND}(\Theta; \mathcal{T}\Theta)$ such that $g_1 = \sum_i \rho_i^1 f_i^1$ and $g_2 = \sum_j \rho_j^2 f_j^2$. Let us define $f_{ij}^1 = f_i^1$ and $f_{ij}^2 = f_j^2$, for $i = 1, \ldots, i_0$, $j = 1, \ldots, j_0$, and use equation (8.76) to obtain

$$g_1 = \sum_{i,j} \rho_i^1 \rho_j^2 f_{ij}^1 \quad \text{and} \quad g_2 = \sum_{i,j} \rho_i^1 \rho_j^2 f_{ij}^2.$$

Then, using the equations in display (8.79), we have

$$\mathcal{P}(g_1 + g_2) = \sum_{i,j} \rho_i^1 \rho_j^2 (\mathcal{P}_{ND} f_{ij}^1 + \mathcal{P}_{ND} f_{ij}^2) = \mathcal{P}g_1 + \mathcal{P}g_2,$$

as required. $\square$

We conclude this subsection with a remark on the absence of nontrivial spectral components of $\sigma(\Gamma)$ for the space $C_{ND}(\Theta; \mathcal{T}\Theta)$. For simplicity, let us assume that the flow $\{\varphi^t\}_{t \in \mathbb{R}}$ is aperiodic and Φ^t is the differential of φ^t. Consider first the situation where Γ, the generator of the semigroup $E^t f = \Phi^t \circ f \circ \varphi^{-t}$, acts on the space $C(\Theta; \mathcal{T}\Theta)$. After some inessential modifications, we can obtain a dichotomic (no spectrum on $i\mathbb{R}$) operator Γ. For example, starting with an Anosov flow, as usual in Mather's theory, such an operator can be obtained by "moding out" the vector field $\mathbf{v}$ that is tangent to the flow $\{\varphi^t\}_{t\in\mathbb{R}}$. However, we will show that "moding out" the flow direction field does not change the spectrum $\sigma(\Gamma)$ on the space $C_{ND}(\Theta;\mathcal{T}\Theta)$. Indeed, the spectrum of Γ, on the direct sum of the quotient space $C_{ND}^{\mathbf{v}} = C_{ND}(\Theta;\mathcal{T}\Theta/[\mathbf{v}])$ and the space of sections generated by $\mathbf{v}$, is the union of the respective spectra. An element of the second space must be a divergence free vector field of the form $\alpha\mathbf{v}$ where α is a function on the manifold Θ. In particular, grad $\alpha = 0$, and it follows that α is constant along the trajectories of $\{\varphi^t\}_{t\in\mathbb{R}}$. Using this fact, we have that $\Gamma\alpha\mathbf{v} = 0$ and, as a result, the spectrum of Γ on this subspace is $\{0\}$. Since the spectrum $\sigma(\Gamma;C_{ND})$ is invariant under vertical translations, $i\mathbb{R} \subset \sigma(\Gamma;C_{ND})$ and, in particular, $i\mathbb{R} \setminus \{0\} \subset \sigma(\Gamma;C_{ND}^{\mathbf{v}})$. Hence, since the spectrum is closed, $\{0\} \subset \sigma(\Gamma;C_{ND}^{\mathbf{v}})$, and it follows that $\sigma(\Gamma;C_{ND}) = \sigma(\Gamma;C_{ND}^{\mathbf{v}})$.

8.2.3. The fast dynamo problem. We will give a brief introduction to the fast dynamo problem of magneto-hydrodynamics. In particular, we will connect the results of the previous two sections to this problem and give an example of a fast dynamo that is generated by the geodesic flow over a two dimensional Riemannian manifold.

8.2.3.1. *The Euclidean setting.* Consider Euler's equations for an ideal fluid with pressure $\mathbf{p}$ on $\Theta = \mathbb{R}^3$, namely,

$$(8.80) \qquad \frac{d\mathbf{v}}{dt} + \langle \mathbf{v}, \nabla \rangle \mathbf{v} = -\text{grad } \mathbf{p}, \quad \text{div } \mathbf{v} = 0.$$

Also, let $\mathbf{v}$ denote a steady state solution with flow $\{\varphi^t\}_{t\in\mathbb{R}}$. If the fluid is a conductor, then, by a model of electro-magnetic induction, the induced magnetic field $\mathbf{u} = \mathbf{u}(t,\theta)$ for $\theta \in \mathbb{R}^3$, is a solution of the equations

$$(8.81) \qquad \frac{d\mathbf{u}}{dt} = \epsilon\Delta\mathbf{u} + \text{curl}(\mathbf{v} \times \mathbf{u}), \quad \text{div } \mathbf{u} = 0$$

where $\epsilon = \text{Re}_m^{-1}$ and Re_m is the magnetic Reynolds number.

Using the fact that div $\mathbf{u} = 0$ and some elementary formulas from vector analysis, the equations (8.81) can be rewritten as

$$(8.82) \qquad \frac{d\mathbf{u}}{dt} = \epsilon\Delta\mathbf{u} - \langle \mathbf{v}, \nabla \rangle \mathbf{u} + \langle \mathbf{u}, \nabla \rangle \mathbf{v}, \quad \text{div } \mathbf{u} = 0.$$

In particular, for an ideally conducting fluid (that is, for $\epsilon = 0$), these equations become

$$(8.83) \qquad \frac{d\mathbf{u}}{dt} = -\langle \mathbf{v}, \nabla \rangle \mathbf{u} + \langle \mathbf{u}, \nabla \rangle \mathbf{v}, \quad \text{div } \mathbf{u} = 0.$$

To be consistent with the notation previously used in this book, let $f(\theta) := \mathbf{u}(0,\theta)$. Also, define the operator Γ_ϵ for each $\epsilon \geq 0$ and $\Gamma = \Gamma_0$ by

$$\Gamma_\epsilon f = \epsilon\Delta f - \langle \mathbf{v}, \nabla \rangle f + \langle f, \nabla \rangle \mathbf{v}, \quad \Gamma f = -\langle \mathbf{v}, \nabla \rangle f + \langle f, \nabla \rangle \mathbf{v}.$$

Then, the solutions of the evolution equations (8.82)–(8.83) are given by

$$\mathbf{u}(\theta,t) = (e^{t\Gamma_\epsilon}f)(\theta)$$

where each operator $e^{t\Gamma_\epsilon}$, for $\epsilon \geq 0$, acts on a space of divergence free vector fields. For the special case of the ideally conducting fluid, the group $\{e^{t\Gamma}\}_{t\in\mathbb{R}}$ is the

evolution group generated by the flow $\{\varphi^t\}_{t \in \mathbb{R}}$ and its differential $D\varphi^t$. In fact, the magnetic field solution
$$\mathbf{u}(t,\theta) = D\varphi^t(\varphi^{-t}\theta)f(\varphi^{-t}\theta) = (e^{t\Gamma}f)(\theta)$$
is called the Alfven solution in the literature on hydrodynamics, see for example Moffatt [**Mo**].

Fix a space of divergence free vector fields and, for $\epsilon \geq 0$, let $\omega_\epsilon = \omega(\Gamma_\epsilon)$ and $s_\epsilon = s(\Gamma_\epsilon)$ denote the growth and the spectral bounds for the semigroup $\{e^{t\Gamma_\epsilon}\}_{t \geq 0}$, see Subsection 2.1.3 for the definitions. Because of the ellipticity of the partial differential operator Γ_ϵ for $\epsilon > 0$, the spectral mapping property $\sigma(e^{t\Gamma_\epsilon}) \setminus \{0\} = e^{t\sigma(\Gamma_\epsilon)}$, for $t > 0$, is valid as long as $\epsilon > 0$. In particular, $s_\epsilon = \omega_\epsilon$ for $\epsilon > 0$ on every "reasonable" space of divergence free vector fields where the semigroup $\{e^{t\Gamma_\epsilon}\}_{t \geq 0}$ is analytic, for example, on a subspace of $L^2(\mathbb{R}^3; \mathbb{R}^3)$.

If $\omega_\epsilon > 0$ for all sufficiently small $\epsilon > 0$, then $\mathbf{v}$ is called a *kinematic dynamo*. The dynamo is called *fast* if, in addition, $\limsup_{\epsilon \to 0} \omega_\epsilon > 0$. A dynamo that is not fast is called slow. Finally, the operator Γ_ϵ, for $\epsilon \geq 0$, is called the kinematic dynamo operator.

The fast dynamo problem, see for example Arnold, Zel'dovich, Rasumaikin, and Sokolov [**AZRS**], is to determine the conditions on the velocity field $\mathbf{v}$ that imply it is a fast dynamo. There is a vast literature on this problem, see the bibliographical comments below. We mention that Vishik [**Vs2**] proved the estimate
$$\limsup_{\epsilon \to 0} \omega_\epsilon \leq \omega_0 = \omega(\Gamma).$$
This gives an "antidynamo" theorem: If $\omega_0 \leq 0$, then the dynamo is slow.

For an ideally conducting fluid where $\epsilon = 0$, the equation (8.83) loses its ellipticity. However, in the setting of Subsection 8.2.1, Corollary 8.24 shows that $s_0 = \omega_0$. Thus, Vishik's result can be reformulated as follows:
$$\limsup_{\epsilon \to 0} s_\epsilon \leq s_0 = s(\Gamma).$$
Vishik [**Vs2**] also found a sufficient condition for the nonexistence of a fast kinematic dynamo in terms of the Lyapunov numbers
$$\lambda_s^+(\theta, x) = \limsup_{t \to \infty} t^{-1} \log |D\varphi^t(\theta)x|,$$
see Definition 6.47. He proves the following theorem: If $\sup\{\lambda_s^+(\theta, x) : \theta \in \Theta, x \in T\Theta\}$ $(= s(\Gamma))$ is nonpositive, then the dynamo is slow. From this point of view, our Corollary 8.25 gives an alternative formulation of this antidynamo theorem in terms of the exact Oseledets-Lyapunov exponents.

Since $\Gamma_\epsilon = \Gamma + \epsilon\Delta$ is a perturbation of the first order differential operator Γ by a "small" higher order operator $\epsilon\Delta$, the determination of the spectrum $\sigma(\Gamma_\epsilon)$, for $\epsilon > 0$, from the spectrum $\sigma(\Gamma)$ is a singular perturbation problem. Therefore, it is not surprising that the set $\sigma(\Gamma_\epsilon)$, for $\epsilon > 0$, is drastically different from the spectrum at $\epsilon = 0$. In fact, the pure point spectrum of the elliptic operator Γ_ϵ, $\epsilon > 0$, becomes a "solid" vertical strip, see Theorem 8.35 and Theorems 8.22–8.23.

8.2.3.2. *A fast dynamo for a geodesic flow.* We will show that the geodesic flow for a two dimensional manifold of constant negative Riemannian curvature, see Subsection 7.5.2.2, is generated by a fast dynamo. Moreover, we give an explicit formula for the leading eigenvalue of the corresponding dynamo operator Γ_ϵ.

Let M denote a smooth two dimensional Riemannian manifold without boundary that has constant negative curvature κ. Let Θ denote its unit tangent bundle and **v** the generator of the geodesic flow $\{\varphi^t\}_{t\in\mathbb{R}}$ on Θ, see Subsection 7.5.2.2. Recall, from Subsection 7.5.2.2, the geodesic vector field **v**, the orthogonal geodesic vector field **y**, and the generator of the fiber rotation flow **z**. For each $\theta \in \Theta$, the vectors $\mathbf{v}(\theta)$, $\mathbf{y}(\theta)$, and $\mathbf{z}(\theta)$ are a basis of $\mathcal{T}_\theta\Theta$. Moreover, there is a unique Riemannian metric on Θ defined by declaring the set $\{\mathbf{v}(\theta), \mathbf{y}(\theta), \mathbf{z}(\theta)\}$ to be an orthonormal basis for $\mathcal{T}_\theta\Theta$. Let us denote the induced scalar product in $\mathcal{T}_\theta\Theta$ by $\langle\cdot,\cdot\rangle$, and the corresponding, normalized volume element on Θ by μ. In particular, μ is chosen such that $\mu(\mathbf{v},\mathbf{y},\mathbf{z}) = 1$.

On the space $L^2_{ND}(\Theta;\mu;\mathcal{T}\Theta)$ of L^2-sections of $\mathcal{T}\Theta$ with zero divergence, cf. (8.26)–(8.27), consider the operator Γ_ϵ, corresponding to the right hand side of (8.81) or (8.82), that is defined by $\Gamma_\epsilon \mathbf{u} = \epsilon\Delta\mathbf{u} + \mathrm{curl}(\mathbf{v}\times\mathbf{u})$ for $\epsilon > 0$. Our objective is to prove the following result.

THEOREM 8.36. *If **v** is the vector field that generates the geodesic flow for a closed two dimensional Riemannian manifold M, then **v** is a steady state solution of Euler's equations on M. In addition, if M has constant negative curvature κ, then for each magnetic Reynolds number $Re_m > \sqrt{-\kappa}$, the corresponding dynamo operator Γ_ϵ has the positive eigenvalue*

$$(8.84) \qquad \lambda_\epsilon := \frac{1}{2}\left[-\epsilon(1+\kappa^2) + \sqrt{\epsilon^2(1-\kappa^2)^2 - 4\kappa}\right]$$

*and $\limsup_{\epsilon\to 0} s_\epsilon \geq \lim_{\epsilon\to 0}\lambda_\epsilon = \sqrt{-\kappa} > 0$. In particular, **v** is a fast kinematic dynamo.*

PROOF. For the proof, we will need some information concerning the vector fields **v**, **y**, and **z**. We use the notation $[\mathbf{u},\mathbf{v}] = \mathbf{u}\mathbf{v} - \mathbf{v}\mathbf{u}$ for the Lie bracket of the vector fields **u** and **v** on Θ, and we let κ also denote the lift of the Gauss curvature of M to Θ. The following bracket relations, as reported by Green [**Gr2, Gr**], are valid:

$$(8.85) \qquad [\mathbf{v},\mathbf{y}] = \kappa\mathbf{z}, \quad [\mathbf{v},\mathbf{z}] = -\mathbf{y}, \quad [\mathbf{y},\mathbf{z}] = \mathbf{v}.$$

For each one-form $\boldsymbol{\xi}$ on Θ, let $\omega_{\mathbf{u}}(\boldsymbol{\xi}) := \langle\boldsymbol{\xi},\mathbf{u}\rangle$. Using the standard notations $\mathbf{d}$ and $\mathbf{i}$ for the exterior and the interior derivatives as in Abraham, Marsden, and Ratiu [**AMR**], recall that the maps $\mathbf{u}\mapsto\omega_{\mathbf{u}}$ and $\mathbf{u}\mapsto\mathbf{i}_{\mathbf{u}}\mu$ are isomorphisms and all the operations of vector analysis on Θ correspond to operations of the exterior differential calculus on the target differential forms. For example, the operators curl and div are defined by

$$(8.86) \qquad \mathbf{d}\omega_{\mathbf{u}} = \mathbf{i}_{\mathrm{curl}\,\mathbf{u}}\mu, \quad \mathbf{d}\mathbf{i}_{\mathbf{u}}\mu = \mathrm{div}\,\mathbf{u}\,\mu.$$

LEMMA 8.37. *For the vector fields **v**, **y**, and **z** on Θ defined above,*

$$(8.87) \qquad \mathrm{curl}\,\mathbf{v} = -\mathbf{v}, \quad \mathrm{curl}\,\mathbf{y} = -\mathbf{y}, \quad \mathrm{curl}\,\mathbf{z} = -\kappa\mathbf{z},$$
$$\mathrm{div}\,\mathbf{v} = 0, \quad \mathrm{div}\,\mathbf{y} = 0, \quad \mathrm{div}\,\mathbf{z} = 0.$$

PROOF. Let $\mathcal{L}_{\boldsymbol{\xi}}$ denote Lie differentiation in the direction $\boldsymbol{\xi}$ and recall the formula (see for example Abraham, Marsden, and Ratiu [**AMR**, n°6.4.11(ii)])

$$(8.88) \qquad \mathbf{d}\omega_{\mathbf{u}}(\boldsymbol{\xi},\boldsymbol{\eta}) = \mathcal{L}_{\boldsymbol{\xi}}(\omega_{\mathbf{u}}(\boldsymbol{\eta})) - \mathcal{L}_{\boldsymbol{\eta}}(\omega_{\mathbf{u}}(\boldsymbol{\xi})) - \omega_{\mathbf{u}}([\boldsymbol{\xi},\boldsymbol{\eta}]).$$

Since $\{\mathbf{v},\mathbf{y},\mathbf{z}\}$ is an orthonormal basis, by an application of formula (8.88), we have that

$$\mathbf{d}\omega_{\mathbf{v}}(\mathbf{y},\mathbf{z}) = \mathcal{L}_{\mathbf{y}}(\omega_{\mathbf{v}}(\mathbf{z})) - \mathcal{L}_{\mathbf{z}}(\omega_{\mathbf{v}}(\mathbf{y})) - \omega_{\mathbf{v}}([\mathbf{y},\mathbf{z}]) = -\omega_{\mathbf{v}}(\mathbf{v}) = -\|\mathbf{v}\|^2 = -1.$$

On the other hand, by the formulas in display (8.86),
$$\mathbf{d}\omega_{\mathbf{v}}(\mathbf{y},\mathbf{z}) = \mathbf{i}_{\mathrm{curl}\ \mathbf{v}}\mu(\mathbf{y},\mathbf{z}) = \mu(\mathrm{curl}\ \mathbf{v},\mathbf{y},\mathbf{z});$$
and, as a result, $\mu(\mathrm{curl}\ \mathbf{v},\mathbf{y},\mathbf{z}) = -1$.

We conclude from the previous formula that $\mathrm{curl}\ \mathbf{v} = -\mathbf{v}$. Indeed, since the vectors $\mathbf{v}$, $\mathbf{y}$, and $\mathbf{z}$ are of unit length and mutually orthogonal with respect to the scalar product $\langle \cdot, \cdot \rangle$, a similar application of formula (8.88) yields $\mathbf{d}\omega_{\mathbf{v}}(\mathbf{v},\mathbf{z}) = 0$ and $\mathbf{d}\omega_{\mathbf{v}}(\mathbf{v},\mathbf{y}) = 0$. Moreover, there are functions α, β, and δ such that $\mathrm{curl}\ \mathbf{v} = \alpha \mathbf{v} + \beta \mathbf{y} + \delta \mathbf{z}$. Using the identities

$$-1 = \mathbf{d}\omega_{\mathbf{v}}(\mathbf{y},\mathbf{z}) = \mu(\mathrm{curl}\ \mathbf{v},\mathbf{y},\mathbf{z}) = \alpha\mu(\mathbf{v},\mathbf{y},\mathbf{z}) + \beta\mu(\mathbf{y},\mathbf{y},\mathbf{z}) + \delta\mu(\mathbf{z},\mathbf{y},\mathbf{z}),$$
$$0 = \mathbf{d}\omega_{\mathbf{v}}(\mathbf{v},\mathbf{z}) = \mu(\mathrm{curl}\ \mathbf{v},\mathbf{v},\mathbf{z}) = \alpha\mu(\mathbf{v},\mathbf{v},\mathbf{z}) + \beta\mu(\mathbf{y},\mathbf{v},\mathbf{z}) + \delta\mu(\mathbf{z},\mathbf{v},\mathbf{z}),$$
$$0 = \mathbf{d}\omega_{\mathbf{v}}(\mathbf{v},\mathbf{y}) = \mu(\mathrm{curl}\ \mathbf{v},\mathbf{v},\mathbf{y}) = \alpha\mu(\mathbf{v},\mathbf{v},\mathbf{y}) + \beta\mu(\mathbf{y},\mathbf{v},\mathbf{y}) + \delta\mu(\mathbf{z},\mathbf{v},\mathbf{y}),$$

we have $\alpha = -1$, $\beta = 0$, and $\delta = 0$, as required.

To prove that $\mathrm{div}\ \mathbf{v} = 0$, use the formulas (8.86) and the equation $\mathrm{curl}\ \mathbf{v} = -\mathbf{v}$ to compute
$$\mathrm{div}\ \mathbf{v}\ \mu = \mathbf{di}_{\mathbf{v}}\mu = \mathbf{di}_{-\mathrm{curl}\ \mathbf{v}}\mu = \mathbf{dd}\omega_{\mathbf{v}} = 0.$$

The remaining formulas in (8.87) are proved similarly. For example, to prove $\mathrm{curl}\ \mathbf{z} = -\kappa \mathbf{z}$, first show that
$$\mathbf{d}\omega_{\mathbf{z}}(\mathbf{v},\mathbf{y}) = \mathcal{L}_{\mathbf{v}}(\omega_{\mathbf{z}}(\mathbf{y})) - \mathcal{L}_{\mathbf{y}}(\omega_{\mathbf{z}}(\mathbf{v})) - \omega_{\mathbf{z}}([\mathbf{v},\mathbf{y}]) = -\kappa\|\mathbf{z}\|^2 = -\kappa.$$

Then, assume that $\mathrm{curl}\ \mathbf{z} = \alpha \mathbf{v} + \beta \mathbf{y} + \delta \mathbf{z}$ and, using the identities

$$-\kappa = \mathbf{d}\omega_{\mathbf{z}}(\mathbf{v},\mathbf{y}) = \mu(\mathrm{curl}\ \mathbf{z},\mathbf{v},\mathbf{y}) = \alpha\mu(\mathbf{v},\mathbf{v},\mathbf{y}) + \beta\mu(\mathbf{y},\mathbf{v},\mathbf{y}) + \delta\mu(\mathbf{z},\mathbf{v},\mathbf{y}),$$
$$0 = \mathbf{d}\omega_{\mathbf{z}}(\mathbf{z},\mathbf{y}) = \mu(\mathrm{curl}\ \mathbf{z},\mathbf{z},\mathbf{y}) = \alpha\mu(\mathbf{v},\mathbf{z},\mathbf{y}) + \beta\mu(\mathbf{y},\mathbf{z},\mathbf{y}) + \delta\mu(\mathbf{z},\mathbf{z},\mathbf{y}),$$
$$0 = \mathbf{d}\omega_{\mathbf{z}}(\mathbf{z},\mathbf{v}) = \mu(\mathrm{curl}\ \mathbf{z},\mathbf{z},\mathbf{v}) = \alpha\mu(\mathbf{v},\mathbf{z},\mathbf{v}) + \beta\mu(\mathbf{y},\mathbf{z},\mathbf{v}) + \delta\mu(\mathbf{z},\mathbf{z},\mathbf{v}),$$

conclude that $\alpha = 0$, $\beta = 0$, and $\delta = -\kappa$. □

Using the identity
$$\langle \boldsymbol{\xi}, \nabla \rangle \boldsymbol{\xi} = \mathrm{grad}\ \frac{\boldsymbol{\xi}^2}{2} - \boldsymbol{\xi} \times \mathrm{curl}\ \boldsymbol{\xi},$$
which depends on the Riemannian metric and Riemannian volume on Θ, we obtain the following coordinate free interpretation of the Euler equations (8.80):

$$(8.89) \qquad \frac{d\mathbf{v}}{dt} - \mathbf{v} \times \mathrm{curl}\ \mathbf{v} + \mathrm{grad}\ \frac{\|\mathbf{v}\|^2}{2} = -\mathrm{grad}\ \mathbf{p}, \quad \mathrm{div}\ \mathbf{v} = 0.$$

Since, for the geodesic vector field, we have the identities $\|\mathbf{v}\|^2 = 1$, $\mathrm{div}\ \mathbf{v} = 0$, and $\mathrm{curl}\ \mathbf{v} = -\mathbf{v}$, we see that $\mathbf{v}$ is a solution of (8.89) with $\mathrm{grad}\ \mathbf{p} = 0$; that is, $\mathbf{v}$ and $\mathbf{p} = 0$ is a steady state solution of Euler's equations.

The Laplacian in the kinematic dynamo equation (8.81) must also be defined on Θ. Again, using standard formulas as in Abraham, Marsden, and Ratiu [**AMR**, p. 605], the Laplace-Beltrami-deRham operator Δ that appears in (8.81) may be defined as the differential operator on the divergence free vector field $\boldsymbol{\xi}$ by $\Delta \boldsymbol{\xi} = -\mathrm{curl}\ \mathrm{curl}\ \boldsymbol{\xi}$. This operator depends on the choice of the volume and the Riemannian metric. However, the operator defined in this manner on Euclidean three dimensional space, with its usual volume, agrees with the definition of Δ as componentwise application of the usual Laplacian. At any rate, using this definition and Lemma 8.37, we have

(8.90) $$\Delta \mathbf{v} = -\mathbf{v}, \quad \Delta \mathbf{y} = -\mathbf{y}, \quad \Delta \mathbf{z} = -\kappa^2 \mathbf{z}.$$

If $\operatorname{div} \boldsymbol{\xi} = 0$ and $\operatorname{div} \boldsymbol{\eta} = 0$, then, by an easy computation, we have the identity $\operatorname{curl}(\boldsymbol{\xi} \times \boldsymbol{\eta}) = -[\boldsymbol{\xi}, \boldsymbol{\eta}]$. Thus, the kinematic dynamo operator Γ_ϵ is given by $\Gamma_\epsilon \mathbf{u} = \epsilon \Delta \mathbf{u} - [\mathbf{v}, \mathbf{u}]$. Using this representation together with the formulas (8.90) and (8.85), we compute

$$\Gamma_\epsilon \mathbf{y} = -\kappa \mathbf{z} - \epsilon \mathbf{y}, \quad \Gamma_\epsilon \mathbf{z} = \mathbf{y} - \epsilon \kappa^2 \mathbf{z}.$$

Observe that Γ_ϵ preserves the two dimensional subspace

$$\mathcal{L} := \{a\mathbf{y} + b\mathbf{z} : a \in \mathbb{R},\ b \in \mathbb{R}\}$$

of divergence free vector fields in $L^2(\Theta, \mu, \mathcal{T}\Theta)$ spanned by $\mathbf{y}$ and $\mathbf{z}$, and the operator Γ_ϵ on $\mathcal{L}$ is represented by the matrix

$$D = \begin{bmatrix} -\epsilon & 1 \\ -\kappa & -\epsilon \kappa^2 \end{bmatrix}.$$

Each eigenvalue λ_ϵ of D is an eigenvalues of Γ_ϵ. Hence, the spectral bound $s_\epsilon = s(\Gamma_\epsilon)$ of Γ_ϵ must satisfy the estimate

$$s_\epsilon \geq \lambda_\epsilon := \frac{1}{2}\left[-\epsilon(1+\kappa^2) + \sqrt{\epsilon^2(1-\kappa^2)^2 - 4\kappa}\right].$$

A corresponding eigenfunction $\mathbf{u}_\epsilon$ for Γ_ϵ is given by $\mathbf{u}_\epsilon = \mathbf{y} + (\epsilon + \lambda_\epsilon)\mathbf{z}$. If, in particular, $\kappa = -1$, then $\lambda_\epsilon = -\epsilon + \sqrt{-\kappa}$ and $\mathbf{u}_\epsilon = \mathbf{y} + \sqrt{-\kappa}\,\mathbf{z}$.

If $\kappa < 0$ and $\epsilon \in (0, 1/\sqrt{-\kappa})$, then $\lambda_\epsilon > 0$. Hence, $\mathbf{v}$ induces a kinematic dynamo. Moreover, since

(8.91) $$\limsup_{\epsilon \to 0} s_\epsilon \geq \sqrt{-\kappa} > 0,$$

the dynamo action is fast. This completes the proof of Theorem 8.36. $\square$

For the case of negative nonconstant curvature $\kappa(\theta) \leq \kappa_0 < 0$ and sufficiently small $\epsilon > 0$, we can construct explicitly a vector field $\mathbf{u}_\epsilon \in L^2_{ND}(\Theta, \mu, \mathcal{T}\Theta)$ with $\|\mathbf{u}_\epsilon\| = 1$ such that

$$\operatorname{Re}\langle \Gamma_\epsilon \mathbf{u}_\epsilon, \mathbf{u}_\epsilon \rangle_{L_2} > 0, \quad \text{and} \quad \limsup_{\epsilon \to 0} \operatorname{Re}\langle \Gamma_\epsilon \mathbf{u}_\epsilon, \mathbf{u}_\epsilon \rangle_{L_2} > 0$$

where $\langle \cdot, \cdot \rangle_{L^2}$ denotes the scalar product in $L^2(\Theta, \mu, \mathcal{T}\Theta)$ induced by the Riemannian metric $\langle \cdot, \cdot \rangle$.

By an elementary calculation, as above, we find that

$$\Gamma_\epsilon \mathbf{v} = -\epsilon \mathbf{v}, \quad \Gamma_\epsilon \mathbf{y} = -\epsilon \mathbf{y} - \kappa \mathbf{z}, \quad \Gamma_\epsilon \mathbf{z} = \epsilon \kappa_2 \mathbf{v} + (1 - \epsilon \kappa_1)\mathbf{y} - \epsilon \kappa^2 \mathbf{z}$$

where $\kappa_1 = \kappa_1(\theta)$ and $\kappa_2 = \kappa_2(\theta)$ are defined such that $\operatorname{grad} \kappa = \kappa_1 \mathbf{v} + \kappa_2 \mathbf{y}$. Hence, the operator Γ_ϵ acts on a vector field $\mathbf{u} = a\mathbf{v} + b\mathbf{y} + c\mathbf{z}$ in the three dimensional subspace

$$\mathcal{M} := \{a\mathbf{v} + b\mathbf{y} + c\mathbf{z} : a \in \mathbb{R},\ b \in \mathbb{R},\ c \in \mathbb{R}\}$$

by $\Gamma_\epsilon \mathbf{u} = a'\mathbf{v} + b'\mathbf{y} + c'\mathbf{z}$. Here, the vector function $\mathbf{q}'(\theta) = (a'(\theta), b'(\theta), c'(\theta))$ is obtained from the vector $\mathbf{q} = (a, b, c)$ by the formula $\mathbf{q}'(\theta) = D(\theta)\mathbf{q}$ where

$$D(\theta) = [D_{ij}(\theta)]_{i,j=1}^3 := \begin{bmatrix} -\epsilon & 0 & 0 \\ 0 & -\epsilon & -\kappa \\ \epsilon \kappa_2 & 1 - \epsilon \kappa_1 & -\epsilon \kappa^2 \end{bmatrix}.$$

Consider the following matrix:
$$d_\epsilon = [d_{ij}]_{i,j=1}^3 \text{ where } d_{ij} := \frac{1}{\mu(\Theta)} \int_\Theta D_{ij}(\theta)\, d\mu(\theta).$$

The eigenvalues of d_ϵ are easy to compute. By this computation and the inequality $\kappa(\theta) \le \kappa_0 < 0$, it follows that if $\epsilon > 0$ is sufficiently small, then the matrix d_ϵ has a positive eigenvalue ν_ϵ. Moreover, for this ν_ϵ, we have that $\limsup_{\epsilon \to 0} \nu_\epsilon > 0$. If $\mathbf{q} = (a,b,c)$ is an eigenvector, with unit norm in $\mathbb{R}^3$, that corresponds to the eigenvalue ν_ϵ, then, for $\mathbf{u}_\epsilon := (\mu(\Theta))^{-1/2}(a\mathbf{v} + b\mathbf{y} + c\mathbf{z})$, we have $\|\mathbf{u}_\epsilon\| = 1$ and
$$\operatorname{Re} \langle \Gamma_\epsilon \mathbf{u}_\epsilon, \mathbf{u}_\epsilon \rangle_{L_2} = \operatorname{Re}\, (d_\epsilon \mathbf{q}, \mathbf{q})_{\mathbb{R}^3} = \nu_\epsilon.$$

8.3. The Ruelle transfer operator

In this section we consider spectral properties of the matrix-coefficient Ruelle transfer operator, $\mathcal{R}$, acting on the spaces of continuous or smooth sections of a finite dimensional bundle. The transfer operator is a generalization of the Mather evolution operator, E, studied in Chapters 6 and 7, and it is one of the central objects in dynamical systems theory, statistical mechanics, etc. A detailed exposition of all these connections is beyond the scope of this book. Instead, our goal in this section is to give some applications of the techniques used in Sections 6.2 and 7.2 (in particular, Mather localization) and Section 8.1 (formulas for the spectral radius that use Oseledets-Lyapunov exponents) to the transfer operator.

We set the stage for our analysis in Subsection 8.3.1 where we define the transfer operator, Markov partitions, topological pressure, etc. In Subsection 8.3.2 we evaluate the spectral radius of the matrix-coefficient transfer operator on the space of continuous sections and construct its approximate eigenfunctions for a partial description of its spectrum. In Subsection 8.3.3 the transfer operator $\mathcal{R}$, which acts on the space of smooth sections of the bundle, is associated with a new transfer operator $\mathcal{K}$ that acts on the space of continuous sections of a certain extended bundle. We will then prove that the essential spectral radius of $\mathcal{R}$ coincides with the spectral radius of $\mathcal{K}$. This will allow us to apply the results of Subsection 8.3.2 to estimate the essential spectral radius of $\mathcal{R}$.

8.3.1. Setting and preliminaries. Let $\varphi : \Theta \to \Theta$ be a continuous surjective covering map on the compact metric space Θ with a metric d. *In this section we will assume that φ is not injective.* We will also assume that the cardinality of the set $\varphi^{-1}(\theta)$ is finite for each $\theta \in \Theta$. In addition, we will often assume that φ is an expanding map; that is, there exists some positive $\delta > 0$ and an *expansion constant* $\rho > 1$ such that $d(\varphi(\theta), \varphi(\eta)) \ge \rho d(\theta, \eta)$ whenever $d(\theta, \eta) < \delta$. For convenience, we will also often assume that φ is topologically mixing; that is, for every open set $U \subset \Theta$ there exists a positive integer n such that $\varphi^n(U) = \Theta$. Finally, we mention that the main emphasis here is on the case where Θ is a compact finite dimensional manifold.

Let $\mathcal{E}$ be a real ℓ-dimensional vector bundle over Θ with projection p and let $\mathcal{E}_\theta$ denote the fiber of $\mathcal{E}$ with base point $\theta \in \Theta$. By the definition of a vector bundle, $\mathcal{E}_\theta$ is linearly isomorphic to $\mathbb{R}^\ell$. Let $\hat{\varphi}$ be a bundle map of $\mathcal{E}$ over φ (a linear skew-product map); that is, $\hat{\varphi} : \mathcal{E} \to \mathcal{E}$, the map $\hat{\varphi}$ is linear when restricted to fibers, and $p \circ \hat{\varphi} = \varphi \circ p$. Note that, for each $\theta \in \Theta$, there is a linear map $\Phi(\theta) : \mathcal{E}_\theta \to \mathcal{E}_{\varphi\theta}$ defined by the restriction $\Phi(\theta) := \hat{\varphi}|_{\mathcal{E}_\theta}$. In this section, we will assume that $\Phi(\theta)$ is invertible for each $\theta \in \Theta$. Finally, we let $C^0 := C^0(\mathcal{E})$ denote the complexified space of continuous sections of $\mathcal{E}$ equipped with the supremum norm. Let us note that the topology of the bundle $\mathcal{E}$ plays no role in the analysis to follow. For technical convenience, we will usually work locally. However, our results are valid for nontrivial bundles.

The matrix-coefficient Ruelle transfer operator, $\mathcal{R}$, is defined on the space C^0 by

$$(8.92) \qquad (\mathcal{R}f)(\theta) = \sum_{\eta \in \varphi^{-1}(\theta)} \Phi(\eta) f(\eta), \quad \theta \in \Theta.$$

Note that, for each $\theta \in \Theta$, all the summands are in the fiber $\mathcal{E}_\theta$. Clearly, the transfer operator, given by $\mathcal{R}f = \sum_i \hat{\varphi} \circ f \circ \varphi_i^{-1}$ where the φ_i^{-1} are branches of the "inverse" of φ, is a natural generalization of the evolution operator defined by $Ef = \hat{\varphi} \circ f \circ \varphi^{-1}$. In Subsection 8.3.3, the transfer operator (8.92) will be considered on the (complexified) space $C^{\mathbf{r}} = C^{\mathbf{r}}(\mathcal{E})$ of $\mathbf{r}$ times continuously differentiable sections of a smooth bundle $\mathcal{E}$ over a smooth compact manifold Θ. In this setting, the maps φ and Φ are also assumed to be smooth.

Spectral properties of the transfer operator give valuable information about the dynamics of the map φ. We will recall several relevant notions in the rest of this section.

The ℓ-dimensional matrix transfer operator was introduced by Ruelle (see [**Ru4**]). It plays a key role in many questions of dynamical system theory and statistical physics even for the case $\ell = 1$ where Φ is usually assumed to have *positive* values. In this case, the line bundle is assumed to be trivial so that the transfer operator can be viewed as an operator acting on the space of scalar valued functions on the metric space Θ. If this is the case, then the transfer operator $\mathcal{R}$ in (8.92) is given by

$$(8.93) \qquad (\mathcal{R}f)(\theta) = \sum_{\eta \in \varphi^{-1}\theta} e^{g(\eta)} f(\eta)$$

where $g(\theta) = \log \Phi(\theta)$. If φ is a smooth map of a smooth manifold Θ, and $\Phi(\theta) := |\det D\varphi(\theta)|^{-1}$, then the transfer operator $\mathcal{R}$, again defined on the space of scalar valued functions on Θ, is the Perron-Frobenius operator given by

$$(8.94) \qquad (\mathcal{R}_{PF} f)(\theta) = \sum_{\eta \in \varphi^{-1}\theta} \frac{f(\eta)}{|\det D\varphi(\eta)|}.$$

EXAMPLE 8.38. As a simple, but important, example, let Θ be the unit circle $\mathbb{T}$ and φ the map given by $\theta \mapsto \theta^N$ for some $N > 1$. Fix a continuous function $g : \Theta \to \mathbb{R}$ and let $\Phi(\theta) = e^{g(\theta)}$. Then, the transfer operator on $C(\mathbb{T})$ is defined by

$$(\mathcal{R}f)(\theta) = \sum_{\{\eta : \eta^N = \theta\}} e^{g(\eta)} f(\eta).$$

In particular, the Perron-Frobenius operator (8.94) for φ is given by

$$(\mathcal{R}_{PF} f)(\theta) = N^{-1} \sum_{\{\eta : \eta^N = \theta\}} f(\eta).$$

$\diamond$

The following example, the subshift of finite type $(\overline{\varphi}, \overline{\Theta})$, is of great importance in dynamical systems theory.

EXAMPLE 8.39. Fix an alphabet $\{1,\ldots,s\}$ of s symbols, and let $\Pi = (\pi_{ij})_{i,j=1}^s$ be a (transition) matrix such that $\pi_{ij} \in \{0,1\}$ for $i,j = 1,\ldots,s$. Let $\overline{\Theta}$ denote the subset of $\{1,\ldots,s\}^{\mathbb{Z}_+}$ consisting of admissible sequences $\overline{\theta} = (\overline{\theta}_j)_{j=0}^\infty$ such that $\overline{\theta}_j \in \{1,\ldots,s\}$ and $\pi_{\overline{\theta}_j \overline{\theta}_{j+1}} = 1$. Then, the set $\overline{\Theta}$ is a compact metric space with respect to the metric

$$d((\overline{\theta}_j)_{j=0}^\infty, (\overline{\eta}_j)_{j=0}^\infty) := \sum_{j=0}^\infty \frac{1 - \delta_{\overline{\theta}_j \overline{\eta}_j}}{\rho^j}$$

where $\rho > 1$ is some fixed real number and $\delta_{\theta_j \eta_j}$ is the Kronecker delta. Define $\overline{\varphi} : \overline{\Theta} \to \overline{\Theta}$ by $\overline{\varphi} : (\overline{\theta}_j)_{j=0}^\infty \mapsto (\overline{\theta}_{j+1})_{j=0}^\infty$. The map $\overline{\varphi}$ on $\overline{\Theta}$ is called a *subshift of finite type*. Here, $\overline{\varphi}$ is an expanding map with expansion constant ρ. Moreover, the subshift of finite type is topologically mixing provided that Π^n has strictly positive entries for some n. ◇

Let us consider an expanding map $\varphi : \Theta \to \Theta$ of a smooth manifold Θ. We will describe the notion of a Markov partition associated with φ. The existence of a Markov partition for an expanding map is a result of Sinai [**Si, Si2**] and Bowen [**Bo**], see [**Ru4**] for a brief summary, and also Krzyzewski [**Kz**]. This result guarantees, in particular, that each expanding map is "conjugate" to a subshift of finite type as described in Example 8.39. We will not use the full power of this result, but rather we will apply it later as a convenient notational tool to keep track for the preimages of points in Θ under iterations of the map φ.

REMARK 8.40. If $\varphi : \Theta \to \Theta$ is an expanding map, then φ has a Markov partition on Θ; that is: For every $\delta > 0$, there are finitely many closed subsets $\Theta_1, \ldots \Theta_s$ of Θ, each with diameter less than δ, such that $\Theta = \cup_{j=1}^s \Theta_j$, the interior Int Θ_j is dense in Θ_j, Int $\Theta_j \cap$ Int $\Theta_i = \emptyset$ for $i \neq j$, and $\varphi(\Theta_j)$ is a union of the Θ_j, (see, e. g., Propositions 2.1 and 2.2 in [**Ru4**]). Define $\pi_{ij} = 1$ if $\varphi(\Theta_i) \supset \Theta_j$ and $\pi_{ij} = 0$ otherwise. Let $\Pi = (\pi_{ij})_{i,j=1}^s$ be the transition matrix, cf. Example 8.39, and consider the associated subshift of finite type $\overline{\varphi}$ on $\overline{\Theta}$; that is,

$$\overline{\Theta} = \{\overline{\theta} := (\overline{\theta}_j)_{j=0}^\infty : \overline{\theta}_j \in \{1,\ldots,s\}, \pi_{\overline{\theta}_j \overline{\theta}_{j+1}} = 1\}$$

and $\overline{\varphi} : (\overline{\theta}_j) \mapsto (\overline{\theta}_{j+1})$ is the shift map. It can be proved that, for each $\overline{\theta} = (\overline{\theta}_j)_{j=0}^\infty \in \overline{\Theta}$, the intersection $\cap_{j \geq 0} \varphi^{-j}(\Theta_{\overline{\theta}_j})$ consists of a single point that we will denote by $\tau(\overline{\theta})$. Also, the map $\tau : \overline{\Theta} \mapsto \Theta$ is a continuous surjection such that $\tau\overline{\varphi} = \varphi\tau$ and the inverse map τ^{-1} is uniquely defined on the set $\Theta \setminus \cup_{j \geq 0} \varphi^j(\cup_{i=1}^s (\Theta_i \setminus \text{Int } \Theta_i))$. If φ is assumed to be topologically mixing, then some power of the matrix Π has all entries strictly positive and $\overline{\varphi}$ is also topologically mixing. ◇

In case φ has a Markov partition, there are "slightly larger" open sets containing each of the subsets of the partition. More precisely, we have the following proposition.

PROPOSITION 8.41. [**Ru4**, Prop.2.3] *Suppose that $\varphi : \Theta \mapsto \Theta$ is an expanding map with the expansion constant ρ and $\{\Theta_1, \ldots, \Theta_s\}$ is a Markov partition. If $\delta > 0$, then, for each $j = 1, \ldots, s$, there is an open set $U_j \supset \Theta_j$ with diam $U_j < \delta$ such that $\varphi(U_i)$ contains the closure of U_j if and only if $\pi_{ij} = 1$, and*

$$\Theta_{i_0} \cap \Theta_{i_1} \cap \ldots \cap \Theta_{i_k} = \emptyset \quad \text{implies} \quad U_{i_0} \cap U_{i_1} \cap \ldots \cap U_{i_k} = \emptyset.$$

For $\pi_{ij} = 1$, there is a unique local inverse $\varphi_{ij}^{-1} : U_j \to U_i$ of the map φ and
$$d(\varphi_{ij}^{-1}\theta, \varphi_{ij}^{-1}\eta) \leq \rho^{-1} d(\theta, \eta)$$
whenever $\theta, \eta \in U_j$. Moreover, if $\overline{\theta} = (\overline{\theta}_j)_{j=0}^{\infty} \in \overline{\Theta}$, then $\cap_{j \geq 0} \varphi^{-j}(U_{\overline{\theta}_j}) = \tau(\overline{\theta})$.

Note that if φ is a $C^{\mathbf{r}}$-map of a smooth manifold, then each map φ_{ij}^{-1} is $C^{\mathbf{r}}$. If Θ is an n-dimensional manifold, then we will assume that the number δ in Proposition 8.41 is so small that each open set U_i belongs to a single coordinate chart. Hence, each $U_i \subset \Theta$ can be viewed, in local coordinates, as a subset of $\mathbb{R}^n$.

Using Proposition 8.41, we will write $i\theta$ to denote the φ-preimage of $\theta \in \Theta$ that is in U_i. In symbols, we have $i\theta \in \varphi^{-1}(\theta) \cap U_i$. Also, as a new notation, let $i^{(m)} = (i_1, \ldots, i_m)$ denote an m-tuple of length m; that is, $i^{(m)} \in \{1, \ldots, s\}^m$. We will say that an m-tuple $i^{(m)} = (i_1, \ldots, i_m)$ is *admissible* if
$$\pi_{i_1 i_2} = \pi_{i_2 i_3} = \cdots = \pi_{i_{m-1} i_m} = 1$$
where π_{ij} is defined in Remark 8.40. For an admissible m-tuple $i^{(m)}$, with $m > 1$, let
$$(8.95) \qquad U_{i^{(m)}} := \varphi_{i_1 i_2}^{-1} \circ \varphi_{i_2 i_3}^{-1} \circ \cdots \circ \varphi_{i_{m-1} i_m}^{-1}(U_{i_m}).$$
Also, let $i^{(m)}\theta := i_1 \ldots i_m \theta \in U_{i_1 \ldots i_m j}$ denote the corresponding preimage of $\theta \in U_j$ under φ^m. Using this notation, we will sometimes write
$$(\mathcal{R}f)(\theta) = \sum_i \Phi(i\theta) f(i\theta) \quad \text{and} \quad (\mathcal{R}^m f)(\theta)) = \sum_{i^{(m)}} \Phi^m(i^{(m)}\theta) f(i^{(m)}\theta)$$
where, as usual, $\{\Phi^m\}_{m \in \mathbb{Z}_+}$, defined by $\Phi^m(\theta) = \Phi(\varphi^{m-1}\theta) \cdots \Phi(\theta)$, is the cocycle over $\{\varphi^m\}_{m \in \mathbb{Z}_+}$ induced by Φ. Finally, let us note that
$$(8.96) \qquad \operatorname{diam} U_{i^{(m)}} \leq \delta \rho^{-(m-1)}$$
for each positive integer m.

We will briefly discuss the notion of topological pressure, see for example Walters [**Wa2**]. Let (Θ, d) be a compact metric space, $\varphi : \Theta \to \Theta$ a continuous map, and $g : \Theta \to \mathbb{R}$ a given continuous function.

Fix $\epsilon > 0$ and $k \in \mathbb{N}$. A set $F \subset \Theta$ is called (k, ϵ)-*separated* if
$$\max_{0 \leq j \leq k-1} d(\varphi^j \theta, \varphi^j \eta) > \epsilon$$
whenever $\theta, \eta \in F$ and $\theta \neq \eta$. A set $G \subset \Theta$ is called a (k, ϵ)-*net* if, for each $\eta \in \Theta$, there exists $\theta \in G$ such that $\max_{0 \leq j \leq k-1} d(\varphi^j \theta, \varphi^j \eta) \leq \epsilon$. Define
$$(8.97) \quad P_k(\varphi, g, \epsilon) := \sup \left\{ \sum_{\theta \in F} \exp \Big(\sum_{j=0}^{k-1} g(\varphi^j \theta) \Big) : F \text{ is a } (k, \epsilon) - \text{separated set} \right\},$$
$$(8.98) \quad Q_k(\varphi, g, \epsilon) := \inf \left\{ \sum_{\theta \in G} \exp \Big(\sum_{j=0}^{k-1} g(\varphi^j \theta) \Big) : G \text{ is a } (k, \epsilon) - \text{net} \right\},$$
and note that, for $g(\theta) \equiv 0$, the numbers $P_k(\varphi, 0, \epsilon)$ and $Q_k(\varphi, 0, \epsilon)$ measure the cardinality of the (k, ϵ)-separated sets and (k, ϵ)-nets, respectively. For a general g, this measurement takes into account the values of the "weight" $\exp g$.

The topological pressure $P(\varphi, g)$ is defined to be
$$(8.99) \qquad P(\varphi, g) = \lim_{\epsilon \to 0} \overline{\lim_{k \to \infty}} k^{-1} \log P_k(\varphi, g, \epsilon) = \lim_{\epsilon \to 0} \overline{\lim_{k \to \infty}} k^{-1} \log Q_k(\varphi, g, \epsilon).$$

To be more precise, the above limits exist and are equal, and $P(\varphi, g)$ is defined to be their common value, see for example [**Wa2**].

As we have mention above, the cardinality of each set $\varphi^{-1}(\theta)$ is assumed to be finite. Recall that a continuous surjective map $\varphi : \Theta \to \Theta$ is a *covering* if, for each point $\theta \in \Theta$, there is an open neighborhood U such that $\varphi^{-1}(U)$ is the disjoint union of finitely many open sets and, for each V in this union, the restriction $\varphi|_V$ is a homeomorphism of V onto U. In particular, the cardinality of $\varphi^{-1}(\theta)$ is constant on each connected component of Θ.

Clearly, for a covering map φ, each $\theta \in \Theta$ has an open neighborhood $V \ni \theta$ such that $\varphi|_V$ is a homeomorphism of V on $\varphi(V)$. Thus, by the compactness of Θ, there are finitely many of the neighborhoods V with this property that cover Θ. By the Lebesgue Covering Lemma (see, e.g. [**Wa2**, Thm. 0.20]), there is some $\epsilon_0 > 0$ such that every ball $B_{\epsilon_0}(\theta)$ with radius ϵ_0 belongs to one of the sets in the covering.

We will need the following examples of (k, ϵ)–separated sets and (k, ϵ)–nets.

EXAMPLE 8.42. Fix $\theta \in \Theta$ and $\epsilon < \epsilon_0$ as above. Then, for each $k \in \mathbb{N}$, the set $F_k(\theta) = \varphi^{-k}(\theta)$ is (k, ϵ)-separated. Indeed, if $\theta_1 \neq \theta_2 \in \varphi^{-k}(\theta)$, then $\varphi^{j+1}(\theta_1) = \varphi^{j+1}(\theta_2)$ and $\varphi^j(\theta_1) \neq \varphi^j(\theta_2)$ for some $0 \leq j \leq k-1$. However, if $d(\varphi^j(\theta_1), \varphi^j(\theta_2)) < \epsilon$, then $\varphi^j(\theta_1) \in B_{\epsilon_0}(\varphi^j(\theta_2)) \subset V$. But, $\varphi|_V$ is a homeomorphism, and $\varphi(\varphi^j(\theta_1)) = \varphi(\varphi^j(\theta_2))$ yields a contradiction. ◇

EXAMPLE 8.43. Assume that φ is a topologically mixing, expanding map. We claim that for each $\epsilon > 0$, there exists a positive integer k_ϵ such that, for an arbitrary $\theta \in \Theta$, the set $G_k(\theta) := \varphi^{-(k+k_\epsilon)}(\theta)$ is a (k, ϵ)-net for each positive integer k.

Indeed, since φ is topologically mixing, there exists k_1 such that $\varphi^{-k_1}(\theta) \cap U_i \neq 0$, for each U_i, as in Proposition 8.41. For the given $\epsilon > 0$, and a corresponding δ as in Proposition 8.41, select $k_2 = k_2(\epsilon)$ such that $\rho^{-k_2} \delta < \epsilon$ for the expansion constant $\rho > 1$. Also, let $k_\epsilon = k_1 + k_2$.

Fix $\theta \in \Theta$ and $k \in \mathbb{N}$. We will show that if $\eta \in \Theta$, then there exists $\theta_0 \in \varphi^{-(k+k_\epsilon)}(\theta)$ such that $d(\varphi^j \theta_0, \varphi^j \eta) \leq \epsilon$ for $j = 0, 1, \ldots, k-1$. First, consider the open set U_i, from Proposition 8.41, that contains the point $\eta_1 := \varphi^{k+k_2}\eta$. There is an admissible $(k+k_2)$-tuple $i^{(k+k_2)} = i_1 \ldots i_{k+k_2}$ such that $\eta = i^{(k+k_2)} \eta_1 \in U_{i_{(k+k_2)i}}$. Moreover, if $\theta_1 \in \varphi^{-k_1}(\theta) \cap U_i$ and $\theta_0 := i^{(k+k_2)}\theta_1$, then $\theta_0 \in \varphi^{-(k+k_\epsilon)}(\theta) \cap U_{i_{(k+k_2)i}}$. Now, for $j = 0, \ldots, k-1$, if we use the choice of k_2 and the inequality (8.96), then we have that

$$d(\varphi^j \theta_0, \varphi^j \eta) = d(\varphi^j(i^{(k+k_2)}\theta_1), \varphi^j(i^{(k+k_2)}\eta_1)) = d(i_{j+1}\ldots i_{k+k_2}\theta_1, i_{j+1}\ldots i_{k+k_2}\eta_1)$$
$$\leq \operatorname{diam}(U_{i_{j+1}\ldots i_{k+k_2}i}) \leq \rho^{-(k+k_2-j)}\delta < \epsilon.$$

◇

Let $\operatorname{Erg}(\varphi)$ denote, as usual, the set of φ-invariant ergodic Borel probability measures on Θ, and, for each $\nu \in \operatorname{Erg}(\varphi)$, let $h_\nu = h_\nu(\varphi)$ denote the measure-theoretic entropy of φ; that is,

$$h_\nu(\varphi) = \sup_{\mathcal{A}} h_\nu(\varphi; \mathcal{A}) \quad \text{where} \quad h_\nu(\varphi; \mathcal{A}) = \lim_{m \to \infty} m^{-1} H_\nu\left(\bigvee_{i=0}^{m-1} \varphi^{-i}\mathcal{A}\right),$$

see for example Bowen [**Bo2**] or Walters [**Wa2**]. Here, $\mathcal{A} = \{A_1, \ldots, A_{|\mathcal{A}|}\}$ is a finite Borel partition of Θ, the partition $\bigvee_{i=0}^{m-1} \varphi^{-i} \mathcal{A}$ is formed by all possible

intersections
$$A_{j_0} \cap \varphi^{-1}(A_{j_1}) \cap \ldots \cap \varphi^{-(m-1)} A_{j_{m-1}}$$
with $A_{j_k} \in \mathcal{A}$, and $H_\nu(\mathcal{A})$ is defined by the formula
$$H_\nu(\mathcal{A}) = \sum_{i=1}^{|\mathcal{A}|} \bigl[-\nu(A_i) \log \nu(A_i)\bigr].$$

The following fundamental *variational principle* relates the topological pressure $P(\varphi, g)$ to the measure-theoretic entropies of φ and the space averages of g, see [**Wa2**] for more details.

THEOREM 8.44 (Variational Principle). *If φ is a continuous map on Θ and $g : \Theta \to \mathbb{R}$ is a continuous function, then*
$$P(\varphi, g) = \sup \Bigl\{ h_\nu(\varphi) + \int_\Theta g \, d\nu : \nu \in \mathrm{Erg}\, (\varphi) \Bigr\}.$$

Let us note that, if $g = 0$, then $P(\varphi, 0) = \sup\{h_\nu : \nu \in \mathrm{Erg}\,(\varphi)\}$ where the quantity on the right hand side of the last equality is called the *topological entropy* of φ.

As we will see in Subsection 8.3.2, if φ is an expanding map, then the spectrum of the scalar-coefficient transfer operator (8.93) on the space $C(\Theta)$ of continuous functions is the disc centered at the origin with radius $\exp P(\varphi, g)$.

In what follows, see Theorem 8.51 and Proposition 8.54, we will use the "scalarization" trick, cf. Theorem 8.15, where we replace the map $\varphi : \Theta \to \Theta$ by its extension $\tilde{\varphi} : \Theta_S \to \Theta_S$, and the matrix-coefficient Φ by a scalar function $\tilde{\Phi}$ defined on Θ_S. More precisely, let us define

(8.100) $\qquad \tilde{\varphi} : (\theta, x) \mapsto \left(\varphi\theta, \dfrac{\Phi(\theta)x}{|\Phi(\theta)x|} \right), \quad \tilde{\Phi}(\theta, x) = \log |\Phi(\theta)x|$

where $(\theta, x) \in \Theta_S := \Theta \times S$ with $S := S^{\ell-1} = \{x \in \mathbb{R}^\ell : |x|_{\mathbb{R}^\ell} = 1\}$. Also, we will use the metric $\tilde{d}$ on Θ_S defined by
$$\tilde{d}((\theta, x), (\eta, y)) = [d^2(\theta, \eta) + |x - y|^2]^{1/2}.$$

The following lemma relates the entropy of φ to the entropy of $\tilde{\varphi}$. To state it, let μ be a Borel $\tilde{\varphi}$-invariant probability measure on Θ_S and $\nu = \mathrm{proj}\,\mu$ its projection on Θ; that is, $\nu(\Theta') = \mu(\pi^{-1}(\Theta'))$ where $\Theta' \subset \Theta$ and the natural projection is given by $\pi : (\theta, x) \mapsto \theta$.

LEMMA 8.45. *If $\nu = \mathrm{proj}\,\mu$, then $h_\mu(\tilde{\varphi}) = h_\nu(\varphi)$.*

PROOF. By the Abramov-Rokhlin formula [**AR**], $h_\mu(\tilde{\varphi}) = h_\nu(\varphi) + h_\mu(\tilde{\varphi}|\varphi)$ where $h_\mu(\tilde{\varphi}|\varphi)$ is the relative entropy. Let $\Omega \subset \Theta_S$ denote the set of nonwandering points for $\tilde{\varphi}$, and define $\pi_\Omega := \pi|\Omega$, $\tilde{\varphi}_\Omega := \tilde{\varphi}|\Omega$, and $\varphi_\Omega := \varphi|\pi(\Omega)$. Since μ is $\tilde{\varphi}$-invariant, we have that $\mu(\Omega) = 1$ and $h_\mu(\tilde{\varphi}_\Omega|\varphi_\Omega) = h_\mu(\tilde{\varphi}|\varphi)$. To prove the equality $h_\mu(\tilde{\varphi}_\Omega|\varphi_\Omega) = 0$, we will use the relative variational principle by Ledrappier and Walters [**LW**, **Wa3**]. As in [**Wa3**, p.3], for $\theta \in \pi(\Omega)$, $n \in \mathbb{N}$, and $\delta > 0$, let
$$Q_n(\tilde{\varphi}, 0, \delta)(\theta) := \inf\{\mathrm{card}\,G : G \subset \pi^{-1}(\theta) \text{ is a } (n, \delta)\text{-net in } \pi^{-1}(\theta) \text{ for } \tilde{\varphi}\},$$
cf. (8.98). The relative pressure $P_\Omega(\theta) := P(\tilde{\varphi}_\Omega, \pi_\Omega, 0)(\theta)$ is defined in [**Wa3**] to be
$$P_\Omega(\theta) = P(\tilde{\varphi}_\Omega, \pi_\Omega, 0)(\theta) = \lim_{\delta \to 0} \overline{\lim_{n \to \infty}}\, n^{-1} \log Q_n(\tilde{\varphi}, 0, \delta)(\theta).$$

Also, the relative variational principle states:

$$\sup_{\mu}\{h_\mu(\tilde\varphi_\Omega|\varphi_\Omega) : \mu \circ \pi^{-1} = \nu\} = \int_{\pi(\Omega)} P_\Omega(\theta)\,d\nu(\theta),$$

see [**LW**, Thm 2.1] or [**Wa3**, p. 3].

Recall from Subsection 8.1.1 that $\Theta_\nu \subset \Theta$ denotes the Oseledets set of full ν-measure such that, for each $\theta \in \Theta_\nu$, the conclusions of the Multiplicative Ergodic Theorems 8.1–8.4 hold. In particular, each subspace $W_{\theta,\nu}^j$, for $j = 1,\ldots,s(\leq \ell)$, is defined for each $\theta \in \Theta_\nu$ (see Theorem 8.4). We will prove that $P_\Omega(\theta) = 0$ for each $\theta \in \Theta_\nu \cap \pi(\Omega)$, as required in the lemma. Let $\mathcal{S}^j := S \cap W_{\theta,\nu}^j$, for $j = 1,\ldots,s$, and let $\hat{\mathcal{S}} := \cup_{j=1}^s \mathcal{S}^j$ denote the "trace" of the Oseledets subbundles on the unit sphere.

Fix some point (θ,x) in the nonwandering set Ω for $\tilde\varphi$. We claim that x is in $\hat{\mathcal{S}}$. To prove the claim, let us first select a sequence $\{n_k\}_{k=1}^\infty$ of integers such that $\lim_{k\to\infty}\tilde\varphi^{n_k}(\theta,x) = (\theta,x)$. Using the equality (8.3), there is some j such that $x = w + u$ with $x \in V_\theta^j$, $u \in V_\theta^{j+1}$, and $w \in W_{\theta,\nu}^j$. Define

$$x_k = \frac{\Phi^{n_k}(\theta)x}{|\Phi^{n_k}(\theta)x|},\quad u_k = \frac{\Phi^{n_k}(\theta)u}{|\Phi^{n_k}(\theta)x|},\quad w_k = \frac{\Phi^{n_k}(\theta)w}{|\Phi^{n_k}(\theta)x|},$$

and note that $x_k = w_k + u_k$. Moreover, $\lim_{k\to\infty} x_k = x$ by the definition of $\tilde\varphi^{n_k}$. On the other hand, for the Lyapunov exponents, $\lambda(\theta,x) = \lambda_\nu^j > \lambda_\nu^{j+1} = \lambda(\theta,u)$; and therefore, $\lim_{k\to\infty}|u_k| = 0$. Finally, $w_k \in W_{\varphi^{n_k}\theta,\nu}^j$ because of the $\Phi^{n_k}(\theta)$-invariance of the subbundle W^j. As a result, $w_k = x_k - u_k \to x$, and so $\lim_{k\to\infty}|w_k| = |x| = 1$. Thus, for $w_k/|w_k| \in \mathcal{S}^j$, we have that $\lim_{k\to\infty} w_k/|w_k| = x \in \hat{\mathcal{S}}$, as claimed.

Since Ω belongs to the Oseledets subbundles, when working with $\tilde\varphi_\Omega$, we can take advantage of the existence of *exact* Oseledets-Lyapunov exponents. Recall that the limit in formula (8.4), for the exact Oseledets-Lyapunov exponent $\lambda(\theta,x)$, exists *uniformly* for $x \in \mathcal{S}^j$, see Oseledets [**Os**, p. 213, Thm. 4]. Therefore, for every $\delta > 0$, there exist constants $C_1 = C_1(\delta)$ and $C_2 = C_2(\delta)$ such that, for each positive integer n,

$$(8.101)\qquad \left|\Phi^n(\theta)\frac{x-y}{|x-y|}\right| \leq C_1 e^{n(\lambda_\nu^j + \delta)}, \quad |\Phi^n(\theta)y| \geq C_2 e^{n(\lambda_\nu^j - \delta)}$$

uniformly for all $x,y \in \mathcal{S}^j$. Using the definition of $\tilde\varphi$, an elementary calculation yields the estimate

$$\tilde d(\tilde\varphi^n(\theta,x),\tilde\varphi^n(\theta,y)) \leq 2\frac{\left|\Phi^n(\theta)\frac{x-y}{|x-y|}\right|}{|\Phi^n(\theta)y|}|x-y|.$$

Moreover, in view of this fact and the estimates (8.101), we find that, for every $\delta > 0$, there exists a constant C_δ such that, for each positive integer n,

$$(8.102)\qquad \tilde d(\tilde\varphi^n(\theta,x),\tilde\varphi^n(\theta,y)) \leq C_\delta e^{2n\delta}|x-y|$$

whenever $x,y \in \hat{\mathcal{S}}$.

For fixed $n \in N$ and $\delta > 0$, let $\epsilon = \delta C_\delta^{-1} e^{-2n\delta}$. Assume that $E_\epsilon = \{x_k\}_{k=1}^q$ is an ϵ-net for S; that is, for each $x \in S$, there exists $x_k \in E_\epsilon$ such that $|x - x_k| < \epsilon$. Of course, $q = q(n,\delta)$ depends on n and δ. Using the inequality (8.102), we see that, for each $\theta \in \pi(\Omega)$, the set $G_n(\theta) := \{(\theta,x_k) : k = 1,\ldots q\}$ is an (n,δ)-net in $\pi_\Omega^{-1}(\theta)$ for the map $\tilde\varphi_\Omega$. Indeed, for $(\theta,x) \in \pi_\Omega^{-1}(\theta)$, take $x_k \in E_\epsilon$ such that

$|x - x_k| \leq \delta C_\delta^{-1} e^{-2n\delta}$. Then, for $j = 0, \ldots, n-1$, use the inequality (8.102) to obtain the estimate

$$\tilde{d}(\tilde{\varphi}^j(\theta, x), \tilde{\varphi}^j(\theta, x_k)) \leq C_\delta e^{2j\delta} |x - x_k| < \delta.$$

As a result, if $\theta \in \pi(\Omega)$ and $\epsilon = \delta C_\delta^{-1} e^{-2n\delta}$, then

$$Q_n(\tilde{\varphi}, 0, \delta)(\theta) \leq q_\epsilon := \inf\{\text{card } E_\epsilon : E_\epsilon \text{ is an } \epsilon - \text{net in } S\}.$$

Obviously, $q_\epsilon \leq C(\ell)\epsilon^{-1}$ where ℓ is the dimension of $\mathcal{E}$. Therefore,

$$P_\Omega(\Theta) \leq \lim_{\delta \to 0} \overline{\lim}_{n \to \infty} n^{-1} \log q_\epsilon = \lim_{\delta \to 0} \overline{\lim}_{n \to \infty} n^{-1} \log[C(\ell)^{-1} \delta^{-1} C_\delta e^{2n\delta}] = 0,$$

as required. $\square$

8.3.2. Continuous sections. Let us consider the transfer operator $\mathcal{R}$, defined in display (8.92), on the (complexified) space $C^0 = C(\mathcal{E})$ of continuous sections of a vector bundle $\mathcal{E}$ over a compact metric space Θ. The spectrum of this transfer operator in the scalar case (Theorem 8.47) is described in Subsection 8.3.2.1, while, in Subsection 8.3.2.2, we study the spectral radius of the matrix-coefficient transfer operator. For the spectral radius, we give an estimate (Theorem 8.51) that holds for an arbitrary covering map φ and an exact formula (Theorem 8.53) that holds for an expanding map. These results are generalizations of the formula for the spectral radius of the evolution operator E given in Theorem 8.15. In Subsection 8.3.2.3, see Theorem 8.55, we specify an annulus that belongs to the approximate point spectrum of $\mathcal{R}$ and give a construction of the approximate eigenfunctions for $\mathcal{R}$ using a construction similar to the Mather localization used in Subsection 6.2.2 for evolution operators.

8.3.2.1. *The scalar case.* Let us consider the transfer operator (8.93) where $g : \Theta \to \mathbb{R}$ is a continuous function and $\Phi = \exp \circ g$. In this case, the transfer operator acts on the space $C(\Theta)$ of continuous scalar functions on Θ. Also, let us use $\Phi^k(\theta)$ to denote $\exp \sum_{j=0}^{k-1} g(\varphi^j \theta)$ for $\theta \in \Theta$. The following proposition is used in the proof of Theorem 8.47 below.

PROPOSITION 8.46. *In the scalar case where $\Phi = \exp \circ g$, if φ is a topologically mixing, expanding map, then*

$$\overline{\lim}_{k \to \infty} k^{-1} \log \min_{\theta \in \Theta} \sum_{\eta \in \varphi^{-k}\theta} \Phi^k(\eta) \geq P(\varphi, g).$$

PROOF. Fix $\epsilon > 0$ and take k_ϵ as in Example 8.43. For $k = m + k_\epsilon$, where m is a positive integer, use the cocycle property for $\{\Phi^k\}_{k \geq 0}$ to compute

$$k^{-1} \log \min_{\theta \in \Theta} \sum_{\eta \in \varphi^{-k}\theta} \Phi^k(\eta) = (m + k_\epsilon)^{-1} \log \min_{\theta \in \Theta} \sum_{\eta \in \varphi^{-k}\theta} \Phi^{k_\epsilon}(\varphi^m \eta) \Phi^m(\eta)$$

$$\geq (m + k_\epsilon)^{-1} \log \min_{\theta \in \Theta} \sum_{\eta \in \varphi^{-k}\theta} \left(\min_{\theta \in \Theta} \Phi^{k_\epsilon}(\theta) \right) \Phi^m(\eta)$$

$$= C(k_\epsilon)(m + k_\epsilon)^{-1}$$

(8.103)
$$+ m(m + k_\epsilon)^{-1} \min_{\theta \in \Theta} m^{-1} \log \sum_{\eta \in \varphi^{-(m+k_\epsilon)}\theta} \Phi^m(\eta)$$

where $C(k_\epsilon) = \log \min_\theta \Phi^{k_\epsilon}(\theta)$. By Example 8.43, the set $G_m(\theta) = \varphi^{-(m+k_\epsilon)}(\theta)$ is an (m,ϵ)-net each $\theta \in \Theta$. Using definition (8.98), the estimate (8.103) implies

$$k^{-1} \log \min_{\theta \in \Theta} \sum_{\eta \in \varphi^{-k}\theta} \Phi^k(\eta) \geq \frac{C(k_\epsilon)}{m+k_\epsilon} + \frac{m}{m+k_\epsilon} \frac{1}{m} \log Q_m(\varphi, g, \epsilon).$$

Passing to the limit as $m \to \infty$ and $\epsilon \to 0$ and using the definition of the topological pressure (8.99), we obtain the desired result. $\square$

THEOREM 8.47. *Suppose that $g \in C(\Theta; \mathbb{R})$. If φ is a topologically mixing, expanding map, then the spectrum of the scalar-coefficient transfer operator $\mathcal{R}$, given by*

$$(\mathcal{R}f)(\theta) = \sum_{\eta \in \varphi^{-1}\theta} e^{g(\eta)} f(\eta)$$

on C^0, is the disk of radius

(8.104) $$r(\mathcal{R}) = \exp P(\varphi, g) = \exp \left(\sup \left\{ h_\nu + \int_\Theta g \, d\nu : \nu \in \mathrm{Erg}\,(\varphi) \right\} \right).$$

Moreover, for each $|z| < r(\mathcal{R})$, the operator $z - \mathcal{R}$ is a surjective operator with nontrivial kernel.

PROOF. For the scalar-coefficient transfer operator, we have

(8.105) $$\|\mathcal{R}\| = \max_{\theta \in \Theta} \sum_{\eta \in \varphi^{-1}\theta} \Phi(\eta), \quad \|\mathcal{R}^k\| = \max_{\theta \in \Theta} \sum_{\eta \in \varphi^{-k}\theta} \Phi^k(\eta), \quad k \in \mathbb{N}.$$

Indeed,

$$\|\mathcal{R}f\|_\infty \leq \max_{\theta \in \Theta} \sum_{\eta \in \varphi^{-1}\theta} \Phi(\eta) \|f\|_\infty, \quad \|\mathcal{R}\mathcal{I}\|_\infty = \max_{\theta \in \Theta} \sum_{\eta \in \varphi^{-1}\theta} \Phi(\eta)$$

where $\mathcal{I}(\theta) = 1$ for $\theta \in \Theta$.

Since φ is a covering map, if $\epsilon > 0$ is sufficiently small and $k \in \mathbb{N}$, then the set $F_k(\theta) = \varphi^{-k}\theta$ is (k,ϵ)-separated for each $\theta \in \Theta$ by Example 8.42. Therefore, using (8.97), we have that

(8.106) $$\|\mathcal{R}^k\| = \max_{\theta \in \Theta} \sum_{\eta \in \varphi^{-k}\theta} \Phi^k(\eta) = \max_{\theta \in \Theta} \sum_{\eta \in F_k(\theta)} \exp \sum_{j=0}^{k-1} g(\varphi^j \eta) \leq P_k(\varphi, g, \epsilon).$$

Moreover, using the definition of the pressure (8.99),

$$\log r(\mathcal{R}) = \lim_{k \to \infty} k^{-1} \log \|\mathcal{R}^k\| \leq \lim_{\epsilon \to 0} \overline{\lim}_{k \to \infty} k^{-1} \log P_k(\varphi, g, \epsilon) = P(\varphi, g).$$

Thus, the inequality $\log r(\mathcal{R}) \leq P(\varphi, g)$ holds for each covering map φ.

On the other hand, since φ is a topologically mixing, expanding map, by an application of Proposition 8.46, it follows that

$$\log r(\mathcal{R}) = \lim_{k \to \infty} k^{-1} \log \max_{\theta \in \Theta} \sum_{\eta \in \varphi^{-k}\theta} \Phi^k(\eta)$$

$$\geq \overline{\lim}_{k \to \infty} k^{-1} \log \min_{\theta \in \Theta} \sum_{\eta \in \varphi^{-k}\theta} \Phi^k(\eta) \geq P(\varphi, g).$$

This fact and the Variational Principle, see Theorem 8.44, gives the desired formula for the spectral radius.

To complete the proof, let
$$\Psi_k(\theta) := \sum_{\eta \in \varphi^{-k}\theta} \Phi^k(\eta), \quad \theta \in \Theta,$$
and consider the composition (Koopman) operator $Tf = f \circ \varphi$ induced by φ. Clearly, T is a strict isometry. Also, $\mathcal{R}T = \Psi_1$ and $\mathcal{R}^k T^k = \Psi_k$, where we use the same notation, Ψ_k, to denote the multiplication operator with multiplier $\Psi_k(\cdot)$.

Note that, because $\Phi(\theta) > 0$, the operator $\mathcal{R}$ is invertible from the right, but not invertible from the left. For $|z| < r(\mathcal{R})$, fix $\epsilon > 0$ such that $|z| < r(\mathcal{R}) - \epsilon$ and consider the identity
$$z^k - \mathcal{R}^k = z^k \mathcal{R}^k T^k \Psi_k^{-1} - \mathcal{R}^k = \mathcal{R}^k \big[z^k T^k \Psi_k^{-1} - I\big].$$
Also, using Proposition 8.46 and formula (8.104), choose k such that
$$\min_{\theta \in \Theta} \sum_{\eta \in \varphi^{-k}\theta} \Phi^k(\eta) > (r(\mathcal{R}) - \epsilon)^k > |z|^k.$$
Then,
$$\big\|z^k T^k \Psi_k^{-1}\big\| \leq |z|^k \big\|\Psi_k^{-1}\big\| = |z|^k \big[\min_{\theta \in \Theta} \sum_{\eta \in \varphi^{-k}\theta} \Phi^k(\eta)\big]^{-1} < 1,$$
and the operator $z^k T^k \Psi_k^{-1} - I$ is invertible from both sides. Thus, $z^k - \mathcal{R}^k$ is surjective but not left-invertible, and so is $z - \mathcal{R}$. □

REMARK 8.48. In the course of the proof of Theorem 8.47 we have established the inequality
$$(8.107) \qquad r(\mathcal{R}) \leq \exp P(\varphi, g) = \exp\big(\sup_{\nu \in \mathrm{Erg}} \{h_\nu + \int_\Theta g\, d\nu\}\big)$$
for *covering* maps that are not necessarily expanding. The following example shows that the inequality in (8.107), generally, can be strict. Let φ denote the endomorphism of the torus $\mathbb{T}^2 = \mathbb{R}^2/\mathbb{Z}^2$ generated by the matrix $\begin{bmatrix} 3 & 1 \\ 1 & 1 \end{bmatrix}$, and suppose that $g(\theta) \equiv 0$. The topological entropy of φ is $P(\varphi, 0) = \log(2 + \sqrt{2})$, see for example Walters [**Wa2**, Thm. 8.15, p. 203], while
$$\log r(\mathcal{R}) = \lim_{k \to \infty} k^{-1} \log \max_{\theta \in \Theta} \sum_{\eta \in \varphi^{-k}\theta} 1 = \log 2.$$
◇

8.3.2.2. *The spectral radius on continuous sections.* Our next goal is to evaluate the spectral radius for the matrix-coefficient transfer operator (8.92) on the space C^0 of continuous sections of an ℓ-dimensional vector bundle $\mathcal{E}$. We will first describe a procedure for "triangularization" of the coefficient Φ. It will allow us to give general formulas for the spectral radius of the transfer operator in terms of the norm of the cocycle (Propositions 8.50 and 8.52). Next, for an arbitrary covering map, respectively an expanding map, we give an estimate in Theorem 8.51, respectively an exact formula in Theorem 8.53, for the spectral radius of the matrix-coefficient transfer operator using a formula similar to the right hand side of equation (8.104). Naturally, cf. Theorem 8.15 and Corollary 8.19, the integrals $\int g\, d\nu$ appearing on the right hand side of (8.104) will be replaced by the largest Oseledets-Lyapunov exponents λ_ν^1 for the cocycle $\{\Phi^k\}_{k \in \mathbb{Z}_+}$.

Let $SO(\ell)$ denote the set of orthogonal $\ell \times \ell$ matrices with unit determinant. Fix $\theta \in \Theta$ and $V \in SO(\ell)$, and consider the $(\ell \times \ell)$ matrix $\Phi(\theta)V$. There exists a unique matrix $\hat{V} = \hat{V}(\theta, V) \in SO(\ell)$ such that the matrix $[\hat{V}(\theta, V)]^T \Phi(\theta)V$, where the "$T$" denotes transposition, is lower triangular with positive diagonal entries (see for example Bylov, Vinograd, Grobman, and Nemytski [**BVGN**] or Johnson, Palmer, and Sell [**JPS**]). Let $\hat{\Theta} = \Theta \times SO(\ell)$, define

$$\hat{\varphi}: (\theta, V) \mapsto (\varphi\theta, \hat{V}(\theta, V)), \quad \hat{\Phi}(\theta, V) = [\hat{V}(\theta, V)]^T \Phi(\theta)V, \quad (\theta, V) \in \hat{\Theta},$$

and note that $\hat{\varphi}$ is a covering map on the compact set $\hat{\Theta}$. Also, $\hat{\Phi}$ takes values in the set of lower triangular matrices with positive diagonal entries.

For the bundle $\mathcal{E} = (\mathcal{E}, \Theta, p)$ over Θ with projection p, let $\hat{\mathcal{E}} = (\hat{\mathcal{E}}, \Theta, p)$ be the bundle $\hat{\mathcal{E}} = \mathcal{E} \times SO(\ell)$ over $\hat{\Theta} = \Theta \times SO(\ell)$ with projection $\hat{p} : ((\theta, x), V) \mapsto (\theta, V)$ where $x \in \mathcal{E}_\theta$. Also, consider the natural projections $\pi_1 : \hat{\Theta} \to \Theta : (\theta, V) \mapsto \theta$ and $\pi_2 : \hat{\Theta} \to SO(\ell) : (\theta, V) \mapsto V$.

We will consider two transfer operators induced by the map $\hat{\varphi}$ on the space $\hat{C}^0 = C^0(\hat{\mathcal{E}})$ of continuous sections of $\hat{\mathcal{E}}$. These operators are given by

$$(\widehat{\mathcal{R}}F)(\theta, V) = \sum_{(\eta,U) \in \hat{\varphi}^{-1}(\theta,V)} \hat{\Phi}(\eta, U) F(\eta, U),$$

$$(RF)(\theta, V) = \sum_{(\eta,U) \in \hat{\varphi}^{-1}(\theta,V)} \Phi \circ \pi_1(\eta, U) F(\eta, U)$$

where $F(\theta, V) \in \mathcal{E}_\theta \simeq \mathbb{R}^\ell$.

PROPOSITION 8.49.

(8.108) $$r(\mathcal{R}; C^0) \geq r(R; \hat{C}^0) = r(\widehat{\mathcal{R}}; \hat{C}^0).$$

PROOF. Note that, by our definitions,

$$\hat{\Phi}(\theta, V) = [\pi_2 \circ \hat{\varphi}(\theta, V)]^{-1} \Phi \circ \pi_1(\theta, V) [\pi_2(\theta, V)], \quad (\theta, V) \in \hat{\Theta}.$$

Next, let us define an operator Π on $C^0(\hat{\mathcal{E}})$ by $(\Pi F)(\theta, V) = V \cdot F(\theta, V)$. Then, because

$$(\widehat{\mathcal{R}}F)(\theta, V) = \sum \hat{\Phi}(\eta, U) F(\eta, U)$$
$$= \sum [\pi_2 \circ \hat{\varphi}(\eta, U)]^{-1} \Phi \circ \pi_1(\eta, U) \cdot U F(\eta, U),$$

we have that $\widehat{\mathcal{R}} = \Pi^{-1} R \Pi$. Thus, we have proved the statement $r(R; \hat{C}^0) = r(\widehat{\mathcal{R}}; \hat{C}^0)$ in (8.108).

To complete the proof, we will show the inequality $\|\mathcal{R}\| \geq \|R\|$. Let $\epsilon > 0$ be given and choose $F \in \hat{C}^0$ such that $\|F\| \leq 1$ and $\|RF\| \geq \|R\| - \epsilon$. Then, by compactness, the maximum in

$$\|RF\| = \max_{(\theta,V)} \left| \sum_{(\eta,U) \in \hat{\varphi}^{-1}(\theta,V)} \Phi \circ \pi_1(\eta, U) F(\eta, U) \right|$$

is attained, say for (θ_0, V_0). For each $\eta \in \varphi^{-1}(\theta_0)$, choose $U_\eta \in SO(\ell)$ such that $(\eta, U_\eta) \in \hat{\varphi}^{-1}(\theta_0, V_0)$, define $f(\eta) = F(\eta, U_\eta)$, and observe that $|f(\eta)| \leq 1$. Also, for each η, find an open neighborhood $B(\eta)$ so that if $\eta, \eta' \in \varphi^{-1}(\theta_0)$ are distinct, then $B(\eta) \cap B(\eta') = \emptyset$. Using a partition of unity subordinate to the collection $\{B(\eta)\}_{\eta \in \varphi^{-1}(\theta_0)}$, extend f to all of Θ so that $f \in C^0$ and $\|f\| \leq 1$. Then,

$$\|R\| - \epsilon \leq \|RF\| = \Big| \sum_{\eta \in \varphi^{-1}\theta_0} \Phi(\eta) f(\eta) \Big| \leq \|\mathcal{R}f\| \leq \|\mathcal{R}\|$$

and Proposition 8.49 is proved. $\square$

The triangularization of Φ is used to prove the following fact.

PROPOSITION 8.50. *If φ is a covering map, then*

(8.109) $$r(\mathcal{R}; C^0) = \lim_{k \to \infty} \Big(\max_{\theta \in \Theta} \sum_{\eta \in \varphi^{-k}\theta} \|\Phi^k(\eta)\| \Big)^{1/k}.$$

PROOF. The inequality "$\leq$" in (8.109) is trivial because

$$\|\mathcal{R}\|_{\mathcal{L}(C^0)} \leq \max_{\theta \in \Theta} \sum_{\eta \in \varphi^{-1}\theta} \|\Phi(\eta)\| \quad \text{and} \quad \|\mathcal{R}^k\|_{\mathcal{L}(C^0)} \leq \max_{\theta \in \Theta} \sum_{\eta \in \varphi^{-k}\theta} \|\Phi^k(\eta)\|.$$

To prove the inequality "$\geq$", we claim, first, that it suffices to consider only the case where $\Phi(\theta)$, for $\theta \in \Theta$, is a lower-triangular matrix with positive entries on the main diagonal. Indeed, by (8.108), $R(\mathcal{R}; C^0) \geq r(\widehat{\mathcal{R}}; \widehat{C}^0)$ where $\widehat{\mathcal{R}}$ is a matrix-coefficient transfer operator with a lower triangular coefficient $\widehat{\Phi}$ with positive diagonal entries. Also, by the definition of $\widehat{\Phi}(\theta, V)$, we have that

$$\|\Phi^k(\theta)\| = \|\widehat{\Phi}^k(\theta, V)\|, \quad (\theta, V) \in \widehat{\Theta} \quad \text{for} \quad \widehat{\Phi}^k(\theta, V) = \widehat{\Phi}(\widehat{\varphi}^{k-1}(\theta, V)) \cdots \widehat{\Phi}(\theta, V).$$

Thus, if the proposition is proved for the triangular case, then

$$r(\mathcal{R}) \geq r(\widehat{\mathcal{R}}) = \lim_{k \to \infty} \max_{(\theta, V)} \Big(\sum_{(\eta, U) \in \widehat{\varphi}^{-k}(\theta, V)} \|\widehat{\Phi}^k(\eta, U)\| \Big)^{1/k}.$$

Because the right hand side of the last equation is equal to the right hand side of (8.109), we have the required result.

To prove the inequality "$\geq$" in the proposition for $(\ell \times \ell)$-lower triangular maps $\Phi(\theta)$ with positive diagonal entries, we use induction on the dimension ℓ. For the scalar case, where $\ell = 1$, the desired inequality follows from (8.105). For the case where $\mathbb{R}^\ell = \mathbb{R}^{\ell_1} \oplus \mathbb{R}^{\ell_2}$, represent Φ and Φ^k as block-matrices as follows:

$$\Phi(\theta) = \begin{bmatrix} A(\theta) & 0 \\ B(\theta) & D(\theta) \end{bmatrix}, \quad \Phi^k(\theta) = \begin{bmatrix} A^k(\theta) & 0 \\ B_k(\theta) & D^k(\theta) \end{bmatrix}.$$

Here $A^k(\theta) = A(\varphi^{k-1}\theta) \cdots A(\theta)$ and $D^k(\theta) = D(\varphi^{k-1}\theta) \cdots D(\theta)$ for $k \geq 1$, $A^0(\theta) = D^0(\theta) = I$, and

$$B_k(\theta) = \sum_{j=0}^{k-1} D^j(\varphi^{k-j}\theta) B(\varphi^{k-1-j}\theta) A^{k-1-j}(\theta), \quad \theta \in \Theta.$$

The operator $\mathcal{R} = \mathcal{R}_\Phi$ has a similar triangular representation which gives $r(\mathcal{R}_\Phi) \geq \max\{r(\mathcal{R}_A), r(\mathcal{R}_D)\}$. By the induction assumption, the desired proposition holds for both $\mathcal{R}_A$ and $\mathcal{R}_D$. Thus, for each $\epsilon > 0$, there exists a positive constant $C_{\epsilon, A}$ such that

(8.110) $$\max_{\theta \in \Theta} \sum_{\eta \in \varphi^{-k}\theta} \|A^k(\eta)\| \leq C_{\epsilon, A}(r(\mathcal{R}_\Phi) + \epsilon)^k$$

for all positive integers k. A similar result holds for D^k.

Using the triangular representation of $\Phi^k(\theta)$, we have the inequality

$$\max_{\theta\in\Theta}\sum_{\eta\in\varphi^{-k}\theta}\|\Phi^n(\eta)\| \leq \max_{\theta\in\Theta}\sum_{\eta\in\varphi^{-k}\theta}\|A^k(\eta)\| + \max_{\theta\in\Theta}\sum_{\eta\in\varphi^{-k}\theta}\|D^k(\eta)\| + b_k;$$

and, using (8.110), we have that

$$b_k \leq \|B\|_\infty \max_{\theta\in\Theta}\sum_{\eta\in\varphi^{-k}\theta}\sum_{j=0}^{k-1}\|D^j(\varphi^{k-j}\eta)\|\|A^{k-1-j}(\eta)\|$$

$$= \|B\|_\infty \max_{\theta\in\Theta}\sum_{j=0}^{k-1}\sum_{\tilde\theta\in\varphi^{-(j+1)}\theta}\|D^j(\varphi\tilde\theta)\|\sum_{\eta\in\varphi^{-(k-j-1)}\tilde\theta}\|A^{k-1-j}(\eta)\|$$

$$\leq \|B\|_\infty \sum_{j=0}^{k-1}\max_{\theta\in\Theta}\sum_{\theta_1\in\varphi^{-1}\theta}\sum_{\theta_2\in\varphi^{-j}\theta_1}\|D^j(\theta_2)\|\max_{\tilde\theta\in\Theta}\sum_{\eta\in\varphi^{-(k-j-1)}\tilde\theta}\|A^{k-1-j}(\eta)\|$$

$$\leq \|B\|_\infty k\cdot\text{const}\cdot C_{\epsilon,D}(r(\mathcal{R}_\Phi)+\epsilon)^j C_{\epsilon,A}(r(\mathcal{R}_\Phi)+\epsilon)^{k-1-j}.$$

$\square$

We have all of the ingredients to prove the main estimate for the spectral radius of a matrix-coefficient transfer operator induced by a covering map.

THEOREM 8.51. *If φ is a covering map on Θ and $\mathcal{R}$ is a corresponding matrix-coefficient transfer operator on the space of continuous sections of an ℓ-dimensional bundle $\mathcal{E}$, then the spectral radius of $\mathcal{R}$ is bounded above as follows:*

(8.111) $$r(\mathcal{R}) \leq \exp(\sup\{h_\nu + \lambda^1_\nu : \nu \in Erg\,(\varphi)\}).$$

PROOF. We will apply a "scalarization" trick similar to the method used in the proof of Theorem 8.15. First, recall the definitions given in display (8.100), and then note that

$$\|\mathcal{R}^k\|_{\mathcal{L}(C^0)} \leq \max_{\theta\in\Theta}\sum_{\eta\in\varphi^{-k}\theta}\|\Phi^k(\eta)\| = \sum_{\eta\in\varphi^{-k}\theta_k}|\Phi^k(\eta)x_\eta|$$

(8.112) $$= \sum_{\eta\in\varphi^{-k}\theta_k}\exp\sum_{j=0}^{k-1}\tilde\Phi(\tilde\varphi^j(\eta,x_\eta))$$

where the points $\theta_k \in \Theta$ and the vectors $x_\eta \in S$, using the compactness, are selected such that

$$\sum_{\eta\in\varphi^{-k}\theta_k}\|\Phi^k(\eta)\| = \max_{\theta\in\Theta}\sum_{\eta\in\varphi^{-k}\theta}\|\Phi^k(\eta)\|, \quad \|\Phi^k(\eta)\| = |\Phi^k(\eta)x_\eta|, \quad \eta \in \varphi^{-k}(\theta_k).$$

Since $\tilde\varphi$ is a covering map, the last sum in (8.112) can be estimated using the topological pressure for $\tilde\varphi$ as in display (8.106). Indeed, for a sufficiently small $\epsilon > 0$ and each θ_k, since $\varphi^{-k}\theta_k$ is (k,ϵ)-separated for φ, the set $F_k := \{(\eta,x_\eta) : \eta \in \varphi^{-k}\theta_k\} \subset \Theta_S$ is (k,ϵ)-separated for $\tilde\varphi$, see Example 8.42. Hence, we have the inequality

$$\sum_{\eta\in\varphi^{-k}\theta_k}\exp\sum_{j=0}^{k-1}\tilde\Phi(\tilde\varphi^j(\eta,x_\eta)) \leq P_k(\tilde\varphi,\tilde\Phi,\epsilon)$$

and, using the Variational Principle in Theorem 8.44,
$$\log r(\mathcal{R}) \leq P(\tilde{\varphi}, \tilde{\Phi}) = \sup\{h_\mu(\tilde{\varphi}) + \int_{\Theta_S} \tilde{\Phi}\, d\mu : \mu \in \mathrm{Erg}\,(\tilde{\varphi}, \Theta_S)\}.$$
Recall that $h_\nu(\tilde{\varphi}) = h_\nu(\varphi)$ for $\nu = \mathrm{proj}\,\mu$ by Lemma 8.45. Then, apply the Birkhoff-Khinchine Pointwise Ergodic Theorem to obtain a full μ-measure subset $\Theta_{\mu,S} \subset \Theta_S$ such that
$$\int_{\Theta_S} \tilde{\Phi}\, d\mu = \lim_{k\to\infty} k^{-1} \sum_{j=0}^{k-1} \tilde{\Phi}(\tilde{\varphi}^j(\theta, x)) = \lim_{k\to\infty} k^{-1} \log|\Phi^k(\theta) x|$$
for each $(\theta, x) \in \Theta_{\mu,S}$. Select $\theta \in \Theta_\nu \subset \Theta$ where Θ_ν is the Oseledets subset for the measure $\nu = \mathrm{proj}\,\mu$, and note that the last limit does not exceed λ_ν^1. $\square$

Next, for an expanding map φ, we will give an exact formula for the spectral radius of the matrix-coefficient transfer operator on the space of continuous sections in terms of the entropy and Oseledets-Lyapunov exponents. We start with the following improvement of Proposition 8.50 for the case of expanding maps.

PROPOSITION 8.52. *If φ is a topologically mixing, expanding map, then*
$$(8.113) \qquad r(\mathcal{R}; C^0) = \lim_{k\to\infty} \Big(\sum_{i^{(k)}} \sup_{\theta \in U_{i^{(k)}}} \|\Phi^k(\theta)\|\Big)^{1/k}$$
where the summation is taken over all admissible k-tuples $i^{(k)} = i_1 \ldots i_k$ and the set $U_{i^{(k)}}$ is defined in display (8.95).

PROOF. For an admissible $(m+k)$-tuple $i^{(m+k)} = i^{(k)} i^{(m)}$, we have the equality $U_{i^{(k+m)}} = \varphi_{i^{(k)}}^{-k}(U_{i^{(m)}})$. Using the cocycle property,
$$\sup_{\theta \in U_{i^{(k+m)}}} \|\Phi^{k+m}(\theta)\| = \sup_{\eta \in U_{i^{(m)}}} \|\Phi^{k+m}(i^{(k)} \eta)\| \leq \sup_{\theta \in \Theta} \|\Phi^m(\theta)\| \sup_{\eta \in U_{i^{(m)}}} \|\Phi^k(i^{(k)} \eta)\|;$$
and therefore,
$$\sum_{i^{(k+m)}} \sup_{\theta \in U_{i^{(k+m)}}} \|\Phi^{k+m}(\theta)\| \leq \sum_{i^{(m)}} \|\Phi^m\|_\infty \sum_{i^{(k)}} \sup_{\eta \in U_{i^{(m)}}} \|\Phi^k(i^{(k)} \eta)\|$$
$$(8.114) \qquad\qquad = \sum_{i^{(m)}} \|\Phi^m\|_\infty \sum_{i^{(k)}} \|\Phi^k(i^{(k)} \eta_1)\|.$$

Here, using the compactness, for each $i^{(m)}$ and $i^{(k)}$, we select $\eta_1 \in \overline{U}_{i^{(m)}}$ where the supremum is attained. We stress the fact that η_1 depends of $i^{(k)}$. Fix an $\eta_2 \in \overline{U}_{i^{(m)}}$ that is independent of $i^{(k)}$. Since each $\overline{U}_{i^{(m)}}$ belongs to one natural coordinate chart on Θ, see page 307, we will work locally; that is, we will treat Φ as a continuous matrix-valued function.

Fix $l = 1, \ldots, k$ and positive integers $j_1, \ldots, j_l$ such that $j_1 + \ldots + j_l = k$. Also, split each admissible k-tuple $i^{(k)} = i_1 \ldots i_k$ as $i^{(k)} = i^{(j_l)} \ldots i^{(j_2)} i^{(j_1)}$ where $i^{(j_l)} = i_1 \ldots i_{j_l}$, etc., $i^{(j_2)} = i_{1+j_3+\ldots+j_l} \ldots i_{j_2+\ldots+j_l}$, and, finally, $i^{(j_1)} = i_{1+j_2+\ldots+j_l} \ldots i_k$. For each k-tuple $i^{(k)} = i_1 \ldots i_k$ and $j = 1, \ldots, k$, let us use the notation
$$\Delta(i^{(k)}) := \Phi(i^{(k)} \eta_1) - \Phi(i^{(k)} \eta_2), \quad \Delta^j(i^{(k)}) := \Delta(i_j \ldots i_k) \cdots \Delta(i_1 \ldots i_k).$$
For each $\delta > 0$, there is some m in (8.114) so large that $\|\Delta^j(i^{(k)})\| \leq \delta^j$ because Φ is continuous and we have the inequality (see (8.96))
$$d(i_j \ldots i_k \eta_1, i_j \ldots i_k \eta_2) \leq \mathrm{const} \cdot \rho^{-(k-j+m)}, \quad \eta_1, \eta_2 \in \overline{U}_{i^{(m)}}, \quad j = 1, \ldots, k.$$

Let us represent $\Phi^k(i^{(k)}\eta_1) = \Phi(i_k\eta_1)\cdots\Phi(i_1\ldots i_k\eta_1)$ as

$$\Phi^k(i^{(k)}\eta_1) = [\Phi(i_k\eta_2) + \Delta(i_k)]\cdots[\Phi(i_1\ldots i_k\eta_2) + \Delta(i_1\ldots i_k)]$$

and note that the Φ's and Δ's in this product do not commute. Therefore, summing over the set of all possible $j_1,\ldots,j_l$ such that $j_1 + \ldots + j_l = k$, we have the representation

$$\Phi^k(i^{(k)}\eta_1) = \sum_{l=1}^{k}\sum_{j_1+\ldots+j_l=k}[\alpha(j_1,\ldots,j_l) + \beta(j_1,\ldots,j_l)].$$

Here, for each even integer l, let

$$\alpha(j_1,\ldots,j_l) := \Phi^{j_1}(i^{(j_1)}\eta_2)\Delta^{j_2}(i^{(j_2)}i^{(j_1)})\cdots\Delta^{j_l}(i^{(j_l)}\ldots i^{(j_1)}),$$
$$\beta(j_1,\ldots,j_l) := \Delta^{j_1}(i^{(j_1)})\Phi^{j_2}(i^{(j_2)}i^{(j_1)}\eta_2)\cdots\Phi^{j_l}(i^{(j_l)}\ldots i^{(j_1)}\eta_2),$$

and, for odd l,

$$\alpha(j_1,\ldots,j_l) := \Phi^{j_1}(i^{(j_1)}\eta_2)\Delta^{j_2}(i^{(j_2)}i^{(j_1)})\cdots\Phi^{j_l}(i^{(j_l)}\ldots i^{(j_1)}\eta_2),$$
$$\beta(j_1,\ldots,j_l) := \Delta^{j_1}(i^{(j_1)})\Phi^{j_2}(i^{(j_2)}i^{(j_1)}\eta_2)\cdots\Delta^{j_l}(i^{(j_l)}\ldots i^{(j_1)}).$$

Changing the order of summation, using $\|\Delta^j\| \leq \delta^j$, and the fact that η_2 does not depend on $i^{(k)}$, we have

$$\sum_{i^{(k)}}\|\Phi^k(i^{(k)}\eta_1)\|$$
$$\leq \sum_{l=1}^{k}\sum_{j_1+\ldots+j_l=k}\left\{\sum_{i^{(j_1)}}\|\Phi^{j_1}(i^{(j_1)}\eta_2)\|\sum_{i^{(j_2)}}\delta^{j_2}\sum_{i^{(j_3)}}\|\Phi^{j_3}(i^{(j_3)}i^{(j_2)}i^{(j_1)}\eta_2)\|\cdots\right.$$
$$\left.+\sum_{i^{(j_1)}}\delta^{j_1}\sum_{i^{(j_2)}}\|\Phi^{j_2}(i^{(j_2)}i^{(j_1)}\eta_2)\|\sum_{i^{(j_3)}}\delta^{j_3}\cdots\right\}.$$

Let $r := r(\mathcal{R};C^0)$ and fix $\epsilon > 0$. Using Proposition 8.50, there is some $c = c_\epsilon$ such that

$$\max_{\eta\in\Theta}\sum_{i^{(k)}}\|\Phi^k(i^{(k)}\eta)\| \leq c(r+\epsilon)^k$$

for each positive integer k. Also, $\sum_{i^{(k)}}\delta^k = (\sum_{i_1}\delta)\cdots(\sum_{i_k}\delta) \leq \delta_0^k$ for some $\delta_0 < 1$ that can be made as small as we wish provided that δ is small enough. Therefore,

$$\sum_{i^{(k)}}\|\Phi^k(i^{(k)}\eta_1)\| \leq \sum_{l=1}^{k}\sum_{j_1+\ldots+j_l=k}\left\{\left[c(r+\epsilon)^{j_1}\delta_0^{j_2}c(r+\epsilon)^{j_3}\cdots\right]\right.$$
$$\left.+\left[\delta_0^{j_1}c(r+\epsilon)^{j_2}\delta_0^{j_3}\cdots\right]\right\}.$$

According to the definitions of the α's and β's above, the first product in $\{\ \}$ ends with the factor $\delta_0^{j_l}$ for an even l and with $c(r+\epsilon)^{j_l}$ for an odd l, while the second product ends with $c(r+\epsilon)^{j_l}$ for an even l and with $\delta_0^{j_l}$ for an odd l. Assume, for example, that l is even and consider the first product. Since $\delta_0 < 1$, we have the inequality

$$\left[c(r+\epsilon)^{j_1}\delta_0^{j_2}c(r+\epsilon)^{j_3}\cdots c(r+\epsilon)^{j_{l-1}}\delta_0^{j_l}\right]$$
$$\leq (c\sqrt{\delta_0})(r+\epsilon)^{j_1}(\sqrt{\delta_0})^{j_2}\cdots(\sqrt{\delta_0})c(r+\epsilon)^{j_{l-1}}(\sqrt{\delta_0})^{j_l}$$
$$\leq (r+\epsilon)^{j_1}(\sqrt{\delta_0})^{j_2}\cdots(r+\epsilon)^{j_{l-1}}(\sqrt{\delta_0})^{j_l}$$

as long as $c\sqrt{\delta_0} \leq 1$. Similar estimates, up to a constant, hold in all other cases. As a result, we have

$$\sum_{i^{(k)}}\|\Phi^k(i^{(k)}\eta_1)\| \leq c \sum_{l=1}^{k} \sum_{j_1+\ldots+j_l=k}\left\{\left[(r+\epsilon)^{j_1}(\sqrt{\delta_0})^{j_2}(r+\epsilon)^{j_3}\cdots\right]\right.$$
$$\left.+\left[(\sqrt{\delta_0})^{j_1}(r+\epsilon)^{j_2}(\sqrt{\delta_0})^{j_3}\cdots\right]\right\} = c(r+\epsilon+\sqrt{\delta_0})^k \leq c(r+2\epsilon)^k,$$

as soon as δ is chosen such that $\sqrt{\delta_0} \leq \epsilon$. Apply this estimate in (8.114) to get

$$\sum_{i^{(k+m)}} \sup_{\theta \in U_{i^{(k+m)}}} \|\Phi^{k+m}(\theta)\| \leq c(m) \cdot c(r+2\epsilon)^k$$

with a constant $c(m)$ independent of k. □

The main result of this subsection follows.

THEOREM 8.53. *If φ is a topologically mixing, expanding map on Θ and $\mathcal{R}$ is a corresponding matrix-coefficient transfer operator on the space of continuous sections of an ℓ-dimensional bundle $\mathcal{E}$, then the spectral radius of the operator $\mathcal{R}$ is given by the formula*

(8.115) $$r(\mathcal{R}) = \exp(\sup\{h_\nu + \lambda_\nu^1 : \nu \in \text{Erg}(\varphi)\}).$$

PROOF. In the proof of the inequality $\log r(\mathcal{R}) \geq h_\nu + \lambda_\nu^1$ we will follow the strategy of the proof of the inequality "$\geq$" in the Variational Principle (Theorem 8.44) given by Bowen in [**Bo2**, Thm.2.10]. Fix $\nu \in \text{Erg}(\varphi, \Theta)$. By the Furstenberg-Kesten formula (see for example [**Wa2**, Cor.10.1.1]), we have that

(8.116) $$\lambda_\nu^1 = \lim_{n\to\infty} m^{-1} \int_\Theta \log\|\Phi^m(\theta)\| d\nu.$$

Step 1. For each finite Borel partition $\mathcal{A} = \{A_1, \ldots, A_{|\mathcal{A}|}\}$ of Θ such that each point $\theta \in \Theta$ belongs to the closure of no more than M elements of $\mathcal{A}$, we will show that

(8.117) $$h_\nu(\varphi, \mathcal{A}) + \lambda_\nu^1 \leq \log r(\mathcal{R}) + \log M,$$

cf. Lemma 2.11 in Bowen [**Bo2**]. Indeed, see the definition of $h_\nu(\varphi, \mathcal{A})$ and $H_\nu(\bigvee_{i=0}^{m-1} \varphi^{-i}\mathcal{A})$ on page 308,

$$H_\nu\Big(\bigvee_{i=0}^{m-1} \varphi^{-i}\mathcal{A}\Big) + \int_\Theta \log\|\Phi^m(\theta)\| d\nu$$
$$\leq \sum_{B \in \bigvee_{i=0}^{m-1}\varphi^{-i}\mathcal{A}} \nu(B)\Big[-\log\nu(B) + \sup_{\theta\in B}\log\|\Phi^m(\theta)\|\Big]$$
$$\leq \log\Big(\sum_{B \in \bigvee_{i=0}^{m-1}\varphi^{-i}\mathcal{A}} \sup_{\theta\in B}\|\Phi^m(\theta)\|\Big).$$

In the last inequality we have used the standard estimate

$$\sum_{i=1}^{k} p_i(a_i - \log p_i) \leq \log \sum_{i=1}^{k} \exp a_i, \quad p_i \geq 0, \quad \sum_{i=1}^{k} p_i = 1, \quad a_i \in \mathbb{R},$$

see for example Walters [**Wa2**, Lemma 9.9] or Bowen [**Bo2**, Lemma 1.1]. For each $B \in \bigvee_{i=0}^{m-1} \varphi^{-i}\mathcal{A}$, select the point $\theta_B \in \overline{B}$ where $\sup_{\theta \in B} \|\Phi^m(\theta)\|$ is attained. Since Θ is covered by the union of the $U_{i^{(m)}}$ over all m-admissible tuples $i^{(m)}$ (see (8.95)), each θ_B belongs to some $U_{i^{(m)}} =: U_{i^{(m)}}(B)$. If $\theta_{B'} \in U_{i^{(m)}}(B)$ for yet another $B' \in \bigvee_{i=0}^{m-1} \varphi^{-i}\mathcal{A}$, then $\overline{B} \cap \overline{B'} \neq \emptyset$. By the assumption in Step 1, the multiplicity of the function $B \mapsto U_{i^{(m)}}(B)$ does not exceed M^m; and therefore,

$$H_\nu\big(\bigvee_{i=0}^{m-1}\varphi^{-1}\mathcal{A}\big) + \int_\Theta \log\|\Phi^m(\theta)\|\, d\nu \leq \log\Big[M^m \sum_{i^{(m)}} \sup_{\theta \in U_{i^{(m)}}} \|\Phi^m(\theta)\|\Big].$$

The inequality (8.117) follows from the definition of $h_\nu(\varphi, \mathcal{A})$, formula (8.116), and Proposition 8.52.

Step 2. Let $\mathcal{A}$ be a finite Borel partition of Θ, fix $n \in \mathbb{N}$, and let $\mathcal{B} = \bigvee_{i=0}^{n-1} \varphi^{-i}\mathcal{A}$. Also, note that

$$h_\nu(\varphi, \mathcal{A}) = n^{-1} h_\nu(\varphi^n, \mathcal{B}) \quad \text{and} \quad \lambda_\nu^1(\varphi, \Phi) = n^{-1}\lambda_\nu^1(\varphi^n, \Phi^n)$$

where $\lambda_\nu^1(\varphi, \Phi)$ is the largest Oseledets-Lyapunov exponent of the cocycle $\{\Phi^m\}_{m \geq 0}$ over φ and $\lambda_\nu^1(\varphi^n, \Phi^n)$ is the largest exponent for the cocycle $\{(\Phi^n)^m\}_{m \geq 0}$ over φ^n. Indeed, the first equality is proved in Lemma 2.6 of [**Bo2**]; the second follows from the definition of λ_ν^1.

Fix $\epsilon > 0$ and choose an open covering $\mathcal{V} = \{V_j\}_{j=1}^{|\mathcal{V}|}$ of Θ with the following property: If $\mathcal{C} = \{C_i\}$ is a finite Borel partition subordinate to $\mathcal{V}$; that is, for each C_i there is a set V_j with $C_i \subset V_j$, then $H_\nu(\mathcal{A}\,|\,\mathcal{C}) < \epsilon$. The existence of such a $\mathcal{V}$ is proved in Lemma 2.3 of [**Bo2**], see also Lemma 2.1 and Lemma 2.2 of [**Bo2**] for properties of the conditional entropy

$$H_\nu(\mathcal{A}\,|\,\mathcal{C}) := H_\nu(\mathcal{A} \vee \mathcal{C}) - H_\nu(\mathcal{C}).$$

By Lemma 2.12 [**Bo2**], for the open covering $\mathcal{V}$ and fixed n, there exists a Borel partition $\mathcal{D}_n$ such that

(a) each $D \in \mathcal{D}_n$ is contained in some element of $\varphi^{-i}\mathcal{V}$ for $i = 0, \ldots, n-1$, and
(b) no more than $n|\mathcal{V}|$ elements of $\mathcal{D}_n$ have a point in common that belongs to each of their closures.

Using [**Bo2**, Lem. 2.2(b)], it follows that

$$h_\nu(\varphi, \mathcal{A}) + \lambda_\nu^1(\varphi, \Phi) = n^{-1}[h_\nu(\varphi^n, \mathcal{B}) + \lambda_\nu^1(\varphi^n, \Phi^n)]$$
$$\leq n^{-1}[h_\nu(\varphi^n, \mathcal{D}_n) + H_\nu(\mathcal{B}\,|\,\mathcal{D}_n) + \lambda_\nu^1(\varphi^n, \Phi^n)].$$

Step 3. Apply (8.117) in Step 1 above to the partition $\mathcal{D}_n$ in place of $\mathcal{A}$, the map φ^n in place of φ, and the transfer operator $\mathcal{R}(\varphi^n, \Phi^n)$ generated by φ^n and Φ^n. Then,

$$h_\nu(\varphi^n, \mathcal{D}_n) + \lambda_\nu^1(\varphi^n, \Phi^n) \leq \log r(\mathcal{R}(\varphi^n, \Phi^n)) + \log(n|\mathcal{V}|).$$

On the other hand, $\mathcal{R}(\varphi^n, \Phi^n) = \mathcal{R}^n$ implies

$$\log r(\mathcal{R}(\varphi^n, \Phi^n)) = \lim_{k \to \infty} k^{-1} \log \|\mathcal{R}^{nk}\| = n \log r(\mathcal{R}).$$

Also, as in the proof of Theorem 2.10 in [**Bo2**], we have
$$H_\nu(\mathcal{B}|\mathcal{D}_n) \leq \sum_{j=0}^{n-1} H_\nu(\varphi^{-i}\mathcal{A}|\mathcal{D}_n) < n\epsilon.$$
Indeed, by (a) above, elements of the partition $\mathcal{D}_n$ are contained in elements of the covering $\varphi^{-i}(\mathcal{V})$ for all $i = 0, \ldots, n-1$, and so $H_\nu(\varphi^{-i}\mathcal{A}\,|\,\mathcal{D}_n) < \epsilon$ for $i = 0, \ldots, n-1$. Combining all of these statements, we have the estimate
$$h_\nu(\varphi, \mathcal{A}) + \lambda_\nu^1(\varphi, \Phi) \leq \log r(\mathcal{R}) + n^{-1} \log(n|\mathcal{V}|) + \epsilon.$$
Finally, the proof is completed by passing to the limit as $n \to \infty$ and $\epsilon \to 0$. □

8.3.2.3. *An annulus in the spectrum.* We will specify a certain annulus that belongs to the approximate point spectrum of the matrix-coefficient transfer operator on the space of continuous sections. The main tool is a variant of Mather localization applied to the transfer operator, cf. Section 6.2.2.

We will use the following lower estimate for the spectral radius $r(\mathcal{R}; C^0)$, cf. Theorem 8.15.

PROPOSITION 8.54. *If φ is a covering map, then*
$$(8.118) \qquad r(\mathcal{R}; C^0) \geq \lim_{k \to \infty} \max_{\theta \in \Theta} \|\Phi^k(\theta)\|^{\frac{1}{k}} = \exp(\sup\{\lambda_\nu^1 : \nu \in \mathrm{Erg}\,(\Theta, \varphi)\}).$$

PROOF. The left inequality in (8.118) follows from Proposition 8.50.

If $\nu \in \mathrm{Erg}\,(\varphi)$ and θ is in the Oseledets set Θ_ν, then
$$(8.119) \qquad \lambda_\nu^1 = \lim_{k \to \infty} k^{-1} \log |\Phi^k(\theta)x| \leq \lim_{k \to \infty} k^{-1} \max_{\theta \in \Theta} \log \|\Phi^k(\theta)\|,$$
see Theorems 8.1–8.4 and Remark 8.5. This proves that the right equality in (8.118) is valid if the equal sign is replaced by "$\geq$".

To prove the reverse inequality, we use the "scalarization" idea introduced in the proof of Theorem 8.15. Recall the notations in display (8.100), and note that
$$\max_{\theta \in \Theta} \|\Phi^k(\theta)\| = \exp \max_{(\theta, x) \in \Theta_S} \frac{1}{k} \sum_{j=0}^{k-1} \tilde{\Phi}^j(\tilde{\varphi}(\theta, x)).$$
The last maximum is attained at some $(\theta_k, x_k) \in \Theta_S = \Theta \times S^{\ell-1}$. Consider the Borel probability measures μ_k on Θ_S defined by
$$\mu_k(h) = k^{-1} \sum_{j=0}^{k-1} h(\tilde{\varphi}^j(\theta_k, x_k)), \quad h \in C^0(\Theta_S), \quad k \in \mathbb{N},$$
and a w^*-limit point μ of the sequence $\{\mu_k\}_{k \in \mathbb{N}}$. Then,
$$\lim_{k \to \infty} \max_{(\theta, x) \in \Theta_S} k^{-1} \sum_{j=0}^{k-1} \tilde{\Phi}(\tilde{\varphi}^j(\theta, x)) = \int_{\Theta_S} \tilde{\Phi}(\theta, x)\, d\mu.$$
Let $\nu = \mathrm{proj}\,\mu$ be the projection of μ on Θ. By the Birkhoff-Khinchine Ergodic Theorem for μ and the Multiplicative Ergodic Theorem for ν, for every $\epsilon > 0$ there exists $(\theta, x) \in \Theta_S$ such that
$$\int_{\Theta_S} \tilde{\Phi}(\theta, x) d\mu - \epsilon \leq \lim_{k \to \infty} k^{-1} \sum_{j=0}^{k-1} \tilde{\Phi}(\tilde{\varphi}^j(\theta, x))$$
$$= \lim_{k \to \infty} k^{-1} \log |\Phi^k(\theta)x| \leq \lambda_\nu^1 \leq \sup\{\lambda_\nu^1 : \nu \in \mathrm{Erg}\,(\varphi)\}. \quad \square$$

Here, let us say a covering φ is *aperiodic*, cf. Subsection 6.2.3, provided that the set of aperiodic points for φ is dense in Θ. In particular, a topologically mixing, expanding map is aperiodic. Also, let

$$R := \lim_{k \to \infty} \max_{\theta \in \Theta} \|\Phi^k(\theta)\|^{\frac{1}{k}} = \exp(\sup\{\lambda_\nu^1 : \nu \in \mathrm{Erg}\,(\varphi)\}).$$

THEOREM 8.55. *If φ is an aperiodic covering map and $\mathcal{R}$ is a corresponding matrix-coefficient transfer operator on C^0, then the approximate point spectrum of the operator $\mathcal{R}$ contains the annulus $\{z \in \mathbb{C} : R \leq |z| \leq r(\mathcal{R}; C^0)\}$.*

PROOF. By rescaling, we may assume that $r(\mathcal{R}; C^0) = 1$ and $1 \in \sigma_{ap}(\mathcal{R})$. Fix $N \in \mathbb{N}$. Using Proposition 6.27, choose $f \in C^0$ with $\|f\| = 1$ such that

(8.120) $\quad \|(\mathcal{R}^N - I)f\|_\infty \leq 1/8 \quad \text{and} \quad \|\mathcal{R}^j f\|_\infty \leq 2, \quad j = 0, \ldots, 2N.$

Select a nonperiodic point $\theta_0 \in \Theta$ such that $|f(\theta_0)| \geq 3/4$ and define $\gamma : \mathbb{R} \to [0, 1]$ by

$$\gamma(t) = t(2N - t)/N^2 \quad \text{for} \quad t \in [0, 2N] \quad \text{and} \quad \gamma(t) = 0 \quad \text{otherwise}.$$

Then, $\gamma(0) = \gamma(2N) = 0$, $\gamma(N) = 1$, and

(8.121) $\quad |\gamma(t+1) - \gamma(t)| \leq 4/N \quad \text{for} \quad t \in \mathbb{R}.$

We will consider two cases: $R < 1$ and $R = 1$.

The Case $R < 1$. Recall that φ is a covering map and, for a sufficiently small neighborhood B of θ_0, the set $\varphi^{-1}(B)$ has disjoint components. With this in mind, choose a neighborhood B of θ_0 and a bump-function $\beta : \Theta \to [0, 1]$ such that

  (i) all components of the set $\cup_{j=0}^{2N} \varphi^{j-N}(B)$ are disjoint;
  (ii) some branch of φ^{N-j}, for $j = 0, \ldots, 2N$, maps each of these components homeomorphically onto B; and
  (iii) the function β is continuous, supported in B, and $\beta(\theta_0) = 1$.

Then, $\alpha = \beta \circ \varphi^N$ is supported in $\varphi^{-N}(B)$.

Fix z with $R < |z| \leq 1$ and define a function $g \in C^0$ by

(8.122) $\quad g(\theta) = \sum_{j=0}^{2N} z^{N-j} \gamma(j) \mathcal{R}^j(\alpha f)(\theta), \quad \theta \in \Theta,$

cf. formula (6.22). We will show that the supremum norm of g is large, $\|g\|_\infty \geq 1/2$, and that $\|\mathcal{R}g - zg\|_\infty$ is $O(1/N)$ as $N \to \infty$.

A calculation shows that

(8.123) $\quad zg(\theta) - (\mathcal{R}g)(\theta) = \sum_{j=1}^{2N} z^{N-j+1} [\gamma(j) - \gamma(j-1)](\mathcal{R}^j \alpha f)(\theta).$

Moreover, the summands on the right hand side of formula (8.123) have disjoint support:

$$\operatorname{supp} \mathcal{R}^j(\alpha f) \subset \varphi^{j-N}(B) \quad \text{for} \quad j = 0, \ldots, 2N.$$

For each $j = 0, 1, \ldots, 2N$, fix some $\theta \in \varphi^{j-N}(B)$. Note that $\alpha(\eta) \neq 0$ for $\eta \in \varphi^{-j}(\theta)$ only if $\eta \in \varphi^{-j}(\theta) \cap \varphi^{-N}(B)$. If $\eta \in \varphi^{-j}(\theta) \cap \varphi^{-N}(B)$, then $\varphi^N(\eta) \in \varphi^{N-j}(\theta \cap B)$. Thus, for each $\theta \in \varphi^{j-N}(B)$, there exists $\tilde{\theta} \in B$, independent of η, such that $\varphi^N(\eta) = \tilde{\theta}$. Hence, $\alpha(\eta) = \beta(\varphi^N(\eta)) = \beta(\tilde{\theta})$, and, as a result, for each

$j = 0, 1, \ldots N$, we have $\mathcal{R}^j(\alpha f)(\theta) = \beta(\tilde{\theta})(\mathcal{R}^j f)(\theta)$. Also, for $j = N+1, \ldots, 2N$, we have $\Phi^j(\eta) = \Phi^{j-N}(\tilde{\theta})\Phi^N(\eta)$; and therefore, $(\mathcal{R}^j \alpha f)(\theta) = \beta(\tilde{\theta})\Phi^{j-N}(\tilde{\theta})(\mathcal{R}^N f)(\tilde{\theta})$.

The inequality (8.121) can be used to estimate the norm in formula (8.123). In fact, we find that

(8.124) $$\|zg - \mathcal{R}g\|_\infty \leq 4M/N$$

where M is the maximum of the following two numbers:

(8.125) $$\max_{1 \leq j \leq N} \max_{\theta \in \varphi^{j-N}(B)} |z^{N-j+1}\beta(\tilde{\theta})(\mathcal{R}^j f)(\theta)|,$$

(8.126) $$\max_{N+1 \leq j \leq 2N} \max_{\theta \in \varphi^{j-N}(B)} |z^{N-j+1}\beta(\tilde{\theta})\Phi^{j-N}(\tilde{\theta})(\mathcal{R}^N f)(\tilde{\theta})|.$$

Since $|z| \leq 1$ and $\|\beta\|_\infty \leq 1$, we can use (8.120), for $j = 1, \ldots, N$, to see that (8.125) is less than or equal to 2. Recall that $R = \exp \sup \lambda_\nu^1 < |z|$ and fix $\epsilon > 0$ such that $|z|^{-1} \exp(\sup \lambda_\nu^1 + \epsilon) < 1$. Now, use the second inequality (8.120) with $j = N$ and Proposition 8.54 to show that (8.126) is less than or equal to

$$\max_{N+1 \leq j \leq 2N} 2|z|^{N-j+1} \max_{\theta \in \Theta} \|\Phi^{j-N}(\theta)\|$$

(8.127) $$\leq \max_{N+1 \leq j \leq 2N} 2|z|^{N-j} C_\epsilon \exp\{(j-N)(\sup \lambda_\nu^1 + \epsilon)\}$$

where the constant C_ϵ may depend upon $\epsilon > 0$, but is independent of N. By the choice of ϵ, the right hand side of inequality (8.127) is bounded by a constant independent of N. Thus, M in the estimate (8.124) does not depend on N.

To estimate $\|g\|_\infty$ from below, let us recall that $\theta_0 \in B$. Then, from (8.123) and (8.120), we have the following inequalities:

$$\|g\|_\infty \geq |g(\theta_0)| = |(\mathcal{R}^N \alpha f)(\theta_0)| \geq |(\mathcal{R}^N f)(\theta_0)| \geq |f(\theta_0)| - 1/8 \geq 1/2.$$

This result, combined with the estimate (8.124), implies the set $\{z : R < |z| \leq 1\}$ is contained in $\sigma_{ap}(\mathcal{R})$. Hence, its closure $\{z : R \leq |z| \leq 1\}$ is also contained in $\sigma_{ap}(\mathcal{R})$.

The Case $R = 1$. We give a slight modification of the construction of g. Given N, choose f and θ_0 as above. Also, find a neighborhood B of $\varphi^n(\theta_0)$ and a continuous bump-function $\beta : \Theta \to [0, 1]$ such that

(i) all components of the set $\cup_{j=0}^{2N} \varphi^{-j}(B)$ are disjoint;
(ii) some branch of φ^j, for $j = 0, 1, \ldots, 2N$, maps each of these components homeomorphically onto B; and
(iii) $\operatorname{supp} \beta \subset B$ and $\beta(\varphi^N \theta_0) = 1$.

Then, $\alpha = \beta \circ \varphi^{2N}$ is supported on $\varphi^{-2N}(B)$. With γ as above and a fixed z with $|z| = 1$, define a function $g \in C^0$ by

$$g(\theta) = \sum_{j=0}^{2N} z^{2N-j} \gamma(j)(\mathcal{R}^j \alpha f)(\theta), \quad \theta \in \Theta.$$

Since $\mathcal{R}^j(\alpha f)(\theta) = \beta \circ \varphi^{2N-j}(\theta)(\mathcal{R}^j f)(\theta)$, the terms in the sum defining g have disjoint supports. Hence, $\|g\|_\infty$ is at least as big as the norm of

$$g(\theta_0) = z^N \gamma(N) \beta \circ \varphi^N(\theta_0)(\mathcal{R}^N f)(\theta_0).$$

Since $|z| = 1 = \gamma(N) = \beta \circ \varphi^N(\theta_0)$, from (8.120) we have

$$\|g\|_\infty \geq |g(\theta_0)| = |(\mathcal{R}^N \alpha f)(\theta_0)| = |(\mathcal{R}^N f)(\theta_0)| = |f(\theta_0)| - 1/8 \geq 1/2.$$

On the other hand, $\|zg - \mathcal{R}g\|_\infty$ is again $O(1/N)$. To see this fact, note first that

$$zg(\theta) - \mathcal{R}g(\theta) = \sum_{j=1}^{2N} z^{2N-j+1}[\gamma(j) - \gamma(j-1)]\mathcal{R}^j(\beta \circ \varphi^{2N} f)(\theta)$$

$$= \sum_{j=1}^{2N} z^{2N-j+1}[\gamma(j) - \gamma(j-1)]\beta \circ \varphi^{2N-j}(\theta)(\mathcal{R}^j f)(\theta).$$

Since the terms in the sum have disjoint support, it suffices to estimate the supremum norm of each term and take the maximum. But $|\gamma(j) - \gamma(j-1)| \leq c/N$, $|z| = 1$, and $\|\beta\|_\infty \leq 1$, so we only need to estimate the maximum of $\|\mathcal{R}^j f\|$ over the integers $j = 1, \ldots, 2N$. But these are each bounded as in the second inequality of display (8.120). □

8.3.3. The transfer operator on smooth sections. Let us consider the matrix-coefficient transfer operator $\mathcal{R}$, as in definition (8.92), acting on the (complexified) space $C^{\mathbf{r}}$ of smooth sections of a smooth ℓ-dimensional bundle $\mathcal{E}$ over a smooth n-dimensional compact manifold Θ. Our main objective is to relate $\mathcal{R}$ to yet another transfer operator, $\mathcal{K}$, that acts on the space of continuous sections of a certain bundle $\mathcal{F}$ over an extended compact space $\Theta_{\mathbf{r}}$. In Subsection 8.3.3.1 we give the definitions of $\Theta_{\mathbf{r}}$ and $\mathcal{K}$. Then, in Subsection 8.3.3.2, we prove that the essential spectral radius $r_{\text{ess}}(\mathcal{R}; C^{\mathbf{r}}(\Theta))$ of $\mathcal{R}$ on $C^{\mathbf{r}}$ is equal to the spectral radius $r(\mathcal{K}; C^0(\Theta_{\mathbf{r}}))$ of $\mathcal{K}$ on $C^0(\Theta_{\mathbf{r}})$. Finally, in subsection 8.3.3.3, we give estimates for $r_{\text{ess}}(\mathcal{R})$ using the Oseledets-Lyapunov exponents for $\{\Phi^k\}$ and $\{D\varphi^k\}$.

8.3.3.1. *The extended transfer operator.* In this subsection we extend Θ to a larger compact space $\Theta_{\mathbf{r}}$, construct a transfer operator $\mathcal{K}$ on $C^0(\Theta_{\mathbf{r}})$, and an intertwining operator $\mathcal{D} : C^{\mathbf{r}}(\Theta) \to C^0(\Theta_{\mathbf{r}})$ such that the difference $\mathcal{DR} - \mathcal{KD}$ is a compact operator.

To fix notations, recall that $\mathcal{E} = (\mathcal{E}, \Theta, p)$ is an ℓ-dimensional vector bundle with the projection $p : \mathcal{E} \to \Theta$, fibers $\mathcal{E}_\theta \simeq \mathbb{R}^\ell$, and base Θ. Here Θ is an n-dimensional smooth compact manifold with the tangent bundle $\mathcal{T}\Theta = (\mathcal{T}\Theta, \Theta, \pi)$ and natural projection $\pi : \mathcal{T}\Theta \to \Theta$. Let $\mathcal{T}^1\Theta = (\mathcal{T}^1\Theta, \Theta, \pi)$ be the unit tangent bundle with the fibers $\mathcal{T}_\theta^1\Theta$ and projection $\pi : w \mapsto \theta$. Here, if in local coordinates $w = (\theta, v)$, then $v \in \mathcal{T}_\theta^1\Theta$ and $|v|_{\mathbb{R}^n} = 1$. For $\mathbf{r} = 1, 2, \ldots$, we use boldface to denote $\mathbf{r}$-tuples, for example, $\mathbf{w} = (w_1, \ldots, w_{\mathbf{r}})$, $\mathbf{v} = (v_1, \ldots, v_{\mathbf{r}})$, etc.

For each positive integer $\mathbf{r}$, define the bundle $\mathcal{T}^1\Theta^{(\mathbf{r})} = (\mathcal{T}^1\Theta^{(\mathbf{r})}, \Theta, \bar{\pi})$ to be

$$\mathcal{T}^1\Theta^{(\mathbf{r})} = \{\mathbf{w} = (w_1, \ldots, w_{\mathbf{r}}) \in \mathcal{T}^1\Theta \times \ldots \times \mathcal{T}^1\Theta :$$
$$w_i \in \mathcal{T}^1\Theta, \quad i = 1, \ldots, \mathbf{r}, \quad \pi(w_1) = \ldots = \pi(w_{\mathbf{r}})\}$$

with fibers $\mathcal{T}^1\Theta_\theta^{(\mathbf{r})} = \mathcal{T}_\theta^1\Theta \times \ldots \times \mathcal{T}_\theta^1\Theta$ and the projection $\bar{\pi} : \mathbf{w} = (\theta, \mathbf{v}) \mapsto \theta = \pi(w_i)$. Here $\mathbf{w} = (w_1 \ldots, w_{\mathbf{r}})$, $w_i = (\theta, v_i)$, and $\mathbf{v} \in \mathcal{T}^1\Theta_\theta^{(\mathbf{r})}$; that is, $\mathbf{v} = (v_1, \ldots, v_{\mathbf{r}})$, $v_i \in \mathcal{T}_\theta^1\Theta$, and $|v_i|_{\mathbb{R}^n} = 1$, for $i = 1, \ldots, \mathbf{r}$. We identify $\mathcal{T}^1\Theta^{(\mathbf{r})}$ with the compact manifold

$$\Theta_{\mathbf{r}} = \{(\theta, \mathbf{v}) : \theta \in \Theta, \mathbf{v} = (v_1, \ldots, v_{\mathbf{r}}), v_i \in \mathcal{T}_\theta^1\Theta, i = 1, \ldots, \mathbf{r}\}$$

by the map $q : \mathcal{T}^1\Theta^{(\mathbf{r})} \to \Theta_{\mathbf{r}} : \mathbf{w} \mapsto (\bar{\pi}(\mathbf{w}), \mathbf{v})$ with $\mathbf{w} = (\theta, \mathbf{v})$, $w_i = (\theta, v_i)$, $i = 1, \ldots \mathbf{r}$, $\mathbf{v} \in \mathcal{T}^1\Theta_\theta^{(\mathbf{r})}$. Locally, $\Theta_{\mathbf{r}}$ is a product of a coordinate neighborhood in Θ with $\mathbf{r}$-copies of the sphere $S^{n-1} = \{v \in \mathbb{R}^n : |v|_{\mathbb{R}^n} = 1\}$.

8.3. THE RUELLE TRANSFER OPERATOR

Consider the bundle $\mathcal{F} = \{(e, \mathbf{w}) \in \mathcal{E} \times T^1\Theta^{(\mathbf{r})} : p(e) = \bar{\pi}(\mathbf{w})\}$ over $\Theta_{\mathbf{r}}$ with projection $\mathcal{F} \to \Theta_{\mathbf{r}} : (e, \mathbf{w}) \mapsto q(\mathbf{w})$ and fibers $\mathcal{F}_{(\theta, \mathbf{v})} \simeq \mathbb{R}^{\ell}$. Locally,

$$\mathcal{F} = \{(\theta, \mathbf{v}, x) : \theta \in \Theta, \mathbf{v} \in S^{n-1} \times \ldots \times S^{n-1} \text{ (}\mathbf{r}\text{ times)}, x \in \mathbb{R}^{\ell}\}.$$

Define $\psi : \Theta_{\mathbf{r}} \to \Theta_{\mathbf{r}}$ and a map $\Psi(\theta, \mathbf{v}) : \mathcal{F}_{(\theta, \mathbf{v})} \to \mathcal{F}_{\psi(\theta, \mathbf{v})}$ as follows:

$$\psi : (\theta, \mathbf{v}) \mapsto \left(\varphi\theta, \frac{D\varphi(\theta)v_1}{|D\varphi(\theta)v_1|}, \ldots, \frac{D\varphi(\theta)v_{\mathbf{r}}}{|D\varphi(\theta)v_{\mathbf{r}}|}\right),$$

(8.128)
$$\Psi(\theta, \mathbf{v}) = \Phi(\theta) \prod_{j=1}^{\mathbf{r}} |D\varphi(\theta)v_j|^{-1}$$

where $D\varphi(\theta)$ is differential of φ and $\mathbf{v} = (v_1, \ldots, v_{\mathbf{r}}) \in T^1\Theta_{\theta}^{(\mathbf{r})}$. Note that for each fixed $\theta \in \Theta$ the map

$$\psi(\theta, \cdot) : \mathbf{v} \mapsto \left(\varphi\theta, D\varphi(\theta)v_1/|D\varphi(\theta)v_1|, \ldots, D\varphi(\theta)v_{\mathbf{r}}/|D\varphi(\theta)v_{\mathbf{r}}|\right)$$

is a homeomorphism of $T^1\Theta_{\theta}^{(\mathbf{r})}$ on $T^1\Theta_{\varphi\theta}^{(\mathbf{r})}$. Therefore, ψ is a covering since φ is a covering.

The maps ψ and Ψ give rise to a transfer operator $\mathcal{K}$ on the space $C^0(\Theta_{\mathbf{r}}) = C^0_{\Theta_{\mathbf{r}}}(\mathcal{F})$ of continuous sections of $\mathcal{F}$ defined by

(8.129)
$$(\mathcal{K}F)(\theta, \mathbf{v}) = \sum_{(\eta, \mathbf{u}) \in \psi^{-1}(\theta, \mathbf{v})} \Psi(\eta, \mathbf{u})F(\eta, \mathbf{u}), \quad (\theta, \mathbf{v}) \in \Theta_{\mathbf{r}}.$$

In other words,

$$(\mathcal{K}F)(\theta, \mathbf{v}) = \sum_i \Phi(i\theta) \prod_{j=1}^{\mathbf{r}} |[D\varphi(i\theta)]^{-1}v_j| F(i\theta, \mathbf{u}),$$

where $\mathbf{v} = (v_1, \ldots, v_{\mathbf{r}})$ and $\mathbf{u} = (u_1, \ldots, u_{\mathbf{r}})$ with

$$u_j = [D\varphi(i\theta)]^{-1}v_j / |[D\varphi(i\theta)]^{-1}v_j|, \quad j = 1, \ldots, \mathbf{r}.$$

Here, we use the notation $i\theta$ introduced on page 307, see also Proposition 8.41.

By Proposition 8.50, the spectral radius of $\mathcal{K}$ on $C^0(\Theta_{\mathbf{r}})$ can be computed as follows:

$$r(\mathcal{K}; C^0(\Theta_{\mathbf{r}})) = \lim_{k \to \infty} \max_{(\theta, \mathbf{u})} \left(\sum_{(\eta, \mathbf{u}) \in \psi^{-k}(\theta, \mathbf{v})} \|\Psi^k(\eta, \mathbf{u})\|\right)^{1/k}$$

where

$$\Psi^k(\theta, \mathbf{v}) = \Phi^k(\theta) \prod_{j=1}^{\mathbf{r}} |D\varphi^k(\theta)v_j|^{-1}, \quad \mathbf{v} = (v_1, \ldots, v_{\mathbf{r}}).$$

The main goal of this subsection is to prove the following result.

THEOREM 8.56. *If φ is an aperiodic covering map, then*

(8.130)
$$r_{ess}(\mathcal{R}; C^{\mathbf{r}}(\Theta)) = r(\mathcal{K}; C^0(\Theta_{\mathbf{r}})).$$

We remark that the difficult part of the proof is to show the inequality "$\geq$"; this is done in Subsection 8.3.3.2. The reverse inequality "$\leq$" is proved in the remaining part of this subsection. In fact, this inequality is proved using general operator-theoretical arguments and the following auxiliary construction. Consider the operator $\mathcal{D} : C^{\mathbf{r}}(\Theta) \to C^0(\Theta_{\mathbf{r}})$ defined by

$$(\mathcal{D}f)(\theta, \mathbf{v}) = (D^{\mathbf{r}}f)(\theta)(\mathbf{v}), \quad \mathbf{v} = (v_1, \ldots, v_{\mathbf{r}}), \quad v_i \in T^1_{\theta}\Theta,$$

and recall that the norm on $C^{\mathbf{r}}$ is defined as

$$\|f\|_{C^{\mathbf{r}}} = \max\{\|f\|_{C(\Theta)}, \|\mathcal{D}^k f\|_{C(\Theta_k)}, k = 1, \ldots, \mathbf{r}\}. \tag{8.131}$$

Here, $\|f\|_{C(\Theta)} := \max_{\theta \in \Theta} |f(\theta)|_{\mathbb{R}^\ell}$, and the norm $\|\mathcal{D}^k f\|_{C(\Theta_k)}$ is defined to be

$$\max\{|(D^k f)(\theta)(v_1, \ldots, v_k)|_{\mathbb{R}^\ell} : \theta \in \Theta, v_j \in T_\theta \Theta, |v_j|_{\mathbb{R}^n} = 1, j = 1, \ldots k\}$$

where D is the differentiation operator. Let us note that the operator $\mathcal{D}$ has closed range and $\dim \operatorname{Ker} \mathcal{D} = n\mathbf{r} < \infty$. Thus, $\mathcal{D}$ is a Φ_+-operator, see for example Gohberg and Fel'dman [**GF**, Sect. I. 11]. In other words, there is an operator $\mathcal{D}^{-1}$ that is bounded as an operator from $C^0(\Theta_{\mathbf{r}})$ to $C^{\mathbf{r}}(\Theta)$ and such that $\mathcal{D}^{-1}\mathcal{D} - I$ is a compact operator on $C^{\mathbf{r}}(\Theta)$.

As the following result shows, the appearance of the operator $\mathcal{K}$ as in (8.129) is quite natural.

LEMMA 8.57. *There is a compact operator L from $C^{\mathbf{r}}(\Theta)$ to $C^0(\Theta_{\mathbf{r}})$ such that $\mathcal{D}\mathcal{R}f = \mathcal{K}\mathcal{D}f + Lf$ for every $f \in C^{\mathbf{r}}(\Theta)$.*

PROOF. Choose a smooth finite partition of unity $\{\rho_i\}$ for Θ; that is,

$$\sum_i \rho_i(\theta) = 1, \quad \theta \in \Theta, \quad \operatorname{supp} \rho_i \setminus (\cup_{j \neq i} \operatorname{supp} \rho_j) \neq \emptyset.$$

By representing the functions f that appear in the lemma in the form $f = \sum_i \rho_i f$, we can assume with no loss of generality that $\operatorname{supp} f$ is arbitrarily small. Thus, we can compute in local coordinates.

For the case where $\mathbf{r} = 1$, use the product and chain rules to obtain

$$(D\mathcal{R}f)(\theta, v) = \sum_{\eta \in \varphi^{-1}\theta} \Phi(\eta)(Df)(\eta)[D\varphi(\eta)]^{-1}(v) + (Lf)(\theta, v) \tag{8.132}$$

where $(Lf)(\theta, v) = \sum_{\eta \in \varphi^{-1}\theta} A(\eta, \theta, v) f(\eta)$ for some $(\ell \times \ell)$-matrices $A(\eta, \theta, v)$.

The second term, Lf, in (8.132) does not contain any differentiations. As a result, L is a compact operator. Indeed, note that the natural embedding $\operatorname{Id} : C^1(\mathcal{E}) \to C^0(\mathcal{E})$ is compact. Also, define a bounded operator on $C^0(\Theta_1)$ by $(\mathcal{A}F)(\theta, v) = \sum_\eta A(\eta, \theta, v) F(\eta, v)$ and a bounded embedding $J : C^0(\Theta) \to C^0(\Theta_1)$ by $(Jf)(\theta, v) \equiv f(\theta)$. Then, the operator $L = \mathcal{A} \circ J \circ \operatorname{Id}$ is compact.

To handle the first term on the right hand side of identity (8.132), use the definition of ψ to see that $u = [D\varphi(\eta)]^{-1}(v)/|[D\varphi(\eta)]^{-1}(v)|$ for $(\eta, u) \in \psi^{-1}(\theta, v)$ and, also using the definition of Ψ, rewrite this term in the form

$$\sum_{\eta \in \varphi^{-1}\theta} \Phi(\eta) |[D\varphi(\eta)]^{-1}(v)| \cdot \left| Df(\eta) \left(\frac{[D\varphi(\eta)]^{-1} v}{|[D\varphi(\eta)]^{-1} v|} \right) \right|_{\mathbb{R}^n}$$

$$= \sum_{(\eta, u) \in \psi^{-1}(\theta, v)} \Psi(\eta, u)(Df)(\eta, u) = (\mathcal{K}\mathcal{D}f)(\theta, v).$$

For the case where $\mathbf{r} \geq 2$ we use the chain rule, see for example Abraham, Marsden, and Ratiu [**AMR**, p. 97],

$$D^{\mathbf{r}}(g \circ h)(\theta)(v_1, \ldots, v_{\mathbf{r}}) = D^{\mathbf{r}} g(h(\theta))(D^1 h(\theta)(v_1), \ldots, D^1 h(\theta)(v_{\mathbf{r}}))$$
$$+ \text{lower order terms}$$

with $g(\cdot) = \Phi(\cdot)f(\cdot)$ and $h = \varphi_i^{-1}$ to compute

$$(D^{\mathbf{r}}\mathcal{R}f)(\theta)(\mathbf{v}) = D^{\mathbf{r}}\left(\sum_i \Phi(\varphi_i^{-1}(\cdot))f(\varphi_i^{-1}(\cdot))\right)(\theta)(v_1,\ldots,v_{\mathbf{r}})$$

$$= \sum_i \Phi(\varphi_i^{-1}\theta)\{D^{\mathbf{r}}f(\varphi_i^{-1}\theta)(D\varphi_i^{-1}(\theta)(v_1),\ldots,D\varphi_i^{-1}(\theta)(v_{\mathbf{r}}))$$

$$+ \text{ lower order terms with derivatives of } f \text{ and } \Phi\}$$

$$= \sum_i \Phi(\varphi_i^{-1}\theta) \prod_{j=1}^{\mathbf{r}} |[D\varphi(\varphi_i^{-1}\theta)]^{-1}v_j| \cdot D^{\mathbf{r}}f(\varphi_i^{-1}\theta)$$

$$\left(\frac{[D\varphi(\varphi_i^{-1}\theta)]^{-1}v_1}{|[D\varphi(\varphi_i^{-1}\theta)]^{-1}v_1|},\ldots,\frac{[D\varphi(\varphi_i^{-1}\theta)]^{-1}v_{\mathbf{r}}}{|[D\varphi(\varphi_i^{-1}\theta)]^{-1}v_{\mathbf{r}}|}\right) + (Lf)(\theta,\mathbf{v})$$

where $Lf(\theta,\mathbf{v})$ contains all terms with derivatives of f of order $\leq \mathbf{r}-1$. This completes the proof of Lemma 8.57. □

Since the constructions of the operators $\mathcal{K}$ and $\mathcal{D}$ are, in general, rather cumbersome, let us illustrate them with the following examples.

EXAMPLE 8.58. Suppose that $\mathbf{r} = 1$, $n = 1$, and $\Theta = [0,1]$, and consider the scalar case, $\ell = 1$, where $\mathcal{D}$ is just the differentiation operator $\mathcal{D}f = f'$. Clearly, $\Theta_1 = [0,1]$ and $\psi = \varphi$. Also,

$$(\mathcal{R}f)' = \mathcal{K}f' + \sum_{\eta \in \varphi^{-1}\theta} [\Phi(\eta)]'f(\eta) \quad \text{where} \quad \mathcal{K}f' = \sum_\eta \frac{\Phi(\eta)}{|\varphi'(\eta)|}f'(\eta). \quad \diamond$$

EXAMPLE 8.59. For $\mathbf{r} = 1$, $n > 1$, $\Theta \subset \mathbb{R}^n$, and $\ell = 1$, we have $(\mathcal{D}f)(\theta,v) = (\nabla f)(\theta) \cdot v$ for the gradient ∇ and $v \in S^{n-1}$. Here we view v as an $(n \times 1)$-vector and $\nabla f(\theta)$ as a $(1 \times n)$-covector. Also,

$$\Theta_{\mathbf{r}} = \Theta \times S^{n-1} \quad \text{and} \quad \psi(\theta,v) = \left(\varphi\theta, \frac{D\varphi(\theta)v}{|D\varphi(\theta)v|}\right).$$

Compute the gradient of $\mathcal{R}f$ and collect the terms with no derivatives of f as the value Lf, to obtain the result in Lemma 8.57. For a fixed $\eta \in \varphi^{-1}\theta$, use the chain rule to obtain the identities

$$\nabla[f(\eta)] = (\nabla f)(\eta) \cdot [D\varphi(\eta)]^{-1},$$
$$(\mathcal{K}f)(\theta,v) = \sum_{\eta \in \varphi^{-1}\theta} \Phi(\eta)(\nabla f)(\eta) \cdot [D\varphi(\eta)]^{-1}v.$$

Let us divide and multiply the last term by $|[D\varphi(\eta)]^{-1}v|$ and note that

$$\psi^{-1}(\theta,v) = \{(\eta,u) : \theta = \varphi(\eta) \quad \text{and} \quad v = D\varphi(\eta)u/|D\varphi(\eta)u|\}.$$

If $\Psi(\theta,v)$ is defined as $\Psi(\theta,v) = \Phi(\theta)/|D\varphi(\theta)v|$, then, for $(\eta,u) \in \psi^{-1}(\theta,v)$, we have that $\Phi(\eta)|[D\varphi(\eta)]^{-1}v| = \Psi(\eta,u)$ and

$$(\mathcal{K}f)(\theta,v) = \sum_{(\eta,u) \in \psi^{-1}(\theta,v)} \Psi(\eta,u)(\nabla f)(\eta) \cdot u.$$

$\diamond$

Using Lemma 8.57, the inequality "$\leq$" in statement (8.130) of Theorem 8.56 is easy to prove. Indeed, recall the Nussbaum formula for the essential spectral radius of an operator A; namely,

$$r_{\text{ess}}(A) = \lim_{m \to \infty} \inf_{L \in \mathcal{C}} \|A^m - L\|^{1/m}$$

where the infimum is taken over the set $\mathcal{C}$ of compact operators, see for example Parry and Pollicott [**PPo**]. Using Lemma 8.57 and, for $\mathcal{D}$, the operator $\mathcal{D}^{-1}$ such that $\mathcal{D}^{-1}\mathcal{D} - I \in \mathcal{C}$, we have

$$\mathcal{R} = \mathcal{D}^{-1}\mathcal{K}\mathcal{D} + L_1, \quad \mathcal{R}^m = \mathcal{D}^{-1}\mathcal{K}^m\mathcal{D} + L_m,$$

for some $L_m \in \mathcal{C}(C^{\mathbf{r}}(\Theta))$. Thus,

$$\inf_{L \in \mathcal{C}} \|\mathcal{R}^m - L\| \leq \|\mathcal{D}^{-1}\mathcal{K}^m\mathcal{D}\| \leq c\|\mathcal{K}^m\|,$$

and the inequality $r_{\text{ess}}(\mathcal{R}; C^{\mathbf{r}}(\Theta)) \leq r(\mathcal{K}; C^0(\Theta_{\mathbf{r}}))$ is proved.

8.3.3.2. *The lower estimate.* We will finish the proof of Theorem 8.56 by showing that

(8.133) $$r_{\text{ess}}(\mathcal{R}; C^{\mathbf{r}}(\Theta)) \geq r(\mathcal{K}; C^0(\Theta_{\mathbf{r}})).$$

Throughout this subsection we will let $r := r(\mathcal{K}; C^0(\Theta_{\mathbf{r}}))$. Using the rescaling $\mathcal{K} \mapsto r^{-1}\mathcal{K}$, we can and will assume that $r = 1$. Since $\mathcal{K}$ acts on C^0-sections, Theorem 8.55 shows that the entire unit circle is contained in the approximate point spectrum $\sigma_{ap}(\mathcal{K}; C^0(\Theta_{\mathbf{r}}))$. In fact, we will show in a moment that if $|z| = 1$, then $z \in \sigma_{ap}(\mathcal{R}; C^{\mathbf{r}}(\Theta))$. In particular, each $z \in \sigma_{ap}(\mathcal{R})$ with $|z| = 1$ is not an isolated eigenvalue. Therefore, $r_{\text{ess}}(\mathcal{R}; C^{\mathbf{r}}(\Theta)) \geq 1$ and we have the desired estimate (8.133).

To prove that $|z| = 1$ implies $z \in \sigma_{ap}(\mathcal{R})$, it suffices to show the implication

(8.134) $$1 \in \sigma_{ap}(\mathcal{K}; C^0(\Theta_{\mathbf{r}})) \quad \text{implies} \quad 1 \in \sigma_{ap}(\mathcal{R}; C^{\mathbf{r}}(\Theta)).$$

Indeed, suppose $\mathcal{K}_z$ and $\mathcal{R}_z$ are (rescaled) transfer operators related to the cocycle $\{z^{-m}\Phi^m\}$ instead of $\{\Phi^m\}$. For $|z| = 1$, the inclusion $z \in \sigma_{ap}(\mathcal{K}; C^0(\Theta_{\mathbf{r}}))$ implies $1 \in \sigma_{ap}(\mathcal{K}_z; C^0(\Theta_{\mathbf{r}}))$. By the implication (8.134) for the operators $\mathcal{R}_z$ and $\mathcal{K}_z$, it follows that $1 \in \sigma_{ap}(\mathcal{R}_z; C^{\mathbf{r}}(\Theta))$ or $z \in \sigma_{ap}(\mathcal{R}; C^{\mathbf{r}}(\Theta))$.

Therefore, in this subsection, under the assumption that $1 \in \sigma_{ap}(\mathcal{K}; C^0(\Theta_{\mathbf{r}}))$ and the spectral radius r of $\mathcal{K}$ on $C^0(\Theta_{\mathbf{r}})$ is $r = 1$, we will show that $1 \in \sigma_{ap}(\mathcal{R}; C^{\mathbf{r}}(\Theta))$. The idea is, again, to use a variant of Mather localization, cf. Subsection 6.2.2 and the proof of Theorem 8.55, to construct an approximate eigenfunction, g, for $\mathcal{R}$ in $C^{\mathbf{r}}(\Theta)$ under the assumption that $1 \in \sigma_{ap}(\mathcal{K}; C^0(\Theta_{\mathbf{r}}))$.

Fix $\epsilon < 1/8$. We will construct $g = g_\epsilon \in C^{\mathbf{r}}(\Theta)$ such that

(8.135) $$\|g_\epsilon - \mathcal{R}g_\epsilon\|_{C^{\mathbf{r}}} \leq O(\epsilon)\|g_\epsilon\|_{C^{\mathbf{r}}} \quad \text{as} \quad \epsilon \to 0.$$

Since $r(\mathcal{K}; C^0(\Theta_{\mathbf{r}})) = 1$, there exists a constant $C = C_{\epsilon^2}$ such that, for each positive integer k, we have

(8.136) $$\|\mathcal{K}^k\|_{\mathcal{L}(C^0(\Theta_{\mathbf{r}}))} \leq C_{\epsilon^2}(1 + \epsilon^2)^k.$$

Choose $N = N(\epsilon)$ sufficiently large (we will specify the exact values later, see (8.145)). Using the assumption that $1 \in \sigma_{ap}(\mathcal{K}; C^0(\Theta_{\mathbf{r}}))$ and Proposition 6.27, pick $F = F_{\epsilon,N}$ from $C^0(\Theta_{\mathbf{r}})$ such that

(8.137) $$\|F\|_{C^0(\Theta_{\mathbf{r}})} = 1, \quad \|\mathcal{K}^N F - F\|_{C^0(\Theta_{\mathbf{r}})} \leq \epsilon,$$

(8.138) $$\|\mathcal{K}^k F\|_{C^0(\Theta_{\mathbf{r}})} \leq 2, \quad k = 0, \ldots, 2N.$$

Choose $\delta = \delta(\epsilon, N)$ sufficiently small (see (8.144) for the exact values).

Fix $(\theta_0, \mathbf{v}^0) \in \Theta_{\mathbf{r}}$, where $\mathbf{v}^0 = (v_1^0, \ldots, v_{\mathbf{r}}^0)$, such that $\theta_0 \in \Theta$ is φ-aperiodic and

(8.139) $$|F(\theta_0, \mathbf{v}^0)| \geq 1/2.$$

In what follows we denote the preimages of $(\theta_0, \mathbf{v}^0)$ under ψ^N by

$$(\eta_0, \mathbf{u}^0) = (\eta_0, (u_1^0, \ldots, u_{\mathbf{r}}^0))$$

where

$$u_j^0 = [D\varphi^N(\eta_0)]^{-1} v_j^0 / |[D\varphi^N(\eta_0)]^{-1} v_j^0|, \quad j = 1, \ldots, \mathbf{r}.$$

Choose a small open set $B \subset \Theta$ such that $\theta_0 \in B$ and the components $\varphi^{N-k}(B)$, for $k = 0, \ldots, 2N+1$, are disjoint. Also, choose a bump-function $\beta : \Theta \to [0, 1]$ such that

(i) $\operatorname{supp} \beta \subset \varphi^{-N}(B)$,
(ii) $\|\beta\|_{C^{\mathbf{r}-1}} \leq \delta$,
(iii) $\|\beta\|_{C^{\mathbf{r}}} \leq 1$, and
(iv) $D^{\mathbf{r}}\beta(\eta)(\mathbf{u}^0) = 1$ for each $\eta \in \varphi^{-N}(\theta_0)$.

Define $h : \Theta \to \mathbb{R}^\ell$ with $\operatorname{supp} h \subset \varphi^{-N}(B)$ by

(8.140) $$h(\eta) = \beta(\eta) F(\eta_0, \mathbf{u}^0)$$

for those $\eta \in \varphi_{i(N)}^{-N}(B)$ that belongs to the same component of $\varphi^{-N}(B)$ that contains $\eta_0 \in \varphi_{i(N)}^{-N}(B)$. Here, $\varphi_{i(N)}^{-N}$ denotes a branch of the inverse map for φ^N, see page 307. Also, define $g_\epsilon \in C^{\mathbf{r}}(\Theta)$ according to the formula (cf. (6.22) and (8.122))

(8.141) $$g_\epsilon(\theta) = (1-\epsilon)^{|N-k|}(\mathcal{R}^k h)(\theta), \quad \theta \in \varphi^{k-N}(B), \quad k = 0, \ldots, 2N.$$

We claim that the inequality (8.135) holds for this g_ϵ.

Note that the supports of the functions $\mathcal{R}^k h$, $k = 0, \ldots, 2N$, in (8.141) are disjoint. Thus, from the definition of g_ϵ it follows that (cf. (6.22) and (8.122))

$$g_\epsilon = \sum_{k=0}^{2N} (1-\epsilon)^{|N-k|} \mathcal{R}^k h \quad \text{and} \quad \mathcal{R} g_\epsilon = \sum_{k=1}^{2N+1} (1-\epsilon)^{|N-k+1|} \mathcal{R}^k h.$$

Hence,

$$g_\epsilon - \mathcal{R} g_\epsilon = (1-\epsilon)^{N+1} h + g_{1,\epsilon} - (1-\epsilon)^N \mathcal{R}^{2N+1} h$$

where

$$g_{1,\epsilon} := \sum_{k=0}^{2N} [(1-\epsilon)^{|N-k|} - (1-\epsilon)^{|N-k+1|}] \mathcal{R}^k h.$$

Due to the fact that the functions $\mathcal{R}^k h$, for $k = 0, \ldots, 2N$, have disjoint supports, we have that

$$\|g_{1,\epsilon}\|_{C^{\mathbf{r}}} = \sup_{0 \leq k \leq 2N} \|[(1-\epsilon)^{|N-k|} - (1-\epsilon)^{|N-k+1|}] \mathcal{R}^k h\|$$

$$\leq \sup_{0 \leq k \leq 2N} |1 - (1-\epsilon)^{|N-k+1|-|N-k|}| \sup_{0 \leq k \leq 2N} \|(1-\epsilon)^{|N-k|} \mathcal{R}^k h\|_{C^{\mathbf{r}}}$$

$$\leq 2\epsilon \|g_\epsilon\|_{C^{\mathbf{r}}};$$

and consequently,

(8.142) $$\|g_\epsilon - \mathcal{R} g_\epsilon\|_{C^{\mathbf{r}}} \leq (1-\epsilon)^{N+1} \|h\|_{C^{\mathbf{r}}} + 2\epsilon \|g_\epsilon\|_{C^{\mathbf{r}}} + (1-\epsilon)^N \|R^{2N+1} h\|_{C^{\mathbf{r}}}.$$

In addition, $\|g_\epsilon\|_{C^\mathbf{r}} \geq \|\mathcal{R}^N h\|_{C^\mathbf{r}}$. Since
$$\|h\|_{C^\mathbf{r}} \leq \|\beta\|_{C^\mathbf{r}} |F(\eta_0, \mathbf{u}^0)| \leq 1,$$
the inequality (8.135) will be established as soon as we are able to estimate $\|\mathcal{R}^N h\|_{C^\mathbf{r}}$ from below and $\|\mathcal{R}^{2N+1} h\|_{C^\mathbf{r}}$ from above. We give the estimates for the **r**-th derivative, lower order derivatives can be treated similarly.

First, we show that

(8.143) $$\|g_\epsilon\|_{C^\mathbf{r}} \geq 1/4$$

for g_ϵ defined in display (8.141), provided that $\delta = \delta(\epsilon, N)$ in (ii) on page 329 is chosen sufficiently small. Since $\theta_0 \in B$, formula (8.141) shows that $g_\epsilon(\theta) = (\mathcal{R}^N h)(\theta)$ for $\theta \in B$. Recall, see (8.131), that $\|g_\epsilon\|_{C^\mathbf{r}} \geq \|\mathcal{D}g_\epsilon\|_{C^0(\Theta_\mathbf{r})} \geq |(\mathcal{D}R^N h)(\theta_0, \mathbf{v}^0)|$, and apply Lemma 8.57 N times to get

$$\mathcal{D}\mathcal{R}^N h = \mathcal{K}^N \mathcal{D}h + L_N h$$

where $L_N h$ contains the derivatives of h of order not exceeding $\mathbf{r} - 1$. Since h is defined as in (8.140) and $\|\beta\|_{C^{\mathbf{r}-1}}$ is bounded by δ according to (ii) on page 329, we conclude that $\|L_N h\|_{C^0(\Theta_\mathbf{r})} \leq C(N)\delta$ with a constant $C(N)$ depending on $\|L_N\|$. As a result, by the triangle inequality,

$$\|g_\epsilon\|_C \geq |(\mathcal{K}^N \mathcal{D}h)(\theta_0, \mathbf{v}^0)| - C(N)\delta.$$

However, by the identity (8.140), $(D^\mathbf{r} h)(\eta)(\mathbf{u}^0) = F(\eta, \mathbf{u}^0) D^\mathbf{r}\beta(\eta)(\mathbf{u}^0)$ and

$$(\mathcal{K}^N \mathcal{D}h)(\theta_0, \mathbf{v}^0) = \sum_{(\eta, \mathbf{u}^0) \in \psi^{-N}(\theta_0, \mathbf{V}^0)} \Psi^N(\eta, \mathbf{u}^0) F(\eta, \mathbf{u}^0) D^\mathbf{r}\beta(\eta)(\mathbf{u}^0)$$
$$= (\mathcal{K}^N f)(\theta_0, \mathbf{v}^0).$$

Indeed, $D^\mathbf{r}\beta(\eta)(\mathbf{u}^0) = 1$ by the choice of β in (iv) on page 329. Therefore, by the inequality (8.139), the desired estimate (8.143) holds as soon as δ is selected such that $C(N)\delta \leq 1/4$.

Next, we estimate $\|\mathcal{R}^{2N+1}h\|_{C^\mathbf{r}}$ from above. Again, we give the estimate for the **r**-th derivative; lower order derivatives are treated similarly. Indeed, apply Lemma 8.57 $2N+1$ times for $f = h$ to get $\mathcal{D}\mathcal{R}^{2N+1} h = \mathcal{K}^{2N+1} \mathcal{D}h + L_{2N+1} h$ where L_{2N+1} does not contain **r**-th derivatives of h. Therefore,

$$\|L_{2N+1} h\|_{C^0(\Theta_\mathbf{r})} \leq C(N)\|h\|_{C^{\mathbf{r}-1}} \leq C(N)\delta$$

by the choice of h in (8.140) and β in (ii) on page 329. Also, by statement (iii) on page 329, it follows that $\|\mathcal{D}h\|_{C^0(\Theta_\mathbf{r})} \leq 1$. Since $\|\mathcal{K}^{2N+1}\|$ satisfies (8.136), we have

$$\|\mathcal{R}^{2N+1}h\|_{C^\mathbf{r}} \leq C_{\epsilon^2}(1+\epsilon^2)^{2N+1} + C(N)\delta.$$

For a given $\epsilon > 0$ and $N = N(\epsilon)$, select $\delta > 0$ so small that

(8.144) $$C(N)\delta \leq C_{\epsilon^2}(1+\epsilon^2)^{2N+1} \quad \text{and} \quad C(N)\delta \leq 1/4.$$

Then (8.143) holds, and $\|\mathcal{R}^{2N+1}\|_{C^\mathbf{r}} \leq 2C_{\epsilon^2}(1+\epsilon^2)^{2N+1}$.

To complete the proof of (8.135), we return to the estimate (8.142). Using (8.143), we derive from (8.143) the inequality

$$\|g_\epsilon - \mathcal{R}g_\epsilon\|_{C^\mathbf{r}} \leq (1-\epsilon)^{N+1} \cdot 4\|g_\epsilon\|_{C^\mathbf{r}} + 2\epsilon\|g_\epsilon\|_{C^\mathbf{r}}$$
$$+ (1-\epsilon)^N \cdot 2C_{\epsilon^2}(1+\epsilon^2)^{2N+1} \cdot 4\|g_\epsilon\|_{C^\mathbf{r}}.$$

For a given $\epsilon > 0$, choose $N = N(\epsilon)$ so large that
$$\max\{4(1-\epsilon)^{N+1}, 8C_{\epsilon^2}(1-\epsilon)^N(1+\epsilon^2)^{2N+1}\} \leq \epsilon. \tag{8.145}$$
Then, inequality (8.135) holds, and the proof of Theorem 8.56 is completed.

8.3.3.3. *Estimates via Oseledets-Lyapunov exponents.* This subsection contains estimates for the spectral radius $r(\mathcal{K})$ on $C^0(\Theta_{\mathbf{r}})$ in terms of the Oseledets-Lyapunov exponents for the cocycles $\{\Phi^k\}_{k\geq 0}$ and $\{D\varphi^k\}_{k\geq 0}$. Theorem 8.56 implies the corresponding estimate for the essential spectral radius $r_{\text{ess}}(\mathcal{R})$ on $C^{\mathbf{r}}(\Theta)$.

By the Multiplicative Ergodic Theorem, for each ergodic measure ν on Θ, there exists a full ν-measure Oseledets subset $\Theta_\nu \subset \Theta$ such that, if $\theta \in \Theta_\nu$, then there exists a filtration of the tangent space
$$\mathcal{T}_\theta \Theta = V_\theta^1 \supset \cdots \supset V_\theta^{n'} \supset V_\theta^{n'+1} = \{0\}, \quad n' \leq n = \dim \Theta,$$
such that the exact Oseledets-Lyapunov exponent
$$\chi(\theta, v) = \lim_{k\to\infty} k^{-1} \log |D\varphi^k(\theta)v|, \quad v \in \mathcal{T}_\theta \Theta,$$
for the cocycle $\{D\varphi^k\}_{k\geq 0}$ exists and is equal to χ_ν^j if and only if $v \in V_\theta^j \setminus V_\theta^{j+1}$. Let $\chi_\nu = \chi_\nu^{n'}$ denote the smallest Oseledets-Lyapunov exponent for the cocycle $\{D\varphi^k\}_{k\geq 0}$ relative to the measure $\nu \in \text{Erg}(\varphi)$ on Θ. Similarly, by the Multiplicative Ergodic Theorem applied to the cocycle $\{\Phi^k\}_{k\geq 0}$, if $\nu \in \text{Erg}(\varphi)$, then there exists a full ν-measure subset $\Theta_\nu \subset \Theta$ and corresponding exact Oseledets-Lyapunov exponents $\lambda_\nu(\theta, x)$, $x \in \mathcal{E}_\theta$. Let $\lambda_\nu = \lambda_\nu^1$ denote the largest Oseledets-Lyapunov exponent for the cocycle $\{\Phi^k\}_{k\geq 0}$; that is, $\lambda_\nu = \lim_{k\to\infty} k^{-1} \log \|\Phi^k(\theta)\|$, $\theta \in \Theta_\nu$. Also, recall that $h_\nu(\varphi)$ denotes the entropy of φ with respect to $\nu \in \text{Erg}(\varphi)$.

THEOREM 8.60. *Suppose that φ is a topologically mixing, expanding map and $\mathcal{R}$ the matrix-coefficient transfer operator, given by formula (8.92) on the space $C^{\mathbf{r}}(\Theta)$ of $\mathbf{r}$ times differentiable sections of an ℓ-dimensional bundle over an n-dimensional manifold Θ. Then, for each positive integer $\mathbf{r}$, the essential spectral radius of $\mathcal{R}$ is bounded above as follows:*
$$r_{ess}(\mathcal{R}; C^{\mathbf{r}}(\Theta)) \leq \exp(\sup\{h_\nu(\varphi) + \lambda_\nu - \mathbf{r}\chi_\nu : \nu \in \text{Erg}(\varphi)\}). \tag{8.146}$$

PROOF. By Theorem 8.56,
$$r_{\text{ess}}(\mathcal{R}; C^{\mathbf{r}}(\Theta)) = r(\mathcal{K}; C^0(\Theta_{\mathbf{r}}))$$
where $\mathcal{K}$, as in definition (8.129), is the transfer operator generated by the covering ψ and the cocycle $\{\Psi^k\}_{k\geq 0}$ acting on the space of continuous sections, as defined in display (8.128). Therefore, Theorem 8.51 implies that
$$r_{\text{ess}}(\mathcal{R}; C^{\mathbf{r}}(\Theta)) \leq \exp\sup\{h_\mu(\psi) + \Lambda_\mu : \mu \in \text{Erg}(\psi)\}.$$
Here $\mu \in \text{Erg}(\psi)$ is a ψ-ergodic measure on $\Theta_{\mathbf{r}}$, $h_\mu(\psi)$ is the entropy of ψ, and
$$\Lambda_\mu = \lim_{k\to\infty} k^{-1} \log \|\Psi^k(\theta)\|, \quad \theta \in \Theta_{\mathbf{r},\mu},$$
is the largest Oseledets-Lyapunov exponent for the cocycle $\{\Psi^k\}_{k\geq 0}$ over $\{\psi^k\}_{k\geq 0}$ on $\Theta_{\mathbf{r}}$.

To express the Oseledets-Lyapunov exponent Λ_μ in terms of λ_ν and χ_ν, we will use a description of the ψ-ergodic measures on $\Theta_{\mathbf{r}}$. This description states that every $\mu \in \text{Erg}(\psi)$ has a form $\mu = \nu \times \mu_1 \times \ldots \times \mu_{\mathbf{r}}$ where each μ_j is supported on one Oseledets subbundle $\mathcal{S}^i$, for some $i \in \{1, \ldots, n'\}$, with the fibers $\mathcal{S}_\theta^i = S^{n-1} \cap V_\theta^i$;

that is, on the trace of an Oseledets subbundle V^i on the unit sphere in the tangent space $\mathcal{T}_\theta \Theta$, see for example Arnold [**A**]. We claim that

(8.147) $\quad \sup\{h_\mu(\psi) + \Lambda_\mu : \mu \in \mathrm{Erg}(\psi)\} = \sup\{h_\nu(\varphi) + \lambda_\nu - \mathbf{r}\chi_\nu : \nu \in \mathrm{Erg}(\varphi)\}.$

To prove the inequality "$\leq$" in (8.147), fix $\mu \in \mathrm{Erg}(\psi)$. As in Lemma 8.45, we have that $h_\mu(\psi) = h_\nu(\varphi)$ where $\nu = \mathrm{proj}\,\mu$ is the projection of μ on Θ. Consider the Oseledets subset $\Theta_{\mathbf{r},\mu}$ such that, for each $(\theta, \mathbf{v}) \in \Theta_{\mathbf{r},\mu}$, there exists the exact Oseledets-Lyapunov exponent $\Lambda_\mu = \lim_{k\to\infty} k^{-1} \log \|\Psi^k(\theta, \mathbf{v})\|$. Pick a point $(\theta, \mathbf{v}) \in \Theta_{\mathbf{r},\mu}$ such that, at the same time, $\theta \in \Theta_\nu$; that is, such that there exist the exact Oseledets-Lyapunov exponents for both cocycles $\{\Phi^k\}_{k\geq 0}$ and $\{D\varphi^k\}_{k\geq 0}$. Then, by the definition of $\{\Psi^k\}_{k\geq 0}$, see (8.128),

$$\Lambda_\mu = \lim_{k\to\infty} k^{-1} \log \|\Psi^k(\theta, \mathbf{v})\|$$

$$= \lim_{k\to\infty} k^{-1} \log \|\Phi^k(\theta)\| - \sum_{j=1}^{\mathbf{r}} \lim_{k\to\infty} k^{-1} \log |D\varphi^k(\theta) v_j|$$

(8.148) $\qquad \leq \lambda_\nu - \mathbf{r}\chi_\nu, \quad \mathbf{v} = (v_1, \ldots, v_\mathbf{r}),$

since χ_ν is the smallest Oseledets-Lyapunov exponent for $\{D\varphi^k\}_{k\geq 0}$.

To prove the inequality "$\geq$" in (8.147), fix $\nu \in \mathrm{Erg}(\varphi)$. Pick a measure μ_0 supported on $\mathcal{S}^{n'}$, the trace of the Oseledets subbundle that corresponds to the smallest exponent $\chi_\nu = \chi_\nu^{n'}$. Form $\mu = \nu \times \mu_0 \times \ldots \times \mu_0$ on $\Theta_\mathbf{r}$. Then, $h_\mu(\psi) = h_\nu(\varphi)$ as before. Also, for each $(\theta, \mathbf{v}) \in \Theta_{\mathbf{r},\mu}$ such that the exact limits Λ_μ, λ_ν, and $\chi_\nu(\theta, v_j)$ exist, we have equality in (8.148) because each vectors v_j is a vector from $\mathcal{S}_\theta^{n'}$. $\qquad\square$

Since $\chi_\nu \geq \log \rho$, where ρ is the expansion constant, we have the following corollary.

COROLLARY 8.61 (Ruelle). *Suppose that φ is a topologically mixing, expanding map with the expansion constant ρ, and $\mathcal{R}$ is the matrix-coefficient transfer operator defined by (8.92) on the space $C^\mathbf{r}(\Theta)$ of $\mathbf{r}$ times differentiable sections of an ℓ-dimensional bundle over an n-dimensional manifold Θ. Then, for each positive integer $\mathbf{r}$, the essential spectral radius of $\mathcal{R}$ is bounded above as follows:*

$$r_{ess}(\mathcal{R}; C^\mathbf{r}(\Theta)) \leq \rho^{-\mathbf{r}} \exp(\sup\{h_\nu(\varphi) + \lambda_\nu : \nu \in \mathrm{Erg}\,(\varphi)\}).$$

8.4. Bibliography and remarks

Oseledets' Theorem and linear skew-product flows. In this section we follow the papers of Latushkin and Stepin [**LSt2, LSt3**].

The Multiplicative Ergodic Theorem. There exists a vast literature on the Multiplicative Ergodic Theorem, see for example the book by Walters [**Wa2**], the book by Krengel [**Kr**], or the review article by Cornfeld and Sinai [**CSi**]. Originally, this theorem was proved by Oseledets in [**Os**]. An excellent exposition and additional references are in the monograph by Arnold [**A**]. The infinite dimensional case described in this subsection was considered by Ruelle in [**Ru3**]. For the Banach space case, see, for example, the papers by Mañe [**Mn3**] and by Thieullen [**Th**]. For the finite dimensional case, see the paper by Johnson, Palmer, and Sell [**JPS**] which is close in spirit to our exposition. They also discuss applications to stochastic differential equations, almost-periodic Schrödinger operators, etc. We remark that the ergodicity of the measure ν is sufficient but not necessary for the Multiplicative

Ergodic Theorem to hold. Similarly, in general the measure ν does not need to be Borel.

Spectral and measurable subbundles. The main results of this subsection, Theorems 8.12–8.14 are taken from Latushkin and Stepin [**LSt2, LSt3**]. The fact that Oseledets subbundles are the "traces" of Sacker-Sell subbundles is also proved by Johnson, Palmer, and Sell [**JPS**] for the finite dimensional case. In addition to Proposition 8.10, we mention the paper of Pliss [**Pl**], where, under the assumption that $0 \notin \Sigma_L$, he proves that, for every $\epsilon > 0$, there exists a set $\Theta_{\nu,\epsilon} \subset \Theta_\nu$ with $\nu(\Theta_{\nu,\epsilon}) = 1 - \epsilon$ such that $\hat{\varphi}|\Theta_{\nu,\epsilon}$ has an exponential dichotomy.

Statement (1) of Theorem 8.12 together with results in Section 6.3 give the description of the Sacker-Sell spectrum for $\dim \mathcal{H} < \infty$, see [**SS3**]. On the other hand, Theorems 8.12–8.14 can be considered as generalizations of the corresponding theorems of Antonevich [**An2**] on the structure of the spectrum and spectral subspaces for evolution operators, see also Antonevich and Lebedev [**An3, AL2**]

Calculation of the spectrum. The formula $\log r(E) = \sup \lambda_\nu^1$ in Theorem 8.15 is proved in [**LSt3**]. For the finite dimensional case, this result was known to Oseledets [**Os2**] and Ya. Pesin (private communication). Also, for the finite dimensional case, this formula follows from results of Johnson, Palmer, and Sell in [**JPS**]. In fact, they prove that the boundary of the Sacker-Sell spectrum is contained in the Oseledets-Lyapunov spectrum. On the other hand, Johnson's paper [**Jo5**] contains an example of an almost periodic system such that a point of the Oseledets-Lyapunov spectrum is contained in the interior of the Sacker-Sell spectrum.

Since, see Corollary 6.45, the general Lyapunov exponent for the cocycle given by $\{\Phi^m\}_{m \in \mathbb{N}}$ coincides with $\log r(E)$, Theorem 8.15 is related to a classical result of Millionshchikov [**Ml**, p. 144]: The general Lyapunov exponent of a linear system belongs to its probability spectrum. On the other hand, the formula $\log r(E) = \sup \lambda_\nu^1$ is a generalization of the scalar result recorded in Corollary 8.19. This corollary was independently obtained by Antonevich and Lebedev [**AL**], Chicone and Swanson [**CS3**], and Kitover [**Ki2**].

The "scalarization" trick used in the proof of Theorem 8.15 goes back to Furstenberg and Kesten [**FKe**], see also Arnold [**A**], and the paper of Furstenberg and Kiefer [**FKi**]. See also the finite dimensional Lemma 8.54, where a result similar to Theorem 8.15 is proved without the complications caused by the facts that the cocycle $\{\Phi^m\}_{m \geq 0}$ is, generally, infinite dimensional and degenerate. Corollary 8.20 is used in Section 8.2 to describe the spectrum of the kinematic dynamo operator.

We also mention an important series of papers by Colonius and Kliemann [**CK2, CK3**] where the connections of Sacker-Sell, Oseledets-Lyapunov, and Morse spectra are described in the setting of control theory. A connection between local and global Lyapunov exponents is studied by Eden, Foias, and Temam in [**EFT**].

We conclude the current comments to Section 8.1 with a brief application to exponentially separated cocycles; for more details see Latushkin and Stepin [**Lt, LSt2**].

Assume that $\dim \mathcal{H} < \infty$, the map $\Phi : \Theta \to \mathcal{GL}(\mathcal{H})$ is continuous and takes invertible values, and Θ is a compact metric space. We say the cocycle $\{\Phi^m\}_{m \in \mathbb{Z}}$ is *exponentially separated* if there exists a continuous projection valued function $\tilde{P} : \Theta \to \text{proj}(\mathcal{H})$ such that, for each $\theta \in \Theta$, we have $\tilde{P}(\varphi^m \theta)\Phi^m(\theta) = \Phi^m(\theta)\tilde{P}(\theta)$, $\theta \in \Theta$ and, for the restrictions $\Phi_s^m(\theta) = \Phi^m(\theta)|\operatorname{Im}\tilde{P}(\theta)$ and $\Phi_u^m(\theta) = \Phi^m(\theta)|\operatorname{Im}\tilde{Q}(\theta)$

where $\tilde{Q}(\theta) = I - \tilde{P}(\theta)$, there exist positive constants C and β such that
$$\|\Phi_u^m(\theta)\| \geq Ce^{\beta m}\|\Phi_s^m(\theta)\| \tag{8.149}$$
for each nonnegative integer m. Clearly, cf. Definition 6.13, an exponentially dichotomic cocycle is trivially exponentially separated. The book by Bronshtein [**Br**] has more details and contains further references. We will need two facts from this book. First, see [**Br**, Corol.6.31], there exists a metric in $\Theta \times \mathcal{H}$ (the Lyapunov metric) for which $C = 1$ in (8.149). So, from now on we will assume $C = 1$. Let us define a scalar function by
$$F(\theta) = (\|\Phi_s(\theta)\| \cdot \|\Phi_u(\theta)\|_\bullet)^{-1}$$
where, as usual, $\|\Phi\|_\bullet := \inf\{|\Phi x| : |x| = 1\}$. For the scalar cocycle defined by
$$F^m(\theta) = F \circ \varphi^{m-1}(\theta) \cdots F(\theta),$$
consider the cocycle defined by $\Psi^m = F^m \Phi^m$. The second result from [**Br**, Thm. 6.30] is the following proposition:

PROPOSITION 8.62. *The cocycle $\{\Phi^m\}_{m\in\mathbb{Z}}$ is exponentially separated if and only if the cocycle $\{\Psi^m\}_{m\in\mathbb{Z}}$ has an exponential dichotomy on Θ.*

For a description of exponentially separated cocycles we will need the Oseledets-Lyapunov spectrum $\Sigma_L = \{\lambda_\nu^j : j = 1, \ldots, s_\nu, \nu \in \text{Erg}(\varphi)\}$ and the measurable Oseledets subbundles W_ν^j, see Theorem 8.4.

THEOREM 8.63. *The cocycle $\{\Phi^m\}_{m\in\mathbb{Z}}$ is exponentially separated if and only if*
(i) *for every measure $\nu \in \text{Erg}(\varphi)$, there exists a number j_ν such that the measurable subbundles $W_\nu^1 + \cdots + W_\nu^{j_\nu}$ and $W_\nu^{j_\nu+1} + \ldots + W_\nu^{s_\nu}$, defined on the Oseledets set $\Theta_\nu \subset \Theta$, have $\hat{\varphi}^m$-invariant continuous extensions $\tilde{\mathbb{U}}$ and $\tilde{\mathbb{S}}$ from Θ_ν to Θ, and*
(ii) *the set*
$$\Sigma_{L,F} = \{\lambda_\nu^j + \int_\Theta F d\nu : j = 1, \ldots, s_\nu, \nu \in \text{Erg}(\varphi)\}$$
is separated from zero.

PROOF. Since $\{F^m\}_{m\geq 0}$ is a scalar cocycle, the Oseledets-Lyapunov spectrum for $\{\Psi^m\}_{m\geq 0}$ is equal to $\Sigma_{L,F}$. If $\{\Phi^m\}_{m\in\mathbb{Z}}$ is exponentially separated, then $\{\Psi^m\}_{m\in\mathbb{Z}}$ has an exponential dichotomy by Proposition 8.62. Hence statement (i) follows from Proposition 8.10 and statement (ii) in Remark 8.9. Conversely, assume that statements (i) and (ii) hold. To prove the hyperbolicity of the evolution operator defined by $E_\Psi f = \Psi f \circ \varphi^{-1}$ on $C(\Theta; \mathcal{H})$, consider the decomposition $C(\Theta; \mathcal{H}) = C(\tilde{\mathbb{S}}) \dotplus C(\tilde{\mathbb{U}})$ where $C(\tilde{\mathbb{S}})$ and $C(\tilde{\mathbb{U}})$ are the spaces of continuous sections of $\tilde{\mathbb{S}}$ and $\tilde{\mathbb{U}}$ defined in statement (i). Accordingly, $E_\Psi = E_{\Psi,S} \dotplus E_{\Psi,U}$ for the corresponding restrictions of E_Ψ. Using statement (ii) and Theorem 8.15, we conclude that $r(E_{\Psi,S}) < 1$ and $r([E_{\Psi,U}]^{-1}) < 1$. □

The kinematic dynamo operator. This section is based on the papers by Chicone, Latushkin, and Montgomery-Smith [**CLM, CLM2**] and [**ChLt3**].

The Spectral Mapping and Annular Hull Theorems. The Spectral Mapping Theorem 8.22 is proved in [**CLM**]. The proof is similar to the proof of the Spectral Mapping Theorem 6.30. Indeed, compare formulas (6.34) and (8.67) for the approximate eigenfunctions of Γ. Thus, the approach taken in this subsection is a generalization of the Mather localization discussed in Subsection 6.2.2. The Annular Hull Theorem 8.23 is proved in [**CLM2**]. Note that formula (8.52) for the approximate eigenfunction for Γ is quite different from formula (7.51) used in the proof of the Annular Hull Theorem 7.25. The Spectral Mapping Theorem 8.22 and the Annular Hull Theorem 8.23 generalize the corresponding results by Chicone and Swanson [**CS, CS2, CS3**] to the case of the space of *divergence free* vector fields. However, the approach taken in [**CS, CS2, CS3**] was developed in Subsection 7.2.2, while in Subsection 8.2.1 we employed an integral variant of the Mather localization formula (6.22). A reason for this is that we need to produce a *divergence free* approximate eigenfunction for Γ. Formulas (7.48) and (7.51) do not guarantee this. However, for the space L_{ND}^2 of divergence free L^2-vector fields on $\Theta = \mathbb{T}^3$, Latushkin and Vishik [**LV**] give the following formula for the divergence free approximate eigenfunction for Γ:

$$f = \varepsilon \nabla \times \left(\nabla s \times \frac{e^{is/\varepsilon}}{|\nabla s|^2} g \right).$$

Here ε is a small parameter, $s : \Theta \to \mathbb{R}$ is a phase, and $g \in L^2$ is an approximate eigenfunction for Γ on L^2 with *no* divergence free condition, cf. formula (8.152) below for the approximative eigenfunction of the linearized Euler operator.

Corollary 8.25 is taken from [**CLM, CLM2**]. We remark that the spectral radius of an evolution operator can be computed in terms of the Oseledets-Lyapunov exponents under very general assumptions, see Theorem 8.15. However, in the calculation of the spectrum the main difficulty is to compute the radius of the next spectral subset, see the related discussion in Subsection 8.1.3. The advantage of the, perhaps more technically challenging, space of divergence free vector fields is that there is no "next" spectral component; that is, the spectrum (or its annular hull) on this space is one annulus.

The spectrum. The effect of "filling the gaps" for the spectrum of the evolution operator on the space C_{ND} versus C was discovered by de la Llave [**dL**]. Theorem 8.35 is proved in [**CLM, CLM2**]. We also mention a recent preprint by de la Llave [**dL2**] where jets are used to study the Mather evolution operators on spaces of smooth vector fields. The paper [**dL2**] also contains a further discussion of evolution operators on the spaces of divergence free vector fields and an interesting development of Mather's original approach in [**Ma**].

The fast dynamo problem. The standard references for the fast dynamo problem are Arnold, Zel'dovich, Rasumaikin, and Sokolov [**Arn, AZRS**], Bayly and Childress [**BaCh**], Finn and Ott [**FO**], Friedlander and Vishik [**FV2, FV, Vs2**], and Soward [**So**], see also a recent book of Arnold and Khesin [**AKh**], their extensive bibliography, and the special issue of *Geophys. Astrophys. Fluid Dynamics*, Vol. 73 (1993). The content of Subsection 8.2.3.1 is taken from Chicone, Latushkin, and Montgomery-Smith [**CLM, CLM2**].

The concept of the fast dynamo was introduced by Zel'dovich and Ruzmaikin [**ZR**], see also the important papers by Arnold [**Arn, AKo**] and Molchanov, Ruzmaikin, and Sokoloff [**MRS**]. As far as we know, the first example of a fast dynamo for the steady-state solution of the Euler equation is given by Soward [**So**]. An example of the fast dynamo in a special hyperbolic flow, as well as some antidynamo theorems are given by Arnold [**Arn**].

The following antidynamo theorem was proved by Vishik [**Vs2**].

THEOREM 8.64. *If there is some exponential stretching in the flow $\{\varphi^t\}_{t\in\mathbb{R}}$ of a conducting fluid, then only slow dynamo action is possible: $\overline{\lim}_{\epsilon\to 0}\omega_\epsilon = 0$.*

The strategy of the proof in [**Vs2**] involves an asymptotic expansion for the Green's kernel for (8.82). In further developments of these methods, Friedlander and Vishik [**FV, FV2**] and Vishik [**Vs3**] studied hydrodynamic instability.

Consider the linearized Navier-Stokes equation for a small perturbation $\mathbf{w}$ about its steady state $\mathbf{v}$, namely,

$$(8.150) \qquad \frac{d\mathbf{w}}{dt} = \epsilon \Delta \mathbf{w} - \langle \mathbf{v}, \nabla \rangle \mathbf{w} - \langle \mathbf{w}, \nabla \rangle \mathbf{v} - \nabla \mathbf{p}, \quad \operatorname{div} \mathbf{w} = 0.$$

It can be shown that the evolution of the vorticity, curl $\mathbf{w}$, satisfies the differential equation

$$\operatorname{curl} \dot{\mathbf{w}} = \epsilon \Delta \operatorname{curl} \mathbf{w} + \operatorname{curl}\,(\mathbf{w} \times \operatorname{curl} \mathbf{v}) + \operatorname{curl}\,(\mathbf{v} \times \operatorname{curl} \mathbf{w}).$$

There is a certain similarity between this last equation and (8.81). Using this similarity, Friedlander and Vishik [**FV2**] give a necessary condition for linearized instability and a sharp estimate on the growth rate of a perturbation for the Navier-Stokes equations in the limit as $\epsilon \to 0$.

We cite the following results from [**FV, FV2, Vs3**] for the linearized Euler equation; that is, $\epsilon = 0$ in equation (8.150). Let $\Theta = \mathbb{T}^3$ and, as above, let $\mathbf{v} : \Theta \to \mathbb{R}^3$ denote a smooth divergence free steady state solution of the nonlinear Euler equation on the space $L^2_{ND} = L^2_{ND}(\Theta; \mathbb{R}^3)$ of L^2-divergence free vector fields. The linearized Euler operator L is given by

$$L\mathbf{w} = -\langle \mathbf{w}, \nabla \rangle \mathbf{v} - \langle \mathbf{v}, \nabla \rangle \mathbf{w} - \nabla \mathbf{p}$$

where $\mathbf{p} : \Theta \to \mathbb{R}$ is the pressure. As in [**FV, FV2, Vs3**], consider the following system of *ordinary* differential equations:

$$(8.151) \qquad \begin{aligned} \dot{\theta} &= \mathbf{v}(\theta), \quad \dot{\xi} = \left[-\frac{\partial \mathbf{v}}{\partial \theta}(\theta)\right]^T \xi, \\ \dot{b} &= -\frac{\partial \mathbf{v}}{\partial \theta}(\theta) b + 2 \left\langle \frac{\partial \mathbf{v}}{\partial \theta}(\theta) b, \frac{\xi}{|\xi|} \right\rangle \frac{\xi}{|\xi|}. \end{aligned}$$

Here $\xi = \xi(\theta, t) \in (\mathbb{R}^3)^*$ is a covector, $b = b(\theta, \xi, t) \in \mathbb{R}^3$, and $\xi \perp b$. Let μ denote the general Lyapunov exponent for this system; that is,

$$\mu = \lim_{t\to\infty} t^{-1} \log \sup\{|b(t; \theta_0, \xi_0, b_0)| :$$

$$(\theta_0, \xi_0, b_0) \in \Theta \times (\mathbb{R}^3)^* \times \mathbb{R}^3, |\xi_0| = |b_0| = 1, \xi_0 \perp b_0\}$$

where $b(t; \theta_0, \xi_0, b_0)$ is the solution for the b-equation in the system (8.151) above with the initial condition (θ_0, ξ_0, b_0).

Using methods from the theory of pseudo-differential operators, Friedlander and Vishik prove in [**FV2**] that $\mu \leq \omega(L)$. Later, Vishik [**Vs3**] proved that $\mu = t^{-1} \log r_{\operatorname{ess}}(e^{tL})$ for $t \neq 0$. Here, $r_{\operatorname{ess}}(e^{tL})$ is the essential spectral radius of e^{tL}.

The beauty of these results is that the study of stability questions for the partial differential operator L and the group $\{e^{tL}\}_{t\in\mathbb{R}}$ is reduced to the study of the finite dimensional Friedlander-Vishik system (8.151)! In particular, the following impressive corollary holds:

COROLLARY 8.65 (S. Friedlander-M. Vishik). *If the flow $\{\varphi^t\}_{t\in\mathbb{R}}$ generated by $\mathbf{v}$ has a positive Lyapunov exponent at some point θ_0, then the vector field $\mathbf{v}$ is linearly unstable as a steady state solution of the Euler equation.*

Latushkin and Vishik [**LV**] use the Friedlander-Vishik system (8.151) and give an explicit formula for the approximate eigenfunctions of L:

$$(8.152) \qquad \mathbf{w} = \varepsilon \nabla \times \left(\nabla s \times \frac{e^{is/\varepsilon}}{|\nabla s|^2} g \right), \quad \mathbf{p} = 2\varepsilon e^{is/\varepsilon} \left\langle \frac{\partial \mathbf{v}}{\partial \theta} g, \frac{\nabla s}{|\nabla s|^2} \right\rangle.$$

Here, under certain conditions, the phase $s : \Theta \to \mathbb{R}$, the amplitude $g : \Theta \to \mathbb{R}^3$, and the value of the small parameter $\varepsilon > 0$ can be determined for each $n \in \mathbb{N}$ such that $\|(L - \mu')\mathbf{w}\| = O(n^{-1})\|\mathbf{w}\|$ for some $\mu' \in \mu + i\mathbb{R}$. The construction of g uses the Mañe point and vector for (8.151). In fact, this construction is very close to the construction of the approximate eigenfunction for Γ presented in Subsection 8.2.1.

To conclude our comments on fast dynamos, we mention the papers by Oseledets [**Os2, Os3**] and Klapper and Young [**KY**]. In particular, the following sufficient condition for the slow dynamo is proved in [**KY**]: If $\Theta \subset \mathbb{R}^n$, $\mathbf{v}$ is of class $C^{\mathbf{r}+1}$, and f is $C^{\mathbf{r}}$ for $\mathbf{r} \geq 2$, then, for the solution $\mathbf{u}(\theta, t) = (e^{t\Gamma_\epsilon} f)(\theta)$ of (8.82), the following estimate holds:

$$\varlimsup_{\epsilon \to 0} \varlimsup_{n \to \infty} n^{-1} \log \int_\Theta |(e^{n\Gamma_\epsilon} f)(\theta)| \, d(\text{mes})(\theta) \leq h(\varphi) + \frac{r(\varphi)}{\mathbf{r}}$$

where $h(\varphi)$ is the topological entropy of $\varphi = \varphi^1$ and

$$r(\varphi) = \lim_{n \to \infty} n^{-1} \log \max_{\theta \in \Theta} |D\varphi^n(\theta)|.$$

The content of Subsection 8.2.3.2 is taken from the paper of Chicone and Latushkin [**ChLt3**]. The fact that the geodesic flow on the unit tangent bundle of a two dimensional manifold with constant negative curvature gives an example of a fast dynamo is probably in [**AZRS**], and more general results can be found in Vishik [**Vs2**] and [**Vs**, Section 9]. For instance, Vishik [**Vs2**] proved that every Anosov flow with smooth foliations on a compact three dimensional manifold produces a fast dynamo. The Metric μ defined in this subsection, is the so-called Sasaki metric, see Chicone [**Ch**] for the definition. Formulas (8.85) for the Lie brackets, used in the proof of Theorem 8.36, were obtained by Green in [**Gr**], see also the related papers of Beem, Chicone, and Ehrlich [**BCE, Ch2, CE**] and the paper by Green [**Gr2**].

The Ruelle transfer operator. Ruelle's transfer operator has its origin in the theory of statistical mechanics ([**Ru**]) where it is used to construct Gibbs measures and equilibrium states. This is also the reason why it was introduced to the theory of dynamical systems, see the work of Bowen and Sinai [**Bo2, Si3**]. The standard references on the transfer operator are the work by Bowen [**Bo2**], Ruelle [**Ru2**], and Sinai [**Si3**]. See also the books by Knauf and Sinai [**KS**], Mayer [**My**], Parry and Pollicott [**PPo**], and Ruelle [**Ru7**], the excellent surveys by Baladi [**Ba**] and Hurt [**Hr**], and the electronic book by P. Cvitanovic at http://www.nbi.dk/ predrag/QCcourse. Also, we mention the paper [**Wa**] by Walters and a recent book [**Ke**] by Keller. The transfer operator is particularly helpful for studying the mixing and statistical properties of measures (see [**Bo2, PPo**] and Young [**Yo2**]). It is also used in the investigation of zeta functions, see the work of the above mentioned authors in [**Ba, PPo, Po2, Ru7**] as well as Haydn [**Hy**] and Tangerman [**Tn**]. Therefore, the transfer operator has stimulated a lot of interest in the theory of piecewise monotone transformations (see papers by Baladi, Isola, Schmitt [**BIS**], Baladi, Jiang, and Lanford [**BJL**], Baladi and Keller [**BK**], Hofbauer and Keller [**HK**]), and in the theory of Fredholm determinants (see for example papers by Ruelle [**Ru5**] and Rugh [**Rg2, Rg**]). There are recent applications to wavelets found by Cohen and Daubechies [**CD**] and Holschneider [**Hs**]. We also mention the references [**CI, My, My2, KK, JMS, FR, KLi, BH, Ru6, Po, KN**]. Finally, we mention some of a large body of work on the transfer operator for the random setting, see for example Baladi and Young [**BY**], and Bogenschütz and Gundlach [**BGu**], and the literature cited therein.

Setting and preliminaries. The setting in Subsection 8.3.1 is taken from the celebrated paper by Ruelle [**Ru4**] that contains an excellent overview of the topics related to the transfer operator, see also [**Ru7**], the literature cited above about the Perron-Frobenius operator, and the, now classic, papers by Lasota and Yorke [**LY**] and by Young [**Yo**]. The definition of topological pressure and the proof of the variational principle can be found, for example, in Walters [**Wa2**]. An equilibrium state is a measure where the supremum in Theorem 8.44 is attained. Gibbs measures are eigenvectors of the operator adjoint of the scalar-coefficient transfer operator. A connection between both measures was studied by Ledrappier [**Ld**]. See Bowen [**Bo2**] for an excellent explanation of the physical meaning of Gibbs measures. Lemma 8.45 is taken from Latushkin and Stepin [**LSt2**] where another type of transfer operators is studied.

We mention the fundamental Ruelle-Perron-Frobenius Theorem. To state it, let us assume that φ is a topologically mixing expanding map, g is α-Hölder continuous, and let us consider the transfer operator (8.93) defined on $C^0(\Theta)$.

THEOREM 8.66. *There is some $\lambda > 0$, a positive function $h \in C(\Theta)$, and a φ-invariant Borel probability measure μ such that $\mathcal{R}h = \lambda h$, $\mathcal{R}^*\mu = \lambda\mu$, $\mu(h) = 1$, and*

$$\lim_{n \to \infty} \|\lambda^{-n}\mathcal{R}^n f - \mu(f)h\|_{C(\Theta)} = 0$$

for all $f \in C(\Theta)$.

The essential element of the proof is the fact that $\mathcal{R}$ preserves the cone of positive functions in $C(\Theta)$, see Bowen [**Bo2**, Sec.1.7]. A general function-analytic setup of the Perron-Frobenius theory is given in the important papers by Yu. Lyubich and M. Lyubich [**LL, LL2**].

Continuous sections. The scalar transfer operator on the space of continuous sections is a classical object. Thus, Theorem 8.47 has been known at least since Ruelle's paper [**Ru2**]. This theorem can be proved using the Ruelle-Perron-Frobenius Theorem 8.66, or it can be derived from general results in [**LL, LL2**]. Formula (8.104) is "similar" to the formula $r(E) = \exp(\sup\{\int_\Theta g d\nu : \nu \in \text{Erg}(\varphi)\})$ given in Corollary 8.19 for the spectral radius of the evolution operator defined by $Ef = e^g f \circ \varphi^{-1}$, see also [**LSt**]. Theorems 8.51 and 8.53 are taken from Campbell and Latushkin [**CL**]; again, compare the RHS of (8.115) and Theorem 8.15 where the spectral radius of the evolution operator $Ef = \Phi f \circ \varphi^{-1}$ is computed. The "triangularization" used in the proof of Proposition 8.50 goes back to the original proof of the Oseledets Theorem in [**Os**], see also Johnson, Palmer, and Sell [**JPS**]. This trick was used in [**LSt2, CL**]. Lemma 8.54 and Theorem 8.55 are taken from [**CL**].

The transfer operator on smooth sections. The exposition of this section is based on the papers [**CL, GL**]. Note that Theorem 8.60 holds, in fact, for every aperiodic covering map φ. However, for a nonexpanding φ, the right hand side of (8.146) might exceed the spectral radius and, thus, does not give any new information.

Corollary 8.61 is due to Ruelle [**Ru4**]. See the work by Antonevich [**An, An3**] for a description of the spectrum of evolution (weighted composition) scalar operators on the space of smooth functions. Important papers on this subject are due to Kitover, see [**Ki2, Ki3, Ki4**]. In particular, the paper [**Ki3**] contains the idea to construct the extended compact $\Theta_\mathbf{r}$ used in Subsection 8.3.3. See Collet and Isola [**CI**] for a formula for the essential spectral radius of the scalar transfer operator, and the recent work of de la Llave [**dL2**] for a study of transfer operators on smooth functions using jets. The results of this subsection are valid also for the random setting, see Gundlach and Latushkin [**GL**].

The eigenvalue of the Perron-Frobenius operator on $C^\mathbf{r}$ with the *second largest* absolute value is of special interest since it determines the rate of decay of correlations for pairs of test functions in $C^\mathbf{r}$. The decay of correlation for $f_1, f_2 \in C^\mathbf{r}$ is defined by

$$\tau(f_1, f_2) := \overline{\lim}_{m \to \infty} m^{-1} \log | \int_\theta (f_1 \circ \varphi^m) f_2 d\mu - \int_\theta f_1 d\mu \cdot \int_\theta f_2 d\mu |$$

where μ is the Gibbs measure for $\log |D\varphi(\theta)|^{-1}$. For this topic, we refer to the beautiful discussion in Baladi and Young [**BY**], and recent advances by Froyland [**Fr**] and Liverani [**Lv**].

Bibliography

[AMR] R. Abraham, J. Marsden, and T. Ratiu, *Manifolds, tensor analysis, and applications*, Springer-Verlag, New York, 1988.

[AR] L. M. Abramov and V. A. Rokhlin, *Entropy of skew-product of a transformation with invariant measure*, Vestnik LGU **17** (1962), 5–13.

[AAK] Yu. Abramovich, E. Arenson, and A. Kitover, *Banach $C(K)$-modules and operators preserving disjointness*, Pitman Res. Notes, vol. 277, Longman, New York, 1992.

[Al] G. Allan, *On one-sided inverses in Banach algebras of holomorphic vector-valued functions*, J. London Math. Soc. **42** (1967), 463–470.

[Ad] B. D. O. Anderson, *External and internal stability of linear time-varying systems*, SIAM J. Contr. Opt. **20**(3) (1982), 408–413.

[An] A. Antonevich, *The spectrum of weighted translation operators on $W_p^l(X)$*, Dokl. Akad. Nauk USSR **264**(5) (1982), 1033–1035.

[An2] A. Antonevich, *Two methods for investigating the invertibility of operators from C^*-algebras generated by dynamical systems*, Math. USSR-Sb **52** (1985), 1–20.

[An3] A. Antonevich, *Linear functional equations: operator approach*, Oper. Theory Adv. Appl. **83**, Birkhäuser-Verlag, Basel, 1996.

[AL] A. Antonevich and A. Lebedev, *Spectral properties of weighted translation operators*, Izvestia AN USSR, Mathem. **47** (1983), 915–941.

[AL2] A. Antonevich and A. Lebedev, *Functional differential equations I: C^*-theory*, Pitman Monogr. Surv. Pure Appl. Math., vol. 70, Longman, New York, 1994.

[ALMZ] J. Appel, V. Lakshmikantham, Nguyen Van Minh, P. Zabreiko, *A general model of evolutionary processes: Exponential dichotomy-I*, Nonlinear Anal. Theory Meth. Appl. **21** (1993), 207–218.

[Ar] W. Arendt, *Spectrum and growth of positive semigroups*, in *Evolution Equations*, G. Ferreyra, G. Ruiz Goldstein and F. Neubrander (Eds.), Lect. Notes Pure Appl. Math., vol. 168, Marcel Dekker, 1995, pp. 21–28.

[Ar2] W. Arendt, *Spectral bound and exponential growth: survey and open problems*, in *Ulmer Seminare-1996 über Funktionanalysis und Differentialgleichungen*, Ulm, 1996, pp. 397–401.

[AB] W. Arendt and C. Batty, *Tauberian theorems and stability of one-parameter semigroups*, Trans. Amer. Math. Soc. **306** (1988), 837–852.

[AB2] W. Arendt and C. Batty, *Almost periodic solutions of first and second-order Cauchy problems*, J. Diff. Eqns. **137** (1997), 363–383.

[AG] W. Arendt and G. Greiner, *The spectral mapping theorem for one-parameter groups of positive operators on $C_0(X)$*, Semigroup Forum **30** (1984), 297–330.

[AEMH] W. Arendt, O. El Mennaoui, and M. Hieber, *Boundary values of holomorphic semigroups*, Proc. Amer. Math. Soc. **125** (1997), 635–647.

[A] L. Arnold, *Random dynamical systems*, Springer-Verlag, New York, 1998.

[Arn] V. I. Arnold, *Some remarks on the antidynamo theorem*, Moscow Univ. Math. Bull. **6** (1982) 50–57.

[AKh] V. I. Arnold and B. A. Khesin, *Topological methods in hydrodynamics*, Springer-Verlag, New York, 1998.

[AKo] V. I. Arnold and E. I. Korkina, *The growth of a magnetic field in a three-dimensional steady incompressible flow*, Vestnik Moscow State Univ. **3** (1983), 43–46.

[AZRS] V. I. Arnold, Ya. B. Zel'dovich, A. A. Rasumaikin, and D. D. Sokolov, *Magnetic field in a stationary flow with stretching in Riemannian space*, Sov. Phys. JETP, **54**(6) (1981), 1083–1086.

[AJ] W. B. Arveson and K. B. Josephson, *Operator algebras and measure preserving automorphisms. II*, J. Funct. Anal. **4**(1) (1969), 100–334.

[AGa] B. Aulbach and B. Garay, *Discretization of semilinear differential equations with an exponential dichotomy*, Advances in Difference Eqns., Comput. Math. Appl. **28** (1994), 23–35.

[AM] B. Aulbach and Nguyen Van Minh, *Semigroups and exponential stability of nonautonomous linear differential equations on the half-line*, in *Dynamical systems and applications*, R. P. Agrawal (Ed.), World Scientific, Singapore, 1995, pp. 45–61.

[AW] B. Aulbach and T. Wanner, *Integral manifolds for Caratheodory type differential equations in Banach spaces*, in *Six lectures on dynamical systems*, B. Aulbach and F. Colonius (Eds.), World Scientific, Singapore, 1996, pp. 45–119.

[Ba] V. Baladi, *Dynamical zeta functions*, in *Real and complex dynamical systems*, NATO Adv. Inst. Ser. C: Math. Phys. Sci., vol. 464, Kluwer Academic Publishers, Dordrecht, 1995, pp. 1–26.

[BIS] V. Baladi, S. Isola, and B. Schmitt, *Transfer operator for piecewise affine approximations for interval maps*, Ann. Inst. H. Poincare Phys. Theor. **62** (1995), 251–265.

[BJL] V. Baladi, Y. Jiang, and O. Lanford III, *Transfer operator acting on Zygmund functions*, Trans. Amer. Math. Soc. **348** (1996), 1599-1615.

[BH] V. Baladi and M. Holschneider, *Approximation of nonessential spectrum of transfer operator*, Preprint, 1998.

[BK] V. Baladi and G. Keller, *Zeta functions and transfer operators for piecewise monotone transformations*, Comm. Math. Phys. **127** (1990), 459–477.

[BY] V. Baladi and L.-S. Young, *On the spectra of randomly perturbed expanding maps*, Comm. Math. Phys. **156** (1993), 355–385.

[BrGK] H. Bart, I. Gohberg, and M. A. Kaashoek, *Wiener-Hopf factorization, inverse Fourier transform and exponentially dichotomous operators*, J. Funct. Anal. **68** (1986), 1–42.

[BaGK] J. Ball, I. Gohberg and M. A. Kaashoek, *A frequency response function for linear, time-varying systems*, Math. Control Signal Syst. **8** (1995), 334–351.

[Bs] A. G. Baskakov, *Some conditions for invertibility of linear differential and difference operators*, Russian Acad. Sci. Dokl. Math. **48**(3) (1994), 498–501.

[Bs2] A. G. Baskakov, *Spectral analysis of linear differential operators and semigroups of difference operators*, Doklady Mathem. **52**(1) (1995), 30–32.

[Bs3] A. G. Baskakov, *Semigroups of difference operators in spectral analysis of linear differential operators*, Funct. Anal. Appl. **30**(3) (1996), 149–157.

[Bs4] A. G. Baskakov, *The spectral analysis of difference operators*, Doklady Mathem. **55**(2) (1997), 237–239.

[BYu] A. G. Baskakov and V. V. Yurgelas, *Indefinite dissipativity and invertibility of linear differential operators*, Ukrainskii Math. Zhurnal **41** (1989), 1613–1618.

[BJ] P. Bates and C. Jones, *Invariant manifolds for semilinear partial differential equations*, in *Dynamics Reported*, vol. 2, 1989, pp. 1–38.

[BC] C. Batty and R. Chill, *Bounded convolutions and solutions of inhomogeneous Cauchy problems*, Preprint.

[BvNR] C. Batty, J. van Neerven, and F. Räbiger, *Tauberian theorems and stability of solutions of the Cauchy problem*, Trans. Amer. Math. Soc. **350** (1998), 2087–2103.

[BvNR2] C. Batty, J. van Neerven, and F. Räbiger, *Local spectra and individual stability of uniformly bounded C_0-semigroups*, Trans. Amer. Math. Soc. **350** (1998), 2071–2085.

[BaCh] B. J. Bayly and S. Childress, *Fast-dynamo action in unsteady flows and maps in three dimensions*, Phys. Rev. Let. **59**(14) (1987), 1573–1576.

[BCE] J. Beem, C. Chicone, and P. Ehrlich, *The geodesic flow and sectional curvature of pseudo-Riemannian manifolds*, Geom. Dedicata **12** (1982), 111–118.

[BG] A. Ben-Artzi and I. Gohberg, *Fredholm properties of band matrices and dichotomy*, Oper. Theory Adv. Appl. **32** (1988), 37–52.

[BG2] A. Ben-Artzi and I. Gohberg, *Dichotomy, discrete Bohl exponents, and spectrum of block weighted shifts*, Integr. Eqns. Oper. Th. **14** (1991), 613–677.

[BG3] A. Ben-Artzi and I. Gohberg, *Dichotomy of systems and invertibility of linear ordinary differential operators*, Oper. Theory Adv. Appl. **56** (1992), 90–119.

[BG4] A. Ben-Artzi and I. Gohberg, *Dichotomies of perturbed time-varying systems and the power method*, Indiana Univ. Math. J. **42**(3) (1993), 699–720.

[BGK]   A. Ben-Artzi, I. Gohberg, and M. A. Kaashoek, *Invertibility and dichotomy of differential operators on a half line*, J. Dynam. Diff. Eqns. **5** (1993), 1–36.

[BGK2] A. Ben-Artzi, I. Gohberg, and M. A. Kaashoek, *A time varying generalization of the canonical factorization theorem for Toeplitz operators*, Indag. Mathem **4** (1993), 385–405.

[BPDM] A. Bensoussan, G. Da Prato, M. Delfour, and S. Mitter, *Representation and control of infinite-dimensional systems*. Vols. 1–2, Birkhäuser-Verlag, Basel, 1992–1993.

[Be]   E. Berkson, *One-parameter semigroups of isometries into H^p*, Pacific J. Math. **86** (1980), 403–413.

[BP]   E. Berkson, and H. Porta, *Semigroups of analytic functions and composition operators*, Michigan Math. J. **25** (1978), 101–115.

[BS]   N. P. Bhatia and G. P. Szegö, *Stability theory of dynamical systems*, Springer-Verlag, New York, 1970.

[Bl]   M. D. Blake, *A spectral bound for asymptotically norm-continuous semigroups*, Preprint, 1998.

[BPh]   S. Bochner and R. S. Phillips, *Absolutely convergent Fourier expansions for noncommutative normed rings*, Annals of Math. **43**(3) (1942), 409–418.

[BGu]   T. Bogenschütz and V. Gundlach, *Ruelle's transfer operator for random subshifts of finite type*, Erg. Th. Dynam. Syst. **15**(3) (1995), 413–447.

[Bo]   R. Bowen, *Markov partitions for Axiom A diffeomorphism*, Trans. Amer. Math. Soc. **154** (1971), 377–397.

[Bo2]   R. Bowen, *Equilibrium states and the ergodic theory of Anosov diffeomorphisms*, Lect. Notes Math., vol. 470, Springer-Verlag, New York, 1975.

[BR]   O. Bratteli and D. Robinson, *Operator algebras and quantum statistical mechanics*, Springer-Verlag, New York, 1979.

[Br]   I. U. Bronshtein, *Nonautonomous dynamical systems*, (Russian), Shtiintza, Kishinev, 1984.

[BKo]   I. U. Bronshtein and A. Ya. Kopanskii, *Smooth invariant manifolds and normal forms*, World Scientific, Singapore, 1994.

[Bu]   C. Buşe, *On the Perron-Bellman theorem for evolutionary processes with exponential growth in Banach spaces*, New Zealand J. Math., to appear.

[BVGN] B. Bylov, R. Vinograd, D. Grobman, and B. Nemytski, *The theory of Lyapunov exponents*, (Russian), Moscow, Nauka, 1966.

[CL]   J. Campbell and Y. Latushkin, *Sharp estimates in Ruelle theorems for matrix transfer operators*, Comm. Math. Phys. **185** (1997), 379–396.

[Ch]   C. Chicone, *Tangent bundle connections and the geodesic flow*, Rocky Mountain J. Math. **11**(2) (1981), 305–317.

[Ch2]   C. Chicone, *The topology of stationary curl parallel solutions of Euler's equations*, Israel J. Math. **39** (1981), 161–166.

[CE]   C. Chicone and P. Ehrlich, *Line integration of Ricci curvature and conjugate points in Lorentzian and Riemannian manifolds*, Manuscr. Math. **31** (1980), 297–316.

[ChLt] C. Chicone and Y. Latushkin, *Quadratic Lyapunov functions for linear skew-product flows and weighted composition operators*, J. Integr. Diff. Eqns. **8** (1995), 289–307.

[ChLt2] C. Chicone and Y. Latushkin, *Hyperbolicity and dissipativity*, in Evolution equations, G. Ferreyra, G. Goldstein, and F. Neubrander (Eds.), Lect. Notes Pure Appl. Math., vol. 168, 1995, pp. 95–106.

[ChLt3] C. Chicone and Y. Latushkin, *The geodesic flow generates a fast dynamo: an elementary proof*, Proc. Amer. Math. Soc. **125** (1997), 3391–3396.

[CLM]   C. Chicone, Y. Latushkin, and S. Montgomery-Smith, *The spectrum of the kinematic dynamo operator for an ideally conducting fluid*, Comm. Math. Phys. **173** (1995), 379–400.

[CLM2] C. Chicone, Y. Latushkin, and S. Montgomery-Smith, *The annular hull theorems for the kinematic dynamo operator for an ideally conducting fluid*, Indiana Univ. Math. J. **45** (1996), 361–379.

[CS]   C. Chicone and R. C. Swanson, *The spectrum of the adjoint representation and the hyperbolicity of dynamical systems*, J. Diff. Eqns. **36** (1980), 28–39.

[CS2]   C. Chicone and R. C. Swanson, *A generalized Poincaré stability criterion*, Proc. Amer. Math. Soc. **81** (1981), 495–500.

[CS3]   C. Chicone and R. Swanson, *Spectral theory for linearization of dynamical systems*, J. Diff. Eqns. **40** (1981), 155–167.

[CLe]   S.-N. Chow and H. Leiva, *Dynamical spectrum for time dependent linear systems in Banach spaces*, Japan J. Industr. Appl. Math. **11**(3) (1994), 379–415.

[CLe2] S.-N. Chow and H. Leiva, *Existence and roughness of the exponential dichotomy for linear skew-product semiflows in Banach spaces*, J. Diff. Eqns. **120** (1995), 429–477.

[CLe3] S.-N. Chow and H. Leiva, *Two definitions of the exponential dichotomy for skew-product semiflow in Banach spaces*, Proc. Amer. Math. Soc. **124** (1996), 1071–1081.

[CLe4] S.-N. Chow and H. Leiva, *Unbounded perturbations of the exponential dichotomy for evolution equation*, J. Diff. Eqns. **129** (1996), 509–531.

[CLL]   S.-N. Chow, X.-B. Lin, and K. Lu, *Smooth invariant foliations in infinite dimensional spaces*, J. Diff. Eqns. **94** (1991), 266–291.

[CLu]   S.-N. Chow and K. Lu, C^k *centre unstable manifolds*, Proc. Royal Soc. Edinburgh **108A** (1988), 303–320.

[CY]   S.-N. Chow and Y. Yi, *Center manifold and stability for skew-product flows*, J. Dynam. Diff. Eqns. **6** (1994), 543–582.

[CFS]   R. Churchill, J. Franks, and J. Selgrade, *A geometric criterion of hyperbolicity of flows*, Proc. Amer. Math. Soc. **62** (1977), 137–143.

[CLRM] S. Clark, Y. Latushkin, T. Randolph, and S. Montgomery-Smith, *Stability radius and internal versus external stability in Banach spaces: an evolution semigroups approach*, Preprint, 1999.

[CG]   P. Clement and S. Guerre-Delabriere, *On the regularity of abstract Cauchy problems of order one and boundary value problems of order two*, Publ. Math. de l'Univers. P. et M. Curie **119**, Oct. 1997.

[CD]   A. Cohen and I. Daubechies, *A new technique to estimate the regularity of refinable functions*, Rev. Math. Iberoamericana **12** (1996), 527–591.

[CI]   P. Collet and S. Isola, *On the essential spectrum of the transfer operator for expanding Markov maps*, Comm. Math. Phys. **139** (1991), 551–557.

[CK]   F. Colonius and W. Kliemann, *Stability radii and Lyapunov exponents*, in *Control of uncertain systems*, D. Hinrichsen and B. Martensson (Eds.), Progr. Syst. Control Theory, vol. 6, Birkhäuser-Verlag, Basel, 1990, pp. 19–55.

[CK2]   F. Colonius and W. Kliemann, *The Morse spectrum of linear flows on vector bundles*, Trans. Amer. Math. Soc. **348** (1996), 4355–4388.

[CK3]   F. Colonius and W. Kliemann, *The Lyapunov spectrum of families of time-varying matrices*, Trans. Amer. Math. Soc. **348** (1996), 4389–4408.

[CSi]   I. P. Cornfeld and Ya. G. Sinai, *General ergodic theory of groups transformations with invariant measure*, in *Dynamical systems. II. Ergodic theory with applications to dynamical systems and statistical mechanics*, Ya. B. Sinai (Ed.), Encycl. Math. Sci., vol. 2, Springer-Verlag, Berlin, 1989.

[CF]   P. Constantin and C. Foias, *Navier-Stokes equations*, Chicago Lect. Math., U. of Chicago Press, Chicago, 1988.

[Co]   W. A. Coppel, *Dichotomies in stability theory*, Lect. Notes Math., vol. 629, Springer-Verlag, New York, 1978.

[Co2]   W. Coppel, *Dichotomies and Lyapunov functions*, J. Diff. Eqns. **52** (1984), 58–65.

[Cw]   C. Cowen, *Composition operators on Hilbert spaces of analytic functions: a status report*, Proc. Symp. Pure Math. **51** (1990), 131–145.

[Cu]   R. Curtain, *Equivalence of input-output stability and exponential stability for infinite-dimensional systems*, Math. Systems Theory **21** (1988), 19–48.

[CLTZ] R. Curtain, H. Logemann, S. Townley, and H. Zwart, *Well-posedness, stabilizability, and admissibility for Pritchard-Salamon systems*, J. Math. Syst. Estim. Control **7** (1997), 439–476.

[CP]   R. Curtain and A. J. Pritchard, *Infinite dimensional linear system theory*, Lect. Notes Control Infor. Sci., vol. 8, Springer-Verlag, New York, 1978.

[CZ]   R. Curtain and H. J. Zwart, *An introduction to infinite-dimensional linear control systems theory*, Springer-Verlag, New York, 1995.

[DK]   J. Daleckij and M. Krein, *Stability of differential equations in Banach space*, Amer. Math. Soc., Providence, RI, 1974.

[DKo] D. Daners and P. Koch Medina, *Abstract evolution equations, periodic problems and applications*, Pitman Res. Notes Math., no. 279, Longman/Wiley, New York, 1992.

[Da] R. Datko, *Extending a theorem of A. M. Lyapunov to Hilbert space*, J. Math. Anal. Appl. **32** (1970), 610–616.

[Da2] R. Datko, *Uniform asymptotic stability of evolutionary processes in a Banach space*, SIAM J. Math. Anal. **3** (1972), 428–445.

[Dv] E. B. Davies, *One-parameter semigroups*, Acad. Press, London, 1980.

[Dy] C. Day, *Spectral mapping theorem for integrated semigroups*, Semigroup Forum **47** (1993), 359–372.

[dL] R. de la Llave, *Hyperbolic dynamical systems and generation of magnetic fields by perfectly conducting fluids*, Geophys. Astrophys. Fluid Dynam. **73** (1993), 123–131.

[dL2] R. de la Llave, *The theory of Mather's spectrum and the spectrum of transfer operators acting on smooth functions, specifically for Anosov systems*, Preprint, 1998.

[De] R. deLaubenfels, *Existence families, functional calculi and evolution equations*, Lect. Notes Math., vol. 1570, Springer-Verlag, Berlin, 1994.

[DU] J. Diestel and J. J. Uhl, *Vector measures*, Math. Surv., no. 15, Amer. Math. Soc., Providence, 1977.

[Do] G. Dore, L_p *regularity for abstract differential equations*, in *Functional analysis and related topics*, Lect. Notes Math., vol. 1540, Springer-Verlag, Berlin, 1993, pp. 25–38.

[DN] J. R. Dorroh and J. W. Neuberger, *A theory of strongly continuous semigroups in terms of Lie generators*, J. Funct. Anal. **136** (1996), 114–126.

[EFT] A. Eden, C. Foias, and R. Temam, *Local and global Lyapunov exponents*, J. Dynam. Diff. Eqns. **3** (1991), 133–177.

[EH] S. Elaydi and O. Hajen, *Exponential dichotomy and trichotomy of nonlinear differential equations*, J. Integr. Diff. Eqns. **3** (1990), 1201–1224.

[El] J. Elliot, *The equation of evolution in a Banach space*, Trans. Amer. Math. Soc. **103** (1962), 470–483.

[EJ] R. Ellis, and R. Johnson, *Topological dynamics and linear differential equations*, J. Diff. Eqns. **44** (1982), 21–39.

[EN] K. Engel and R. Nagel, *One-parameter semigroups for linear evolution equations*, Springer-Verlag, New York, to appear.

[En] R. Engelking, *General topology*, Polish Scientific Publisher, Warsaw, 1977.

[Ev] D. E. Evans, *Time dependent perturbations and scattering of strongly continuous groups on Banach spaces*, Math. Ann. **221** (1976), 275–290.

[Fa] H. O. Fattorini, *The Cauchy problem*, Encycl. Math. Appl., no. 18, Addison Wesley, Reading Mass., 1983.

[FO] J. M. Finn and E. Ott, *Chaotic flows and magnetic dynamos*, Phys. Rev. Lett. **60**(9) (1988), 760–763.

[FvN] A. Fischer and J. M. A. M. van Neerven, *Robust stability of C_0-semigroups and an application to stability of delay equations*, J. Math. Anal. Appl. **226** (1998), 82–100.

[FR] T. Fischer and H. Rugh, *Transfer operators for coupled analytic maps*, Chao-dyn/9711018, Preprint, 1997.

[FRo] J. Franks and C. Robinson, *A quasi-Anosov diffeomorphism that is not Anosov*, Trans. Amer. Math. Soc. **233** (1976), 267–278.

[FV] S. Friedlander and M. Vishik, *Instability criteria for steady flows of a perfect fluid*, Chaos **2**(3) (1992), 455–460.

[FV2] S. Friedlander and M. Vishik, *Dynamo theory methods for hydrodynamic stability*, J. Math. Pures Appl. **72** (1993), 145–180.

[Fr] G. Froyland, *Computer-assisted bounds for the rate of decay of correlations*, Comm. Math. Phys. **189** (1997), 237–257.

[FKe] H. Furstenberg and H. Kesten, *Products of random matrices*, Annals of Math. **31** (1960), 457–469.

[FKi] H. Furstenberg and Y. Kiefer, *Random matrix products and measures on projective spaces*, Israel. J. Math. **46**(1–2) (1983), 12–32.

[Ge] L. Gearhart, *Spectral theory for contraction semigroups on Hilbert spaces*, Trans. Amer. Math. Soc. **236** (1978), 385–394.

[GF] I. Gohberg and I. Fel'dman, *Convolution equations and projection methods for their solutions*, Amer. Math. Soc. Transl., Providence, RI, 1974.

[GL]   I. Gohberg and J. Leiterer, *Factorization of operator functions with respect to a contour. II. Factorization of operator functions close to the identity*, Math. Nachr. **54** (1973), 41–74.

[Go]   J. Goldstein, *On the operator equation $AX + BX = Q$*, Proc. Amer. Math. Soc. **70** (1978), 31–34.

[Go2] J. Goldstein, *Semigroups of linear operators and applications*, Oxford Univ. Press, New York, 1985.

[Go3] J. Goldstein, *Asymptotics for bounded semigroups on Hilbert space*, in *Aspects of positivity in functional analysis*, North-Holland Math. Stud., vol. 122, 1986, pp. 49-62.

[Gr]   L. Green, *Geodesic flows*, Lect. Notes Math., vol. 200, 1971, pp. 25–27.

[Gr2] L. Green, *When is an Anosov flow geodesic?* Erg. Th. Dynam. Syst. **12** (1992), 227–232.

[GM]   G. Greiner and M. Müller, *The spectral mapping theorem for integrated semigroups*, Semigroup Forum **47** (1993), 115–122.

[GS]   G. Greiner and M. Schwarz, *Weak spectral mapping theorems for functional differential equations*, J. Diff. Eqns. **94** (1991), 205–216.

[GVW] G. Greiner, J. Voigt, and M. Wolff, *On the spectral bound of the generator of semigroups of positive operators*, J. Oper. Theory **5** (1981), 245–256.

[GH]   J. Guckenheimer and P. Holmes, *Nonlinear oscillations, dynamical systems, and bifurcations of vector fields*, Springer-Verlag, New York, 1983.

[GL]   V.M. Gundlach and Y. Latushkin, *Essential spectral radius of Ruelle's operator on smooth and Hölder spaces*, Comp. Rend. Acad. Sci. Paris **325**, Serie I (1997), 889–894.

[HH]   D. Hadwin and T. Hoover, *Operator algebras and the conjugacy of transformations*, J. Funct. Anal. **77** (1988), 112–122.

[HH2] D. Hadwin and T. Hoover, *Operator algebras and the conjugacy of transformations II*, Proc. Amer. Math. Soc. **106** (1989), 365–369.

[HH3] D. Hadwin and T. Hoover, *Representations of weighted translation algebras*, Houston J. Math. **18** (1992), 295–318.

[Ha]   J. Hale, *Ordinary Differential Equations*, Wiley, New York, 1969.

[Ha2] J. Hale, *Asymptotic behavior of dissipative systems*, Math. Surv. Monogr., vol. 25, Amer. Math. Soc., Providence, RI, 1988.

[HL]   J. Hale and S. M. Verduyn Lunel, *Introduction to functional-differential equations*, Appl. Math. Sci., vol. 99, Springer-Verlag, New York, 1993.

[Hm]   P. Halmos, *A Hilbert space problem book*, Van Nostrand, Princeton, 1967.

[Hy]   N. Haydn, *Meromorphic extension of the zeta function for Axiom A flows*, Erg. Th. Dynam. Syst. **10** (1990), 347–360.

[He]   D. Henry, *Geometric theory of nonlinear parabolic equations*, Lect. Notes Math., vol. 840, Springer-Verlag, New York, 1981.

[He2] D. Henry, *Topics in analysis*, Publ. Mat. **31**(1) (1987), 29–84.

[He3] D. Henry, *Evolution equations in Banach spaces*, Course notes, Inst. de Mat. e Estat., Univ. São Paulo; unpublished

[Hr]   I. Herbst, *Contraction semigroups and the spectrum of $A_1 \otimes I + I \otimes A_2$*, J. Oper. Theory **7** (1982), 61–78.

[Hil] E. Hille, *Functional analysis and semi-groups*, Amer. Math. Soc., New York 1948.

[HPh] E. Hille and R. Phillips, *Functional analysis and semi-groups*, Amer. Math. Soc., Coll. Publ., vol. 31, Providence, 1957.

[HIP] D. Hinrichsen, A. Ilchmann, and A. J. Pritchard, *Robustness of stability of time-varying linear systems*, J. Diff. Eqns. **82** (1989), 219–250.

[HP]   D. Hinrichsen and A. J. Pritchard, *Stability radius for structured perturbations and the algebraic Riccati equation*, Syst. Control Lett. **8** (1986), 105–113.

[HP2] D. Hinrichsen and A. J. Pritchard, *Real and complex stability radii: a survey*, in *Control of uncertain systems*, D. Hinrichsen and B. Martensson (Eds.), Progr. Syst. Control Theory, vol. 6, Birkhäuser-Verlag, Basel, 1990, pp. 119–162.

[HP3] D. Hinrichsen and A. J. Pritchard, *Robust stability of linear evolution operators on Banach spaces*, SIAM J. Contr. Opt. **32**(6) (1994), 1503–1541.

[HPS] M. Hirsch, C. Pugh, and M. Shub, *Invariant manifolds*, Lect. Notes Math., vol. 583, Springer-Verlag, New York, 1977.

[HK]   F. Hofbauer and G. Keller, *Zeta-functions and transfer operators for piecewise linear transformations*, J. Reine Angew. Math. **352** (1984), 100–113.

[Hs]   M. Holschneider, *Wavelet analysis of transfer operators acting on n-dimensional Hölder, Zygmund, Triebel spaces*, Preprint CPT-96/P3337, Centre de Phys. Theor. CNRS, 1996.

[HLQ] T. Hoover, A. Lambert, and J. Quinn, *The Markov process determined by a weighted composition operator*, Studia Math. **72** (1982), 225–235.

[Ho]   J. S. Howland, *Stationary scattering theory for time-dependent Hamiltonians*, Math. Ann. **207** (1974), 315–335.

[Ho2] J. Howland, *On a theorem of Gearhart*, Integr. Eqns. Oper. Th. **7** (1984), 138–142.

[Hu]   F. Huang, *Characteristic conditions for exponential stability of linear dynamical systems in Hilbert spaces*, Ann. Diff. Eqns. **1** (1985), 45–53.

[Hn]   S. Huang, *Characterizing spectra of closed operators through existence of slowly growing solutions of their Cauchy problems*, Studia Math. **116** (1995), 23–41.

[Hr]   N. E. Hurt, *Zeta functions and periodic orbit theory: a review*, Results in Math. **23** (1993), 55–120.

[Ht]   W. Hutter, *Spectral theory and almost periodicity of mild solutions of nonautonomous Cauchy problems*, Dissertation, Tübingen, 1997.

[Ja]   B. Jacob, *Time-varying infinite dimensional state-space systems*, Dissertation, Universität Bremen, 1995.

[Ja1] B. Jacob, *A formula for the stability radius of time-varying systems*, J. Diff. Eqns. **142** (1998), 167–187.

[JDP] B. Jacob, V. Dragon, and A. J. Pritchard, *Robust stability of infinite dimensional time-varying systems with respect to nonlinear perturbations*, Integr. Eqns. Oper. Th. **22** (1995), 440–462.

[JN]   C. A. Jacobson and C. N. Nett, *Linear state space systems in infinite-dimensional space: the role and characterization of joint stabilizability/detectability*, IEEE Trans. Automat. Control **33**(6) (1988), 541–550.

[JMS] Y. Jiang, T. Morita, and D. Sullivan, *Expanding direction of the period doubling operator*, Comm. Math. Phys. **144** (1992), 509–520.

[Jo]   R. Johnson, *Concerning a theorem of Sell*, J. Diff. Eqns. **30** (1978), 324–329.

[Jo2] R. Johnson, *Analyticity of spectral subbundles*, J. Diff. Eqns. **35** (1980), 366–387.

[Jo3] R. Johnson, *A review of recent works on almost periodic differential and difference operators*, Acta. Applic. Math. **1** (1983), 241–261.

[Jo4] R. Johnson, *Exponential dichotomy, rotation numbers, and linear differential operators with bounded coefficients*, J. Diff. Eqns. **61** (1986), 54–78.

[Jo5] R. Johnson, *The Oseledets and Sacker-Sell spectra for almost periodic linear systems: an example*, Proc. Amer. Math. Soc. **99** (1987), 261–267.

[JN]   R. Johnson and M. Nerurkar, *Exponential dichotomy and rotation number for linear Hamiltonian systems*, J. Diff. Eqns. **108** (1994), 201–216.

[JPS] R. Johnson, K. Palmer, and G. Sell, *Ergodic properties of linear dynamical systems*, SIAM J. Math. Anal. **1**(1) (1987), 1–33.

[JK]   L. K. Jones and U. Krengel, *On transformations without finite invariant measure*, Adv. Math. **12** (1974), 275–295.

[Ka]   Yu. I. Karlovich, *The local-trajectory method of studying invertibility in C^*-algebras of operators with discrete groups of shifts*, Soviet Math. Dokl. **37** (1988), 407–412.

[KKL] Yu. Karlovich, V. Kravchenko, and G. Litvinchuk, *The Noether theory of singular integral operators with a shift*, Izv. Vuzov Matem. **4** (1983), 3–27.

[KL]   M.A. Kaashoek and S. M. Verduyn Lunel, *An integrability condition on the resolvent for hyperbolicity of the semigroup*, J. Diff. Eqns. **112** (1994), 374–406.

[Ka]   T. Kato, *Perturbation theory for linear operators*, Springer-Verlag, New York, 1966.

[KH]   A. Katok and B. Hasselblatt, *Introduction to the modern theory of dynamical systems*, Encycl. Math. Appl., vol. 54, Cambridge Univ. Press, Cambridge, 1995.

[Ke]   G. Keller, *Equilibrium states in ergodic theory*, London Mathematical Society Student Press, Cambridge, 1998.

[KK]   G. Keller and M. Künzle, *Transfer operators for coupled map lattices*, Erg. Th. Dynam. Syst. **12** (1992), 297–318.

[KLi] G. Keller and C. Liverani, *Stability of the spectrum for transfer operators*, Preprint, 1998.

[KN]   G. Keller and T. Nowicki, *Spectral theory, zeta functions and the distribution of periodic points for Collet-Eckmann maps*, Comm. Math. Phys. **149** (1992), 31–69.

[KP]   U. Kirchgraber and K. Palmer, *Geometry in the neighborhood of invariant manifolds of maps and flows and linearization*, Pitman, London, 1991.

[Ki]   A. Kitover, *On spectra of operators in ideal spaces*, Zapiski Nauchn. Sem. Leningrad. Otdel. Mat. Inst. Steklov (LOMI), **65** (1976), 196–198.

[Ki2] A. Kitover, *The spectrum of weighted automorphisms and Kamowitz-Sheinberg theorem*, Funct. Anal. Appl. **13** (1979), 70–71.

[Ki3] A. Kitover, *Operators on C^1 induced by smooth mappings*, Funct. Anal. Appl. **16** (1982), 61–62.

[Ki4] A. Kitover, *Spectral properties of weighted endomorphisms in commutative Banach algebras*, Funct. Theory Funct. Anal. Appl. **41** (1984), 70–77.

[KY]   I. Klapper and L. S. Young, *Rigorous bounds on the fast dynamo growth rate involving topological entropy*, Comm. Math. Phys. **173** (1995), 623–646.

[KS]   A. Knauf and Ya. Sinai, *Classical nonintegrability, quantum chaos*, Birkhäuser-Verlag, Basel, 1997.

[Ko]   W. Konig, *Semicocycles and weighted composition semigroups on H^p*, Michigan Math. J. **37** (1990), 469–476.

[KLc] V. G. Kravchenko and G. S. Litvinchuk, *Introduction to the theory of singular integral operators with shift*, Math. Appl., vol. 289, Kluwer, Dordrecht, 1994.

[Kr]   U. Krengel, *Ergodic theorems*, de Gruyter Stud. in Math., vol. 6, Walter de Gruyter, New York, 1985.

[Kz]   K. Krzyzewski, *On connection between expanding mappings and Markov chains*, Bull. Acad. Polon. Sci. Sr. Sci. Math. Astronom. Phys. **19** (1971), 291–293.

[Ku]   V. G. Kurbatov *Lyneinye differentsial'no-rasnostnye uravneniya (Linear differential-difference equations)* (Russian), Voronez University, Voronez, 1990.

[KPp] J. Kurzweil and G. Papaschinopoulos, *Structural stability of linear discrete systems via the exponential dichotomy*, Czechoslov. Math. J. **38** (1988), 280–284.

[La]   S. Laederich, *Boundary value problems for partial differential equations with exponential dichotomies*, J. Diff. Eqns. **100** (1992), 1–21.

[LlM] G. Lancien and C. Le Merdy, *A generalized H^∞ functional calculus for operators on subspaces of L^p and application to maximal regularity*, Illinois J. Math. **42** (1998), 470–480.

[LT]   I. Lasiecka and R. Triggiani, *Feedback semigroups and cosine operators for boundary feedback parabolic and hyperbolic equations*, J. Diff. Eqns. **47** (1983), 246–272.

[LY]   A. Lasota and J. York, *Exact dynamical systems and the Frobenius-Perron operator*, Trans. Amer. Math. Soc. **273** (1982), 375–384.

[Lt]   Y. Latushkin, *Exact Lyapunov exponents and exponentially separated cocycles*, in *Partial differential equations*, J. Hale and J. Wiener (Eds.), Pitman Res. Notes Math., vol. 273, Longman Scientific, 1992, pp. 91–95.

[Lt2] Y. Latushkin, *Green's function, continual weighted composition operators along trajectories, and hyperbolicity of linear extensions for dynamical systems*, J. Dynam. Diff. Eqns. **6**(1) (1994), 1–21.

[LM]   Y. Latushkin and S. Montgomery-Smith, *Evolutionary semigroups and Lyapunov theorems in Banach spaces*, J. Funct. Anal. **127** (1995), 173–197.

[LM2] Y. Latushkin and S. Montgomery-Smith, *Lyapunov theorems for Banach spaces*, Bull. Amer. Math. Soc. **31**(1) (1994), 44–49.

[LM3] Y. Latushkin and S. Montgomery-Smith, *Spectral mapping theorems for evolution semigroups*, in *Seminar Notes in Funct. Anal. and PDE*, Tiger Notes, LSU, 1993-94, pp. 246-255.

[LMR] Y. Latushkin, S. Montgomery-Smith, and T. Randolph, *Evolutionary semigroups and dichotomy of linear skew-product flows on locally compact spaces with Banach fibers*, J. Diff. Eqns. **125** (1996), 73–116.

[LR]   Y. Latushkin and T. Randolph, *Dichotomy of differential equations on Banach spaces and an algebra of weighted composition operators*, Integr. Eqns. Oper. Th. **23** (1995), 472–500.

[LRS] Y. Latushkin, T. Randolph, and R. Schnaubelt, *Exponential dichotomy and mild solutions of nonautonomous equations in Banach spaces*, J. Dynam. Diff. Eqns. **10** (1998), 489–510.

[LS] Y. Latushkin and R. Schnaubelt, *Exponential dichotomy of cocycles, evolution semigroups and translation algebras*, J. Diff. Eqns., to appear.

[LS2] Y. Latushkin and R. Schnaubelt, *The spectral mapping theorem for evolution semigroups on L^p associated with strongly continuous cocycles*, Semigroup Forum, to appear.

[LSt] Y. Latushkin and A. M. Stepin, *Weighted composition operator on topological Markov chains*, Funct. Anal. Appl. **22** (1988), 86–87.

[LSt2] Y. Latushkin and A. M. Stepin, *Weighted composition operators and linear extensions of dynamical systems*, Russian Math. Surveys **46**(2) (1992), 95–165.

[LSt3] Y. Latushkin and A. M. Stepin, *Weighted shift operators, spectral theory of linear extensions and the multiplicative ergodic theorem*, Math. USSR Sbornik **70**(1) (1991), 143–163.

[LSt4] Y. Latushkin and A. M. Stepin, *On the perturbation theorem for the dynamical spectrum*, unpublished.

[LV] Y. Latushkin and M. Vishik, *A characterization of linear stability for Euler equation*, Preprint.

[Ld] F. Ledrappier, *Principe variationnel et systeèmes dynamiques symboliques*, Z. Wahr. Verw. Geb. **30** (1974), 185–202.

[LW] F. Ledrappier and P. Walters, *A relativised variational principle for continuous transformations*, J. London Math. Soc. **16** (1977), 568–576.

[Le] H. Leiva, *Sacker-Sell spectrum for scalar parabolic equations*, Preprint, 1997.

[Le2] H. Leiva, *Exponential dichotomy for a non-autonomous system of parabolic equations*, J. Dynam. Diff. Eqns. **10** (1998), 475–488.

[Lv] V. B. Levenshtam, *Averaging of quasilinear parabolic equations with a rapidly oscillating main part. Exponential dichotomy*, Russian Acad. Sci. Izv. Math. **41** (1993), 95–132.

[LZ] B. M. Levitan and V. V. Zhikov, *Almost periodic functions and differential equations*, Cambridge Univ. Press, Cambridge, 1982.

[Lw] J. Lewowicz, *Lyapunov functions and topological stability*, J. Diff. Eqns. **38** (1980), 192–209.

[Li] X.-B. Lin, *Exponential dichotomies in intermediate spaces with applications to a diffusively perturbed predator-prey model*, J. Diff. Eqns. **108** (1994), 36–63.

[Lc] G. S. Litvinchuk, *Boundary value problems and singular integral equations with a shift*, Nauka Publ., Moscow, 1977.

[Lv] C. Liverani, *Decay of correlations*, Ann. Math. **142** (1995), 239–301.

[Lo] D. Lovelady, *On the generation of linear evolution operators*, Duke Math. J. **42** (1975), 57–69.

[LL] M. Lyubich and Yu. Lyubich, *Perron-Frobenius theory for almost periodic operators and semigroup representations*, Funct. Theory Funct. Anal. Appl. **46** (1986), 54–72.

[LL2] M. Lyubich and Yu. Lyubich, *On splitting off the boundary spectrum for almost periodic operators and semigroup representations*, Funct. Theory Funct. Anal. Appl. **45** (1986), 69–84.

[LP] Yu. I. Lyubich and Vũ Quoc Phong, *Asymptotic stability of linear differential equations on Banach spaces*, Studia Math. **88** (1988), 37–42.

[LP2] Yu. L. Lyubich and Vũ Quoc Phong, *On the spectral mapping theorem for one-parameter groups of operators*, Zapiski Nauchn. Sem. Leningrad. Otdel. Mat. Inst. Steklov (LOMI) **178** (1989), 146–150.

[Lu] G. Lumer, *Équations d'évolution, semigroupes en espace-temps et perturbations*, C. R. Acad. Sci. Paris Série I **300** (1985), 169–172.

[Lu2] G. Lumer, *Opérateurs d'évolution, comparaison de solutions, perturbations et approximations*, C. R. Acad. Sci. Paris Série I **301** (1985), 351–354.

[Lu3] G. Lumer, *Equations de diffusion dans le domaines (x,t) non-cylindriques et semigroupes "espace-temps"*, Lect. Notes Math., vol. 1393, Springer-Verlag, New York, 1989, pp. 161–179.

[LSc] G. Lumer and R. Schnaubelt, *Local operator methods and time dependent parabolic equations on non-cylindrical domains*, Preprint, 1998.

[Ln] A. Lunardi, *Analytic semigroups and optimal regularity in parabolic problems*, Progr. Nonlin. Diff. Eqns. Appl., vol. 16, Birkhäuser, Basel, 1995.

[Mg] L. T. Magalhães, *Persistence and smoothness of hyperbolic invariant manifold for functional differential equations*, SIAM J. Math. Anal. **18**(3) (1987), 670–693.

[Mg2]   L. Magalhães, *The spectrum of invariant sets for dissipative semiflows*, in *Dynamics of infinite-dimensional systems (Lisbon, 1986)* NATO Adv. Sci. Inst. Ser. F Comput. Systems Sci., vol. 37, 1987, pp. 161–168.

[MSe]   J. Mallet-Paret and G. Sell, *Inertial manifolds for reaction diffusion equations in higher space dimensions*. J. Amer. Math. Soc. **1** (1988), 805–866.

[Mn]   R. Mãné, *Persistent manifolds are normally hyperbolic*, Bull. Amer. Math. Soc. **80** (1974), 90–91.

[Mn2]   R. Mãné, *Quasi-Anosov diffeomorphisms and hyperbolic manifolds*, Trans. Amer. Math. Soc. **229** (1977), 351–370.

[Mn3]   R. Mãné, *Lyapunov exponents and stable manifolds for compact transformations*, in *Geometric Dynamics*, J. Palis (Ed.), Lect. Notes Math., vol. 1007, Springer-Verlag, New York, 1983, pp. 522–577.

[Mr]   A. Marchesi, *Exponential dichotomy and strong solutions for abstract parabolic non-autonomous equations*, Preprint, 1997.

[MM]   J. Martinez and J. Mazon, C_0-*semigroups norm continuous at infinity*, Semigroup Forum **52** (1996), 213–224.

[Ma]   J. Mather, *Characterization of Anosov diffeomorphisms*, Indag. Math. **30** (1968), 479–483.

[MS]   J. Massera and J. Schäffer, *Linear differential equations and functional analysis. III. Lyapunov's second method in the case of conditional stability*, Ann. Math. **69** (1959), 535–574.

[MS2]   J. Massera and J. Schäffer, *Linear differential equations and function spaces*, Academic Press, New York, 1966.

[My]   D. Mayer, *The Ruelle-Araki transfer operator in classical statistical mechanics*, Lect. Notes Phys., vol. 123, Springer-Verlag, New York, 1980.

[My2]   D. Mayer, *Continued fractions and related transformations*, in *Ergodic theory, symbolic dynamics and hyperbolic spaces*, T. Bedfrod, M. Keane, and C. Series (Eds.), Oxford Univ. Press, 1991, pp. 175–222.

[Mi]   Nguyen Van Minh, *Semigroups and stability of nonautonomous differential equations in Banach spaces*, Trans. Amer. Math. Soc., **345**(1) (1994), 223–241.

[Mi2]   Nguyen Van Minh, *Spectral theory for linear non-autonomous differential equations*, J. Math. Anal. Appl. **187** (1994), 339–351.

[Mi3]   Nguyen Van Minh, *On the proof of characterizations of the exponential dichotomy*, Proc. Amer. Math. Soc. **127** (1999), 779–782.

[MRS]   Nguyen Van Minh, F. Räbiger, and R. Schnaubelt, *Exponential stability, exponential expansiveness, and exponential dichotomy of evolution equations on the half-line*, Integr. Eqns. Oper. Th. **32** (1998), 332–353.

[ML]   M. Megan and R. Latcu, *Exponential dichotomy of evolution operators in Banach spaces*, Intern. Ser. Numer. Math. **107**, Birkhäuser-Verlag, Basel, 1992, pp. 47–62.

[Ml]   V. M. Millionshchikov, *Statistically regular systems*, Matem. Sbornik **75** (1968), 140–151, English transl. MathUSSR Sb. **4** (1968), 125–135.

[MST]   K. Mischaikow, H. Smith, and H. Thieme, *Asymptotically autonomous semiflows: chain recurrence and Lyapunov functions*, Trans. Amer. Math. Soc. **347** (1995), 1669–1685.

[Mo]   H. K. Moffatt, *Magnetic field generation in electrically conducting fluids*, Cambridge Univ. Press, Cambridge, 1978.

[MRS]   S. A. Molchanov, A. A. Ruzmaikin, and D. D. Sokolov, *Kinematic dynamo in random flow*, Sov. Phys. Usp. **28**(4) (1985), 307–327.

[Mt]   S. Montgomery-Smith, *Stability and dichotomy of positive semigroups on L_p*, Proc. Amer. Math. Soc. **124** (1996), 2433–2437.

[MSK]   Y. A. Mitropolskij, A. M. Samojlenko, and V. L. Kulik, *Issledovanija dichotomii lineinyh sistem differencial'nyh uravnenij (Dichotomy of systems of linear differential equations)*, (Russian), Naukova Dumka, Kiev, 1990.

[Na]   R. Nagel (Ed.) *One parameter semigroups of positive operators*, Lect. Notes Math., vol. 1184, Springer-Verlag, Berlin, 1984.

[Na2]   R. Nagel, *Semigroup methods for non-autonomous Cauchy problems*, in *Evolution Equations (Baton Rouge, LA, 1992)* Lect. Notes Pure Appl. Math., vol. 168, 1995, pp. 301–316.

[NP]   R. Nagel and J. Poland, *The critical spectrum of a strongly continuous semigroup*, Preprint, 1998.

[NR]   R. Nagel and A. Rhandi, *Positivity and Lyapunov conditions for linear systems*, Adv. Math. Sci. Appl. **3** (1993-94), 33–41.

[NR2] R. Nagel and A. Rhandi, *A characterization of Lipschitz continuous evolution families on Banach spaces*, Oper. Th. Adv. Appl. **75** (1995), 275–288.

[NS]   R. Nagel and E. Sinestrari, *Inhomogeneous Volterra integrodifferential equations for Hille-Yosida operators*, in *Functional Analysis (Essen, 1991)*, Lect. Notes Pure Appl. Math., vol. 150, 1994, pp. 51–70.

[Ne]   H. Neidhardt, *On abstract linear evolution equations, I*, Math. Nachr. **103** (1981), 283–298.

[NRL] A. F. Neves, H. S. Ribeiro, and D. Lopes, *On the spectrum of evolution operators generated by hyperbolic systems*, J. Funct. Anal. **67** (1986), 320–344.

[Ni]   G. Nickel, *On evolution semigroups and wellposedness of nonautonomous Cauchy problems*, Dissertation, Tübingen, 1996.

[No]   E. A. Nordgren, *Composition operators in Hilbert spaces*, Lect. Notes Math., vol. 693, 1978, pp. 37–64.

[Nu]   M. Nuñez, *Localized eigenmodes of the induction equation*, SIAM J. Appl. Math. **54**(5) (1994), 1254–1267.

[Os]   V. Oseledets, *A multiplicative ergodic theorem. Lyapunov characteristic numbers for dynamical systems*, Trans. Moscow Math. Soc. **19** (1968), 197–231.

[Os2] V.I. Oseledets, *Λ-entropy and the antidynamo theorem*, Proc. 6th Intern. Symp. Inform. Theory, Part III, Tashkent, 1984, pp. 162–163.

[Os3] V.I. Oseledets, *Fast dynamo problem for a smooth map on a two-torus*, Geophys. Astrophys. Fluid Dynam. **73** (1993), 133–145.

[Ot]   N. Otsuki, *A characterization of Anosov flows for geodesic flows*, Hiroshima Math. J. **4** (1974), 397–412.

[PP]   J. Palis and C. Pugh, *Fifty problems in dynamical systems*, Lect. Notes Math., vol. 468, 1974, pp. 345–353.

[Pa]   K. Palmer, *The structurally stable linear systems on the half-line are those with exponential dichotomies*, J. Diff. Eqns. **33** (1979), 16–25.

[Pa2] K. Palmer, *Two linear systems criteria for exponential dichotomy*, Ann. Mat. Pura Appl. **124** (1980), 199–216.

[Pa3] K. Palmer, *Exponential dichotomy and Fredholm operators*, Proc. Amer. Math. Soc. **104** (1988), 149–156.

[Pa4] K. Palmer, *Exponential dichotomies, the shadowing lemma and transversal homoclinic points*, in *Dynamics Reported*, vol. 1, 1988, pp. 256–306.

[Pf]   L. Pandolfi, *A Lyapunov theorem for semigroups of operators*, Syst. Control Lett. **15** (1990), 147–151.

[Pq]   L. Paquet, *Semigroups géneralisés et équations d'évolution*, (French) in *Séminaire de Théorie du Potentiel (Paris 1977/1978)*, vol. 4, Lect. Notes. Math., vol. 713, Springer-Verlag, New York, 1979, pp. 243–263.

[Pz]   A. Pazy, *Semigroups of linear operators and applications to partial differential equations*, Springer-Verlag, New York, 1983.

[Pz2] A. Pazy, *On the applicability of Lyapunov theorem in Hilbert space*, SIAM J. Math. Anal. **3** (1972), 291–294.

[PPo] W. Parry and M. Pollicott, *Zeta functions and the periodic orbit structure of hyperbolic dynamics*, Soc. Math. de France, Asterisque, vols. 187-188, Paris, 1990.

[Pe]   G. K. Pedersen, *C^*-algebras and their automorphisms groups*, London Math. Soc. Monogr., vol. 14, Academic Press, New York, 1979.

[Pn]   O. Perron, *Die stabilitätsfrage dei differentialgleichungen*, Math. Z. **32** (1930), 703–728.

[Ps]   Ya. B. Pesin, *Hyperbolic theory*, Encycl. Math. Sci. Dynam. Syst., vol. 2, 1988.

[Ph]   R. S. Phillips, *Perturbation theory for semi-groups of linear operators*, Trans. Amer. Math. Soc. **74** (1953), 199–221.

[PR]   Vũ Quoc Phong and W. Ruess, *Asymptotically almost periodic solutions of evolution equations in Banach spaces*, J. Diff. Eqns. **122** (1995), 282–301.

[PS]   Vũ Quoc Phong and E. Schüler, *The operator equation $AX - XB = C$, admissibility, and asymptotic behavior of differential equations*, J. Diff. Eqns. **145** (1998), 394–419.

[Pl] V. A. Pliss, *On hyperbolicity of smooth cocycles over flows with an invariant ergodic measure*, Časpois Pro Péstorani Matem. **3** (1986), 146–154.

[Po] M. Pollicott, *A complex Ruelle-Perron-Frobenius theorem and two counterexamples*, Erg. Th. Dynam. Syst. **4** (1984), 135–146.

[Po2] M. Pollicott, *Meromorphic extensions of generalized zeta functions*, Invent. Math. **85** (1986), 147–164.

[PM] P. Preda and M. Megan, *Exponential dichotomy of evolutionary processes in Banach spaces*, Czechoslov. Math. J. **35** (1985), 312–323.

[PT] A. J. Pritchard and S. Townley, *Robustness of linear systems*, J. Diff. Eqns. **77** (1989), 254–286.

[Pr] J. Prüss, *On the spectrum of C_0-semigroups*, Trans. Amer. Math. Soc. **284**(2) (1984), 847–857.

[Pr2] J. Prüss, *Evolutionary integral equations and applications*, Monogr. Math., vol. 87, Birkhäuser-Verlag, Basel, 1993.

[RS] F. Räbiger and R. Schnaubelt, *A spectral characterization of exponentially dichotomic and hyperbolic evolution families*, Tübingen Berichte zur Funktionalanalysis, Heft 3, Jahrang 1993/94, 1994, pp. 204–221.

[RS2] F. Räbiger and R. Schnaubelt, *The spectral mapping theorem for evolution semigroups on spaces of vector-valued functions*, Semigroup Forum **52** (1996), 225–239.

[RRS] F. Räbiger, A. Rhandi, and R. Schnaubelt, *Perturbation and abstract characterization of evolution semigroups*, J. Math. Anal. Appl. **198** (1996), 516–533.

[RRSV] F. Räbiger, A. Rhandi, R. Schnaubelt, and J. Voigt, *Non-autonomous Miyadera perturbation*, Preprint.

[RLC] T. Randolph, Y. Latushkin and S. Clark, *Evolution semigroups and stability of time-varying systems on Banach spaces*, Proc. 36th IEEE Conference on Decision and Control, San Diego, CA, December 1997, pp. 3932–3937.

[Ra] R. Rau, *Hyperbolic evolution semigroups*, Dissertation, Tübingen, 1992.

[Ra2] R. Rau, *Hyperbolic evolutionary semigroups on vector-valued function spaces*, Semigroup Forum **48** (1994), 107–118.

[Ra3] R. Rau, *Hyperbolic linear skew-product semiflows*, Z. Anal. Anwendungen **15** (1996), 865–880.

[RSe] G. Raugel and G. Sell, *Navier-Stokes equations on thin 3D domains. I: global attractors and global regularity of solutions*, J. Amer. Math. Soc. **6** (1993), 503–568.

[Re] R. Rebarber, *Frequency domain methods for proving the uniform stability of vibrating systems*, in *Analysis and optimization of systems: state and frequency domain approaches for infinite-dimensional systems (Sophia-Antipolis, 1992)*, R. F. Curtain, A. Bensoussan, and J.-L. Lions (Eds.), Lect. Notes in Control Inform. Sci., vol. 185, Springer-Verlag, New York, 1993, pp. 366–377.

[Re2] R. Rebarber, *Conditions for the equivalence of internal and external stability for distributed parameter systems*, IEEE Trans. on Automat. Control **31**(6) (1993), 994–998.

[Rn] M. Renardy, *On the linear stability of hyperbolic PDEs and viscoelastic flows*, Z. Angew Math. Phys. **45** (1994), 854–865.

[Rn2] M. Renardy, *On the stability of differentiability of semigroups*, Semigroup Forum **51** (1995), 343–346.

[Rn3] M. Renardy, *Spectrally determined growth is generic*, Proc. Amer. Math. Soc. **124** (1996), 2451–2453.

[Rh] A. Rhandi, *Lipschitz stetige evolutionsfamilen und die exponentielle dichotomie*, Dissertation, Tübingen, 1994.

[RR] H. M. Rodrigues and J. G. Ruas-Filho, *Evolution equations: dichotomies and the Fredholm alternative for bounded solutions*, J. Diff. Eqns. **119** (1995), 263–283.

[Ru] D. Ruelle, *Statistical mechanics for a one-dimensional lattice*, Comm. Math. Phys. **9** (1968), 267–278.

[Ru2] D. Ruelle, *Thermodynamic formalism*, Addison-Wesley, New York, 1978.

[Ru3] D. Ruelle, *Characteristics exponents and invariant manifolds in Hilbert space*, Ann. Math. **115** (1982), 243–290.

[Ru4] D. Ruelle, *The thermodynamic formalism for expanding maps*, Comm. Math. Phys. **125** (1989), 239–262.

[Ru5] D. Ruelle, *An extension of the theory of Fredholm determinants*, Publ. Math IHES **72** (1990), 175–193.

[Ru6] D. Ruelle, *Spectral properties of a class of operators associated with maps in one dimension*, Erg. Th. Dynam. Syst. **11** (1991), 757–767.

[Ru7] D. Ruelle, *Dynamical zeta functions for piecewise monotone maps of the interval*, CRM Monograph Series, vol. 4, Centre Res. Math. Universite de Montreal, Amer. Math. Soc., Providence, 1991.

[Rg] H. Rugh, *On the asymptotic form and the reality of spectra of Perron-Frobenius operators*, Nonlinearity **7** (1994), 1055-1066.

[Rg2] H. Rugh, *Generalized Fredholm determinants and Selberg zeta function for Axiom A dynamical systems*, Erg. Th. Dynam. Syst. **16** (1996), 805–819.

[Sa] R. Sacker, *Existence of dichotomies and invariant splittings for linear differential systems. IV*, J. Diff. Eqns. **27** (1978), 106–137.

[Sa2] R. Sacker, *The splitting index for linear differential systems*, J. Diff. Eqns, **33** (1979), 368–405.

[SS] R. Sacker and G. Sell, *A note of Anosov diffeomorphisms*, Bull. Amer. Math. Soc. **80** (1974), 278–280.

[SS2] R. Sacker and G. Sell, *Existence of dichotomies and invariant splitting for linear differential systems, I,II,III*, J. Diff. Eqns. **15, 22** (1974, 1976), 429–458, 478–522.

[SS3] R. Sacker and G. Sell, *A spectral theory for linear differential systems*, J. Diff. Eqns. **27** (1978), 320–358.

[SS4] R. Sacker and G. Sell, *The spectrum of an invariant submanifold*, J. Diff. Eqns. **38** (1980), 135–160.

[SS5] R. Sacker and G. Sell, *Dichotomies for linear evolutionary equations in Banach spaces*, J. Diff. Eqns. **113** (1994), 17–67.

[Se] G. Sell, *Linearization and global dynamics*, Proc. Intern. Congress of Math., Aug. 16-24, 1983, Warsaw, vol. 2, pp. 1283-1296.

[Se2] G. Sell, *References on dynamical systems*, University of Minnesota Preprint, 1991.

[SK] R. Saeks and G. Knowles, *The Arveson frequency response and systems theory*, Int. J. Control **42**(3) (1985), 639–650.

[Sm] A. M. Samojlenko, *Elements of the mathematical theory of multifrequency oscillations. Invariant tori*, Nauka Publ., Moscow, 1987.

[Sc] R. Schnaubelt, *Exponential bounds and hyperbolicity of evolution families*, Dissertation, Tübingen, 1996.

[Sg] J. Selgrade, *Isolated invariant sets for flows on vector bundles*, Trans. Amer. Math. Soc. **203** (1975), 359–390.

[Sy] V. Semenyuta, *On singular operator equations with a shift on the circle*, Dokl. AN SSSR **237**(6) (1977), 1301-1302.

[Sh] J. Shapiro, *The essential norm of a composition operator*, Ann. Math. **125** (1987), 375–404.

[Sh2] J. Shapiro, *Composition operators and classical function theory*, Springer-Verlag, New York, 1993.

[SY] W. Shen and Y. Yi, *On minimal sets of scalar parabolic equations with skew product structures*, Trans. Amer. Math. Soc. **347** (1995), 4413–4431.

[SY2] W. Shen and Y. Yi, *Almost automorphic and almost periodic dynamics in skew product semiflows*, Memoirs Amer. Math. Soc. **136** (1998), no. 647, 93 pp.

[Sb] M. Shubin, *Pseudo-difference operators and their inversion*, DAN SSSR **276**(3) (1984), 567–570.

[Sb2] M. Shubin, *Pseudo-difference operators and their Green function*, Izv. AN SSSR, Ser. Mat. **49** (1985), 652–671.

[Si] Ya. G. Sinai, *Markov partitions and C-diffeomorphism*, Funct. Anal. Appl. **2** (1968), 61–82.

[Si2] Ya. G. Sinai, *Construction of Markov partitions*, Funct. Anal. Appl. **2** (1968), 245–253.

[Si3] Ya. G. Sinai, *Gibbs measures in ergodic theory*, Russian Math. Surv. **27** (1972), 21–69.

[Ss] A. G. Siskakis, *Weighted composition semigroups on Hardy spaces*, Linear Algebr. Appl. **84** (1986), 359–371.

[So] A.M. Soward, *Fast dynamo actions in a steady flow*, J. Fluid Mech. **180** (1987), 267–295.

[Sn]   R. Swanson, *The spectrum of vector bundle flows with invariant subbundles*, Proc. Amer. Math. Soc. **83** (1981), 141–145.

[Sn2] R. Swanson, *The spectral characterization of normal hyperbolicity*, Proc. Amer. Math. Soc. **89** (1983), 503–509.

[Ta]   H. Tanabe, *Equations of evolution*, Pitman, London, 1979.

[Tn]   F. Tangerman, *Meromorphic continuation of Ruelle zeta function*, Dissertation, Boston University, 1986.

[Te]   R. Temam, *Infinite-dimensional dynamical systems in mechanics and physics*, Springer-Verlag, New York, 1997.

[Th]   P. Thieullen, *Fibres dynamiques asumptotiquement compact, exposants de Lyapunov. Entropie. Dimension*, (French) Ann. Inst. H. Poincaré Anal. Nonlin. **4**(1) (1987), 49–97.

[Ty]   V. M. Tyurin, *Invertibility of linear differential operators in some function spaces*, Siberian Math. J. **32** (1991), 485–490.

[vK]   B. van Keulen, *$\mathcal{H}_\infty$-control for distributed parameter systems: a state-space approach*, Birkhäuser-Verlag, Basel, 1993.

[vN]   J. M. A. M. van Neerven, *The asymptotic behavior of semigroups of linear operators*, Oper. Theory Adv. Appl., vol. 88, Birkhäuser-Verlag, 1996.

[vN2] J. M. A. M. van Neerven, *Characterization of exponential stability of a semigroup of operators in terms of its action by convolution on vector-valued function spaces over $\mathbb{R}_+$*, J. Diff. Eqns. **124** (1996), 324–342.

[VI]   A. Vanderbauwhede and G. Iooss, *Center manifold theory in infinite dimensions*, in *Dynamics Reported*, vol. 1 (new ser.), 1992, pp. 125–163.

[VG]   A. Vanderbauwhede and S. A. Van Gils, *Center manifolds and contractions on a scale of Banach spaces*, J. Funct. Anal. **72** (1987), 209–224.

[Vi]   J. Vilms, *Connections of tangent bundles*, J. Diff. Geom. **16** (1967), 234–243.

[Vn]   R. Vinograd, *Exact bounds for exponential dichotomy roughness*, J. Diff. Eqns. **90** (1991), 203–210; **91** (1991), 245–267; **71** (1988), 63–71.

[Vs]   M. M. Vishik, *On a system of equations arising in magneto-hydrodynamics*, Soviet Math. Dokl. **29**(2) (1984), 372–376.

[Vs2] M. M. Vishik, *Magnetic field generation by the motion of a highly conducting fluid*, Geophys. Astrophys. Fluid Dynam. **48** (1989), 151–167.

[Vs3] M. M. Vishik, *Spectrum of small oscillations of an ideal fluid and Lyapunov exponents*, J. Math. Pures et Appl. **75**(6) (1996), 531–558.

[Vo]   J. Voigt, *On the perturbation theory for strongly continuous semigroups*, Math. Ann. **229** (1977), 163–171.

[VS]   L. R. Volevich and A. R. Shirikyan, *Exponential dichotomy and exponential splitting for hyperbolic equations*, Trans. Moscow Math. Soc. **58** (1998), 95–133.

[Wa]   P. Walters, *Invariant measures and equilibrium states for some mappings which expand distances*, Trans. Amer. Math. Soc. **236** (1978), 121–153.

[Wa2] P. Walters, *An introduction to ergodic theory*, Grad. Texts in Math., vol. 79, Springer-Verlag, New York, 1982.

[Wa3] P. Walters, *Relative pressure, relative equilibrium states, compensation functions and many-to-one codes between subshifts*, Trans. Amer. Math. Soc. **296** (1986), 1–31.

[Way] C. E. Wayne, *Invariant manifolds for parabolic partial differential equations on unbounded domains*, Arch. Rational Mech. Anal. **138** (1997), 279–306.

[Wb]   G. F. Webb, *Theory of nonlinear age-dependent population dynamics*, Monogr. Text. Pure Appl. Math., vol. 89, Marcel Dekker, New York, 1985.

[We]   L. Weis, *The stability of positive semigroups on L_p spaces*, Proc. Amer. Math. Soc. **123** (1995), 3089–3094.

[We2] L. Weis, *Stability theorems for semigroups via multiplier theorems*, in *Differential equations, asymptotic analysis, and mathematical physics (Potsdam, 1996)* Math. Res., vol. 100, 1997, pp. 407–411.

[We3] L. Weis, *A short proof for the stability theorem for positive semigroups on $L^p(\mu)$*, Proc. Amer. Math. Soc. **126** (1998), 3253–3256.

[WW]   L. Weis and V. Wrobel, *Asymptotic behavior of C_0-semigroups in Banach spaces*, Proc. Amer. Math. Soc. **124** (1996), 3663–3671.

[Ws]   G. Weiss, *Weak L^p-stability of a linear semigroup on a Hilbert space implies exponential stability*, J. Diff. Eqns. **76** (1988), 269–285.

[Ws2]   G. Weiss, *Representation of shift invariant operators on L^2 by H^∞ transfer functions: an elementary proof, a generalization to L^p and a counterexample for L^∞*, Math. Control Signals Syst. **4** (1991), 193–203.

[Ws3]   G. Weiss, *Transfer functions of regular linear systems, part I: Characterizations of regularity*, Trans. Amer. Math. Soc. **342**(2) (1994), 827–854.

[Wn]   Z. Weinian, *Generalized exponential dichotomies and invariant manifolds for differential equations*, Adv. Math. **22**(1), (1993), 1–44.

[Wm]   J. Willems, *Topological classification and structural stability of linear systems*, J. Diff. Eqns. **35** (1980), 306–318.

[Wi]   F. R. Wirth, *Robust stability of discrete-time systems under time-varying perturbations*, Dissertation, Universität Bremen, 1995.

[Wo]   M. Wolff, *A remark on the spectral bound of the generator of a semigroup of positive operators with applications to stability theory*, in *Functional analysis and approximation*, P. L. Butzer, et al. (Eds.), Birkhäuser-Verlag, Basel, 1981, pp. 39–50.

[Wr]   V. Wrobel, *Asymptotic behavior of C_0-semigroups on B-convex spaces*, Indiana Univ. Math. J. **38** (1989), 101–114.

[Wr2]   V. Wrobel, *Stability and spectra of C_0-semigroups*, Math. Ann. **285** (1989), 201–219.

[Yi]   Y. Yi, *A generalized integral manifold theorem*, J. Diff. Eqns. **102** (1993), 153–187.

[Yo]   L.-S. Young, *Decay of correlations for certain quadratic maps*. Comm. Math. Phys. **146** (1992), 123–138.

[Yo2]   L.-S. Young, *Statistical properties of dynamical systems with some hyperbolicity*. Ann. of Math. **147** (1998), 585–650.

[Za]   J. Zabczyk, *A note on C_0-semigroups*, Bull. Acad. Polon. Sci. **23** (1975), 895–898.

[Za2]   J. Zabczyk, *Mathematical control theory: an introduction*, Birkhäuser-Verlag, Basel, 1995.

[ZM]   P. P. Zabreiko and Nguyen Van Minh, *Group of characteristic operators and its applications in the theory of linear differential equations*, Russian Acad. Sci. Dokl. Math. **45**(3) (1992), 517–521.

[ZR]   Ya. B. Zel'dovich and A.A. Ruzmaikin, *The magnetic field in a conducting fluid in two-dimensional motion*, J. Exper. Theor. Phys. **78** (1980), 980–985.

[Ze]   W. Zeng, *Exponential dichotomies and transversal homoclinic orbits in degenerate cases*, J. Dynam. Diff. Eqns. **7** (1995), 521–548.

[Zh]   W. Zhang, *The Fredholm alternative and exponential dichotomies for parabolic equations*, J. Math. Anal. Appl. **191** (1995), 180–201.

List of Notations

$\mathbb{N}$ – the natural numbers;
$\mathbb{Z}$ – the integers;
$\mathbb{R}$ – the real numbers;
$\mathbb{C}$ – the complex numbers;
$\mathbb{Z}_+, \mathbb{R}_+$ – the nonnegative integers, the nonnegative real numbers;
$\mathbb{T}$ – unit circle;
$\mathbb{T}_\lambda = \{z \in \mathbb{C} : |z| = e^\lambda\}$, $\lambda \in \mathbb{R}$;
$\mathbb{D}$ – open unit disk;
$\mathbb{C}_+ = \{z \in \mathbb{C} : \operatorname{Re} z > 0\}$;
$\mathbb{C}_- = \{z \in \mathbb{C} : \operatorname{Re} z < 0\}$;
X – Banach space with a norm $|\cdot|$;
X^* – dual space;
$\mathcal{H}$ – Hilbert space;
$\langle x, y \rangle$ – scalar product;
$\mathcal{L}(X)$ – linear bounded operators on X;
$\mathcal{GL}(X)$ – invertible operators in $\mathcal{L}(X)$;
$\mathcal{SL}(\mathcal{H})$ – self-adjoint operators;
$\mathcal{C}(X)$ – compact operators;
$SO(\ell)$ – orthogonal $\ell \times \ell$ matrices with unit determinant;
$C_0(\mathbb{R}; X)$ – continuous X-valued functions f on $\mathbb{R}$ such that $\lim_{t \to \pm\infty} f(t) = 0$;
$C_{00}(\mathbb{R}_+; X)$ – continuous X-valued functions f on $\mathbb{R}_+$ such that $f(0) = 0$
        and $\lim_{t \to \infty} f(t) = 0$;
$C_c^1(\mathbb{R})$ – compactly supported and continuously differentiable scalar functions;
$c_0(\mathbb{Z}; X)$ – sequences $(x_n)_{n \in \mathbb{Z}} \in \ell^\infty(\mathbb{Z}; X)$ such that $\lim_{n \to \pm\infty} |x_n| = 0$;
$\mathcal{S} = \mathcal{S}(\mathbb{R}; X)$ – Schwartz class;
$C^0, C^{\mathbf{r}}$ – continuous, respectively $\mathbf{r}$-smooth, sections of a vector bundle;
$\rho(A)$, $R(z; A) = (z - A)^{-1}$ – resolvent set, resolvent for the operator A;
$\mathcal{D}(A)$ – domain of the operator A;
$\sigma(A)$, $\sigma_{\mathrm{ap}}(A)$ – spectrum, approximate point spectrum of the operator A;
$\sigma_p(A)$, $\sigma_r(A)$ – point spectrum, residual spectrum of the operator A;
$\operatorname{Im} A$, $\operatorname{Ker} A$, A^* – range, kernel, and adjoint of the operator A;
$\|A\|_\bullet = \|A\|_{\bullet, X} = \inf\{|Ax| : x \in \mathcal{D}(A), |x| = 1\}$ – lower bound of A on X;
$A|Y$ – restriction of A to a subspace Y;
$|x|_A = |x| + |Ax|$, $x \in \mathcal{D}(A)$ – graph norm;
$r(A)$, $r_{\mathrm{ess}}(A)$ – spectral radius, essential spectral radius for an operator A;
$s(A) = s(\{e^{tA}\}) = \sup\{\operatorname{Re} z : z \in \sigma(A)\}$ – spectral bound;
$s_0(A)$ – the abscissa of uniform boundedness of the resolvent;
$\omega(A) = \omega(\{e^{tA}\}) = \lim_{t \to \infty} t^{-1} \log \|e^{tA}\|$ – growth bound;
$\mathcal{AH}(S)$ – annular hull of the set S;

LIST OF NOTATIONS

$\mathcal{A}$, $\mathcal{B}$ – multiplication operators with operator-valued multipliers $A(\cdot)$, $B(\cdot)$;
P, $Q = I - P$ – complementary projections;
$P(\cdot)$, $Q(\cdot) = I - P(\cdot)$ – dichotomy projections;
$\mathcal{P}$ – Riesz projection;
$\mathfrak{A}$ – algebra of multiplication operators;
$\mathfrak{B}$ – algebra of weighted translations;
$\mathfrak{C}$ – algebra of weighted shifts;
$\mathfrak{D}$ – algebra of diagonal operators;
$\{U(\theta, \tau)\}_{\theta \geq \tau}$ – evolution family;
$\{E^t\}_{t \geq 0}$ – evolution semigroup;
Γ – generator of the evolution semigroup;
$E = E^1 = e^{\Gamma}$ – evolution operator;
$(V^t f)(\theta) = f(\theta - t)$ – translation operator;
$S : (x_n)_{n \in \mathbb{Z}} \mapsto (x_n)_{n \in \mathbb{Z}}$ – shift operator;
Θ – a locally compact metric space or a compact manifold;
$B(\theta, r)$ – ball in Θ of radius r centered at θ;
$\mathcal{E}$ – ℓ-dimensional vector bundle over the n-dimensional manifold Θ;
$\{\varphi^t\}_{t \in \mathbb{R}}$– continuous flow on Θ;
$\mathcal{O}(\theta)$– orbit through θ;
$p(\theta) = \inf\{T > 0 : \varphi^T(\theta) = \theta\}$– prime period of $\theta \in \Theta$;
$\mathcal{B}(\Theta) = \{\theta \in \Theta : p \text{ is bounded in a neighborhood of } \theta\}$;
$\{\Phi^t\}_{t \geq 0} = \{\Phi^t(\theta) : t \geq 0, \theta \in \Theta\}$ – strongly continuous cocycle over a flow $\{\varphi^t\}_{t \in \mathbb{R}}$;
$\varphi := \varphi^1$, $\Phi(\theta) := \Phi^1(\theta)$;
$(\mathbf{d}f)(\theta) = \frac{d}{dt} f(\varphi^t \theta)\big|_{t=0}$;
$L_{\mathbf{v}}$ – Lie derivative in the direction of the vector field $\mathbf{v}$;
$\mathcal{T}\Theta$ – tangent bundle of the manifold Θ;
$\mathcal{T}_\theta \Theta$ – tangent space at $\theta \in \Theta$;
Erg (Θ, φ) – Borel probability φ-ergodic measures on Θ;
Θ_ν – Oseledets set;
λ^j_ν – Oseledets-Lyapunov exponents with respect of measure ν;
λ^1_ν – largest Oseledets-Lyapunov exponent;
$h_\nu(\varphi)$ – measure-theoretic entropy;
$P(\varphi, g)$ – topological pressure;
$\mathcal{R}, \mathcal{K}$ – transfer operators;
$i^{(m)} = (i_1 \ldots i_m)$ – admissible m-tuple;
$i^{(m)}\theta$ – preimage of θ under φ^m.

Index

C^*-dynamical system, 244

abscissa of uniform boundedness of the resolvent, 27
abstract Cauchy problem, 58
admissible m-tuple, 307
Alfven solution, 300
algebra
 convolution, 206
 cross-product, 242, 245
 inverse-closed, 88, 92, 99, 127, 243
 operators of multiplication, 93, 206, 241
 weighted shifts, 88, 127, 208
 weighted translations, 93, 127, 207, 243
amenable group, 244
annular hull, 227
Anosov system, 6, 255
aperiodic
 flow, 177
 map, 177
 point, 177
approximate eigenfunction
 for evolution semigroup, 39, 46, 177, 226
approximate eigenvalue and eigenvector, 22

Bohl spectrum, 62
bounded solutions, 107, 127

center manifold, 113
Cesàro mean, 30
change-of-variables, 45, 65, 74, 77, 215
classical solution, 23, 58, 107
cocycle, 164
 eventually norm continuous, 217
 smooth, 166
 strongly continuous, 164
Condition (M_λ), 230
Condition (M_{C_0}), 111
Condition (M_{L^p}), 111
Condition $(M_{\mathcal{F}_\alpha})$, 112
Condition (M), 107
continuous flow, 164
continuous subbundles, 195, 265
control space, 132
control system, 132
convolution operator, 48
core, 24

covering map, 308
cross-section, 186, 224
curvature, 255, 303

decay of correlations, 339
dichotomy, 61, 84, 170
 at a point, 170, 214
 bound, 62, 170
 constants, 170
 on Θ, 170
 projection, 170
 radius, 150
Dichotomy Theorem, 70
differential, 169
Discrete Dichotomy Theorem, 100, 213
divergence free
 extension, 279
 vector field, 274

entropy, 308, 309, 319, 331
equation
 Euler, 293, 299, 302, 336
 Hamiltonian, 253
 Lyapunov, 55, 128
 mild, 59, 230
 Navier-Stokes, 168, 336
 parabolic, 167
 Riccati, 119, 159, 253
 Schrödinger, 252
 variational, 165, 221
evolution family, 57
evolution semigroup
 along trajectories, 214
 autonomous case, 38, 42, 47
 cocycle case, 173, 176, 185
 nonautonomous case, 62, 63, 73
 norm of the generator, 52
 spectral symmetry, 38, 43, 48, 73, 178, 180, 188
exact Lyapunov exponents, 263
example
 Arendt, 33, 54, 153
 Greiner-Voigt-Wolff, 32, 54
 Hille-Phillips-Henry, 34, 54
 Montgomery-Smith, 36, 54
 Renardy, 35, 54
 Zabczyk, 32, 54

360 INDEX

expanding map, 304, 317, 319
expansion constant, 304
exponential dichotomy, *see also* dichotomy
exponentially bounded solutions, 111, 128
exponentially separated cocycle, 333

fast dynamo, 300, 301
flow-box, 186, 224
Fourier
 multiplier, 54, 159
 series, 92
 transform, 51
frequency response function, 146, 147

geodesic flow, 254, 300
global dichotomy, 170
Green's function, 103, 104, 106, 127, 129, 231, 238
growth bound, 26, 46, 54, 60, 142

Howland semigroup, 62
hull, 165
hyperbolic
 cocycle, 171
 operator, 28
 semigroup, 28, 54
hyperbolicity
 and Gearhart's Theorem, 30
 and Lyapunov function, 118, 246

kinematic dynamo, 300

Laplace transform formula, 23
largest Oseledets-Lyapunov exponent, 264
lemma
 Commutation, 67, 192
 Halmos-Rokhlin, 186
 Mañe, 179, 229
 Räbiger-Schnaubelt, 68
Lie derivative, 176
linear skew-product flow, 166, 169, 201
lower bound, 22
LSPF, 164
Lyapunov exponent, 195, 199, 263
Lyapunov function, 118, 129, 246, 247
Lyapunov numbers, 196, 300
Lyapunov Stability Theorem, 2

Mañe sequence, 222, 223, 225
Mañe point and vector, 180, 222, 277
Markov partition, 306
Mather localization, 177
Mather semigroup, 173, 185
Mather's Hyperbolicity Theorem, 3
matrix transfer operator, 305, 317, 319, 331
maximal regularity, 55
measurable subbundles, 263, 265, 268, 273
mild solution, 59, 107
 for autonomous problem, 23
 for cocycles, 230

 for nonautonomous problem, 107, 127
minimal set, 204
Morse set, 204
Multiplicative Ergodic Theorem, 262, 332
multiplicities, 264

negative continuation, 202
net, 307
Neumann map, 242
nominal system, 132

operator
 convolution, 41, 48, 76, 103
 discrete, 93, 94, 127, 185, 209
 dissipative, 118
 extended transfer, 324
 finite-diagonal, 90
 Green's, 103, 106, 108, 112, 231, 238, 239
 input-output, 132, 134
 kinematic dynamo, 300
 Koopman, 313
 multiplication, 22
 Perron-Frobenius, 305
 push-forward, 176
 Schrödinger, 252
 transfer, 305, 311, 316, 317, 331
 two-diagonal, 89, 91
 uniformly W-dissipative, 119
 uniformly dissipative, 118
 uniformly injective, 22
 uniformly negative, 118
 uniformly strongly elliptic, 60
 weighted shift, 88, 126, 185, 209, 213
 weighted translation, 87, 209
Oseledets
 Lyapunov exponents, 263, 268, 273, 276, 300, 319, 331
 set, 263
 subbundles, 263
output space, 132

period function, 177
periodic point, 177
perturbation
 bounded, 161
 Miyadera-type, 157, 161
pointwise dichotomy, 170, 214
prime period function, 177
projection, 22
 dichotomy, 100, 170, 193, 194
 Riesz, 28, 70, 100, 102, 193, 195
 splitting, 29, 103, 105, 238
projectivization, 269
pseudo-orbit, 218

quasi-Anosov system, 6, 171, 202

Radon-Nikodým derivative, 185, 241
regular admissibility, 128
regular representation, 245

representation, 243
rescaling, 24, 171
Ruelle transfer operator, 305

scalar-coefficient transfer operator, 305, 311
Schwartz class, 30, 50
semigroup
    analytic, 25
    eventually compact, 25
    eventually differentiable, 25
    eventually norm continuous, 25
    hyperbolic, 28, 119
    integrated, 53
    of contractions, 119
    translation, 97
separated set, 307
smooth cocycle, 166
solution
    classical, 58
    mild, 59, 107
spectral bound, 27, 46
Spectral Inclusion Theorem, 25
spectral mapping property, 24, 46, 48, 65, 73, 227
    weak, 53
spectral mapping theorem
    for evolution semigroups, 40, 44, 46, 48, 73, 180, 188, 227, 275
    for point spectrum, 25
    for residual spectrum, 25
    Gearhart's, for Hilbert space, 26, 53
    Gearhart's-type, for Banach space, 40, 53, 55, 148
spectral radius, 26
spectral subbundles, 195, 196, 198, 204
spectrum
    approximate point, 21, 177
    Bohl, 62, 72, 80, 83
    continuous, 21
    critical, 53
    dynamical, 172
    normal, 199
    Oseledets-Lyapunov, 264
    point, 21
    residual, 21
    Sacker-Sell, 172, 194, 196, 199, 204, 266, 268, 273
    tangential, 199
splitting projection, 238
    for evolution families, 105
    for hyperbolic semigroup, 29, 31, 103
stability
    external, 133
    individual, 54
    input-output, 133
    internal, 132
    uniform exponential, 26, 50, 53, 60, 77
stability radius, 144, 159
    constant, 144

pointwise, 149
stable family of generators, 123
state space, 132
strongly continuous cocycle, 164
subshift of finite type, 305
system
    detectable, 133
    Friedlander-Vishik, 336
    stabilizable, 133

theorem
    Annular Hull, 227, 276
    antidynamo, 300, 335
    Baskakov, 116
    Birkhoff-Khinchine Ergodic, 271
    Bochner-Phillips, 92, 102
    Datko-Pazy, 49, 75
    Datko-van Neerven, 49
    Dichotomy, 70, 79, 193
    Discrete Dichotomy, 100, 116, 213
    Friedlander-Vishik, 336
    Gearhart, 26, 142
    Generalized Lyapunov, 118, 129
    Hille-Yosida, 23
    Huang, 53
    Isomorphism, 242, 244
    Kaashoek-Verduyn Lunel, 31, 54, 143
    Lumer-Phillips, 119
    Lyapunov Stability, 2, 27
    Mañe, 202
    Multiplicative Ergodic, 262, 332
    Perron-Daleckij-Krein, 107, 127
    Prüss, 53
    Ruelle, 332
    Ruelle-Perron-Frobenius, 338
    Sacker-Sell Perturbation, 218
    Spectral Inclusion, 25
    Spectral Mapping, 180, 227, 275
    Spectral Projection, 66, 79, 191, 217
    Wiener, 92, 102
topological
    entropy, 309
    free action, 244
    mixing, 304
    pressure, 307, 313
transfer function, 133, 144, 146, 159
    for autonomous system, 132
    for nonautonomous system, 147
triangularization, 313

unit tangent bundle, 254

variational principle, 309

weak dichotomy, 171, 202

Selected Titles in This Series

(*Continued from the front of this publication*)

- 40.3 **Daniel Gorenstein, Richard Lyons, and Ronald Solomon,** The classification of the finite simple groups, number 3, 1998
- 40.2 **Daniel Gorenstein, Richard Lyons, and Ronald Solomon,** The classification of the finite simple groups, number 2, 1995
- 40.1 **Daniel Gorenstein, Richard Lyons, and Ronald Solomon,** The classification of the finite simple groups, number 1, 1994
- 39 **Sigurdur Helgason,** Geometric analysis on symmetric spaces, 1994
- 38 **Guy David and Stephen Semmes,** Analysis of and on uniformly rectifiable sets, 1993
- 37 **Leonard Lewin, Editor,** Structural properties of polylogarithms, 1991
- 36 **John B. Conway,** The theory of subnormal operators, 1991
- 35 **Shreeram S. Abhyankar,** Algebraic geometry for scientists and engineers, 1990
- 34 **Victor Isakov,** Inverse source problems, 1990
- 33 **Vladimir G. Berkovich,** Spectral theory and analytic geometry over non-Archimedean fields, 1990
- 32 **Howard Jacobowitz,** An introduction to CR structures, 1990
- 31 **Paul J. Sally, Jr. and David A. Vogan, Jr., Editors,** Representation theory and harmonic analysis on semisimple Lie groups, 1989
- 30 **Thomas W. Cusick and Mary E. Flahive,** The Markoff and Lagrange spectra, 1989
- 29 **Alan L. T. Paterson,** Amenability, 1988
- 28 **Richard Beals, Percy Deift, and Carlos Tomei,** Direct and inverse scattering on the line, 1988
- 27 **Nathan J. Fine,** Basic hypergeometric series and applications, 1988
- 26 **Hari Bercovici,** Operator theory and arithmetic in H^∞, 1988
- 25 **Jack K. Hale,** Asymptotic behavior of dissipative systems, 1988
- 24 **Lance W. Small, Editor,** Noetherian rings and their applications, 1987
- 23 **E. H. Rothe,** Introduction to various aspects of degree theory in Banach spaces, 1986
- 22 **Michael E. Taylor,** Noncommutative harmonic analysis, 1986
- 21 **Albert Baernstein, David Drasin, Peter Duren, and Albert Marden, Editors,** The Bieberbach conjecture: Proceedings of the symposium on the occasion of the proof, 1986
- 20 **Kenneth R. Goodearl,** Partially ordered abelian groups with interpolation, 1986
- 19 **Gregory V. Chudnovsky,** Contributions to the theory of transcendental numbers, 1984
- 18 **Frank B. Knight,** Essentials of Brownian motion and diffusion, 1981
- 17 **Le Baron O. Ferguson,** Approximation by polynomials with integral coefficients, 1980
- 16 **O. Timothy O'Meara,** Symplectic groups, 1978
- 15 **J. Diestel and J. J. Uhl, Jr.,** Vector measures, 1977
- 14 **V. Guillemin and S. Sternberg,** Geometric asymptotics, 1977
- 13 **C. Pearcy, Editor,** Topics in operator theory, 1974
- 12 **J. R. Isbell,** Uniform spaces, 1964
- 11 **J. Cronin,** Fixed points and topological degree in nonlinear analysis, 1964
- 10 **R. Ayoub,** An introduction to the analytic theory of numbers, 1963
- 9 **Arthur Sard,** Linear approximation, 1963
- 8 **J. Lehner,** Discontinuous groups and automorphic functions, 1964
- 7.2 **A. H. Clifford and G. B. Preston,** The algebraic theory of semigroups, Volume II, 1961

For a complete list of titles in this series, visit the
AMS Bookstore at **www.ams.org/bookstore/**.